BAYESIAN METHODS
for Statistical Analysis

BAYESIAN METHODS
for Statistical Analysis

BY BOREK PUZA

Australian
National
University

eVIEW

Published by ANU eView
The Australian National University
Acton ACT 2601, Australia
Email: enquiries.eview@anu.edu.au
This title is also available online at http://eview.anu.edu.au

National Library of Australia Cataloguing-in-Publication entry

Creator:	Puza, Borek, author.
Title:	Bayesian methods for statistical analysis / Borek Puza.
ISBN:	9781921934254 (paperback) 9781921934261 (ebook)
Subjects:	Bayesian statistical decision theory. Statistical decision.
Dewey Number:	519.542

Cover design and layout by ANU Press

Contents

Abstract

'Bayesian Methods for Statistical Analysis' is a book on statistical methods for analysing a wide variety of data. The book consists of 12 chapters, starting with basic concepts and covering numerous topics, including Bayesian estimation, decision theory, prediction, hypothesis testing, hierarchical models, Markov chain Monte Carlo methods, finite population inference, biased sampling and nonignorable nonresponse. The book contains many exercises, all with worked solutions, including complete computer code. It is suitable for self-study or a semester-long course, with three hours of lectures and one tutorial per week for 13 weeks.

Acknowledgements

'Bayesian Methods for Statistical Analysis' derives from the lecture notes for a four-day course titled 'Bayesian Methods', which was presented to staff of the Australian Bureau of Statistics, at ABS House in Canberra, in 2013. Lectures of three hours each were held in the mornings of 11, 18 and 25 November and 9 December, and three-hour tutorials were held in the mornings of 14, 20 and 27 November and 11 December.

Of the 30-odd participants, some of whom attended via video link from regional ABS offices, special thanks go to Anura Amarasinghe, Rachel Barker, Geoffrey Brent, Joseph Chien, Alexander Hanysz, Sebastien Lucie, Peter Radisich and Anthony Russo, who asked insightful questions, pointed out errors, and contributed to an improved second edition of the lecture notes. Thanks also to Siu-Ming Tam, First Australian Statistician of the Methodology and Data Management Division at ABS, for useful comments, and for inviting the author to present the course in the first place, after having read Puza (1995). Last but not least, special thanks go to Kylie Johnson for her excellent work as the course administrator.

Further modifications to 'Bayesian Methods' led to the present work, which is published as an eView textbook by the ANU Press, Canberra. Many thanks to David Gardiner, Brian Kennett, Teresa Prowse, Emily Tinker and two anonymous referees for useful comments and suggestions which helped to further improve the quality of the book. Thanks also to Yi (James) Zhang for his proofreading of the book whilst learning the material as part of his Honours studies.

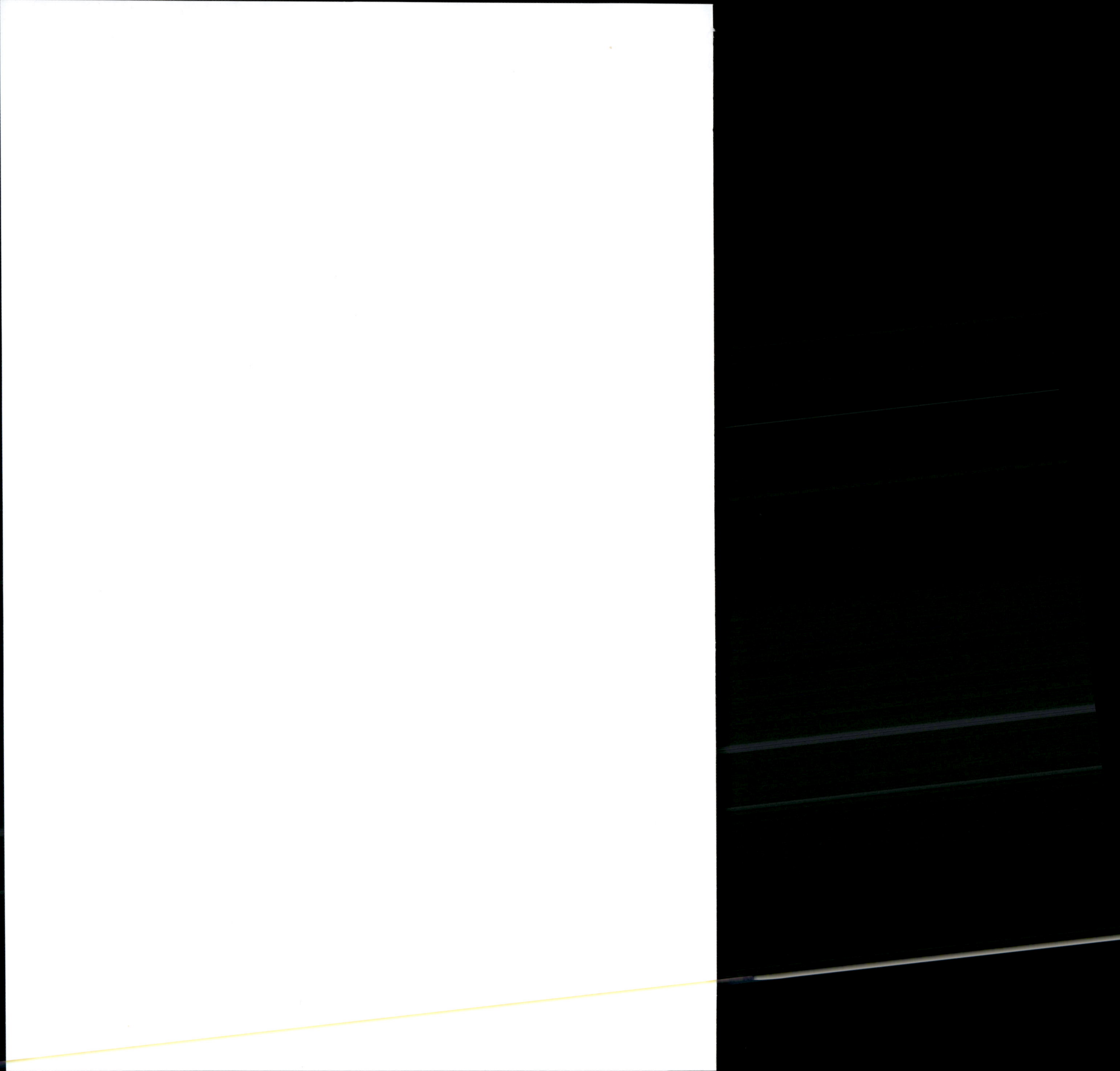

Preface

'Bayesian Methods for Statistical Analysis' is a book which can be used as the text for a semester-long course and is suitable for anyone who is familiar with statistics at the level of 'Mathematical Statistics with Applications' by Wackerly, Mendenhall and Scheaffer (2008). The book does not attempt to cover all aspects of Bayesian methods but to provide a 'guided tour' through the subject matter, one which naturally reflects the author's particular interests gained over years of research and teaching.

For a more comprehensive account of Bayesian methods, the reader is referred to the very extensive literature on this subject, including 'Theory of Probability' by Jeffreys (1961), 'Bayesian Inference in Statistical Analysis' by Box and Tiao (1973), 'Markov Chain Monte Carlo in Practice' by Gilks et al. (1996), 'Bayesian Statistics: An Introduction' by Lee (1997), 'Bayesian Methods: An Analysis for Statisticians and Interdisciplinary Researchers' by Leonard and Hsu (1999), 'Bayesian Data Analysis' by Gelman et al. (2004), 'Computational Bayesian Statistics' by Bolstad (2009) and 'Handbook of Markov Chain Monte Carlo' by Brooks et al. (2011). See also Smith and Gelfand (1992) and O'Hagan and Forster (2004).

The software packages which feature in this book are R and WinBUGS.

R is a general software environment for statistical computing and graphics which compiles and runs on UNIX platforms, Windows and MacOS. This software is available for free at www.r-project.org/ Two useful guides to R are 'Bayesian Computation With R' by Albert (2009) and 'Data Analysis and Graphics Using R: An Example-Based Approach' by Maindonald and Braun (2010).

BUGS stands for 'Bayesian Inference Using Gibbs Sampling' and is a specialised software environment for the Bayesian analysis of complex statistical models using Markov chain Monte Carlo methods. WinBUGS, a version of BUGS for Microsoft Windows, is available for free at www.mrc-bsu.cam.ac.uk/software/bugs/ Two useful guides to WinBUGS are 'Bayesian Modeling Using WinBUGS' by Ntzoufras (2009) and 'Bayesian Population Analysis Using WinBUGS' by Kéry and Schaub (2012).

The present book includes a large number of exercises, interspersed throughout and each followed by a detailed solution, including complete computer code. A student should be able to reproduce all of the numerical and graphical results in the book by running the provided code. Although many of the exercises are straightforward, some are fairly involved, and a few will be of interest only to the particularly keen or advanced student. All of the code in this book is also available in the form of an electronic text document which can be obtained from the same website as the book.

This book is in the form of an Adobe PDF file saved from Microsoft Word 2013 documents, with the equations as MathType 6.9 objects. The figures in the book were created using Microsoft Paint, the Snipping Tool in Windows, WinBUGS and R. In the few instances where color is used, this is only for additional clarity. Thus, the book can be printed in black and white with no loss of essential information.

The following chapter provides an overview of the book. Appendix A contains several additional exercises with worked solutions, Appendix B has selected distributions and notation, and Appendix C lists some abbreviations and acronyms. Following the appendices is a bibliography for the entire book.

The last four of the 12 chapters in this book constitute a practical companion to 'Monte Carlo Methods for Finite Population Inference', a largely theoretical manuscript written by the author (Puza, 1995) during the last year of his employment at the Australian Bureau of Statistics in Canberra.

Overview

Chapter 1: Bayesian Basics Part 1 (pages 1–60)

Introduces Bayes' rule, Bayes factors, Bayesian models, posterior distributions, and the proportionality formula. Also covered are the binomial-beta model, the Jeffreys' famous tramcar problem, the distinction between finite population inference and superpopulation inference, conjugacy, point and interval estimation, inference on functions of parameters, credibility estimation, the normal-normal model, and the normal-gamma model.

Chapter 2: Bayesian Basics Part 2 (pages 61–108)

Covers the frequentist characteristics of Bayesian estimators including bias and coverage probabilities, mixture priors, uninformative priors including the Jeffreys prior, and Bayesian decision theory including the posterior expected loss and Bayes risk.

Chapter 3: Bayesian Basics Part 3 (pages 109–152)

Covers inference based on functions of the data including censoring and rounded data, predictive inference, posterior predictive p-values, multiple-parameter models, and the normal-normal-gamma model including an example of Bayesian finite population inference.

Chapter 4: Computational Tools (pages 153–200)

Covers the Newton-Raphson (NR) algorithm including its multivariate version, the expectation-maximisation (EM) algorithm, hybrid search algorithms, integration techniques including double integration, optimisation in R, and specification of prior distributions.

Chapter 5: Monte Carlo Basics (pages 201–262)

Covers Monte Carlo integration, importance sampling, the method of composition, Buffon's needle problem, testing the coverage of Monte Carlo confidence intervals, random number generation including the inversion technique, rejection sampling, and applications to Bayesian inference including prediction in the normal-normal-gamma model, Rao-Blackwell estimation, and estimation of posterior predictive p-values.

Chapter 6: MCMC Methods Part 1 (pages 263–320)

Covers Markov chain Monte Carlo (MCMC) methods including the Metropolis-Hastings algorithm, the Gibbs sampler, specification of tuning parameters, the batch means method, computational issues, and applications to the normal-normal-gamma model.

Chapter 7: MCMC Methods Part 2 (pages 321–364)

Covers stochastic data augmentation, a comparison of classical and Bayesian methods for linear regression and logistic regression, respectively, and a Bayesian model for correlated Bernoulli data.

Chapter 8: MCMC Inference via WinBUGS (pages 365–406)

Provides a detailed tutorial in the WinBUGS computer package including running WinBUGS within R, and shows how WinBUGS can be used for linear regression, logistic regression and ARIMA time series analysis.

Chapter 9: Bayesian Finite Population Theory (pages 407–466)

Introduces notation and terminology for Bayesian finite population inference in the survey sampling context, and discusses ignorable and nonignorable sampling mechanisms. These concepts are illustrated by way of examples and exercises, some of which involve MCMC methods.

Chapter 10: Normal Finite Population Models (pages 467–514)

Contains a generalisation of the normal-normal-gamma model to the finite population context with covariates. Useful vector and matrix formulae are provided, special cases such as ratio estimation are treated in detail, and it is shown how MCMC methods can be used for both descriptive and analytic inferences.

Chapter 11: Transformations and Other Topics (pages 515–558)

Shows how MCMC methods can be used for inference on complicated functions of superpopulation and finite population quantities, as well for inference based on transformed data. Frequentist characteristics of Bayesian estimators are discussed in the finite population context, with examples of how Monte Carlo methods can be used to estimate model bias, design bias, model coverage and design coverage.

Chapter 12: Biased Sampling and Nonresponse (pages 559–608)

Discusses and provides examples of ignorable and nonignorable response mechanisms, with an exercise involving follow-up data. The topic of self-selection bias in volunteer surveys is studied from a frequentist perspective, then treated using Bayesian methods, and finally extended to the finite population context.

Appendix A: Additional Exercises (pages 609–666)

Provides practice at applying concepts in the last four chapters.

Appendix B: Distributions and Notation (pages 667–672)

Provides details of some distributions which feature in the book.

Appendix C: Abbreviations and Acronyms (pages 673–676)

Catalogues many of the simplified expressions used throughout.

Computer Code in Bayesian Methods for Statistical Analysis

Combines all of the R and WinBUGS code interspersed throughout the 679-page book. This separate 126-page PDF file is available online at: http://eview.anu.edu.au/bayesian_methods/pdf/computer_code.pdf.

CHAPTER 1

Bayesian Basics Part 1

1.1 Introduction

Bayesian methods is a term which may be used to refer to any mathematical tools that are useful and relevant in some way to *Bayesian inference*, an approach to statistics based on the work of Thomas Bayes (1701–1761). Bayes was an English mathematician and Presbyterian minister who is best known for having formulated a basic version of the well-known *Bayes' Theorem*.

Figure 1.1 (page 3) shows part of the Wikipedia article for Thomas Bayes. Bayes' ideas were later developed and generalised by many others, most notably the French mathematician Pierre-Simon Laplace (1749–1827) and the British astronomer Harold Jeffreys (1891–1989).

Bayesian inference is different to *classical inference* (or *frequentist inference*) mainly in that it treats model parameters as *random variables* rather than as *constants*. The Bayesian framework (or paradigm) allows for prior information to be formally taken into account. It can also be useful for formulating a complicated statistical model that presents a challenge to classical methods.

One drawback of Bayesian inference is that it invariably requires a prior distribution to be specified, even in the absence of any prior information. However, suitable *uninformative* prior distributions (also known as *noninformative*, *objective* or *reference* priors) have been developed which address this issue, and in many cases a nice feature of Bayesian inference is that these priors lead to exactly the same point and interval estimates as does classical inference. The issue becomes even less important when there is at least a moderate amount of data available. As sample size increases, the Bayesian approach typically converges to the same inferential results, irrespective of the specified prior distribution.

Another issue with Bayesian inference is that, although it may easily lead to suitable formulations of a challenging statistical problem, the types of calculation needed for inference can themselves be very complicated. Often, these calculations take on the form of multiple integrals (or summations) which are intractable and difficult (or impossible) to solve, even with the aid of advanced numerical techniques.

In such situations, the desired solutions can typically be approximated to any degree of precision using *Monte Carlo* (MC) methods. The idea is to make clever use of a large sample of values generated from a suitable probability distribution.

How to generate this sample presents another problem, but one which can typically be solved easily via *Markov chain Monte Carlo* (MCMC) *methods*. Both MC and MCMC methods will feature in later chapters of the course.

1.2 Bayes' rule

The starting point for Bayesian inference is *Bayes' rule*. The simplest form of this is

$$P(A \mid B) = \frac{P(A)P(B \mid A)}{P(A)P(B \mid A) + P(\overline{A})P(B \mid \overline{A})},$$

where A and B are events such that $P(B) > 0$. This is easily proven by considering that:

$$P(A \mid B) = \frac{P(AB)}{P(B)} \quad \text{by the definition of conditional probability}$$

$$P(AB) = P(A)P(B \mid A) \quad \text{by the multiplicative law of probability}$$

$$P(B) = P(AB) + P(\overline{A}B) = P(A)P(B \mid A) + P(\overline{A})P(B \mid \overline{A})$$
$$\text{by the law of total probability.}$$

We see that the posterior probability $P(A \mid B)$ is equal to the prior probability $P(A)$ multiplied by a factor, where this factor is given by $P(B \mid A) / P(B)$.

As regards terminology, we call $P(A)$ the *prior* probability of A (meaning the probability of A *before* B is known to have occurred), and we call $P(A \mid B)$ the *posterior* probability of A *given* B (meaning the probability of A *after* B is known to have occurred). We may also say that $P(A)$ represents our *a priori* beliefs regarding A, and $P(A \mid B)$ represents our *a posteriori* beliefs regarding A.

Figure 1.1 Beginning of the Wikipedia article on Thomas Bayes

Source: en.wikipedia.org/wiki/Thomas_Bayes, 29/10/2014

Thomas Bayes

From Wikipedia, the free encyclopedia

Thomas Bayes (/ˈbeɪz/; c. 1701 – 7 April 1761)[1][2][note a] was an English statistician, philosopher and Presbyterian minister, known for having formulated a specific case of the theorem that bears his name: Bayes' theorem. Bayes never published what would eventually become his most famous accomplishment; his notes were edited and published after his death by Richard Price.[3]

Thomas Bayes

Portrait used of Bayes in the 1936 book *History of Life Insurance*; it is dubious whether it actually depicts Bayes.[1] No earlier portrait or claimed portrait survived.

Born	c. 1701 London, England
Died	7 April 1761 (aged 59) Tunbridge Wells, Kent, England
Residence	Tunbridge Wells, Kent, England
Nationality	English

Signature

T. Bayes.

Contents [hide]

Biography [edit]

Thomas Bayes was the son of London Presbyterian minister Joshua Bayes,[4] and was possibly born in Hertfordshire.[5] He came from a prominent nonconformist family from Sheffield. In 1719, he enrolled at the University of Edinburgh to study logic and theology. On his return around 1722, he assisted his father at the latter's chapel in London before moving to Tunbridge Wells, Kent, around 1734. There he was minister of the Mount Sion chapel, until 1752.[6]

More generally, we may consider any event B such that $P(B) > 0$ and $k > 1$ events $A_1, ..., A_k$ which form a partition of any superset of B (such as the entire sample space S). Then, for any $i = 1, ..., k$, it is true that

$$P(A_i \mid B) = \frac{P(A_i B)}{P(B)},$$

where $P(B) = \sum_{j=1}^{n} P(A_j B)$ and $P(A_j B) = P(A_j)P(B \mid A_j)$.

Exercise 1.1 Medical testing

The incidence of a disease in the population is 1%. A medical test for the disease is 90% accurate in the sense that it produces a false reading 10% of the time, both: (a) when the test is applied to a person with the disease; and (b) when the test is applied to a person without the disease.

A person is randomly selected from population and given the test. The test result is positive (i.e. it indicates that the person has the disease).

What is the probability that the person actually has the disease?

Solution to Exercise 1.1

Let A be the event that the person has the disease, and let B be the event that they test positive for the disease. Then:

$P(A) = 0.01$ (the *prior* probability of the person having the disease)

$P(B \mid A) = 0.9$ (the true positive rate, also called

          the *sensitivity* of the test)

$P(\overline{B} \mid \overline{A}) = 0.9$ (the true negative rate, also called

          the *specificity* of the test).

So: $P(AB) = P(A)P(B \mid A) = 0.01 \times 0.9 = 0.009$

         $P(\overline{A}B) = P(\overline{A})P(B \mid \overline{A}) = 0.99 \times 0.1 = 0.099$.

So the unconditional (or prior) probability of the person testing positive is $P(B) = P(AB) + P(\overline{A}B) = 0.009 + 0.099 = 0.108$.

So the required *posterior* probability of the person having the disease is

$$P(A \mid B) = \frac{P(AB)}{P(B)} = \frac{0.009}{0.108} = \frac{1}{12} = 0.08333.$$

Figure 1.2 is a Venn diagram which illustrates how B may be considered as the union of AB and $\overline{A}B$. The required posterior probability of A given B is simply the probability of AB divided by the probability of B.

Figure 1.2 Venn diagram for Exercise 1.1

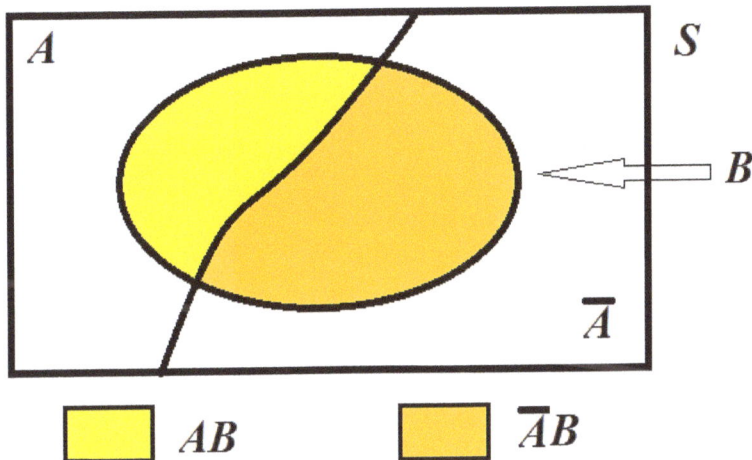

Discussion

It may seem the posterior probability that the person has the disease (1/12) is rather low, considering the high accuracy of the test (namely $P(B\,|\,A) = P(\overline{B}\,|\,\overline{A}) = 0.9$).

This may be explained by considering 1,000 random persons in the population and applying the test to each one. About 10 persons will have the disease, and of these, 9 will test positive. Of the 990 who do not have the disease, 99 will test positive. So the total number of persons testing positive will be $9 + 99 = 108$, and the proportion of these 108 who actually have the disease will be $9/108 = 1/12$. This heuristic derivation of the answer shows it to be small on account of the large number of false positives (99) amongst the overall number of positives (108).

On the other hand, it may be noted that the posterior probability of the person having the disease is actually very *high* relative to the prior probability of them having the disease ($P(A) = 0.01$). The positive test result has greatly increased the person's chance of having the disease (increased it by more than 700%, since $0.01 + 7.333 \times 0.01 = 0.08333$).

It is instructive to generalise the answer (1/12) as a function of the prevalence (i.e. proportion) of the disease in the population, $p = P(A)$, and the common accuracy rate of the test, $q = P(B | A) = P(\overline{B} | \overline{A})$.

We find that

$$P(A|B) = \frac{P(A)P(B|A)}{P(A)P(B|A) + P(\overline{A})P(B|\overline{A})} = \frac{pq}{pq + (1-p)(1-q)}.$$

Figure 1.3 shows the posterior probability of the person having the disease $(P(A|B))$ as a function of p with q fixed at 0.9 and 0.95, respectively (subplot (a)), and as a function of q with p fixed at 0.01 and 0.05, respectively (subplot (b)). In each case, the answer (1/12) is represented as a dot corresponding to $p = 0.01$ and $q = 0.9$.

Figure 1.3 Posterior probability of disease as functions of p and q

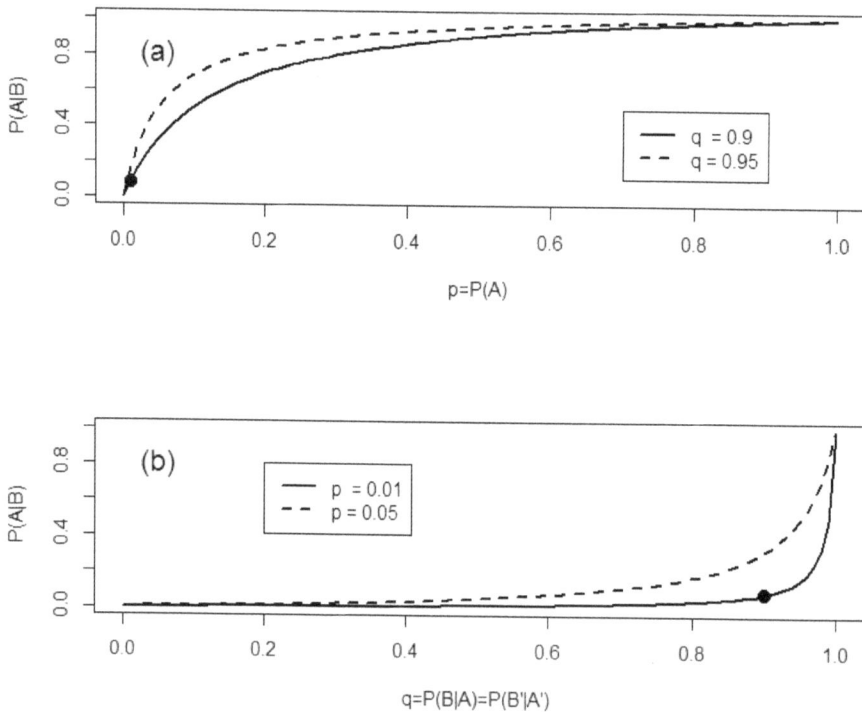

R Code for Exercise 1.1

```
PAgBfun=function(p=0.01,q=0.9){  p*q  / (p*q+(1-p)*(1-q))  }
PAgBfun() # 0.08333333

pvec=seq(0,1,0.01); Pveca=PAgBfun(p=pvec,q=0.9)
        Pveca2=PAgBfun(p=pvec,q=0.95)
qvec=seq(0,1,0.01); Pvecb=PAgBfun(p=0.01,q=qvec)
        Pvecb2=PAgBfun(p=0.05,q=qvec)

X11(w=8,h=7); par(mfrow=c(2,1));

plot(pvec,Pveca,type="l",xlab="p=P(A)",ylab="P(A|B)",lwd=2)
points(0.01,1/12,pch=16,cex=1.5); text(0.05,0.8,"(a)",cex=1.5)
lines(pvec,Pveca2,lty=2,lwd=2)
legend(0.7,0.5,c("q = 0.9","q = 0.95"),lty=c(1,2),lwd=c(2,2))

plot(qvec,Pvecb,type="l",xlab="q=P(B|A)=P(B'|A')",ylab="P(A|B)",lwd=2)
points(0.9,1/12,pch=16,cex=1.5); text(0.05,0.8,"(b)",cex=1.5)
lines(qvec,Pvecb2,lty=2,lwd=2)
legend(0.2,0.8,c("p = 0.01","p = 0.05"),lty=c(1,2),lwd=c(2,2))

# Technical note: The graph here was copied from R as 'bitmap' and then
# pasted into a Word document, which was then saved as a PDF. If the graph
# is copied from R as 'metafile', it appears correct in the Word document,
# but becomes corrupted in the PDF, with axis legends slightly off-centre.
# So, all graphs in this book created in R were copied into Word as 'bitmap'.
```

Exercise 1.2 Blood types

In a particular population:

        10% of persons have Type 1 blood,

                and of these, 2% have a particular disease;

        30% of persons have Type 2 blood,

                and of these, 4% have the disease;

        60% of persons have Type 3 blood,

                and of these, 3% have the disease.

A person is randomly selected from the population and found to have the disease.

What is the probability that this person has Type 3 blood?

Solution to Exercise 1.2

Let: A = 'The person has Type 1 blood'
B = 'The person has Type 2 blood'
C = 'The person has Type 3 blood'
D = 'The person has the disease'.

Then: $P(A) = 0.1$, $\quad P(D|A) = 0.02$
$P(B) = 0.3$, $\quad P(D|B) = 0.04$
$P(C) = 0.6$, $\quad P(D|C) = 0.03$.

So:
$$P(D) = P(AD) + P(BD) + P(CD)$$
$$= P(A)P(D|A) + P(B)P(D|B) + P(C)P(D|C)$$
$$= 0.1 \times 0.02 + 0.3 \times 0.04 + 0.6 \times 0.03$$
$$= 0.002 + 0.012 + 0.018 = 0.032.$$

Hence: $P(C|D) = \dfrac{P(CD)}{P(D)} = \dfrac{0.018}{0.032} = \dfrac{9}{16} = 56.25\%.$

Figure 1.4 is a Venn diagram showing how D may be considered as the union of AD, BD and CD. The required posterior probability of C given D is simply the probability of CD divided by the probability of D.

Figure 1.4 Venn diagram for Exercise 1.2

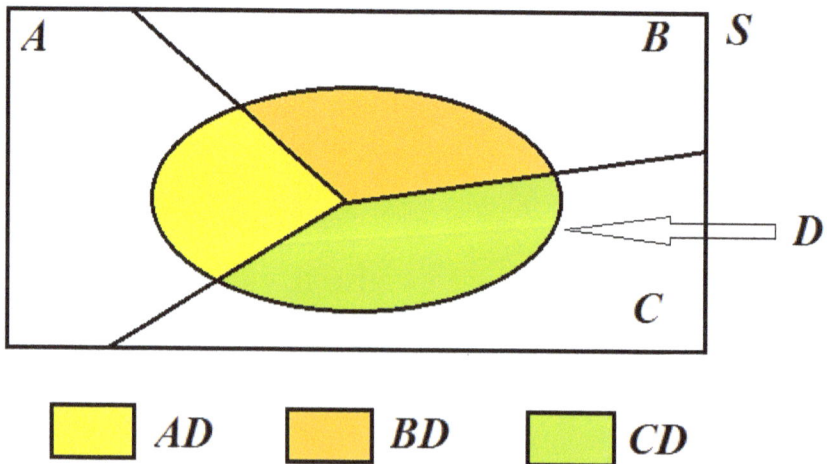

| AD | BD | CD |

1.3 Bayes factors

One way to perform hypothesis testing in the Bayesian framework is via the theory of *Bayes factors*. Suppose that on the basis of an observed event D (standing for *data*) we wish to test a *null hypothesis*

$$H_0 : E_0$$

versus an *alternative hypothesis*

$$H_1 : E_1,$$

where E_0 and E_1 are two *events* (which are not necessarily mutually exclusive or even exhaustive of the event space).

Then we calculate:

$\pi_0 = P(E_0)$ = the prior probability of the null hypothesis

$\pi_1 = P(E_1)$ = the prior probability of the alternative hypothesis

$PRO = \pi_0 / \pi_1$ = the prior odds in favour of the null hypothesis

$p_0 = P(E_0 \mid D)$ = the posterior probability of the null hypothesis

$p_1 = P(E_1 \mid D)$ = the posterior probability of the alternative hypothesis

$POO = p_0 / p_1$ = the posterior odds in favour of the null hypothesis.

The *Bayes factor* is then defined as $BF = POO / PRO$. This may be interpreted as the factor by which the data have multiplied the odds in favour of the null hypothesis relative to the alternative hypothesis. If $BF > 1$ then the data has *increased* the relative likelihood of the null, and if $BF < 1$ then the data has *decreased* that relative likelihood. The magnitude of BF tells us how much effect the data has had on the relative likelihood.

Note 1: Another way to express the Bayes factor is as

$$BF = \frac{p_0 / p_1}{\pi_0 / \pi_1} = \frac{P(E_0 \mid D) / P(E_1 \mid D)}{P(E_0) / P(E_1)} = \frac{P(D)P(E_0 \mid D) / P(E_0)}{P(D)P(E_1 \mid D) / P(E_1)}$$

$$= \frac{P(D \mid E_0)}{P(D \mid E_1)}.$$

Thus, the Bayes factor may also be interpreted as the ratio of the likelihood of the data given the null hypothesis to the likelihood of the data given the alternative hypothesis.

Note 2: The idea of a Bayes factor extends to situations where the null and alternative hypotheses are *statistical models* rather than *events*. This idea may be taken up later.

Exercise 1.3 Bayes factor in disease testing

The incidence of a disease in the population is 1%. A medical test for the disease is 90% accurate in the sense that it produces a false reading 10% of the time, both: (a) when the test is applied to a person with the disease; and (b) when the test is applied to a person without the disease.

A person is randomly selected from population and given the test. The test result is positive (i.e. it indicates that the person has the disease).

Calculate the Bayes factor for testing that the person has the disease versus that they do not have the disease.

Solution to Exercise 1.3

Recall in Exercise 1.1, where A = 'Person has disease' and B = 'Person tests positive', the relevant probabilities are $P(A) = 0.01$, $P(B \mid A) = 0.9$ and $P(\bar{B} \mid \bar{A}) = 0.9$, from which can be deduced that $P(A \mid B) = 1/12$.

We now wish to test $H_0 : A$ vs $H_1 : \bar{A}$. So we calculate:

$\pi_0 = P(A) = 0.01$, $\pi_1 = P(\bar{A}) = 0.99$, $PRO = \pi_0 / \pi_1 = 1/99$,

$p_0 = P(A \mid B) = 1/12$, $p_1 = P(\bar{A} \mid B) = 11/12$, $POO = p_0 / p_1 = 1/11$.

Hence the required Bayes factor is $BF = POO/PRO = (1/11)/(1/99) = 9$.

This means the positive test result has multiplied the odds of the person having the disease relative to not having it by a factor of 9 or 900%. Another way to say this is that those odds have increased by 800%.

Note: We could also work out the Bayes factor here as

$$BF = \frac{P(B \mid A)}{P(B \mid \bar{A})} = \frac{0.9}{0.1} = 9,$$

namely as the ratio of the probability that the person tests positive given they have the disease to the probability that they test positive given they do not have the disease.

1.4 Bayesian models

Bayes' formula extends naturally to statistical models. A *Bayesian model* is a parametric model in the classical (or frequentist) sense, but with the addition of a *prior probability distribution* for the model parameter, which is treated as a *random variable* rather than an *unknown constant*. The basic components of a Bayesian model may be listed as:
 • the *data*, denoted by y
 • the *parameter*, denoted by θ
 • the *model distribution,* given by a specification of
     $f(y|\theta)$ or $F(y|\theta)$ or the distribution of $(y|\theta)$
 • the *prior distribution,* given by a specification of
     $f(\theta)$ or $F(\theta)$ or the distribution of θ.

Here, F is a generic symbol which denotes *cumulative distribution function* (cdf), and f is a generic symbol which denotes *probability density function* (pdf) (when applied to a continuous random variable) or *probability mass function* (pmf) (when applied to a discrete random variable). For simplicity, we will avoid the term pmf and use the term pdf or density for all types of random variable, including the mixed type.

Note 1: A *mixed distribution* is defined by a cdf which exhibits at least one discontinuity (or jump) and is strictly increasing over at least one interval of values.

Note 2: The prior may be specified by writing a statement of the form '$\theta \sim ...$', where the symbol '$\sim$' means 'is distributed as', and where '...'denotes the relevant distribution. Likewise, the model for the data may be specified by writing a statement of the form '$(y|\theta) \sim ...$'.

Note 3: At this stage we will not usually distinguish between y as a random variable and y as a value of that random variable; but sometimes we may use Y for the former. Each of y and θ may be a scalar, vector, matrix or array. Also, each component of y and θ may have a discrete distribution, a continuous distribution, or a mixed distribution.

In the first few examples below, we will focus on the simplest case where both y and θ are scalar and discrete.

1.5 The posterior distribution

Bayesian inference requires determination of the *posterior probability distribution* of θ. This task is equivalent to finding the *posterior pdf* of θ, which may be done using the equation

$$f(\theta \mid y) = \frac{f(\theta)f(y \mid \theta)}{f(y)}.$$

Here, $f(y)$ is the *unconditional* (or *prior*) pdf of y, as given by

$$f(y) = \int f(y \mid \theta)dF(\theta) = \begin{cases} \int f(\theta)f(y \mid \theta)d\theta & \text{if } \theta \text{ is continuous} \\ \sum_{\theta} f(\theta)f(y \mid \theta) & \text{if } \theta \text{ is discrete.} \end{cases}$$

Note: Here, $\int f(y \mid \theta)dF(\theta)$ is a *Lebesgue-Stieltjes integral*, which may need evaluating by breaking the integral into two parts in the case where θ has a mixed distribution. In the continuous case, think of $dF(\theta)$ as $\dfrac{dF(\theta)}{d\theta}d\theta = f(\theta)d\theta$.

Exercise 1.4 Loaded dice

Consider six loaded dice with the following properties. Die A has probability 0.1 of coming up 6, each of Dice B and C has probability 0.2 of coming up 6, and each of Dice D, E and F has probability 0.3 of coming up 6.

A die is chosen randomly from the six dice and rolled twice. On both occasions, 6 comes up.

What is the posterior probability distribution of θ, the probability of 6 coming up on the chosen die.

Solution to Exercise 1.4

Let y be the number of times that 6 comes up on the two rolls of the chosen die, and let θ be the probability of 6 coming up on a single roll of that die. Then the Bayesian model is:

$$(y \mid \theta) \sim Bin(2, \theta)$$

$$f(\theta) = \begin{cases} 1/6, & \theta = 0.1 \\ 2/6, & \theta = 0.2 \\ 3/6, & \theta = 0.3. \end{cases}$$

In this case $y = 2$ and so

$$f(y \mid \theta) = \binom{2}{y} \theta^y (1-\theta)^{2-y} = \binom{2}{2} \theta^2 (1-\theta)^{2-2} = \theta^2.$$

So $f(y) = \sum_\theta f(\theta) f(y \mid \theta) = \dfrac{1}{6}(0.1)^2 + \dfrac{2}{6}(0.2)^2 + \dfrac{3}{6}(0.3)^2 = 0.06.$

So $f(\theta \mid y) = \dfrac{f(\theta) f(y \mid \theta)}{f(y)} = \begin{cases} (1/6)0.1^2 / 0.06 = 0.02778, & \theta = 0.1 \\ (2/6)0.2^2 / 0.06 = 0.22222, & \theta = 0.2 \\ (3/6)0.3^2 / 0.06 = 0.75, & \theta = 0.3. \end{cases}$

Note: This result means that if the chosen die were to be tossed again a large number of times (say 10,000) then there is a 75% chance that 6 would come up about 30% of the time, a 22.2% chance that 6 would come up about 20% of the time, and a 2.8% chance that 6 would come up about 10% of the time.

1.6 The proportionality formula

Observe that $f(y)$ is a constant with respect to θ in the Bayesian equation

$$f(\theta \mid y) = f(\theta) f(y \mid \theta) / f(y),$$

which means that we may also write the equation as

$$f(\theta \mid y) = \frac{f(\theta) f(y \mid \theta)}{k},$$

or as

$$f(\theta \mid y) = c f(\theta) f(y \mid \theta),$$

where $k = f(y)$ and $c = 1/k$.

We may also write

$$f(\theta \mid y) \propto f(\theta) f(y \mid \theta),$$

where $\propto$ is the proportionality sign.

Equivalently, we may write

$$f(\theta|y) \overset{\theta}{\propto} f(\theta)f(y|\theta)$$

to emphasise that the proportionality is specifically with respect to θ.

Another way to express the last equation is

$$f(\theta|y) \propto f(\theta) \times L(\theta|y),$$

where $L(\theta|y)$ is the *likelihood function* (defined as the model density $f(y|\theta)$ multiplied by any constant with respect to θ, and viewed as a function of θ rather than of y).

The last equation may also be stated in words as:

The posterior is proportional to the prior times the likelihood.

These observations indicate a shortcut method for determining the required posterior distribution which obviates the need for calculating $f(y)$ (which may be difficult).

This method is to multiply the prior density (or the kernel of that density) by the likelihood function and try to identify the resulting function of θ as the density of a well-known or common distribution.

Once the posterior distribution has been identified, $f(y)$ may then be obtained easily as the associated normalising constant.

Exercise 1.5 Loaded dice with solution via the proportionality formula

As in Exercise 1.4, suppose that Die A has probability 0.1 of coming up 6, each of Dice B and C has probability 0.2 of coming up 6, and each of Dice D, E and F has probability 0.3 of coming up 6.

A die is chosen randomly from the six dice and rolled twice. On both occasions, 6 comes up.

Using the proportionality formula, find the posterior probability distribution of θ, the probability of 6 coming up on the chosen die.

Solution to Exercise 1.5

With y denoting the number of times 6 comes up, the Bayesian model may be written:

$$f(y|\theta) = \binom{2}{y}\theta^y(1-\theta)^{2-y}, \; y = 0,1,2$$

$$f(\theta) = 10\theta/6, \; \theta = 0.1, 0.2, 0.3.$$

Note: $10\theta/6 = 1/6$, 2/6 and 3/6 for $\theta = 0.1$, 0.2 and 0.3, respectively.

Hence $f(\theta|y) \propto f(\theta)f(y|\theta)$

$$= \frac{10\theta}{6} \times \binom{2}{y}\theta^y(1-\theta)^{2-y}$$

$$\propto \theta \times \theta^2 \quad \text{since } y = 2.$$

Thus $f(\theta|y) \propto \theta^3 = \begin{cases} 0.1^3 = 1/1000, \theta = 0.1 \\ 0.2^3 = 8/1000, \theta = 0.2 \\ 0.3^3 = 27/1000, \theta = 0.3 \end{cases} \propto \begin{cases} 1, \theta = 0.1 \\ 8, \theta = 0.2 \\ 27, \theta = 0.3. \end{cases}$

Now, $1 + 8 + 27 = 36$, and so $f(\theta|y) = \begin{cases} 1^3/36 = 0.02778, \theta = 0.1 \\ 2^3/36 = 0.22222, \theta = 0.2 \\ 3^3/36 = 0.75, \theta = 0.3, \end{cases}$

which is the same result as obtained earlier in Exercise 1.4.

Exercise 1.6 Buses

You are visiting a town with buses whose licence plates show their numbers consecutively from 1 up to however many there are. In your mind the number of buses could be anything from one to five, with all possibilities equally likely.

Whilst touring the town you first happen to see Bus 3.

Assuming that at any point in time you are equally likely to see any of the buses in the town, how likely is it that the town has at least four buses?

Solution to Exercise 1.6

Let θ be the number of buses in the town and let y be the number of the bus that you happen to first see. Then an appropriate Bayesian model is:
$$f(y|\theta) = 1/\theta, \ y = 1,...,\theta$$
$$f(\theta) = 1/5, \ \theta = 1,...,5 \quad \text{(prior)}.$$

Note: We could also write this model as:
$$(y|\theta) \sim DU(1,...,\theta)$$
$$\theta \sim DU(1,...,5),$$
where *DU* denotes the *discrete uniform distribution*. (See Appendix B.9 for details regarding this distribution. Appendix B also provides details regarding some other important distributions that feature in this book.)

So the posterior density of θ is
$$f(\theta|y) \propto f(\theta)f(y|\theta)$$
$$\propto 1 \times 1/\theta, \ \ \theta = y,...,5 .$$

Noting that $y = 3$, we have that
$$f(\theta|y) \propto \begin{cases} 1/3, \theta = 3 \\ 1/4, \theta = 4 \\ 1/5, \theta = 5. \end{cases}$$

Now, $1/3 + 1/4 + 1/5 = (20 + 15 + 12)/60 = 47/60$, and so
$$f(\theta|y) = \begin{cases} \dfrac{1/3}{47/60} = \dfrac{20}{47}, \theta = 3 \\[2mm] \dfrac{1/4}{47/60} = \dfrac{15}{47}, \theta = 4 \\[2mm] \dfrac{1/5}{47/60} = \dfrac{12}{47}, \theta = 5. \end{cases}$$

So the posterior probability that the town has at least four buses is
$$P(\theta \geq 4|y) = \sum_{\theta:\theta \geq 4} f(\theta|y) = f(\theta = 4|y) + f(\theta = 5|y)$$
$$= 1 - f(\theta = 3|y) = 1 - \frac{20}{47} = \frac{27}{47} = 0.5745.$$

Discussion

This exercise is a variant of the famous 'tramcar problem' considered by Harold Jeffreys in his book *Theory of Probability* and previously suggested to him by M.H.A. Newman (see Jeffreys, 1961, page 238). Suppose that before entering the town you had *absolutely no idea* about the number of buses θ. Then, according to Jeffreys' logic, a prior which may be considered as suitably uninformative (or noninformative) in this situation is given by $f(\theta) \propto 1/\theta$, $\theta = 1, 2, 3, \ldots$.

Now, this prior density is problematic because it is *improper* (since $\sum_{\theta=1}^{\infty} 1/\theta = \infty$). However, it leads to a *proper* posterior density given by

$$f(\theta \mid y) = \frac{1}{c\theta^2}, \ \theta = 3, 4, 5, \ldots,$$

where $c = \dfrac{1}{3^2} + \dfrac{1}{4^2} + \dfrac{1}{5^2} + \ldots = \dfrac{\pi^2}{6} - \left(\dfrac{1}{1^2} + \dfrac{1}{2^2}\right) = 0.394934.$

So, under this alternative prior, the probability of there being at least four buses in the town (given that you have seen Bus 3) works out as

$$P(\theta \geq 4 \mid y) = 1 - P(\theta = 3 \mid y) = 1 - \frac{1}{9c} = 0.7187.$$

The logic which Jeffreys used to come up with the prior $f(\theta) \propto 1/\theta$ in relation to the tramcar problem will be discussed further in Chapter 2.

R Code for Exercise 1.6

```
options(digits=6); c=(1/6)*(pi^2)-5/4; c # 0.394934
1- (1/3^2)/c # 0.718659
```

Exercise 1.7 Balls in a box

In each of nine indistinguishable boxes there are nine balls, the ith box having i red balls and $9 - i$ white balls ($i = 1, \ldots, 9$).

One box is selected randomly from the nine, and then three balls are chosen randomly from the selected box (without replacement and without looking at the remaining balls in the box).

Exactly two of the three chosen balls are red. Find the probability that the selected box has at least four red balls remaining in it.

Solution to Exercise 1.7

Let: N = the number of balls in each box (9)

n = the number of balls chosen from the selected box (3)

θ = the number of red balls initially in the selected box

(1,2,...,8 or 9)

y = the number of red balls amongst the n chosen balls (2).

Then an appropriate Bayesian model is:

$(y \mid \theta) \sim Hyp(N, \theta, n)$ (Hypergeometric with parameters

N, θ and n, and having mean $n\theta / N$)

$\theta \sim DU(1, ..., N)$ (discrete uniform over the integers 1,2,...,N).

For this model, the posterior density of θ is

$$f(\theta \mid y) \propto f(\theta) f(y \mid \theta) = \frac{1}{N} \times \binom{\theta}{y}\binom{N-\theta}{n-y} \bigg/ \binom{N}{n}$$

$$\propto \frac{\theta!(N-\theta)!}{(\theta-y)!(N-\theta-(n-y))!}, \quad \theta = y, ..., N-(n-y).$$

In our case,

$$f(\theta \mid y) \propto \frac{\theta!(9-\theta)!}{(\theta-2)!(9-\theta-(3-2))!}, \quad \theta = 2, ..., 9-(3-2),$$

or more simply,

$$f(\theta \mid y) \propto \theta(\theta-1)(9-\theta), \quad \theta = 2, ..., 8.$$

Thus $f(\theta \mid y) \propto \begin{cases} 14, \theta = 2 \\ 36, \theta = 3 \\ 60, \theta = 4 \\ 80, \theta = 5 \\ 90, \theta = 6 \\ 84, \theta = 7 \\ 56, \theta = 8 \end{cases} \equiv k(\theta),$

where

$$c \equiv \sum_{\theta=1}^{8} k(\theta) = 14 + 36 + ... + 56 = 420.$$

$$\text{So } f(\theta \mid y) = \frac{k(\theta)}{c} = \begin{cases} 14 / 420 = 0.03333, \theta = 2 \\ 36 / 420 = 0.08571, \theta = 3 \\ 60 / 420 = 0.14286, \theta = 4 \\ 80 / 420 = 0.19048, \theta = 5 \\ 90 / 420 = 0.21429, \theta = 6 \\ 84 / 420 = 0.20000, \theta = 7 \\ 56 / 420 = 0.13333, \theta = 8. \end{cases}$$

The probability that the selected box has at least four red balls remaining is the posterior probability that θ (the number of red balls initially in the box) is at least 6 (since two red balls have already been taken out of the box). So the required probability is

$$P(\theta \geq 6 \mid y) = \frac{90 + 84 + 56}{420} = \frac{23}{42} = 0.5476.$$

R Code for Exercise 1.7

```
tv=2:8; kv=tv*(tv-1)*(9-tv); c=sum(kv); c # 420
options(digits=4);  cbind(tv,kv,kv/c,cumsum(kv/c))
# [1,] 2  14  0.03333  0.03333
# [2,] 3  36  0.08571  0.11905
# [3,] 4  60  0.14286  0.26190
# [4,] 5  80  0.19048  0.45238
# [5,] 6  90  0.21429  0.66667
# [6,] 7  84  0.20000  0.86667
# [7,] 8  56  0.13333  1.00000

23/42 # 0.5476
1-0.45238 # 0.5476  (alternative calculation of the required probability)
sum((kv/c)[tv>=6]) #  0.5476
                # (yet another calculation of the required probability)
```

1.7 Continuous parameters

The examples above have all featured a target parameter which is *discrete*. The following example illustrates Bayesian inference involving a *continuous* parameter. This case presents no new problems, except that the prior and posterior densities of the parameter may no longer be interpreted directly as probabilities.

Exercise 1.8 The *binomial-beta model* (or *beta-binomial model*)

Consider the following Bayesian model:
$$(y \mid \theta) \sim Binomial(n, \theta)$$
$$\theta \sim Beta(\alpha, \beta) \quad \text{(prior)}.$$

Find the posterior distribution of θ.

Solution to Exercise 1.8

The posterior density is
$$f(\theta \mid y) \propto f(\theta) f(y \mid \theta)$$
$$= \frac{\theta^{\alpha-1}(1-\theta)^{\beta-1}}{B(\alpha, \beta)} \times \binom{n}{y} \theta^{y}(1-\theta)^{n-y}$$
$$\propto \theta^{\alpha-1}(1-\theta)^{\beta-1} \times \theta^{y}(1-\theta)^{n-y} \quad \text{(ignoring constants which}$$
$$\text{do not depend on } \theta)$$
$$= \theta^{(\alpha+y)-1}(1-\theta)^{(\beta+n-y)-1}, \ 0 < \theta < 1.$$

This is the kernel of the beta density with parameters $\alpha + y$ and $\beta + n - y$. It follows that the posterior distribution of θ is given by
$$(\theta \mid y) \sim Beta(\alpha + y, \beta + n - y),$$
and the posterior density of θ is (exactly)
$$f(\theta \mid y) = \frac{\theta^{(\alpha+y)-1}(1-\theta)^{(\beta+n-y)-1}}{B(\alpha + y, \beta + n - y)}, \ 0 < \theta < 1.$$

For example, suppose that $\alpha = \beta = 1$, that is, $\theta \sim Beta(1,1)$.

Then the prior density is $f(\theta) = \dfrac{\theta^{1-1}(1-\theta)^{1-1}}{B(1,1)} = 1, \ 0 < \theta < 1$.

Thus the prior may also be expressed by writing $\theta \sim U(0,1)$.

Also, suppose that $n = 2$. Then there are three possible values of y, namely 0, 1 and 2, and these lead to the following three posteriors, respectively:
$$(\theta \mid y) \sim Beta(1+0, 1+2-0) = Beta(1,3)$$
$$(\theta \mid y) \sim Beta(1+1, 1+2-1) = Beta(2,2)$$
$$(\theta \mid y) \sim Beta(1+2, 1+2-2) = Beta(3,1).$$

These three posteriors and the prior are illustrated in Figure 1.5.

Note: The prior here may be considered uninformative because it is 'flat' over the entire range of possible values for θ, namely 0 to 1. This prior was originally used by Thomas Bayes and is often called the *Bayes prior*. However, other uninformative priors have been proposed for the binomial parameter θ. These will be discussed later, in Chapter 2.

Figure 1.5 The prior and three posteriors in Exercise 1.8

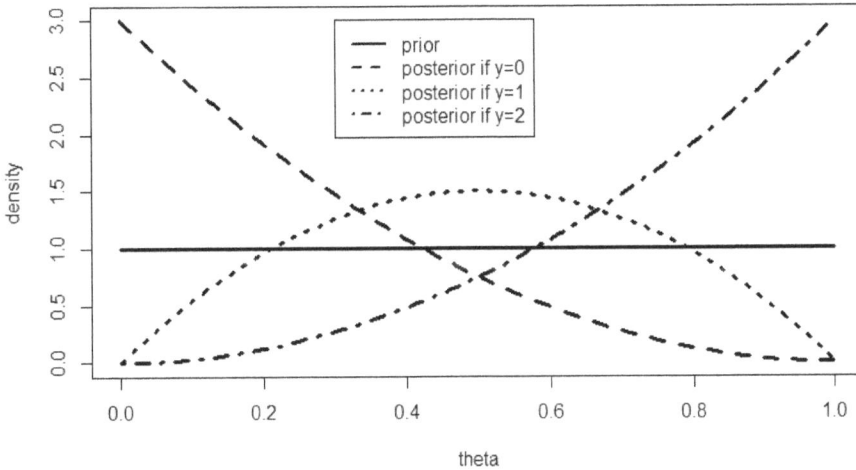

R Code for Exercise 1.8

```
X11(w=8,h=5); par(mfrow=c(1,1));

plot(c(0,1),c(0,3),type="n",xlab="theta",ylab="density")

lines(c(0,1),c(1,1),lty=1,lwd=3); tv=seq(0,1,0.01)
lines(tv,3*(1-tv)^2,lty=2,lwd=3)
lines(tv,3*2*tv*(1-tv),lty=3,lwd=3)
lines(tv,3*tv^2,lty=4,lwd=3)

legend(0.3,3,c("prior","posterior if y=0","posterior if y=1","posterior if y=2"),
    lty=c(1,2,3,4),lwd=rep(2,4))
```

1.8 Finite and infinite population inference

In the last example (Exercise 1.8), with the model:

$(y|\theta) \sim Binomial(n,\theta)$

$\theta \sim Beta(\alpha,\beta)$,

the quantity of interest θ is the probability of success on a single Bernoulli trial.

This quantity may be thought of as the average of a hypothetically *infinite* number of Bernoulli trials. For that reason we may refer to derivation of the posterior distribution,

$(\theta|y) \sim Beta(\alpha+y,\beta+n-y)$,

as *infinite population inference*.

In contrast, for the 'buses' example further above (Exercise 1.6), which involves the model:

$f(y|\theta) = 1/\theta, \ y = 1,...,\theta$

$f(\theta) = 1/5, \ \theta = 1,...,5$,

the quantity of interest θ represents the number of buses in a population of buses, which of course is *finite*.

Therefore derivation of the posterior,

$$f(\theta|y) = \begin{cases} 20/47, \ \theta = 3 \\ 15/47, \ \theta = 4 \\ 12/47, \ \theta = 5, \end{cases}$$

may be termed *finite population inference*.

Another example of finite population inference is the 'balls in a box' example (Exercise 1.7), where the model is:

$(y|\theta) \sim Hyp(N,\theta,n)$

$\theta \sim DU(1,...,N)$,

and where the quantity of interest θ is the number of red balls initially in the selected box (1,2,...,8 or 9).

And another example of *infinite* population inference is the 'loaded dice' example (Exercises 1.4 and 1.5), where the model is:

$$f(y|\theta) = \binom{2}{y} \theta^y (1-\theta)^{2-y}, \ y = 0,1,2$$

$f(\theta) = 10\theta/6, \ \theta = 0.1, 0.2, 0.3$,

and where the quantity of interest θ is the probability of 6 coming up on a single roll of the chosen die (i.e. the average number of 6s that come up on a hypothetically infinite number of rolls of that particular die).

Generally, finite population inference may also be thought of in terms of *prediction* (e.g. in the 'buses' example, we are *predicting* the total number of buses in the town). For that reason, finite population inference may also be referred to as *predictive inference*. Yet another term for finite population inference is *descriptive inference*. In contrast, infinite population inference may also be called *analytic inference*. More will be said on finite population/predictive/descriptive inference in later chapters of the course.

1.9 Continuous data

So far, all the Bayesian models considered have featured data which is modelled using a *discrete* distribution. (Some of these models have a discrete parameter and some have a continuous parameter.) The following is an example with data that follows a *continuous* probability distribution. (This example also has a continuous parameter.)

Exercise 1.9 The *exponential-exponential model*

Suppose θ has the standard exponential distribution, and the conditional distribution of y given θ is exponential with mean $1/\theta$. Find the posterior density of θ given y.

Solution to Exercise 1.9

The Bayesian model here is:
$$f(y \mid \theta) = \theta e^{-\theta y}, \, y > 0$$
$$f(\theta) = e^{-\theta}, \theta > 0.$$

So $f(\theta \mid y) \propto f(\theta) f(y \mid \theta) \propto e^{-\theta} \times \theta e^{-\theta y} = \theta^{2-1} e^{-\theta(y+1)}, \, y > 0.$

This is the kernel of a gamma distribution with parameters 2 and $y + 1$, as per the definitions in Appendix B.2. Thus we may write
$$(\theta \mid y) \sim Gamma(2, y + 1),$$
from which it follows that the posterior density of θ is
$$f(\theta \mid y) = \frac{(y+1)^2 \theta^{2-1} e^{-\theta(y+1)}}{\Gamma(2)}, \theta > 0.$$

Exercise 1.10 The *uniform-uniform model*

Consider the Bayesian model given by:
$$(y \mid \theta) \sim U(0, \theta)$$
$$\theta \sim U(0, 1).$$

Find the posterior density of θ given y.

Solution to Exercise 1.10

Noting that $0 < y < \theta < 1$, we see that the posterior density is
$$f(\theta \mid y) = \frac{f(\theta)f(y \mid \theta)}{f(y)} = \frac{1 \times (1/\theta)}{\int\limits_{y}^{1} 1 \times (1/\theta)d\theta}$$

$$= \frac{1/\theta}{\log 1 - \log y} = \frac{-1}{\theta \log y}, \; y < \theta < 1.$$

Note: This is a 'non-standard' density and strictly decreasing. To give a physical example, a stick of length 1 metre is cut at a point randomly located along its length. The part to the right of the cut is discarded and then another cut is made randomly along the stick which remains. Then the part to the right of that second cut is likewise discarded. The length of the stick remaining after the first cut is a random variable with density as given above, with y being the length of the finally remaining stick.

1.10 Conjugacy

When the prior and posterior distributions are members of the same class of distributions, we say that they form a *conjugate pair*, or that the prior is *conjugate*. For example, consider the binomial-beta model:
$$(y \mid \theta) \sim Binomial(n, \theta)$$
$$\theta \sim Beta(\alpha, \beta) \qquad \text{(prior)}$$
$$\Rightarrow \; (\theta \mid y) \sim Beta(\alpha + y, \beta + n - y) \qquad \text{(posterior)}.$$
Since both prior and posterior are beta, the prior is conjugate.

Likewise, consider the exponential-exponential model:
$$f(y \mid \theta) = \theta e^{-\theta y}, \; y > 0$$
$$f(\theta) = e^{-\theta}, \theta > 0 \quad \text{(i.e. } \theta \sim Gamma(1, 1)) \quad \text{(prior)}$$
$$\Rightarrow \; (\theta \mid y) \sim Gamma(2, y + 1) \qquad \text{(posterior)}.$$

Since both prior and posterior are gamma, the prior is conjugate.

On the other hand, consider the model in the buses example:
$$(y \mid \theta) \sim DU(1,...,\theta)$$
$$\theta \sim DU(1,...,5) \qquad \text{(prior)}$$
$$\Rightarrow \quad f(\theta \mid y = 3) = \begin{cases} 20/47, \ \theta = 3 \\ 15/47, \ \theta = 4 \\ 12/47, \ \theta = 5 \end{cases} \qquad \text{(posterior)}.$$

The prior is discrete uniform but the posterior is not. So in this case the prior is not conjugate.

Specifying a Bayesian model using a conjugate prior is generally desirable because it can simplify the calculations required.

1.11 Bayesian point estimation

Once the posterior distribution or density $f(\theta \mid y)$ has been obtained, Bayesian point estimates of the model parameter θ can be calculated. The three most commonly used point estimates are as follows.

- The *posterior mean* of θ is

$$E(\theta \mid y) = \int \theta \, dF(\theta \mid y) = \begin{cases} \int \theta f(\theta \mid y) d\theta & \text{if } \theta \text{ is continuous} \\ \sum_{\theta} \theta f(\theta \mid y) & \text{if } \theta \text{ is discrete.} \end{cases}$$

- The *posterior mode* of θ is
  $Mode(\theta \mid y)$ = any value $m \in \Re$ which satisfies
  $$f(\theta = m \mid x) = \max_{\theta} f(\theta \mid x)$$
  $$\text{or } \lim_{\theta \to m} f(\theta \mid x) = \sup f(\theta \mid x),$$
  or the set of all such values.

- The *posterior median* of θ is
  $Median(\theta \mid y)$ = any value m of θ such that
  $$P(\theta \le m \mid y) \ge 1/2$$
  $$\text{and } P(\theta \ge m \mid y) \ge 1/2,$$
  or the set of all such values.

Note 1: In some cases, the posterior mean does not exist or it is equal to infinity or minus infinity.

Note 2: Typically, the posterior mode and posterior median are unique. The above definitions are given for completeness.

Note 3: The integral $\int \theta dF(\theta \mid y)$ is a Lebesgue-Stieltje's integral. This may need to be evaluated as the sum of two separate parts in the case where θ has a mixed distribution. In the continuous case, it is useful to think of $dF(\theta \mid y)$ as $\dfrac{dF(\theta \mid y)}{d\theta} d\theta = f(\theta \mid y)d\theta$.

Note 4: The above three Bayesian point estimates may be interpreted in an intuitive manner. For example, θ's posterior mode is the value of θ which is 'made most likely by the data'. They may also be understood in the context of *Bayesian decision theory* (discussed later).

1.12 Bayesian interval estimation

There are many ways to construct a Bayesian interval estimate, but the two most common ways are defined as follows. The $1-\alpha$ (or $100(1-\alpha)\%$) *highest posterior density region* (HPDR) for θ is the smallest set S such that:

$$P(\theta \in S \mid y) \geq 1-\alpha$$

and $f(\theta_1 \mid y) \geq f(\theta_2 \mid y)$ if $\theta_1 \in S$ and $\theta_2 \notin S$.

Figure 1.6 illustrates the idea of the HPDR. In the very common situation where θ is scalar, continuous and has a posterior density which is unimodal with no local modes (i.e. has the form of a single 'mound'), the $1-\alpha$ HPDR takes on the form of a single interval defined by two points at which the posterior density has the same value. When the HPDR is a single interval, it is the shortest possible single interval over which the area under the posterior density is $1-\alpha$.

The $1-\alpha$ *central posterior density region* (CPDR) for a scalar parameter θ may be defined as the shortest single interval $[a,b]$ such that:

$$P(\theta < a \mid y) \leq \alpha/2$$

and $P(\theta > b \mid y) \leq \alpha/2$.

Figure 1.6 An 80% HPDR

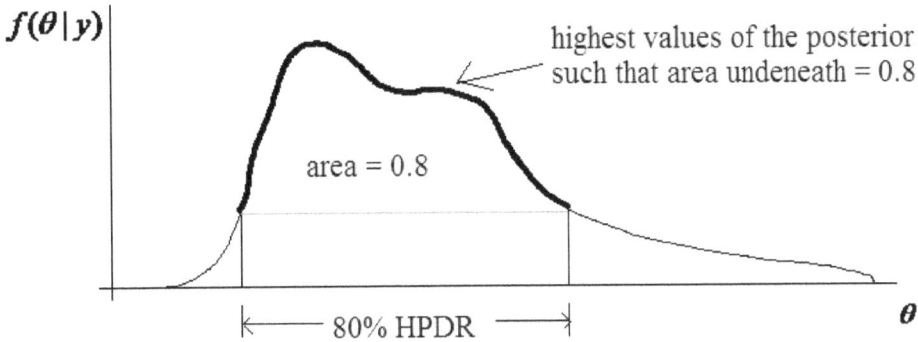

Figure 1.7 illustrates the idea of the CPDR. One drawback of the CPDR is that it is only defined for a *scalar* parameter. Another drawback is that some values *inside* the CPDR may be less likely *a posteriori* than some values *outside* it (which is not the case with the HPDR). For example, in Figure 1.7, a value *just below the upper bound* of the 80% CPDR has a smaller posterior density than a value *just below the lower bound* of that CPDR. However, CPDRs are typically easier to calculate than HPDRs.

In the common case of a continuous parameter with a posterior density in the form of a single 'mound' which is furthermore symmetric, the CPDR and HPDR are identical.

Note 1: The $1-\alpha$ CPDR for θ may alternatively be defined as the shortest single *open* interval (a,b) such that:
$$P(\theta \leq a \mid y) \leq \alpha/2$$
and $P(\theta \geq b \mid y) \leq \alpha/2$.

Other variations are possible (of the form $[a,b]$ and $(a,b]$); but when the parameter of interest θ is continuous these definitions are all equivalent. Yet another definition of the $1-\alpha$ CPDR is any of the CPDRs as defined above but with all *a posteriori* impossible values of θ excluded.

Note 2: As regards terminology, whenever the HPDR is a single interval, it may also be called the *highest posterior density interval* (HPDI). Likewise, the CPDR, which is always a single interval, may also be called the *central posterior density interval* (CPDI).

Figure 1.7 An 80% CPDR

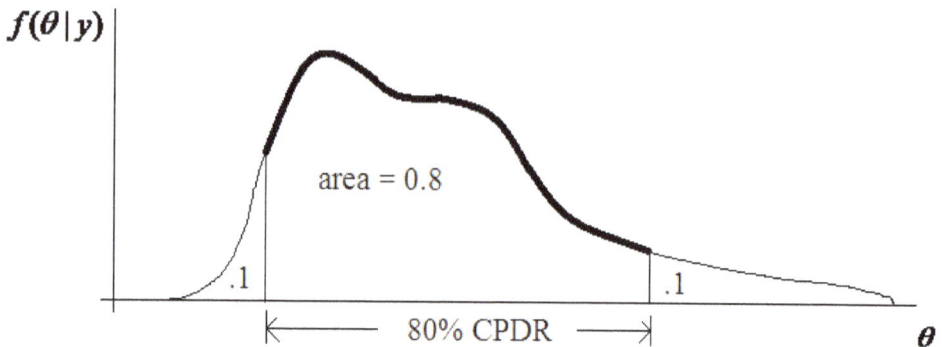

Exercise 1.11 A bent coin

We have a bent coin, for which θ, the probability of heads coming up, is unknown. Our prior beliefs regarding θ may be described by a standard uniform distribution. Thus no value of θ is deemed more or less likely than any other.

We toss the coin $n = 5$ times (independently), and heads come up every time.

Find the posterior mean, mode and median of θ. Also find the 80% HPDR and CPDR for θ.

Solution to Exercise 1.11

Recall the binomial-beta model:
$$(y \,|\, \theta) \sim Binomial(n, \theta)$$
$$\theta \sim Beta(\alpha, \beta),$$
for which $(\theta \,|\, y) \sim Beta(\alpha + y, \beta + n - y)$.

We now apply this result with $n = y = 5$ and $\alpha = \beta = 1$ (corresponding to $\theta \sim U(0,1)$), and find that:
$$(\theta \,|\, y) \sim Beta(1 + 5, 5 - 5 + 1) = Beta(6, 1)$$
$$f(\theta \,|\, y) = \frac{\theta^{6-1}(1-\theta)^{1-1}}{B(6,1)} = 6\theta^5, \quad 0 < \theta < 1$$
$$F(\theta \,|\, y) = \int_0^\theta 6t^5 \, dt = \theta^6, \quad 0 < \theta < 1.$$

Therefore: $\quad E(\theta \mid y) = \dfrac{6}{6+1} = \dfrac{6}{7} = 0.8571$

$$Mode(\theta \mid y) = \frac{6-1}{(6-1)+(1-1)} = 1$$

$Median(\theta \mid y) = $ solution in θ of $F(\theta \mid y) = 1/2$, i.e. $\theta^6 = 0.5$
$$= (0.5)^{1/6} = 0.8909.$$

Also, the 80% HPDR is $(0.2^{1/6}, 1) = (0.7647, 1)$ (since $f(\theta \mid y)$ is strictly increasing), and the 80% CPDR is $(0.1^{1/6}, 0.9^{1/6}) = (0.6813, 0.9826)$. The three point estimate and two interval estimates just derived are shown in Figure 1.8.

Figure 1.8 Inference in Exercise 1.11

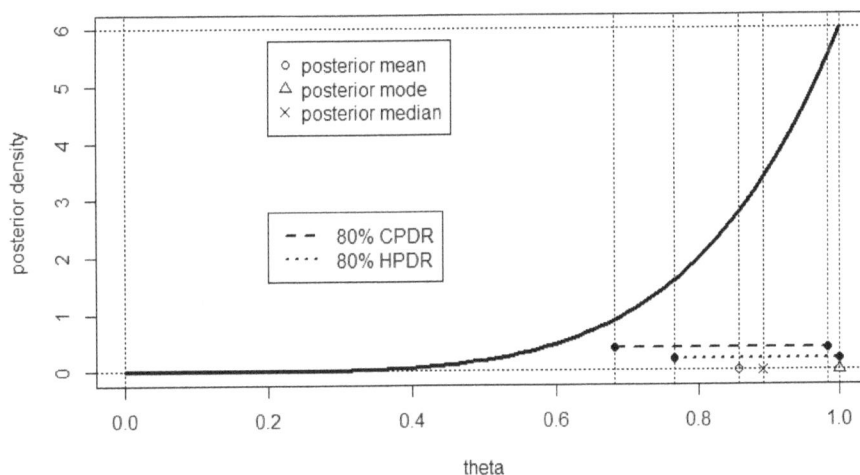

R Code for Exercise 1.11

```
options(digits=4); postmean=6/7; postmode=1; postmedian=0.5^(1/6)
c(postmean,postmode,postmedian) # 0.8571 1.0000 0.8909
hpdr=c(0.2^(1/6),1); cpdr=c(0.1,0.9)^(1/6)
c(hpdr,cpdr) # 0.7647 1.0000 0.6813 0.9826

X11(w=8,h=5); par(mfrow=c(1,1)); tv=seq(0,1,0.01); fv=dbeta(tv,6,1)
plot(tv,fv,type="l",lwd=3,xlab="theta",ylab="posterior density")
points(c(postmean,postmode,postmedian),c(0,0,0),pch=c(1,2,4))
points(hpdr,rep(0.2,2),pch=16); lines(hpdr,rep(0.2,2),lty=3,lwd=2)
```

```
points(cpdr,rep(0.4,2),pch=16); lines(cpdr,rep(0.4,2),lty=2,lwd=2)
abline(v=c(postmean,postmode,postmedian),lty=3)
abline(v=c(0,hpdr,cpdr),lty=3); abline(h=c(0,6),lty=3)
legend(0.2,5.8,c("posterior mean","posterior mode",
        "posterior median"),pch=c(1,2,4))
legend(0.2,2.8,c("80% CPDR","80% HPDR"),lty=c(2,3),lwd=c(2,2))
```

Exercise 1.12 HPDR and CPDR for a discrete parameter

Consider the posterior distribution from Exercise 1.7 (Balls in a box):

$$f(\theta \mid y) = \begin{cases} 14/420 = 0.03333, \ \theta = 2 \\ 36/420 = 0.08571, \ \theta = 3 \\ 60/420 = 0.14286, \ \theta = 4 \\ 80/420 = 0.19048, \ \theta = 5 \\ 90/420 = 0.21429, \ \theta = 6 \\ 84/420 = 0.20000, \ \theta = 7 \\ 56/420 = 0.13333, \ \theta = 8. \end{cases}$$

Find the 90% HPDR and 90% CPDR for θ. Also find the 50% HPDR and 50% CPDR for θ. For each region, calculate the associated exact coverage probability.

Solution to Exercise 1.12

The 90% HPDR is the set $\{3,4,5,6,7,8\}$;
this has exact coverage $1 - 14/420 = 0.9667$.

The 90% CPDR is the closed interval $[3, 8]$;
this likewise has exact coverage 0.9667.

The 50% HPDR is $\{5,6,7\}$;
this has exact coverage $(80 + 90 + 84)/420 = 0.6047$.

The 50% CPDR is $[4, 7]$;
this has exact coverage $(60 + 80 + 90 + 84)/420 = 0.7476$.

Note: The lower bound of the 50% CPDR cannot be equal to 5. This is because $P(\theta < 5 \mid y) = (14+36+60)/420 = 0.2619$, which is not less than or equal to $\alpha/2 = 0.25$, as required by the definition of CPDR.

Exercise 1.13 Illustration of the definition of HPDR

Suppose that the posterior probabilities of a parameter θ given data y are exactly 10%, 40% and 50% for values 1, 2 and 3, respectively. Find S, the 40% HPDR for θ.

Solution to Exercise 1.13

The smallest set S such that $P(\theta \in S \mid y) \geq 0.4$ is $\{2\}$ or $\{3\}$. With the additional requirement that $f(\theta_1 \mid y) \geq f(\theta_2 \mid y)$ if $\theta_1 \in S$ and $\theta_2 \notin S$, we see that $S = \{3\}$ (only). That is, the 40% HPDR is the singleton set $\{3\}$.

1.13 Inference on functions of the model parameter

So far we have examined Bayesian models with a single parameter θ and described how to perform posterior inference on that parameter. Sometimes there may also be interest in some *function* of the model parameter, denoted by (say)

$$\psi = g(\theta).$$

Then the posterior density of ψ can be derived using distribution theory, for example by applying the transformation rule,

$$f(\psi \mid y) = f(\theta \mid y) \left| \frac{d\theta}{d\psi} \right|,$$

in cases where $\psi = g(\theta)$ is strictly increasing or strictly decreasing.

Point and interval estimates of ψ can then be calculated in the usual way, using $f(\psi \mid y)$. For example, the posterior mean of ψ equals

$$E(\psi \mid y) = \int \psi f(\psi \mid y) d\psi.$$

Sometimes it is more practical to calculate point and interval estimates another way, without first deriving $f(\psi \mid y)$.

For example, another expression for the posterior mean is

$$E(\psi \mid y) = E(g(\theta) \mid y) = \int g(\theta) f(\theta \mid y) d\theta.$$

Also, the posterior median of ψ, call this M, can typically be obtained by simply calculating
$$M = g(m),$$
where m is the posterior median of θ.

Note: To see why this works, we write
$$P(\psi < M \mid y) = P(g(\theta) < M \mid y)$$
$$= P(g(\theta) < g(m) \mid y) = P(\theta < m \mid y) = 1/2.$$

Exercise 1.14 Estimation of an exponential mean

Suppose that θ has the standard exponential distribution, and y given θ is exponential with mean $1/\theta$. Find the posterior density and posterior mean of the model mean, $\psi = E(y \mid \theta) = 1/\theta$, given the data y.

Solution to Exercise 1.14

Recall that the Bayesian model
$$f(y \mid \theta) = \theta e^{-\theta y}, y > 0$$
$$f(\theta) = e^{-\theta}, \theta > 0$$
implies the posterior $(\theta \mid y) \sim Gamma(2, y+1)$.

So, by definition, $(\psi \mid y) \sim InverseGamma(2, y+1)$,

with density $f(\psi \mid y) = \dfrac{(y+1)^2 \psi^{-(2+1)} e^{-(y+1)/\psi}}{\Gamma(2)} = \dfrac{(y+1)^2}{\psi^3 e^{(y+1)/\psi}}, \psi > 0$,

and mean $E(\psi \mid y) = \dfrac{y+1}{2-1} = y+1$.

Note: This mean could also be obtained as follows:
$$E(\psi \mid y) = E\left(\frac{1}{\theta}\Big|\, y\right) = \int_0^\infty \frac{1}{\theta} f(\theta \mid y) d\theta$$

$$= \int_0^\infty \frac{1}{\theta} \times \frac{(y+1)^2 \theta^{2-1} e^{-\theta(y+1)}}{\Gamma(2)} d\theta$$

$$= \frac{\Gamma(1)(y+1)^2}{\Gamma(2)(y+1)^1} \int_0^\infty \frac{1}{\theta} \times \frac{(y+1)^1 \theta^{1-1} e^{-\theta(y+1)}}{\Gamma(1)} d\theta$$

$$= y+1 \quad \text{(using the fact that the last integral equals 1).}$$

Exercise 1.15 Inference on a function of the binomial parameter

Recall the binomial-beta model given by:

$$(y \mid \theta) \sim Binomial(n, \theta)$$
$$\theta \sim Beta(\alpha, \beta),$$

for which $(\theta \mid y) \sim Beta(\alpha + y, \beta + n - y)$.

Find the posterior mean, density function and distribution function of $\psi = \theta^2$ in the case where $n = 5$, $y = 5$, and $\alpha = \beta = 1$.

Note: In the context where we toss a bent coin five times and get heads every time (and the prior on the probability of heads is standard uniform), the quantity ψ may be interpreted as the *probability of the next two tosses both coming up heads*, or equivalently, as the proportion of times heads will come up twice if the coin is repeatedly tossed in groups of two tosses a hypothetically infinite number of times.

Solution to Exercise 1.15

Here, $(\theta \mid y) \sim Beta(1 + 5, 1 + 5 - 5) \sim Beta(6, 1)$
with pdf $f(\theta \mid y) = 6\theta^5, 0 < \theta < 1$.

Now $\theta = \psi^{1/2}$ and so, by the transformation method, the posterior density function of ψ is

$$f(\psi \mid y) = f(\theta \mid y) \left| \frac{d\theta}{d\psi} \right| = 6\psi^{5/2} \left| -\frac{1}{2}\psi^{-\frac{1}{2}} \right| = 3\psi^2, 0 < \psi < 1.$$

It follows that the posterior mean of ψ is

$$\hat{\psi} = E(\psi \mid y) = \int_0^1 \psi\left(3\psi^2\right) d\psi = 0.75,$$

and the posterior distribution function of ψ is

$$F(\psi \mid y) = \int_0^\psi f(\psi = t \mid y) dt = \int_0^\psi 3t^2 dt = \psi^3, 0 < \psi < 1.$$

Note 1: The posterior mean of $\psi = \theta^2$ can also be obtained by writing

$$\hat{\psi} = E(\theta^2 \mid y) = \int_0^1 \theta^2 \left(6\theta^5\right) d\theta = 0.75$$

or $\quad \hat{\psi} = E(\theta^2 \mid y) = V(\theta \mid y) + \{E(\theta \mid y)\}^2$

$$= \frac{6 \times 1}{(6+1)^2(6+1+1)} + \left(\frac{6}{6+1}\right)^2 = 0.75$$

or $\quad (\psi \mid y) \sim Beta(3,1) \implies \hat{\psi} = E(\psi \mid y) = 3/(3+1) = 0.75.$

Note 2: The distribution function of $\psi = \theta^2$ can also be obtained by writing

$$F(\psi = v \mid y) = P(\psi \leq v \mid y) = P(\theta^2 \leq v \mid y) = P(\theta \leq v^{1/2} \mid y)$$

$$= F(\theta = v^{1/2} \mid y) = \left[\theta^6 \Big|_{\theta = v^{1/2}} \right] = v^3, \, 0 < v < 1.$$

Note 3: In the above, $f(\psi = t \mid y)$ denotes the pdf of ψ given y, but evaluated at t. This pdf could also be written as $f_\psi(t \mid y)$ or as $\left[f(\psi \mid y) \big|_{\psi = t} \right]$. Likewise, $F(\psi = v \mid y) \equiv F_\psi(v \mid y) \equiv \left[F(\psi \mid y) \big|_{\psi = v} \right]$.

1.14 Credibility estimates

In actuarial studies, a *credibility estimate* is one which can be expressed as a weighted average of the form
$$C = (1 - k)A + kB,$$
where:

A is the *subjective estimate* (or the *collateral data estimate*)
B is the *objective estimate* (or the *direct data estimate*)
k is the *credibility factor*, a number that is between 0 and 1 (inclusive) and represents the weight assigned to the objective estimate.

A high value of k implies $C \cong B$, representing a situation where the objective estimate is assigned 'high credibility'. A primary aim of *credibility theory* is to determine an appropriate value or formula for k, as is done, for example, in the theory of the *Bühlmann model* (Bühlmann, 1967). Many Bayesian models lead to a point estimate which can be expressed as an intuitively appealing credibility estimate.

Exercise 1.16 Credibility estimation in the binomial-beta model

Consider the binomial-beta model: $(y \mid \theta) \sim Binomial(n, \theta)$
$$\theta \sim Beta(\alpha, \beta).$$

Express the posterior mean of θ as a credibility estimate and discuss.

Solution to Exercise 1.16

Earlier we showed that
$$(\theta \mid y) \sim Beta(\alpha + y, \beta + n - y),$$
and hence that the posterior mean of θ is
$$\hat{\theta} = E(\theta \mid y) = \frac{(\alpha + y)}{(\alpha + y) + (\beta + n - y)} = \frac{\alpha + y}{\alpha + \beta + n}.$$

Observe that the prior mean of θ is $E\theta = \alpha / (\alpha + \beta)$, and the maximum likelihood estimate (MLE) of θ is y/n. This suggests that we write
$$\hat{\theta} = \frac{\alpha}{\alpha + \beta + n} + \frac{y}{\alpha + \beta + n}$$
$$= \frac{\alpha}{\alpha + \beta + n}\left(\frac{\alpha + \beta}{\alpha}\right)\left(\frac{\alpha}{\alpha + \beta}\right) + \frac{n}{\alpha + \beta + n}\left(\frac{y}{n}\right)$$
$$= \frac{\alpha + \beta}{\alpha + \beta + n}\left(\frac{\alpha}{\alpha + \beta}\right) + \frac{n}{\alpha + \beta + n}\left(\frac{y}{n}\right).$$

Thus $\hat{\theta} = (1 - k)A + kB$

where: $A = \frac{\alpha}{\alpha + \beta}$, $B = \frac{y}{n}$, $k = \frac{n}{\alpha + \beta + n}$.

We see that the posterior mean $\hat{\theta}$ is a credibility estimate in the form of a weighted average of the prior mean $A = E\theta = \alpha / (\alpha + \beta)$ and the MLE $B = y / n$, where the weight assigned to the MLE is the credibility factor given by $k = n / (n + \alpha + \beta)$. Observe that as n increases, the credibility factor k approaches 1. This makes sense: if there is a lot of data then the prior should not have much influence on the estimation.

Figure 1.9 illustrates this idea by showing relevant densities, likelihoods and estimates for the following two cases, respectively:

(a) $n = 5$, $y = 4$, $\alpha = 2$, $\beta = 6$

(b) $n = 20$, $y = 16$, $\alpha = 2$, $\beta = 6$.

In both cases, the prior mean is the same ($A = 2/(2 + 6) = 0.25$), as is the MLE ($B = 4/5 = 16/20 = 0.8$). However, due to n being larger in case (b) (i.e. there being more direct data), case (b) leads to a larger credibility factor (0.714 compared to 0.385) and hence a posterior mean closer to the MLE (0.643 compared to 0.462).

Note: Each likelihood function in Figure 1.9 has been normalised so that the area underneath it is exactly 1. This means that in each case (a) and (b), the likelihood function $L(\theta)$ as shown is identical to the posterior density which would be implied by the standard uniform prior, i.e. under $f_{U(0,1)}(\theta) = f_{Beta(1,1)}(\theta)$. Thus, $L(\theta) = f_{Beta(1+y,1+n-y)}(\theta)$.

Figure 1.9 Illustration for Exercise 1.16

Legend: solid line = prior, dashed line = likelihood, dotted line = posterior, circle = prior mean, triangle = MLE, cross = posterior mean

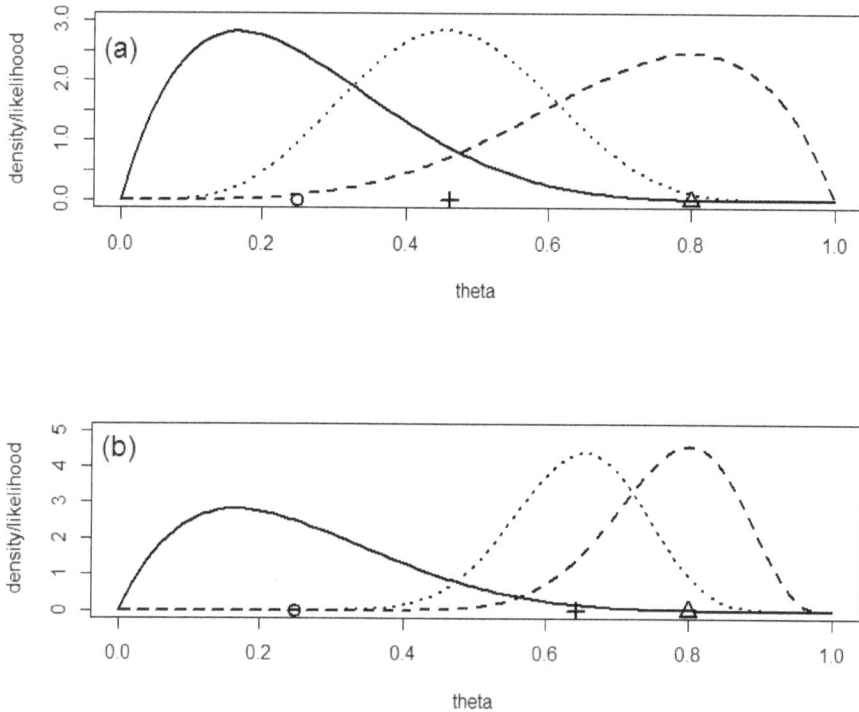

R Code for Exercise 1.16

```
X11(w=8,h=7); par(mfrow=c(2,1))

alp=2; bet=6; n = 5; y = 4; pvec=seq(0,1,0.01)
plot(c(0,1),c(0,3),type="n",xlab="theta",ylab="density/likelihood")
lines(pvec,dbeta(pvec,alp,bet),lty=1,lwd=2)
lines(pvec,dbeta(pvec,1+y,n-y+1),lty=2,lwd=2)
lines(pvec,dbeta(pvec,alp+y,n-y+bet),lty=3,lwd=2)

points(c(alp/(alp+bet), y/n,(alp+y)/(alp+bet+n)),c(0,0,0),pch=c(1,2,3),
       cex=rep(1.5,3),lwd=2);  text(0,2.5,"(a)",cex=1.5)
c(alp/(alp+bet), y/n,(alp+y)/(alp+bet+n)) # 0.2500000 0.8000000 0.4615385
n/(alp+bet+n) # 0.3846154

alp=2; bet=6; n = 20; y = 16; pvec=seq(0,1,0.01)
plot(c(0,1),c(0,5),type="n",xlab="theta",ylab="density/likelihood")
lines(pvec,dbeta(pvec,alp,bet),lty=1,lwd=2)
lines(pvec,dbeta(pvec,1+y,n-y+1),lty=2,lwd=2)
lines(pvec,dbeta(pvec,alp+y,n-y+bet),lty=3,lwd=2)

points(c(alp/(alp+bet), y/n,(alp+y)/(alp+bet+n)),c(0,0,0),pch=c(1,2,3),
       cex=rep(1.5,3),lwd=2);  text(0,4.5,"(b)",cex=1.5)
c(alp/(alp+bet), y/n,(alp+y)/(alp+bet+n)) # 0.2500000 0.8000000 0.6428571
n/(alp+bet+n) # 0.7142857
```

Exercise 1.17 Further credibility estimation in the binomial-beta model

Consider the binomial-beta model:
$$(Y \mid \theta) \sim Binomial(n, \theta)$$
$$\theta \sim Beta(\alpha, \beta).$$

If possible, express the posterior *mode* of θ as a credibility estimate.

Solution to Exercise 1.17

Since $(\theta \mid y) \sim Beta(\alpha + y, \beta + n - y)$, the posterior mode of θ is
$$Mode(\theta \mid y) = \frac{(\alpha + y - 1)}{(\alpha + y - 1) + (\beta + n - y - 1)} = \frac{\alpha + y - 1}{\alpha + \beta + n - 2}.$$

Now, the prior mode of θ is $Mode(\theta) = \dfrac{(\alpha-1)}{(\alpha-1)+(\beta-1)} = \dfrac{\alpha-1}{\alpha+\beta-2}$.

So we write $Mode(\theta \mid y) = \dfrac{\alpha-1}{\alpha+\beta+n-2} + \dfrac{y}{\alpha+\beta+n-2}$

$$= \frac{\alpha-1}{\alpha+\beta+n-2}\left(\frac{\alpha+\beta-2}{\alpha-1}\right)\left(\frac{\alpha-1}{\alpha+\beta-2}\right) + \frac{n}{\alpha+\beta+n-2}\left(\frac{y}{n}\right).$$

We see that the posterior mode is a credibility estimate of the form
$$Mode(\theta \mid y) = (1-c)\,Mode(\theta) + c\hat{\theta},$$

where: $Mode(\theta) = \dfrac{\alpha-1}{\alpha+\beta-2}$ is the prior mode

$\hat{\theta} = \dfrac{y}{n}$ is the maximum likelihood estimate

(mode of the likelihood function)

$c = \dfrac{n}{n+\alpha+\beta-2}$ is the credibility factor

(assigned to the direct data estimate, $\hat{\theta}$).

Exercise 1.18 The *normal-normal model*

Consider the following Bayesian model:
$$(y_1,\ldots,y_n \mid \mu) \sim iid\ N(\mu,\sigma^2)$$
$$\mu \sim N(\mu_0,\sigma_0^2),$$
where σ^2, μ_0 and σ_0^2 are known or specified constants.

Find the posterior distribution of μ given data in the form of the vector $y = (y_1,\ldots,y_n)$.

Solution to Exercise 1.18

The posterior density of μ is
$$f(\mu \mid y) \propto f(\mu)f(y \mid \mu)$$

$$\overset{\mu}{\propto} \exp\left\{-\frac{1}{2}\left(\frac{\mu-\mu_0}{\sigma_0}\right)^2\right\} \times \prod_{i=1}^{n}\exp\left\{-\frac{1}{2}\left(\frac{y_i-\mu}{\sigma}\right)^2\right\}$$

$$= \exp\left(-\frac{1}{2}\left[\frac{1}{\sigma_0^2}\left(\mu^2 - 2\mu\mu_0 + \mu_0^2\right) + \frac{1}{\sigma^2}\left(\sum_{i=1}^{n} y_i^2 - 2\mu n\bar{y} + n\mu^2\right)\right]\right), \quad (1.1)$$

where $\bar{y} = (y_1 + \ldots + y_n)/n$ is the sample mean.

We see that the posterior density of μ is proportional to the exponent of a quadratic in μ. That is,

$$f(\mu \mid y) \propto \exp\left(-\frac{1}{2\sigma_*^2}\left(\mu - \mu_*\right)^2\right), \quad (1.2)$$

which then implies that

$$(\mu \mid y) \sim N(\mu_*, \sigma_*^2),$$

for some constants μ_* and σ_*^2.

It remains to find the normal mean and variance parameters, μ_* and σ_*^2. (These must be functions of the known quantities n, $\bar{y}$, σ, μ_0 and σ_0.)

One way to obtain these parameters which completely define μ's posterior distribution is to complete the square in the exponent of (1.2). To this end we write

$$f(\mu \mid y) \propto \exp\left(-\frac{1}{2}q\right),$$

where

$$q = \frac{1}{\sigma_0^2}\left(\mu^2 - 2\mu\mu_0\right) + \frac{1}{\sigma^2}\left(-2\mu n\bar{y} + n\mu^2\right)$$

(ignoring constants with respect to μ)

$$= \mu^2\left(\frac{1}{\sigma_0^2} + \frac{n}{\sigma^2}\right) - 2\mu\left(\frac{\mu_0}{\sigma_0^2} + \frac{n\bar{y}}{\sigma^2}\right) + c$$

(where c is a constant with respect to μ)

$$= a\mu^2 - 2b\mu + c \quad \text{where} \quad a = \frac{1}{\sigma_0^2} + \frac{n}{\sigma^2} \quad \text{and} \quad b = \frac{\mu_0}{\sigma_0^2} + \frac{n\bar{y}}{\sigma^2}$$

$$= a\left(\mu^2 - 2\frac{b}{a}\mu\right) + c \quad = a\left(\mu^2 - 2\left(\frac{b}{a}\right)\mu + \left(\frac{b}{a}\right)^2\right) + c'$$

(where c' is a constant with respect to μ)

$$= \frac{1}{1/a}\left(\mu - \frac{b}{a}\right)^2 + c'.$$

Thus, $\quad f(\mu \mid y) \propto \exp\left(-\frac{1}{2(1/a)}\left(\mu - \frac{b}{a}\right)^2\right).$ $\qquad$ (1.3)

So, equating (1.2) and (1.3), we obtain:

$$\sigma_*^2 = \frac{1}{a} = \frac{1}{\dfrac{1}{\sigma_0^2} + \dfrac{n}{\sigma^2}} = \frac{\sigma^2 \sigma_0^2}{\sigma^2 + n\sigma_0^2}$$

$$\mu_* = \frac{b}{a} = \frac{\dfrac{\mu_0}{\sigma_0^2} + \dfrac{n\bar{y}}{\sigma^2}}{\dfrac{1}{\sigma_0^2} + \dfrac{n}{\sigma^2}} = \frac{\sigma^2 \mu_0 + n\sigma_0^2 \bar{y}}{\sigma^2 + n\sigma_0^2}. \qquad (1.4)$$

Note 1: A little algebra (left as an additional exercise) shows that the posterior mean can also be written as

$$\mu_* = (1-k)\mu_0 + k\bar{y},$$

and the posterior variance can be written as

$$\sigma_*^2 = k\frac{\sigma^2}{n},$$

where

$$k = \frac{n}{n + \dfrac{\sigma^2}{\sigma_0^2}}.$$

We see that μ's posterior mean is a credibility estimate in the form of a weighted average of the prior mean μ_0 and the sample mean $\bar{y}$ (which is also the maximum likelihood estimate), with the weight assigned to $\bar{y}$ being the credibility factor, k. More will be said on this further down.

Note 2: Another way to derive μ_* and σ_*^2 is to write (1.2) as

$$f(\mu \mid y) \propto \exp\left(-\frac{1}{2\sigma_*^2}\left(\mu^2 - 2\mu\mu_* + \mu_*^2\right)\right) \qquad (1.5)$$

and then equate coefficients of powers of μ in (1.1) and (1.5). This logic leads to $\dfrac{1}{\sigma_*^2} = \dfrac{1}{\sigma_0^2} + \dfrac{n}{\sigma^2}$ and $\dfrac{\mu_*}{\sigma_*^2} = \dfrac{\mu_0}{\sigma_0^2} + \dfrac{n\bar{y}}{\sigma^2}$ and ultimately the same formulae for μ_* and σ_*^2 as given by (1.4).

Note 3: Since both prior and posterior are normal, the prior is *conjugate*.

Note 4: The posterior mean, mode and median of μ are the same and equal to μ_*. The $1 - \alpha$ CPDR and $1 - \alpha$ HPDR for μ are the same and equal to $(\mu_* \pm z_{\alpha/2}\sigma_*)$.

Note 5: The posterior distribution of μ depends on the data vector $y = (y_1, \ldots, y_n)$ only by way of the sample mean, i.e. $\bar{y} = (y_1 + \ldots + y_n)/n$. Therefore, the main result, $(\mu \mid y) \sim N(\mu_*, \sigma_*^2)$, also implies that $(\mu \mid \bar{y}) \sim N(\mu_*, \sigma_*^2)$.

That is, if we know only the sample mean $\bar{y}$, the posterior distribution of μ is the same as if we know y, i.e. all n sample values. Knowing the individual y_i values makes no difference to the inference.

Note 6: The formula for the credibility factor in Note 1, namely
$$k = \frac{n}{n + \dfrac{\sigma^2}{\sigma_0^2}} = \frac{1}{1 + \dfrac{\sigma^2/n}{\sigma_0^2}},$$
makes sense in the following ways:

(i) If the prior standard deviation σ_0 is *small* then $k \approx 0$, so that $\mu_* \approx \mu_0$ and $\sigma_* \approx \sigma_0$. Therefore $(\mu \mid y) \sim N(\mu_0, \sigma_0^2)$.

That is, if the prior information is very 'precise' or 'definite', the data has little influence on the posterior. So the posterior is approximately equal to the prior; i.e. $f(\mu \mid y) \approx f(\mu)$, or equivalently, $(\mu \mid y) \sim \mu$. In this case the posterior mean, mode and median of μ are approximately equal to μ_0. Also, the $1 - \alpha$ CPDR and $1 - \alpha$ HPDR for μ are approximately equal to $(\mu_0 \pm z_{\alpha/2}\sigma_0)$.

(ii) If σ_0 is *large* then $k \approx 1$, so that $\mu_* \approx \bar{y}$, $\sigma_*^2 \approx \sigma^2/n$, and so $(\mu \mid y) \sim N(\bar{y}, \sigma^2/n)$.

That is, a large σ_0 corresponds to a highly disperse prior, reflecting little prior information and so little influence of the prior distribution (as specified by μ_0 and σ_0) on the inference. In this case the posterior mean, mode and median of μ are approximately equal to $\bar{y}$. Also, the $1-\alpha$ CPDR and $1-\alpha$ HPDR for μ are approximately equal to $(\bar{y} \pm z_{\alpha/2}\sigma/\sqrt{n})$. Thus, inference is almost the same as implied by the classical approach.

(iii) If the sample size n is *large* then $k \approx 1$, so that $\mu_* \approx \bar{y}$ and $\sigma_*^2 \approx \sigma^2/n$. Therefore $(\mu \mid y) \dot\sim N(\bar{y}, \sigma^2/n)$.

So, in this case, just as when σ_0 is large, the prior distribution has very little influence on the posterior, and the ensuing inference is almost the same as that implied by the classical approach.

Note 7: In the case of *a priori* ignorance (meaning no prior information at all) it is customary to take $\sigma_0 = \infty$, which implies that
$$\mu \sim N(0, \infty).$$

This prior on μ appears to be problematic, because it is *improper*. However, it meaningfully leads to a *proper* posterior, namely
$$(\mu \mid y) \sim N(\bar{y}, \sigma^2/n),$$
which then leads to the same point and interval estimates implied by the classical approach, namely the MLE $\bar{y}$ and $1-\alpha$ CI $(\bar{y} \pm z_{\alpha/2}\sigma/\sqrt{n})$.

The improper prior $\mu \sim N(0, \infty)$ may be described as 'flat' or 'uniform over the whole real line' and can also be written as
$$\mu \sim U(-\infty, \infty)$$
or $f(\mu) \propto 1, \mu \in \Re$.

In some cases (more complicated models not considered here), using an improper prior may lead to an improper posterior, which then becomes problematic. For more information on this topic, see Hobert and Casella (1996).

Summary: For the *normal-normal model*, defined by:
$$(y_1, \ldots, y_n \mid \mu) \sim iid \; N(\mu, \sigma^2)$$
$$\mu \sim N(\mu_0, \sigma_0^2),$$
the posterior distribution of the normal mean μ is given by
$$(\mu \mid y) \sim N(\mu_*, \sigma_*^2),$$
where: $\mu_* = (1-k)\mu_0 + k\bar{y}$

$$\sigma_*^2 = k\frac{\sigma^2}{n}$$

$$k = \frac{n}{n + \sigma^2 / \sigma_0^2} \quad \text{(the \textit{normal-normal model credibility factor}).}$$

The posterior mean, mode and median of μ are all equal to μ_*, and the $1-\alpha$ CPDR and HPDR for μ are both $(\mu_* \pm z_{\alpha/2}\sigma_*)$.

In the case of *a priori* ignorance it is appropriate to set $\sigma_0 = \infty$.

This defines an improper prior
$$f(\mu) \propto 1, \mu \in \Re$$
and the proper posterior
$$(\mu \mid y) \sim N(\bar{y}, \sigma^2 / n).$$

Exercise 1.19 Practice with the normal-normal model

In the context of the normal-normal model, given by:
$$(y_1, \ldots, y_n \mid \mu) \sim iid \; N(\mu, \sigma^2)$$
$$\mu \sim N(\mu_0, \sigma_0^2),$$
suppose that $y = (8.4, 10.1, 9.4)$, $\sigma = 1$, $\mu_0 = 5$ and $\sigma_0 = 1/2$.

Calculate the posterior mean, mode and median of μ.

Also calculate the 95% CPDR and 95% HPDR for μ.

Create a graph which shows these estimates as well as the prior density, prior mean, likelihood, MLE and posterior density.

Solution to Exercise 1.19

Here: $n = 3$,

$$\bar{y} = (8.4 + 10.1 + 9.4)/3 = 9.3$$

$$k = \frac{1}{1 + \dfrac{1^2/3}{(1/2)^2}} = \frac{3}{7} = 0.4285714$$

$$\mu_* = \left(1 - \frac{3}{7}\right)5 + \frac{3}{7} \times 9.3 = 6.8428571$$

$$\sigma_*^2 = \frac{3}{7} \times \frac{1^2}{3} = \frac{1}{7} = 0.1428571.$$

So the posterior mean/mode/median is

$$\mu_* = 6.84286,$$

and the 95% CPDR/HPDR is

$$(\mu_* \pm z_{0.025}\sigma_*) = (6.84286 \pm 1.96\sqrt{0.14286})$$
$$= (6.102, 7.584).$$

Figure 1.10 shows the various densities and estimates here, as well as the normalised likelihood. Note that the likelihood function as shown is also the posterior density if the prior is taken to be uniform over the whole real line, i.e. $\mu \sim U(-\infty, \infty)$.

Discussion

If we change σ_0 from 0.5 to 2 we get $k = 0.923$ and results as illustrated in Figure 1.11.

If we change σ_0 from 0.5 to 0.25 we get $k = 0.158$ and results as illustrated in Figure 1.12 (page 46).

If we keep σ_0 as 0.5 but change σ from 1 to 2 we get $k = 0.158$ and results as illustrated in Figure 1.13 (page 46).

Note that the posteriors in Figures 1.12 and 1.13 have the same mean but different variances.

Figure 1.10 Results if $\sigma_0 = 0.5$, $\sigma = 1$, $k = n/(n + \sigma^2/\sigma_0^2) = 0.429$

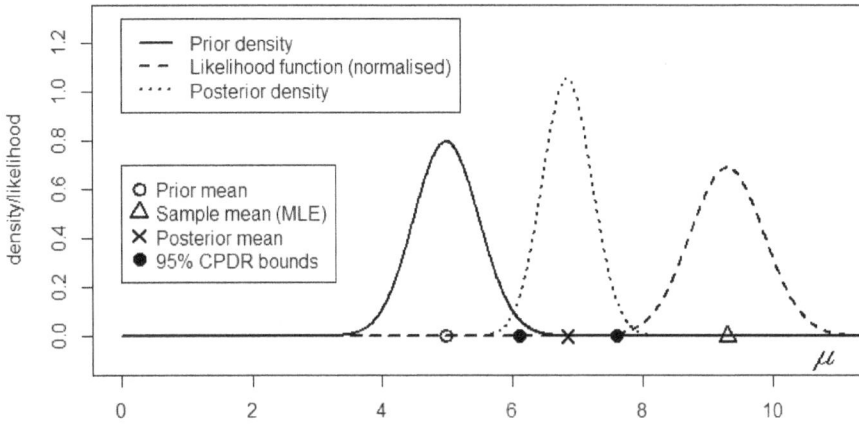

Figure 1.11 Results if $\sigma_0 = 2$, $\sigma = 1$, $k = n/(n + \sigma^2/\sigma_0^2) = 0.9223$

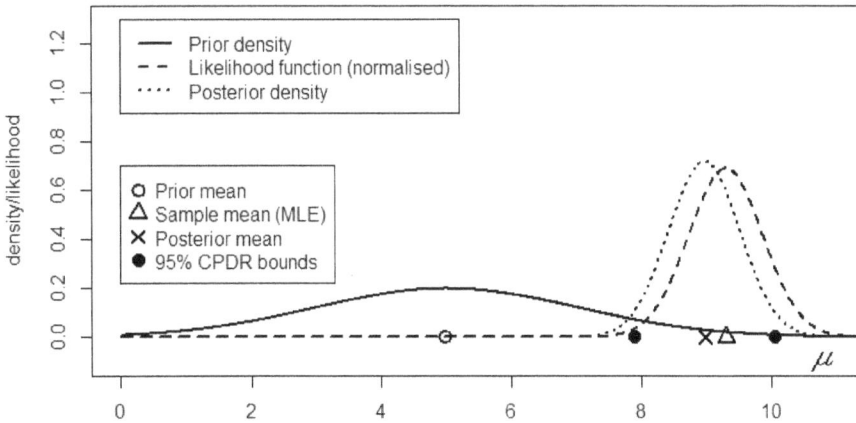

Figure 1.12 Results if $\sigma_0 = 0.25$, $\sigma = 1$, $k = n/(n + \sigma^2/\sigma_0^2) = 0.158$

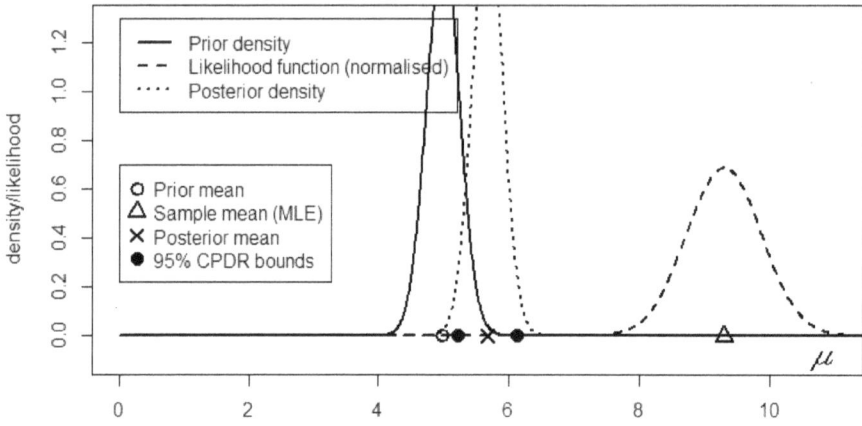

Figure 1.13 Results if $\sigma_0 = 0.5$, $\sigma = 2$, $k = n/(n + \sigma^2/\sigma_0^2) = 0.158$

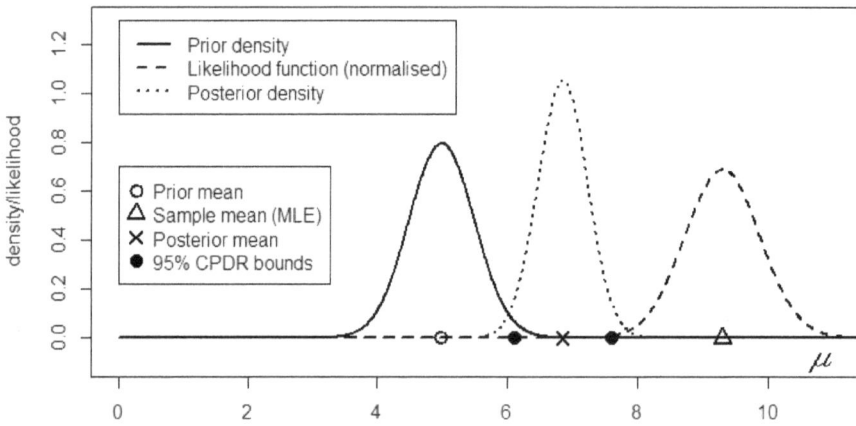

R Code for Exercise 1.19

```
X11(w=8,h=5); par(mfrow=c(1,1)); mu0=5; sig0=0.5; sig=1

y = c(8.4, 10.1, 9.4); n = length(y); k=1/(1+(sig^2/n)/sig0^2); k # 0.4285714
ybar=mean(y); ybar # 9.3
mus = (1-k)*mu0 + k*ybar; sigs2=k*sig^2/n
c(mus,sigs2) # 6.8428571 0.1428571
muv=seq(0,15,0.01)
prior = dnorm(muv,mu0,sig0); post=dnorm(muv,mus,sqrt(sigs2))
like = dnorm(muv,ybar,sig/sqrt(n))
cpdr=mus+c(-1,1)*qnorm(0.975)*sqrt(sigs2)
cpdr # 6.102060 7.583654

plot(c(0,11),c(-0.1,1.3),type="n",xlab="",ylab="density/likelihood")
lines(muv,prior,lty=1,lwd=2); lines(muv,like,lty=2,lwd=2)
lines(muv,post,lty=3,lwd=2)
points(c(mu0,ybar,mus),c(0,0,0),pch=c(1,2,4),cex=rep(1.5,3),lwd=2)
points(cpdr,c(0,0),pch=rep(16,2),cex=rep(1.5,2))
legend(0,1.3,
    c("Prior density","Likelihood function (normalised)","Posterior density"),
        lty=c(1,2,3),lwd=c(2,2,2))
legend(0,0.7,c("Prior mean","Sample mean (MLE)","Posterior mean",
    "95% CPDR bounds"), pch=c(1,2,4,16),pt.cex=rep(1.5,4),pt.lwd=rep(2,4))
text(10.8,-0.075,"m", vfont=c("serif symbol","italic"), cex=1.5)

# Repeat above with sig0=2 to obtain Figure 1.11
# Repeat above with sig0=0.25 to obtain Figure 1.12
# Repeat above with sig0=0.5 and sig=2 to obtain Figure 1.13
```

Exercise 1.20 The *normal-gamma model*

Consider the following Bayesian model:

$$(y_1,\ldots,y_n \mid \lambda) \sim iid\ N(\mu, 1/\lambda)$$
$$\lambda \sim G(\alpha, \beta)\,.$$

Find the posterior distribution of λ given $y = (y_1,\ldots,y_n)$.

Note 1: In the normal-normal model, the normal mean μ is unknown and the normal variance σ^2 is known. Now we consider the same Bayesian model but with those roles reversed, i.e. with μ known and σ^2 unknown. For an example of where this kind of situation might arise, see Byrne and Dracoulis (1985).

Note 2: For reasons of mathematical convenience and conjugacy, we parameterise the normal distribution here via the *precision parameter*
$$\lambda = 1/\sigma^2$$
rather than using σ^2 directly as before in the normal-normal model.

Note 3: An equivalent formulation of the *normal-gamma model* being considered here is:
$$(y_1,\ldots,y_n \mid \sigma^2) \sim iid\ N(\mu,\sigma^2)$$
$$\sigma^2 \sim IG(\alpha,\beta),$$
where this may be called the *normal-inverse-gamma* model.

Solution to Exercise 1.20

The posterior density of λ is
$$f(\lambda \mid y) \propto f(\lambda) f(y \mid \lambda)$$

$$\propto \lambda^{\alpha-1} e^{-\beta\lambda} \times \prod_{i=1}^{n} \frac{1}{1/\sqrt{\lambda}} \exp\left\{-\frac{1}{2}\left(\frac{y_i - \mu}{1/\sqrt{\lambda}}\right)^2\right\}$$

$$= \lambda^{\alpha-1} e^{-\beta\lambda} \times \lambda^{n/2} \exp\left\{-\frac{\lambda}{2}\sum_{i=1}^{n}(y_i - \mu)^2\right\}$$

$$= \lambda^{a-1} e^{-b\lambda} \quad \text{for some } a \text{ and } b.$$

We see that
$$(\lambda \mid y) \sim G(a,b),$$

where: $\quad a = \alpha + \dfrac{n}{2}$

$$b = \beta + \frac{n}{2}s_\mu^2$$

$$s_\mu^2 = \frac{1}{n}\sum_{i=1}^{n}(y_i - \mu)^2.$$

Note 1: The posterior mean of λ, namely

$$E(\lambda \mid y) = \frac{a}{b} = \frac{\alpha + n/2}{\beta + ns_{\mu}^2/2},$$

converges to $\hat{\lambda} = \dfrac{1}{s_{\mu}^2}$ (the MLE of λ) as $n \to \infty$.

If $\alpha = \beta = 0$ then $E(\lambda \mid y) = \hat{\lambda}$ exactly for all n.

Note 2: Unlike the posterior mean of μ in the normal-normal model, the posterior mean of λ cannot be expressed as a credibility estimate of the form

$$(1 - c)\lambda_0 + c\hat{\lambda},$$

where: $\lambda_0 = E\lambda = \dfrac{\alpha}{\beta}$ (the prior mean of λ)

$\hat{\lambda} = \dfrac{1}{s_{\mu}^2}$ (the MLE of λ).

Note 3: We may write the posterior as

$$(\lambda \mid y) \sim G\left(\frac{2\alpha + n}{2}, \frac{2\beta + ns_{\mu}^2}{2}\right).$$

It can then be shown via the method of transformations that

$$(u \mid y) \sim G\left(\frac{2\alpha + n}{2}, \frac{1}{2}\right) \sim \chi^2(2\alpha + n),$$

where $u = (2\beta + ns_{\mu}^2)\lambda$.

So the $1 - A$ CPDR for u is $\left(\chi_{1-A/2}^2(2\alpha + n), \chi_{A/2}^2(2\alpha + n)\right)$.

So the $1 - A$ CPDR for $\lambda = \dfrac{u}{2\beta + ns_{\mu}^2}$ is $\left(\dfrac{\chi_{1-A/2}^2(2\alpha + n)}{2\beta + ns_{\mu}^2}, \dfrac{\chi_{A/2}^2(2\alpha + n)}{2\beta + ns_{\mu}^2}\right)$.

So the $1 - A$ CPDR for $\sigma^2 = \dfrac{1}{\lambda}$ is $\left(\dfrac{2\beta + ns_{\mu}^2}{\chi_{A/2}^2(2\alpha + n)}, \dfrac{2\beta + ns_{\mu}^2}{\chi_{1-A/2}^2(2\alpha + n)}\right)$.

If $\alpha = \beta = 0$, this is exactly the same as the classical $1 - A$ CI for σ^2.

Note 4: The classical $1 - A$ CI for σ^2 may be derived as follows. First consider all parameters fixed as constants. Then

$$\frac{y_1 - \mu}{\sigma}, ..., \frac{y_n - \mu}{\sigma} \sim iid\ N(0,1).$$

So

$$\left(\frac{y_1 - \mu}{\sigma}\right)^2, ..., \left(\frac{y_n - \mu}{\sigma}\right)^2 \sim iid\ \chi^2(1).$$

So

$$\sum_{i=1}^{n}\left(\frac{y_i - \mu}{\sigma}\right)^2 = \frac{ns_\mu^2}{\sigma^2} \sim \chi^2(n).$$

So

$$1 - A = P\left(\chi^2_{1-A/2}(n) < \frac{ns_\mu^2}{\sigma^2} < \chi^2_{A/2}(n)\right)$$

$$= P\left(\frac{ns_\mu^2}{\chi^2_{A/2}(n)} < \sigma^2 < \frac{ns_\mu^2}{\chi^2_{1-A/2}(n)}\right).$$

Note 5: Notes 1 to 3 indicate that in the case of *a priori* ignorance, a reasonable specification is

$$\alpha = \beta = 0,$$

or equivalently,

$$f(\lambda) \propto 1/\lambda,\ \lambda > 0.$$

This improper prior may be thought of as the limiting case as $\varepsilon \to 0$ of the proper prior

$$\lambda \sim Gam(\varepsilon, \varepsilon),$$

where $\varepsilon \approx 0$.

Observe that

$$E\lambda = \varepsilon / \varepsilon = 1$$

for all ε, and

$$V\lambda = \varepsilon / \varepsilon^2 \to \infty$$

as $\varepsilon \to 0$.

Summary: For the *normal-gamma model*, defined by:
$$(y_1, \ldots, y_n \mid \lambda) \sim iid \ N(\mu, 1/\lambda)$$
$$\lambda \sim G(\alpha, \beta),$$
the posterior distribution of λ is given by
$$(\lambda \mid y) \sim G(a, b),$$
where: $\quad a = \alpha + \dfrac{n}{2}, \qquad b = \beta + \dfrac{n}{2} s_\mu^2, \qquad s_\mu^2 = \dfrac{1}{n} \sum_{i=1}^{n} (y_i - \mu)^2.$

The posterior mean of λ is a/b. The posterior median is $F_{G(a,b)}^{-1}(1/2)$.

The posterior mode of λ is $(a-1)/b$ if $a > 1$; otherwise that mode is 0.

The $1 - A$ CPDR for λ is $\left(F_{G(a,b)}^{-1}(A/2), F_{G(a,b)}^{-1}(1 - A/2) \right)$

and may also be written as $\left(\dfrac{\chi_{1-A/2}^2(2\alpha + n)}{2\beta + ns_\mu^2}, \dfrac{\chi_{A/2}^2(2\alpha + n)}{2\beta + ns_\mu^2} \right).$

The $1 - A$ CPDR for $\sigma^2 = 1/\lambda$ is $\left(\dfrac{2\beta + ns_\mu^2}{\chi_{A/2}^2(2\alpha + n)}, \dfrac{2\beta + ns_\mu^2}{\chi_{1-A/2}^2(2\alpha + n)} \right).$

In the case of *a priori* ignorance it is appropriate to set $\alpha = \beta = 0$.
This defines an improper prior with density
$$f(\lambda) \propto 1/\lambda, \ \lambda > 0,$$
and a proper posterior distribution given by
$$(ns_\mu^2 \lambda \mid y) \sim \chi^2(n).$$

Exercise 1.21 Practice with the normal-gamma model

In the context of the normal-gamma model, given by:
$$(y_1, \ldots, y_n \mid \lambda) \sim iid \ N(\mu, 1/\lambda)$$
$$\lambda \sim Gamma(\alpha, \beta),$$
suppose that $y = (8.4, 10.1, 9.4)$, $\mu = 8$, $\alpha = 3$ and $\beta = 2$.

(a) Calculate the posterior mean, mode and median of the model precision λ. Also calculate the 95% CPDR for λ. Create a graph which shows these estimates as well as the prior density, prior mean, likelihood, MLE and posterior density.

(b) Calculate the posterior mean, mode and median of the model variance $\sigma^2 = 1/\lambda$. Also calculate the 95% CPDR for σ^2. Create a graph which shows these estimates as well as the prior density, prior mean, likelihood, MLE and posterior density.

(c) Calculate the posterior mean, mode and median of the model standard deviation σ. Also calculate the 95% CPDR for σ. Create a graph which shows these estimates as well as the prior density, prior mean, likelihood, MLE and posterior density.

(d) Examine each of the point estimates in (a), (b) and (c) and determine which ones, if any, can be easily expressed in the form of a credibility estimate.

Solution to Exercise 1.21

(a) The required posterior distribution is $(\lambda \mid y) \sim Gamma(a,b)$, where:

$$a = \alpha + \frac{n}{2} = 4.5, \; b = \beta + \frac{n}{2}s_\mu^2 = 5.265, \; s_\mu^2 = \frac{1}{n}\sum_{i=1}^{n}(y_i - \mu)^2 = 2.177.$$

So:
- the posterior mean of λ is $E(\lambda \mid y) = a/b = 0.8547$
- the posterior mode is $Mode(\lambda \mid y) = (a-1)/b = 0.6648$
- the posterior median is the 0.5 quantile of the $G(a,b)$ distribution and works out as $Median(\lambda \mid y) = 0.7923$
    (as obtained using the qgamma() function in R; see below)
- the 95% CPDR for λ is (0.2564, 1.8065) (where the bounds are the 0.025 and 0.975 quantiles of the $G(a,b)$ distribution).

Also:
- the prior mean is $E\lambda = \alpha/\beta = 1.5$
- the prior mode is $Mode(\lambda) = (\alpha-1)/\beta = 1$
- the prior median is $Median(\lambda) = 1.3370$
- the MLE of λ is $\hat{\lambda} = 1/s_\mu^2 = 0.4594$
    (note that this estimate is biased).

Figure 1.14 shows the various densities and estimates here, as well as the normalised likelihood function.

Note: The normalised likelihood function (with area below equal to 1) is the same as the posterior density of λ if the prior is taken to be uniform over the positive real line, i.e. $\lambda \sim U(0,\infty)$. This prior is specified by taking $\alpha = 1$ and $\beta = 0$, because then $f(\lambda) \propto \lambda^{1-1}e^{-0\lambda} \propto 1$.

Figure 1.14 Results for Exercise 1.21(a)

Inference on the model precision parameter

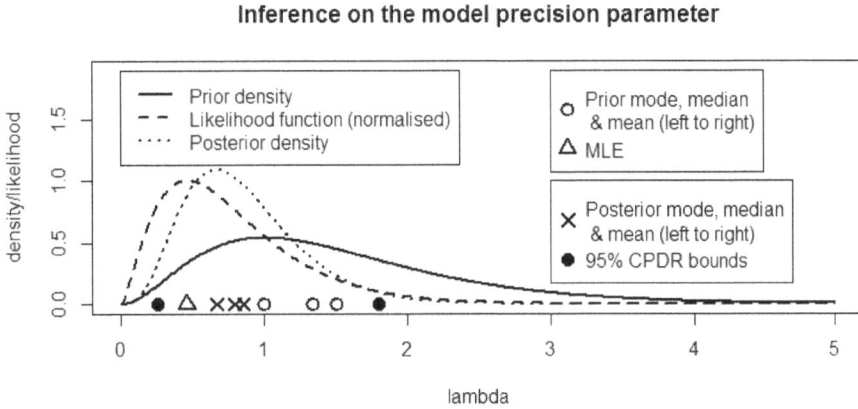

(b) As regards the model variance $\sigma^2 = 1/\lambda$ we note that $\sigma^2 \sim IG(\alpha,\beta)$ with density

$$f(\sigma^2) = f(\lambda)\left|\frac{d\lambda}{d\sigma^2}\right| \quad \text{where } \lambda = \left(\sigma^2\right)^{-1}$$

$$= \frac{\beta^\alpha[(\sigma^2)^{-1}]^{\alpha-1}e^{-\beta(\sigma^2)^{-1}}}{\Gamma(\alpha)}\left|-(\sigma^2)^{-2}\right|$$

$$= \frac{\beta^\alpha}{\Gamma(\alpha)}(\sigma^2)^{-\alpha-1}e^{-\beta/\sigma^2}, \sigma^2 > 0. \tag{1.6}$$

Then, by well-known properties of the inverse gamma distribution and maximum likelihood theory:

- the prior mean of σ^2 is $E\sigma^2 = \beta/(\alpha-1) = 1$
- the prior mode is $Mode(\sigma^2) = \beta/(\alpha+1) = 0.5$
- the prior median is $Median(\sigma^2) = 1/Median(\lambda) = 0.7479$
- the MLE of σ^2 is $\hat{\sigma}^2 = 1/\hat{\lambda} = s_\mu^2 = 2.1767$

   (note that this estimate is unbiased).

By analogy with the prior (1.6), we find that $(\sigma^2 \mid y) \sim IG(a,b)$ with density

$$f(\sigma^2 \mid y) = \frac{b^a}{\Gamma(a)}(\sigma^2)^{-a-1}e^{-b/\sigma^2}, \sigma^2 > 0,$$

and hence that:

- the posterior mean of σ^2 is $E(\sigma^2 \mid y) = b/(a-1) = 1.5043$
- the posterior mode is $Mode(\sigma^2 \mid y) = b/(a+1) = 0.9573$
- the posterior median is
    $$Median(\sigma^2 \mid y) = 1/Median(\lambda \mid y) = 1.2622$$
    (since $1/2 = P(\sigma^2 < m \mid y) = P(1/\lambda < m \mid y) = P(1/m < \lambda \mid y)$)
- the 95% CPDR for σ^2 is (0.5535, 3.8994)   (where the lower and upper bounds are the inverses of the 0.975 and 0.025 quantiles of the $G(a,b)$ distribution, respectively).

Figure 1.15 shows the various densities and estimates here, as well as the normalised likelihood function.

Note: The normalised likelihood function is the same as the posterior density of σ^2 if the prior on σ^2 is taken to be uniform over the positive real line, i.e. $f(\sigma^2) \propto 1, \sigma^2 > 0$. This prior is specified by $\lambda \sim G(-1,0)$, i.e. by $\alpha = -1$ and $\beta = 0$ as is evident from (1.6) above.

Figure 1.15 Results for Exercise 1.21(b)

Inference on the model variance parameter

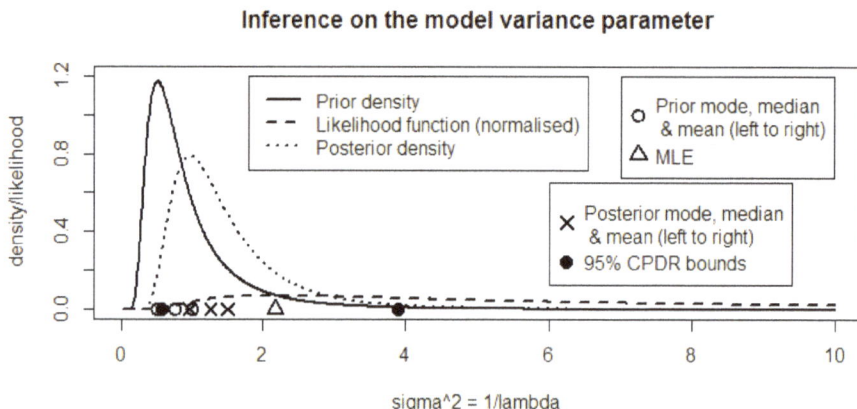

sigma^2 = 1/lambda

(c) As regards the model standard deviation $\sigma = 1/\sqrt{\lambda}$, observe that the prior density of this quantity is

$$f(\sigma) = f(\lambda)\left|\frac{d\lambda}{d\sigma}\right| \quad \text{where } \lambda = \sigma^{-2}$$

$$= \frac{\beta^{\alpha}(\sigma^{-2})^{\alpha-1}e^{-\beta\sigma^{-2}}}{\Gamma(\alpha)}\left|-2\sigma^{-3}\right| = \frac{2\beta^{\alpha}}{\Gamma(\alpha)}\sigma^{-2\alpha-1}e^{-\beta/\sigma^2}, \sigma > 0. \quad (1.7)$$

We find that:

- the prior mean of σ is

$$E\sigma = E\lambda^{-1/2} = \int_0^\infty \lambda^{-1/2}\frac{\beta^{\alpha}\lambda^{\alpha-1}e^{-\beta\lambda}}{\Gamma(\alpha)}d\lambda$$

$$= \frac{\beta^{\alpha}\Gamma(\alpha-1/2)}{\beta^{\alpha-1/2}\Gamma(\alpha)}\int_0^\infty \frac{\beta^{\alpha-\frac{1}{2}}\lambda^{\alpha-\frac{1}{2}-1}e^{-\beta\lambda}}{\Gamma(\alpha-1/2)}d\lambda$$

$$= \beta^{1/2}\frac{\Gamma(\alpha-1/2)}{\Gamma(\alpha)} = 0.9400$$

- the prior mode of σ is $Mode(\sigma) = \sqrt{\frac{2\beta}{2\alpha+1}} = 0.7559$

  (obtained by setting the derivative of the logarithm of (1.7) to zero, where that derivative is derived as follows:

$$l(\sigma) = \log f(\sigma) = -(2\alpha+1)\log\sigma - \beta\sigma^{-2} + \text{constant}$$

$$\Rightarrow l'(\sigma) = -\frac{2\alpha+1}{\sigma} + 2\beta\sigma^{-3} \overset{set}{=} 0 \Rightarrow \sigma^2 = \frac{2\beta}{2\alpha+1})$$

- the prior median of σ is $Median(\sigma) = \sqrt{Median(\sigma^2)} = 0.8648$
- the MLE of σ is $\hat{\sigma} = \sqrt{s_\mu^2} = 1.4754$ (which is biased).

By analogy with the above, $f(\sigma \mid y) = \frac{2b^a}{\Gamma(a)}\sigma^{-2a-1}e^{-b/\sigma^2}, \sigma > 0$.

So we find that:

- the posterior mean of σ is $E(\sigma \mid y) = b^{1/2}\frac{\Gamma(a-1/2)}{\Gamma(a)} = 1.1836$

- the posterior mode is $Mode(\sigma \mid y) = \sqrt{\frac{2b}{2a+1}} = 1.0262$

- the posterior median is

$$Median(\sigma \mid y) = \sqrt{Median(\sigma^2 \mid y)} = 1.1235$$

$$(\text{since } 1/2 = P(\sigma^2 < m \mid y) = P(\sigma < \sqrt{m} \mid y))$$

- the 95% CPDR for σ is $(0.7440, 1.9747)$ (where these bounds are the square roots of the bounds of the 95% CPDR for σ^2).

Figure 1.16 shows the various densities and estimates here, as well as the normalised likelihood function.

Note: The normalised likelihood function is the same as the posterior density of σ if the prior on σ is taken to be uniform over the positive real line, i.e. $f(\sigma) \propto 1, \sigma > 0$. This prior is specified by $\lambda \sim G(-1/2, 0)$, i.e. by $\alpha = -1/2$ and $\beta = 0$, as is evident from (1.7) above.

Figure 1.16 Results for Exercise 1.21(c)

Inference on the model standard deviation parameter

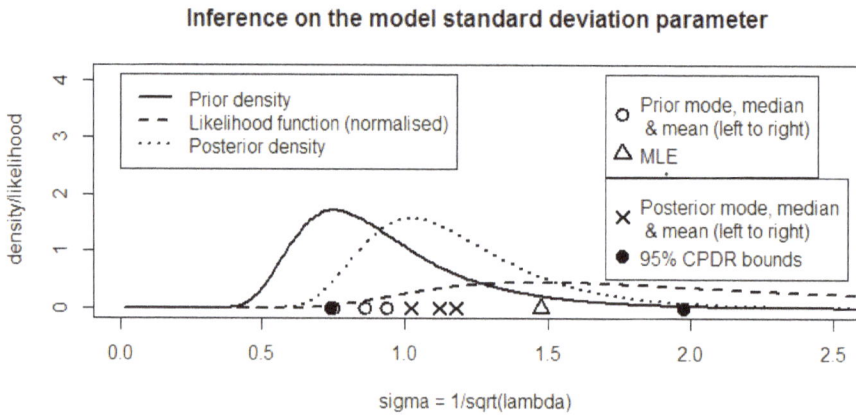

sigma = 1/sqrt(lambda)

(d) Considering the various point estimates of λ, σ^2 and σ derived above, we find that two of them can easily be expressed as credibility estimates, as follows. First, observe that

$$E(\sigma^2 \mid y) = \frac{b}{a-1} = \frac{\beta + ns_\mu^2/2}{\alpha + (n/2) - 1} = \frac{2\beta + ns_\mu^2}{2\alpha + n - 2}$$

$$= \left(\frac{n}{n+2\alpha-2}\right)s_\mu^2 + \frac{2\beta}{n+2\alpha-2},$$

where

$$\frac{2\beta}{n+2\alpha-2} = \frac{2\beta}{n+2\alpha-2} \times \frac{\alpha-1}{\beta} \times \frac{\beta}{\alpha-1} = \frac{2\alpha-2}{n+2\alpha-2} \times E\sigma^2.$$

We see that the posterior *mean* of σ^2 is a credibility estimate of the form

$$E(\sigma^2 \mid y) = (1-c)E\sigma^2 + cs_\mu^2,$$

where:

$E\sigma^2 = \dfrac{\beta}{\alpha-1}$ is the prior mean of σ^2

$s_\mu^2 = \dfrac{1}{n}\displaystyle\sum_{i=1}^{n}(y_i - \mu)^2$ is the MLE of σ^2

$c = \dfrac{n}{n+2\alpha-2}$ is the credibility factor (assigned to the MLE).

Likewise,

$$Mode(\sigma^2 \mid y) = \frac{b}{a+1} = \frac{\beta+ns_\mu^2/2}{\alpha+(n/2)+1} = \frac{2\beta+ns_\mu^2}{2\alpha+n+2}$$

$$= \left(\frac{n}{n+2\alpha+2}\right)s_\mu^2 + \frac{2\beta}{n+2\alpha+2},$$

where

$$\frac{2\beta}{n+2\alpha+2} = \frac{2\cancel{\beta}}{n+2\alpha+2} \times \frac{\alpha+1}{\cancel{\beta}} \times \frac{\beta}{\alpha+1}$$

$$= \frac{2\alpha+2}{n+2\alpha+2} \times Mode(\sigma^2)$$

$$= \left(1 - \frac{n}{n+2\alpha+2}\right)Mode(\sigma^2).$$

We see that the posterior *mode* of σ^2 is a credibility estimate of the form

$$Mode(\sigma^2 \mid y) = (1-d)Mode(\sigma^2) + ds_\mu^2,$$

where:

$Mode(\sigma^2) = \dfrac{\beta}{\alpha+1}$ is the prior mode of σ^2

$s_\mu^2 = \dfrac{1}{n}\displaystyle\sum_{i=1}^{n}(y_i - \mu)^2$ is the MLE of σ^2

(i.e. mode of the likelihood function)

$d = \dfrac{n}{n+2\alpha+2}$ is the credibility factor (assigned to the MLE).

R Code for Exercise 1.21

```
# (a) Inference on lambda ----------------------------------------------

y = c(8.4, 10.1, 9.4); n = length(y); mu=8;  alp=3; bet=2; options(digits=4)
a=alp+n/2; sigmu2=mean((y-mu)^2); b=bet+(n/2)*sigmu2

c(a,sigmu2,b) # 4.500 2.177 5.265

lampriormean=alp/bet; lamlikemode=1/sigmu2; lampriormode=(alp-1)/bet
lampriormedian= qgamma(0.5,alp,bet)
lampostmean=a/b; lampostmode=(a-1)/b; lampostmedian=qgamma(0.5,a,b)
lamcpdr=qgamma(c(0.025,0.975),a,b)

c(lampriormean,lamlikemode,lampriormode,lampriormedian,
        lampostmode,lampostmedian, lampostmean,lamcpdr)
  # 1.5000 0.4594 1.0000 1.3370 0.6648 0.7923 0.8547 0.2564 1.8065

lamv=seq(0,5,0.01); prior=dgamma(lamv,alp,bet)
post=dgamma(lamv,a,b);  like=dgamma(lamv,a-alp+1,b-bet+0)

X11(w=8,h=4); par(mfrow=c(1,1))

plot(c(0,5),c(0,1.9),type="n",
        main="Inference on the model precision parameter",
        xlab="lambda",ylab="density/likelihood")
lines(lamv,prior,lty=1,lwd=2); lines(lamv,like,lty=2,lwd=2);
lines(lamv,post,lty=3,lwd=2)
points(c(lampriormean,lampriormode, lampriormedian,
        lamlikemode,lampostmode,lampostmedian,lampostmean),
        rep(0,7),pch=c(1,1,1,2,4,4,4),cex=rep(1.5,7),lwd=2)
points(lamcpdr,c(0,0),pch=rep(16,2),cex=rep(1.5,2))

legend(0,1.9,
   c("Prior density","Likelihood function (normalised)","Posterior density"),
        lty=c(1,2,3),lwd=c(2,2,2))
legend(3,1.9,c("Prior mode, median\n & mean (left to right)",
        "MLE"), pch=c(1,2),pt.cex=rep(1.5,4),pt.lwd=rep(2,4))
legend(3,1,c("Posterior mode, median\n & mean (left to right)",
        "95% CPDR bounds"), pch=c(4,16),pt.cex=rep(1.5,4),pt.lwd=rep(2,4))
```

```
# (b)  Inference on sigma2 = 1/lambda --------------------------------------------------

sig2priormean=bet/(alp-1); sig2likemode=sigmu2; sig2priormode=bet/(alp+1)
sig2postmean=b/(a-1); sig2postmode=b/(a+1);
sig2postmedian=1/lampostmedian
sig2cpdr=1/qgamma(c(0.975,0.025),a,b); sig2priormedian= 1/lampriormedian

c(sig2priormean, sig2likemode, sig2priormode, sig2priormedian,
       sig2postmode, sig2postmedian, sig2postmean, sig2cpdr)
  # 1.0000 2.1767 0.5000 0.7479 0.9573 1.2622 1.5043 0.5535 3.8994

sig2v=seq(0.01,10,0.01); prior=dgamma(1/sig2v,alp,bet)/sig2v^2
post=dgamma(1/sig2v,a,b)/sig2v^2;
like=dgamma(1/sig2v,a-alp-1,b-bet+0)/sig2v^2

plot(c(0,10),c(0,1.2),type="n",
       main="Inference on the model variance parameter",
       xlab="sigma^2 = 1/lambda",ylab="density/likelihood")
lines(sig2v,prior,lty=1,lwd=2); lines(sig2v,like,lty=2,lwd=2)
lines(sig2v,post,lty=3,lwd=2)

points(c(sig2priormean, sig2priormode, sig2priormedian, sig2likemode,
       sig2postmode, sig2postmedian,sig2postmean),
       rep(0,7),pch=c(1,1,1,2,4,4,4),cex=rep(1.5,7),lwd=2)
points(sig2cpdr,c(0,0),pch=rep(16,2),cex=rep(1.5,2))

legend(1.8,1.2,
   c("Prior density","Likelihood function (normalised)","Posterior density"),
       lty=c(1,2,3),lwd=c(2,2,2))
legend(7,1.2,c("Prior mode, median\n & mean (left to right)",
       "MLE"), pch=c(1,2),pt.cex=rep(1.5,4),pt.lwd=rep(2,4))
legend(6,0.65,c("Posterior mode, median\n & mean (left to right)",
       "95% CPDR bounds"), pch=c(4,16),pt.cex=rep(1.5,4),pt.lwd=rep(2,4))

# abline(h=max(like),lty=3)  # Checking likelihood and MLE are consistent
# fun=function(t){ dgamma(1/t,a-alp-1,b-bet+0)/t^2  }
# integrate(f=fun,lower=0,upper=Inf)$value
       # 1  Checking likelihood is normalised
```

```
# (c)  Inference on sigma = 1/sqrt(lambda) ---------------------------------------------

sigpriormean=sqrt(bet)*gamma(alp-1/2)/gamma(alp);
siglikemode=sqrt(sigmu2); sigpriormode=sqrt(2*bet/(2*alp+1))
sigpostmean= sqrt(b)*gamma(a-1/2)/gamma(a)
sigpostmode= sqrt(2*b/(2*a+1)); sigpostmedian=sqrt(sig2postmedian)
sigcpdr=sqrt(sig2cpdr); sigpriormedian= sqrt(sig2priormedian)

c(sigpriormean, siglikemode, sigpriormode, sigpriormedian,
        sigpostmode, sigpostmedian, sigpostmean, sigcpdr)
  # 0.9400 1.4754 0.7559 0.8648 1.0262 1.1235 1.1836 0.7440 1.9747

sigv=seq(0.01,3,0.01); prior=dgamma(1/sigv^2,alp,bet)*2/sigv^3
post=dgamma(1/sigv^2,a,b)*2/sigv^3;
like=dgamma(1/sigv^2,a-alp-1/2,b-bet+0)*2/sigv^3

plot(c(0,2.5),c(0,4.1),type="n",
        main="Inference on the model standard deviation parameter",
        xlab="sigma = 1/sqrt(lambda)",ylab="density/likelihood")
lines(sigv,prior,lty=1,lwd=2)
lines(sigv,like,lty=2,lwd=2)
lines(sigv,post,lty=3,lwd=2)
points(c(sigpriormean, sigpriormode, sigpriormedian, siglikemode,
        sigpostmode, sigpostmedian,sigpostmean),
        rep(0,7),pch=c(1,1,1,2,4,4,4),cex=rep(1.5,7),lwd=2)
points(sigcpdr,c(0,0),pch=rep(16,2),cex=rep(1.5,2))

legend(0,4.1,
    c("Prior density","Likelihood function (normalised)","Posterior density"),
        lty=c(1,2,3),lwd=c(2,2,2))
legend(1.7,4.1,c("Prior mode, median\n & mean (left to right)",
        "MLE"), pch=c(1,2),pt.cex=rep(1.5,4),pt.lwd=rep(2,4))
legend(1.7,2.3,c("Posterior mode, median\n & mean (left to right)",
        "95% CPDR bounds"), pch=c(4,16),pt.cex=rep(1.5,4),pt.lwd=rep(2,4))
```

CHAPTER 2

Bayesian Basics Part 2

2.1 Frequentist characteristics of Bayesian estimators

Consider a Bayesian model defined by a likelihood $f(y|\theta)$ and a prior $f(\theta)$, leading to the posterior

$$f(\theta|y) = \frac{f(\theta)f(y|\theta)}{f(y)}.$$

Suppose that we choose to perform inference on θ by constructing a point estimate $\hat{\theta}$ (such as the posterior mean, mode or median) and a $(1-\alpha)$-level interval estimate $I = (L, U)$ (such as the CPDR or HPDR).

Then $\hat{\theta}$, I, L and U are functions of the data y and may be written $\hat{\theta}(y)$, $I(y)$, $L(y)$ and $U(y)$. Once these functions are defined, the estimates which they define stand on their own, so to speak, and may be studied from many different perspectives.

Naturally, the characteristics of these estimates may be seen in the context of the Bayesian framework in which they were constructed. More will be said on this below when we come to discuss *Bayesian decision theory*.

However, another important use of Bayesian estimates is as a *proxy* for *classical* estimates. We have already mentioned this in relation to the normal-normal model:

$$(y_1, \ldots, y_n | \mu) \sim iid \ N(\mu, \sigma^2)$$
$$\mu \sim N(\mu_0, \sigma_0^2),$$

where the use of a particular prior, namely the one specified by $\sigma_0 = \infty$, led to the point estimate $\hat{\mu} = \hat{\mu}(y) = \bar{y}$ and the interval estimate

$$I(y) = (L(y), U(y)) = (\bar{y} \pm z_{\alpha/2}\sigma / \sqrt{n}).$$

As we noted earlier, these estimates are exactly the same as the usual estimates used in the context of the corresponding classical model,

$$y_1, \ldots, y_n \sim iid \ N(\mu, \sigma^2),$$

where μ is an unknown constant and σ^2 is given.

Therefore, the frequentist operating characteristics of the Bayesian estimates are immediately known. In particular, we refer to the fact that the frequentist bias of $\hat{\mu}$ is zero, and the frequentist coverage probability of I is exactly $1 - \alpha$. These statements mean that the expected value of $\overline{y}$ given μ is μ for all possible values of μ, and that the probability of μ being inside I given μ is $1 - \alpha$ for all possible values of μ.

More generally, in the context of a Bayesian model as above, we may define the *frequentist bias* of a Bayesian point estimate
$$\hat{\theta} = \hat{\theta}(y)$$
as
$$B_\theta = E\{\hat{\theta}(y) - \theta \,|\, \theta\}.$$

Also, we may define the *frequentist relative bias* of $\hat{\theta}$ as
$$R_\theta = E\left(\left. \frac{\hat{\theta}(y) - \theta}{\theta} \right| \theta \right) = \frac{B_\theta}{\theta} \quad (\theta \neq 0).$$

Furthermore, we may define the *frequentist coverage probability* (FCP) of a Bayesian interval estimate
$$I(y) = (L(y),\ U(y))$$
as
$$C_\theta = P\{\theta \in I(y) \,|\, \theta\}.$$

Thus, for the normal-normal model with $\sigma_0 = \infty$, we may write:
$$B_\mu = E\{\hat{\mu}(y) - \mu \,|\, \mu\} = E(\overline{y} \,|\, \mu) - \mu = \mu - \mu = 0 \ \ \forall\ \mu \in \Re$$
$$R_\mu = \frac{0}{\mu} = 0 \quad (\mu \neq 0)$$
$$C_\mu = P\{\mu \in I(y) \,|\, \mu\}$$
$$= P\left(\overline{y} - z_{\alpha/2} \frac{\sigma}{\sqrt{n}} < \mu < \overline{y} + z_{\alpha/2} \frac{\sigma}{\sqrt{n}} \,\middle|\, \mu \right) = 1 - \alpha \ \ \forall\ \mu \in \Re.$$

The above analysis is straightforward enough. However, in the case of an *informative* prior (one with $\sigma_0 < \infty$), or in the context of other

Bayesian models, the frequentist bias of a Bayesian point estimate (B_θ) and the frequentist coverage probability of a Bayesian interval estimate (C_θ) may not be so obvious. Working out these functions may be useful for adding insight to the estimation process as well as for deciding whether or not to use a set of Bayesian estimates as frequentist proxies.

Exercise 2.1 Frequentist characteristics of estimators in the normal-normal model

Consider the normal-normal model:

$$(y_1,\ldots,y_n \mid \mu) \sim iid\ N(\mu,\sigma^2)$$
$$\mu \sim N(\mu_0,\sigma_0^2).$$

Work out general formulae for the frequentist and relative bias of the posterior mean of μ, and for the frequentist coverage probability of the $1-\alpha$ HPDR for μ.

Produce graphs showing a number of examples of each of these three functions.

Solution to Exercise 2.1

Recall that

$$(\mu \mid y) \sim N(\mu_*, \sigma_*^2),$$

where:

$$\mu_* = (1-k)\mu_0 + k\overline{y} \quad \text{is } \mu\text{'s posterior mean}$$

$$\sigma_*^2 = k\frac{\sigma^2}{n} \quad \text{is } \mu\text{'s posterior variance}$$

$$k = \frac{n}{n+\sigma^2/\sigma_0^2} \quad \text{is a credibility factor.}$$

Also, recall that μ's HPDR (and CPDR) is

$$(\mu_* \pm z_{\alpha/2}\sigma_*).$$

Using these results, we find that the frequentist bias of the posterior mean of μ is

$$B_\mu = E(\mu_* - \mu \mid \mu) = (1-k)\mu_0 + kE(\overline{y} \mid \mu) - \mu$$
$$= (1-k)\mu_0 + k\mu - \mu$$
$$= (1-k)(\mu_0 - \mu).$$

Also, the frequentist relative bias of that mean is

$$R_\mu = \frac{R_\mu}{\mu} = \frac{(1-k)(\mu_0 - \mu)}{\mu}$$

$$= (1-k)\left(\frac{\mu_0}{\mu} - 1\right) \quad (\mu \neq 0).$$

Further, the frequentist coverage probability of the $1 - \alpha$ HPDR for μ is

$$C_\mu = P\{\mu \in (\mu_* \pm z_{\alpha/2}\sigma_*)|\mu\}$$

$$= P\left(\mu_* - z_{\alpha/2}\sigma_* < \mu < \mu_* + z_{\alpha/2}\sigma_*|\mu\right)$$

$$= P\left(\mu_* - z_{\alpha/2}\sigma_* < \mu, \mu < \mu_* + z_{\alpha/2}\sigma_*|\mu\right)$$

$$= P\left((1-k)\mu_0 + k\bar{y} - z_{\alpha/2}\sigma_* < \mu, \mu < (1-k)\mu_0 + k\bar{y} + z_{\alpha/2}\sigma_*|\mu\right)$$

$$= P\left(\bar{y} < \frac{\mu - (1-k)\mu_0 + z_{\alpha/2}\sigma_*}{k}, \frac{\mu - (1-k)\mu_0 - z_{\alpha/2}\sigma_*}{k} < \bar{y}|\mu\right)$$

$$= P\left(\bar{y} < b(\mu), a(\mu) < \bar{y}|\mu\right),$$

where:

$$b(\mu) = \frac{\mu - (1-k)\mu_0 + z_{\alpha/2}\sigma_*}{k}$$

$$a(\mu) = \frac{\mu - (1-k)\mu_0 - z_{\alpha/2}\sigma_*}{k}.$$

Thus, we find that

$$C_\mu = P\left(a(\mu) < \bar{y} < b(\mu)|\mu\right)$$

$$= P\left(\frac{a(\mu) - \mu}{\sigma/\sqrt{n}} < \frac{\bar{y} - \mu}{\sigma/\sqrt{n}} < \frac{b(\mu) - \mu}{\sigma/\sqrt{n}}\bigg|\mu\right)$$

$$= P\left(\frac{a(\mu) - \mu}{\sigma/\sqrt{n}} < Z < \frac{b(\mu) - \mu}{\sigma/\sqrt{n}}\right)$$

where $Z \sim N(0,1)$, since $\left(\frac{\bar{y} - \mu}{\sigma/\sqrt{n}}\bigg|\mu\right) \sim N(0,1)$

$$= \Phi\left(\frac{b(\mu) - \mu}{\sigma/\sqrt{n}}\right) - \Phi\left(\frac{a(\mu) - \mu}{\sigma/\sqrt{n}}\right).$$

Note: Here, Φ denotes the standard normal cdf.

Figures 2.1, 2.2 and 2.3 (pages 66 and 67) show B_μ, R_μ and C_μ for selected values of σ_0, with $n = 10$, $\mu_0 = 1$, $\sigma = 1$ and $\alpha = 0.05$ in each case. The strength of the prior belief is represented by σ_0, with large values of this parameter indicating relative ignorance.

In Figure 2.1, we see that, for any given value of μ, the frequentist bias B_μ of the posterior mean $\mu_* = E(\mu \mid y)$ converges to zero as the prior belief tends to total ignorance, that is, in the limit as $\sigma_0 \to \infty$.

Also, $B_\mu \to \mu_0 - \mu$ as the prior belief tends to complete certainty, that is, in the limit as $\sigma_0 \to 0$.

Note: One of the thin dotted guidelines in Figure 2.1 shows the function $B_\mu = \mu_0 - \mu$ in this latter extreme case of 'absolute' prior belief that $\mu = \mu_0$. In all of the examples, $\mu_0 = 1$.

In Figure 2.2, we see that, for any given value of μ, the frequentist relative bias R_μ of the posterior mean $\mu_* = E(\mu \mid y)$ converges to zero as $\sigma_0 \to \infty$. Also, $R_\mu \to (\mu_0 / \mu) - 1$ as $\sigma_0 \to 0$.

Note: The curved thin dotted guidelines in Figure 2.2 shows the function $R_\mu = (\mu_0 / \mu) - 1$ in this latter extreme case of 'absolute' prior belief that $\mu = \mu_0$.

In Figure 2.3, we see that, for any given value of μ, the frequentist coverage probability C_μ of the $1 - \alpha$ (i.e. 0.95 or 95%) HPDR, namely $(\mu_* \pm z_{\alpha/2}\sigma_*)$, converges to $1 - \alpha$ as $\sigma_0 \to \infty$.

Also, $C_\mu \to 0$ as $\sigma_0 \to 0$, except at exactly $\mu = \mu_0$ where $C_\mu \to 1$; thus, $C_\mu \to I(\mu = \mu_0)$ as $\sigma_0 \to 0$ (where I denotes the standard indicator function).

Note: In Figure 2.3, the thin dotted horizontal guidelines show the values 0, 0.95 and 1.

Figure 2.1 Frequentist bias in Exercise 2.1

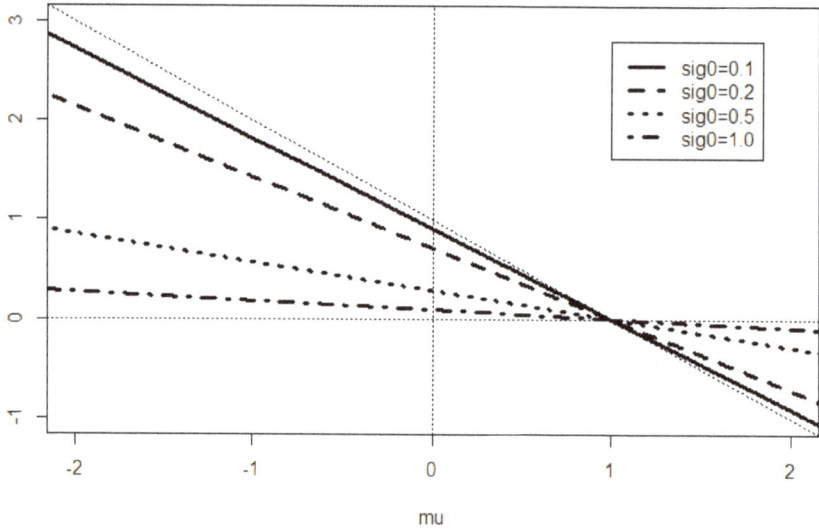

Figure 2.2 Frequentist relative bias in Exercise 2.1

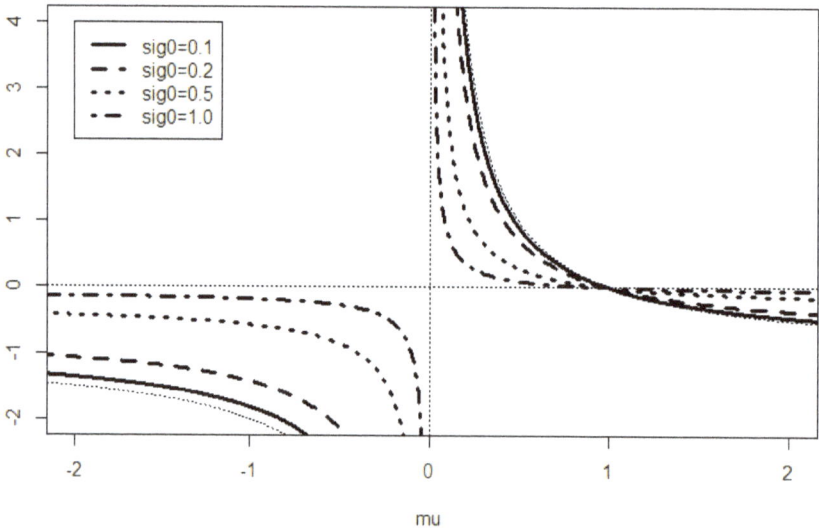

Figure 2.3 Frequentist coverage probability in Exercise 2.1

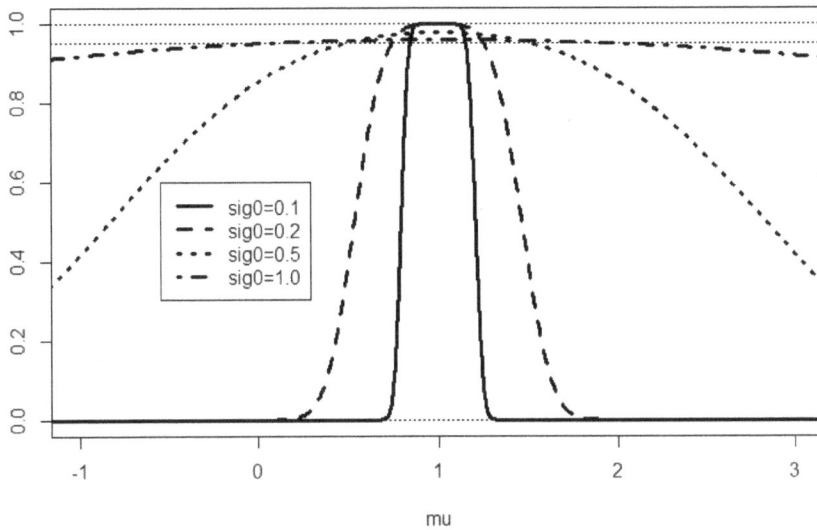

R Code for Exercise 2.1

```
biasfun = function(mu,n,sig,mu0,sig0){
    k = n/(n+(sig/sig0)^2)
    (1-k)*mu0-mu*(1-k)      }

coverfun = function(mu,n,sig,mu0,sig0,alp=0.05){
    k = n/(n + (sig/sig0)^2)
    sigstar = sig*sqrt(k/n); z=qnorm(1-alp/2)
    a= (  mu-(1-k)*mu0-z*sigstar  ) / k
    b= (  mu-(1-k)*mu0+z*sigstar  ) / k
    u= pnorm((b-mu)/(sig/sqrt(n)))
    l= pnorm((a-mu)/(sig/sqrt(n)))
    u-l      }

X11(w=8,h=5.5); par(mfrow=c(1,1))
muvec=seq(-5,5,0.01); mu0=1; sig=1; n=10;  sig0v=c(0.1,0.2,0.5,1)

plot(c(-2,2),c(-1,3),type="n",xlab="mu",ylab="",main=" ")
abline(1,-1,lty=3); abline(v=0,lty=3); abline(h=0,lty=3)
lines(muvec,biasfun(mu=muvec,n=n,sig=sig,mu0=mu0,sig0=sig0v[1]),
    lty=1,lwd=3)
```

```
lines(muvec,biasfun(mu=muvec,n=n,sig=sig, mu0=mu0,sig0=sig0v[2]),
    lty=2,lwd=3)
lines(muvec,biasfun(mu=muvec,n=n,sig=sig, mu0=mu0,sig0=sig0v[3]),
    lty=3,lwd=3)
lines(muvec,biasfun(mu=muvec,n=n,sig=sig, mu0=mu0,sig0=sig0v[4]),
    lty=4,lwd=3)
legend(1,2.8,c("sig0=0.1","sig0=0.2","sig0=0.5","sig0=1.0"),
    lty=1:4,lwd=rep(3,4))


plot(c(-2,2),c(-2,4),type="n",xlab="mu",ylab="",main=" ")
abline(v=0,lty=3); abline(h=0,lty=3); lines(muvec, mu0/muvec-1,lty=3)
lines(muvec, biasfun(mu=muvec,n=n,sig=sig,mu0=mu0, sig0=sig0v[1])/muvec,
    lty=1,lwd=3)
lines(muvec, biasfun(mu=muvec,n=n,sig=sig,mu0=mu0, sig0=sig0v[2])/muvec,
    lty=2,lwd=3)
lines(muvec, biasfun(mu=muvec,n=n,sig=sig,mu0=mu0, sig0=sig0v[3])/muvec,
    lty=3,lwd=3)
lines(muvec, biasfun(mu=muvec,n=n,sig=sig,mu0=mu0, sig0=sig0v[4])/muvec,
    lty=4,lwd=3)
legend(-2,4,c("sig0=0.1","sig0=0.2","sig0=0.5","sig0=1.0"),
    lty=1:4,lwd=rep(3,4))


plot(c(-1,3),c(0,1),type="n",xlab="mu",ylab="",main=" ")
abline(h=c(0,0.95,1),lty=3)
lines(muvec, coverfun(mu=muvec,n=n,sig=sig,mu0=mu0,sig0=sig0v[1]),
    lty=1,lwd=3)
lines(muvec, coverfun(mu=muvec,n=n,sig=sig,mu0=mu0,sig0=sig0v[2]),
    lty=2,lwd=3)
lines(muvec, coverfun(mu=muvec,n=n,sig=sig,mu0=mu0,sig0=sig0v[3]),
    lty=3,lwd=3)
lines(muvec, coverfun(mu=muvec,n=n,sig=sig,mu0=mu0,sig0=sig0v[4]),
    lty=4,lwd=3)
legend(-0.55,0.6,c("sig0=0.1","sig0=0.2","sig0=0.5","sig0=1.0"),
    lty=1:4,lwd=rep(3,4))
```

Exercise 2.2 Frequentist characteristics of estimators in the normal-gamma model

Consider the normal-gamma model given by:

$$(y_1,\ldots,y_n \mid \lambda) \sim iid\ N(\mu, 1/\lambda)$$
$$\lambda \sim Gamma(\alpha, \beta).$$

(a) Work out general formulae for the frequentist bias and relative bias of the posterior mean of $\sigma^2 = 1/\lambda$, and for the frequentist coverage probability of the $1 - \alpha$ CPDR for σ^2.

Produce graphs showing examples of each of these three functions.

(b) Attempt to find a *single* prior under this model (that is, a single suitable pair of values α, β) which results in *both*:

> **(i)** a Bayesian posterior mean of σ^2 that is unbiased (in the frequentist sense) for all possible values of σ^2; *and*
>
> **(ii)** a CPDR for σ^2 that has frequentist coverage probabilities exactly equal to the desired coverage for all possible values of σ^2.

Solution to Exercise 2.2

(a) Recall that the posterior mean of σ^2 is

$$\hat{\sigma}^2 = E(\sigma^2 \mid y) = \frac{b}{a-1},$$

where: $\quad a = \alpha + \dfrac{n}{2}, \quad b = \beta + \dfrac{n}{2}s_\mu^2, \quad s_\mu^2 = \dfrac{1}{n}\sum_{i=1}^{n}(y_i - \mu)^2.$

Thus, $\hat{\sigma}^2 = \dfrac{\beta + (n/2)s_\mu^2}{\alpha + (n/2) - 1} = \dfrac{2\beta + ns_\mu^2}{2\alpha + n - 2}.$

So the frequentist bias of $\hat{\sigma}^2$ is

$$B_{\sigma^2} = E(\hat{\sigma}^2 - \sigma^2 \mid \sigma^2) = \frac{2\beta + nE(s_\mu^2 \mid \sigma^2)}{2\alpha + n - 2} - \sigma^2 = \frac{2\beta + n\sigma^2}{2\alpha + n - 2} - \sigma^2.$$

Note: This follows because, conditional on σ^2, it is true that

$$\frac{ns_\mu^2}{\sigma^2} = \sum_{i=1}^{n}\left(\frac{y_i - \mu}{\sigma}\right)^2 \sim \chi^2(n) \quad \text{(with mean } n\text{)}.$$

Therefore the frequentist relative bias of $\hat{\sigma}^2$ is

$$R_{\sigma^2} = \frac{B_{\sigma^2}}{\sigma^2} = \frac{(2\beta/\sigma^2) + n}{2\alpha + n - 2} - 1.$$

Note: We see that for any fixed σ^2, α and β it is true that

$$B_{\sigma^2}, R_{\sigma^2} \to 0 \text{ as } n \to \infty.$$

Thus the posterior mean of σ^2 is *asymptotically unbiased*, in the frequentist sense.

Next, recall that the $1 - A$ CPDR for $\sigma^2 = 1/\lambda$ is

$$I = I(y) = \left(\frac{2\beta + ns_\mu^2}{v}, \frac{2\beta + ns_\mu^2}{u} \right),$$

where: $v = \chi^2_{A/2}(2\alpha + n) = F^{-1}_{\chi^2(2\alpha+n)}(1 - A/2)$

$u = \chi^2_{1-A/2}(2\alpha + n) = F^{-1}_{\chi^2(2\alpha+n)}(A/2).$

So the frequentist coverage probability of I is

$$C_{\sigma^2} = P\left\{ \sigma^2 \in I(y) \middle| \sigma^2 \right\}$$

$$= P\left(\frac{2\beta + ns_\mu^2}{v} < \sigma^2 < \frac{2\beta + ns_\mu^2}{u} \middle| \sigma^2 \right)$$

$$= P\left\{ \sigma^2 \in I(y) \middle| \sigma^2 \right\}$$

$$= P\left(\frac{ns_\mu^2}{\sigma^2} < v - \frac{2\beta}{\sigma^2}, u - \frac{2\beta}{\sigma^2} < \frac{ns_\mu^2}{\sigma^2} \middle| \sigma^2 \right)$$

$$= F_{\chi^2(n)}\left(v - \frac{2\beta}{\sigma^2} \right) - F_{\chi^2(n)}\left(u - \frac{2\beta}{\sigma^2} \right).$$

Figures 2.4, 2.5 and 2.6 (pages 72 and 73) show B_{σ^2}, R_{σ^2} and C_{σ^2} for selected values of α and β, with $n = 10$ and $A = 0.05$ in each case.

(b) Observe that under the prior given by $\alpha = 1$ and $\beta = 0$
(that is, $f(\lambda) \propto \lambda^{1-1} e^{-0\lambda} \propto 1$), it is true that:

- the posterior mean of σ^2 equals the MLE, namely s_μ^2, and so is unbiased

- the $1 - A$ CPDR for σ^2 is $\left(\frac{ns_\mu^2}{\chi^2_{A/2}(n+2)}, \frac{ns_\mu^2}{\chi^2_{1-A/2}(n+2)} \right),$

  which has coverage probability less than $1 - A$ for all σ^2.

Also, under the prior given by $\alpha = \beta = 0$ (i.e. $f(\lambda) \propto \lambda^{0-1} e^{-0\lambda} \propto 1/\lambda$), it is true that:

- the posterior mean of σ^2 equals $s_\mu^2 / (1 - 2/n)$ and so is biased

- the $1 - A$ CPDR for σ^2 is the same as the classical CI, namely

$$\left(\frac{n s_\mu^2}{\chi_{A/2}^2(n)}, \frac{n s_\mu^2}{\chi_{1-A/2}^2(n)} \right), \text{ and so has coverage exactly } 1 - A \text{ for all}$$

σ^2.

We see that there is no single gamma prior for $\lambda = 1/\sigma^2$ which results in *both*:

(i) a Bayesian posterior mean of σ^2 that is unbiased (in the frequentist sense) for all possible values of σ^2; *and*

(ii) a CPDR for σ^2 that has frequentist coverage probabilities exactly equal to the desired coverage for all possible values of σ^2.

Note: It is easy to modify or 'correct' the posterior mean under $\alpha = \beta = 0$ so that it becomes unbiased. Explictly, if $\alpha = \beta = 0$, then

$$E(\hat{\sigma}^2 \mid \sigma^2) = \frac{n\sigma^2}{n-2}.$$

So an unbiased estimate of σ^2 is

$$\tilde{\sigma}^2 = \frac{n-2}{n} \hat{\sigma}^2 = \frac{n-2}{n} \times \frac{0 + (n/2) s_\mu^2}{0 + (n/2) - 1} = s_\mu^2 \text{ (i.e. the MLE).}$$

Figure 2.4 Frequentist bias in Exercise 2.2

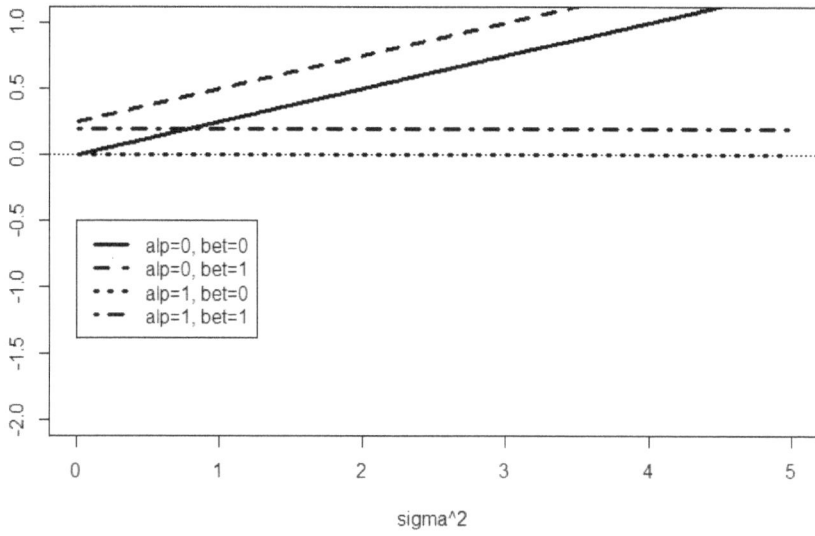

Figure 2.5 Frequentist relative bias in Exercise 2.2

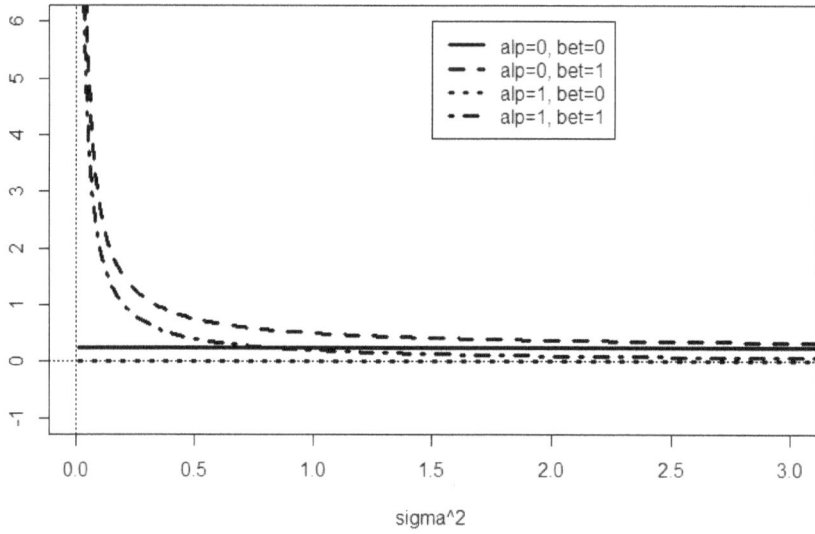

Figure 2.6 Frequentist coverage probability in Exercise 2.2

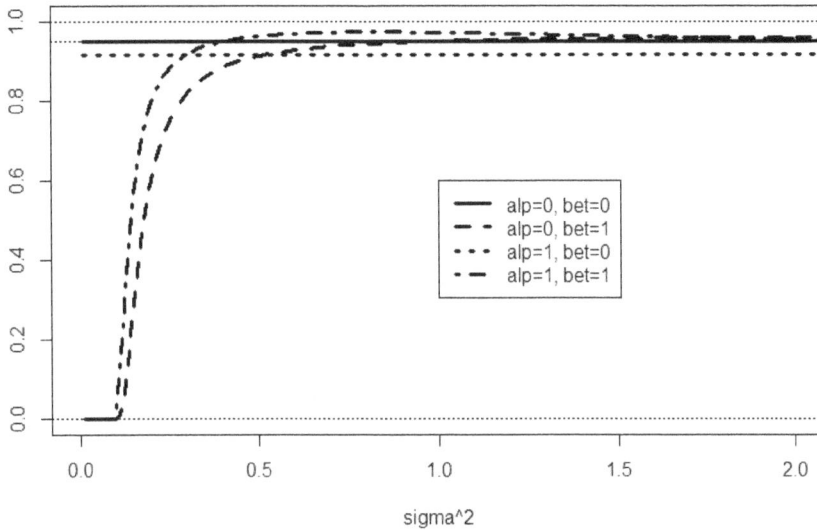

R Code for Exercise 2.2

```
biasfun = function(sig2,n=10,alp=0,bet=0){ (2*bet+n*sig2)/(2*alp+n-2)-sig2 }

coverfun = function(sig2,n=10,alp=0,bet=0,A=0.05){
    u = qchisq(A/2,2*alp+n); v = qchisq(1-A/2,2*alp+n)
    pchisq(v-2*bet/sig2, n) - pchisq(u-2*bet/sig2, n)   }

X11(w=8,h=5.5); par(mfrow=c(1,1))
sig2vec=seq(0.01,5,0.01); n=10;  alpv=c(0.1,1,5); betv=c(0.1,1,5)

plot(c(0,5),c(-2,1),type="n",xlab="sigma^2",ylab="",main=" ")
abline(h=0,lty=3)

lines(sig2vec,biasfun(sig2=sig2vec,alp=0,bet=0), lty=1,lwd=3)
lines(sig2vec,biasfun(sig2=sig2vec,alp=0,bet=1), lty=2,lwd=3)
lines(sig2vec,biasfun(sig2=sig2vec,alp=1,bet=0), lty=3,lwd=3)
lines(sig2vec,biasfun(sig2=sig2vec,alp=1,bet=1), lty=4,lwd=3)
legend(0,-0.5,c("alp=0, bet=0","alp=0, bet=1","alp=1, bet=0","alp=1, bet=1"),
        lty=1:4,lwd=rep(3,4))
```

```
plot(c(0,3),c(-1,6),type="n",xlab="sigma^2",ylab="",main=" ")
abline(h=0,lty=3); abline(v=0,lty=3)

lines(sig2vec,biasfun(sig2=sig2vec,alp=0,bet=0)/ sig2vec, lty=1,lwd=3)
lines(sig2vec,biasfun(sig2=sig2vec,alp=0,bet=1)/ sig2vec, lty=2,lwd=3)
lines(sig2vec,biasfun(sig2=sig2vec,alp=1,bet=0)/ sig2vec, lty=3,lwd=3)
lines(sig2vec,biasfun(sig2=sig2vec,alp=1,bet=1)/ sig2vec, lty=4,lwd=3)
legend(1.5,6,c("alp=0, bet=0","alp=0, bet=1","alp=1, bet=0","alp=1, bet=1"),
       lty=1:4,lwd=rep(3,4))

plot(c(0,2),c(0,1),type="n",xlab="sigma^2",ylab="",main=" ")
abline(h=c(0,0.95,1),lty=3)

lines(sig2vec, coverfun(sig2=sig2vec,n=10,alp=0,bet=0,A=0.05), lty=1,lwd=3)
lines(sig2vec, coverfun(sig2=sig2vec,n=10,alp=0,bet=1,A=0.05), lty=2,lwd=3)
lines(sig2vec, coverfun(sig2=sig2vec,n=10,alp=1,bet=0,A=0.05), lty=3,lwd=3)
lines(sig2vec, coverfun(sig2=sig2vec,n=10,alp=1,bet=1,A=0.05), lty=4,lwd=3)
legend(1,0.6,c("alp=0, bet=0","alp=0, bet=1","alp=1, bet=0","alp=1, bet=1"),
       lty=1:4,lwd=rep(3,4))
```

2.2 Mixture prior distributions

So far we have considered Bayesian models with priors that are limited in the types of prior information that they can represent. For example, the normal-normal model does not allow a prior for the normal mean which has two or more modes. If a non-normal class of prior is used to represent one's complicated prior beliefs regarding the normal mean, then that prior will not be conjugate, and this will lead to difficulties down the track when making inferences based on the nonstandard posterior distribution.

Fortunately, this problem can be addressed in any Bayesian model for which a conjugate class of prior exists by specifying the prior as a *mixture* of members of that class.

Generally, a random variable X with a *mixture distribution* has a density of the form

$$f(x) = \sum_{m=1}^{M} c_m f_m(x),$$

where each $f_m(x)$ is a proper density and the c_m values are positive and sum to 1.

If our prior beliefs regarding a parameter θ do not follow any single well-known distribution, those beliefs can in that case be conveniently *approximated* to any degree of precision by a suitable *mixture prior distribution* with a density having the form

$$f(\theta) = \sum_{m=1}^{M} c_m f_m(\theta).$$

It can be shown (see Exercise 2.3 below) that if each component prior $f_m(\theta)$ is conjugate then $f(\theta)$ is also conjugate. This means that θ's posterior distribution is also a mixture with density of the form

$$f(\theta \mid y) = \sum_{m=1}^{M} c'_m f_m(\theta \mid y), \tag{2.1}$$

where $f_m(\theta \mid y)$ is the posterior implied by the mth prior $f_m(\theta)$ and is from the same family of distributions as that prior.

Exercise 2.3 Binomial-beta model with a mixture prior

(a) Consider the following Bayesian model:
$$(y \mid \theta) \sim Bin(n, \theta)$$
$$f(\theta) = k f_{Beta(a_1,b_1)}(\theta) + (1-k) f_{Beta(a_2,b_2)}(\theta),$$
where n, k and the a_i, b_i are specified constants.

Note: Here, $f_{Beta(a,b)}(t)$ denotes the density at t of the beta distribution with parameters a and b (and mean $a/(a+b)$).

Find the posterior distribution of θ and shows that θ's prior is conjugate. Then create a figure showing the prior, likelihood and posterior for the situation defined by:
$$n = 5, \; k = 3/4, \; a_1 = 8, \; b_1 = 25, \; a_2 = 20, \; b_2 = 20 \text{ and } y = 4.$$

Also calculate the prior mean of θ, the posterior mean of θ and the MLE of θ. Then mark these three points in the figure.

(b) Show that any mixture of conjugate priors is also conjugate and derive a general formula which could be used to calculate the mixture weights c'_m in (2.1) above.

Solution to Exercise 2.3

(a) The posterior density is

$$f(\theta \mid y) \propto f(\theta) f(y \mid \theta)$$

$$= \left(k \frac{\theta^{a_1-1}(1-\theta)^{b_1-1}}{B(a_1,b_1)} + (1-k) \frac{\theta^{a_2-1}(1-\theta)^{b_2-1}}{B(a_2,b_2)} \right) \times \theta^y (1-\theta)^{n-y}$$

$$= \left[k \frac{B(a_1+y,b_1+n-y)}{B(a_1,b_1)} \right] \left(\frac{\theta^{(a_1+y)-1}(1-\theta)^{(b_1+n-y)-1}}{B(a_1+y,b_1+n-y)} \right)$$

$$+ \left\{ (1-k) \frac{B(a_2+y,b_2+n-y)}{B(a_2,b_2)} \right\} \left(\frac{\theta^{(a_2+y)-1}(1-\theta)^{(b_2+n-y)-1}}{B(a_2+y,b_2+n-y)} \right).$$

Thus
$$f(\theta \mid y) \propto c_1 f_1(\theta \mid y) + c_2 f_2(\theta \mid y),$$
where:
$$c_1 = k \frac{B(a_1+y,b_1+n-y)}{B(a_1,b_1)}$$

$$c_2 = (1-k) \frac{B(a_2+y,b_2+n-y)}{B(a_2,b_2)}$$

$$f_i(\theta \mid y) = f_{Beta(a_i+y,b_i+n-y)}(\theta) = \frac{\theta^{(a_i+y)-1}(1-\theta)^{(b_i+n-y)-1}}{B(a_i+y,b_i+n-y)}, \quad 0 < \theta < 1$$

(the posterior density corresponding to $\theta \sim Beta(a_i,b_i)$ as prior).

Now,
$$\int f(\theta \mid y)d\theta = 1,$$
and so
$$f(\theta \mid y) = c \, f_{Beta(a_1+y,b_1+n-y)}(\theta) + (1-c) \, f_{Beta(a_2+y,b_2+n-y)}(\theta),$$
where
$$c = \frac{c_1}{c_1+c_2}.$$

Note: This ensures that $\int f(\theta \mid y)d\theta = c \times 1 + (1-c) \times 1 = 1$.

We see that the prior $f(\theta)$ and posterior $f(\theta \mid y)$ are in the same family, namely the *family of mixtures of two beta distributions*. Therefore the mixture prior is *conjugate*.

For the situation where
$$n = 5, \, k = 3/4, \, a_1 = 8, \, b_1 = 25, \, a_2 = 20, \, b_2 = 20 \text{ and } y = 4,$$
we find that:

- the prior mean is
$$E\theta = k\left(\frac{a_1}{a_1 + b_1}\right) + (1-k)\left(\frac{a_2}{a_2 + b_2}\right) = 0.3068$$

- the maximum likelihood estimate is
$$y/n = 0.8$$

- the posterior mean is
$$E(\theta \mid y) = c\left(\frac{a_1 + y}{a_1 + b_1 + n}\right) + (1-c)\left(\frac{a_2 + y}{a_2 + b_2 + n}\right) = 0.4772.$$

Figure 2.7 shows the prior density $f(\theta)$, the likelihood function $L(\theta)$, and the posterior density $f(\theta \mid y)$, as well as the prior mean, the MLE and the posterior mean.

Note: The likelihood function in Figure 2.7 has been normalised so that the area underneath it is exactly 1. This means that this likelihood function is identical to the posterior density under the standard uniform prior, i.e. under $f_{U(0,1)}(\theta) = f_{Beta(1,1)}(\theta)$. Thus, $L(\theta) = f_{Beta(1+y,1+n-y)}(\theta)$.

Figure 2.7 also shows the two component prior densities and the two component posterior densities. It may be observed that, whereas the lower component prior has the highest weight, 0.8, the opposite is the case regarding the component posteriors. For these, the weight associated with the lower posterior is only 0.2583. This is because the inference is being 'pulled up' in the direction of the likelihood (with the posterior mean being between the prior mean and the MLE, 0.8).

Figure 2.7 Densities and likelihood in Exercise 2.3

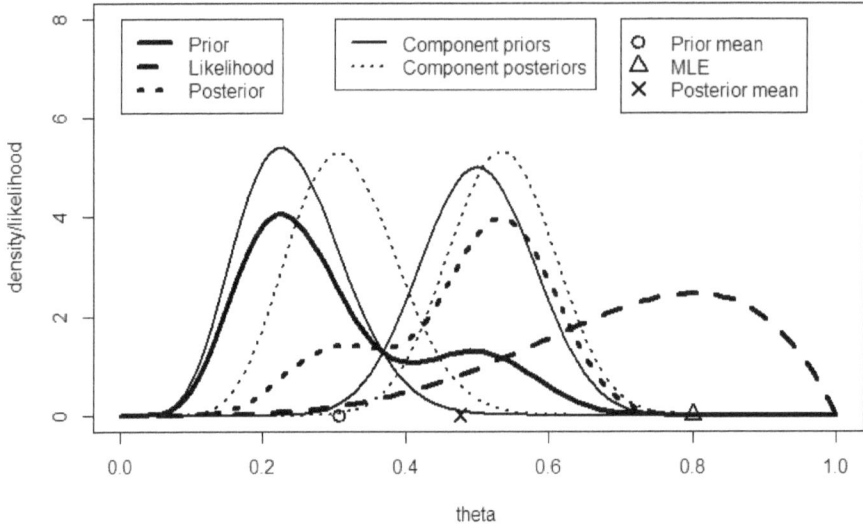

(b) Suppose that θ has a mixture prior of the general form

$$f(\theta) = \sum_{m=1}^{M} c_m f_m(\theta),$$

where each $f_m(\theta)$ is conjugate for the data model.

Then the posterior density is

$$f(\theta \mid y) \propto f(\theta) f(y \mid \theta) = \left(\sum_{m=1}^{M} c_m f_m(\theta) \right) f(y \mid \theta)$$

$$= \sum_{m=1}^{M} \{ c_m f_m(\theta) f(y \mid \theta) \} = \sum_{m=1}^{M} \left\{ (c_m f_m(y)) \left(\frac{f_m(\theta) f(y \mid \theta)}{f_m(y)} \right) \right\},$$

where $f_m(y) = \int f_m(\theta) f(y \mid \theta) d\theta$ is the unconditional density of the data under the mth prior, $f_m(\theta)$. Thus

$$f(\theta \mid y) \propto \sum_{m=1}^{M} k_m f_m(\theta \mid y),$$

where

$$k_m = c_m f_m(y) \text{ and } f_m(\theta \mid y) = \frac{f_m(\theta) f(y \mid \theta)}{f_m(y)}$$

is the posterior density of θ under the mth prior, $f_m(\theta)$.

It follows that

$$f(\theta \mid y) = \sum_{m=1}^{M} c'_m f_m(\theta \mid y),$$

where $c'_m = k_m / (k_1 + \ldots + k_M)$.

Thus θ's posterior is a mixture of distributions from the same families to which the components of θ's mixture prior belong, respectively. This shows that θ's mixture prior is conjugate. Note that the component prior distributions can be from different classes, so long as each is conjugate in relation to its own class.

R Code for Exercise 2.3

```
n=5; k=3/4; a1=8; b1=25; a2=20; b2=20; y=4; thetav=seq(0,1,0.01)
prior1=dbeta(thetav,a1,b1); prior2=dbeta(thetav,a2,b2)
post1=dbeta(thetav,a1+y,b1+n-y); post2=dbeta(thetav,a2+y,b2+n-y)
prior = k*prior1 + (1-k)*prior2

c1=k*beta(a1+y,b1+n-y)/beta(a1,b1); c2=(1-k)*beta(a2+y,b2+n-y)/beta(a2,b2)
c=c1/(c1+c2); post=c*post1 + (1-c)*post2; options(digits=4); c # 0.2583
like=dbeta(thetav,1+y,1+n-y) # likelihood = post. under U(0,1)=beta(1,1) prior

X11(w=8,h=5.5)
plot(c(0,1),c(0,8),type="n",xlab="theta",ylab="density/likelihood")
lines(thetav,prior,lty=1,lwd=4)
lines(thetav,like,lty=2,lwd=4)
lines(thetav,post,lty=3,lwd=4)
legend(0,8,c("Prior","Likelihood","Posterior"),lty=c(1,2,3),lwd=c(4,4,4))
lines(thetav,prior1,lty=1,lwd=2)
lines(thetav,prior2,lty=1,lwd=2)
lines(thetav,post1,lty=3,lwd=2)
lines(thetav,post2,lty=3,lwd=2)
legend(0.3,8,c("Component priors","Component posteriors"),
       lty=c(1,3),lwd=c(2,2))

mle=y/n; priormean=k*a1/(a1+b1)+(1-k)*a2/(a2+b2)
postmean=c*(a1+y)/(a1+b1+n) + (1-c)*(a2+y)/(a2+b2+n)
points(c(priormean,mle,postmean),c(0,0,0),pch=c(1,2,4),cex=c(1.5,1.5,1.5),
       lwd=c(2,2,2))
c(priormean,mle,postmean) # 0.3068 0.8000 0.4772
legend(0.7,8,c("  Prior mean","  MLE","  Posterior mean"),
       pch=c(1,2,4),pt.cex=c(1.5,1.5,1.5),pt.lwd=c(2,2,2))
```

2.3 Dealing with a priori ignorance

The Bayesian approach requires a prior distribution to be specified even when there is complete (or total) *a priori* ignorance (meaning no prior information at all). This feature presents a general and philosophical problem with the Bayesian paradigm, one for which several theoretical solutions have been advanced but which does not yet have a universally accepted solution. We have already discussed finding an uninformative prior in relation to particular Bayesian models, as follows.

For the *normal-normal model* defined by $(y_1,\ldots,y_n \mid \mu) \sim iid\ N(\mu,\sigma^2)$ and $\mu \sim N(\mu_0,\sigma_0^2)$, an uninformative prior is given by $\sigma_0 = \infty$, that is, $f(\mu) \propto 1, \mu \in \Re$.

For the *normal-gamma model* defined by $(y_1,\ldots,y_n \mid \mu) \sim iid\ N(\mu,1/\lambda)$ and $\lambda \sim Gamma(\alpha,\beta)$, an uninformative prior is given by $\alpha = \beta = 0$, that is, $f(\lambda) \propto 1/\lambda, \lambda > 0$.

For the *binomial-beta model* defined by $(y \mid \theta) \sim Binomial(n,\theta)$ and $\theta \sim Beta(\alpha,\beta)$ (having the posterior $(\theta \mid y) \sim Beta(\alpha+y,\beta+n-y)$), an uninformative prior is the *Bayes prior* given by $\alpha = \beta = 1$, that is, $f(\theta) = 1, 0 < \theta < 1$. This is the prior that was originally advocated by Thomas Bayes.

Unlike for the normal-normal and normal-gamma models, more than one uninformative prior specification has been proposed as reasonable in the context of the binomial-beta model.

One of these is the improper *Haldane prior*, defined by $\alpha = \beta = 0$, or

$$f(\theta) \propto \frac{1}{\theta(1-\theta)}, \quad 0 < \theta < 1.$$

Under the prior $\theta \sim Beta(\alpha,\beta)$ generally, the posterior mean of θ is

$$\hat{\theta} = E(\theta \mid y) = \frac{(\alpha+y)}{(\alpha+y)+(\beta+n-y)} = \frac{\alpha+y}{\alpha+\beta+n}.$$

This reduces to the MLE y/n under the Haldane prior but not under the Bayes prior. In contrast, the Bayes prior leads to a posterior *mode* which is equal to the MLE.

The Haldane prior may be considered as being most appropriate for allowing the data to 'speak for itself' in cases of *a priori* ignorance.

However, the Haldane prior leads to an improper and degenerate posterior if the data y happens to be either 0 or n. Specifically:
$$y = 0 \implies (\theta \mid y) \sim Beta(0, n), \text{ or equivalently, } P(\theta = 0 \mid y) = 1$$
$$y = n \implies (\theta \mid y) \sim Beta(n, 0), \text{ or equivalently, } P(\theta = 1 \mid y) = 1.$$

So in each case, point estimation is possible but not interval estimation.

No such problems occur using the Bayes prior. This is because that prior is proper and so *cannot* lead to an improper posterior, whatever the data may be. Interestingly, there is a third choice which provides a kind of compromise between the Bayes and Haldane priors, as described below.

2.4 The Jeffreys prior

The statistician Harold Jeffreys devised a rule for finding a suitable uninformative prior in a wide variety of situations. His idea was to construct a prior which is invariant under reparameterisation. For the case of a univariate model parameter θ, the *Jeffreys prior* is given by the following equation (also known as *Jeffreys' rule*):
$$f(\theta) \propto \sqrt{I(\theta)},$$
where $I(\theta)$ is the *Fisher information* defined by
$$I(\theta) = E\left\{ \left(\frac{\partial}{\partial \theta} \log f(y \mid \theta) \right)^2 \middle| \theta \right\}.$$

Note 1: If $\log f(y \mid \theta)$ is twice differentiable with respect to θ, and certain regularity conditions hold, then
$$I(\theta) = -E\left\{ \frac{\partial^2}{\partial \theta^2} \log f(y \mid \theta) \middle| \theta \right\}.$$

Note 2: Jeffreys' rule also extends to the multi-parameter case (not considered here).

The significance of Jeffreys' rule may be described as follows. Consider a prior given by $f(\theta) \propto \sqrt{I(\theta)}$ and the transformed parameter $\phi = g(\theta)$, where g is a strictly increasing or decreasing function. (For simplicity, we only consider this case.) Then the prior density for ϕ is

$$f(\phi) \propto f(\theta) \left| \frac{\partial \theta}{\partial \phi} \right| \quad \text{by the transformation rule}$$

$$\propto \sqrt{I(\theta) \left(\frac{\partial \theta}{\partial \phi} \right)^2} = \sqrt{E \left\{ \left(\frac{\partial}{\partial \theta} \log f(y \mid \theta) \right)^2 \middle| \theta \right\} \left(\frac{\partial \theta}{\partial \phi} \right)^2}$$

$$= \sqrt{E \left\{ \left(\frac{\partial}{\partial \theta} \log f(y \mid \theta) \frac{\partial \theta}{\partial \phi} \right)^2 \middle| \theta \right\}}$$

$$= \sqrt{E \left\{ \left(\frac{\partial}{\partial \phi} \log f(y \mid \phi) \right)^2 \middle| \phi \right\}}$$

$$= \sqrt{I(\phi)} .$$

Thus, Jeffreys' rule is 'invariant under reparameterisation', in the sense that if a prior is constructed according to

$$f(\theta) \propto \sqrt{I(\theta)} ,$$

then, for another parameter $\phi = g(\theta)$, it is also true that

$$f(\phi) \propto \sqrt{I(\phi)} .$$

Exercise 2.4 Jeffreys prior for the normal-normal model

Find the Jeffreys prior for μ if $(y_1, \ldots, y_n \mid \mu) \sim iid\ N(\mu, \sigma^2)$, where σ is known.

Solution to Exercise 2.4

Here: $f(y \mid \mu) \overset{\mu}{\propto} \prod_{i=1}^{n} \exp \left\{ -\frac{1}{2\sigma^2} (y_i - \mu)^2 \right\} = \exp \left\{ -\frac{1}{2\sigma^2} \sum_{i=1}^{n} (y_i - \mu)^2 \right\}$

$\log f(y \mid \mu) = -\frac{1}{2\sigma^2} \sum_{i=1}^{n} (y_i - \mu)^2 + c$ (where c is a constant)

$\frac{\partial}{\partial \mu} \log f(y \mid \mu) = -\frac{1}{2\sigma^2} \sum_{i=1}^{n} 2(y_i - \mu)^1 (-1) = \frac{n}{\sigma^2} (\bar{y} - \mu)$

$\left(\frac{\partial}{\partial \mu} \log f(y \mid \mu) \right)^2 = \frac{n^2}{\sigma^4} (\bar{y} - \mu)^2 .$

Therefore the Fisher information is

$$I(\mu) = E\left\{\left(\frac{\partial}{\partial\mu}\log f(y\mid\mu)\right)^2\Bigg|\mu\right\} = E\left\{\frac{n^2}{\sigma^4}(\bar{y}-\mu)^2\Big|\mu\right\}$$

$$= \frac{n^2}{\sigma^4}V(\bar{y}\mid\mu) = \frac{n^2}{\sigma^4}\frac{\sigma^2}{n} = \frac{n}{\sigma^2}.$$

It follows that the Jeffreys prior is $f(\mu) \propto \sqrt{I(\mu)} = \sqrt{\dfrac{n}{\sigma^2}} \overset{\mu}{\propto} 1,\ \mu\in\Re$.

Note 1: This is the same prior as used earlier in the uninformative case.

Note 2: The Fisher information here can also be derived as follows:

$$\frac{\partial^2}{\partial\mu^2}\log f(y\mid\mu) = -\frac{n}{\sigma^2}$$

$$\Rightarrow I(\mu) = -E\left\{\frac{\partial^2}{\partial\theta^2}\log f(y\mid\theta)\Big|\theta\right\} = -E\left(-\frac{n}{\sigma^2}\right) = \frac{n}{\sigma^2}.$$

Exercise 2.5 Jeffreys prior for the normal-gamma model

Find the Jeffreys prior for λ if $(y_1,\ldots,y_n\mid\lambda)\sim iid\ N(\mu,1/\lambda)$, where μ is known.

Solution to Exercise 2.5

Here: $f(y\mid\lambda) \overset{\lambda}{\propto} \prod_{i=1}^{n}\lambda^{1/2}\exp\left\{-\frac{\lambda}{2}(y_i-\mu)^2\right\} = \lambda^{n/2}\exp\left\{-\frac{\lambda}{2}\sum_{i=1}^{n}(y_i-\mu)^2\right\}$

$\log f(y\mid\lambda) = \frac{n}{2}\log\lambda - \frac{\lambda}{2}\sum_{i=1}^{n}(y_i-\mu)^2 + c$ (where c is a constant)

$\dfrac{\partial\log f(y\mid\lambda)}{\partial\lambda} = \dfrac{n}{2\lambda} - \dfrac{1}{2}\sum_{i=1}^{n}(y_i-\mu)^2,\quad \dfrac{\partial^2\log f(y\mid\lambda)}{\partial\lambda^2} = -\dfrac{n}{2\lambda^2}.$

So the Fisher information is

$$I(\lambda) = -E\left\{\frac{\partial^2\log f(y\mid\lambda)}{\partial\lambda^2}\Big|\lambda\right\} = -E\left\{-\frac{n}{2\lambda^2}\Big|\lambda\right\} = \frac{n}{2\lambda^2}.$$

So the Jeffreys prior is $f(\lambda) \propto \sqrt{I(\lambda)} = \sqrt{\dfrac{n}{2\lambda^2}} \overset{\lambda}{\propto} \dfrac{1}{\lambda}, \ \lambda \in \Re$.

Note 1: This is the same prior as used earlier in the uninformative case.

Note 2: Another way to obtain the the Fisher information is to first write

$$\frac{\partial \log f(y \mid \lambda)}{\partial \lambda} = \frac{n}{2\lambda} - \frac{1}{2\lambda}\left\{ \lambda \sum_{i=1}^{n} (y_i - \mu)^2 \right\} = \frac{1}{2\lambda}(n-q),$$

where: $q = \sum_{i=1}^{n} \left(\dfrac{y_i - \mu}{1/\sqrt{\lambda}} \right)^2$, $(q \mid \lambda) \sim \chi^2(n)$, $E(q \mid \lambda) = n$, $V(q \mid \lambda) = 2n$.

We may then write $\left(\dfrac{\partial \log f(y \mid \lambda)}{\partial \lambda} \right)^2 = \dfrac{1}{4\lambda^2}(n^2 - 2nq + q^2)$,

and so the Fisher information is $I(\lambda) = E\left\{ \left(\dfrac{\partial \log f(y \mid \lambda)}{\partial \lambda} \right)^2 \Bigg| \lambda \right\}$

$= \dfrac{1}{4\lambda^2}\{n^2 - 2nE(q \mid \lambda) + E(q^2 \mid \lambda)\} = \dfrac{1}{4\lambda^2}\{n^2 - 2nn + [2n + n^2]\} = \dfrac{n}{2\lambda^2}$.

Exercise 2.6 Jeffreys prior for the binomial-beta model

Find the Jeffreys prior for θ if $(y \mid \theta) \sim Binomial(n, \theta)$, where n is known.

Solution to Exercise 2.6

Here: $f(y \mid \theta) = \dbinom{n}{y} \theta^y (1-\theta)^{n-y}$

$\log f(y \mid \theta) = \log \dbinom{n}{y} + y \log \theta + (n-y) \log(1-\theta)$

$\dfrac{\partial}{\partial \theta} \log f(y \mid \theta) = 0 + y\theta^{-1} - (n-y)(1-\theta)^{-1}$

$\dfrac{\partial^2}{\partial \theta^2} \log f(y \mid \theta) = -y\theta^{-2} - (n-y)(1-\theta)^{-2}$.

So the Fisher information is

$$I(\theta) = -E\left\{\frac{\partial^2}{\partial\theta^2}\log f(y\,|\,\theta)\Big|\theta\right\}$$

$$= -E\left\{-y\theta^{-2} - (n-y)(1-\theta)^{-2}\Big|\theta\right\}$$

$$= (n\theta)\theta^{-2} + (n-n\theta)(1-\theta)^{-2}$$

$$= n\left(\frac{1}{\theta} + \frac{1}{1-\theta}\right) = n\frac{(1-\theta+\theta)}{\theta(1-\theta)} = \frac{n}{\theta(1-\theta)}.$$

It follows that the Jeffreys prior is given by

$$f(\theta) \propto \sqrt{I(\theta)} = \sqrt{\frac{n}{\theta(1-\theta)}} \overset{\theta}{\propto} \frac{1}{\sqrt{\theta(1-\theta)}}, \quad 0 < \theta < 1.$$

Note: We may also write the Jeffreys prior density as

$$f(\theta) \propto \theta^{\frac{1}{2}-1}(1-\theta)^{\frac{1}{2}-1}, \quad 0 < \theta < 1.$$

Thus the Jeffreys prior can be specified by writing
$$\theta \sim Beta(\alpha, \beta)$$
with $\alpha = \beta = 1/2$.

We see that the Jeffreys prior may be thought of as 'half-way' between:
- the Bayes prior, defined by $\alpha = \beta = 1$; and
- the Haldane prior, defined by $\alpha = \beta = 0$.

Exercise 2.7 Jeffreys prior for the tramcar problem

Recall the discussion of the tramcar problem following Exercise 1.6, in relation to the model $(y\,|\,\theta) \sim DU(1,...,\theta)$. Find the Jeffreys prior for θ.

Solution to Exercise 2.7

Here,

$$f(y\,|\,\theta) = 1/\theta = \theta^{-1}$$
$$\Rightarrow \quad \log f(y\,|\,\theta) = -\log\theta$$
$$\Rightarrow \quad \frac{\partial}{\partial\theta}\log f(y\,|\,\theta) = -\frac{1}{\theta}$$

$$\Rightarrow \quad \left(\frac{\partial}{\partial\theta}\log f(y\,|\,\theta)\right)^2 = \frac{1}{\theta^2}$$

$$\Rightarrow \quad I(\theta) = E\left\{\left(\frac{\partial}{\partial\theta}\log f(y\,|\,\theta)\right)^2 \bigg|\, \theta\right\} = \frac{1}{\theta^2}\,.$$

It follows that the Jeffreys prior for θ is given by
$$f(\theta) \propto \sqrt{I(\theta)} \propto 1/\theta,\ \theta = 1,2,3,\dots$$

2.5 Bayesian decision theory

The posterior mean, mode and median, as well as other Bayesian point estimates, can all be derived and interpreted using the principles and theory of *decision theory*. Suppose we wish to choose an estimate of θ which minimises costs in some sense. To this end, let $L(\hat{\theta},\theta)$ denote generally a *loss function* (LF) associated with an estimate $\hat{\theta}$.

Note: The estimator $\hat{\theta}$ is a function of the data y and so could also be written $\hat{\theta}(y)$. For example, in the context where $(y\,|\,\theta) \sim Bin(n,\theta)$, the sample proportion or MLE is the function given by $\hat{\theta} = \hat{\theta}(y) = y/n$.

The loss function L represents the cost incurred when the true value θ is estimated by $\hat{\theta}$ and usually satisfies the property $L(\theta,\theta) = 0$.

The three most commonly used loss functions are defined as follows:

$$L(\hat{\theta},\theta) = |\hat{\theta}-\theta| \qquad \text{the } \textit{absolute error loss function } \text{(AELF)}$$

$$L(\hat{\theta},\theta) = (\hat{\theta}-\theta)^2 \qquad \text{the } \textit{quadratic error loss function } \text{(QELF)}$$

$$L(\hat{\theta},\theta) = I(\hat{\theta} \neq \theta) = \begin{cases} 0 & \text{if } \hat{\theta}=\theta \\ 1 & \text{if } \hat{\theta} \neq \theta \end{cases} \qquad \text{the } \textit{indicator error loss}$$

$\qquad\qquad$ *function* (IELF), also known as the *zero-one loss function* (ZOLF) or the *all-or-nothing* error loss function (ANLF).

Figures 2.8 and 2.9 illustrate these three basic loss functions.

Figure 2.8 The three most important loss functions

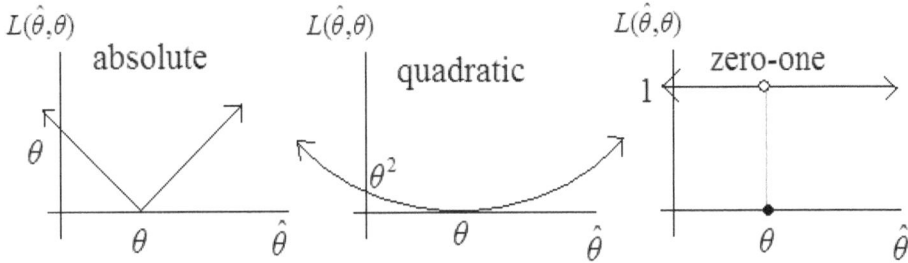

Figure 2.9 Alternative representation of the absolute error loss function
(The other two loss functions can be represented similarly)

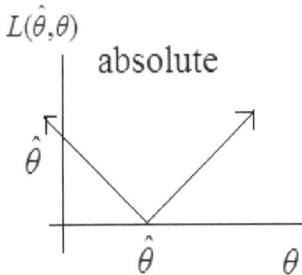

Given a Bayesian model, loss function and estimator, we would like to quantify what the loss is likely to be. However, this loss depends on θ and y, which complicates things. An idea of the expected loss may be provided by the *risk function*, defined as the conditional expectation

$$R(\theta) = E(L(\hat{\theta}, \theta) \mid \theta) = \int L(\hat{\theta}(y), \theta) f(y \mid \theta) dy \,.$$

The risk function $R(\theta)$ provides us with an idea of the expected loss given any particular value of θ. Figure 2.10 illustrates the idea.

Figure 2.10 The idea of a risk function

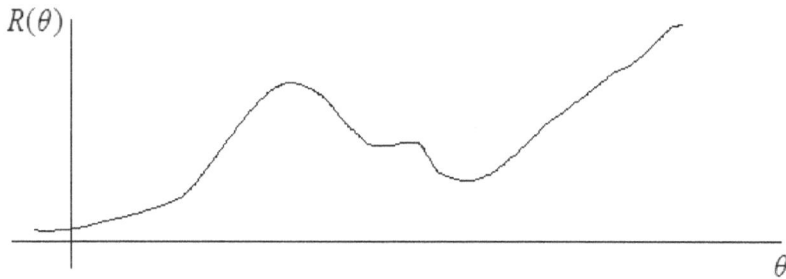

To obtain the *overall* expected loss we need to average the risk function over all possible values of θ. This overall expected loss is called the *Bayes risk* and may be defined as

$$r = EL(\hat{\theta},\theta) = EE\{L(\hat{\theta},\theta)\,|\,\theta\} = ER(\theta) = \int R(\theta)f(\theta)d\theta \,.$$

Exercise 2.8 Examples of the risk function and Bayes risk

Consider the normal-normal model: $\quad (y_1,\ldots,y_n\,|\,\mu) \sim iid\ N(\mu,\sigma^2)$
$$\mu \sim N(\mu_0,\sigma_0^2)\,.$$

For each of the following estimators, derive a formulae for the risk function under the quadratic error loss function:

(a) $\hat{\mu} = \bar{y} = \dfrac{1}{n}(y_1 + \ldots + y_n)$ (the sample mean)

(b) $\hat{\mu} = |\bar{y}|$ (the absolute value of the sample mean).

In each case, use the derived risk function to determine the Bayes risk.

Solution to Exercise 2.8

For both parts of this exercise, the loss function is given by
$$L(\hat{\mu},\mu) = (\hat{\mu} - \mu)^2\,.$$

(a) If $\hat{\mu} = \bar{y}$ then the risk function is
$$R(\mu) = E\{L(\hat{\mu},\mu)\,|\,\mu\} = E\{(\bar{y} - \mu)^2\,|\,\mu\} = V(\bar{y}\,|\,\mu)$$
$$= \sigma^2/n \ \ (\text{a constant}).$$

So the Bayes risk is simply

$$r = ER(\mu) = E(\sigma^2 / n) = \sigma^2 / n \quad \text{(i.e. the same constant)}.$$

(b) If $\hat{\mu} = |\bar{y}|$ then the risk function is

$$R(\mu) = E\left\{(|\bar{y}| - \mu)^2 \mid \mu\right\} = E\left\{|\bar{y}|^2 - 2\mu|\bar{y}| + \mu^2 \mid \mu\right\}$$

$$= E(\bar{y}^2 \mid \mu) - 2\mu E\left(|\bar{y}| \mid \mu\right) + \mu^2$$

$$= \left(\frac{\sigma^2}{n} + \mu^2\right) - 2\mu m + \mu^2, \quad \text{where } m = E\left(|\bar{y}| \mid \mu\right).$$

Now,

$$m = \int_{-\infty}^{0} (-\bar{y}) f(\bar{y} \mid \mu) d\bar{y} + \int_{0}^{\infty} (+\bar{y}) f(\bar{y} \mid \mu) d\bar{y}$$

$$= -\int_{-\infty}^{0} \bar{y} f(\bar{y} \mid \mu) d\bar{y} + \int_{0}^{\infty} \bar{y} f(\bar{y} \mid \mu) d\bar{y} + \int_{-\infty}^{0} \bar{y} f(\bar{y} \mid \mu) d\bar{y} - \int_{-\infty}^{0} \bar{y} f(\bar{y} \mid \mu) d\bar{y}$$

$$= -2\int_{-\infty}^{0} \bar{y} f(\bar{y} \mid \mu) d\bar{y} + \int_{-\infty}^{\infty} \bar{y} f(\bar{y} \mid \mu) d\bar{y}$$

$$= \mu - 2I, \quad \text{where } I = \int_{-\infty}^{0} \bar{y} f(\bar{y} \mid \mu) d\bar{y}.$$

Here,

$$I = \int_{-\infty}^{-\mu/c} (\mu + cz) \phi(z) dz \quad \text{after putting } z = \frac{\bar{y} - \mu}{\sigma / \sqrt{n}} \text{ with } c = \frac{\sigma}{\sqrt{n}}$$

$$= \mu \int_{-\infty}^{-\mu/c} \phi(z) dz + c \int_{-\infty}^{-\mu/c} z \phi(z) dz$$

$$= \mu \Phi\left(-\frac{\mu}{c}\right) + cJ, \quad \text{where } J = \int_{-\infty}^{-\mu/c} z \phi(z) dz.$$

Note: Here, $\phi(z) = \dfrac{1}{\sqrt{2\pi}} e^{-\frac{1}{2}z^2}$ and $\Phi(z) = \displaystyle\int_{-\infty}^{z} \phi(t) dt$ are the standard normal pdf and cdf, respectively.

Now, $\quad J = \int\limits_{-\infty}^{-\mu/c} z \dfrac{1}{\sqrt{2\pi}} e^{-\frac{1}{2}z^2} dz$

$\qquad = \int\limits_{\infty}^{\frac{\mu^2}{2c^2}} \dfrac{1}{\sqrt{2\pi}} e^{-w} dw \qquad$ after substituting $w = \dfrac{1}{2}z^2$

$\qquad = \dfrac{1}{\sqrt{2\pi}}\left[-e^{-w}\Big|_{w=\infty}^{\frac{\mu^2}{2c^2}} \right] = \dfrac{1}{\sqrt{2\pi}}\left(-e^{-\frac{\mu^2}{2c^2}} + e^{-\infty} \right) = -\phi\!\left(\dfrac{\mu}{c}\right).$

Hence $\quad I = \mu\Phi\!\left(-\dfrac{\mu}{c}\right) + cJ = \mu\Phi\!\left(-\dfrac{\mu}{c}\right) - c\phi\!\left(\dfrac{\mu}{c}\right),$

and so $\quad m = \mu - 2I = \mu - 2\left[\mu\Phi\!\left(-\dfrac{\mu}{c}\right) - c\phi\!\left(\dfrac{\mu}{c}\right) \right].$

Therefore

$$R(\mu) = \dfrac{\sigma^2}{n} + 2\mu^2 - 2\mu m = \dfrac{\sigma^2}{n} + 2\mu^2 - 2\mu\left\{ \mu - 2\left[\mu\Phi\!\left(-\dfrac{\mu}{c}\right) - c\phi\!\left(\dfrac{\mu}{c}\right) \right] \right\}.$$

Thereby we obtain:

$$R(\mu) = \dfrac{\sigma^2}{n} + 4\mu^2\Phi\!\left(-\dfrac{\mu}{\sigma/\sqrt{n}}\right) - 4\mu\dfrac{\sigma}{\sqrt{n}}\phi\!\left(\dfrac{\mu}{\sigma/\sqrt{n}}\right), \quad \mu \in \Re.$$

The Bayes risk is then given by

$$r = ER(\mu) = \int R(\mu) f(\mu) d\mu = \int\limits_{-\infty}^{\infty} g(\mu) d\mu,$$

where

$$g(\mu) = \left\{ \dfrac{\sigma^2}{n} + 4\mu^2\Phi\!\left(-\dfrac{\mu}{\sigma/\sqrt{n}}\right) - 4\mu\dfrac{\sigma}{\sqrt{n}}\phi\!\left(\dfrac{\mu}{\sigma/\sqrt{n}}\right) \right\} \times \dfrac{1}{\sigma_0}\phi\!\left(\dfrac{\mu - \mu_0}{\sigma_0}\right).$$

We see that the Bayes risk r is an intractable integral equal to the area under the integrand, $g(\mu) = R(\mu)f(\mu)$. However, this area can be evaluated numerically (using techniques discussed later). Figures 2.11 and 2.12 show examples of the risk function $R(\mu)$ and the integrand function $g(\mu)$. For the case $n = \sigma = \mu_0 = \sigma_0 = 1$, we find that $r = 1.16$.

Figure 2.11 Some risk functions in Exercise 2.8

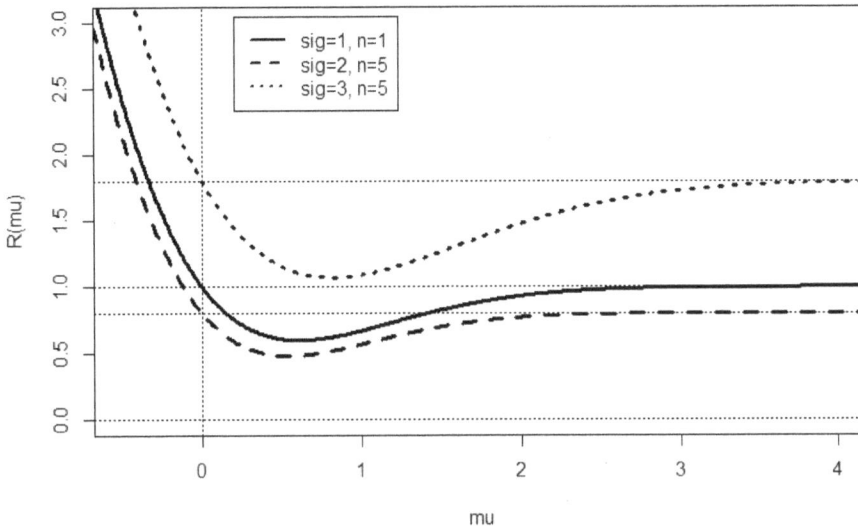

Figure 2.12 Some integrand functions used to calculate the Bayes risk

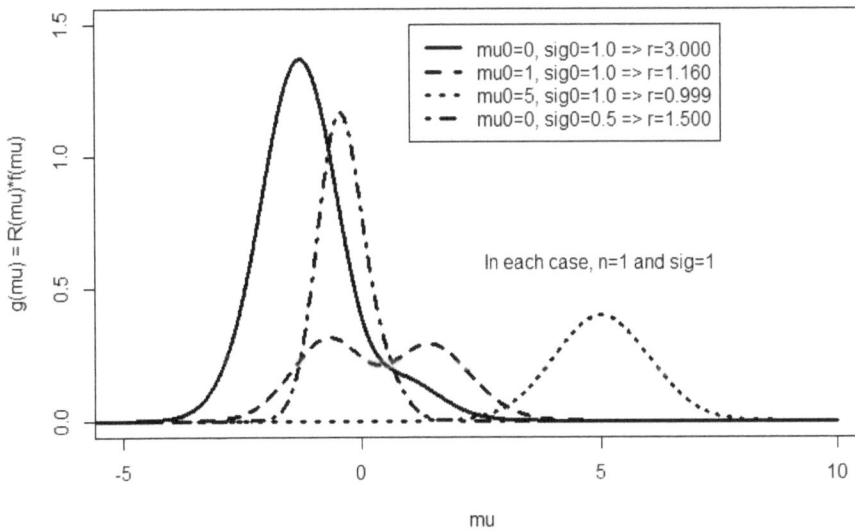

R Code for Exercise 2.8

```
Rfun=function(mu,sig,n){  sig^2/n+4*mu*(  mu*pnorm(-mu/(sig/sqrt(n))) -
                  (sig/sqrt(n))*dnorm(mu/(sig/sqrt(n)))  )  }
muvec=seq(-10,10,0.01); options(digits=4)

X11(w=8,h=5.5); par(mfrow=c(1,1));

plot(c(-0.5,4),c(0,3),type="n",xlab="mu",ylab="R(mu)",main=" ")

n=1; sig=1; lines(muvec,Rfun(muvec,sig=sig,n=n),lty=1,lwd=3);
      abline(v=0,lty=3); abline(h=c(0,sig^2/n),lty=3)
n=5; sig=2; lines(muvec,Rfun(muvec,sig=sig,n=n),lty=2,lwd=3);
      abline(h= sig^2/n,lty=3)
n=5; sig=3; lines(muvec,Rfun(muvec,sig=sig,n=n),lty=3,lwd=3);
      abline(h= sig^2/n,lty=3)
legend(0.2,3.05,c("sig=1, n=1","sig=2, n=5","sig=3, n=5"),
      lty=c(1,2,3),lwd=c(2,2,2))

Ifun = function(mu,sig,n,mu0,sig0){
      Rfun(mu=mu,sig=sig,n=n)*dnorm(mu,mu0,sig0)    }

plot(c(-5,10),c(0,1.5),type="n", xlab="mu",ylab="g(mu) = R(mu)*f(mu)",
      main=" ")
n=1; sig=1; mu0=0; sig0=1
lines(muvec, Ifun(mu=muvec,sig=sig,n=n,mu0=mu0, sig0=sig0),lty=1,lwd=3)
      # Check range over which to integrate the integrand
integrate(f=Ifun,lower=-7,upper=7, sig=sig,n=n,mu0=mu0, sig0=sig0)$value
      # 3

n=1; sig=1; mu0=1; sig0=1
lines(muvec, Ifun(mu=muvec,sig=sig,n=n,mu0=mu0, sig0=sig0),lty=2,lwd=3)
      # Check range over which to integrate the integrand
integrate(f=Ifun,lower=-7,upper=7, sig=sig,n=n,mu0=mu0, sig0=sig0)$value
      # 1.16

n=1; sig=1; mu0=5; sig0=1
lines(muvec, Ifun(mu=muvec,sig=sig,n=n,mu0=mu0, sig0=sig0),lty=3,lwd=3)
      # Check range over which to integrate the integrand
integrate(f=Ifun,lower=0,upper=10, sig=sig,n=n,mu0=mu0, sig0=sig0)$value
      # 0.9994
```

n=1; sig=1; mu0=0; sig0=0.5
lines(muvec, lfun(mu=muvec,sig=sig,n=n,mu0=mu0, sig0=sig0),lty=4,lwd=3)
        # Check range over which to integrate the integrand
integrate(f=lfun,lower=-5,upper=5, sig=sig,n=n,mu0=mu0, sig0=sig0)$value
        # 1.5

legend(1,1.5,c("mu0=0, sig0=1.0 => r=3.000", "mu0=1, sig0=1.0 => r=1.160",
        "mu0=5, sig0=1.0 => r=0.999","mu0=0, sig0=0.5 => r=1.500"),
        lty=c(1,2,3,4),lwd=c(3,3,3,3)); text(5,0.6,"In each case, n=1 and sig=1")

2.6 The posterior expected loss

We have defined the risk function as the expectation of the loss function
given the parameter, namely

$$R(\theta) = E(L(\hat{\theta},\theta)\,|\,\theta) = \int L(\hat{\theta}(y),\theta)f(y\,|\,\theta)dy\,.$$

Conversely, we now define the *posterior expected loss* (PEL) as the
expectation of the loss function given the *data*, and we denote this
function by

$$PEL(y) = E\{L(\hat{\theta},\theta)\,|\,y\} = \int L(\hat{\theta}(y),\theta)f(\theta\,|\,y)d\theta\,.$$

Then, just as the risk function can be used to compute the Bayes risk
according to

$$r = EL(\hat{\theta},\theta) = EE\{L(\hat{\theta},\theta)\,|\,\theta\} = ER(\theta) = \int R(\theta)f(\theta)d\theta\,,$$

so also can the PEL be used, but with the formula

$$r = EL(\hat{\theta},\theta) = EE\{L(\hat{\theta},\theta)\,|\,y\} = E\{PEL(y)\} = \int PEL(y)f(y)dy\,.$$

Note: Both of these formulae for the Bayes risk use the law of iterated
expectation, but with different conditionings.

Exercise 2.9   Examples of the PEL and Bayes risk

Consider the normal-normal model:

$$(y_1,\ldots,y_n\,|\,\mu) \sim iid\ N(\mu,\sigma^2)$$
$$\mu \sim N(\mu_0,\sigma_0^2)\,.$$

93

For each of the following estimators, derive a formula for the posterior expected loss under the quadratic error loss function:

(a) $\hat{\mu} = \bar{y} = \dfrac{1}{n}(y_1 + ... + y_n)$ (the sample mean)

(b) $\hat{\mu} = |\bar{y}|$ (the absolute value of the sample mean).

In each case, use the derived PEL to obtain the Bayes risk.

Note: This exercise is an extension of Exercise 2.8.

Solution to Exercise 2.9

(a) If $\hat{\mu} = \bar{y}$ then the PEL function is

$$
\begin{aligned}
PEL(y) &= E\{L(\hat{\mu}, \mu) \mid y\} \\
&= E\{(\bar{y} - \mu)^2 \mid y\} \\
&= \bar{y}^2 - 2\bar{y}E(\mu \mid y) + E(\mu^2 \mid y),
\end{aligned}
$$

where:

$$
E(\mu \mid y) = \mu_*
$$
$$
E(\mu^2 \mid y) = V(\mu \mid y) + \{E(\mu \mid y)\}^2
$$
$$
= \sigma_*^2 + \mu_*^2
$$
$$
\mu_* = (1-k)\mu_0 + k\bar{y}, \quad \sigma_*^2 = k\frac{\sigma^2}{n}, \quad k = \frac{n}{n + \sigma^2/\sigma_0^2}.
$$

Thus, more explicitly,

$$
PEL(y) = \bar{y}^2 - 2\bar{y}\{(1-k)\mu_0 + k\bar{y}\} + \sigma_*^2 + \{(1-k)\mu_0 + k\bar{y}\}^2
$$

$$
= \bar{y}^2 - 2(1-k)\mu_0\bar{y} - 2k\bar{y}^2 + \sigma_*^2 + (1-k)^2\mu_0^2 + 2(1-k)\mu_0 k\bar{y} + k^2\bar{y}^2
$$

$$
= \bar{y}^2(1-k)^2 - \bar{y}(1-k)^2 2\mu_0 + \sigma_*^2 + (1-k)^2\mu_0^2
$$

$$
= \sigma_*^2 + (1-k)^2(\bar{y} - \mu_0)^2.
$$

Note: This is a quadratic in $\bar{y}$ with a minimum of σ_*^2 at $\bar{y} = \mu_0$.

The Bayes risk is then

$$r = E\{PEL(y)\}$$
$$= \sigma_*^2 + (1-k)^2 E\{(\bar{y}-\mu_0)^2\},$$

where

$$E\{(\bar{y}-\mu_0)^2\} = V\bar{y}$$
$$= EV(\bar{y}\mid\mu) + VE(\bar{y}\mid\mu)$$
$$= E\left(\frac{\sigma^2}{n}\right) + V\mu$$
$$= \frac{\sigma^2}{n} + \sigma_0^2.$$

Thus $\quad r = \sigma_*^2 + (1-k)^2\left(\dfrac{\sigma^2}{n} + \sigma_0^2\right)$

$$= k\frac{\sigma^2}{n} + (1-k)^2\left(\frac{\sigma^2}{n} + \sigma_0^2\right) \qquad \text{(where } k = \frac{n}{n + \sigma^2/\sigma_0^2}\text{)}$$

$$= \frac{\sigma^2}{n} \quad \text{(after a little algebra)}.$$

Note: This is in agreement with Exercise 2.8, where the result was obtained much more easily by taking the mean of the risk function, as follows:

$$r = ER(\mu) = E(\sigma^2/n) = \sigma^2/n.$$

(b) If $\hat{\mu} = |\bar{y}|$ then the posterior expected loss function is

$$PEL(y) = E\left\{(|\bar{y}|-\mu)^2 \mid y\right\}$$
$$= \bar{y}^2 - 2|\bar{y}|E(\mu\mid y) + E(\mu^2\mid y)$$
$$= \bar{y}^2 - 2|\bar{y}|\mu_* + \sigma_*^2 + \mu_*^2$$
$$= \bar{y}^2 - 2|\bar{y}|\{(1-k)\mu_0 + k\bar{y}\} + \sigma_*^2 + \{(1-k)\mu_0 + k\bar{y}\}^2.$$

Some examples of this PEL function are shown in Figure 2.13. In all these examples, $n = \sigma = 1$.

Figure 2.13 Some posterior expected loss functions

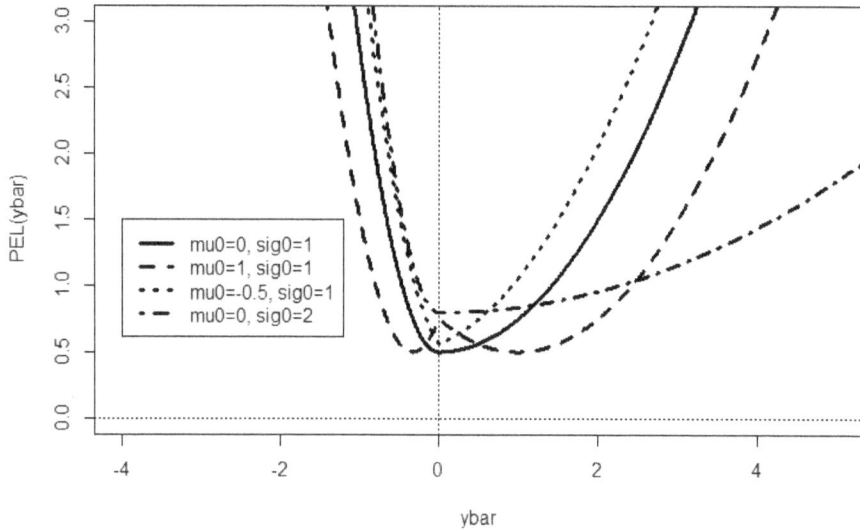

In terms of the PEL function, the Bayes risk can be expressed as

$$r = E\{PEL(y)\} = \int_{-\infty}^{\infty} PEL(\bar{y})f(\bar{y})d\bar{y},$$

where

$$f(\bar{y}) = f_{N\left(\mu_0, \sigma_0^2 + \sigma^2/n\right)}(\bar{y}),$$

since

$$\bar{y} \sim N\left(\mu_0, \sigma_0^2 + \frac{\sigma^2}{n}\right).$$

As an example, we consider the case $n = \sigma = \mu_0 = \sigma_0 = 1$. Figure 2.14 shows the integrand function $PEL(\bar{y})f(\bar{y})$. The area under this function works out as 1.16, in agreement with an alternative working for the Bayes risk in Exercise 2.8 (taking an expectation of the risk function).

Figure 2.14 An integrand function with area underneath equal to 1.16

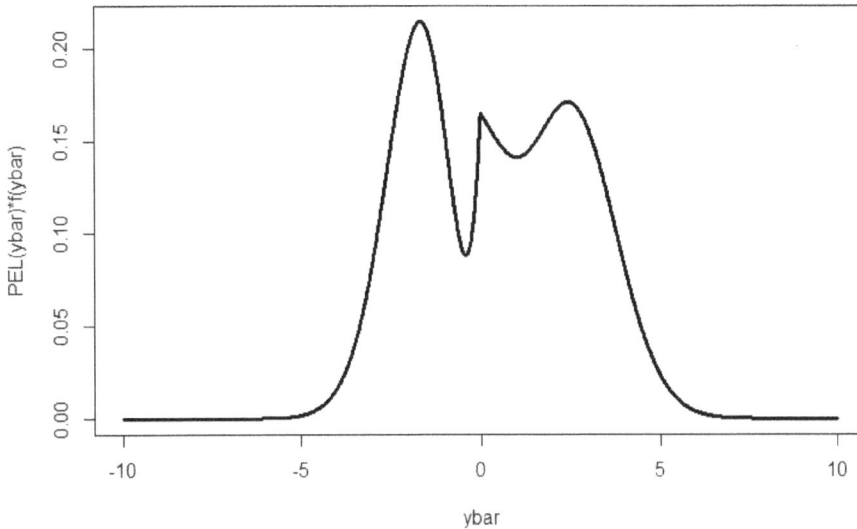

R Code for Exercise 2.9

```
PELfun=function(ybar,sig,n,sig0,mu0){
        k=n/(n+sig^2/sig0^2)
        mustar=(1-k)*mu0+k*ybar
        sigstar2=k*sig^2/n
        ybar^2-2*abs(ybar)*mustar+sigstar2 + mustar^2
        }

ybarvec=seq(-10,10,0.01); options(digits=4)
X11(w=8,h=5.5); par(mfrow=c(1,1));

plot(c(-4,5),c(0,3),type="n",xlab="ybar",ylab="PEL(ybar)", main=" ")
abline(v=0,lty=3); abline(h=0,lty=3)

n=1; sig=1; mu0=0; sig0=1
lines(ybarvec,PELfun(ybarvec,sig=sig,n=n,sig0=sig0,mu0=mu0),lty=1,lwd=3);

n=1; sig=1; mu0=1; sig0=1
lines(ybarvec,PELfun(ybarvec,sig=sig,n=n,sig0=sig0,mu0=mu0),lty=2,lwd=3);
```

```
n=1; sig=1; mu0=-0.5; sig0=1
lines(ybarvec,PELfun(ybarvec,sig=sig,n=n,sig0=sig0,mu0=mu0),lty=3,lwd=3);

n=1; sig=1; mu0=0; sig0=2
lines(ybarvec,PELfun(ybarvec,sig=sig,n=n,sig0=sig0,mu0=mu0),lty=4,lwd=3);

legend(-4,1.5,c("mu0=0, sig0=1","mu0=1, sig0=1","mu0=-0.5, sig0=1",
        "mu0=0, sig0=2"), lty=c(1,2,3,4), lwd=c(3,3,3,3))

# Calculate r when n=1, sig=1, mu0=1, sig0=1 (should get 1.16 as before)

Jfun = function(ybar,sig,n,sig0,mu0){
        PELfun(ybar=ybar,sig=sig,n=n,sig0=sig0,mu0=mu0)*
        dnorm(ybar,mu0,sqrt(sig0^2+sig^2/n))
        }

n=1; sig=1; mu0=1; sig0=1

plot(ybarvec,    PELfun(ybar=ybarvec,sig=sig,n=n,sig0=sig0,mu0=mu0)*
        dnorm(ybarvec,mu0,sqrt(sig0^2+sig^2/n)),
        type="l", xlab="ybar",ylab="PEL(ybar)*f(ybar)", lwd=3)

integrate(f=Jfun,lower=-10,upper=10, sig=sig,n=n,mu0=mu0, sig0=sig0)$value
        # 1.16   Correct (same as in last exercise)
```

2.7 The Bayes estimate

The *Bayes estimate* (or *estimator*) is defined to be the choice of the function $\hat{\theta} = \hat{\theta}(y)$ for which the Bayes risk $r = EL(\hat{\theta},\theta)$ is minimised. This estimator has the smallest overall expected loss over all estimators under the specified loss function $L(\hat{\theta},\theta)$.

In many cases, the procedure for finding a Bayes estimate can be considerably simplified by considering which estimate minimises the posterior expected loss function, $PEL(y) = E\{L(\hat{\theta},\theta)\,|\,y\}$.

If we can find an estimate $\hat{\theta} = \hat{\theta}(y)$ which minimises $PEL(y)$ for *all* possible values of the data y, then that estimate must also minimise the Bayes risk.

This is because the Bayes risk may be written as a *weighted average of the PEL*, namely

$$r = EL(\hat{\theta}, \theta) = EE\{L(\hat{\theta}, \theta) \mid y\} = E\{PEL(y)\} = \int PEL(y)f(y)dy.$$

Exercise 2.10 Bayes estimate under the QELF

Find the Bayes estimate under the quadratic error loss function.

Solution to Exercise 2.10

Observe that
$$\begin{aligned}
PEL(y) &= E\{(\hat{\theta} - \theta)^2 \mid y\} = E\{\hat{\theta}^2 - 2\hat{\theta}\theta + \theta^2 \mid y\} \\
&= \hat{\theta}^2 - 2\hat{\theta}E(\theta \mid y) + E(\theta^2 \mid y) \\
&= \left[\hat{\theta} - E(\theta \mid y)\right]^2 - \{E(\theta \mid y)\}^2 + E(\theta^2 \mid y).
\end{aligned}$$

Note: We have completed the square in $\hat{\theta}$.

We see that the PEL is a quadratic function of $\hat{\theta}$ which is clearly minimised at the posterior mean, $\hat{\theta} = E(\theta \mid y)$. So the Bayes estimate under the QELF is that posterior mean.

Note 1: This result can also be obtained using *Leibniz's rule for differentiating an integral*, which is generally

$$\frac{d}{dx}\int_a^b G(u, x)du = \int_a^b \frac{\partial G(u, x)}{\partial x}du + G(b, x)\frac{db}{dx} - G(a, x)\frac{da}{dx}$$

and which reduces to $\displaystyle\int_a^b \frac{\partial G(u, x)}{\partial x}du + 0 - 0$ if a and b are constants.

Thus we may write
$$\begin{aligned}
\frac{\partial}{\partial \hat{\theta}}PEL(y) &= \frac{\partial}{\partial \hat{\theta}}\int (\hat{\theta} - \theta)^2 f(\theta \mid y)d\theta \\
&= \int \frac{\partial}{\partial \hat{\theta}}\{(\hat{\theta} - \theta)^2 f(\theta \mid y)\}d\theta + 0 - 0 \\
&= \int 2(\hat{\theta} - \theta)^1 f(\theta \mid y)d\theta = 2\{\hat{\theta} - \int \theta f(\theta \mid y)d\theta\}.
\end{aligned}$$

Setting this to zero yields $\hat{\theta} = \int \theta f(\theta \mid y)d\theta = E(\theta \mid y)$.

Note 2: To check that this *minimises* the PEL (rather than *maximises* it) we may further calculate

$$\frac{\partial^2}{\partial \hat{\theta}^2} PEL(y) = 2 \frac{\partial}{\partial \hat{\theta}} \left\{ \hat{\theta} - \int \theta f(\theta \mid y) d\theta \right\} = 2\{1 - 0\} > 0.$$

Thus the slope of the PEL ($\partial PEL(y) / \partial \hat{\theta}$) is *increasing* with $\hat{\theta}$, implying that $PEL(y)$ is indeed *minimised* at $\hat{\theta} = \hat{\theta}(y) = E(\theta \mid y)$.

Exercise 2.11 Bayes estimate under the AELF

Find the Bayesian estimate under the absolute error loss function.

Solution to Exercise 2.11

Suppose that the parameter θ is *continuous*, and let t denote $\hat{\theta} = \hat{\theta}(y)$.

Then
$$PEL(y) = \int_{-\infty}^{\infty} |t - \theta| f(\theta \mid y) d\theta$$

$$= \int_{-\infty}^{t} (t - \theta) f(\theta \mid y) d\theta + \int_{t}^{\infty} (\theta - t) f(\theta \mid y) d\theta.$$

So, by Leibniz's rule for differentiation of an integral (in Exercise 2.10),

$$\frac{\partial}{\partial t} PEL(y) = \left[\int_{-\infty}^{t} \frac{\partial (t - \theta)}{\partial t} f(\theta \mid y) d\theta + \{(t - t) f(\theta = t \mid y)\} \frac{dt}{dt} - (\cdot) \frac{d(-\infty)}{dt} \right]$$

$$+ \left[\int_{t}^{\infty} \frac{\partial (\theta - t)}{\partial t} f(\theta \mid y) d\theta + (\cdot) \frac{d(\infty)}{dt} - \{(t - t) f(\theta = t \mid y)\} \frac{dt}{dt} \right]$$

$$= \left[\int_{-\infty}^{t} f(\theta \mid y) d\theta + 0 - 0 \right] + \left[\int_{t}^{\infty} (-1) f(\theta \mid y) d\theta + 0 - 0 \right]$$

$$= P(\theta < t \mid y) - P(\theta > t \mid y).$$

Setting this to zero implies $P(\theta < t \mid y) = P(\theta > t \mid y)$ which yields t as the posterior median. So the Bayes estimate under the AELF is the posterior median. This argument can easily be adapted to the case where θ is *discrete*. The idea is to approximate θ's discrete prior distribution with a continuous distribution and then apply the result already proved.

Exercise 2.12 Bayes estimate under the IELF

Find the Bayes estimate under the indicator error loss function.

Solution to Exercise 2.12

Let t denote $\hat{\theta} = \hat{\theta}(y)$ and first suppose that the parameter θ is *discrete*. The indicator error loss function is $L(t, \theta) = I(t \neq \theta) = 1 - I(t = \theta)$. Therefore

$$\begin{aligned} PEL(y) = E\{L(t,\theta) \mid y\} = E\{1 - I(t = \theta) \mid y\} &= 1 - E\{I(t = \theta) \mid y\} \\ &= 1 - P(t = \theta \mid y) \\ &= 1 - f(\theta = t \mid y). \end{aligned}$$

Thus $PEL(y)$ is *minimised* at the value of t which *maximises* the posterior density $f(\theta \mid y)$. So, when θ is discrete, the Bayes estimate under the IELF is the posterior mode, $Mode(\theta \mid y)$.

Now suppose that θ is *continuous*. In that case, consider the approximating loss function

$$L_\varepsilon(t, \theta) = 1 - I(t - \varepsilon < \theta < t + \varepsilon),$$

where $\varepsilon > 0$, and observe that

$$\lim_{\varepsilon \to 0} L_\varepsilon(t, \theta) = 1 - I(t = \theta) = L(t, \theta).$$

The posterior expected loss under the loss function $L_\varepsilon(t, \theta)$ is

$$\begin{aligned} PEL_\varepsilon(y) = E\{L_\varepsilon(t,\theta) \mid y\} &= 1 - E\{I(t - \varepsilon < \theta < t + \varepsilon) \mid y\} \\ &= 1 - P(t - \varepsilon < \theta < t + \varepsilon \mid y). \end{aligned}$$

The value of t which minimises the $PEL_\varepsilon(y)$ is the value which maximises the area $P(t - \varepsilon < \theta < t + \varepsilon \mid y)$. But in the limit as $\varepsilon \to 0$, that value is the posterior mode. So, when θ is continuous, the Bayes estimate under the IELF is (as before) the posterior mode, $Mode(\theta \mid y)$.

Note: To clarify the above argument, observe that if ε is small then

$$PEL_\varepsilon(t) \approx 1 - 2\varepsilon f_\theta(t \mid y).$$

This function of t is minimised at *approximately $t = Mode(\theta \mid y)$* and at *exactly $t = Mode(\theta \mid y)$* in the limit as $\varepsilon \to 0$. Figure 2.15 illustrates.

Figure 2.15 Illustration for the continuous case in Exercise 2.12

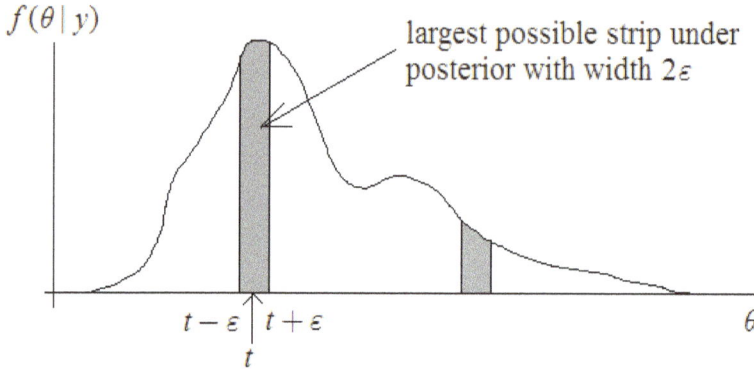

Exercise 2.13 Bayesian decision theory in the Poisson-gamma model

Consider a random sample $y_1, ..., y_n$ from the Poisson distribution with parameter λ whose prior density is gamma with parameters α and β.

(a) Find the risk function, Bayes risk and posterior expected loss implied by the estimator $\hat{\lambda} = 2\bar{y}$ under the quadratic error loss function.

(b) Assuming quadratic error loss, find an estimator of λ with a smaller Bayes risk than the one in (a).

Solution to Exercise 2.13

(a) The risk function is

$$
\begin{aligned}
R(\lambda) &= E\{L(\hat{\lambda}, \lambda) \mid \lambda\} \\
&= E\left\{(2\bar{y} - \lambda)^2 \big| \lambda\right\} \\
&= E\left\{4\bar{y}^2 - 4\bar{y}\lambda + \lambda^2 \big| \lambda\right\} \\
&= 4E\left\{\bar{y}^2 \big| \lambda\right\} - 4\lambda E\left\{\bar{y} \big| \lambda\right\} + \lambda^2 \\
&= 4\left[V\left\{\bar{y} \big| \lambda\right\} + E\left\{\bar{y} \big| \lambda\right\}^2\right] - 4\lambda E\left\{\bar{y} \big| \lambda\right\} + \lambda^2 \\
&= 4\left(\frac{\lambda}{n} + \lambda^2\right) - 4\lambda\lambda + \lambda^2 \\
&= \lambda^2 + 4\lambda / n, \ \lambda > 0 \quad \text{(an increasing quadratic).}
\end{aligned}
$$

So the Bayes risk is

$$r = ER(\hat{\lambda}, \lambda)$$
$$= E(\lambda^2 + 4\lambda/n)$$
$$= \{V\lambda + (E\lambda)^2\} + 4(E\lambda)/n$$
$$= \frac{\alpha}{\beta^2} + \left(\frac{\alpha}{\beta}\right)^2 + \frac{4\alpha}{n\beta}.$$

To find the posterior expected loss, we first derive λ's posterior density:

$$f(\lambda \mid y) \propto f(\lambda) f(y \mid \lambda)$$
$$= \frac{\beta^\alpha \lambda^{\alpha-1} e^{-\beta\lambda}}{\Gamma(\alpha)} \times \prod_{i=1}^{n} \frac{e^{-\lambda}\lambda^{y_i}}{y_i!}$$
$$\overset{\lambda}{\propto} \lambda^{\alpha+n\bar{y}-1} e^{-\lambda(\beta+n)}.$$

We see that

$$f(\lambda \mid y) \sim \text{Gam}(\alpha + n\bar{y}, \beta + n).$$

It follows that

$$PEL(y) = E\{L(\hat{\lambda}, \lambda) \mid y\}$$
$$= E\left\{(2\bar{y} - \lambda)^2 \middle| y\right\}$$
$$= E\left\{4\bar{y}^2 - 4\bar{y}\lambda + \lambda^2 \middle| y\right\}$$
$$= 4\bar{y}^2 - 4\bar{y}E(\lambda \mid y) + E(\lambda^2 \mid y)$$
$$= 4\bar{y}^2 - 4\bar{y}\left(\frac{\alpha + n\bar{y}}{\beta + n}\right) + \left\{\frac{\alpha + n\bar{y}}{(\beta + n)^2} + \left(\frac{\alpha + n\bar{y}}{\beta + n}\right)^2\right\}.$$

Note: The Bayes risk could also be computed using an argument which begins as follows:

$$r = E\{PEL(y)\}$$
$$= E\left\{4\bar{y}^2 - 4\bar{y}\left(\frac{\alpha + n\bar{y}}{\beta + n}\right) + \frac{\alpha + n\bar{y}}{(\beta + n)^2} + \left(\frac{\alpha + n\bar{y}}{\beta + n}\right)^2\right\},$$

where, for example,

$$E\bar{y} = EE(\bar{y} \mid \lambda) = EE(y_1 \mid \lambda) = E\lambda = \alpha/\beta.$$

(b) The Bayes estimate under the QELF is the posterior mean,

$$E(\lambda \mid y) = \frac{\alpha + n\bar{y}}{\beta + n}.$$

This estimator has the smallest Bayes risk amongst all possible estimators, including the one in (a), which is different. So $E(\lambda \mid y)$ must have a smaller Bayes risk than the estimator in (a).

Discussion

The last statement could be verified by calculating r according to

$$E\left(\frac{\alpha + n\bar{Y}}{\beta + n} - \lambda\right)^2.$$

The result should be an expression for r which is smaller than

$$\frac{\alpha}{\beta^2} + \left(\frac{\alpha}{\beta}\right)^2 + \frac{4\alpha}{n\beta},$$

for all $n = 1, 2, 3, \dots$, and all $\alpha, \beta > 0$.

We leave the required working as an additional exercise.

Exercise 2.14 A non-standard loss function

Consider the Bayesian model given by:
$$(y \mid \mu) \sim N(\mu, 1)$$
$$\mu \sim N(0, 1).$$

Then suppose that the loss function is
$$L(t, \mu) = \begin{cases} 0 & \text{if } 0 < \mu < t < 2\mu \\ 1 & \text{otherwise.} \end{cases}$$

(a) Find the risk function and Bayes risk for the estimator $\hat{\mu} = y$.

Sketch the risk function.

(b) Find the Bayes estimate and sketch it as a function of the data y.

Explicitly calculate the Bayes estimate at $y = -1$, 0 and 1, respectively.

Solution to Exercise 2.14

(a) For convenience we will sometimes denote $\hat{\mu} = y$ by t. Then, the loss function may be written as

$$L(t,\mu) = \begin{cases} 1 - I(\mu < t < 2\mu), \mu > 0 \\ \qquad 1, \qquad \mu \leq 0. \end{cases}$$

Now, for $\mu \leq 0$ the risk function is simply
$$R(\mu) = E\{L(y,\mu) \,|\, \mu\} = 1.$$

For $\mu > 0$, the risk function is
$$\begin{aligned} R(\mu) &= 1 - P(\mu < y < 2\mu \,|\, \mu) \;\; = 1 - P(0 < y - \mu < \mu \,|\, \mu) \\ &= 1 - P(0 < Z < \mu) \quad \text{where } Z \sim N(0,1) \\ &= 1 - (\Phi(\mu) - 1/2) \;\; = 1.5 - \Phi(\mu). \end{aligned}$$

In summary, $R(\mu) = \begin{cases} 1, & \mu \leq 0 \\ 1.5 - \Phi(\mu), & \mu > 0 \end{cases}$, as shown in Figure 2.16.

Figure 2.16 Risk function in Exercise 2.14

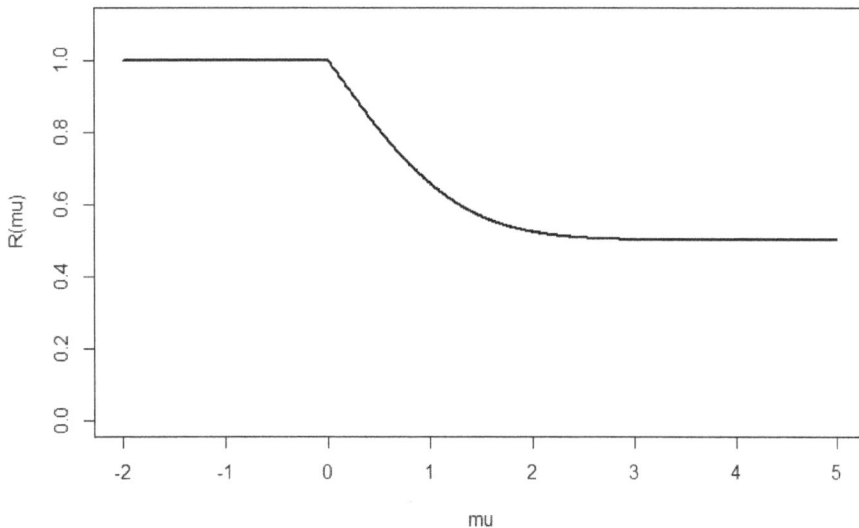

The associated Bayes risk is

$$r = ER(\mu) = \int_{-\infty}^{0} \phi(\mu)d\mu + \frac{3}{2}\int_{0}^{\infty} \phi(\mu)d\mu - \int_{0}^{\infty} \Phi(\mu)\phi(\mu)d\mu$$

$$= \frac{1}{2} + \frac{3}{2} \times \frac{1}{2} - I,$$

where $I = \int_{1/2}^{1} wdw$ = 3/8, after putting $w = \Phi(\mu)$ with $\dfrac{dw}{d\mu} = \phi(\mu)$.

So, for the estimator $\hat{\mu} = y$, the Bayes risk is

$$r = \frac{1}{2} + \frac{3}{2} \times \frac{1}{2} - \frac{3}{8} = \frac{7}{8}.$$

(b) Here, by the theory of the normal-normal model we have that
$$(\mu \mid y) \sim N(\mu_*, \sigma_*^2),$$
where:

$$\mu_* = (1-k)\mu_0 + k\bar{y}, \quad \sigma_*^2 = k\sigma^2/n, \quad k = 1/(1 + \sigma^2/(n\sigma_0^2))$$
$$n = 1, \quad \mu_0 = 0, \quad \sigma_0 = 1, \quad \bar{y} = y.$$

Thus $k = 1/2$, $\mu_* = y/2$ and $\sigma_*^2 = 1/2$, and so
$$(\mu \mid y) \sim N(y/2, 1/2).$$

The posterior expected loss is
$$PEL(y) = E\{L(t,\mu) \mid y\},$$
where t is a function of y (i.e. $t = t(y)$).

Now
$$L(t,\mu) = 1 - I(0 < \mu < t < 2\mu),$$
and so
$$PEL(y) = E\{1 - I(0 < \mu < t < 2\mu) \mid y\}$$
$$= 1 - P(0 < \mu < t < 2\mu \mid y).$$

We see that if $t = t(y) \le 0$ then
$$PEL(y) = 1.$$

Also, if $t > 0$ then
$$PEL(t) = 1 - E\{I(0 < \mu < t < 2\mu) \mid y\}.$$
$$= 1 - P(0 < \mu < t < 2\mu \mid y)$$
$$= 1 - P(t/2 < \mu < t \mid y)$$
$$= 1 - \psi(t),$$

where
$$\psi(t) = F(\mu = t \mid y) - F(\mu = t/2 \mid y)$$
is to be maximised.

Now, $\psi'(t) = f(\mu = t \mid y) - f(\mu = t/2 \mid y) \times 1/2$
$$= \frac{1}{(1/\sqrt{2})\sqrt{2\pi}} e^{-\frac{1}{2(1/2)}(t-y/2)^2} - \frac{1}{2} \times \frac{1}{(1/\sqrt{2})\sqrt{2\pi}} e^{-\frac{1}{2(1/2)}((t/2)-y/2)^2}.$$

Setting $\psi'(t)$ to zero we obtain
$$2e^{-(t-y/2)^2} = e^{-((t/2)-y/2)^2}$$
$$\Rightarrow \log 2 - \left\{t^2 - 2t\left(\frac{y}{2}\right) + \left(\frac{y}{2}\right)^2\right\} = -\left\{\left(\frac{t}{2}\right)^2 + 2\left(\frac{t}{2}\right)\left(\frac{y}{2}\right) - \left(\frac{y}{2}\right)^2\right\}$$
$$\Rightarrow \frac{3}{4}t^2 - \frac{1}{2}ty - \log 2 = 0$$
$$\Rightarrow t = \frac{\frac{y}{2} \pm \sqrt{\frac{y^2}{4} + 4 \times \frac{3}{4}\log 2}}{2(3/4)}.$$

Hence we find that the Bayes estimate of μ is given by
$$\hat{\mu} = \hat{\mu}(y) = \frac{1}{3}\left(y + \sqrt{y^2 + 12\log 2}\right),$$
as shown in Figure 2.17.

We see that the Bayes estimate is a strictly increasing function of y and converges to zero as y tends to negative infinity. The required values of the Bayes estimate are:
$$\hat{\mu}(-1) = \frac{1}{3}\left(-1 + \sqrt{1 + 12\log 2}\right) = 0.6842$$
$$\hat{\mu}(0) = \frac{1}{3}\left(0 + \sqrt{0 + 12\log 2}\right) = 0.9614$$
$$\hat{\mu}(1) = \frac{1}{3}\left(1 + \sqrt{1 + 12\log 2}\right) = 1.3508.$$

Figure 2.17 Bayes estimate in Exercise 2.14

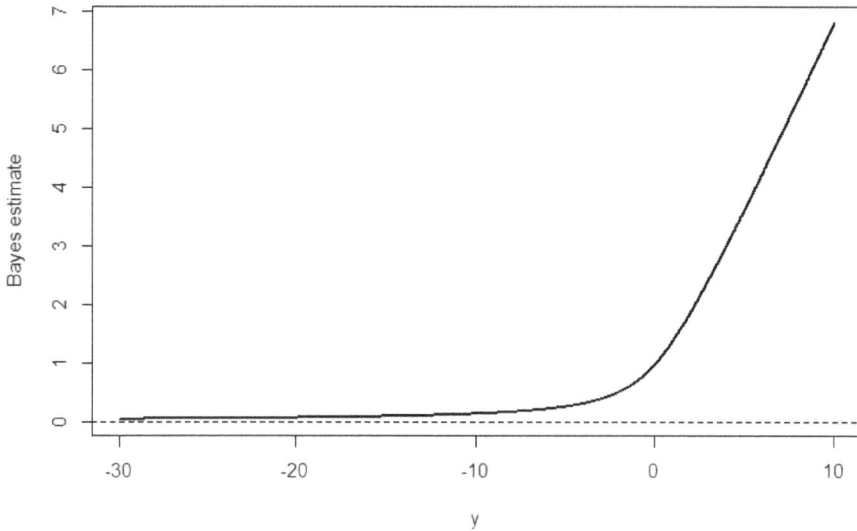

R Code for Exercise 2.14

```
X11(w=8,h=5.5)

muvec <- seq(0,5,0.01) ; Rvec <- 1.5-pnorm(muvec);
plot(c(-2,5),c(0,1.1),type= "n",xlab="mu",ylab="R(mu)",cex=1.5)
lines(muvec,Rvec,lwd=2) ; lines(c(-2,0),c(1,1),lwd=2)

yvec <- seq(-30,10,0.01); muhatvec <- (1/3)*(yvec+sqrt(yvec^2 + 12*log(2)))
plot(yvec,muhatvec,type="l",xlab="y",ylab="Bayes estimate",cex=1.5,lwd=2)
abline(h=0,lty=2)

(1/3)*(c(-1,0,1)+sqrt(c(-1,0,1)^2 + 12*log(2)))
# 0.6841672 0.9613513 1.3508339
```

CHAPTER 3

Bayesian Basics Part 3

3.1 Inference given functions of the data

Sometimes we observe a *function* of the data rather than the data itself. In such cases the function typically *degrades* the information available in some way. An example is *censoring*, where we observe a value only if that value is less than some cut-off point (right censoring) or greater than some cut-off value (left censoring). It is also possible to have censoring on the left and right simultaneously. Another example is *rounding*, where we only observe values to the nearest multiple of 0.1, 1 or 5, etc.

Exercise 3.1 Right censoring of exponential observations

Each light bulb of a certain type has a life which is conditionally exponential with mean $m = 1/c$, where c has a prior distribution which is standard exponential. We observe $n = 5$ light bulbs of this type for 6 units of time, and the lifetimes are:

$$2.6, 3.2, *, 1.2, *,$$

where $*$ indicates a right-censored value which is greater than 6. (Only values less than or equal to 6 could be observed.)

Find the posterior distribution and mean of the average light bulb lifetime, m.

Solution to Exercise 3.1

The data here is

$$D = \{ y_1 = 2.6, y_2 = 3.2, y_3 > 6, y_4 = 1.2, y_5 > 6 \},$$

and the probability of censoring is

$$P(y_i > 6 \mid c) = \int_6^\infty c e^{-cy_i} dy_i = e^{-6c}.$$

Therefore the posterior density of c is

$$f(c \mid D) \propto f(c) f(D \mid c)$$

$$\propto f(c) f(y_1 \mid c) f(y_2 \mid c) P(y_3 > 6 \mid c) f(y_4 \mid c) P(y_5 > 6 \mid c)$$

$$\propto e^{-c} \left(ce^{-cy_1} \right) \left(ce^{-cy_2} \right) \left(e^{-6c} \right) \left(ce^{-cy_4} \right) \left(e^{-6c} \right)$$
$$= c^3 \exp\{-c(1 + y_1 + y_2 + 6 + y_4 + 6)\}$$
$$= c^{4-1} \exp\{-c(1 + 2.6 + 3.2 + 6 + 1.2 + 6)\}$$
$$= c^{4-1} \exp(-20c).$$

Hence: $(c \mid D) \sim G(4, 20)$

$\qquad (m \mid D) \sim IG(4, 20)$

$\qquad f(m \mid D) = 20^4 m^{-(4+1)} e^{-20/m} / \Gamma(4), \, m > 0$

$\qquad E(m \mid D) = 20 / (4 - 1) = 6.667.$

It will be observed that this estimate of m is appropriately higher than the estimate obtained by simply averaging the observed values, namely
$\qquad (1/3)(2.6 + 3.2 + 1.2) = 2.333.$

The estimate 6.667 is also higher than the estimate obtained by simply replacing the censored values with 6, namely
$\qquad (1/3)(2.6 + 3.2 + 6 + 1.2 + 6) = 3.8.$

Exercise 3.2 A uniform-uniform model with rounded data

Suppose that:
$\qquad (y \mid \theta) \sim U(0, \theta)$
$\qquad \theta \sim U(0, 2),$
where the data is
$\qquad x = g(y)$ = the value of y rounded to the nearest integer.

Find the posterior density and mean of θ if we observe $x = 1$.

Solution to Exercise 3.2

Observe that:
$\qquad x = 0 \quad \text{if } 0 < y < 1/2$
$\qquad x = 1 \quad \text{if } 1/2 < y < 3/2$
$\qquad x = 2 \quad \text{if } 3/2 < y < 2.$

Therefore, considering y and θ on a number line from 0 to 2 in each case, we have that:

$$P(x=0\,|\,\theta) = P\left(0 < y < \frac{1}{2}\Big|\theta\right) = \begin{cases} 1 & \text{if } \theta < 1/2 \\[2mm] \dfrac{1/2}{\theta} & \text{if } \theta > 1/2 \end{cases}$$

$$P(x=1\,|\,\theta) = P\left(\frac{1}{2} < y < \frac{3}{2}\Big|\theta\right) = \begin{cases} 0 & \text{if } 0 < \theta < 1/2 \\[2mm] \dfrac{\theta-1/2}{\theta} & \text{if } \dfrac{1}{2} < \theta < \dfrac{3}{2} \\[2mm] \dfrac{1}{\theta} & \text{if } \dfrac{3}{2} < \theta < 2 \end{cases}$$

$$P(x=2\,|\,\theta) = P\left(\frac{3}{2} < y < 2\Big|\theta\right) = \begin{cases} 0 & \text{if } 0 < \theta < 3/2 \\[2mm] \dfrac{\theta-3/2}{\theta} & \text{if } \dfrac{3}{2} < \theta < 2. \end{cases}$$

Since we observe $x = 1$, the posterior density of θ is

$$f(\theta\,|\,x=1) \propto f(\theta)f(x\,|\,\theta) \propto \begin{cases} 1 \times \dfrac{\theta-1/2}{\theta}, & \dfrac{1}{2} < \theta < \dfrac{3}{2} \\[3mm] 1 \times \dfrac{1}{\theta}, & \dfrac{3}{2} < \theta < 2. \end{cases}$$

Now, the area under this function is

$$B = \int_{1/2}^{3/2} \frac{\theta-1/2}{\theta}\,d\theta + \int_{3/2}^{2} \frac{1}{\theta}\,d\theta$$

$$= \left[\theta - \frac{1}{2}\log\theta\,\Big|_{1/2}^{3/2}\right] + \left[\log\theta\,|_{3/2}^{2}\right]$$

$$= \left[\frac{3}{2} - \frac{1}{2}\log\frac{3}{2} - \frac{1}{2} + \frac{1}{2}\log\frac{1}{2}\right] + \left[\log 2 - \log\frac{3}{2}\right]$$

$$= 0.7383759.$$

So the required posterior density is

$$f(\theta\,|\,x=1) = \begin{cases} \dfrac{\theta-1/2}{B\theta}, & \dfrac{1}{2} < \theta < \dfrac{3}{2} \\[3mm] \dfrac{1}{B\theta}, & \dfrac{3}{2} < \theta < 2, \end{cases}$$

and the associated posterior mean of θ is

$$E_1 = E(\theta \mid x = 1) = \int_{1/2}^{3/2} \theta\left(\frac{\theta - 1/2}{B\theta}\right)d\theta + \int_{3/2}^{2} \theta\left(\frac{1}{B\theta}\right)d\theta$$

$$= \frac{1}{B} = 1.354 \quad \text{(after some working).}$$

Discussion

In contrast to $f(\theta \mid x)$, the posterior density of θ given the original data y is

$$f(\theta \mid y) = \frac{f(\theta)f(y \mid \theta)}{f(y)} = \frac{(1/2)(1/\theta)}{\int_y^2 (1/2)(1/\theta)d\theta} = \frac{1}{\theta(\log 2 - \log y)}, \; y < \theta < 2,$$

and the corresponding posterior mean is

$$\hat{\theta} = E(\theta \mid y) = \int_y^2 \theta\left(\frac{1}{\theta(\log 2 - \log y)}\right)d\theta = \frac{2 - y}{\log 2 - \log y}.$$

Figure 3.1 shows $f(\theta \mid x = 1)$ and examples of $f(\theta \mid y)$ which are consistent with $x = 1$.

Figure 3.1 Posteriors given x = 1 and given y = 0.6, 1, 1.1, 1.4

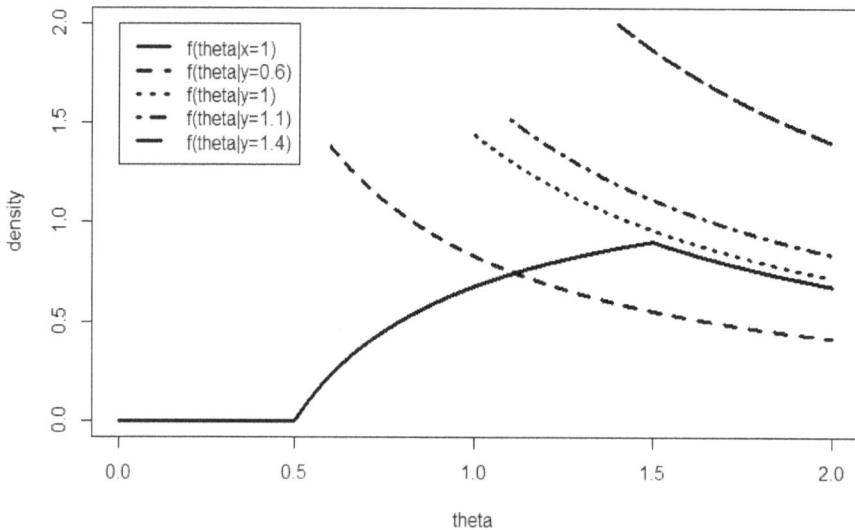

It is now of interest to also calculate $f(\theta \mid x)$ for the other two possible values of x, namely 0 and 2. We find that:

$$f(\theta \mid x = 0) = \begin{cases} \dfrac{1}{A}, & 0 < \theta < \dfrac{1}{2} \\[2ex] \dfrac{1}{2A\theta}, & \dfrac{1}{2} < \theta < 2 \end{cases}$$

where $A = \dfrac{1}{2} + \dfrac{1}{2}\log 2 - \dfrac{1}{2}\log\dfrac{1}{2} = 1.1931$

$$f(\theta \mid x = 2) = \dfrac{1}{C}\left(1 - \dfrac{3}{2\theta}\right), \dfrac{3}{2} < \theta < 2$$

where $C = 2 - \dfrac{3}{2}\log 2 - \dfrac{3}{2} + \dfrac{3}{2}\log\dfrac{3}{2} = 0.068477$.

Figure 3.2 shows these two posteriors, and further examples of $f(\theta \mid y)$.

Figure 3.2 Posteriors given x = 0, 1, 2, and given y = 0.1, ..., 1.9

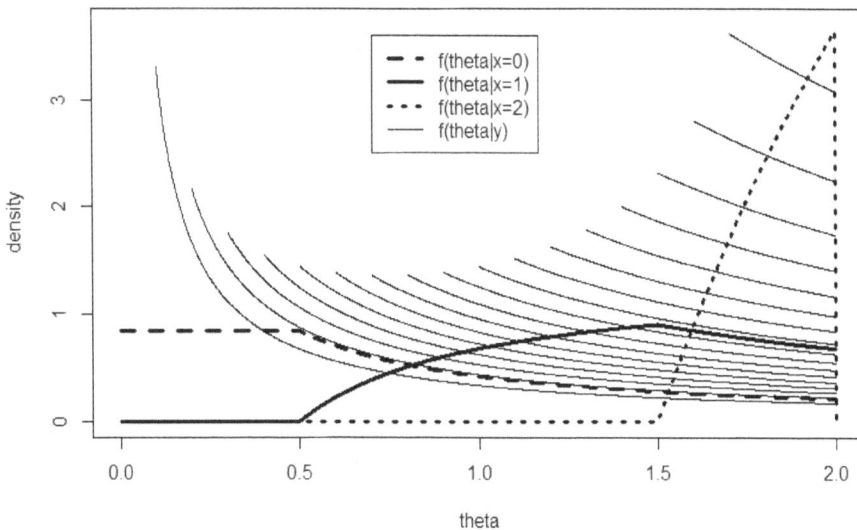

For completeness and checking we now also calculate the other two posterior means:

$$E_0 = E(\theta \mid x = 0) = \dfrac{7}{8A} = 0.7334$$

$$E_2 = E(\theta \mid x = 2) = \dfrac{1}{8C} = 1.8254,$$

as well as the unconditional probabilities of the data:

$$P_0 = P(x = 0) = P\left(y < \frac{1}{2}\right) = EP\left(y < \frac{1}{2}\middle|\theta\right) = \int P\left(y < \frac{1}{2}\middle|\theta\right) f(\theta) d\theta$$

$$= \int_0^{1/2} 1 \times \frac{1}{2} d\theta + \int_{1/2}^2 \frac{1/2}{\theta} \times \frac{1}{2} d\theta = \frac{1}{4}\left(1 + \log 2 - \log \frac{1}{2}\right) = 0.5966$$

$$P_1 = P(x = 1) = 0.3692$$
$$P_2 = P(x = 2) = 0.0342.$$

As a check on our calculations, we note that
$$P_0 + P_1 + P_2 = 1 \quad \text{(which is correct)}.$$

We may also calculate the prior mean of θ (which is obviously 1) as
$$E\theta = EE(\theta \mid x)$$
$$= E(\theta \mid x = 0)P(x = 0) + E(\theta \mid x = 1)P(x = 1) + E(\theta \mid x = 2)P(x = 2)$$
$$= E_0 P_0 + E_1 P_1 + E_2 P_2$$
$$= 0.7334 \times 0.5966 + 1.354 \times 0.3692 + 1.825 \times 0.03424$$
$$= 1.000 \quad \text{(correct)}.$$

R Code for Exercise 3.2

```
X11(w=8,h=5.5); par(mfrow=c(1,1)); options(digits=7)
B=1.5-0.5*log(3/2)-0.5+0.5*log(0.5)+log(2)-log(1.5); c(B,1/B)
        # 0.7383759 1.3543237
postfunB= function(theta,B=0.7383759){ res=0;
        if((theta>=1/2)&&(theta<3/2)) res=1-1/(2*theta)
        if((theta>=3/2)&&(theta<=2)) res=1/theta
        res/B   }

thetavec = seq(0,2,0.001); postvecB=thetavec;
for(i in 1:length(thetavec)) postvecB[i]=postfunB(theta=thetavec[i])
plot(c(0,2),c(0,2),type="n",xlab="theta",ylab="density", main=" ")
lines(thetavec, postvecB,lwd=3)
y=0.6; k=1/(log(2)-log(y))
        lines(thetavec[thetavec>y],k/ thetavec[thetavec>y], lty=2,lwd=3)
y=1; k=1/(log(2)-log(y))
        lines(thetavec[thetavec>y],k/ thetavec[thetavec>y], lty=3,lwd=3)
y=1.1; k=1/(log(2)-log(y))
        lines(thetavec[thetavec>y],k/ thetavec[thetavec>y], lty=4,lwd=3)
y=1.4; k=1/(log(2)-log(y))
        lines(thetavec[thetavec>y],k/ thetavec[thetavec>y], lty=5,lwd=3)
```

```
legend(0,2,c("f(theta|x=1)","f(theta|y=0.6)","f(theta|y=1)","f(theta|y=1.1)",
        "f(theta|y=1.4)"), lty=c(1,2,3,4,5), lwd=c(3,3,3,3,3))


C=2-1.5*log(2)-1.5+1.5*log(1.5)
A=0.5+0.5*log(2)-0.5*log(0.5)
options(digits=7); c(A,B,C) # 1.19314718 0.73837593 0.06847689
E0=7/(8*A); E1=1/B; E2=1/(8*C); c(E0,E1,E2)
        # 0.7333546 1.3543237 1.8254333


P0=1/4+(1/4)*(log(2)-log(1/2))
P1=0.5*(1.5-0.5*log(1.5)-0.5+0.5*log(0.5)) +0.5*(log(2)-log(1.5))
P2=0.5*(2-1.5*log(2)-1.5+1.5*log(1.5))


P0+P1+P2 # 1  Correct
c(P0,P1,P2) # 0.59657359 0.36918796 0.03423845
E0*P0 + E1*P1 + E2*P2 # 1  Correct


postfunA= function(theta,A=1.19314718){ res=0;
        if((theta>=0)&&(theta<1/2)) res=1
        if((theta>=1/2)&&(theta<=2))  res=1/(2*theta)
        res/A   }
postfunC= function(theta,C=0.06847689){ res=0;
        if((theta>=3/2)&&(theta<2)) res=1-3/(2*theta)
        res/C   }


postvecA=thetavec; postvecC=thetavec;
for(i in 1:length(thetavec)){      postvecA[i]=postfunA(theta=thetavec[i])
                                   postvecC[i]=postfunC(theta=thetavec[i])  }
plot(c(0,2),c(0,3.7),type="n",xlab="theta",ylab="density", main=" ")
lines(thetavec, postvecA,lty=2,lwd=3)
lines(thetavec, postvecB,lty=1,lwd=3)
lines(thetavec, postvecC,lty=3,lwd=3)
for(y in seq(0.1,1.9,0.1)){ k=1/(log(2)-log(y))
        lines(thetavec[thetavec>y],k/ thetavec[thetavec>y], lty=1,lwd=1)  }


legend(0.7,3.6,c("f(theta|x=0)","f(theta|x=1)","f(theta|x=2)","f(theta|y)"),
        lty=c(2,1,3,1), lwd=c(3,3,3,1))
```

3.2 Bayesian predictive inference

In addition to estimating model parameters (and functions of those parameters) there is often interest in predicting some future data (or some other quantity which is not just a function of the model parameters).

Consider a Bayesian model specified by $f(y|\theta)$ and $f(\theta)$, with posterior as derived in ways already discussed and given by $f(\theta|y)$.

Now consider any other quantity x whose distribution is defined by a density of the form $f(x|y,\theta)$.

The *posterior predictive distribution* of x is given by the *posterior predictive density* $f(x|y)$. This can typically be derived using the following equation:

$$f(x|y) = \int f(x,\theta|y)d\theta$$
$$= \int f(x|y,\theta)f(\theta|y)d\theta.$$

Note: For the case where θ is discrete, a summation needs to be performed rather than an integral.

The posterior predictive density $f(x|y)$ forms a basis for making probability statements about the quantity x given the observed data y.

Point and interval estimation for future values x can be performed in very much the same way as that for model parameters, except with a slightly different terminology.

Now, instead of referring to $\hat{x} = E(x|y)$ as the *posterior mean* of x, we may instead use the term *predictive mean*.

Also, the 'P' in HPDR, and CPDR may be read as *predictive* rather than as *posterior*. For example, the CPDR for x is now the *central predictive density region* for x.

As an example of point prediction, the predictive mean of x is

$$\hat{x} = E(x|y) = \int x f(x|y)dx.$$

Often it is easier to obtain the predictive mean of x using the equation
$$\hat{x} = E(x\,|\,y) = E\{E(x\,|\,y,\theta)\,|\,y\}$$
$$= \int E(x\,|\,y,\theta) f(\theta\,|\,y) d\theta.$$

Note: The basic *law of iterated expectation* (LIE) implies that $E(x) = EE(x\,|\,\theta)$. This equation must also be true after conditioning throughout on y. We thereby obtain $E(x\,|\,y) = E\{E(x\,|\,y,\theta)\,|\,y\}$.

Likewise, the predictive variance of x can be calculated via the equation
$$V(x\,|\,y) = E\{V(x\,|\,y,\theta)\,|\,y\} + V\{E(x\,|\,y,\theta)\,|\,y\}.$$

Note: This follows from the basic *law of iterated variance* (LIV), $Vx = EV(x\,|\,\theta) + VE(x\,|\,\theta)$, after conditioning throughout on y.

An important special case of Bayesian predictive inference is where the quantity of interest x is *an independent future replicate* of y.

This means that $(x\,|\,y,\theta)$ has exactly the same distribution as $(y\,|\,\theta)$, which in turn may be expressed mathematically as
$$(x\,|\,y,\theta) \sim (y\,|\,\theta)$$
or equivalently as
$$f(x\,|\,y,\theta) = f(y = x\,|\,\theta) = \left[f(y\,|\,\theta)\big|_{y=x} \right].$$

Note: The last equation indicates that the pdf of $(x\,|\,y,\theta)$ is the same as the pdf of $(y\,|\,\theta)$ but with y changed to x in the density formula.

In the case where x is an independent future replicate of y, we may write $f(x\,|\,y,\theta)$ as $f(x\,|\,\theta)$, and this then implies that
$$f(x\,|\,y) = \int f(x\,|\,\theta) f(\theta\,|\,y) d\theta.$$

Exercise 3.3 Prediction in the exponential-exponential model

Suppose that θ has the standard exponential distribution, and the conditional distribution of y given θ is exponential with mean $1/\theta$.

Find the posterior predictive density of x, a future independent replicate of y.

Then, for $y = 2.0$, find the predictive mean, mode and median of x, and also the 80% central predictive density region and 80% highest predictive density region for x.

Solution to Exercise 3.3

Recall that the Bayesian model given by:
$$f(y|\theta) = \theta e^{-\theta y}, y > 0$$
$$f(\theta) = e^{-\theta}, \theta > 0$$
implies the posterior $(\theta|y) \sim Gamma(2, y+1)$.

Now let x be a future independent replicate of the data y, so that
$$f(x|y,\theta) = f(x|\theta) = f(y = x|\theta) = \theta e^{-\theta x}, x > 0.$$

Then the posterior predictive density of x is
$$f(x|y) = \int f(x|y,\theta) f(\theta|y) d\theta$$
$$= \int_0^\infty (\theta e^{-\theta x}) \left(\frac{(y+1)^2 \theta^{2-1} e^{-\theta(y+1)}}{\Gamma(2)} \right) d\theta$$
$$= \frac{\Gamma(3)(y+1)^2}{\Gamma(2)(x+y+1)^3} \int_0^\infty \frac{(x+y+1)^3 \theta^{3-1} e^{-\theta(x+y+1)}}{\Gamma(3)} d\theta$$
$$= \frac{2(y+1)^2}{(x+y+1)^3}, x > 0.$$

Check: $\int f(x|y)dx = \int_0^\infty \frac{2(y+1)^2}{(x+y+1)^3} dx$
$$= 2(y+1)^2 \int_{0+y+1}^{\infty+y+1} u^{-3} du \quad \text{(where } u = x+y+1\text{)}$$
$$= 2(y+1)^2 \left[\frac{u^{-2}}{-2} \right]_{u=y+1}^{\infty} = -(y+1)^2 \left[\frac{1}{\infty^2} - \frac{1}{(y+1)^2} \right] = 1 \quad \text{(correct)}.$$

Next, suppose that $y = 2$. Then
$$f(x|y) = 18(x+3)^{-3}, x > 0.$$

This is a strictly decreasing function, and so the predictive mode is zero.

The predictive mean can be calculated according to the equation

$$E(x \mid y) = \int_0^\infty x 18(x+3)^{-3} dx.$$

An easier way to find the predictive mean is to note that
$$(\theta \mid y) \sim Gamma(2,3)$$
and then write

$$E(x \mid y) = E\{E(x \mid y,\theta) \mid y\} = E(\theta^{-1} \mid y) = \int_0^\infty \theta^{-1} \left(\frac{3^2 \theta^{2-1} e^{-3\theta}}{\Gamma(2)} \right) d\theta$$

$$= \frac{3^2 \Gamma(1)}{3^1 \Gamma(2)} \int_0^\infty \frac{3^1 \theta^{1-1} e^{-3\theta}}{\Gamma(1)} d\theta = 3.$$

An even easier way to do the calculation is to recall a previous exercise where it was shown that the posterior mean of $\psi = 1/\theta$ is given by
$$E(\psi \mid y) = y+1.$$

Thus, $E(x \mid y) = E\{E(x \mid y,\theta) \mid y\} = E(\psi \mid y) = y+1 = 3$ when $y = 2$.

One way to find the predictive median of x is to solve $F(x \mid y) = 1/2$ for x, where $F(x \mid y)$ is the predictive cdf of x, or equivalently, to calculate $Q(1/2)$, where $Q(p) = F^{-1}(p \mid y)$ is the predictive quantile function of x.

Now, the predictive cdf of x is

$$F(x \mid y) = \int_0^x 18(t+3)^{-3} dt = \int_3^{3+x} 18u^{-3} dt \qquad \text{where } u = 3 + t$$

$$= 18 \left[\frac{u^{-2}}{-2} \right]_{u=3}^{3+x} = -9 \left[\frac{1}{(3+x)^2} - \frac{1}{3^2} \right] = 1 - \frac{9}{(3+x)^2}.$$

Setting this to p and solving for x yields the predictive quantile function,

$$Q(p) = F^{-1}(p \mid y) = 3 \left(\frac{1}{\sqrt{1-p}} - 1 \right).$$

So the predictive median is $Q\left(\frac{1}{2} \right) = 3 \left(\frac{1}{\sqrt{1-1/2}} - 1 \right) = 1.2426.$

The predictive quantile function can now also be used to calculate the 80% CPDR for x,

$$(Q(0.1), Q(0.9)) = (0.1623,\ 6.4868),$$

and the 80% HPDR for x,

$$(0, Q(0.8)) = (0,\ 3.7082).$$

Another way to calculate the predictive median of x is as the solution in q of

$$1/2 = P(x < q \mid y)$$

after noting that the right hand side of this equation also equals

$$E\{P(x < q \mid y, \theta) \mid y\} = E(1 - e^{-\theta q} \mid y)$$
$$= 1 - m(-q),$$

where $m(t)$ is the posterior moment generating function (mgf) of θ.

But $(\theta \mid y) \sim Gamma(2, y+1)$, and so $m(t) = (1 - t/(y+1))^{-2}$.

So we need to solve $1/2 = (1 - (-q)/(y+1))^{-2}$ for q. The result is $q = (y+1)(\sqrt{2} - 1) = 1.2426$ when $y = 2$ (same as before).

R Code for Exercise 3.3

```
Qfun=function(p){ 3*(-1+1/sqrt(1-p)) }; Qfun(0.5)  # 1.242641
c(Qfun(0.1),Qfun(0.9))  # 0.1622777 6.4868330
c(0,Qfun(0.8))  # 0.000000 3.708204
```

Exercise 3.4 Predicting a bus number (Extension of Exercise 1.6)

You are visiting a small town with buses whose license plates show their numbers consecutively from 1 up to however many there are. In your mind the number of buses could be anything from 1 to 5, with all possibilities equally likely. Whilst touring the town you first happen to see Bus 3.

Assuming that at any point in time you are equally likely to see any of the buses in the town, how likely is it that the *next* bus number you see will be at least 4?

Also, what is the expected value of the bus number that you will *next* see?

Solution to Exercise 3.4

As in Exercise 1.6, let θ be the number of buses in the town and let y be the number of the bus you happen to first see. Recall that a suitable Bayesian model is:

$$f(y|\theta) = 1/\theta, \ y = 1,...,\theta$$
$$f(\theta) = 1/5, \ \theta = 1,...,5 \quad \text{(prior)},$$

and that the posterior density of θ works out as

$$f(\theta|y) = \begin{cases} 20/47, & \theta = 3 \\ 15/47, & \theta = 4 \\ 12/47, & \theta = 5. \end{cases}$$

Now let x be the number on the *next* bus that you happen to see in the town. Then

$$f(x|y,\theta) = \frac{1}{\theta}, \ x = 1,...,\theta \quad \text{(same distribution as that of } (y|\theta)).$$

This may also be written

$$f(x|y,\theta) = I(x \le \theta)/\theta, \ x = 1, 2, 3,...,$$

and so the posterior predictive density of x is

$$f(x|y) = \sum_{\theta} f(x,\theta|y) = \sum_{\theta} f(x|y,\theta)f(\theta|y) = \sum_{\theta=y}^{5} \frac{I(x \le \theta)}{\theta} f(\theta|y).$$

In our case, the observed value of y is 3 and so:

$$f(x=1|y) = \frac{1}{3} \times \frac{20}{47} + \frac{1}{4} \times \frac{15}{47} + \frac{1}{5} \times \frac{12}{47} = 0.27270$$

$$f(x=2|y) = \frac{1}{3} \times \frac{20}{47} + \frac{1}{4} \times \frac{15}{47} + \frac{1}{5} \times \frac{12}{47} = 0.27270$$

$$f(x=3|y) = \frac{1}{3} \times \frac{20}{47} + \frac{1}{4} \times \frac{15}{47} + \frac{1}{5} \times \frac{12}{47} = 0.27270$$

$$f(x=4|y) = \frac{1}{4} \times \frac{15}{47} + \frac{1}{5} \times \frac{12}{47} = 0.13085$$

$$f(x=5|y) = \frac{1}{5} \times \frac{12}{47} = 0.05106.$$

Check: $\sum_{x=1}^{5} f(x|y) = 0.27270 \times 3 + 0.13085 + 0.05106 = 1$ (correct).

In summary, for $y = 3$, we have that $f(x \mid y) = \begin{cases} 0.27270, & x = 1, 2, 3 \\ 0.13085, & x = 4 \\ 0.05106, & x = 5. \end{cases}$

So the probability that the next bus you see will have a number on it which is at least 4 equals

$$P(x \geq 4 \mid y) = \sum_{x : x \geq 4} f(x \mid y) = f(x = 4 \mid y) + f(x = 5 \mid y)$$

$$= 0.13085 + 0.05106 = 18.2\%.$$

Also, the expected value of the bus number you will next see is
$$E(x \mid y) = 1(0.27270) + 2(0.27270) + 3(0.27270)$$
$$+ 4(0.13085) + 5(0.05106) = 2.4149.$$

Alternatively, $E(x \mid y) = E\{E(x \mid y, \theta) \mid y\} = E\left(\frac{1 + \theta}{2} \middle| y\right) = \frac{1}{2} E(\theta \mid y)$

$$= \left\{ \frac{1 + 3(20/47) + 4(15/47) + 5(12/47)}{2} \right\} = \frac{1 + 180/47}{2} = \frac{227}{94} = 2.4149.$$

R Code for Problem 3.4

```
fv=rep(NA,5);  fv[1] = (1/3)*(20/47)+(1/4)*(15/47)+(1/5)*(12/47)
fv[2] = fv[1]; fv[3] = fv[1];  fv[4] = (1/4)*(15/47)+(1/5)*(12/47)
fv[5] = (1/5)*(12/47); options(digits=5)
fv  # 0.272695 0.272695 0.272695 0.130851 0.051064
sum(fv) # 1  (OK)
sum(fv[4:5]) # 0.18191
sum((1:5)*fv) # 2.4149
227/94 #  2.4149
```

Exercise 3.5 Prediction in the binomial-beta model

(a) For the Bayesian model given by $(Y \mid \theta) \sim Bin(n, \theta)$ and the prior $\theta \sim Beta(\alpha, \beta)$, find the posterior predictive density of a future data value x, whose distribution is defined by $(x \mid y, \theta) \sim Bin(m, \theta)$.

(b) A bent coin is tossed 20 times and 6 heads come up. Assuming a flat prior on the probability of heads on a single toss, what is the probability that exactly one head will come up on the next two tosses of the same coin? Answer this using results in (a).

(c) A bent coin is tossed 20 times and 6 heads come up. Assume a *Beta*(20.3,20.3) prior on the probability of heads.

Find the expected number of times you will have to toss the same coin again repeatedly until the next head comes up.

(d) A bent coin is tossed 20 times and 6 heads come up. Assume a *Beta*(20.3,20.3) prior on the probability of heads.

Now consider tossing the coin repeatedly until the next head, writing down the number of tosses, and then doing all of this again repeatedly, again and again.

The result will be a sequence of natural numbers (for example 3, 1, 1, 4, 2, 2, 1, 5, 1, ...), where each number represents a number of tails in a row within the sequence, plus one.

Next define ψ to be the average of a very long sequence like this (e.g. one of length 1,000,000). Find the posterior predictive density and mean of ψ (approximately).

Note: In parts (c) and (d) the parameters of the beta distribution (both 20.3) represent a prior belief that the probability of heads is about 1/2, is equally likely to be on either side of 1/2, and is 80% likely to be between 0.4 and 0.6. See the R Code below for details.

Solution to Exercise 3.5

(a) First note that x is *not* a future independent *replicate* of the observed data y, except in the special case where $m = n$.

Next recall that $(\theta \mid y) \sim Beta(a,b)$, where:
$$a = \alpha + y, \qquad b = \beta + n - y.$$

Thus the posterior predictive density of x is
$$f(x \mid y) = \int f(x,\theta \mid y)d\theta$$
$$= \int f(x \mid y,\theta)f(\theta \mid y)d\theta$$
$$= \int_0^1 \binom{m}{x} \theta^x (1-\theta)^{m-x} \frac{\theta^{a-1}(1-\theta)^{b-1}}{B(a,b)} d\theta$$

$$= \binom{m}{x} \frac{B(x+a, m-x+b)}{B(a,b)} \int_0^1 \frac{\theta^{x+a-1}(1-\theta)^{m-x+b-1}}{B(x+a, m-x+b)} d\theta$$

$$= \binom{m}{x} \frac{B(x+\alpha+y, m-x+\beta+n-y)}{B(\alpha+y, \beta+n-y)}, \quad x = 0, \ldots, m.$$

Note: The distribution of $(x \mid y)$ here may be called the *beta-binomial*.

(b) Here, we consider the situation in (a) with $n = 20$, $y = 6$, $m = 2$, $\alpha = 1$, $\beta = 1$ and $x = 0$, 1 or 2. So, specifically,

$$f(x \mid y) = \binom{2}{x} \frac{B(x+1+6, 2-x+1+20-6)}{B(1+6, 1+20-6)}$$

$$= \binom{2}{x} \frac{\Gamma(7+x)\Gamma(17-x)/\Gamma(24)}{\Gamma(7)\Gamma(15)/\Gamma(22)}$$

$$= \frac{2!}{x!(2-x)!} \frac{(6+x)!(16-x)!/23!}{6!14!/21!}$$

$$= \begin{cases} 0.4743, & x = 0 \\ 0.4150, & x = 1 \\ 0.1107, & x = 2. \end{cases}$$

Check: $0.4743 + 0.4150 + 0.1107 = 1$ (correct).

So the (posterior predictive) probability that heads will come up on exactly one of the next two tosses is $f(x=1 \mid y=6) = 41.5\%$.

Note: An alternative way to do the working here is to see that if $y = 6$ then

$$(\theta \mid y) \sim Beta(1+6, 1+20-6) \sim Beta(7, 15),$$

so that:

$$E(\theta \mid y) = \frac{7}{7+15} = \frac{7}{22}$$

$$V(\theta \mid y) = \frac{7 \times 15}{(7+15)^2(7+15+1)} = 0.009432.$$

Also, $(x \mid y, \theta) \sim Bin(2, \theta)$ (if $y = 6$).

It follows that

$$P(x=1 \mid y) = E\{P(x=1 \mid y, \theta) \mid y\}$$
$$= E\{2\theta(1-\theta) \mid y\}$$
$$= 2\{E(\theta \mid y) - E(\theta^2 \mid y)\}$$
$$= 2\{E(\theta \mid y) - [V(\theta \mid y) + (E(\theta \mid y))^2]\}$$
$$= 2\left[\frac{7}{22} - \left[0.009432 + \left(\frac{7}{22}\right)^2\right]\right] = 0.415.$$

(c) Let z be the number of tosses until the next head. Then
$$(z \mid y, \theta) \sim Geometric(\theta)$$
with pdf
$$f(z \mid y, \theta) = (1-\theta)^{z-1}\theta, z = 1,2,3,....$$

So the posterior predictive density of z can be obtained via the equation

$$f(z \mid y) = \int f(z, \theta \mid y)d\theta = \int f(z \mid y, \theta)f(\theta \mid y)d\theta.$$

It will be noted that $(z \mid y)$ has a density with a similar form to that of $(x \mid y)$ in (a), but with an infinite range ($z = 1,2,3,...$). If we were to write down $f(z \mid y)$, we could then evaluate the expected number of tosses until the next head according to the equation

$$E(z \mid y) = \sum_{z=1}^{\infty} zf(z \mid y).$$

More easily, the posterior predictive mean of z can be obtained as

$$E(z \mid y) = E\{E(z \mid y, \theta) \mid y\} = E\left(\frac{1}{\theta}\Big|y\right) = \int_0^1 \frac{1}{\theta} \times \frac{\theta^{a-1}(1-\theta)^{b-1}}{B(a,b)} d\theta$$
$$= \frac{B(a-1,b)}{B(a,b)} \int_0^1 \frac{\theta^{(a-1)-1}(1-\theta)^{b-1}}{B(a-1,b)} d\theta$$
$$= \frac{\Gamma(a-1)\Gamma(b)/\Gamma(a-1+b)}{\Gamma(a)\Gamma(b)/\Gamma(a+b)} \times 1 = \frac{a+b-1}{a-1}$$
$$= \frac{(\alpha+y)+(\beta+n-y)-1}{(\alpha+y)-1} = \frac{\alpha+\beta+n-1}{\alpha+y-1}.$$

For $n = 20$, $y = 6$ and $\alpha = \beta = 20.3$, we find that $E(z \mid y) = 2.356$.

(d) Here, ψ represents the average of a very large number of independent realisations of the random variable z in (c). Therefore (approximately), $\psi = E(z \mid y, \theta) = 1/\theta$.

It follows that the posterior predictive density of ψ is

$$f(\psi \mid y) = f(\theta \mid y)\left|\frac{d\theta}{d\psi}\right|,$$

where $\theta = \psi^{-1}$ and $d\theta / d\psi = -\psi^{-2}$. Thus

$$f(\psi \mid y) = \frac{(1/\psi)^{a-1}(1-1/\psi)^{b-1}}{B(a,b)}\left|\frac{-1}{\psi^2}\right|$$

$$= \frac{(\psi-1)^{b-1}}{\psi^{a+b}B(a,b)}, \psi > 1.$$

So the posterior predictive mean of ψ is

$$E(\psi \mid y) = \int_1^\infty \psi \frac{(\psi-1)^{b-1}}{\psi^{a+b}B(a,b)}d\psi$$

$$= \frac{B(a-1,b)}{B(a,b)}\int_1^\infty \frac{(\psi-1)^{b-1}}{\psi^{(a-1)+b}B(a-1,b)}d\psi.$$

The last integral is 1, by analogy of its integrand with $f(\psi \mid y)$. Thus we obtain the same expression as for $E(z \mid y)$ and $E(1/\theta \mid y)$ in (c), namely

$$E(\psi \mid y) = \frac{\alpha + \beta + n - 1}{\alpha + y - 1}.$$

R Code for Exercise 3.5

```
options(digits=4); pbeta(0.4,20.3,20.3) # 0.1004
pbeta(0.6,20.3,20.3) - pbeta(0.4,20.3,20.3) # 0.7993

x=0:2
( 2*factorial(6+x)*factorial(16-x)/factorial(23)   )/
       ( factorial(x)*factorial(2-x) * factorial(6)*factorial(14)/factorial(21)  )
                  # 0.4743 0.4150 0.1107

7*15/(22^2*23) # 0.009432
2 * (7/22 - ( 0.009432267 + (7/22)^2 ) ) # 0.415
(20.3+20.3+20-1)/(20.3+6-1) # 2.356
```

Exercise 3.6 Prediction in the normal-normal model (with variance known)

Consider the Bayesian model given by:
$$(y_1,...,y_n \mid \mu) \sim iid \ N(\mu,\sigma^2)$$
$$\mu \sim N(\mu_0,\sigma_0^2),$$
and suppose we have data in the form of the vector $y = (y_1,...,y_n)$.

Also suppose there is interest in m future values:
$$(x_1,...,x_m \mid y,\mu) \sim iid \ N(\mu,\sigma^2).$$

Find the posterior predictive distribution of
$$\overline{x} = (x_1 +...+ x_m)/m,$$
both generally and in the case of *a priori* ignorance regarding μ.

Solution to Exercise 3.6

By Exercise 1.18 the posterior distribution of μ is given by
$$(\mu \mid y) \sim N(\mu_*,\sigma_*^2),$$

where: $\mu_* = (1-k)\mu_0 + k\overline{y}$, $\sigma_*^2 = k\dfrac{\sigma^2}{n}$, $k = \left(1 + \dfrac{\sigma^2/n}{\sigma_0^2}\right)^{-1}$.

Now, $(\overline{x} \mid y,\mu) \sim N(\mu,\sigma^2/m)$, and therefore
$$f(\overline{x} \mid y) = \int f(\overline{x} \mid y,\mu)f(\mu \mid y)d\mu$$

$$\propto \int_{-\infty}^{\infty} \exp\left\{-\frac{(\overline{x} - \mu)^2}{2\sigma^2/m}\right\} \exp\left\{-\frac{(\mu - \mu_*)^2}{2\sigma_*^2}\right\}d\mu.$$

This is the integral of the exponent of a quadratic in both $\overline{x}$ and μ and so must equal the exponent of a quadratic in $\overline{x}$. It follows that
$$(\overline{x} \mid y) \sim N(\eta,\delta^2),$$
where η and δ^2 are to be determined. This final step is easily achieved as follows:
$$\eta = E(\overline{x} \mid y)$$
$$= E\{E(\overline{x} \mid y,\mu) \mid y\}$$
$$= E\{\mu \mid y\} = \mu_*$$

$$\delta^2 = V(\bar{x} \mid y)$$
$$= E\{V(\bar{x} \mid y, \mu) \mid y\} + V\{E(\bar{x} \mid y, \mu) \mid y\}$$
$$= E\left\{\frac{\sigma^2}{m} \middle| y\right\} + V\{\mu \mid y\} = \frac{\sigma^2}{m} + \sigma_*^2.$$

Thus generally we have that

$$(\bar{x} \mid y) \sim N\left(\mu_*, \sigma_*^2 + \frac{\sigma^2}{m}\right) \sim N\left((1-k)\mu_0 + k\bar{y}, k\frac{\sigma^2}{n} + \frac{\sigma^2}{m}\right).$$

A special case is where there is no prior information regarding the normal mean μ. In this case, assuming it is appropriate to set $\sigma_0 = \infty$ (so that $f(\mu) \propto 1$, $\mu \in \Re$), we have that $k = 1$ and hence

$$(\bar{x} \mid y) \sim N\left(\bar{y}, \frac{\sigma^2}{n} + \frac{\sigma^2}{m}\right).$$

Exercise 3.7 Prediction in the normal-gamma model (with a known mean)

Consider the Bayesian model given by :
$$(y_1, \ldots, y_n \mid \lambda) \sim iid\ N(\mu, 1/\lambda)$$
$$\lambda \sim G(\alpha, \beta),$$
and suppose we have data in the form of the vector $y = (y_1, \ldots, y_n)$.

Also, suppose we are interested in m future values:
$$(x_1, \ldots, x_m \mid y, \lambda) \sim iid\ N(\mu, 1/\lambda).$$

Find the posterior predictive distribution of
$$\bar{x} = (x_1 + \ldots + x_m)/m,$$
both generally and in the case of *a priori* ignorance regarding λ.

Solution to Exercise 3.7

By Exercise 1.20 the posterior distribution of λ is given by
$$(\lambda \mid y) \sim Gamma(a, b),$$
where: $a = \alpha + \dfrac{n}{2}$, $b = \beta + \dfrac{n}{2}s_{y\mu}^2$, $s_{y\mu}^2 = \dfrac{1}{n}\sum_{i=1}^{n}(y_i - \mu)^2$.

Now, $(\bar{x} \mid y, \lambda) \sim N(\mu, 1/(m\lambda))$, and therefore

$$f(\overline{x} \mid y) = \int f(\overline{x} \mid y, \lambda) f(\lambda \mid y) d\lambda$$

$$\propto \int_0^\infty \sqrt{\lambda} \exp\left\{-\frac{m\lambda}{2}(\overline{x} - \mu)^2\right\} \lambda^{a-1} \exp\left(-\lambda b\right) d\lambda$$

$$= \int_0^\infty \lambda^{\left(a + \frac{1}{2}\right) - 1} \exp\left\{-\lambda\left[b + \frac{m}{2}(\overline{x} - \mu)^2\right]\right\} d\lambda$$

$$\propto \left[b + \frac{m}{2}(\overline{x} - \mu)^2\right]^{-\left(a + \frac{1}{2}\right)} \propto \left(1 + \frac{\left[\dfrac{m2a(\overline{x} - \mu)^2}{2b}\right]}{2a}\right)^{-\frac{1}{2}(2a+1)}.$$

Now let $Q = \sqrt{\dfrac{m2a(\overline{x} - \mu)^2}{2b}} = \dfrac{\overline{x} - \mu}{\left(\sqrt{b/a}\right)/\sqrt{m}}$, so that $\overline{x} = \mu + Q\dfrac{\sqrt{b/a}}{\sqrt{m}}$.

Then by the transformation rule,

$$f(Q \mid y) = f(\overline{x} \mid y)\left|\frac{d\overline{x}}{dQ}\right| \propto \left(1 + \frac{Q^2}{2a}\right)^{-\frac{1}{2}(2a+1)} \left|\frac{\sqrt{b/a}}{\sqrt{m}}\right| \propto \left(1 + \frac{Q^2}{2a}\right)^{-\frac{1}{2}(2a+1)}.$$

This implies that $(Q \mid y) \sim t(2a)$, or equivalently,

$$\left(\left.\frac{\overline{x} - \mu}{\sqrt{\dfrac{s_{y\mu}^2 + 2\beta/n}{1 + 2\alpha/n}} \Big/ \sqrt{m}}\right| y\right) \sim t(n + 2\alpha).$$

A special case of this general result is when there is no prior information regarding the precision parameter λ. In that case, and assuming it is then appropriate to set $\alpha = \beta = 0$ (so that $f(\lambda) \propto 1/\lambda, \lambda > 0$), we have that

$$\left(\left.\frac{\overline{x} - \mu}{s_{y\mu}/\sqrt{m}}\right| y\right) \sim t(n).$$

3.3 Posterior predictive p-values

Earlier, in Section 1.3, we discussed *Bayes factors* as a form of hypothesis testing within the Bayesian framework. An entirely different way to perform hypothesis testing in that framework is via the theory of *posterior predictive p-values* (Meng, 1994). As in the theory of Bayes factors, this involves first specifying a *null hypothesis*

$$H_0 : E_0$$

and an *alternative hypothesis*

$$H_1 : E_1,$$

where E_0 and E_1 are two events.

Note: As in Section 1.3, E_0 and E_1 may or may not be disjoint. Also, E_0 and E_1 may instead represent two different *models* for the same data.

In the context of a single Bayesian model with data y and parameter θ, the theory of posterior predictive p-values involves the following steps:

   (i) Define a suitable *discrepancy measure* (or *test statistic*), denoted $T(y,\theta)$,

      following careful consideration of both H_0 and H_1 (see below).

   (ii) Define x as an *independent future replicate* of the data y.

   (iii) Calculate the *posterior predictive p-value* (ppp-value), defined as
$$p = P\{T(x,\theta) \ge T(y,\theta) \,|\, y\}.$$

Note 1: The ppp-value is calculated under the implicit assumption that H_0 is true. Thus we could also write $p = P\{T(x,\theta) \ge T(y,\theta) \,|\, y, H_0\}$.

Note 2: The discrepancy measure may or may not depend on the model parameter, θ. Thus in some cases, $T(y,\theta)$ may also be written as $T(y)$.

The underlying idea behind the choice of discrepancy measure T is that if the observed data y is highly inconsistent with H_0 in favour of H_1 then p should likely be *small*. This is the same idea as behind classical hypothesis testing. In fact, the classical theory may be viewed as a special case of the theory of ppp-values. The advantage of the ppp-value framework is that it is far more versatile and can be used in situations where it is not obvious how the classical theory should be applied.

An example of how ppp-value theory can perform well relative to the classical theory is where the null hypothesis is *composite*, meaning that it consists of the specification of multiple values rather than a single value (e.g. $H_0 : |\theta| < \varepsilon$ as compared to $H_0 : \theta = 0$). The next exercise illustrates this feature.

Exercise 3.8 Posterior predictive p-values for testing a composite null hypothesis

Consider the Bayesian model given by:
$$(y \mid \lambda) \sim Poisson(\lambda)$$
$$f(\lambda) = e^{-\lambda}, \lambda > 0,$$
and suppose that we observe $y = 3$.

(a) Find a suitable ppp-value for testing
$$H_0 : \lambda = 1 \quad \text{versus} \quad H_1 : \lambda > 2.$$

(b) Find a suitable ppp-value for testing
$$H_0 : \lambda \in \{1,2\} \quad \text{versus} \quad H_1 : \lambda > 2.$$

Solution to Exercise 3.8

(a) Here, $(x \mid y, \lambda) \sim Poi(\lambda)$, and we may define the test statistic as
$$T(y, \lambda) = y.$$

Then, the posterior predictive p-value is
$$p = P(x \geq y \mid y, \lambda = 1)$$
$$= 1 - F_{Poi(1)}(y - 1),$$
where $y = 3$ and where $F_{Poi(q)}(r)$ is the cumulative distribution function of a Poisson random variable with mean q, evaluated at r.

Thus a suitable ppp-value is
$$p = 1 - \left(\frac{e^{-1}1^0}{0!} + \frac{e^{-1}1^1}{1!} + \frac{e^{-1}1^2}{2!} \right) = 0.08030.$$

Note: This is just the probability that a *Poisson*(1) random variable will take on a value greater than 2, and so is the same as the classical p-value which would be used in this situation.

(b) Here we first observe that

$$f(\lambda \mid y, H_0) \propto f(\lambda \mid H_0) f(y \mid H_0, \lambda)$$

$$= \frac{e^{-\lambda}}{e^{-1} + e^{-2}} \times \frac{e^{-\lambda} \lambda^y}{y!} \propto e^{-2\lambda} \lambda^y, \quad \lambda = 1, 2 \quad \text{(with } y = 3\text{)}.$$

Thus: $P(\lambda = 1 \mid y, H_0) = \dfrac{e^{-2 \times 1} 1^3}{e^{-2 \times 1} 1^3 + e^{-2 \times 2} 2^3} = 0.48015$

$P(\lambda = 2 \mid y, H_0) = 1 - 0.48015 = 0.51985.$

So a suitable ppp-value is

$$p = P(x \geq y \mid y, H_0) = E\{P(x \geq y \mid y, H_0, \lambda) \mid y, H_0\}$$

$$= E\{1 - F_{Poi(\lambda)}(y - 1) \mid y, H_0\}$$

$$= 0.48015 \times (1 - F_{Poi(1)}(2)) + 0.51985 \times (1 - F_{Poi(2)}(2))$$

$$= 0.48015 \left\{ 1 - \left(\frac{e^{-1} 1^0}{0!} + \frac{e^{-1} 1^1}{1!} + \frac{e^{-1} 1^2}{2!} \right) \right\}$$

$$+ 0.51985 \left\{ 1 - \left(\frac{e^{-2} 2^0}{0!} + \frac{e^{-2} 2^1}{1!} + \frac{e^{-2} 2^2}{2!} \right) \right\}$$

$$= 0.20664.$$

R Code for Exercise 3.8

```
options(digits=5); 1-ppois(2,1) # 0.080301
p1=exp(-2)/(exp(-2)+8*exp(-4)); c(p1,1-p1) # 0.48015 0.51985
p1*(1-ppois(2,1))+(1-p1)*(1-ppois(2,2))  # 0.20664
```

Exercise 3.9 Posterior predictive p-values for testing a normal mean

Consider a random sample $y_1, \ldots, y_n$ from a normal distribution with variance σ^2, where the prior on the precision parameter $\lambda = 1/\sigma^2$ is given by $\lambda \sim Gamma(0, 0)$, or equivalently by $f(\lambda) \propto 1/\lambda, \lambda > 0$.

We wish to test the null hypothesis

H_0: that the normal mean equals μ

against the alternative hypothesis

H_1: that the normal mean is greater than μ

(where μ is a specified constant of interest).

Derive a formula for the ppp-value under each of the following three choices of the test statistic:

(a) $T(y, \lambda) = \bar{y}$, (b) $T(y, \lambda) = \dfrac{\bar{y} - \mu}{\sigma / \sqrt{n}}$, (c) $T(y, \lambda) = \dfrac{\bar{y} - \mu}{s_y / \sqrt{n}}$,

where: $\bar{y} = \dfrac{1}{n} \sum_{i=1}^{n} y_i$ (the sample mean)

$s_y^2 = \dfrac{1}{n-1} \sum_{i=1}^{n} (y_i - \bar{y})^2$ (the sample variance).

For each of these choices of test statistic, report the ppp-value for the case where $\mu = 2$ and $y = (2.1, 4.0, 3.7, 5.5, 3.0, \; 4.6, 8.3, 2.2, 4.1, 6.2)$.

Solution to Exercise 3.9

(a) Let $\bar{x} = (x_1 + ... + x_n) / n$ be the mean of an independent replicate of the sample values, defined by $(x_1, ..., x_n \mid y, \lambda) \sim iid \; N(\mu, \sigma^2)$.

Then, by Exercise 3.7, $\left(\left. \dfrac{\bar{x} - \mu}{s_{y\mu} / \sqrt{n}} \right| y \right) \sim t(n)$, where $s_{y\mu}^2 = \dfrac{1}{n} \sum_{i=1}^{n} (y_i - \mu)^2$.

From this, if the test statistic is $T(y, \lambda) = \bar{y}$, then the ppp-value is

$$p = P(\bar{x} > \bar{y} \mid y) = P\left(\left. \dfrac{\bar{x} - \mu}{s_{y\mu} / \sqrt{n}} > \dfrac{\bar{y} - \mu}{s_{y\mu} / \sqrt{n}} \right| y \right) = 1 - F_{t(n)} \left(\dfrac{\bar{y} - \mu}{s_{y\mu} / \sqrt{n}} \right).$$

Here: $\mu = 2$, $n = 10$, $\bar{y} = \dfrac{1}{n} \sum_{i=1}^{n} y_i = 4.370$, $s_{y\mu}^2 = \dfrac{1}{n} \sum_{i=1}^{n} (y_i - \mu)^2 = 2.978$.

Therefore $\dfrac{\bar{y} - \mu}{s_{y\mu} / \sqrt{n}} = 2.51658$, and so $p = 1 - F_{t(10)}(2.51658) = 0.01528$.

(b) If $T(y, \lambda) = \dfrac{\bar{y} - \mu}{\sigma / \sqrt{n}}$ then the ppp-value is

$$p = P\left(\left. \dfrac{\bar{x} - \mu}{\sigma / \sqrt{n}} > \dfrac{\bar{y} - \mu}{\sigma / \sqrt{n}} \right| y \right) = P(\bar{x} > \bar{y} \mid y).$$

We see that the answer here is exactly the same as in (a).

(c) If $T(y,\lambda) = \dfrac{\bar{y} - \mu}{s_y / \sqrt{n}}$ then the ppp-value is

$$p = P\left(\left. \frac{\bar{x} - \mu}{s_x / \sqrt{n}} > \left| \frac{\bar{y} - \mu}{s_y / \sqrt{n}} \right| \right| y \right) \qquad \text{where } s_x^2 = \frac{1}{n-1} \sum_{i=1}^{n} (x_i - \bar{x})^2$$

$$= E\left\{ P\left(\left. \frac{\bar{x} - \mu}{s_x / \sqrt{n}} > \left| \frac{\bar{y} - \mu}{s_y / \sqrt{n}} \right| \right| y, \lambda \right) \middle| y \right\}$$

by the law of iterated expectation

$$= E\left\{ 1 - F_{t(n-1)}\left(\left| \frac{\bar{y} - \mu}{s_y / \sqrt{n}} \right| \right) \middle| y \right\} \qquad \text{since } \left(\left. \frac{\bar{x} - \mu}{s_x / \sqrt{n}} \right| y, \lambda \right) \sim t(n-1)$$

$$= 1 - F_{t(n-1)}\left(\frac{\bar{y} - \mu}{s_y / \sqrt{n}} \right).$$

We see that the ppp-value derived is exactly the same as the classical p-value which would be used in this setting. Numerically, we have that:

$$s_y^2 = \frac{1}{n-1} \sum_{i=1}^{n} (y_i - \bar{y})^2 = 1.901, \qquad \frac{\bar{y} - \mu}{s_y / \sqrt{n}} = 3.942645.$$

Consequently, the ppp-value is $p = 1 - F_{t(9)}(3.942645) = 0.001696$.

Note: A fourth test statistic which makes sense in the present context is

$$T(y,\lambda) = \frac{\bar{y} - \mu}{s_{y\mu} / \sqrt{n}} \qquad \text{where } s_{y\mu}^2 = \frac{1}{n} \sum_{i=1}^{n} (y_i - \mu)^2 \quad \text{(as before).}$$

This implies a ppp-value given by

$$p = P\left(\left. \frac{\bar{x} - \mu}{s_{x\mu} / \sqrt{n}} > \left| \frac{\bar{y} - \mu}{s_{y\mu} / \sqrt{n}} \right| \right| y \right) \qquad \text{where } s_{x\mu}^2 = \frac{1}{n} \sum_{i=1}^{n} (x_i - \mu)^2.$$

This ppp-value is more difficult to calculate, and it cannot be expressed in terms of well-known quantities, e.g. the cdf of a t distribution, as in (a), (b) and (c). (Here, $\bar{x}$ and $s_{x\mu}$ are *not* independent, given y and μ.)

For more details, regarding this exercise specifically and ppp-values generally, see Meng (1994) and Gelman et al. (2004).

R Code for Exercise 3.9

```
options(digits=4); mu=2; y = c(2.1, 4.0, 3.7, 5.5, 3.0,   4.6, 8.3, 2.2, 4.1, 6.2);
n=length(y); ybar=mean(y); s=sd(y); smu=sqrt(mean((y-mu)^2))
c(ybar,s,smu) # 4.370 1.901 2.978
arga=(ybar-mu)/(smu/sqrt(n)); pppa=1-pt(arga,n); c(arga,pppa)
        # 2.51658 0.01528
argc=(ybar-mu)/(s/sqrt(n)); pppc=1-pt(argc,n-1); c(argc,pppc)
        # 3.942645 0.001696
```

3.4 Bayesian models with multiple parameters

So far we have examined Bayesian models involving some data y and a parameter θ, where θ is a strictly *scalar* quantity. We now consider the case of Bayesian models with *multiple* parameters, starting with a focus on just two, say θ_1 and θ_2. In that case, the Bayesian model may be defined by specifying $f(y|\theta)$ and $f(\theta)$ in the same way as previously, but with an understanding that θ is a *vector* of the form $\theta = (\theta_1, \theta_2)$.

The first task now is to find the *joint posterior density* of θ_1 and θ_2, according to
$$f(\theta|y) \propto f(\theta)f(y|\theta),$$
or equivalently
$$f(\theta_1, \theta_2 | y) \propto f(\theta_1, \theta_2)f(y|\theta_1, \theta_2),$$
where
$$f(\theta) = f(\theta_1, \theta_2)$$
is the joint prior density of the two parameters.

Often, this joint prior density is specified as an *unconditional prior* multiplied by a *conditional prior*, for example as
$$f(\theta_1, \theta_2) = f(\theta_1)f(\theta_2 | \theta_1).$$

Once a Bayesian model with two parameters has been defined, one task is to find the *marginal posterior densities* of θ_1 and θ_2, respectively, via the equations:
$$f(\theta_1|y) = \int f(\theta_1, \theta_2 | y)d\theta_2$$
$$f(\theta_2|y) = \int f(\theta_1, \theta_2 | y)d\theta_1.$$

From these two marginal posteriors, one may obtain point and interval estimates of θ_1 and θ_2 in the usual way (treating each parameter separately). For example, the marginal posterior mean of θ_1 is

$$\hat{\theta}_1 = E(\theta_1 \mid y) = \int \theta_1 f(\theta_1 \mid y) d\theta_1 .$$

Another way to do this calculation is via the *law of iterated expectation*, according to

$$\hat{\theta}_1 = E(\theta_1 \mid y) = E\{E(\theta_1 \mid y, \theta_2) \mid y\}$$
$$= \int E(\theta_1 \mid y, \theta_2) f(\theta_2 \mid y) d\theta_2 .$$

Note: The equation $E(\theta_1 \mid y) = E\{E(\theta_1 \mid y, \theta_2) \mid y\}$ follows from the simpler identity $E\theta_1 = EE(\theta_1 \mid \theta_2)$ after conditioning throughout on y.

Here, $E(\theta_1 \mid y, \theta_2)$ is called the *conditional posterior mean* of θ_1 and can be calculated as

$$E(\theta_1 \mid y, \theta_2) = \int \theta_1 f(\theta_1 \mid y, \theta_2) d\theta_1 .$$

Also, $f(\theta_1 \mid y, \theta_2)$ is called the *conditional posterior density* of θ_1 and may be obtained according to

$$f(\theta_1 \mid y, \theta_2) \propto f(\theta_1, \theta_2 \mid y) . \tag{3.1}$$

Note: Equation (3.1) follows after first considering the equation $f(\theta_1 \mid \theta_2) \propto f(\theta_1, \theta_2)$ and then conditioning throughout on y.

The main idea of Equation (3.1) is to examine the joint posterior density
$$f(\theta_1, \theta_2 \mid y)$$
(or any kernel thereof), think of all terms in this as constant except for θ_1, and then try to recognise a well-known density function of θ_1.

This density function will define the *conditional posterior distribution* of θ_1, from which estimates such as the *conditional posterior mean* of θ_1 (i.e. $E(\theta_1 \mid y, \theta_2)$) will hopefully be apparent.

One may also be interested in some function,
$$\psi = g(\theta_1, \theta_2),$$

of the two parameters (possibly of only one). Then advanced distribution theory may be required to obtain the posterior pdf of ψ, i.e. $f(\psi \mid y)$.

This posterior density may then be used to calculate point and interval estimates of ψ. For example, the posterior mean of ψ is

$$\hat{\psi} = E(\psi \mid y) = \int \psi f(\psi \mid y) d\psi .$$

Alternatively, this mean may be obtained using the equation

$$\hat{\psi} = E(g(\theta_1, \theta_2) \mid y) = \int \int g(\theta_1, \theta_2) f(\theta_1, \theta_2 \mid y) d\theta_1 d\theta_2 .$$

Further, one may be interested in predicting some other quantity x, whose model distribution is specified in the form $f(x \mid y, \theta)$.

To obtain the posterior predictive density of x will generally require a double integral (or summation) of the form

$$f(x \mid y) = \int \int f(x \mid y, \theta_1, \theta_2) f(\theta_1, \theta_2 \mid y) d\theta_1 d\theta_2 .$$

Further integrations will then be required to produce point and interval estimates, such as the predictive mean of x,

$$\hat{x} = E(x \mid y) = \int x f(x \mid y) dx .$$

Exercise 3.10 A bent coin which is tossed an unknown number of times

Suppose that five heads have come up on an *unknown* number of tosses of a bent coin.

Before the experiment, we believed the coin was going to be tossed a number of times equal to 1, 2, 3, ..., or 9, with all possibilities equally likely. As regards the probability of heads coming up on a single toss, we deemed no value more or less likely than any other value. We also considered the probability of heads as unrelated to the number of tosses.

Find the marginal posterior distribution and mean of the number of tosses and of the probability of heads, respectively. Also find the number of heads we could expect to come up if the coin were to be tossed again the *same* number of times.

Solution to Exercise 3.10

For this problem it is appropriate to consider the following three-level hierarchical Bayesian model:

$$(y \mid \theta, n) \sim Binomial(n, \theta)$$
$$(\theta \mid n) \sim U(0,1)$$
$$n \sim DU(1,...,k), \quad k = 9 \qquad \text{(i.e. } f(n) = 1/9, \ n = 1,...,9\text{)}.$$

Under this model, the joint posterior density of the two parameters n and θ is

$$
\begin{aligned}
f(n, \theta \mid y) &\propto f(n, \theta) f(y \mid n, \theta) \\
&= f(n) f(\theta \mid n) f(y \mid n, \theta) \\
&= \frac{1}{k} \times 1 \times \binom{n}{y} \theta^y (1-\theta)^{n-y} \\
&\propto \binom{n}{y} \theta^y (1-\theta)^{n-y}, \quad 0 < \theta < 1, \quad n = y, y+1, ..., 9.
\end{aligned}
$$

So the marginal posterior density of n is

$$
\begin{aligned}
f(n \mid y) &= \int f(n, \theta \mid y) d\theta \\
&\propto \int_0^1 \binom{n}{y} \theta^y (1-\theta)^{n-y} d\theta, \quad n = y, y+1, ..., 9 \quad \text{(since } y = 0, ..., n\text{)} \\
&= \binom{n}{y} B(y+1, n-y+1) \int_0^1 \frac{\theta^{y+1-1}(1-\theta)^{n-y+1-1}}{B(y+1, n-y+1)} d\theta, \quad n = 5, 6, 7, 8, 9 \\
&= \binom{n}{y} \frac{\Gamma(y+1)\Gamma(n-y+1)}{\Gamma(y+1+n-y+1)} \times 1 \quad \text{(since the integral equals 1)} \\
&= \frac{n!}{y!(n-y)!} \frac{y!(n-y)!}{(n+1)!} \\
&= \frac{1}{n+1} \\
&= \begin{cases} 1/6, & n=5 \\ 1/7, & n=6 \\ 1/8, & n=7 \\ 1/9, & n=8 \\ 1/10, & n=9. \end{cases}
\end{aligned}
$$

After normalising (i.e. dividing each of these five numbers by their sum, 0.6456), we find that, to four decimals, n's posterior pdf is

$$f(n\,|\,y) = \begin{cases} 0.2581, & n=5 \\ 0.2213, & n=6 \\ 0.1936, & n=7 \\ 0.1721, & n=8 \\ 0.1549, & n=9. \end{cases}$$

Thus, for example, there is a 17.2% chance *a posteriori* the coin was tossed 8 times.

It follows that n's posterior mean is

$$\hat{n} = E(n\,|\,y) = \sum_{n=6}^{9} nf(n\,|\,y)$$
$$= 0.2581 \times 5 + 0.2213 \times 6 + ... + 0.1549 \times 9$$
$$= 6.744.$$

Next, the marginal posterior density of θ is

$$f(\theta\,|\,y) = \sum_{n} f(n,\theta\,|\,y)$$

$$\propto \sum_{n=y}^{9} \binom{n}{y} \theta^y (1-\theta)^{n-y}$$

$$= \sum_{n=5}^{9} \binom{n}{y} B(y+1,n-y+1) \frac{\theta^{y+1-1}(1-\theta)^{n-y+1-1}}{B(y+1,n-y+1)}$$

$$= \sum_{n=y}^{9} \frac{1}{n+1} f_{Beta(y+1,n-y+1)}(\theta).$$

Recall that $f(n\,|\,y) \propto 1/(n+1)$. It follows that θ's marginal posterior density must be exactly

$$f(\theta\,|\,y) = \sum_{n=5}^{9} f(n\,|\,y) f_{Beta(y+1,n-y+1)}(\theta)$$

$$= 0.2581 \frac{\theta^5 (1-\theta)^{5-5}}{5!(5-5)!/(5+1)!} + ... + 0.1549 \frac{\theta^5 (1-\theta)^{9-5}}{5!(9-5)!/(9+1)!}.$$

We see that θ's posterior is a mixture of five beta distributions.

Note: This result can also be obtained, more directly, as follows. By considering the 'ordinary' binomial-beta model (from earlier), we see that in the present context the conditional posterior distribution of θ (given n) is given by

$$(\theta \mid y, n) \sim Beta(y+1, n-y+1).$$

It immediately follows that

$$f(\theta \mid y) = \sum_n f(\theta, n \mid y) = \sum_n f(n \mid y) f(\theta \mid y, n)$$

$$= \sum_{n=5}^{9} f(n \mid y) f_{Beta(y+1, n-y+1)}(\theta).$$

We may now perform inference on θ. The posterior mean of θ is

$$\hat{\theta} = E(\theta \mid y) = E\{E(\theta \mid y, n) \mid y\} = E\left(\frac{y+1}{n+2} \,\middle|\, y\right)$$

$$= (y+1) \sum_{n=5}^{9} \left(\frac{1}{n+2}\right) f(n \mid y)$$

$$= 6\left\{\left(\frac{1}{7}\right)0.2581 + \left(\frac{1}{8}\right)0.2213 + \left(\frac{1}{9}\right)0.1936 + \left(\frac{1}{10}\right)0.1721 + \left(\frac{1}{11}\right)0.1549\right\}$$

$$= 0.7040.$$

Figures 3.3 and 3.4 (page 141) show the marginal posterior densities of n and θ, respectively, with the posterior means $\hat{n} = 6.744$ and $\hat{\theta} = 0.7040$ marked by vertical lines.

Finally, we consider x, the number of heads on the next n tosses.

The distribution of x is defined by $(x \mid y, n, \theta) \sim Bin(n, \theta)$.

So the posterior predictive mean of x is

$$E(x \mid y) = E\{E(x \mid y, n, \theta) \mid y\} = E(n\theta \mid y)$$

$$= E\{E(n\theta \mid y, n) \mid y\} = E\{nE(\theta \mid y, n) \mid y\}$$

$$= E\left(n \times \frac{y+1}{n+2} \,\middle|\, y\right) = (y+1) \sum_{n=5}^{9} \left(\frac{n}{n+2}\right) f(n \mid y)$$

$$= 6\left\{\left(\frac{5}{7}\right)0.2581 + \left(\frac{6}{8}\right)0.2213 + \left(\frac{7}{9}\right)0.1936 + \left(\frac{8}{10}\right)0.1721 + \left(\frac{9}{11}\right)0.1549\right\}$$

$$= 4.592.$$

Figure 3.3 Posterior density of *n*

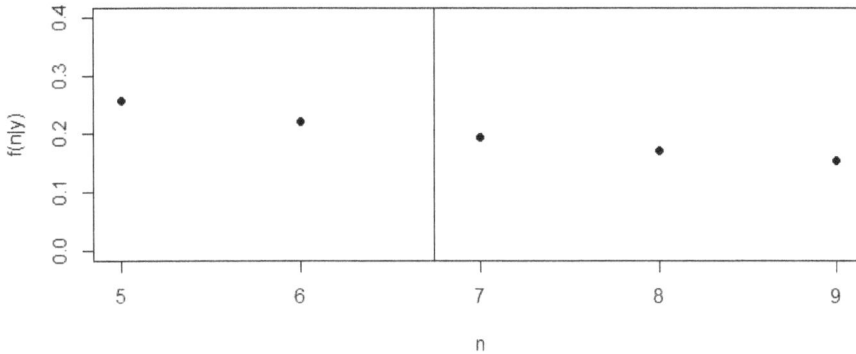

Figure 3.4 Posterior density of θ

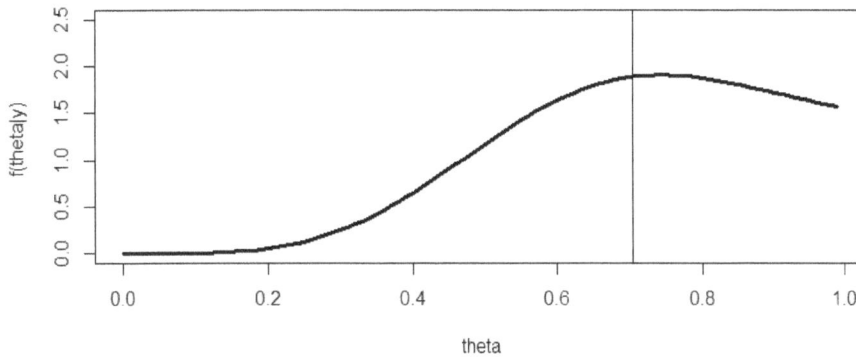

R Code for Exercise 3.10

```
y <- 5; k <- 9; options(digits=4)
nvec <- y:k;   avec <- 1/(nvec+1);   sumavec <- sum(avec); sumavec   # 0.6456
fny <- avec/sumavec;    rbind(nvec,avec,fny)
  # nvec 5.0000 6.0000 7.0000 8.0000 9.0000
  # avec 0.1667 0.1429 0.1250 0.1111 0.1000
  #  fny 0.2581 0.2213 0.1936 0.1721 0.1549
nhat <- sum(nvec*fny); nhat    # 6.744
thhat <- sum( fny * (y+1)/(nvec+2) ); thhat   # 0.704
xhat <- sum( fny * nvec * (y+1)/(nvec+2) ); xhat   # 4.592
```

```
thvec <- seq(0,0.99,0.01);  fthyvec <- thvec
for(i in 1:length(thvec))  fthyvec[i] <- sum( fny * dbeta(thvec[i],y+1,nvec-y+1) )

X11(w=8,h=4); par(mfrow=c(1,1))

plot(nvec,fny,type="n",xlab="n",ylab="f(n|y)",ylim=c(0,0.4))
points(nvec,fny,pch=16,cex=1);   abline(v=nhat)

plot(thvec,fthyvec,type="n",xlab="theta",ylab="f(theta|y) ",ylim=c(0,2.5))
lines(thvec,fthyvec,lwd=3);   abline(v=thhat)
```

Exercise 3.11 The *uninformative normal-normal-gamma model*

Consider the following Bayesian model:
$$(y_1,\ldots,y_n \mid \mu,\lambda) \sim iid\ N(\mu,1/\lambda)$$
$$(\mu \mid \lambda) \sim N(0,\infty)$$
$$\lambda \sim Gamma(0,0),$$
with observed data $y = (y_1,\ldots,y_n)$.

(a) Find the marginal posterior distribution of μ.

(b) Find the marginal posterior distribution of λ.

(c) Find the posterior mean of the *signal to noise ratio*, defined as
$$\gamma = \mu/\sigma = \mu\sqrt{\lambda}.$$

(d) Find the posterior predictive distribution of
$$\bar{x} = (x_1 + \ldots + x_m)/m,$$
where the x_i values have a distribution given by
$$(x_1,\ldots,x_m \mid y,\mu,\lambda) \sim N(\mu,1/\lambda).$$

Note: Both μ and λ are assigned uninformative priors. The joint prior distribution of these two parameters could also be specified by:
$$f(\mu \mid \lambda) \propto 1,\ \mu \in \Re$$
$$f(\lambda) \propto 1/\lambda,\ \lambda > 0,$$
or by the single statement
$$f(\mu,\lambda) \propto 1/\lambda,\ \mu \in \Re, \lambda > 0.$$

Solution to Exercise 3.11

(a) The joint posterior density of the two parameters μ and λ is

$$f(\mu, \lambda \mid y) \propto f(\mu, \lambda) f(y \mid \mu, \lambda) = f(\lambda) f(\mu \mid \lambda) f(y \mid \mu, \lambda)$$

$$\overset{\mu, \lambda}{\propto} \lambda^{-1} \times 1 \times \prod_{i=1}^{n} \frac{1}{\sqrt{1/\lambda}} \exp\left(-\frac{(y_i - \mu)^2}{2(1/\lambda)}\right)$$

$$= \lambda^{\frac{n}{2}-1} \exp\left(-\frac{\lambda}{2} \sum_{i=1}^{n} (y_i - \mu)^2\right).$$

So the marginal posterior density of μ is

$$f(\mu \mid y) = \int f(\mu, \lambda \mid y) d\lambda$$

$$\overset{\mu}{\propto} \int_0^\infty \lambda^{\frac{n}{2}-1} \exp\left(-\lambda \left\{\frac{1}{2}\sum_{i=1}^{n}(y_i-\mu)^2\right\}\right) d\lambda$$

$$= \frac{\Gamma(n/2)}{\left\{\dfrac{1}{2}\displaystyle\sum_{i=1}^{n}(y_i-\mu)^2\right\}^{\frac{n}{2}}} \int_0^\infty \left[\frac{1}{\Gamma(n/2)}\left\{\frac{1}{2}\sum_{i=1}^{n}(y_i-\mu)^2\right\}^{\frac{n}{2}}\right.$$

$$\left. \times \lambda^{\frac{n}{2}-1} \exp\left(-\lambda\left\{\frac{1}{2}\sum_{i=1}^{n}(y_i-\mu)^2\right\}\right)\right] d\lambda$$

$$= \frac{\Gamma(n/2)}{\left\{\dfrac{1}{2}\displaystyle\sum_{i=1}^{n}(y_i-\mu)^2\right\}^{\frac{n}{2}}} \overset{\mu}{\propto} \left\{\sum_{i=1}^{n}(y_i-\mu)^2\right\}^{-\frac{n}{2}}.$$

Note: The last integral is that of a gamma density and so is equal to 1.

Now observe that

$$\sum_{i=1}^{n}(y_i-\mu)^2 = \sum_{i=1}^{n}\left\{(y_i-\overline{y})+(\overline{y}-\mu)\right\}^2$$

$$= \sum_{i=1}^{n}(y_i-\overline{y})^2 + 2(\overline{y}-\mu)\sum_{i=1}^{n}(y_i-\overline{y}) + (\overline{y}-\mu)^2\sum_{i=1}^{n}1$$

$$= (n-1)\left\{\frac{1}{n-1}\sum_{i=1}^{n}(y_i-\overline{y})^2\right\} + 2(\overline{y}-\mu)(n\overline{y}-n\overline{y}) + n(\overline{y}-\mu)^2$$

$$= (n-1)s^2 + n(\mu-\overline{y})^2, \text{ where } s^2 \text{ is the sample variance.}$$

This result implies that

$$f(\mu \mid y) \propto \left\{ (n-1)s^2 + n(\mu - \bar{y})^2 \right\}^{-\frac{n}{2}}$$

$$\overset{\mu}{\propto} \left\{ 1 + \frac{n(\mu - \bar{y})^2}{(n-1)s^2} \right\}^{-\frac{1}{2}\{(n-1)+1\}} = \left\{ 1 + \frac{\left(\dfrac{\mu - \bar{y}}{s/\sqrt{n}} \right)^2}{(n-1)} \right\}^{-\frac{1}{2}\{(n-1)+1\}} .$$

We now define $r = \dfrac{\mu - \bar{y}}{s/\sqrt{n}}$, so that $\mu = \bar{y} + \dfrac{s}{\sqrt{n}} r$ and $\dfrac{d\mu}{dr} = \dfrac{s}{\sqrt{n}}$.

By the transformation rule, we then have that

$$f(r \mid y) = f(\mu \mid y) \left| \frac{d\mu}{dr} \right|$$

$$\overset{r}{\propto} \left\{ 1 + \frac{r^2}{(n-1)} \right\}^{-\frac{1}{2}\{(n-1)+1\}} \times \left| \frac{s}{\sqrt{n}} \right| \overset{r}{\propto} \left\{ 1 + \frac{r^2}{(n-1)} \right\}^{-\frac{1}{2}\{(n-1)+1\}} .$$

By definition of the t distribution, we see that $(r \mid y) \sim t(n-1)$.

It follows that the marginal posterior distribution of μ is given by

$$\left(\left. \frac{\mu - \bar{y}}{s/\sqrt{n}} \right| y \right) \sim t(n-1). \tag{3.2}$$

Note 1: In result (3.2), the data vector y appears only by way of the sample mean $\bar{y}$ and sample standard deviation s. So it is also true that

$$\left(\left. \frac{\mu - \bar{y}}{s/\sqrt{n}} \right| \bar{y}, s \right) \sim t(n-1).$$

Here, s may not be left out of the conditioning. So it is *not* true that

$$\left(\left. \frac{\mu - \bar{y}}{s/\sqrt{n}} \right| \bar{y} \right) \sim t(n-1).$$

Note 2: Result (3.2) implies that the marginal posterior mean, mode and median of μ are all equal to $\bar{y}$, and the $1-\alpha$ CPDR/HPDR for μ is

$$(\bar{y} \pm t_{\alpha/2}(n-1)s/\sqrt{n}).$$

This inference is identical to that obtained via the classical approach and thereby justifies the use of the joint prior

$$f(\mu,\lambda) \propto 1/\lambda, \mu \in \Re, \lambda > 0$$

in cases of *a priori* ignorance regarding both μ and λ.

Note 3: The exact marginal posterior density of μ is

$$f(\mu \mid y) = f(r \mid y)\left|\frac{dr}{d\mu}\right|,$$

where $r = \dfrac{\mu - \bar{y}}{s/\sqrt{n}}$ and $(r \mid y) \sim t(n-1)$.

Thus $\quad f(\mu \mid y) = \dfrac{\Gamma(\{(n-1)+1\}/2)}{\Gamma((n-1)\pi)\Gamma((n-1)/2)}$

$$\times \left[1 + \frac{1}{n-1}\left(\frac{\mu-\bar{y}}{s/\sqrt{n}}\right)^2\right]^{-\frac{1}{2}((n-1)+1)} \frac{\sqrt{n}}{s}, \mu \in \Re.$$

This density can be calculated in R at any point μ by first calculating the corresponding value of r and then returning

```
dt(r,n-1)*sqrt(n)/s
```

(see below for examples).

(b) The marginal posterior density of λ is

$$f(\lambda \mid y) = \int f(\mu,\lambda \mid y)d\mu$$

$$\propto \int_{-\infty}^{\infty} \lambda^{\frac{n}{2}-1} \exp\left(-\frac{\lambda}{2}\{(n-1)s^2 + n(\mu-\bar{y})^2\}\right)d\mu$$

$$= \lambda^{\frac{n}{2}-1} e^{-\lambda\left(\frac{n-1}{2}\right)s^2} \sqrt{(1/(n\lambda))}\sqrt{2\pi}$$

$$\times \int_{-\infty}^{\infty} \frac{1}{\sqrt{(1/(n\lambda))}\sqrt{2\pi}} \exp\left(-\frac{1}{2(1/(n\lambda))}(\mu-\bar{y})^2\right)d\mu$$

$$= \lambda^{\frac{n}{2}-1} e^{-\lambda\left(\frac{n-1}{2}\right)s^2} \sqrt{(1/(n\lambda))}\sqrt{2\pi} \quad \propto \lambda^{\left(\frac{n-1}{2}\right)-1} e^{-\lambda\left(\frac{n-1}{2}\right)s^2}.$$

Note: The last integral is that of a normal density and so equals 1.

It follows that

$$(\lambda \mid y) \sim Gamma\left(\left(\frac{n-1}{2}\right), \left(\frac{n-1}{2}\right)s^2\right),$$
(3.3)

and hence also that

$$((n-1)s^2\lambda \mid y) \sim \chi^2(n-1).$$
(3.4)

Note 1: Result (3.4) can be proved as follows. Let

$$u = (n-1)s^2\lambda,$$

so that $\lambda = \dfrac{u}{(n-1)s^2}$ and $\dfrac{d\lambda}{du} = \dfrac{1}{(n-1)s^2}$.

Then, by the transformation rule,

$$f(u \mid y) = f(\lambda \mid y)\left|\frac{d\lambda}{du}\right|$$

$$\propto \left(\frac{u}{(n-1)s^2}\right)^{\left(\frac{n-1}{2}\right)-1} e^{-\frac{u}{(n-1)s^2}\left(\frac{n-1}{2}\right)s^2} \left|\frac{1}{(n-1)s^2}\right|$$

$$\propto u^{\left(\frac{n-1}{2}\right)-1} e^{-\frac{u}{2}}.$$

Thus $(u \mid y) \sim Gamma\left(\dfrac{n-1}{2}, \dfrac{1}{2}\right) \sim \chi^2(n-1)$, which confirms (3.4).

Note 2: Results (3.3) and (3.4) imply that λ has posterior mean $1/s^2$. This makes sense because $\lambda = 1/\sigma^2$, and s^2 is an unbiased estimator of σ^2. We see that the inverse of the posterior mean of λ provides us with the classical estimator of σ^2.

Also, result (3.4) implies that the $1-\alpha$ CPDR for λ is

$$\left(\frac{\chi^2_{1-\alpha/2}(n-1)}{(n-1)s^2}, \frac{\chi^2_{\alpha/2}(n-1)}{(n-1)s^2}\right).$$

It follows that the $1-\alpha$ CPDR for $\sigma^2 = 1/\lambda$ is

$$\left(\frac{(n-1)s^2}{\chi^2_{\alpha/2}(n-1)}, \frac{(n-1)s^2}{\chi^2_{1-\alpha/2}(n-1)} \right).$$

It will be observed that this is exactly the same as the usual classical $1-\alpha$ CI for σ^2 when the normal mean μ is unknown.

(c) The posterior mean of $\gamma = \mu/\sigma = \mu\sqrt{\lambda}$ could be calculated using the equation

$$\hat{\gamma} = \int \gamma f(\gamma \mid y) d\gamma,$$

where $f(\gamma \mid y)$ is the posterior density of γ.

However, obtaining this density may be difficult. We could use Jacobian theory to find the joint posterior density of μ and γ, and then integrate that joint density with respect to μ. The result would be $f(\gamma \mid y)$.

Another approach is to calculate the mean as

$$\hat{\gamma} = E(\mu\sqrt{\lambda} \mid y) = \int\limits_{\mu=-\infty}^{\infty} \int\limits_{\lambda=0}^{\infty} \mu\sqrt{\lambda} f(\mu, \lambda \mid y) d\mu d\lambda,$$

where: $f(\mu, \lambda \mid y) = \dfrac{k(\mu, \lambda)}{c}$

$$k(\mu, \lambda) = \lambda^{\frac{n}{2}-1} \exp\left(-\frac{\lambda}{2}\sum_{i=1}^{n}(y_i - \mu)^2\right)$$

$$c = \int\limits_{\mu=-\infty}^{\infty} \int\limits_{\lambda=0}^{\infty} h(\mu, \lambda) d\mu d\lambda.$$

More simply, we may use the law of iterated expectation to write

$$\hat{\gamma} = E(\mu\sqrt{\lambda} \mid y) = E\{E(\mu\sqrt{\lambda} \mid y, \lambda) \mid y\} = E\{\sqrt{\lambda}E(\mu \mid y, \lambda) \mid y\}$$

$$= E\{\sqrt{\lambda}\bar{y} \mid y\} = \bar{y}E(\lambda^{1/2} \mid y)$$

$$= \bar{y}\frac{\Gamma\left(\dfrac{n-1}{2}+\dfrac{1}{2}\right)}{\left(\left(\dfrac{n-1}{2}\right)s^2\right)^{1/2}\Gamma\left(\dfrac{n-1}{2}\right)} \quad \text{by (3.3)}$$

$$= \frac{\bar{y}}{s}c_n,$$

147

where

$$c_n = \frac{\Gamma\left(\dfrac{n-1}{2}+\dfrac{1}{2}\right)}{\left(\dfrac{n-1}{2}\right)^{1/2}\Gamma\left(\dfrac{n-1}{2}\right)}.$$

Note 1: By a well-known property of the gamma function, $c_n \to 1$ as $n \to \infty$. So for large n the posterior mean of $\gamma = \mu/\sigma$ is approximately the same as γ's MLE, $\bar{y}/s$.

Note 2: Suppose that we wish to find the posterior median or mode of γ or the 95% CPDR or HPDR for that quantity. Then we first need to determine $f(\gamma|y)$. This and subsequent calculations may be difficult. This points to the need for another strategy. As will be seen later, most of these issues can be easily sidestepped using Monte Carlo methods.

(d) Recall from previous exercises that:

$$(\bar{x}|y,\lambda) \sim N\left(\bar{y}, \frac{1/\lambda}{m}+\frac{1/\lambda}{n}\right) \sim N\left(\bar{y}, \frac{n+m}{nm\lambda}\right)$$

$$(\lambda|y) \sim Gamma\left(\left(\frac{n-1}{2}\right), \left(\frac{n-1}{2}\right)s^2\right).$$

Hence $f(\bar{x}|y) = \int f(\bar{x}|y,\lambda)f(\lambda|y)d\lambda$

$$\propto \int_0^\infty \left[\lambda^{1/2}\exp\left\{-\frac{nm\lambda(\bar{x}-\bar{y})^2}{2(n+m)}\right\}\right]$$

$$\times\left[\lambda^{\left(\frac{n-1}{2}\right)-1}\exp\left\{-\left(\frac{n-1}{2}\right)s^2\lambda\right\}\right]d\lambda$$

$$\propto \int_0^\infty \lambda^{\left(\frac{n}{2}\right)-1}\exp\left\{-\lambda\left[\frac{nm(\bar{x}-\bar{y})^2}{2(n+m)}+\left(\frac{n-1}{2}\right)s^2\right]\right\}d\lambda$$

$$\propto \left[\frac{nm(\bar{x}-\bar{y})^2}{2(n+m)}+\left(\frac{n-1}{2}\right)s^2\right]^{-\left(\frac{n}{2}\right)}$$

$$\propto \left[1 + \frac{nm(\bar{x} - \bar{y})^2}{(n-1)(n+m)s^2} \right]^{-\left(\frac{n}{2}\right)}$$

$$\propto \left[1 + \frac{\left(\dfrac{\bar{x} - \bar{y}}{(s/\sqrt{n})\sqrt{(n+m)/m}} \right)^2}{n-1} \right]^{-\left(\frac{n}{2}\right)}.$$

It follows that

$$\left(\frac{\bar{x} - \bar{y}}{(s/\sqrt{n})\sqrt{(n+m)/m}} \,\middle|\, y \right) \sim t(n-1). \tag{3.5}$$

Note 1: Equation (3.5) can be used to derive the predictive distribution of the average of all $n + m$ values considered (both past and future).

That average may be written

$$a = \frac{1}{n+m} \left(\sum_{i=1}^{n} y_i + \sum_{j=1}^{m} x_i \right) = \frac{n\bar{y} + m\bar{x}}{n+m}.$$

Consequently,

$$\bar{x} = \frac{(n+m)a - n\bar{y}}{m}.$$

It follows that in (3.5),

$$\frac{\bar{x} - \bar{y}}{(s/\sqrt{n})\sqrt{(n+m)/m}} = \frac{\left(\dfrac{(n+m)a - n\bar{y}}{m} \right) - \bar{y}}{(s/\sqrt{n})\sqrt{(n+m)/m}}$$

$$= \frac{(a - \bar{y})(n+m)/m}{(s/\sqrt{n})\sqrt{(n+m)/m}}$$

$$= \frac{a - \bar{y}}{(s/\sqrt{n})\sqrt{m/(n+m)}},$$

and therefore

$$\left(\frac{a - \bar{y}}{(s/\sqrt{n})\sqrt{m/(n+m)}} \,\middle|\, y \right) \sim t(n-1). \tag{3.6}$$

This may look familiar to some readers, the reason being as follows.

Denote the total number of values, $n + m$, as N, and write the average of all these observations, a, as $\bar{Y}$. Then (3.6) is equivalent to the result

$$\left(\left| \frac{\bar{Y} - \bar{y}}{\frac{s}{\sqrt{n}} \sqrt{1 - \frac{n}{N}}} \right| y \right) \sim t(n-1). \tag{3.7}$$

So the posterior predictive mean of $\bar{Y}$ is the observed sample mean $\bar{y}$, and the 95% central (and highest) predictive density region for $\bar{Y}$ is

$$\left(\bar{y} \pm t_{\alpha/2}(n-1) \frac{s}{\sqrt{n}} \sqrt{1 - \frac{n}{N}} \right). \tag{3.8}$$

It will be noted that this inference is exactly the same as implied by the standard approach in the classical survey sampling framework (e.g. see Cochran, 1977).

Recall that in this framework, $\sqrt{1 - n/N}$ is the finite population correction (fpc) factor. As N increases, the fpc factor tends to 1 and (3.8) reduces to

$$\left(\bar{y} \pm t_{\alpha/2}(n-1) \frac{s}{\sqrt{n}} \right),$$

which is the 'standard' CI for a normal mean when the normal variance is unknown.

We have here touched on the topic of *Bayesian finite population inference*. More will be said on this topic later in the book.

Note 2: The exact posterior predictive density of the finite population mean $\bar{Y}$ may be obtained according to

$$f(\bar{Y} \mid y) = f(q \mid y) \left| \frac{dq}{d\bar{Y}} \right|,$$

where: $q = \dfrac{\bar{Y} - \bar{y}}{(s/\sqrt{n}) \sqrt{1 - n/N}}$

$$(q \mid y) \sim t(n-1).$$

We thereby obtain the density $\quad f(\bar{Y} \mid y) = \dfrac{\Gamma(\{(n-1)+1\}/2)}{\Gamma((n-1)\pi)\Gamma((n-1)/2)}$

$$\times \left[1 + \frac{1}{n-1} \left(\frac{\bar{Y}-\bar{y}}{(s/\sqrt{n})\sqrt{1-n/N}} s \right)^2 \right]^{-\frac{1}{2}((n-1)+1)} \frac{\sqrt{n}}{s\sqrt{1-n/N}}, \quad \bar{Y} \in \Re.$$

This density can be calculated in R at any point $\bar{Y}$ by first calculating the corresponding value of q (as defined above) and then returning
    dt(q,n-1)*sqrt(n)/(s*sqrt(1-n/N))
(see below for an example).

Note 3: The posterior predictive density of $\bar{Y}$ converges to the marginal posterior density of μ as N tends to infinity with n fixed. That is,

$$f(\bar{Y}=c \mid y) \to f(\mu=c \mid y) \text{ as } N \to \infty.$$

This is on account of the fpc factor $\sqrt{1-n/N}$ converging to unity. Thus μ may be interpreted as the average of a hypothetically infinite number of values from the underlying superpopulation, $N(\mu, 1/\lambda)$.

Figure 3.5 shows the predictive density $f(\bar{Y} \mid y)$ for various values of N, as well as the posterior density $f(\mu \mid y)$, corresponding to the limiting case $N=\infty$. In each case, the values of n, $\bar{y}$ and s are (arbitrarily) taken as 5, 10 and 2, respectively. Note that $N=\infty \Leftrightarrow m=\infty$ since $m=N-n$.

Note 4: Consider the following Bayesian model:
$$(y_1,\ldots,y_n \mid \mu,\lambda) \sim iid\ N(\mu, 1/\lambda)$$
$$(\mu \mid \lambda) \sim N(\mu_0, \sigma_0^2)$$
$$\lambda \sim Gamma(\alpha,\beta),$$
where σ_0 is not necessarily ∞ and α and β are not necessarily 0.

This may be called the *(general) normal-normal-gamma model,* as distinct from the *uninformative normal-normal-gamma model,* here in Exercise 3.11. In the *general* model, the inferences typically required are much more difficult to perform. Later in the book, it will be shown how to proceed in this—and similarly difficult—situations using *Monte Carlo methods,* including *Markov chain Monte Carlo* (MCMC) *methods.*

Figure 3.5 Predictive density of the finite population mean
(See Note 3 on page 151)

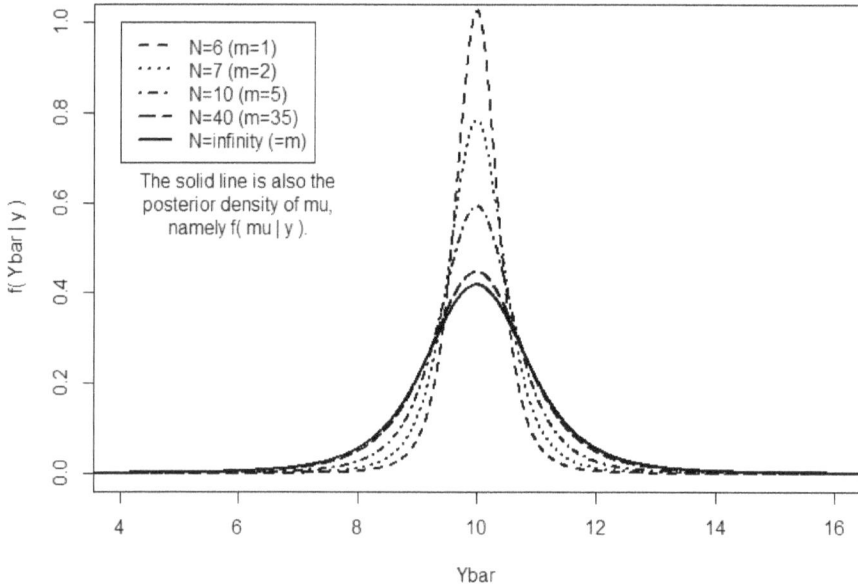

R Code for Exercise 3.11

```
X11(w=8,h=6); par(mfrow=c(1,1))
ybar=10; s=2; cv=seq(0,20,0.005)
plot(c(4,16),c(0,1),type="n",xlab="Ybar",ylab="f( Ybar | y )", main=" ")
n=5; rv=(cv-ybar)*sqrt(n)/s; lines(cv, dt(rv,n-1)*sqrt(n)/s,lty=1,lwd=2)
Nvec=c(6,7,10,40)
for(i in 1:length(Nvec)){ N=Nvec[i];   qv=rv/sqrt(1-n/N)
        lines(cv, dt(qv,n-1)*sqrt(n)/(s*sqrt(1-n/N)),lty=i+1,lwd=2)   }
legend(4,1,
   c("N=6 (m=1)","N=7 (m=2)","N=10 (m=5)","N=40 (m=35)","N=infinity (=m)"),
        lty=c(2:5,1),lwd=2)
text(6,0.6,
   "The solid line is also the\nposterior density of mu,\nnamely f( mu | y ).")
```

CHAPTER 4

Computational Tools

4.1 Solving equations

In most of the Bayesian models so far examined, the calculations required could be done analytically. For example, the model given by:

$$(Y \mid \theta) \sim Binomial(5, \theta)$$

$$\theta \sim U(0,1),$$

together with data $y = 5$, implies the posterior $(\theta \mid y) \sim Beta(6,1)$. So θ has posterior pdf $f(\theta \mid y) = 6\theta^5$ and posterior cdf $F(\theta \mid y) = \theta^6$. Then, setting $F(\theta \mid y) = 1/2$ yields the posterior median, $\theta = 1/2^{1/6} = 0.8909$.

But what if the equation $F(\theta \mid y) = 1/2$ were not so easy to solve? In that case we could employ a number of strategies. One of these is *trial and error*, and another is via special functions in software packages, for example using the qbeta() function in R. This yields the correct answer. Yet another method is the Newton-Raphson algorithm, our next topic.

R Code for Section 4.1

```
qbeta(0.5,6,1)   # 0.8908987
```

4.2 The Newton-Raphson algorithm

The *Newton-Raphson* (NR) *algorithm* is a useful technique for solving equations of the form $g(x) = 0$.

This algorithm involves choosing a suitable starting value x_0 and iteratively applying the equation

$$x_{j+1} = x_j - g'(x_j)^{-1} g(x_j)$$

until convergence had been achieved to a desired degree of precision.

How does the NR algorithm work? Figure 4.1 illustrates the idea.

Figure 4.1 The Newton-Raphson algorithm

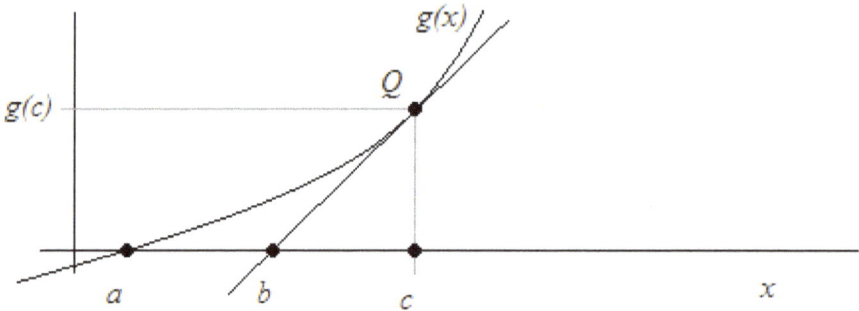

Here, a is the desired solution of the equation $g(x) = 0$, c is a guess at that solution, and b is a better estimate of a. Observe that the slope of the tangent at point Q is equal to both $g'(c)$ and $g(c)/(c-b)$. Equating these two expressions we get $b = c - g(c)/g'(c)$.

Note: Sometimes the NR algorithm takes a long time to converge, and sometimes it converges to the wrong or even impossible value or gets 'stuck' and fails to converge at all. This is a general problem with the NR algorithm, namely its instability and the need to start it off with an initial guess that is sufficiently close to the desired solution.

Exercise 4.1 Calculating a posterior median via the Newton-Raphson algorithm

Suppose that the posterior cdf of a parameter is $F(\theta \mid y) = \theta^6$.

Find the posterior median by solving the equation $F(\theta \mid y) = 1/2$ via the Newton-Raphson algorithm.

Note: The algorithm should converge to the analytical solution, namely $\theta = 1/2^{1/6} = 0.8909$.

Solution to Exercise 4.1

We wish to solve $g(\theta) = 0$, where $g(\theta) = F(\theta \mid y) - 1/2$.

Here, $g'(\theta) = f(\theta \mid y) - 0$, where $f(\theta \mid y) = 6\theta^5$.

So the algorithm is given by

$$\theta_{j+1} = \theta_j - \frac{g(\theta_j)}{g'(\theta_j)} = \theta_j - \frac{F(\theta_j \mid y) - 1/2}{f(\theta_j \mid y)} = \theta_j - \frac{\theta_j^6 - 1/2}{6\theta_j^5}.$$

Starting at the posterior mode, $\theta_0 = 1$ (chosen arbitrarily), we get the sequence shown in Table 4.1.

Table 4.1 NR algorithm starting from 1

j	0	1	2	3	4
θ_j	1.0000	0.9167	0.8926	0.8909	0.8909

So the posterior median is 0.8909. The same result is obtained if we start with $\theta_0 = 0.8$, as shown in table 4.2

Table 4.2 NR algorithm starting from 0.8

j	0	1	2	3	4
θ_j	0.8000	0.9210	0.8933	0.8909	0.8909

Note 1: The median must satisfy

$$\theta = \theta - \frac{\theta^6 - 1/2}{6\theta^5}.$$

This equation is indeed satisfied at the solution $\theta = 0.8909$ (working to four decimals). This illustrates how to check whether or not the NR algorithm has converged properly.

Note 2: In this simple example, one could get the answer by solving the equation $\theta = \theta - g(\theta)/g'(\theta)$ analytically. In general, that won't be possible, and iterating the algorithm will be required. Of course, if it is possible to solve that equation analytically, there is no need to iterate.

R Code for Exercise 4.1

```
NR <- function(th,J=5){
# This function performs the Newton-Raphson algorithm for J iterations
# after starting at the value th. It outputs a vector of th values of length J+1.
thvec <- th; for(j in 1:J){
        num <- th^6-1/2          # theta's posterior cdf minus 1/2 (numerator)
        den <- 6*th^5            # theta's posterior pdf (denominator)
        th <- th - num/den
        thvec <- c(thvec,th) }
thvec   }
```

```
options(digits=4)
NR(th=1,J=6) # 1.0000 0.9167 0.8926 0.8909 0.8909 0.8909 0.8909
NR(th=0.8,J=6) # 0.8000 0.9210 0.8933 0.8909 0.8909 0.8909 0.8909
0.8909-(0.8909^6-0.5)/(6*0.8909^5) # 0.8909 (Check)
```

Exercise 4.2 Further practice with the NR algorithm

Use the Newton-Raphson algorithm to solve the equation $t^2 = e^t$.

Note: In this case there is no analytical solution.

Solution to Exercise 4.2

We wish to solve $g(t) = 0$, where $g(t) = t^2 - e^t$. Now, $g'(t) = 2t - e^t$.

So we iterate according to $t_{j+1} = t_j - \left(\dfrac{t_j^2 - e^{t_j}}{2t_j - e^{t_j}} \right)$.

Let us arbitrarily choose $t_0 = 0$. Then we get:

$$t_1 = 0 - \frac{0^2 - e^0}{2(0) - e^0} = -1.000000, \quad t_2 = (-1) - \frac{(-1)^2 - e^{-1}}{2(-1) - e^{-1}} = -0.733044$$

$$t_3 = (-0.733044) - \frac{(-0.733044)^2 - e^{-0.733044}}{2(-0.733044) - e^{-0.733044}} = -0.703808$$

$$t_4 = (-0.703808) - \frac{(-0.703808)^2 - e^{-0.703808}}{2(-0.703808) - e^{-0.703808}} = -0.703467$$

$$t_5 = (-0.703467) - \frac{(-0.703467)^2 - e^{-0.703467}}{2(-0.703467) - e^{-0.703467}} = -0.703467, \quad \text{etc.}$$

Thus the output of the NR algorithm starting from 0 is:

0.000000, -1.000000, -0.733044, -0.703808, -0.703467, -0.703467, -0.703467, -0.703467, ...

Also, we find that the output of the NR algorithm starting from 1 is:

1.000000, -1.392211, -0.835088, -0.709834, -0.703483, -0.703467, -0.703467, -0.703467, ...

From these results we feel confident that the required solution to 6 decimals is -0.703467. As a check, we calculate

$$g(-0.703467) = (-0.703467)^2 - e^{-0.703467} = 0.000000803508 \approx 0.$$

Figure 4.2 illustrates the function g and the output of the NR algorithm starting from -5, which is:

-5.000000, -2.502357, -1.287421, -0.802834, -0.707162, -0.703473, -0.703467, -0.703467, ...

Figure 4.2 Solution via the NR algorithm starting at −5

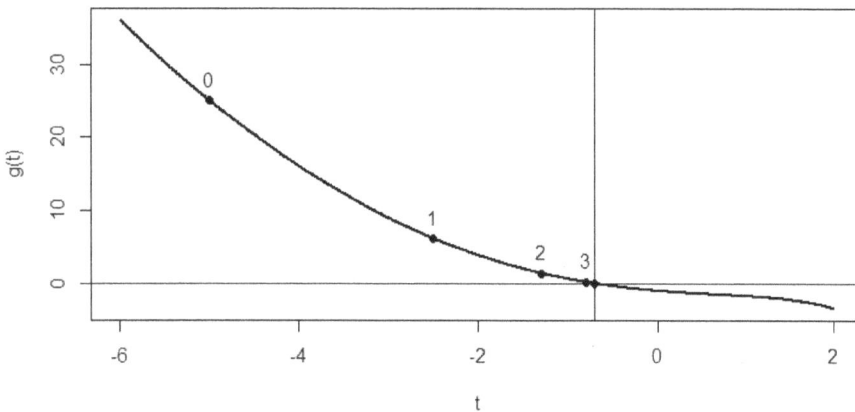

R Code for Exercise 4.2

```
options(digits=6); t=0; tv=t; for(j in 1:7){ t=t-(t^2-exp(t))/(2*t-exp(t))
tv=c(tv,t) }; tv
# 0.000000 -1.000000 -0.733044 -0.703808 -0.703467 -0.703467 -0.703467
# -0.703467
# Check:
  t^2-exp(t)  # 0
  (-0.703467)^2-exp(-0.703467)  # -8.03508e-07
```

```
t=1; tv=t; for(j in 1:7){ t=t-(t^2-exp(t))/(2*t-exp(t)); tv=c(tv,t) }; tv
# 1.000000 -1.392211 -0.835088 -0.709834 -0.703483 -0.703467 -0.703467
# -0.703467

t=-5; tv=t; for(j in 1:7){ t=t-(t^2-exp(t))/(2*t-exp(t)); tv=c(tv,t) }; tv
# -5.000000 -2.502357 -1.287421 -0.802834 -0.707162 -0.703473 -0.703467
# -0.703467

tvec=seq(-6,2,0.01); gvec= tvec^2-exp(tvec)
X11(w=8,h=4.5); par(mfrow=c(1,1))
plot(tvec,gvec,type="l",lwd=2,xlab="t",ylab="g(t)", main="")
abline(h=0,v=t); points(tv, tv^2-exp(tv),pch=16)
text( tv[1:4],  tv[1:4]^2-exp(tv[1:4]) + 3,  0:3)
```

Exercise 4.3 Another example of the NR algorithm

Consider the Bayesian model:
$$(x\,|\,p) \sim Bin(3,p)$$
$$p \sim U(0,1),$$
and suppose the observed value of x is 2. Find the posterior median of p.

Solution to Exercise 4.3

The posterior distribution of p is given by
$$(p\,|\,x) \sim Beta(1+2,1+1),$$
with density
$$f(p\,|\,x) = \frac{p^{3-1}(1-p)^{2-1}}{\Gamma(3)\Gamma(2)/\Gamma(5)} = 12p^2(1-p),\, 0 < p < 1.$$

So, the posterior cdf is
$$F(p\,|\,x) = \int_0^p 12r^2(1-r)\,dr = 12\left(\frac{p^3}{3} - \frac{p^4}{4}\right) = 4p^3 - 3p^4,\; 0 < p < 1.$$

To find the posterior median of p we need to solve $F(p\,|\,x) = 1/2$, or equivalently $g(p) = 0$, where $g(p) = F(p\,|\,x) - 1/2 = 4p^3 - 3p^4 - 1/2$.

Now, $g'(p) = 12p^2 - 12p^3$. So the NR algorithm is defined by iterating
$$p_{j+1} = p_j - \frac{g(p_j)}{g'(p_j)} = p_j - \left(\frac{4p_j^3 - 3p_j^4 - 1/2}{12p_j^2 - 12p_j^3}\right).$$

What's a good starting value here? Let's try the MLE, $p_0 = 2/3$.

Using this value, we get:
  0.666667, 0.614583, 0.614272, 0.614272, 0.614272, 0.614272,
  0.614272, 0.61427, ...

Starting at other values (0.5, 0.9 and 0.1), we get the following (three) sequences (respectively):
  0.500000, 0.625000, 0.614306, 0.614272, 0.614272, 0.614272,
  0.614272, 0.614272, ...

  0.900000, 0.439403, 0.649191, 0.614501, 0.614272, 0.614272,
  0.614272, 0.614272, ...

  0.10000, 4.69537, 3.62690, 2.83403, 2.25146, 1.83195, 1.54254,
  1.36156, ...

The last sequence does not seem to have converged. Let's run this for a bit longer. The result is:
  0.10000, 4.69537, 3.62690, 2.83403, 2.25146, 1.83195, 1.54254,
  1.36156, 1.27282, 1.24913, 1.24749, 1.24748, 1.24748, 1.24748,
  1.24748, 1.24748, 1.24748, 1.24748, 1.24748, 1.24748, ...

Thus if we start at 0.1, the algorithm converges to an *impossible* value of p, namely 1.24748.

It appears that the required posterior median is 0.61427. As a check we may calculate
$$F(p = 0.61427 \mid x) = 4(0.61427)^3 - 3(0.61427)^4 = 0.499999 \approx 0.5.$$

Figures 4.3 and 4.4 show the posterior median 0.61427, as well as the other solution of $g(p) = 0$ (i.e. root of g), namely 1.24748. This is not actually a solution of $F(p \mid x) = 0.5$, because the values of $F(p \mid x)$ for $p < 0$ and $p > 1$ are 0 and 1, respectively.

Thus, the definition of g above is 'deceptive', and a better definition is:
$$g(p) = F(p \mid x) - 1/2 \ = \begin{cases} 0 - 1/2 = -1/2, & p < 0 \\ 4p^3 - 3p^4 - 1/2, & 0 \le p \le 1 \\ 1 - 1/2 = 1/2, & p > 1. \end{cases}$$

Figure 4.3 Posterior cdf and median of *p*

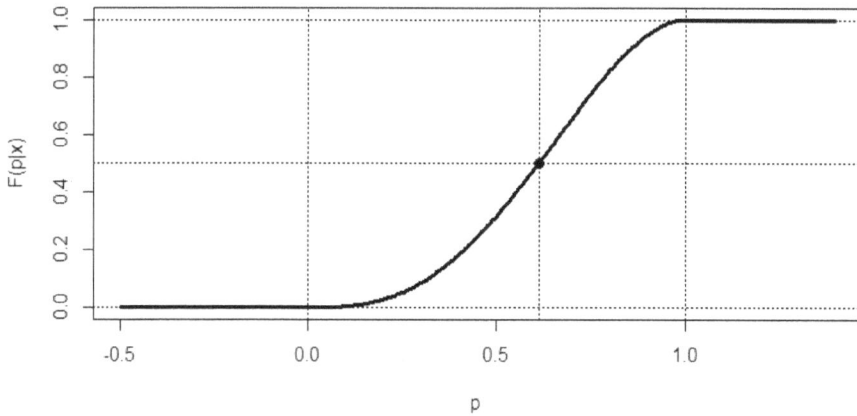

Figure 4.4 Posterior median of *p* and the other root of *g*

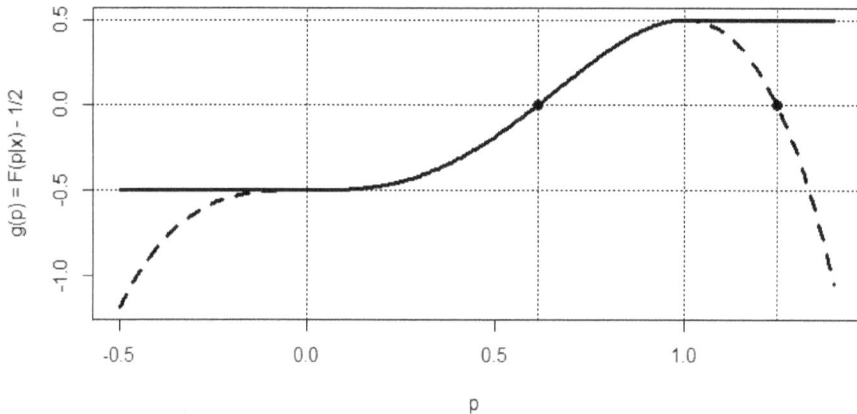

R Code for Exercise 4.3

```
options(digits=6); p=2/3; pv=p; for(j in 1:7){
        p = p - (4*p^3-3*p^4-1/2)/(12*p^2-12*p^3);  pv=c(pv,p) }; pv
# 0.666667 0.614583 0.614272 0.614272 0.614272 0.614272 0.614272
# 0.614272

p=0.5; pv=p; for(j in 1:7){ p = p - (4*p^3-3*p^4-1/2)/(12*p^2-12*p^3);
pv=c(pv,p) }; pv # 0.500000 0.625000 0.614306 0.614272 0.614272 0.614272
# 0.614272 0.614272
```

```
p=0.9; pv=p; for(j in 1:7){ p = p - (4*p^3-3*p^4-1/2)/(12*p^2-12*p^3);
pv=c(pv,p) }; pv  # 0.900000 0.439403 0.649191 0.614501 0.614272 0.614272
# 0.614272 0.614272
```

```
p=0.1; pv=p; for(j in 1:7){ p = p - (4*p^3-3*p^4-1/2)/(12*p^2-12*p^3);
pv=c(pv,p) }; pv
# 0.10000 4.69537 3.62690 2.83403 2.25146 1.83195 1.54254 1.36156
```

```
p=0.1; pv=p; for(j in 1:20){ p = p - (4*p^3-3*p^4-1/2)/(12*p^2-12*p^3);
pv=c(pv,p) }; pv
# 0.10000 4.69537 3.62690 2.83403 2.25146 1.83195 1.54254 1.36156
# 1.27282 1.24913 1.24749 1.24748 1.24748 1.24748 1.24748 1.24748
# 1.24748 1.24748 1.24748 1.24748 1.24748
```

```
4*(0.614272)^3-3*(0.614272)^4  # 0.499999
pvec=seq(-0.5,1.4,0.005); Fvec = 4*pvec^3-3*pvec^4
Fvec[pvec<=0] = 0; Fvec[pvec>=1] = 1
```

```
X11(w=8,h=4.5); par(mfrow=c(1,1))
```

```
plot(pvec,Fvec,type="l",lwd=3,xlab="p",ylab="F(p|x)", main=" ")
abline(h=0.5,v=0.614272,lty=3); points(0.614272,0.5,pch=16, cex=1.2)
abline(h=c(0,1),lty=3); abline(v=c(0,1),lty=3)
gvecwrong=4*pvec^3-3*pvec^4-0.5
```

```
plot(pvec, gvecwrong,type="n",lwd=2,xlab="p",ylab="g(p) = F(p|x) - 1/2",
        main=" ")
lines(pvec,Fvec-0.5,lwd=3)
lines(pvec[pvec<0], gvecwrong[pvec<0],lty=2,lwd=3)
lines(pvec[pvec>1], gvecwrong[pvec>1],lty=2,lwd=3)
abline(v=c(0.614272, 1.24748),lty=3); abline(h=0,lty=3)
points(c(0.614272, 1.24748),c(0,0),pch=16,cex=1.2)
abline(h=c(-0.5,0,0.5),lty=3); abline(v=c(0,1),lty=3)
```

4.3 The multivariate Newton-Raphson algorithm

The Newton-Raphson algorithm can also be used to solve several equations simultaneously, say

$$g_k(x_1,...,x_K) = 0, \; k = 1,...,K.$$

$$\text{Let: } x = \begin{pmatrix} x_1 \\ \vdots \\ x_K \end{pmatrix}, \ g(x) = \begin{pmatrix} g_1(x) \\ \vdots \\ g_K(x) \end{pmatrix}, \ 0 = \begin{pmatrix} 0 \\ \vdots \\ 0 \end{pmatrix} \text{ (a column vector of length } K).$$

Then the system of K equations may be expressed as
$$g(x) = 0,$$
and the NR algorithm involves iterating according to
$$x^{(j+1)} = x^{(j)} - g'(x^{(j)})^{-1} g(x^{(j)}),$$

where: $x^{(j)} = \begin{pmatrix} x_1^{(j)} \\ \vdots \\ x_K^{(j)} \end{pmatrix}$ is the value of x at the jth iteration

$$x^{(j+1)} = \begin{pmatrix} x_1^{(j+1)} \\ \vdots \\ x_K^{(j+1)} \end{pmatrix}, \qquad g(x^{(j)}) = \begin{pmatrix} g_1(x^{(j)}) \\ \vdots \\ g_K(x^{(j)}) \end{pmatrix} = \left[\begin{pmatrix} g_1(x) \\ \vdots \\ g_K(x) \end{pmatrix} \right]_{x=x^{(j)}}$$

$$g'(x^{(j)}) = \left[g'(x) \big|_{x=x^{(j)}} \right]$$

$$g'(x) = \begin{pmatrix} \partial g_1(x)/\partial x^T \\ \vdots \\ \partial g_K(x)/\partial x^T \end{pmatrix} = \begin{pmatrix} \partial g_1(x)/\partial x_1 & \cdots & \partial g_1(x)/\partial x_K \\ \vdots & \ddots & \vdots \\ \partial g_K(x)/\partial x_1 & \cdots & \partial g_K(x)/\partial x_K \end{pmatrix}.$$

Exercise 4.4 Finding a HPDR via the multivariate NR algorithm

Consider the Bayesian model: $(x \mid \lambda) \sim Poisson(\lambda)$
$$f(\lambda) \propto 1, \lambda > 0,$$
and suppose that we observe $x = 1$. Find the 80% HPDR for λ.

Solution to Exercise 4.4

First, $f(\lambda \mid x) \propto f(\lambda) f(x \mid \lambda) = 1 \times e^{-\lambda} \lambda^x / x! = e^{-\lambda} \lambda$, since $x = 1$.

Thus $(\lambda \mid x) \sim Gamma(2,1)$, with $f(\lambda \mid x) = \lambda e^{-\lambda}, \lambda > 0$.

The 80% HPDR for λ is (a,b), where a and b satisfy the two equations:

$$F(b\,|\,x) - F(a\,|\,x) = 0.8 \tag{4.1}$$

$$f(b\,|\,x) = f(a\,|\,x). \tag{4.2}$$

Note: Here, $f(b\,|\,x)$ is the posterior pdf of λ evaluated at b, $F(b\,|\,x)$ is the posterior cdf of λ evaluated at b, etc. Equations (4.1) and (4.2) reflect the requirement that $\lambda \in (a,b)$ with posterior probability 0.8, and that the posterior density of λ must be the same at both a and b, considering that λ's posterior pdf is bell-shaped and unimodal.

Thus we wish to solve the equation

$$g(t) = 0,$$

where:

$$0 = \begin{pmatrix} 0 \\ 0 \end{pmatrix}, \quad t = \begin{pmatrix} a \\ b \end{pmatrix}, \quad g(t) = \begin{pmatrix} g_1(t) \\ g_2(t) \end{pmatrix} = \begin{pmatrix} F(b\,|\,x) - F(a\,|\,x) - 0.8 \\ f(b\,|\,x) - f(a\,|\,x) \end{pmatrix}.$$

The Newton-Raphson algorithm for solving this equation is

$$t^{(j+1)} = t^{(j)} - g'(t^{(j)})^{-1} g(t^{(j)}),$$

where:

$$t^{(j)} = \begin{pmatrix} a_j \\ b_j \end{pmatrix}$$

$$g'(t) = \begin{pmatrix} \partial g_1(t)/\partial a & \partial g_1(t)/\partial b \\ \partial g_2(t)/\partial a & \partial g_2(t)/\partial b \end{pmatrix} = \begin{pmatrix} -ae^{-a} & be^{-b} \\ e^{-a}(a-1) & e^{-b}(1-b) \end{pmatrix}.$$

Starting at

$$t^{(0)} = \begin{pmatrix} a_0 \\ b_0 \end{pmatrix} = \begin{pmatrix} 0.5 \\ 3.0 \end{pmatrix}$$

(based on a visual inspection of the posterior density $f(\lambda\,|\,x) = \lambda e^{-\lambda}$), we obtain results as shown in Table 4.3.

Table 4.3 Multivariate NR algorithm

j	0	1	2	3	4	5
a_j	0.5	0.0776524	0.163185	0.167317	0.16730	0.16730
b_j	3.0	2.7406883	3.025571	3.079274	3.08029	3.08029

It seems that the 80% CPDR for λ is (0.16730, 3.08029). This interval is illustrated in Figure 4.5 and appears to be correct.

As another check on our calculations, we find that:
$$f(\lambda = 3.08029 \,|\, x) - f(\lambda = 0.16730 \,|\, x) = 0.14153 - 0.14153 = 0$$
$$F(\lambda = 3.08029 \,|\, x) - F(\lambda = 0.16730 \,|\, x) = 0.81253 - 0.01253 = 0.8.$$

Figure 4.5 An 80% HPDR

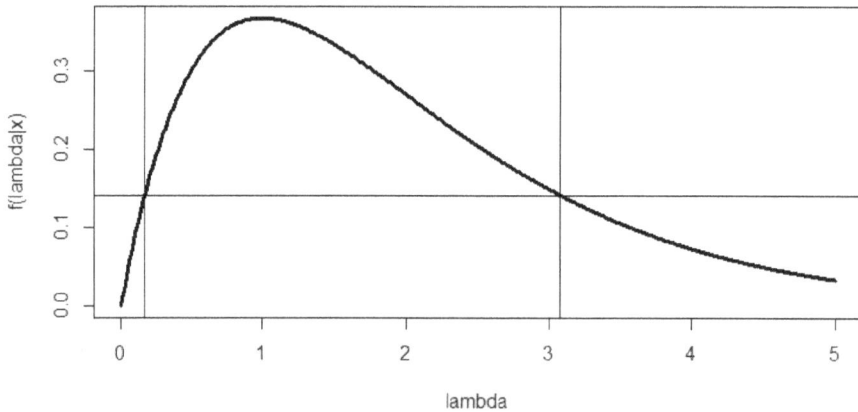

R Code for Exercise 4.4

```
gfun = function(a,b){
    g1=pgamma(b,2,1)-pgamma(a,2,1)-0.8; g2=dgamma(b,2,1)-dgamma(a,2,1);
    c(g1,g2) }

gpfun = function(a,b){   m11=-dgamma(a,2,1);   m12=dgamma(b,2,1)
    m21=exp(-a)*(a-1);   m22=exp(-b)*(1-b)
    matrix(c(m11,m12,m21,m22),nrow=2,byrow=T)   }

gvec=c(0.5,3); gmat=gvec; for(j in 1:7){
        a=gvec[1]; b=gvec[2]
        gvec = gvec - solve(gpfun(a,b)) %*% gfun(a,b)
        gmat = cbind(gmat,gvec)   }

options(digits=6); gmat
# [1,]  0.5 0.0776524 0.163185 0.167317 0.16730 0.16730 0.16730 0.16730
# [2,]  3.0 2.7406883 3.025571 3.079274 3.08029 3.08029 3.08029 3.08029
```

```
lamv=seq(0,5,0.01); fv=dgamma(lamv,2,1)
X11(w=8,h=4.5); par(mfrow=c(1,1))
plot(lamv,fv,type="l",lwd=3,xlab="lambda",ylab="f(lambda|x)", main=" ")
abline(h=c(dgamma(a,2,1)),v=c(a,b),lty=1)

# Checks:
c(a,b,dgamma(c(a,b),2,1))  # 0.167300 3.080291 0.141527 0.141527
c(pgamma(a,2,1), pgamma(b,2,1), pgamma(b,2,1) - pgamma(a,2,1))
#   0.0125275 0.8125275 0.8000000
```

4.4 The Expectation-Maximisation (EM) algorithm

We have shown how the Newton-Raphson algorithm for solving $g(x) = 0$ numerically can be useful for finding the posterior median and the HPDR. That algorithm can also be used for finding the posterior mode, when this is the solution of

$$\frac{\partial f(\theta \mid y)}{\partial \theta} = 0,$$

or equivalently

$$\frac{\partial \log f(\theta \mid y)}{\partial \theta} = 0.$$

In some situations, finding the posterior mode either analytically or via the NR algorithm may be problematic because the posterior density $f(\theta \mid y)$ has a very complicated form. In that case, one may consider applying the *Expectation-Maximisation* (EM) *algorithm*.

This algorithm first requires the specification (i.e. definition by the user) of some suitable *latent data*, which we will denote by z, and then the application of the following two steps iteratively until convergence.

Note: The choice of the latent data z will depend on the particular application.

Step 1. The Expectation Step (E-Step)

Determine the *Q-function*, defined as

$$Q_j(\theta) = E_z\{\log f(\theta \mid y, z) \mid y, \theta_j\}$$

$$= \int \log f(\theta \mid y, z) f(z \mid y, \theta_j) dz, \tag{4.3}$$

or, in words, as

> *the expectation of the log-augmented posterior density with respect to the distribution of the latent data given the observed data and current parameter estimates.*

Step 2. The Maximisation Step (M-Step)

Find the value of θ which maximises the Q-function, for example using the Newton-Raphson algorithm.

This value becomes the current parameter estimate in the next iteration.

Note 1: For mathematical convenience, the Q-function may also be defined as at (4.3) but plus and/or multiplied by any constants which do not depend on the parameter θ. This extended definition allows us to ignore terms which have no impact on the final results. If (4.3) is multiplied by a *negative* constant, the resulting Q-function should be *minimised* at Step 2 rather than maximised.

Note 2: If there is a choice between using the NR algorithm or the EM algorithm, one should consider the fact that the EM algorithm is slower to converge but far more stable. In fact, under certain regularity conditions, the EM algorithm is guaranteed to move closer to the required solution at each iteration. By contrast, the NR algorithm may not converge at all if started at a value far away from the required solution. Thus, one plausible strategy is to use the EM algorithm to obtain an approximate solution which is sufficiently close to the correct answer, and then to obtain a very high precision using just a few iterations of the NR algorithm.

Exercise 4.5 Illustration of the EM algorithm

Consider the Bayesian model given by:

$$(y_1, ..., y_n \mid \lambda) \sim iid \ Gamma(1, \lambda)$$
$$f(\lambda) \propto 1, \lambda > 0.$$

Suppose that the data, denoted D, consists of the *observed* data vector, denoted by

$$y_o = (y_1, ..., y_k),$$

and the partially observed (or *missing*) data vector, denoted by

$$y_m = (y_{k+1}, ..., y_n).$$

We don't know the values in y_m exactly, only that each of those values is greater than some specified constant c.

Suppose that $c = 10$, $n = 5$, $k = 3$ and $y_o = (3.1, 8.2, 6.9)$.

(a) Find the posterior mode of λ by maximising the posterior density directly.

(b) Find the posterior mode of λ using the EM algorithm.

Solution to Exercise 4.5

(a) First, $f(\lambda \mid D) \propto f(\lambda) f(D \mid \lambda)$

$$\propto 1 \times \left(\prod_{i=1}^{k} f(y_i \mid \lambda) \right) \left(\prod_{i=k+1}^{n} P(y_i > c \mid \lambda) \right),$$

where: $f(y_i \mid \lambda) = \lambda e^{-\lambda y_i}$

$$P(y_i > c \mid \lambda) = \int_{c}^{\infty} \lambda e^{-y_i \lambda} dy_i = e^{-c\lambda}.$$

Then $f(\lambda \mid D) \propto \left(\prod_{i=1}^{k} \lambda e^{-\lambda y_i} \right) \left(\prod_{i=k+1}^{n} e^{-c\lambda} \right)$

$$= \lambda^k \exp\{-\lambda[y_{oT} + (n-k)c]\},$$

where $y_{oT} = y_1 + ... + y_k = 18.2$ (the total of the observed values).

So $l(\lambda) \equiv \log f(\lambda \mid D) = k \log \lambda - \lambda[y_{oT} + (n-k)c]$

$$\Rightarrow l'(\lambda) = \frac{k}{\lambda} - [y_{oT} + (n-k)c].$$

Setting $l'(\lambda)$ to zero yields the posterior mode,

$$\frac{k}{y_{oT} + (n-k)c} = 0.078534.$$

(b) The latent data here may be defined as $z = y_m = (y_{k+1}, \ldots, y_n)$.

Then, the augmented posterior density is
$$f(\lambda \mid y_o, y_m) \propto \prod_{i=1}^{n} \lambda e^{-\lambda y_i} = \lambda^n \exp\{-\lambda[y_{oT} + y_{mT}]\},$$
where $y_{mT} = y_{k+1} + \ldots + y_n$ (the total of the missing values).

So the log-augmented density is
$$\log f(\lambda \mid y_o, y_m) = n \log \lambda - \lambda[y_{oT} + y_{mT}] + c_1$$
(where c_1 is a constant with respect to λ).

Now, $f(y_i \mid y_i > c, \lambda) = \dfrac{\lambda e^{-\lambda y_i}}{e^{-\lambda c}} = \lambda e^{-\lambda(y_i - c)}, \quad y_i > c$

(an exponential pdf shifted to the right by c).

Therefore, $\quad E(y_i \mid y_i > c, \lambda) = c + \dfrac{1}{\lambda}.$

It follows that the Q-function is given by
$$Q_j(\lambda) = n \log \lambda - \lambda \left[y_{oT} + (n-k)\left(c + \frac{1}{\lambda_j}\right) \right]$$
(note the distinction here between λ and λ_j).

That concludes the E-Step.

As regards the M-Step, we now calculate the derivative
$$Q'_j(\lambda) = \frac{n}{\lambda} - \left[y_{oT} + (n-k)\left(c + \frac{1}{\lambda_j}\right) \right].$$
Setting this derivative to zero yields a formula for the next value,
$$\lambda_{j+1} = \frac{n}{y_{oT} + (n-k)\left(c + 1/\lambda_j\right)}. \tag{4.4}$$

Implementing the above EM algorithm starting at $\lambda_0 = 1$ we get the following sequence:
  1.000000, 0.124378, 0.092115, 0.083456, 0.080431, 0.079282,
  0.078832, 0.078653, 0.078581, 0.078553, 0.078542, 0.078537,
  0.078535, 0.078535, 0.078534, 0.078534, ...

We see that the EM algorithm has converged correctly to the answer obtained in (a), namely 0.078534.

Note: Writing (4.4) with $\lambda_j = \lambda_{j+1} = \lambda$ (i.e. the limiting value) gives

$$\lambda = \frac{n}{y_{oT} + (n-k)(c+1/\lambda)},$$

and this can be solved easily for the same formula as derived in (a), namely

$$\lambda = \frac{k}{y_{oT} + (n-k)c}.$$

Thus, in this exercise it was not necessary to actually perform any iterations of the EM algorithm.

R Code for Exercise 4.5

```
# (a)
n=5; k=3; c=10;  yo=c(3.1, 8.2, 6.9); yoT=sum(yo); yoT # 18.2
k/(yoT+(n-k)*c) # 0.078534

# (b)
lam = 1; lamv = lam; options(digits=5)
for(j in 1:20){ lam=n/(yoT+(n-k)*(c+1/lam)); lamv=c(lamv,lam) }
lamv
# 1.000000 0.124378 0.092115 0.083456 0.080431 0.079282 0.078832
# 0.078653 0.078581 0.078553 0.078542 0.078537 0.078535 0.078535
# 0.078534 0.078534 0.078534 0.078534 0.078534 0.078534 0.078534
```

Exercise 4.6 EM algorithm for right-censored Gaussian data

Consider the Bayesian model given by:
$$(y_1,...,y_n \mid \lambda) \sim iid\ N(\mu, \sigma^2)$$
$$f(\mu) \propto 1, \mu \in \Re.$$

Suppose that the data, denoted D, consists of the observed data vector
$$y_o = (y_1,...,y_k)$$
and the partially observed (or 'missing') data vector
$$y_m = (y_{k+1},...,y_n).$$

We don't know the values in y_m exactly, but only that each of these values is greater than some specified constant c.

Suppose that $c = 10$, $n = 5$, $k = 3$ and $y_o = (3.1, 8.2, 6.9)$.

(a) Find the log-posterior density of μ and describe how it could be used to find the posterior mode of μ. (Do not actually find that mode in this way.)

(b) Find the posterior mode of μ using the EM algorithm. Then check your answer by showing the mode in plots of the likelihood and log-likelihood functions.

Solution to Exercise 4.6

(a) Observe that $f(\mu \mid D) \propto 1 \times \left(\prod_{i=1}^{k} f(y_i \mid \mu) \right)\left(\prod_{i=k+1}^{n} P(y_i > c \mid \mu) \right)$.

Here, $\displaystyle\prod_{i=1}^{k} f(y_i \mid \mu) \propto \prod_{i=1}^{k} e^{-\frac{1}{2\sigma^2}(y_i - \mu)^2} = \exp\left\{ -\frac{1}{2\sigma^2} \sum_{i=1}^{k} (y_i - \mu)^2 \right\}$

$$= \exp\left\{ -\frac{1}{2\sigma^2}\left[(k-1)s_o^2 + k(\mu - \bar{y}_o)^2 \right] \right\},$$

where: $\displaystyle \bar{y}_o = \frac{1}{k}\sum_{i=1}^{k} y_i$ (the observed sample mean)

$\displaystyle s_o^2 = \frac{1}{k-1}\sum_{i=1}^{k}(y_i - \bar{y}_o)^2$ (the observed sample variance).

Also, $\displaystyle P(y_i > c \mid \mu) = \int_{c}^{\infty} \frac{1}{\sigma\sqrt{2\pi}} e^{-\frac{1}{2}(y_i - \mu)^2} \, dy_i$

$$= P\left(Z > \frac{c - \mu}{\sigma} \right) = 1 - \Phi\left(\frac{c - \mu}{\sigma} \right),$$

where $Z \sim N(0,1)$ and $\Phi(z) = P(Z \le z)$ (the standard normal cdf).

Therefore $\displaystyle f(\mu \mid D) \propto \exp\left\{ -\frac{k}{2\sigma^2}(\mu - \bar{y}_o)^2 \right\}\left(1 - \Phi\left(\frac{c - \mu}{\sigma} \right) \right)^{n-k}$.

So the log-posterior is

$$\log f(\mu \mid D) = -\frac{k}{2\sigma^2}(\mu - \bar{y}_o)^2 + (n-k)\log\left(1 - \Phi\left(\frac{c-\mu}{\sigma}\right)\right) + c_1$$

(where c_1 is a term which does not depend on μ).

To find the posterior mode of μ we could solve the equation

$$l'(\mu) = 0,$$

where $\quad l'(\mu) = \dfrac{\partial \log f(\mu \mid D)}{\partial \mu}$

$$= -\frac{k}{\sigma^2}(\mu - \bar{y}_o) + \frac{(n-k)}{\left(1 - \Phi\left(\dfrac{c-\mu}{\sigma}\right)\right)}\left(-\phi\left(\frac{c-\mu}{\sigma}\right)\right)\left(\frac{-1}{\sigma}\right).$$

This solution could be obtained via the NR algorithm defined by

$$\mu_{j+1} = \mu_j - \frac{l'(\mu_j)}{l''(\mu_j)},$$

where $\quad l''(\mu) = \dfrac{\partial l'(\mu)}{\partial \mu} = -\dfrac{k}{\sigma^2} + \ldots$

As a further exercise, one could complete the formula for $l''(\mu)$ above and actually implement the NR algorithm.

Note: The posterior mode here is also the maximum likelihood estimate, since the prior is proportional to a constant.

(b) With $y_m = (y_{k+1}, \ldots, y_n)$ as the latent data, the augmented posterior is

$$f(\mu \mid y_o, y_m) \propto 1 \times \left(\prod_{i=1}^{k} f(y_i \mid \mu)\right)\left(\prod_{i=k+1}^{n} f(y_i \mid \mu)\right)$$

$$\propto \exp\left\{-\frac{1}{2\sigma^2}\sum_{i=1}^{k}(y_i - \mu)^2\right\}\exp\left\{-\frac{1}{2\sigma^2}\sum_{i=k+1}^{n}(y_i - \mu)^2\right\}.$$

So the log-augmented posterior is

$$\log f(\mu \mid y_o, y_m) = -\frac{1}{2\sigma^2}\sum_{i=1}^{k}(y_i - \mu)^2 - \frac{1}{2\sigma^2}\sum_{i=1}^{k}(y_i - \mu)^2 + c_1$$

$$= -\frac{1}{2\sigma^2}\sum_{i=1}^{k}(y_i^2 - 2\mu y_i + \mu^2) - \frac{1}{2\sigma^2}\sum_{i=k+1}^{n}(y_i^2 - 2\mu y_i + \mu^2) + c_1$$

$$= c_2\left\{\left(k\mu^2 - 2\mu n\bar{y}_o\right) + \left((n-k)\mu^2 - 2\mu(n-k)\bar{y}_m\right)\right\} + c_3,$$

where: $\bar{y}_o = \dfrac{1}{k}\sum_{i=1}^{k}y_i$ (the sample mean of the *observed* values)

$$\bar{y}_m = \frac{1}{n-k}\sum_{i=k+1}^{n}y_i \quad \text{(the sample mean of the *missing* values)}.$$

Thus the Q-function may be taken as

$$Q_j(\mu) = k\mu^2 - 2\mu k\bar{y}_o + (n-k)\mu^2 - 2\mu(n-k)e_j$$

$$= 2n\mu - 2\{k\bar{y}_o + (n-k)e_j\},$$

where $e_j = E(\bar{y}_m \mid D, \mu_j) = E(y_i \mid D, \mu_j) \quad (i > k)$.

We see that $e_j = \left[E(X \mid X > c)\big|_{\mu=\mu_j}\right]$,

where $X \sim N(\mu, \sigma^2)$ (with μ taken as a constant).

Now observe that

$$E(X \mid X > c) = \int_c^{\infty} x\frac{f(x)}{P(X > c)}dx = \frac{I}{P(X > c)},$$

where $P(X > c) = 1 - P(X < c) = 1 - P\left(Z < \frac{c-\mu}{\sigma}\right) = 1 - \Phi\left(\frac{c-\mu}{\sigma}\right)$,

and where

$$I = \int_c^{\infty} x\frac{1}{\sigma\sqrt{2\pi}}e^{-\frac{1}{2\sigma^2}(x-\mu)^2}dx$$

$$= \int_c^{\infty}(x-\mu)\frac{1}{\sigma\sqrt{2\pi}}e^{-\frac{1}{2\sigma^2}(x-\mu)^2}dx + \int_c^{\infty}\mu\frac{1}{\sigma\sqrt{2\pi}}e^{-\frac{1}{2\sigma^2}(x-\mu)^2}dx$$

$$= \int_{(c-\mu)^2/2}^{\infty}\frac{1}{\sigma\sqrt{2\pi}}e^{-\frac{1}{\sigma^2}t}dt + \mu P(X > c)$$

where $t = \dfrac{1}{2}(x-\mu)^2$ and $dt = (x-\mu)dx$

$$= \frac{\sigma}{\sqrt{2\pi}}\int_{(c-\mu)^2/2}^{\infty}\frac{1}{\sigma^2}e^{-\frac{1}{\sigma^2}t}dt + \mu P(X > c)$$

$$= \frac{\sigma}{\sqrt{2\pi}}^{-\frac{1}{\sigma^2}(c-\mu)^2/2} + \mu P(X>c) \quad = \sigma\left(\frac{1}{\sqrt{2\pi}}^{-\frac{1}{2}\left(\frac{c-\mu}{\sigma}\right)^2}\right) + \mu P(X>c)$$

$$= \sigma\phi\left(\frac{c-\mu}{\sigma}\right) + \mu P(X>c) \quad \text{where } \phi(z) \text{ is the standard normal pdf.}$$

Thus $\quad E(X \mid X>c) = \dfrac{1}{P(X>c)}\left\{\sigma\phi\left(\dfrac{c-\mu}{\sigma}\right) + \mu P(X>c)\right\}$

$$= \mu + \sigma\phi\left(\frac{c-\mu}{\sigma}\right)\bigg/\left\{1-\Phi\left(\frac{c-\mu}{\sigma}\right)\right\},$$

and consequently $\quad e_j = \mu_j + \sigma\phi\left(\dfrac{c-\mu_j}{\sigma}\right)\bigg/\left\{1-\Phi\left(\dfrac{c-\mu_j}{\sigma}\right)\right\}.$

That completes the E-Step, which may be summarised by writing
$$Q_j(\mu) = n\mu^2 - 2\mu\{k\overline{y}_o + (n-k)e_j\},$$
where e_j is as given above.

The M-Step then involves calculating
$$Q_j'(\mu) = 2n\mu - 2\{k\overline{y}_o + (n-k)e_j\}$$
and setting this to zero so as to yield the next parameter estimate,
$$\mu_{j+1} = \frac{k\overline{y}_o + (n-k)e_j}{n}$$

$$= \frac{1}{n}\left[k\overline{y}_o + (n-k)\left(\mu_j + \sigma\phi\left(\frac{c-\mu_j}{\sigma}\right)\bigg/\left\{1-\Phi\left(\frac{c-\mu_j}{\sigma}\right)\right\}\right)\right].$$

Implementing the above EM algorithm starting at 5 (arbitrarily), we obtain the sequence:
5.000000, 8.137838, 8.371786, 8.395701, 8.398209, 8.398473, 8.398501, 8.398504, 8.398504, 8.398504, 8.398504, ...

We conclude that the posterior mode of μ is 8.3985.

Figure 4.6 shows the posterior density (top subplot) and the log-posterior density (bottom subplot). Each of these density functions is drawn scaled, meaning correct only up to a constant of proportionality. In each subplot, the posterior mode is indicated by way of a vertical dashed line.

Figure 4.6 Posterior and log-posterior densities (scaled)

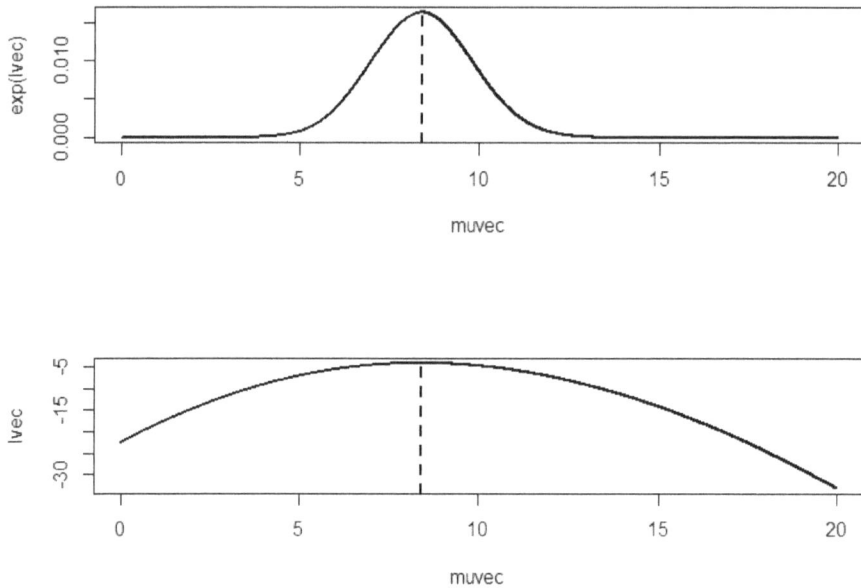

R Code for Exercise 4.6

```
# (b)
options(digits=6); yo = c(3.1, 8.2, 6.9); n=5; k = 3; c= 10; sig=3;
yoT=sum(yo); c(yoT, yoT/3)  # 18.20000  6.06667
mu=5; muv=mu;   for(j in 1:10){
  ej = mu + sig * dnorm((c-mu)/sig) / ( 1-pnorm((c-mu)/sig) )
  mu = ( yoT + (n-k)*ej  )/n
  muv=c(muv,mu)    }
muv      # 5.00000 8.13784 8.37179 8.39570 8.39821 8.39847
         # 8.39850 8.39850 8.39850 8.39850 8.39850
modeval=muv[length(muv)]; modeval # 8.3985

muvec=seq(0,20,0.001); lvec=muvec
for(i in 1:length(muvec)){ muval=muvec[i]
  lvec[i]=(-1/(2*sig^2))*sum((yo-muval)^2) +
                          (n-k)*log(1-pnorm((c-muval)/sig))  }
iopt=(1:length(muvec))[lvec==max(lvec)]; muopt=muvec[iopt]; muopt # 8.399

X11(w=8,h=6); par(mfrow=c(2,1));
plot(muvec,exp(lvec),type="l",lwd=2); abline(v=modeval,lty=2,lwd=2)
plot(muvec,lvec,type="l",lwd=2); abline(v=modeval,lty=2,lwd=2)
```

4.5 Variants of the NR and EM algorithms

The Newton-Raphson and Expectation-Maximisation algorithms can be modified and combined in various ways to produce a number of useful variants or 'hybrids'. For example, the NR algorithm can be used at each M-Step of the EM algorithm to maximise the Q-function.

If the EM algorithm is applied to find the mode of a parameter vector, say $\theta = (\theta_1, \theta_2)$, then the multivariate NR algorithm for doing this may be problematic and one may consider using the *ECM algorithm* (where *C* stands for *Conditional*).

The idea is, at each M-Step, to maximise the Q-function with respect to θ_1, with θ_2 fixed at its current value; and then to maximise the Q-function with respect to θ_2, with θ_1 fixed at its current value.

If each of these conditional maximisations is achieved via the NR algorithm, the procedure can be modified to become the *ECM1 algorithm*. This involves applying only *one step* of each NR algorithm (rather than finding the exact conditional maximum). In many cases the ECM1 algorithm will be more efficient at finding the posterior mode than the ECM algorithm.

Sometimes, when the simultaneous solution of several equations via the multivariate NR algorithm is problematic, a more feasible solution is to apply a suitable *CNR algorithm* (where again *C* stands for *Conditional*).

For example, suppose we wish to solve two equations simultaneously, say:
$$g_1(a,b) = 0$$
$$g_2(a,b) = 0,$$
for a and b. Then it may be convenient to define the function
$$g(a,b) = g_1(a,b)^2 + g_2(a,b)^2,$$
which clearly has a minimum value of zero at the required solutions for a and b.

This suggests that we iterate two steps as follows:

Step 1. Minimise $g(a,b)$ with respect to a, with b held fixed.

Step 2. Minimise $g(a,b)$ with respect to b, with a held fixed.

The first of these two steps involves solving

$$\frac{\partial g(a,b)}{\partial a} = 0,$$

where $\quad \dfrac{\partial g(a,b)}{\partial a} = 2g_1(a,b)\dfrac{\partial g_1(a,b)}{\partial a} + 2g_2(a,b)\dfrac{\partial g_2(a,b)}{\partial a}.$

Assuming the current values of a and b are a_j and b_j, this can be achieved via the NR algorithm by setting $a_0' = a_j$ and iterating until convergence as follows ($k = 0, 1, 2, ...$):

$$a_{k+1}' = a_k' - \frac{\left[\dfrac{\partial g(a,b)}{\partial a}\Big|a=a_k', b=b_j\right]}{\left[\dfrac{\partial^2 g(a,b)}{\partial a^2}\Big|a=a_k', b=b_j\right]},$$

and finally setting

$$a_{j+1} = a_\infty'. \tag{4.5}$$

The second of the two steps involves solving

$$\frac{\partial g(a,b)}{\partial b} = 0,$$

where $\quad \dfrac{\partial g(a,b)}{\partial b} = 2g_1(a,b)\dfrac{\partial g_1(a,b)}{\partial b} + 2g_2(a,b)\dfrac{\partial g_2(a,b)}{\partial b}.$

This can be achieved via the NR algorithm by setting $b_0' = b_j$ and iterating until convergence as follows ($k = 0, 1, 2, ...$):

$$b_{k+1}' = b_k' - \frac{\left[\dfrac{\partial g(a,b)}{\partial b}\Big|a=a_{j+1}, b=b_k\right]}{\left[\dfrac{\partial^2 g(a,b)}{\partial b^2}\Big|a=a_{j+1}, b=b_k\right]},$$

and finally setting

$$b_{j+1} = b_\infty'. \tag{4.6}$$

A variant of the CNR algorithm is the *CNR1 algorithm*. This involves performing only *one step* of each NR algorithm in the CNR algorithm.

In the above example, the CNR1 algorithm implies we set $a_{j+1} = a_1'$ at (4.5) and $b_{j+1} = b_1'$ at (4.6) (rather than $a_{j+1} = a_\infty'$ and $b_{j+1} = b_\infty'$).

This modification will also result in eventual convergence to the solution of $g_1(a,b) = 0$ and $g_2(a,b) = 0$.

One application of the CNR and CNR1 algorithms is to finding the HPDR for a parameter.

For example, in Exercise 4.4 we considered the model given by
$$(x \mid \lambda) \sim Poisson(\lambda)$$
$$f(\lambda) \propto 1, \lambda > 0,$$
with observed data $x = 1$.

The 80% HPDR for λ was shown to be (a,b), where a and b are the simultaneous solutions of the two equations:
$$g_1(a,b) = F(b \mid x) - F(a \mid x) - 0.8$$
$$g_2(a,b) = f(b \mid x) - f(a \mid x).$$

Applying the CNR or CNR1 algorithm as described above should also lead to the same interval as obtained earlier via the multivariate NR algorithm, namely (0.16730, 3.08029).

For further details regarding the EM algorithm, the Newton-Raphson algorithm, and extensions thereof, see Lachlan and Krishnan (2008).

Exercise 4.7 Application of the EM and ECM algorithms to a normal mixture model

Consider the following Bayesian model:
$$(y_i \mid R, \mu, \delta) \sim \perp N(\mu + \delta R_i, \sigma^2), i = 1,...,n$$
$$(R_1,..., R_n \mid \mu, \delta) \sim iid \ Bernoulli(\pi), i = 1,...,n$$
$$f(\mu, \delta) \propto 1, \quad \mu \in \Re, \ \delta > 0.$$

This model says that each value y_i has a common variance σ^2 and one of two means, these being: $\quad \mu \quad$ if $R_i = 0$
$$\mu + \delta \quad \text{if} \ R_i = 1.$$

Each of the 'latent' indicator variables R_i has *known* probability π of being equal to 1, and probability $1 - \pi$ of being equal to 0.

Note: In more advanced models, the quantity π could be treated as unknown and assigned a prior distribution, along with the other two model parameters, μ and δ. The model here provides a 'stepping stone' to understanding and implementing such more complex models.

(a) Consider the situation where $n = 100$, $\pi = 1/3$, $\mu = 20$, $\delta = 10$ and $\sigma = 3$. Generate a data vector $y = (y_1, ..., y_n)$ using these specifications and create a histogram of the simulated values.

(b) Design an EM algorithm for finding the posterior mode of $\theta = (\mu, \delta)$. Then implement the algorithm so as to find that mode.

(c) Modify the EM algorithm in part (b) so that it is an ECM algorithm. Then run the ECM algorithm so as to check your answer to part (b).

(d) Create a plot which shows the routes taken by the algorithms in parts (b) and (c).

Solution to Exercise 4.7

(a) Figure 4.7 shows a histogram of the sampled values which clearly shows the two component normal densities and the mixture density. The sample mean of the data is 23.16. Also, 29 of the 100 R_i values are equal to 1, and 71 of them are equal to 0.

Figure 4.7 Histogram of simulated data

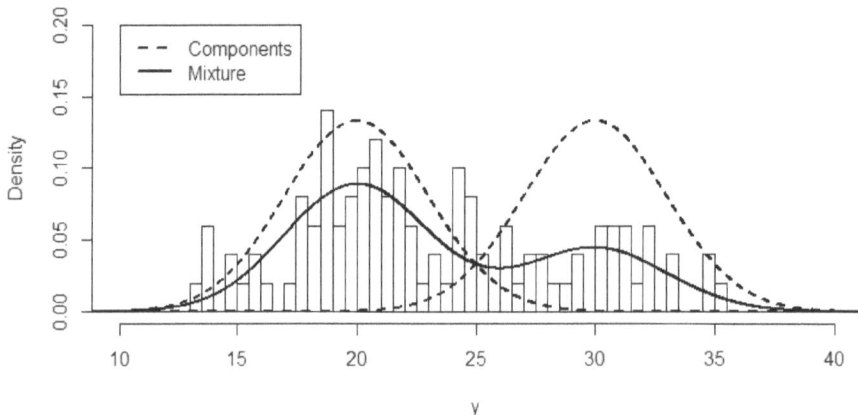

(b) We will here take the vector $R = (R_1, ..., R_n)$ as the latent data. The conditional posterior of μ and δ given this latent data is

$$f(\mu, \delta \mid y, R) \propto f(\mu, \delta, y, R)$$

$$= f(\mu, \delta) f(R \mid \mu, \delta) f(y \mid R, \mu, \delta)$$

$$\propto 1 \times \prod_{i=1}^{n} \pi^{R_i} (1-\pi)^{1-R_i} \times \prod_{i=1}^{n} \exp\left\{ -\frac{1}{2\sigma^2} \left(y_i - [\mu + R_i \delta]\right)^2 \right\}$$

$$\propto 1 \times 1 \times \exp\left\{ -\frac{1}{2\sigma^2} \sum_{i=1}^{n} \left(y_i - [\mu + R_i \delta]\right)^2 \right\}.$$

So the log-augmented posterior density is

$$\log f(\mu, \delta \mid y, R) = -\frac{1}{2\sigma^2} \sum_{i=1}^{n} \left(y_i - [\mu + R_i \delta]\right)^2$$

$$= -\frac{1}{2\sigma^2} \sum_{i=1}^{n} \left(y_i^2 - 2y_i [\mu + R_i \delta] + [\mu + R_i \delta]^2\right)$$

$$= -\frac{1}{2\sigma^2} \left\{ \sum_{i=1}^{n} y_i^2 - 2\sum_{i=1}^{n} y_i [\mu + R_i \delta] + \sum_{i=1}^{n} [\mu + R_i \delta]^2 \right\}$$

$$= -c_1 \left\{ c_2 - 2\mu n \bar{y} - 2\delta \sum_{i=1}^{n} y_i R_i + n\mu^2 + 2\mu\delta \sum_{i=1}^{n} R_i + \delta^2 \sum_{i=1}^{n} R_i^2 \right\},$$

where c_1 and c_2 are positive constants which do not depend on μ or δ in any way. We see that

$$\log f(\mu, \delta \mid y, R) = -c_1 \left\{ c_2 - 2\mu n \bar{y} - 2\delta \sum_{i=1}^{n} y_i R_i + n\mu^2 + 2\mu\delta R_T + \delta^2 R_T \right\},$$

where $R_T = \sum_{i=1}^{n} R_i$.

Note: Each R_i equals 0 or 1, and therefore $R_i^2 = R_i$.

So the Q-function is

$$Q_j(\mu, \delta) = E_R\{\log f(\mu, \delta \mid y, R) \mid y, \mu_j, \delta_j\}$$

$$= -c_1 \left\{ c_2 - 2\mu n \bar{y} - 2\delta \sum_{i=1}^{n} y_i e_{ij} + n\mu^2 + 2\mu\delta e_{Tj} + \delta^2 e_{Tj} \right\},$$

where: $e_{ij} = E(R_i \mid y, \mu_j, \delta_j)$

$$e_{Tj} = E(R_T \mid y, \mu_j, \delta_j) = \sum_{i=1}^{n} e_{ij}.$$

We now need to obtain formulae for the e_{ij} values. Observe that

$$f(R \mid y, \mu, \delta) \propto f(\mu, \delta, y, R)$$

$$\propto 1 \times \prod_{i=1}^{n} \pi^{R_i} (1-\pi)^{1-R_i} \times \prod_{i=1}^{n} \exp\left\{ -\frac{1}{2\sigma^2} (y_i - [\mu + R_i \delta])^2 \right\}.$$

It follows that

$$(R_i \mid y, \mu, \delta) \sim \perp Bernoulli(e_i), \; i = 1, \ldots, n,$$

where

$$e_i = \frac{\pi \exp\left(-\dfrac{1}{2\sigma^2} (y_i - [\mu + \delta])^2 \right)}{\pi \exp\left(-\dfrac{1}{2\sigma^2} (y_i - [\mu + \delta])^2 \right) + (1-\pi) \exp\left(-\dfrac{1}{2\sigma^2} (y_i - \mu)^2 \right)}.$$

Therefore

$$e_{ij} = \frac{\pi \exp\left(-\dfrac{1}{2\sigma^2} (y_i - [\mu_j + \delta_j])^2 \right)}{\pi \exp\left(-\dfrac{1}{2\sigma^2} (y_i - [\mu_j + \delta_j])^2 \right) + (1-\pi) \exp\left(-\dfrac{1}{2\sigma^2} (y_i - \mu_j)^2 \right)}.$$

Thereby the E-Step of the EM algorithm has been defined.

Next, the M-Step requires us to maximise the Q-function. We begin by writing:

$$\frac{\partial Q_j(\mu, \delta)}{\partial \mu} = -c_1 \left\{ 0 - 2n\bar{y} - 0 + 2n\mu + 2\delta e_{Tj} + 0 \right\}$$

$$\frac{\partial Q_j(\mu, \delta)}{\partial \delta} = -c_1 \left\{ 0 - 0 - 2\sum_{i=1}^{n} y_i e_{ij} + 0 + 2\mu e_{Tj} + 2\delta e_{Tj} \right\}.$$

Setting both of these derivatives to zero and solving for μ and δ *simultaneously*, we obtain the next two values in the algorithm:

$$\mu_{j+1} = \frac{\bar{y} - \dfrac{1}{n} \sum_{i=1}^{n} y_i e_{ij}}{1 - \dfrac{1}{n} e_{Tj}}, \quad \delta_{j+1} = \frac{\sum_{i=1}^{n} y_i e_{ij}}{e_{Tj}} - \mu_{j+1}.$$

The EM algorithm is now completely defined.

Starting the algorithm from $(\mu_0, \delta_0) = (10,1)$, we obtain the sequence shown in Table 4.4. We see that the algorithm has converged to what we believe to be the posterior mode, $(\hat{\mu}, \hat{\delta}) = (20.08, 9.72)$.

Running the algorithm from different starting points we obtain the same final results. Unlike the NR algorithm, we find that the EM algorithm always converges, regardless of the point from which it is started.

Table 4.4 Results of an EM algorithm

j	μ_j	δ_j
0	10.000	1.000
1	21.169	3.032
2	20.321	7.07
3	19.843	9.139
4	19.926	9.518
5	20.005	9.626
6	20.046	9.674
7	20.066	9.697
8	20.075	9.708
9	20.08	9.713
10	20.082	9.715
11	20.083	9.717
12	20.084	9.717
13	20.084	9.717
14	20.084	9.718
15	20.084	9.718
16	20.084	9.718
17	20.084	9.718
18	20.084	9.718
19	20.084	9.718
20	20.084	9.718

(c) The ECM requires us to once again examine the Q-function,

$$Q_j(\mu, \delta) = -c_1 \left\{ c_2 - 2\mu n \bar{y} - 2\delta \sum_{i=1}^{n} y_i e_{ij} + n\mu^2 + 2\mu\delta e_{Tj} + \delta^2 e_{Tj} \right\},$$

but now to maximise this function with respect to μ and δ *individually* (rather than *simultaneously* as for the EM algorithm in (c)).

Thus, setting $\dfrac{\partial Q_j(\mu,\delta)}{\partial \mu} = -c_1\left\{0 - 2n\bar{y} - 0 + 2n\mu + 2\delta e_{Tj} + 0\right\}$

to zero we get $\mu_{j+1} = \bar{y} - \delta_j \dfrac{1}{n} e_{Tj}$ (after substituting in $\delta = \delta_j$).

Then, setting $\dfrac{\partial Q_j(\mu,\delta)}{\partial \delta} = -c_1\left\{0 - 0 - 2\sum_{i=1}^{n} y_i e_{ij} + 0 + 2\mu e_{Tj} + 2\delta e_{Tj}\right\}$

to zero we get $\delta_{j+1} = \dfrac{\sum_{i=1}^{n} y_i e_{ij}}{e_{Tj}} - \mu_{j+1}$ (same equation as in (c)).

We see that the ECM algorithm here is fairly similar to the EM algorithm.

Starting the algorithm at $(\mu_0, \delta_0) = (10, 1)$ we obtain the sequence shown in Table 4.5 (page 184). We see that the ECM algorithm has converged to the same values as the EM algorithm, but along a slightly different route.

(d) Figure 4.8 (page 185) shows a contour plot of the log-posterior density $\log f(\mu, \delta \mid y, R)$ and the routes of the EM and ECM algorithms in parts (b) and (c), each from the starting point $(\mu_0, \delta_0) = (10, 1)$ to the mode, $(\hat{\mu}, \hat{\delta}) = (20.08, 9.72)$. Also shown are two other pairs of routes, one pair starting from (5, 30), and the other from (35, 20).

Note 1: In this exercise there is little difference between the EM and ECM algorithms, both as regards complexity and performance. In more complex models we may expect the EM algorithm to converge faster but have an M-Step which is more difficult to complete than the set of separate Conditional Maximisation-Steps (CM-Steps) of the ECM algorithm.

Note 2: The log-posterior density in Figure 4.8 has a formula which can be derived as follows. First, the joint posterior of all unknowns in the model is

$$f(\mu, \delta, R \mid y) \propto f(\mu, \delta, y, R)$$
$$\propto 1 \times \prod_{i=1}^{n} \pi^{R_i}(1-\pi)^{1-R_i} \times \prod_{i=1}^{n} \exp\left\{-\frac{1}{2\sigma^2}\left(y_i - [\mu + R_i\delta]\right)^2\right\}$$

$$= \prod_{i=1}^{n} \pi^{R_i} (1-\pi)^{1-R_i} \exp\left\{ -\frac{1}{2\sigma^2} \left(y_i - [\mu + R_i \delta] \right)^2 \right\}.$$

So the joint posterior density of just μ and δ is

$$f(\mu, \delta \mid y) = \sum_R f(\mu, \delta, R \mid y)$$

$$\propto \prod_{i=1}^{n} \sum_{R_i=0}^{1} \pi^{R_i} (1-\pi)^{1-R_i} \exp\left\{ -\frac{1}{2\sigma^2} \left(y_i - [\mu + R_i \delta] \right)^2 \right\}$$

$$= \prod_{i=1}^{n} \left(\pi \exp\left\{ -\frac{1}{2\sigma^2} \left(y_i - [\mu + \delta] \right)^2 \right\} + (1-\pi) \exp\left\{ -\frac{1}{2\sigma^2} \left(y_i - \mu \right)^2 \right\} \right).$$

So the log-posterior density of μ and δ is

$$l(\mu, \delta) \equiv \log f(\mu, \delta \mid y)$$

$$= c + \sum_{i=1}^{n} \log\left(\pi \exp\left\{ -\frac{1}{2\sigma^2} \left(y_i - [\mu + \delta] \right)^2 \right\} \right.$$

$$\left. + (1-\pi) \exp\left\{ -\frac{1}{2\sigma^2} \left(y_i - \mu \right)^2 \right\} \right),$$

where c is an additive constant and can arbitrarily be set to zero.

Note 3: As an additional exercise (and a check on our calculations above), we could apply the Newton-Raphson algorithm so as to find the mode of $l(\mu, \delta)$. But this would require us to first determine formulae for the following rather complicated partial derivatives:

$$\frac{\partial l(\mu, \delta)}{\partial \mu}, \ \frac{\partial l(\mu, \delta)}{\partial \delta}, \ \frac{\partial^2 l(\mu, \delta)}{\partial \mu^2}, \ \frac{\partial^2 l(\mu, \delta)}{\partial \delta^2}, \ \frac{\partial^2 l(\mu, \delta)}{\partial \delta \partial \mu},$$

and could prove to be unstable. That is, the algorithm might fail to converge if started from a point not very near the required solution.

Another option is to apply the CNR algorithm (the conditional Newton-Raphson algorithm). This would obviate the need for one of the derivatives above, $\dfrac{\partial^2 l(\mu, \delta)}{\partial \delta \partial \mu}$, and might be more stable, albeit at the cost of not converging so quickly as the plain NR algorithm.

As yet another possibility, we could apply the CNR1 algorithm. This is the same as the CNR algorithm, except that at each conditional step we perform just one iteration of the univariate NR algorithm before moving on to the other of the two conditional steps.

Finally, we could use the R function optim() to maximise $l(\mu,\delta)$. Although this function will be formally introduced later, we can report that it does indeed find the posterior mode, $(\hat{\mu},\hat{\delta}) = (20.08, 9.72)$. For details, see the bottom of the R code below.

Table 4.5 Results of an ECM algorithm

j	μ_j	δ_j
0	10.000	1.000
1	22.505	1.696
2	22.566	3.882
3	21.905	6.811
4	21.139	8.729
5	20.611	9.501
6	20.322	9.732
7	20.181	9.774
8	20.118	9.764
9	20.093	9.746
10	20.085	9.732
11	20.083	9.725
12	20.083	9.720
13	20.083	9.719
14	20.084	9.718
15	20.084	9.718
16	20.084	9.718
17	20.084	9.718
18	20.084	9.718
19	20.084	9.718
20	20.084	9.718

Figure 4.8 Routes of the EM and ECM algorithms

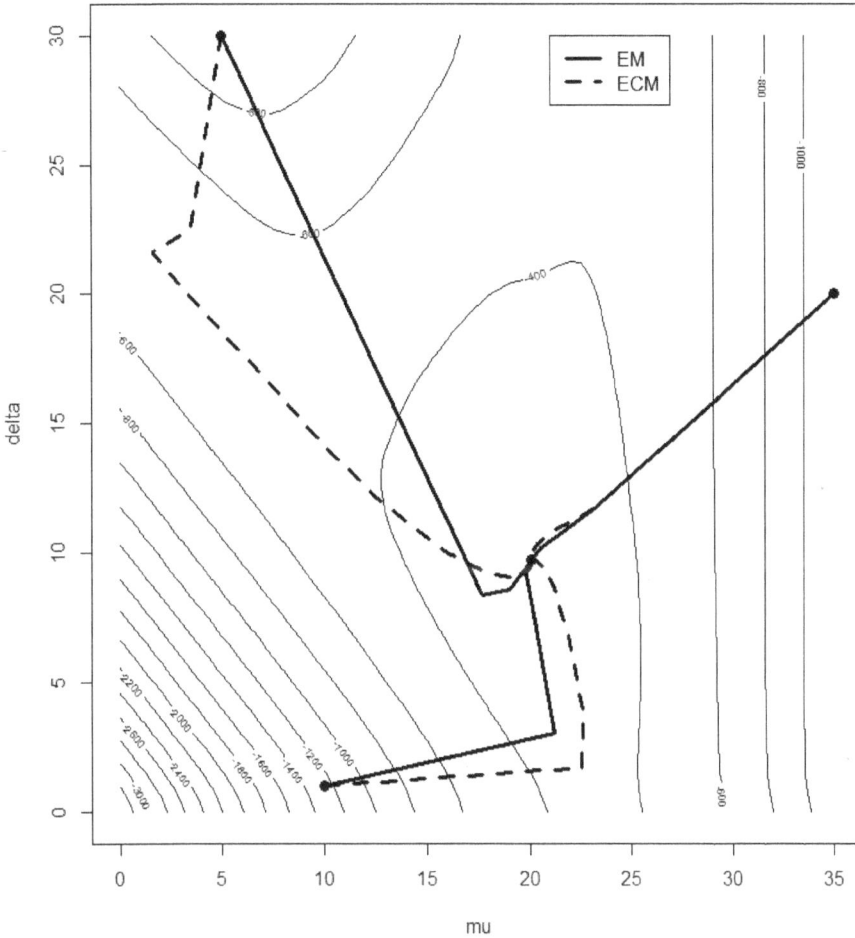

R Code for Exercise 4.7

```
# (a)
X11(w=8,h=4.5); par(mfrow=c(1,1)); options(digits=4)
ntrue=100; pitrue=1/3; mutrue=20; deltrue=10; sigtrue=3

set.seed(512); Rvec=rbinom(ntrue,1,pitrue); sum(Rvec) # 29
yvec=rnorm(ntrue,mutrue+deltrue*Rvec,sigtrue)
ybar=mean(yvec); ybar # 23.16

hist(yvec,prob=T,breaks=seq(0,50,0.5),xlim=c(10,40),ylim=c(0,0.2),
     xlab="y", main=" ")
```

185

```
yv=seq(0,50,0.01); lines(yv,dnorm(yv,mutrue,sigtrue),lty=2,lwd=2)
lines(yv,dnorm(yv,mutrue+deltrue, sigtrue),lty=2,lwd=2)
lines(yv,  (1-pitrue)*dnorm(yv,mutrue,sigtrue)+
          pitrue*dnorm(yv,mutrue+deltrue,sigtrue),    lty=1,lwd=2)
legend(10,0.2,c("Components","Mixture"),lty=c(2,1),lwd=c(2,2))

# (b)
evalsfun= function(y=yvec, pii=pitrue, mu=mutrue,del=deltrue,sig=sigtrue){
# This function outputs (e1,e2,...,en)
        term1vals=pii*dnorm(y,mu+del,sig)
        term0vals=(1-pii)*dnorm(y,mu,sig)
        term1vals/(term1vals+term0vals) }

EMfun=function(J=20, mu=10, del=1, y=yvec, pii=pitrue, sig=sigtrue){
        muv=mu; delv=del;  ybar=mean(y); n=length(y)
        for(j in 1:J){
                evals=evalsfun(y=y, pii=pii, mu=mu, del=del, sig=sig)
                sumyevals = sum(y*evals); sumevals=sum(evals)
                mu=(ybar-sumyevals/n) / (1-sumevals/n)
                del=sumyevals/sumevals - mu
                muv=c(muv,mu); delv=c(delv,del)
                }
        list(muv=muv,delv=delv)
        }
EMres=EMfun(J=20, mu=10, del=1,y=yvec,pii=pitrue,sig=sigtrue)
outmat = cbind(0:20,EMres$muv, EMres$delv)
print.matrix <- function(m){    write.table(format(m, justify="right"),
                                row.names=F, col.names=F, quote=F)  }
print.matrix(outmat)
#  0.000 10.000  1.000
#  1.000 21.169  3.032
#  2.000 20.321  7.070
#  3.000 19.843  9.139
#  4.000 19.926  9.518
#  5.000 20.005  9.626
# .................................
# 16.000 20.084  9.718
# 17.000 20.084  9.718
# 18.000 20.084  9.718
# 19.000 20.084  9.718
# 20.000 20.084  9.718

muhat=EMres$muv[21]; delhat=EMres$delv[21];
c(muhat,delhat) # 20.084  9.718
```

```
# (c)
CEMfun=function(J=20, mu=10, del=1, y=yvec, pii=pitrue, sig=sigtrue){
        muv=mu; delv=del;  ybar=mean(y); n=length(y)
        for(j in 1:J){
                evals=evalsfun(y=y, pii=pii, mu=mu, del=del, sig=sig)
                sumyevals = sum(y*evals); sumevals=sum(evals)
                mu=ybar-del*sumevals/n
                del=sumyevals/sumevals - mu
                muv=c(muv,mu); delv=c(delv,del)
                }
        list(muv=muv,delv=delv)
        }
CEMres=CEMfun(J=20, mu=10, del=1,y=yvec,pii=pitrue,sig=sigtrue)
outmat2 = cbind(0:20, CEMres$muv, CEMres$delv)
print.matrix(outmat2)

#   0.000 10.000  1.000
#   1.000 22.505  1.696
#   2.000 22.566  3.882
#   3.000 21.905  6.811
#   4.000 21.139  8.729
#   5.000 20.611  9.501
# ...................................
# 16.000 20.084  9.718
# 17.000 20.084  9.718
# 18.000 20.084  9.718
# 19.000 20.084  9.718
# 20.000 20.084  9.718


# (d)
X11(w=8,h=9); par(mfrow=c(1,1))
logpostfun=function(mu=10,del=10,y=yvec,pii=pitrue,sig=sigtrue){
        sum(log(pii*dnorm(y,mu+del,sig)+(1-pii)*dnorm(y,mu,sig)))  }
mugrid=seq(0,35,0.5); delgrid=seq(0,30,0.5)
logpostmat=as.matrix(mugrid %*% t(delgrid))
dim(logpostmat)  # 41 21 OK

for(i in 1:length(mugrid)) for(j in 1:length(delgrid))        logpostmat[i,j] =
        logpostfun(mu=mugrid[i],del=delgrid[j],y=yvec,pii=pitrue,sig=sigtrue)

contour(x=mugrid, y=delgrid, z=logpostmat, nlevels=20,
        xlab="mu", ylab="delta"); points(muhat,delhat, pch=16,cex=1.2)

points(10,1,pch=16,cex=1.2)
```

```
EMres=EMfun(J=20, mu=10, del=1,y=yvec,pii=pitrue,sig=sigtrue)
CEMres=CEMfun(J=20, mu=10, del=1,y=yvec,pii=pitrue,sig=sigtrue)
lines(EMres$muv, EMres$delv,lty=1,lwd=3)
lines(CEMres$muv, CEMres$delv,lty=2,lwd=3)


points(5,30,pch=16,cex=1.2)
EMres=EMfun(J=50, mu=5, del=30,y=yvec,pii=pitrue,sig=sigtrue)
CEMres=CEMfun(J=50, mu=5, del=30, y=yvec,pii=pitrue,sig=sigtrue)
lines(EMres$muv, EMres$delv,lty=1,lwd=3)
lines(CEMres$muv, CEMres$delv,lty=2,lwd=3)


points(35,20,pch=16,cex=1.2)
EMres=EMfun(J=50, mu=35, del=20,y=yvec,pii=pitrue,sig=sigtrue)
CEMres=CEMfun(J=50, mu=35, del=20, y=yvec,pii=pitrue,sig=sigtrue)
lines(EMres$muv, EMres$delv,lty=1,lwd=3)
lines(CEMres$muv, CEMres$delv,lty=2,lwd=3)
legend(21,30,c("EM","ECM"),lty=c(1,2),lwd=c(3,3))

# Note 2. Maximisation of the logposterior density of mu and delta using optim()
logpostfun2=function(theta=c(10,1),y=yvec,pii=pitrue,sig=sigtrue){
        -sum(log(pii*dnorm(y,theta[1]+theta[2],sig)+
        (1-pii)*dnorm(y,theta[1],sig)))
        }
res=optim(par=c(10,1),fn= logpostfun2)$par; res  # 20.08  9.72
res=optim(par=c(5,30),fn= logpostfun2)$par; res  # 20.085  9.716
res=optim(par=c(35,20),fn= logpostfun2)$par; res  # 20.084  9.716
res=optim(par=res,fn= logpostfun2)$par; res  # 20.084  9.718
  # Here we fine-tune the answer by starting at the previous solution.
```

4.6 Integration techniques

Bayesian inference typically involves a great deal of integration (and/or summation). For example, consider the posterior density

$$f(\theta \mid y) = 6\theta^5, 0 < \theta < 1$$

(which featured in previous exercise involving the binomial-beta model) and suppose that we wish to find the posterior mean estimate of $\lambda = \theta^2$. This estimate is

$$\hat{\lambda} = E(\theta^2 \mid y) = \int_0^1 \theta^2 \times \left(6\theta^5\right) d\theta = 0.75.$$

But what if this integral did not have a simple analytical solution?

In that case, we could consider a number of other strategies. First, we might re-express the posterior mean as

$$\hat{\lambda} = \int \lambda f(\lambda \mid y) d\lambda,$$

where, using the method of transformation,

$$f(\lambda \mid y) = f(\theta \mid y) \left| \frac{d\theta}{d\lambda} \right| = 6\left(\lambda^{1/2}\right)^5 \left| \frac{1}{2} \lambda^{-1/2} \right| = 3\lambda^2, \quad 0 < \lambda < 1,$$

so that

$$\hat{\lambda} = \int_0^1 \lambda \left(3\lambda^2\right) d\lambda = 0.75.$$

If this strategy does not help, we may then consider using a *numerical integration technique*.

For example, we could apply the integrate() function in R to get $\hat{\lambda} = 0.75$, as follows:

```
gfun = function(t){ 6*t^7 } # Define the function to be integrated
integrate(f=gfun,lower=0,upper=1)$value # 0.75
```

In some cases the function requiring integration is very complicated or does not have a closed form expression. In that case, direct application of the integrate() function may not work or be practicable, and then it may be useful to apply the *trapezoidal rule* or *Simpson's rule* to evaluate the integral.

When working in R, the following is often a convenient strategy:

(i) evaluate $g(\theta) = \theta^2 \times 6\theta^5$ at each θ on the grid
0, 0.1, 0.2, ..., 0.9, 1 (say); then

(ii) create a spline through these points, using the fit() and predict() functions; and then

(iii) find the area under this spline using the integrate() function.

Applying this method (see the R code below for details) yields 0.7558 as an estimate of $\hat{\lambda}$. Repeating, but with the evaluations on the grid 0.01, 0.02, ...,1 yields 0.7500. Repeating again, but with evaluations on the grid 0.001, 0.002, ..., 1 yields 0.7500. It appears that a limit has been reached and that using a finer grid would not result in any improvements to the results of this numerical procedure.

We may conclude that $\hat{\lambda} = 0.7500$ (to 4 decimals).

R Code for Section 4.6

```
gfun = function(t){ 6*t^7 } # Define the function to be integrated
integrate(f=gfun,lower=0,upper=1)$value # 0.75

INTEG <- function(xvec, yvec, a = min(xvec), b = max(xvec)){
        # Integrates numerically under a spline through the
        # points given by the vectors xvec and yvec, from a to b.
        fit <- smooth.spline(xvec, yvec)
        spline.f <- function(x){predict(fit, x)$y }
        integrate(spline.f, a, b)$value  }

gfun=function(t){ 6*t^7 }
tvec <- seq(0,1,0.1);   gvec <- gfun(tvec)
        INTEG(tvec,gvec,0,1) # 0.755803
tvec <- seq(0,1,0.01);   gvec <- gfun(tvec)
        INTEG(tvec,gvec,0,1)  # 0.75
tvec <- seq(0,1,0.001);  gvec <- gfun(tvec)
        INTEG(tvec,gvec,0,1)  # 0.75
```

Exercise 4.8 Numerical integration

Suppose that $X \sim N(\mu, \sigma^2)$ and $Y = (X \mid X > c)$ where $\mu = 8$, $\sigma = 3$ and $c = 10$. Find EY using numerical techniques and compare your answer with the exact value,

$$\mu + \sigma \phi\left(\frac{c-\mu}{\sigma}\right) \bigg/ \left\{ 1 - \Phi\left(\frac{c-\mu}{\sigma}\right)\right\},$$

which was derived analytically in Exercise 4.6.

Solution to Exercise 4.8

The required integral is

$$EY = \int_c^\infty g(x)dx,$$

where:
$$g(x) = \frac{xf(x)}{P(X > 0)} \qquad , \qquad f(x) = \frac{1}{\sigma}\phi\left(\frac{x-\mu}{\sigma}\right) \qquad ,$$

$$P(X > 0) = 1 - \Phi\left(\frac{c-\mu}{\sigma}\right).$$

Applying the integrate() function directly to $g(x)$ we get $EY = 11.7955$.

Applying the INTEG() function (defined in Section 4.6) with coordinates given by $(10,10.1,10.2,...,30)$ and $(g(10),g(10.1),g(10.2),...,g(30))$, we also get $EY = 11.7955$. The exact value of EY is in fact

$$\mu + \sigma \phi\left(\frac{c-\mu}{\sigma}\right) \bigg/ \left\{1 - \Phi\left(\frac{c-\mu}{\sigma}\right)\right\} = 11.7955.$$

Note: If we use the integrate() function with bounds from 10 to 20 rather than 10 to 30, we get 11.7929, which is slightly in error. Exactly the same happens with the INTEG() function. Thus, when using either of these functions, care must be taken to choose a large enough range. Ideally, we will *sketch* the integrand function and make sure the range of integration is sufficiently broad to cover all important regions (where the integrand is significantly positive). In practice, it is useful to gradually increase the range of integration until the answer stops changing. Likewise, it is useful to gradually increase the grid density chosen for the INTEG() function until the answer stops changing.

R Code for Exercise 4.8

```
# First declare the function INTEG() as defined in the previous exercise

mu=8; sig=3; c = 10; options(digits=6)
PXpos = (1-pnorm((c-mu)/sig))
gfun=function(x){ x * dnorm(x,mu,sig) / PXpos }
integrate(gfun,c,20)$value # 11.7929
integrate(gfun,c,30)$value # 11.7955
xvec <- seq(c,20,0.1);   gvec <- gfun(xvec);   INTEG(xvec,gvec,c,20)  # 11.7929
xvec <- seq(c,30,0.1);   gvec <- gfun(xvec);   INTEG(xvec,gvec,c,30)  # 11.7955
true=mu + sig*dnorm((c-mu)/sig)/(1-pnorm((c-mu)/sig)); true # 11.7955
```

Exercise 4.9 Double integration

Use the integrate() and INTEG() functions in at least two different ways so as to calculate the double integral

$$I = \int_{x=0}^{1}\left(\int_{t=0}^{x^3} t^t \, dt\right) dx.$$

Illustrate your calculations with suitable graphs of the relevant functions involved.

Solution to Exercise 4.9

Using the integrate() function alone (and not the INTEG() function), the integral can be worked out as follows:

```
integrate(function(x) {
        sapply(x, function(x) {
        integrate(function(t) {
        sapply(t, function(t) t^t )
        }, 0, x^3)$value }) }, 0, 1)

# 0.192723 with absolute error < 7.8e-10
```

Another approach is as follows. Observe that

$$I = \int_{x=0}^{1} g(x)\, dx,$$

where

$$g(x) = \int_{t=0}^{x^3} h(t)\, dt$$

and

$$h(t) = t^t.$$

We will now use the integrate() function to obtain $g(x)$ for each value of x in the grid 0, 0.01, 0.02, ..., 1. We will then apply the INTEG() function to the resulting coordinates.

Figure 4.9 below displays the two functions $h(t)$ and $g(x)$. The value $g(0.8) = 0.381116$ is the area under $h(t)$ between 0 and 0.8. The total area under $h(t)$ (from 0 to 1) is 0.78343.

The total area under $g(x)$ (from 0 to 1) is estimated as 0.192723. Using the grid 0, 0.001, 0.002, ..., 1 also leads to 0.192723, whereas using the grid 0, 0.1, 0.2, ..., 1 leads to 0.193054.

We conclude that the exact value of the required integral I to 4 decimals is 0.1927, which is in agreement with the first approach above which doesn't make use of the INTEG() function.

One could also adapt the second approach above so as to calculate the double integral using the INTEG() function only (without using the integrate() function directly). This might be useful if the inner integral

$$g(x) = \int\limits_{t=0}^{x^3} h(t)dt \quad \text{where } h(t) = t^t$$

could not be evaluated easily using integrate() directly, for example if $h(t)$ were a very complicated function which could not be expressed in closed form.

Note: The integrate() function is called within the INTEG() function and so is used at least indirectly in all of the approaches considered here.

Figure 4.9 Two functions

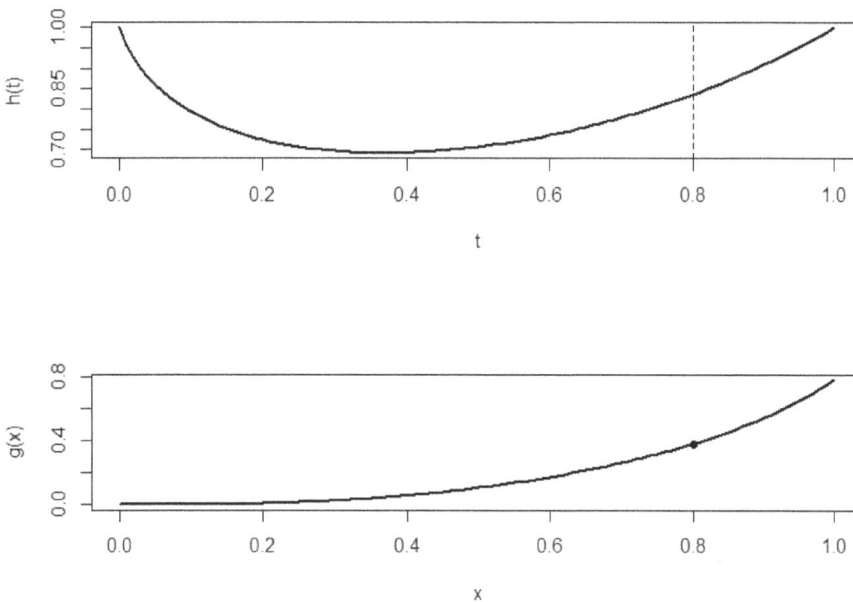

R Code for Exercise 4.9

```
integrate(function(x) {
        sapply(x, function(x) {
        integrate(function(t) {
        sapply(t, function(t) t^t )
        }, 0, x^3)$value }) }, 0, 1)
# 0.192723 with absolute error < 7.8e-10
```

```
# Declare the function INTEG() as defined in the previous exercise
options(digits=6); X11(w=8,h=6); par(mfrow=c(2,1))

hfun= function(t){ t^t }
tvec=seq(0,1,0.01); hvec=hfun(tvec)
plot(tvec,hvec,type="l",xlab="t",ylab="h(t)",lwd=2); abline(v=0.8,lty=2)

integrate(f=hfun,lower=0,upper=0.8^3)$value
        # 0.381116   This is g(0.8) = area under h(t) to left of 0.8
integrate(f=hfun,lower=0,upper=1)$value
        # 0.78343   This is the total areas under h(t) (from 0 to 1)

xvec = seq(0,1,0.01); gvec = rep(NA,length(xvec))
for(i in 1:length(xvec)){  xval = xvec[i]
        gvec[i] = integrate(f=hfun,lower=0,upper=xval^3)$value }
INTEG(xvec,gvec) # 0.192723
plot(xvec,gvec,type="l",xlab="x",ylab="g(x)",lwd=2)
points(0.8, 0.381116 , pch=16, cex=1)

# Apply INTEG() using different grids

xvec = seq(0,1,0.001); gvec = rep(NA,length(xvec))
for(i in 1:length(xvec)){  xval = xvec[i]
        gvec[i] = integrate(f=hfun,lower=0,upper=xval^3)$value }
INTEG(xvec,gvec) # 0.192723

xvec = seq(0,1,0.1); gvec = rep(NA,length(xvec))
for(i in 1:length(xvec)){  xval = xvec[i]
        gvec[i] = integrate(f=hfun,lower=0,upper=xval^3)$value }
INTEG(xvec,gvec) # 0.193053
```

4.7 The optim() function

The function optim() in R is a very useful and versatile tool for maximising or minimising functions, both of one and of several variables.

This R function can also be adapted for solving single or simultaneous equations and provides an alternative to other techniques such as trial and error, the Newton-Raphson algorithm and the EM algorithm.

The second of the next two exercises shows how the optim() function can be used to specify a prior distribution.

Exercise 4.10 Simple examples of the optim() function

Use the optim() function to 'find' the mode of each of the following:

(a) $g(x) = x^2 e^{-5x}$, $x > 0$ (mode = 2/5)

(b) $g(x) = \dfrac{|x|^x e^{-(x-1)^2}}{1+|x|}$, $x \in \Re$ (the mode has no closed form)

(c) $g(x,y) = y^3 e^{-y\{(x-1)^2+(x-3)^2\}}$, $x \in \Re$, $y > 0$

$$\text{(mode} = (x, y) = ((1+3)/2, 3/2)).$$

Solution to Exercise 4.10

In each of these cases, the optim() function (which minimises a function by default) may be applied to the *negative* of the specified function (so as to maximise that function).

(a) The function correctly returns $x = 2/5$. (NB: The warning message may be ignored.)

(b) The function returns a value of 1.5047. (We presume that this is correct; see below for a verification.)

(c) The mode is correctly computed as $(x, y) = (2, 1.5)$. (Note that this solution is obvious by analogy with maximum likelihood estimation of the normal mean and variance.)

Figure 4.10 illustrates these three solutions, with each mode being marked by a dot and vertical line. Subplot (c) shows several examples of the function $g(x, y)$ in part (c) considered as a function of only x, with each line defined by a fixed value of y on the grid 0, 0.5, 1, ...,4.5, 5.

Figure 4.10 Maximisation of function g in parts (a), (b) and (c)

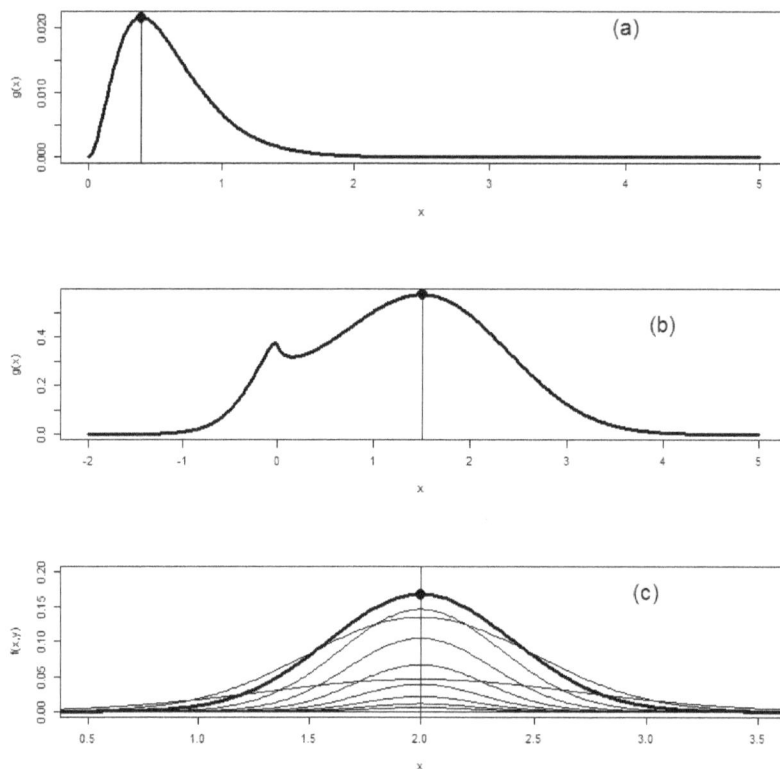

R Code for Exercise 4.10

```
help(optim); options(digits=5);  X11(w=8,h=8); par(mfrow=c(3,1))

# (a)
fun=function(x){ -x^2 * exp(-5*x) }
res0=optim(par=0.5,fn=fun)$par; res0 # 0.4
        # Warning message:
        # In optim(par = 0.5, fn = fun) :
        #   one-diml optimization by Nelder-Mead is unreliable:
        # use "Brent" or optimize() directly
plot(seq(0,5,0.01), -fun(seq(0,5,0.01)),type="l",lwd=3,xlab="x",ylab="g(x)");
abline(v=res0); points(res0, -fun(res0), pch=16, cex=2); text(4,0.02,"(a)",cex=2)

# (b)
fun=function(x){ -exp(-(x-1)^2) * abs(x)^x/(1+abs(x)) }
res0=optim(par=1,fn=fun)$par; res0 # 1.5047
plot(seq(-2,5,0.01), -fun(seq(-2,5,0.01)),type="l",lwd=3, xlab="x",ylab="g(x)");
abline(v=res0); points(res0, -fun(res0), pch=16, cex=2); text(4,0.45,"(b)",cex=2)
```

```
# (c)
fun=function(v){ -v[2]^3 * exp(  -v[2]  * ( (v[1]-1)^2 + (v[1]-3)^2 )  )  }
res0=optim(par=c(2,2),fn=fun, lower = c(-Inf,0), upper = c(Inf,Inf),
        method = "L-BFGS-B")$par; res0 # 2.0 1.5

fun2=function(x,y){ y^3 * exp( -y  * ( (x-1)^2 + (x-3)^2 )  )  }

plot(c(0.5,3.5),c(0,0.2), type="n",xlab="x",ylab="f(x,y)")
for(y in seq(0,5,0.5))
        lines(seq(0,5,0.01),  fun2(x=seq(0,5,0.01),y=y), lty=1)
abline(v=res0[1]); points(res0[1],fun2(res0[1],res0[2]), pch=16, cex=2);
lines(seq(0,5,0.01),fun2(x= seq(0,5,0.01), y=res0[2]),lty=1,lwd=3);
text(3,0.17,"(c)",cex=2)
```

Exercise 4.11 Specification of parameters in a prior distribution using the optim() function

Consider the normal-gamma model given by:

$$(y_1,\ldots,y_n \mid \lambda) \sim iid\ N(\mu, 1/\lambda)$$
$$\lambda \sim G(\eta, \tau).$$

Use the optim() function in R to find the values of η and τ which correspond to a prior belief that the population standard deviation $\sigma = 1/\sqrt{\lambda}$ lies between 0.5 and 1 with 95% probability, and that σ is equally likely to be below 0.5 as it is to be above 1.

Solution to Exercise 4.11

We wish to find the values of η and τ which satisfy the two equations:

$$P(\sigma < a) = \alpha/2 \qquad \text{and} \qquad P(\sigma < b) = 1 - \alpha/2,$$

where $a = 0.5$, $b = 1$ and $\alpha = 0.05$.

These two equations are together equivalent to each of the following five pairs of equations:

$P(\sigma^2 < a^2) = \alpha/2$	and	$P(\sigma^2 < b^2) = 1 - \alpha/2$
$P(1/\lambda < a^2) = \alpha/2$	and	$P(1/\lambda < b^2) = 1 - \alpha/2$
$P(1/a^2 < \lambda) = \alpha/2$	and	$P(1/b^2 < \lambda) = 1 - \alpha/2$
$P(\lambda < 1/a^2) = 1 - \alpha/2$	and	$P(\lambda < 1/b^2) = \alpha/2$
$F_{G(\eta,\tau)}(1/a^2) - (1 - \alpha/2) = 0$	and	$F_{G(\eta,\tau)}(1/b^2) - \alpha/2 = 0.$

We now focus on the last of these pairs of two equations. Two obvious ways to solve these equations are by trial and error and via the multivariate Newton-Raphson algorithm, as illustrated earlier. But the solution can be obtained more easily by using the optim() function to minimise

$$g(\eta,\tau) = \left[F_{G(\eta,\tau)}(1/a^2) - (1-\alpha/2) \right]^2 + \left[F_{G(\eta,\tau)}(1/b^2) - (\alpha/2) \right]^2.$$

Note: Clearly, this function has a value of zero at the required values of η and τ.

With the default settings and starting at $\eta = 0.2$ and $\tau = 6$, optim() produced some warning messages (which we ignored) and provided the solution, $\eta = 8.4764$ and $\tau = 3.7679$.

Now, this solution is not exactly correct, because the probabilities of a *Gamma*(8.4764, 3.7679) random variable lying below $1/b^2 = 1$ and below $1/a^2 = 4$, respectively, are 0.025048 and 0.975104 (i.e. not exactly 0.025 and 0.975 as desired).

However, applying the optim() function *again* but starting at the previous solution, namely $\eta = 8.4764$ and $\tau = 3.7679$, yielded a 'refined' solution, $\eta = 8.4748$ and $\tau = 3.7654$.

This solution may be considered correct, because the probabilities of a *Gamma*(8.4748, 3.7654) random variable being less than $1/b^2 = 1$ and less than $1/a^2 = 4$, respectively, are exactly 0.025 and 0.975.

Discussion

It is instructive to derive and plot the corresponding density of the precision parameter λ, and then to do this also for the variance parameter $\sigma^2 = \lambda^{-1}$ and the standard deviation parameter $\sigma = \lambda^{-1/2}$, respectively.

The three densities are plotted in Figure 4.11 (in the stated order from top to bottom). The vertical lines show the 0.025 and 0.975 quantiles of each distribution. The formulae for the three densities are as follows:

$$f(\lambda) = f_{G(\eta,\tau)}(\lambda) = \frac{\tau^\eta \lambda^{\eta-1} e^{-\tau\lambda}}{\Gamma(\eta)}, \quad \lambda > 0$$

$$f(\sigma^2) = f_{IG(\eta,\tau)}(\sigma^2) = f(\lambda)\left|\frac{d\lambda}{d(\sigma^2)}\right| = f_{G(\eta,\tau)}(\lambda = \sigma^{-2})\left|-(\sigma^2)^{-2}\right|,$$

$$\text{where } \lambda = (\sigma^2)^{-1}$$

$$= \frac{\tau^{\eta}(1/\sigma^2)^{\eta-1}e^{-\tau(1/\sigma^2)}}{\Gamma(\eta)}(\sigma^2)^{-2}, \ \sigma^2 > 0$$

$$f(\sigma) = f(\lambda)\left|\frac{d\lambda}{d\sigma}\right| = f_{G(\eta,\tau)}(\lambda = \sigma^{-2})\left|-2\sigma^{-3}\right| \qquad \text{where } \lambda = (\sigma)^{-2}$$

$$= \frac{\tau^{\eta}(1/\sigma^2)^{\eta-1}e^{-\tau(1/\sigma^2)}}{\Gamma(\eta)}2\sigma^{-3}, \ \sigma > 0.$$

As a check on the last of these three densities, the integrate() function was used to show that the area under that density is exactly 1, and that the areas underneath it to the left of 0.5 and to the right of 1 are both exactly 0.025.

Figure 4.11 Three prior densities

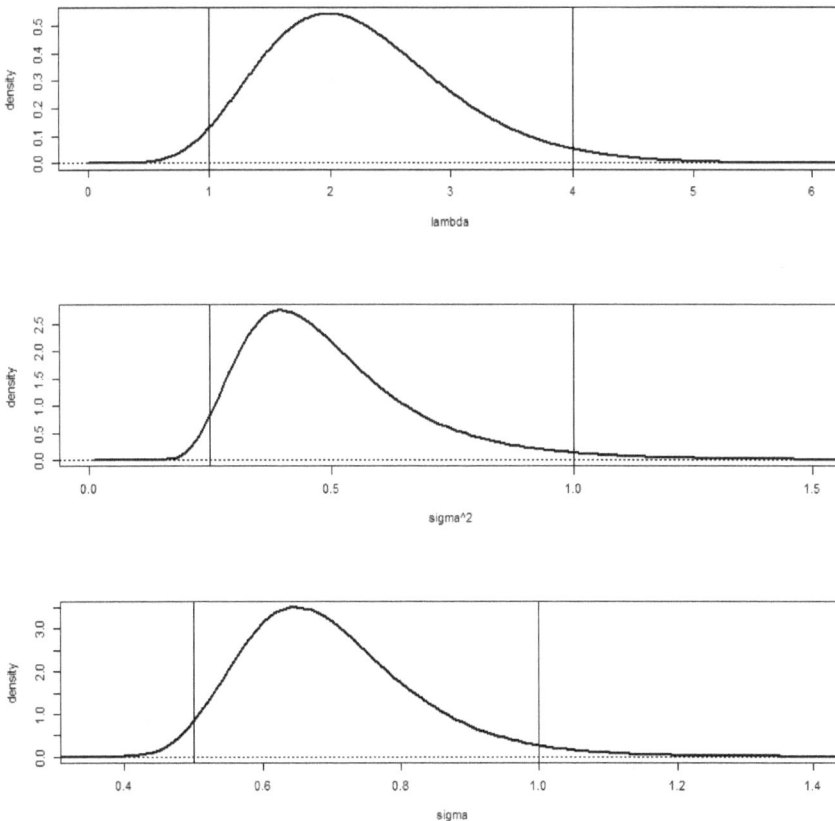

R Code for Exercise 4.11

```
options(digits=5); a=0.5; b=1; alp=0.05;
fun=function(v,alp=0.05,a=0.5,b=1){
    (pgamma(1/a^2,v[1],v[2])-(1-alp/2))^2 +
    (pgamma(1/b^2,v[1],v[2])-(alp/2))^2   }

res0=optim(par=c(0.2,6),fn=fun)$par
res0 # 8.4764 3.7679
pgamma(c(1/b^2,1/a^2),res0[1],res0[2]) # 0.025048 0.975104 Close

res=optim(par=res0,fn=fun)$par; res # 8.4748 3.7654
pgamma(c(1/b^2,1/a^2),res[1],res[2]) # 0.025 0.975   Correct

res2=optim(par=c(6,3),fn=fun)$par; res2 # 8.4753 3.7655
pgamma(c(1/b^2,1/a^2),res2[1],res2[2]) # 0.024992 0.974996   Close

res3=optim(par=res2,fn=fun)$par; res3 # 8.4748 3.7654
pgamma(c(1/b^2,1/a^2),res3[1],res3[2]) # 0.025 0.975   Correct

par(mfrow=c(3,1));  tv=seq(0,10,0.01)

plot(tv, dgamma(tv,res[1],res[2]),type="l",lwd=2, xlim=c(0,6),
        xlab="lambda",ylab="density"); abline(v=c(1/a^2,1/b^2));
abline(h=0,lty=3)

plot(tv,dgamma(1/tv,res[1],res[2])/tv^2, type="l", lwd=2, xlim=c(0,1.5),
        xlab="sigma^2",ylab="density");
abline(v=c(a^2,b^2)); abline(h=0,lty=3)

plot(tv,dgamma(1/tv^2,res[1],res[2])*2/tv^3, type="l", lwd=2,
xlim=c(0.35,1.4), xlab="sigma",ylab="density");
abline(v=c(a,b)); abline(h=0,lty=3)

# Check areas under the last curve
func=function(t){ dgamma(1/t^2,res[1],res[2])*2/t^3  }
integrate(func,lower=0,upper=Inf)$value # 1 Correct
integrate(func,lower=0,upper=0.5)$value # 0.025  Correct
integrate(func,lower=1,upper=Inf)$value # 0.025 Correct
```

CHAPTER 5

Monte Carlo Basics

5.1 Introduction

The term *Monte Carlo* (MC) *methods* refers to a broad collection of tools that are useful for approximating quantities based on artificially generated random samples. These include the *Monte Carlo integration* (for estimating an integral using such a sample), the *inversion technique* (for generating the required sample), and *Markov chain Monte Carlo methods* (an advanced topic in Chapter 6). In principle, the approximation can be made as good as required simply by making the Monte Carlo sample size sufficiently large. As will be seen (further down), Monte Carlo methods are a very useful tool in Bayesian inference.

To illustrate the basic idea of Monte Carlo methods, consider *Buffon's needle problem*, where a needle of length 10 cm (say) is dropped randomly onto a floor with parallel lines being distance 10 cm apart. What is p, the probability of the needle crossing a line? The exact value of p can be worked out *analytically* as $2/\pi = 0.63662$ (this is done in one of the exercises below). But this takes mathematical effort. If this analytical solution were not possible (or just too much work), we could instead estimate p via *Monte Carlo*. The simplest way to do this would be to toss the needle onto the floor 1,000 times (randomly and independently). If the needle crosses a line 641 times (say), then the Monte Carlo estimate of p is just $641/1,000 = 0.641$.

As a variation on this *physical* experiment (which could be rather laborious), we could toss the needle 1,000 times *virtually*, meaning that we simulate each drop (or rather the *parameters* of each drop) on a computer and each time determine whether the virtual needle has crossed a virtual line.

This method will be faster and more accurate; but it will also require at least some mathematical work to identify exactly what the parameters of each drop are and what configuration of those parameters correspond to the needle crossing a line (again, this is done in one of the exercises below).

In this chapter, we will first discuss Monte Carlo methods and their usefulness under the assumption that we have available or can generate the required random samples. As we will see in the exercises and their solutions, such samples can often be obtained very easily using inbuilt R functions, e.g. runif() and rnorm().

After this we will describe special methods for generating a random samples, starting with the simplest, such as the inversion technique and rejection sampling. We reserve the more complicated techniques which involve Markov chain theory to the next and later chapters.

Also, as part of the structure of the present chapter, we will first discuss Monte Carlo methods and random number generation in a fully general setting. Only after we have finished our treatment of these two topics (to a certain level at least) will we discuss their application to Bayesian inference. Hopefully this format will minimise any confusion.

5.2 The method of Monte Carlo integration for estimating means

One of the most important applications of Monte Carlo methods is the estimation of *means*. Suppose we are interested in μ, the mean of some distribution defined by a density $f(x)$ (or by a cumulative distribution function $F(x)$), but we are unable to calculate μ exactly (or easily), for example by applying the formula

$$\mu = Ex = \int xf(x)dx$$

(or $\mu = Ex = \sum_x xf(x)$ or $\mu = Ex = \int xdF(x)$).

Also suppose, however, that we are able to generate (or obtain) a random sample from the distribution in question. Denote this sample as

$$x_1,...,x_J \sim iid \ f(x)$$

(or $x_1,...,x_J \sim iid \ F(x)$).

Then we may use this sample to estimate μ by

$$\bar{x} = \frac{1}{J}\sum_{j=1}^{J} x_j.$$

Also, a $1-\alpha$ confidence interval (CI) for μ given by

$$CI = (\bar{x} \pm z_{\alpha/2} s / \sqrt{J}),$$

where

$$s^2 = \frac{1}{J-1} \sum_{j=1}^{J} (x_j - \bar{x})^2$$

is the sample variance of the random values.

In this context we refer to:

$x_1,...,x_J$	as the *Monte Carlo sample values* or the *Monte Carlo sample*
$\bar{x}$	as the *Monte Carlo sample mean* or the *Monte Carlo estimate*
CI	as the *Monte Carlo $1-\alpha$ confidence interval* for μ
J	as the *Monte Carlo sample size*
s^2	as the *Monte Carlo sample variance*
s	as the *Monte Carlo sample standard deviation*
$s/\sqrt{J}$	as the *Monte Carlo standard error* (SE).

Three important facts here are that:

- $\bar{x}$ is unbiased for μ (i.e. $E\bar{x} = \mu$)
- the CI has coverage approximately $1-\alpha$, by the central limit theorem
- the width of the CI converges to zero as the MC sample size J tends to infinity.

Exercise 5.1 Monte Carlo estimation of a known gamma mean

(a) Use the R function rgamma() to generate a random sample of size $J = 100$ from the *Gamma*(3,2) distribution, whose mean is $\mu = 3/2 = 1.5$. Then use the method of Monte Carlo to produce a point estimate μ and a 95% CI for μ.

(b) Repeat (a) but with MC sample sizes of 1,000 and 10,000, and discuss the results.

Note: In this exercise we are focusing on the integral

$$\mu = \int xf(x)dx = \int_0^\infty x\left(\frac{2^3 x^{3-1} e^{-2x}}{\Gamma(3)}\right)dx,$$

showing how it could be estimated via MC if it were not possible to evaluate analytically. Exactly the same approach could be applied if the integral were impossible to evaluate.

Solution to Exercise 5.1

(a) Applying the above procedure (see the R code below) we estimate μ by $\bar{x} = 1.5170$. The Monte Carlo 95% confidence interval for μ is

$$CI = (\bar{x} \pm z_{0.025} s / \sqrt{J}) = (1.3539, 1.6800).$$

We note that $\bar{x}$ is 'close' to the true value, $\mu = 1.5$, and the CI contains that true value.

(b) Repeating (a) with $J = 1,000$ we obtain the point estimate 1.5199 and the interval estimate (1.4658, 1.5740).

Repeating (a) with $J = 10,000$ we obtain the point estimate 1.4942 and the interval estimate (1.4773, 1.5110).

As in (a) we note in each case that $\bar{x}$ is 'close' to μ, and the CI contains μ. We also note that as J increases the MC point estimate tends to get closer to μ, and the 95% CI tends to get narrower. (The widths of the three CIs are 0.3261, 0.1081 and 0.0337.)

R Code for Exercise 5.1

```
options(digits=4); J = 100; set.seed(221); xv=rgamma(J,3,2)
xbar=mean(xv); s=sd(xv); ci=xbar + c(-1,1)*qnorm(0.975)*s/sqrt(J)
c(xbar,s,s^2,ci,ci[2]-ci[1]) # 1.5170 0.8320 0.6921 1.3539 1.6800 0.3261

J = 1000; set.seed(231); xv=rgamma(J,3,2)
xbar=mean(xv); s=sd(xv); ci=xbar + c(-1,1)*qnorm(0.975)*s/sqrt(J)
c(xbar,s,s^2,ci,ci[2]-ci[1]) # 1.5199 0.8722 0.7607 1.4658 1.5740 0.1081
J = 10000; set.seed(211); xv=rgamma(J,3,2)
xbar=mean(xv); s=sd(xv); ci=xbar + c(-1,1)*qnorm(0.975)*s/sqrt(J)
c(xbar,s,s^2,ci,ci[2]-ci[1]) # 1.4942 0.8597 0.7391 1.4773 1.5110 0.0337
```

5.3 Other uses of the MC sample

Once a Monte Carlo sample $x_1,...,x_J \sim iid \ f(x)$ has been obtained, it can be used for much more than just estimating the mean of the distribution, $\mu = Ex$. For example, suppose we are interested in the (lower) p-quantile of the distribution, namely

$$q_p = F_X^{-1}(p) = \{\text{value of } x \text{ such that } F(x) = p \}.$$

The MC estimate of q_p is simply $\hat{q}_p$, the empirical p-quantile of $x_1,...,x_J$. For instance, the median $q_{1/2}$ can be estimated by the middle number amongst $x_1,...,x_J$ after sorting in increasing order. This assumes that J is odd. If J is even, we estimate $q_{1/2}$ by the average of the two middle numbers. Thus we may write the MC estimate of $q_{1/2}$ as

$$\hat{q}_{1/2} = \begin{cases} x_{((J+1)/2)}, & J \text{ odd} \\ \dfrac{x_{(J/2)} + x_{((J+1)/2)}}{2}, & J \text{ even}, \end{cases}$$

where $x_{(k)}$ is the kth smallest value amongst $x_1,...,x_J$ $(k = 1,...,J)$.

Also, we estimate the $1-\alpha$ *central density region* (CDR) for x, namely $(q_{\alpha/2}, q_{1-\alpha/2})$, by $(\hat{q}_{\alpha/2}, \hat{q}_{1-\alpha/2})$.

Further, suppose we are interested in the expected value of some function of x, say $y = g(x)$. That is, we wish to estimate the quantity/integral

$$\psi = Ey = \int yf(y)dy = Eg(x) = \int g(x)f(x)dx.$$

Then we simply calculate $y_j = g(x_j)$ for each $j = 1,...,J$. The result will be a random sample $y_1,...,y_J \sim iid \ f(y)$ to which the method of Monte Carlo can then be applied in the usual way. Thus, an estimate of ψ is

$$\bar{y} = \frac{1}{J}\sum_{j=1}^{J} y_j \quad \text{(the sample mean of the } y\text{-values)},$$

and a $1-\alpha$ CI for ψ is

$$\left(\bar{y} \pm z_{\alpha/2} \frac{s_y}{\sqrt{J}} \right),$$

where $s_y^2 = \dfrac{1}{J-1}\sum_{j=1}^{J}(y_j - \bar{y})^2$ (the sample variance of the y-values).

This idea applies to even very complicated functions $y = g(x)$ for which the exact or even approximate value of $\psi = Ey$ would otherwise be very difficult to obtain, either analytically or numerically using a deterministic technique such as numerical integration (or quadrature).

Also, the density $f(x)$ can be estimated by smoothing a probability histogram of $x_1,...,x_J$. Likewise, the density $f(y)$ can be estimated by smoothing a probability histogram of $y_1,...,y_J$. (This could be extremely useful if y is a very complicated function of x.)

Note 1: As we will see later, it is often the case that we are able to sample from a distribution without knowing—or being able to derive—the exact form of its density function.

Note 2: Smoothing a histogram requires some arbitrary decisions to be made about the degree of smoothing and other smoothing parameters. So the MC estimate of a density is not uniquely defined.

Exercise 5.2 Monte Carlo estimation of complicated quantities

Suppose that $x \sim G(3,2)$. Use MC methods and a sample of size $J = 1,000$ to estimate:

$\mu = Ex$, the 80% CDR for x, and $f(x)$

$\psi = Ey$, the 80% CDR for y, and $f(y)$, where $y = \dfrac{x^2 e^{-x}}{1+x+1/x}$.

Present your results graphically, and wherever possible show the true values of the quantities being estimated. Then repeat everything but using a Monte Carlo sample size of $J = 10,000$.

Solution to Exercise 5.2

The required graphs are shown in Figures 5.1 to 5.4. See the R code below for more details.

Figure 5.1 Histogram of *x*-value (*J* = 1,000)

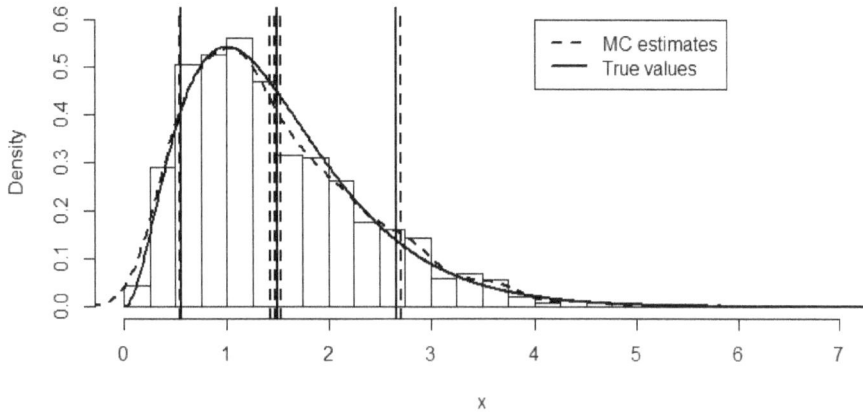

Figure 5.2 Histogram of y-value (*J* = 1,000)

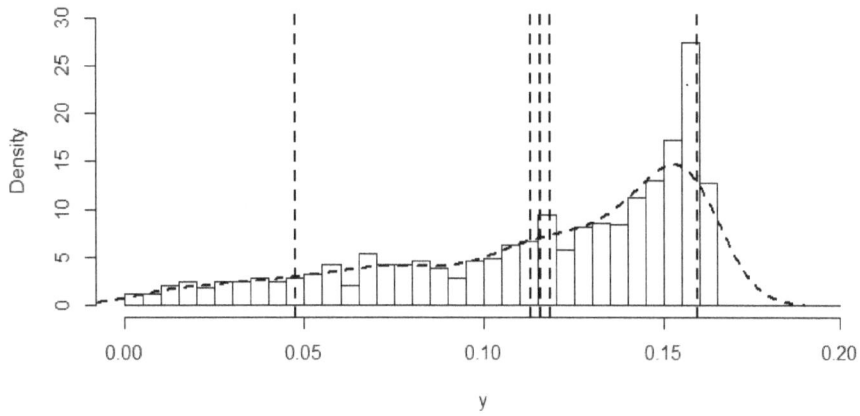

Figure 5.3 Histogram of x-value (*J* = 10,000)

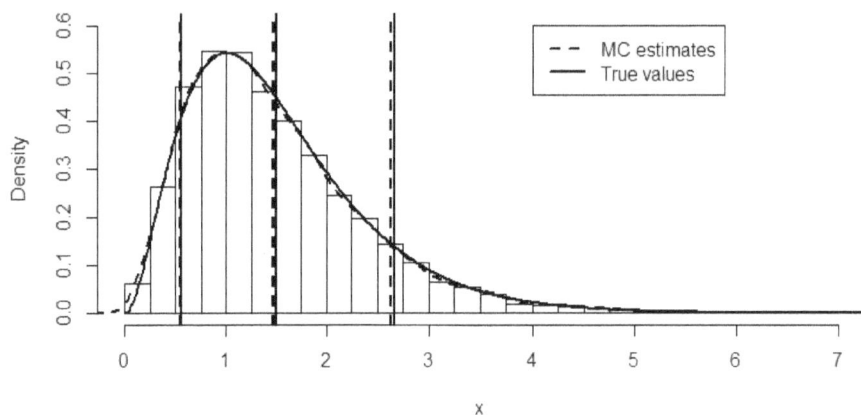

Figure 5.4 Histogram of y-value (*J* = 10,000)

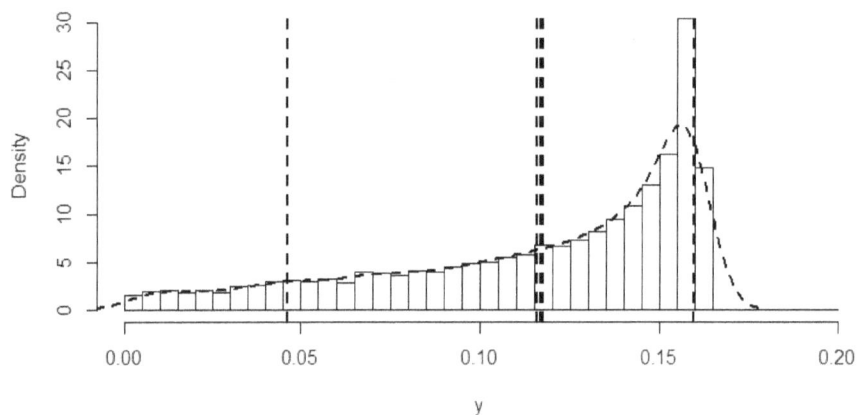

R Code for Exercise 5.2

```
X11(w=8,h=4.5);  par(mfrow=c(1,1)); options(digits=4);
J = 1000; set.seed(221); xv=rgamma(J,3,2)
xbar=mean(xv); xci=xbar + c(-1,1)*qnorm(0.975)*sd(xv)/sqrt(J)
xcdr=quantile(xv,c(0.1,0.9)); xden=density(xv)
yv=xv^2 * exp(-xv) / ( 1 + xv + 1/xv )
ybar=mean(yv); yci=ybar + c(-1,1)*qnorm(0.975)*sd(yv)/sqrt(J)
ycdr=quantile(yv,c(0.1,0.9)); yden=density(yv)
```

```
hist(xv,prob=T,breaks=seq(0,7,0.25),xlim=c(0,7),ylim=c(0,0.6),xlab="x",
   main=""); lines(xden,lty=2,lwd=2)
xvec=seq(0,10,0.01); lines(xvec,dgamma(xvec,3,2),lty=1,lwd=2)
abline(v= c(xbar, xci, xcdr), lty=2, lwd=2)
abline(v=c(3/2,qgamma(c(0.1,0.9),3,2)), lty=1,lwd=2)
legend(4,0.6,c("MC estimates","True values"),lty=c(2,1),lwd=c(2,2))


hist(yv,prob=T,breaks=seq(0,0.2,0.005),xlim=c(0,0.2),ylim=c(0,30),xlab="y",
   main=""); lines(yden,lty=2,lwd=2)
abline(v= c(ybar, yci, ycdr), lty=2, lwd=2)
legend(4,0.6,c("MC estimates","True values"),lty=c(2,1),lwd=c(2,2))


# Repeat with J = 10000 ----------------------------


J = 10000; set.seed(221); xv=rgamma(J,3,2)
xbar=mean(xv); xci=xbar + c(-1,1)*qnorm(0.975)*sd(xv)/sqrt(J)
xcdr=quantile(xv,c(0.1,0.9)); xden=density(xv)
yv=xv^2 * exp(-xv) / ( 1 + xv + 1/xv )
ybar=mean(yv); yci=ybar + c(-1,1)*qnorm(0.975)*sd(yv)/sqrt(J)
ycdr=quantile(yv,c(0.1,0.9)); yden=density(yv)


hist(xv,prob=T,breaks=seq(0,9,0.25),xlim=c(0,7),ylim=c(0,0.6),xlab="x",
   main=""); lines(xden,lty=2,lwd=2)
xvec=seq(0,10,0.01); lines(xvec,dgamma(xvec,3,2),lty=1,lwd=2)
abline(v= c(xbar, xci, xcdr), lty=2, lwd=2)
abline(v=c(3/2,qgamma(c(0.1,0.9),3,2)), lty=1,lwd=2)
legend(4,0.6,c("MC estimates","True values"),lty=c(2,1),lwd=c(2,2))


hist(yv,prob=T,breaks=seq(0,0.2,0.005),xlim=c(0,0.2),ylim=c(0,30),xlab="y",
   main="")
lines(yden,lty=2,lwd=2); abline(v= c(ybar, yci, ycdr), lty=2, lwd=2)
legend(4,0.6,c("MC estimates","True values"),lty=c(2,1),lwd=c(2,2))
```

5.4 Importance sampling

When applying the method of MC to estimate an integral of the form

$$\psi = Eg(x) = \int g(x)f(x)dx,$$

suppose it is impossible (or difficult) to sample from $f(x)$, but it is easy to sample from a distribution/density $h(x)$ which is 'similar' to $f(x)$.

Then we may write

$$\psi = \int \left(g(x) \frac{f(x)}{h(x)} \right) h(x) dx = \int w(x) h(x) dx ,$$

where

$$w(x) = g(x) \frac{f(x)}{h(x)} .$$

This suggests that we sample $x_1, ..., x_J \sim iid\ h(x)$ and use MC to estimate ψ by

$$\hat{\psi} = \bar{w} = \frac{1}{J} \sum_{j=1}^{J} w_j ,$$

where

$$w_j = w(x_j) = g(x_j) \frac{f(x_j)}{h(x_j)} .$$

This techniques is called *importance sampling*, and there are several issues to consider. As already indicated, the method works best if $h(x)$ is chosen to be very similar to $f(x)$.

Another issue is that $f(x)$ may be known only up to a multiplicative constant, i.e. where $f(x) = k(x) / c$, where the kernel $k(x)$ is known exactly but it is too difficult or impossible to evaluate the normalising constant $c = \int k(x) dx$. In that case, we may write

$$\psi = \int g(x) \frac{k(x)}{c} dx = \frac{\int g(x) k(x) dx}{\int k(x) dx}$$

$$= \frac{\int \left(g(x) \frac{k(x)}{h(x)} \right) h(x) dx}{\int \left(\frac{k(x)}{h(x)} \right) h(x) dx} = \frac{\int w(x) h(x) dx}{\int u(x) h(x) dx} ,$$

where:

$$w(x) = g(x) \frac{k(x)}{h(x)}$$

$$u(x) = \frac{k(x)}{h(x)} .$$

This suggests that we sample $x_1,...,x_J \sim iid\ h(x)$ (as before) and apply MC estimation to the means of $w(x)$ and $u(x)$, respectively (each with respect to the distribution defined by density $h(x)$) so as to obtain the estimate

$$\hat{\psi} = \frac{\overline{w}}{\overline{u}} = \frac{\dfrac{1}{J}\sum_{j=1}^{J} w_j}{\dfrac{1}{J}\sum_{j=1}^{J} u_j} = \frac{w_1 + ... + w_J}{u_1 + ... + u_J},$$

where $w_j = w(x_j)$ and $u_j = u(x_j)$.

Exercise 5.3 Example of Monte Carlo with importance sampling

We wish to find $\mu = Ex$ where x has density

$$f(x) \propto \frac{1}{x+1}e^{-x}, x > 0.$$

Use Monte Carlo methods and importance sampling to estimate μ.

Solution to Exercise 5.3

Here, $k(x) = \dfrac{1}{x+1}e^{-x}$, and it is convenient to use $h(x) = e^{-x}, x > 0$ (the standard exponential density, or $Gamma(1,1)$ density). Then,

$$\mu = Ex = \int_0^\infty xf(x)dx = \frac{\int xk(x)dx}{\int k(x)dx}$$

$$= \frac{\int \left(x\dfrac{k(x)}{h(x)}\right)h(x)dx}{\int \left(\dfrac{k(x)}{h(x)}\right)h(x)dx} = \frac{\int \dfrac{x}{x+1}h(x)dx}{\int \dfrac{1}{x+1}h(x)dx}.$$

So a MC estimate of μ is $\hat{\mu} = \dfrac{\dfrac{1}{J}\sum_{j=1}^{J} \dfrac{x_j}{x_j+1}}{\dfrac{1}{J}\sum_{j=1}^{J} \dfrac{1}{x_j+1}}$,

where $x_1,...,x_J \sim iid\ G(1,1)$.

Implementing this with $J = 100,000$, we get $\hat{\mu} = \dfrac{0.40345}{0.59655} = 0.67631$.

Note 1: For interest we use numerical techniques to get the exact answer, $\mu = 0.67687$.

Thus the relative error is -0.084%. Figure 5.5 illustrates.

Note 2: The exact value of the normalising constant is
$$c = \int k(x)dx \quad \text{is } 0.596347.$$

From the above we see that our MC estimate of c is 0.59655 (similar).

Figure 5.5 Illustration of importance sampling

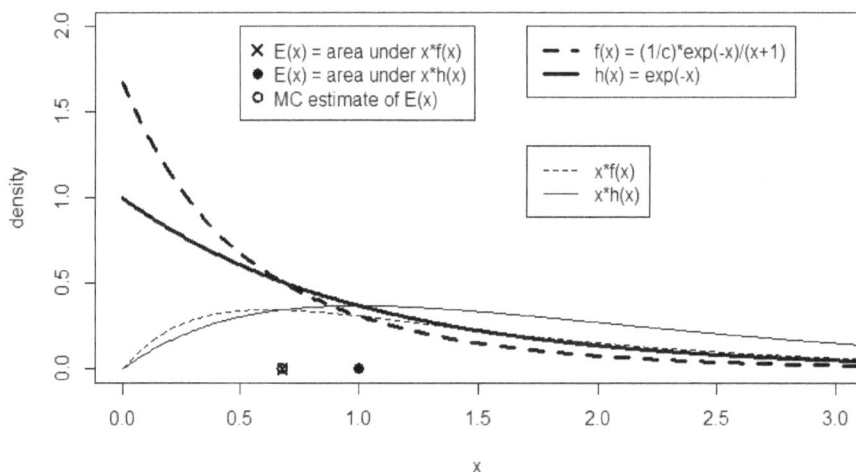

R Code for Exercise 5.3

```
options(digits=10);
kfun=function(x){ exp(-x)/(x+1) }
c=integrate(f=kfun,lower=0,upper=Inf)$value; c # 0.5963473624
ffun=function(x){ (1/ 0.5963473624)*exp(-x)/(x+1) }
integrate(f=ffun,lower=0,upper=Inf)$value; # 0.9999999999
xffun= function(x){ x*(1/0.5963474)*exp(-x)/(x+1) }
mu= integrate(f=xffun,lower=0,upper=Inf)$value; mu # 0.6768749849
```

```
J=100000; set.seed(413); xv=rgamma(J,1,1)
num=mean(xv/(xv+1)); den=mean(1/(xv+1))
est=num/den; c(num, den, est) # 0.4034510685 0.5965489315 0.6763084254
err=100* (est-mu)/mu; err # -0.08370222467

plot(c(0,3),c(0,2),type="n",xlab="x",ylab="density"); xvec=seq(0,5,0.01);
lines(xvec,dgamma(xvec,1,1),lty=1,lwd=3)
lines(xvec,xvec*dgamma(xvec,1,1),lty=1,lwd=1)
lines(xvec,ffun(xvec),lty=2,lwd=3); lines(xvec,xvec*ffun(xvec),lty=2,lwd=1)
points(c(1,mu,est),c(0,0,0),pch=c(16,4,1),lwd=c(2,2,2),cex=c(1.2,1.2,1.2))
legend(1.7,2,c( "f(x) = (1/c)*exp(-x)/(x+1)", "h(x) = exp(-x)" ),
   lty=c(2,1), lwd=c(3,3))
legend(1.7,1.3,c( "x*f(x)", "x*h(x)" ), lty=c(2,1), lwd=c(1,1))
legend(0.5,2,c("E(x) = area under x*f(x)", "E(x) = area under x*h(x)",
   "MC estimate of E(x)"), pch=c(4,16,1),pt.lwd=c(2,2,2),pt.cex=c(1.2,1.2,1.2))
```

5.5 MC estimation involving two or more random variables

All the examples so far have involved only a single random variable x. However, the method of Monte Carlo generalises easily to two or more random variables. In fact, the procedure for MC estimation of the mean of a function, as described above, is already valid in the case where x is a vector. We will now focus on the bivariable case, but the same principles apply when three or more random variables are being considered simultaneously.

Suppose that we have a random sample from the bivariate distribution of two random variables x and y, denoted $(x_1, y_1), ..., (x_J, y_J) \sim iid\ f(x, y)$, and we are interested in some function of x and y, say $r = g(x, y)$. Then we simply calculate $r_j = g(x_j, y_j)$ and perform MC inference on the resulting sample $r_1, ..., r_J \sim iid\ f(r)$.

Note 1: This procedure applies whether or not the random variables x and y are independent. If they are independent then we simply sample $x_j \sim f(x)$ and $y_j \sim f(y)$.

Note 2: If x and y are *dependent*, it may not be obvious how to generate $(x_j, y_j) \sim f(x, y)$.

Then, one approach is to apply the *method of composition*, as detailed below. If that fails, other methods are available, in particular ones which involve Markov chain theory. Much more will be said on these methods later in the course.

5.6 The method of composition

Suppose we wish to sample a vector $(x_j, y_j) \sim f(x, y)$. Often this can be done in two different ways via the *method of composition*, as follows.

One way is to first sample $x_j \sim f(x)$ and then sample $y_j \sim f(y \mid x_j)$. The result will be the desired $(x_j, y_j) \sim f(x, y)$. This follows by the identity (or 'composition')
$$f(x, y) = f(x) f(y \mid x).$$

Note: Having obtained $(x_j, y_j) \sim f(x, y)$ in this manner, suppose we 'discard' x_j. Then this will leave behind a single number, $y_j \sim f(y)$. This could be useful if all we really want is a sample from $f(y)$ but sampling from this distribution/density directly is difficult.

Alternatively, first sample $y_j \sim f(y)$ and then sample $x_j \sim f(x \mid y_j)$. The result will again be $(x_j, y_j) \sim f(x, y)$. This follows by the identity
$$f(x, y) = f(y) f(x \mid y).$$

Note: Having obtained $(x_j, y_j) \sim f(x, y)$ in this second manner, suppose that we 'discard' y_j. This will leave behind a single number, $x_j \sim f(x)$. This could be useful if all we really desire is a sample from $f(x)$ but sampling from this distribution/density directly is difficult.

This idea of composition generalises easily to higher dimensions. For example, one of several different ways to sample a triplet
$$(x_j, y_j, z_j) \sim f(x, y, z)$$
is first sample $y_j \sim f(y)$, then sample $x_j \sim f(x \mid y_j)$ and finally sample $z_j \sim f(z \mid x_j, y_j)$. This works because of the identity
$$f(x, y, z) = f(y) f(x \mid y) f(z \mid x, y).$$

Exercise 5.4

Suppose that we are interested in the distribution of a random variable defined by $r = y/\left(x + \sqrt{|y|}\right)$, where x and y have a joint distribution defined by the pdf $f(x, y) = f(x)f(y \mid x)$, and where $x \sim G(3, 2)$ and $(y \mid x) \sim N(x, x)$.

Use the R functions rgamma() and rnorm() to generate a sample of size $J = 1,000$ from the joint distribution of x and y. Then use the method of MC to estimate $\psi = Er$, and report a 95% CI for ψ. Also estimate the 80% CDR for r and $f(r)$. Present your results both graphically and numerically.

Solution to Exercise 5.4

Numerically, we estimate ψ by 0.4256, and our 95% CI for ψ is (0.4026, 0.4486). We also estimate the 80% CDR for r by (–0.1025, 0.8339). The required graph is shown in Figure 5.6.

Figure 5.6 Histogram of r-values (J = 1,000)

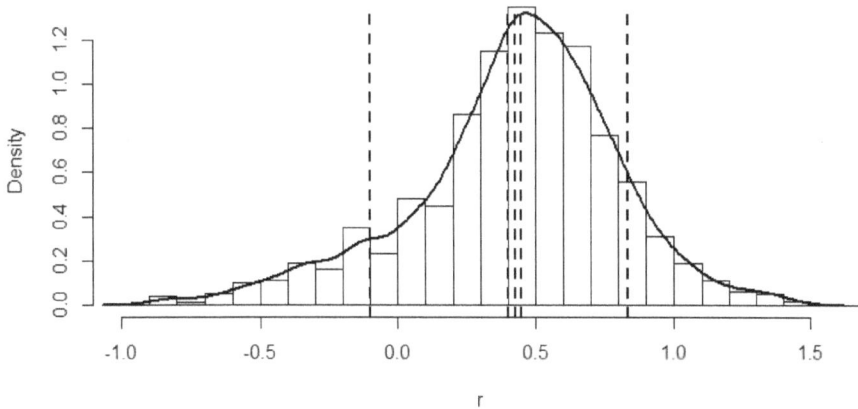

R Code for Exercise 5.4

X11(w=8,h=4.5); par(mfrow=c(1,1)); options(digits=4);

J = 1000; set.seed(221); xv=rgamma(J,3,2); yv = rnorm(xv,sqrt(xv))
rv = yv/(xv+sqrt(abs(yv)))
rbar=mean(rv); rci=rbar + c(-1,1)*qnorm(0.975)*sd(rv)/sqrt(J)
rcdr=quantile(rv,c(0.1,0.9)); rden=density(rv)

c(rbar,rci,rcdr) # 0.4256 0.4026 0.4486 -0.1025 0.8339

hist(rv,prob=T, breaks=seq(-1,1.8,0.1),xlim=c(-1,1.6),ylim=c(0,1.3),xlab="r",
   main=""); lines(rden,lty=1,lwd=2); abline(v= c(rbar, rci, rcdr), lty=2, lwd=2)

5.7 Monte Carlo estimation of a binomial parameter

Suppose we are interested in a binomial proportion (i.e. probability) p but have difficulty calculating this quantity exactly. Then we may interpret p as the mean μ of a Bernoulli distribution and directly apply the method of Monte Carlo in the usual way. In this special case, there are certain simplifications which result in slightly different-looking final formulae.

Explicitly, suppose we are able to generate
$$x_1,...,x_J \sim iid\ Bernoulli(p).$$

Then the MC estimate of p is
$$\bar{x} = \frac{1}{J}\sum_{j=1}^{J}x_j \qquad \text{(the sample proportion of 1s in the sample),}$$
and the MC sample variance is
$$s^2 = \frac{1}{J-1}\left(\sum_{j=1}^{J}x_j^2 - J\bar{x}^2\right)$$
$$= \frac{1}{J-1}\left(J\bar{x} - J\bar{x}^2\right) \quad \text{since } x_j^2 = x_j \text{ (because each } x_j \text{ is 0 or 1)}$$
$$= \frac{J}{J-1}\bar{x}(1-\bar{x}).$$

So the MC SE is $\dfrac{s}{\sqrt{J}} = \dfrac{1}{\sqrt{J}}\sqrt{\dfrac{J}{J-1}\bar{x}(1-\bar{x})} = \sqrt{\dfrac{\bar{x}(1-\bar{x})}{J-1}}$.

It follows that a MC $1-\alpha$ CI for p is

$$\left(\overline{x} \pm z_{\alpha/2}\frac{s}{\sqrt{J}} \right) = \left(\overline{x} \pm z_{\alpha/2}\sqrt{\frac{\overline{x}(1-\overline{x})}{J-1}} \right).$$

The MC estimate $\overline{x}$ is often written as $\hat{p}$, and $J-1$ is often replaced by J (for simplicity). These changes lead to the standard form of the MC $1-\alpha$ confidence interval for p,

$$\left(\hat{p} \pm z_{\alpha/2}\sqrt{\frac{\hat{p}(1-\hat{p})}{J}} \right).$$

Note 1: The above theory is really nothing other than the usual classical theory for estimating a binomial proportion. Thus, there are many other CIs that could be substituted, (e.g. the Wilson CI whose coverage is closer to $1-\alpha$, and the Clopper-Pearson CI whose coverage is always guaranteed to be at least $1-\alpha$ but which is typically wider).

Note 2: The above MC inference depends on the x_j values only by way of the sample mean $\overline{x}$ or, equivalently, by way of the sample total $x_T = x_1 + \ldots + x_J = J\overline{x}$. A consequence of this is that exactly the same Monte Carlo inference can be performed if we observe only a single value of the total x_T, whose distribution is given by $x_T \sim Bin(J, p)$.

Note 3: A common application of the theory here is where the binomial parameter is the probability of some event involving random variables, for example $p = P(x > 1)$ and $p = P(x < y)$.

For the first example here, we generate $x_1 \sim f(x)$, let $r_1 = I(x_1 > 1)$, and then repeat independently many times so as to generate a random sample $r_1, \ldots, r_J \sim iid \; Bern(p)$. That sample can then be used for MC inference on $p = P(x > 1)$.

The procedure for the second example is similar, except that it involves sampling $(x_1, y_1) \sim f(x, y)$ and determining $r_1 = I(x_1 < y_1)$, etc.

Note 4: One use of MC CIs for a binomial proportion is to assess the coverage of MC CIs.

Often, the true coverage probability of a MC CI is not exactly the nominal level, say 95%. This may be due to the MC sample size J being insufficiently large or for some other reason.

If we are concerned about this, we may wish to estimate the true coverage of the MC CI by repeating the entire MC inference procedure itself a large number of times, say M. Each time we record an indicator r for the MC CI containing the quantity of interest.

The result will be a sample $r_1, ..., r_M \sim iid\ Bern(p)$, where p is the true coverage probability, which can then be estimated via MC methods in the usual way.

Exercise 5.5 Estimating a probability via Monte Carlo

Use MC to estimate $p = P\left(\sqrt{\dfrac{x}{x+1}} > 0.3e^x \right)$, where $x \sim Gamma(3,2)$.

Solution to Exercise 5.5

With $J = 20{,}000$, we sample $x_1, ..., x_J \sim iid\ G(3,2)$ and let

$$r_j = I\left(\sqrt{\frac{x_j}{x_j+1}} > 0.3e^{x_j} \right).$$

Thereby we obtain an estimate of p equal to $\hat{p} = \dfrac{1}{J}\sum_{j=1}^{J} r_j = 0.2117$

and a 95% CI for p equal to $\left(\hat{p} \pm 1.96\sqrt{\dfrac{\hat{p}(1-\hat{p})}{200000}} \right) = (0.2060, 0.2173)$.

Note 1: We may also view p as $p = P(y > 0.3)$, where $y = e^{-x}\sqrt{\dfrac{x}{x+1}}$

(for example). In that case, we sample $x_1, ..., x_J \sim iid\ G(3,2)$, calculate

$$y_j = e^{-x_j}\sqrt{\frac{x_j}{x_j+1}} \text{ , and then let } r_j = I\left(y_j > 0.3\right). \text{ This leads to exactly}$$

the same results regarding p. As a by-product of this second approach, we obtain an estimate of the density function of the random variable

$$y = e^{-x}\sqrt{\frac{x}{x+1}} \text{ , namely } f(y), \text{ which would be very difficult to obtain}$$

analytically. Figure 5.7 illustrates.

Note 2: The density() function in R used to smooth the histogram does not adequately capture the upper region of the density $f(y)$, nor the fact that $f(y) = 0$ when $y < 0$.

Figure 5.7 Histogram of 20,000 values of y

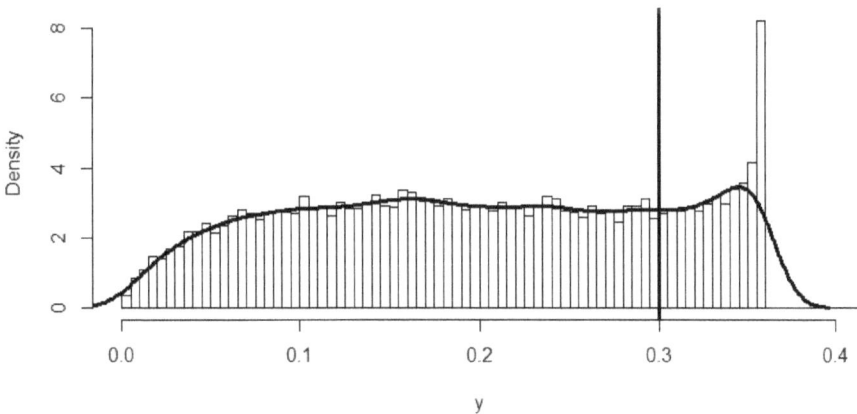

R Code for Exercise 5.5

```
X11(w=8,h=4.5); par(mfrow=c(1,1)); options(digits=4)

J=20000; set.seed(162); xv=rgamma(J,3,2); ct=0
yv= sqrt(xv)*exp(-xv) / sqrt(xv+1)
for(j in 1:J)  if(yv[j] > 0.3) ct=ct+1
phat=ct/J; ci=phat+c(-1,1)*qnorm(0.975)*sqrt(phat*(1-phat)/J)
c(phat,ci) # 0.2117 0.2060 0.2173
hist(yv,prob=T,breaks=seq(0,0.5,0.005),xlim=c(0,0.4),xlab="y",main=" ")
abline(v=0.3,lwd=3); lines(density(yv),lwd=3)
```

Exercise 5.6 Buffon's needle problem

A needle of length 10 cm is dropped randomly onto a floor with lines on it that are parallel and 10 cm apart.

(a) Analytically derive p, the probability that the needle crosses a line.

(b) Now forget that you know p. Estimate p using Monte Carlo methods on a computer and a sample size of 1,000. Also provide a 95% confidence interval for p. Then repeat with a sample size of 10,000 and discuss.

Solution to Exercise 5.6

(a) Let: X = perpendicular distance from centre of needle to nearest line
in units of 5 cm
Y = acute angle between lines and needle in radians
C = 'The needle crosses a line'.

Then: $X \sim U(0,1)$ with density $f(x) = 1, 0 < x < 1$

$Y \sim U\left(0, \dfrac{\pi}{2}\right)$ with density $f(y) = \dfrac{2}{\pi}, 0 < y < \dfrac{\pi}{2}$

$X \perp Y$ (i.e. X and Y are independent, so that

$$f(x,y) = f(x)f(y) = 1 \times \frac{2}{\pi}, \ 0 < x < 1, \ 0 < y < \frac{\pi}{2})$$

$C = \{X < \sin Y\} = \{(x, y) : x < \sin y\}$.

Figure 5.8 illustrates this setup.

It follows that
$$p = P(C) = P(X < \sin Y)$$

$$= \iint\limits_{x < \sin y} f(x,y)dxdy = \frac{2}{\pi} \int\limits_{y=0}^{\pi/2} \left(\int\limits_{x=0}^{\sin y} dx \right) dy = \frac{2}{\pi} \int\limits_{y=0}^{\pi/2} \sin y \, dy$$

$$= \frac{2}{\pi}\left[-\cos y \Big|_0^{\pi/2} \right] = \frac{2}{\pi}\left(-\cos\left(\frac{\pi}{2}\right) - (-\cos 0) \right)$$

$$= \frac{2}{\pi}(-0 - (-1)) = \frac{2}{\pi} = 0.63662.$$

Figure 5.9 illustrates the integration here.

Figure 5.8 Illustration of Buffon's needle problem

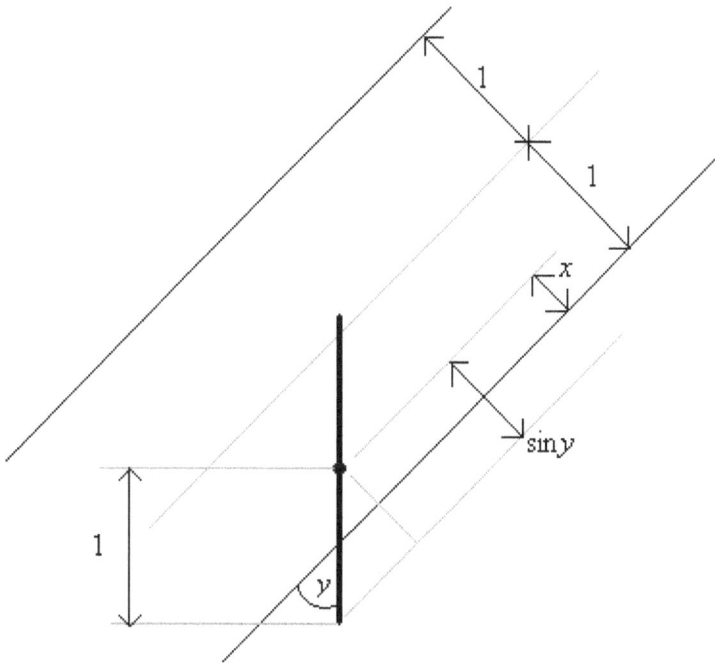

Figure 5.9 Illustration of the solution to Buffon's needle problem

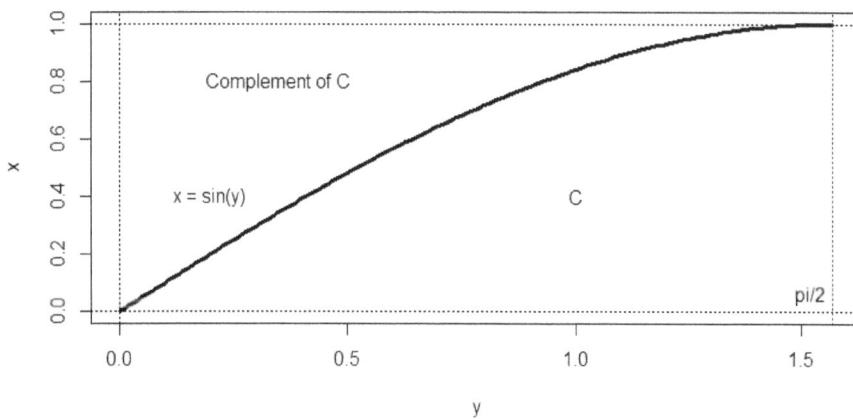

Note 1: Another way to express the above working is to first note that

$$P(C \,|\, y) \equiv P(C \,|\, Y = y) = P(X < \sin y \,|\, y) = P(X < \sin y) = \sin y,$$

since $(X \,|\, y) \sim X \sim U(0,1)$ with cdf $F(x \,|\, y) = F(x) = x, 0 < x < 1$.

It follows that

$$p = P(C) = EP(C \,|\, Y) = E \sin Y = \int_0^{\pi/2} (\sin y) \frac{2}{\pi} dy = \frac{2}{\pi},$$

as before.

Note 2: It can be shown that if the length of the needle is r times the distance between lines, then the probability that the needle will cross a line is given by the formula

$$p = \begin{cases} 2r / \pi, & r \leq 1 \\ 1 - \dfrac{2}{\pi}\left(\sqrt{r^2 - 1} - r + \sin^{-1}\left(\dfrac{1}{r}\right) \right), & r > 1. \end{cases}$$

(b) For this part, we will make use of the analysis in (a) whereby
$$C = \{(x,y) : x < \sin y\},$$
and where:

$$x \sim U(0,1), \quad y \sim U\left(0, \frac{\pi}{2}\right), \quad X \perp Y.$$

Note: We suppose that these facts are understood but that the integration required to then proceed on from these facts to the final answer (as in (a)) is too difficult.

We now sample $x_1, ..., x_J \sim iid \; U(0,1)$ and $y_1, ..., y_J \sim iid \; U(0, \pi/2)$ (all independently of one another). Next, we obtain the indicators defined by
$$r_j = I(x_j < \sin y_j) = \begin{cases} 1 & \text{if } x_j < \sin y_j \\ 0 & \text{otherwise.} \end{cases}$$

The result is the MC sample $r_1, ..., r_J \sim iid \; Bern(p)$ (i.e. a sample of size J to be used for inference on p). (Equivalently, we may obtain $r_T \equiv r_1 + ... + r_J \sim Bin(J, p)$, which will lead to the same final results.)

The MC estimate of p is $\hat{p} = \bar{r} = \dfrac{1}{J}\sum_{j=1}^{J} r_j = \dfrac{r_T}{J}$,

and a 95% CI for p is $CI = \left(\hat{p} \pm z_{\alpha/2} \sqrt{\dfrac{\hat{p}(1-\hat{p})}{J}} \right)$.

Carrying out this experiment in R with $J = 1{,}000$ we get
$\hat{p} = 0.618$ and $CI = (0.588,\ 0.648)$.

Then repeating, but with $J = 10{,}000$ instead, we obtain
$\hat{p} = 0.633$ and $CI = (0.624,\ 0.643)$.

We see that increasing the MC sample size (from 1,000 to 10,000) has reduced the width of the MC CI from 0.060 to 0.019. Both intervals contain the true value, namely $2/\pi = 0.6366$.

R Code for Exercise 5.6

```
# (a)
X11(w=8,h=4.5); par(mfrow=c(1,1))
plot(seq(0,pi/2,0.01),sin(seq(0,pi/2,0.01)), type="l",lwd=3,xlab="y", ylab="x")
abline(v=c(0,pi/2),lty=3); abline(h=c(0,1),lty=3)
text(0.2,0.4,"x = sin(y)"); text(1,0.4,"C"); text(0.35,0.8,"Complement of C")
text(1.52,0.06,"pi/2")

# (b)
J=1000; set.seed(213); xv=runif(J,0,1); yv=runif(J,0,pi/2); rv=rep(0,J)
options(digits=4); for(j in 1:J) if(xv[j]<sin(yv[j])) rv[j]=1

phat=mean(rv); z=qnorm(0.975); pci=phat+c(-1,1)*z*sqrt(phat*(1-phat)/J)
c(phat,pci,pci[2]-pci[1]) # 0.61800 0.58789 0.64811 0.06023

J=10000; set.seed(215); xv=runif(J,0,1); yv=runif(J,0,pi/2); rv=rep(0,J)
for(j in 1:J) if(xv[j]<sin(yv[j])) rv[j]=1

phat=mean(rv); z=qnorm(0.975); pci=phat+c(-1,1)*z*sqrt(phat*(1-phat)/J)
c(phat,pci,pci[2]-pci[1]) # 0.63320 0.62375 0.64265 0.01889
```

Exercise 5.7 MC CIs for the coverage probabilities of MC CIs for a gamma mean

(a) Using the R function rgamma(), generate a random sample of size $J = 100$ from the gamma distribution with parameters 3 and 2 and mean $\mu = 3/2$. Then use the method of Monte Carlo to estimate μ. In your estimation, include a 95% CI for μ and the width of this CI. Also report whether the CI contains the true value of μ.

(b) Repeat (a) but with $J = 200$, 500, 1,000, 10,000 and 100,000, respectively. Report the widths of the resulting CIs and, for each CI, state whether it contains μ. Discuss any patterns that you see.

(c) Repeat (a) $M = 100$ times and report the proportion of the resulting M 95% MC CIs which contain the true value of the mean. (In each case use $J = 100$.) Hence calculate a 95% CI for p, the true coverage probability of the 95% MC CI for μ based on a MC sample of size $J = 100$ from the *Gamma*(3,2) distribution.

(d) Repeat (c), but with $M = 200$, 500, 1,000 and 10,000, respectively. Discuss any patterns that you see.

Solution to Exercise 5.7

(a) Applying the procedure (see the R code below) we estimate μ by $\bar{x} = 1.517$. The Monte Carlo 95% confidence interval for μ is
$$CI = (\bar{x} \pm z_{0.025} s / \sqrt{J}) = (1.354,\ 1.680).$$

We observe that this interval has width 0.326 and contains μ.

(b) Repeating (a) as required, we obtain:
$\bar{x} = 1.471$ and $CI = (1.348, 1.593)$ with width 0.245   for $J = 200$
$\bar{x} = 1.430$ and $CI = (1.358, 1.502)$ with width 0.144   for $J = 500$
$\bar{x} = 1.475$ and $CI = (1.419, 1.530)$ with width 0.111   for $J = 1,000$
$\bar{x} = 1.490$ and $CI = (1.473, 1.508)$ with width 0.0344 for $J = 10,000$
$\bar{x} = 1.502$ and $CI = (1.497, 1.507)$ with width 0.0107 for $J = 100,000$.

We see that $\bar{x}$ appears to be converging towards $\mu = 1.5$. The width of the CI appears to be decreasing as J increases. Each of these five CIs contains μ, just like the CI in (a).

(c) Repeating (a) $M = 100$ times leads to $M = 100$ MC CIs of which 93 contain $\mu = 1.5$. Thus $\hat{p} = 93\%$, which as expected is 'close' to the 95% nominal coverage of the CI.

A 95% CI for p is $\left(0.93 \pm 1.96 \sqrt{\dfrac{0.93(1-0.93)}{100}} \right) = (0.880 \ \ 0.980)$.

This is consistent with the MC 95% CI for μ having coverage 95%.

(d) Repeating (a) $M = 200$ times leads to $\hat{p} = 94.5\%$ of the 200 CIs containing 1.5, with a 95% CI for p,

$$\left(0.945 \pm 1.96 \sqrt{\frac{0.945(1-0.945)}{200}} \right) = (0.913, 0.977).$$

Repeating (a) $M = 500$ times leads to $\hat{p} = 94.2\%$ of the 500 CIs containing 1.5 with a 95% CI for p,

$$\left(0.942 \pm 1.96 \sqrt{\frac{0.942(1-0.942)}{500}} \right) = (0.922, 0.962).$$

Repeating (a) $M = 1,000$ times leads to $\hat{p} = 93.5\%$ of the 1,000 CIs containing 1.5, with a 95% CI for p,

$$\left(0.935 \pm 1.96 \sqrt{\frac{0.935(1-0.935)}{1,000}} \right) = (0.935, 0.963).$$

Repeating (a) $M = 10,000$ times leads to $\hat{p} = 94.4\%$ of the 10,000 CIs containing 1.5, with a 95% CI for p,

$$\left(0.944 \pm 1.96 \sqrt{\frac{0.94(1-0.94)}{10,000}} \right) = (0.940, 0.949).$$

The widths of all five CIs for p are: 0.100, 0.063, 0.041, 0.027 and 0.009. We see that the CI for p becomes narrower as M increases. Also, the proportion of CIs containing 1.5 converges towards 95% as M increases. The convergence does not seem to be uniform. This is because of Monte Carlo error. If we repeated the experiment again, we might find a slightly different pattern.

Each of the CIs for p is consistent with $p = 0.95$, except the one with $M = 10,000$, which is the most reliable. In that case the CI for p is

(0.940, 0.949), which is entirely below 0.95. This suggests that the true coverage probability of the 95% MC CI for μ is slightly less than 95%.

The observed proportions appear to be converging to this limit rather than to 95% exactly. This is explainable by the fact that the MC sample size $J = 100$ is far from infinity. If we repeated (d) with a larger value of J in each case, say $J = 1,000$, we would see the proportion of the M CIs converge towards a limiting value which is even closer to 95%. But then an even larger value of M would be necessary to establish that there is in fact any difference between the limiting value and 95%.

R Code for Exercise 5.7

```
# (a)
options(digits=5); J = 100; set.seed(221); xv=rgamma(J,3,2)
xbar=mean(xv); ci=xbar + c(-1,1)*qnorm(0.975)*sd(xv)/sqrt(J)
c(xbar,ci) # 1.5170 1.3539 1.6800


# (b)
Jvec=c(100,200,500,1000,10000,100000); K = length(Jvec)
xbarvec=rep(NA,K); LBvec= rep(NA,K); UBvec= rep(NA,K);
set.seed(221);
for(k in 1:K){    J=Jvec[k]; xv=rgamma(J,3,2); xbar=mean(xv)
                  ci=xbar + c(-1,1)*qnorm(0.975)*sd(xv)/sqrt(J)
                  xbarvec[k]=xbar; LBvec[k]=ci[1]; UBvec[k]=ci[2]
                  }
Wvec=UBvec-LBvec
print(rbind(Jvec, xbarvec, LBvec,UBvec, Wvec),digits=4)

# Jvec    100.0000 200.0000 500.0000 1000.000 1.000e+04 1.000e+05
# xbarvec 1.5170   1.4705   1.4299   1.475 1.490e+00 1.502e+00
# LBvec   1.3539   1.3480   1.3577   1.419 1.473e+00 1.497e+00
# UBvec   1.6800   1.5930   1.5020   1.530 1.508e+00 1.507e+00
# Wvec    0.3261   0.2451   0.1443   0.111 3.441e-02 1.073e-02


# (c)
J=100; M=100; ct=0; set.seed(442); for(m in 1:M){
        xv=rgamma(J,3,2)
        xbar=mean(xv); ci=xbar + c(-1,1)*qnorm(0.975)*sd(xv)/sqrt(J)
        if((ci[1]<=1.5)&&(1.5<=ci[2])) ct = ct + 1 }
p=ct/M; ci=p+c(-1,1)*qnorm(0.975)*sqrt(p*(1-p)/J)
c(ct,p,ci) # 93.00000 0.93000 0.87999 0.98001
```

```
# (d)
J=100; Mvec=c(200,500,1000,10000); set.seed(651)
  for(M in Mvec){        ct=0
        for(m in 1:M){
                xv=rgamma(J,3,2);        xbar=mean(xv)
                ci=xbar + c(-1,1)*qnorm(0.975)*sd(xv)/sqrt(J)
                if((ci[1]<=1.5)&&(1.5<=ci[2])) ct = ct + 1
                }
        p=ct/M; ci=p+c(-1,1)*qnorm(0.975)*sqrt(p*(1-p)/M)
        print(c(M,p,ci,ci[2]-ci[1]),digits=3)  }

# [1] 200.0000  0.9450  0.9134  0.9766  0.0632
# [1] 500.000  0.942  0.922  0.962  0.041
# [1] 1.00e+03  9.49e-01 9.35e-01 9.63e-01 2.73e-02
# [1] 1.00e+04  9.44e-01 9.40e-01 9.49e-01 9.00e-03
```

5.8 Random number generation

So far we have assumed the availability of the sample required for Monte Carlo estimation, such as $x_1,...,x_J \sim iid\ f(x)$. The issue was skipped over by making use of ready made functions in R such as runif(), rbeta() and rgamma(). However, many applications involve dealing with complicated distributions from which sampling is not straightforward.

So we will next discuss some basic techniques that can be used to generate the required Monte Carlo sample from a given distribution. More advanced techniques will be treated later. We will first treat the discrete case, which is the simplest, and then the continuous case. It will be assumed throughout that we can at least sample easily from the standard uniform distribution, i.e. that we can readily generate $u \sim U(0,1)$.

Note: This sampling is easily achieved using the runif() function in R. Alternatively, it can be done physically by using a hat with 10 cards in it, where these have the numbers 0,1,2,.....,9 written on them. Three cards (say) are drawn out of the hat, randomly and *with replacement.* The three numbers thereby selected are written down in a row, and a decimal point is placed in front of them. The resulting number (e.g. 0.472, 0.000 or 0.970) is an *approximate* draw from the standard uniform distribution. Repeating the entire procedure several times results in a random sample from that distribution. Increasing 'three' above (to 'five', say) improves the approximation (e.g. yielding 0.47207, 0.00029 or 0.97010).

5.9 Sampling from an arbitrary discrete distribution

Suppose we wish to sample a value $x \sim f(x)$ where $f(x)$ is a discrete pdf defined over the possible values $x = x_1,...,x_K$. First define

$$f_k = f(x_k)$$

and

$$F_k = f_1 + ... + f_k \quad (k = 1,...,K),$$

noting that $F_K = 1$.

Then sample $u \sim U(0,1)$, and finally return:

$$x = x_1 \qquad \text{if } 0 \le u \le F_1$$
$$x = x_2 \qquad \text{if } F_1 < u \le F_2$$
$$\dots\dots\dots\dots\dots\dots\dots$$
$$x = x_K \qquad \text{if } F_{K-1} < u \le F_K \, (=1).$$

One way to implement the above is to set $k = 1$, to repeatedly increment k by 1 until $F_{k-1} < u \le F_k$, and then, using the final value of k thereby obtained, to return $x = x_k$.

Note 1: We see that this procedure will work also in the case where K is *infinite*. In that case a practical alternative is to redefine K as a value k for which F_k is very close to 1 (e.g. 0.9999) and then approximate $f(x)$ by zero for all $x > x_K$.

Note 2: In R, an alternative to using $u \sim U(0,1)$ is to apply the function sample() with appropriate specifications of $x_1,...,x_K$ and $f_1,...,f_K$ (as illustrated in an exercise below).

Exercise 5.8 Example of sampling from a simple discrete distribution

Show that the above method works when applied to generating a value x from the *Bin*(2,1/2) distribution, i.e. that it returns $x = 0$, 1 and 2 with probabilities 1/4, 1/2 and 1/4, respectively.

Solution to Exercise 5.8

In this case, $K = 3$ and: $\quad x_1 = 0, \quad F(x_1) = P(x \le 0) = 0.25$
$$x_2 = 1, \quad F(x_2) = P(x \le 1) = 0.75$$
$$x_3 = 2, \quad F(x_3) = P(x \le 2) = 1.00.$$

Let $u \sim U(0,1)$. Then the method returns:
$x = x_1 = 0 \quad$ if $\quad 0 < u < F(x_1) \quad$ i.e. if $\;\; 0.00 < u < 0.25$
$x = x_2 = 1 \quad$ if $\quad F(x_1) < u < F(x_2) \quad$ i.e. if $\;\; 0.25 < u < 0.75$
$x = x_3 = 2 \quad$ if $\quad F(x_2) < u < F(x_3) \quad$ i.e. if $\;\; 0.75 < u < 1.00.$

Thus, x has: $\quad 0.25 - 0.00 = 0.25$ probability of being set to 0
$0.75 - 0.25 = 0.50$ probability of being set to 1
$1.00 - 0.75 = 0.25$ probability of being set to 2 (all correct).

Exercise 5.9 Sampling from a complicated discrete distribution

Consider the discrete distribution defined by the pdf

$$f(x) \propto \frac{x^3 e^{-x}}{1 + \sqrt{x}}, x = 1, 3, 5, \ldots$$

Find the mean of the distribution by performing appropriate summations. Then generate a random sample from this distribution and use it to confirm the mean.

Solution to Exercise 5.9

Using R we calculate $k(x) = \dfrac{x^3 e^{-x}}{1 + \sqrt{x}}, x = 1, 3, 5, \ldots, 41$ (here k stands for

kernel), noting that the last two values of $k(x)$ are tiny (9.455201e-14 and 1.454999e-14).

We then calculate the sum of the kernel values,
$$c = k(1) + k(3) + \ldots + k(41) = 1.051009,$$
and thereby normalise the kernel to obtain
$$f(x) = \frac{k(x)}{c}, x = 1, 3, 5, \ldots, 41.$$

The pdf may also be written as $f(x) = k(x)/c$, $x = x_1,...,x_K$, where: $x_k = 2k - 1$; $k = 1,...,K$; $K = 21$. The exact mean of the distribution is then evaluated numerically as

$$\mu = \sum_{k=1}^{K} x_k f(x_k) = 3.6527.$$

Note: Changing 41 to 101 here changes the approximation to 3.6527, i.e. makes no difference to 4 decimals. This suggests that taking the upper bound as 41 is good enough.

To sample $J = 100,000$ values from the distribution we may write
        sample(x=xvec,size=J,replace=TRUE,prob=fvec)
where xvec is a vector with values 1,3,...,41 and fvec is a vector with the values $f(1)$, $f(3)$,..., $f(41)$ (see the R Code below).

Note: We could also change fvec to kvec here, where kvec is a vector with the values $k(1), k(3),..., k(41)$; both possibilities will work since sample() will automatically normalise the values in its parameter 'prob'.

The Monte Carlo estimate of μ works out as 3.6494 with 95% CI (3.6374, 3.6615). We note that this CI contains the true value, 3.6527.

R Code for Exercise 5.9

```
kfun = function(x){ x^3*exp(-x)/(1 + sqrt(x)) };  options(digits=5)
xvec=seq(1,41,2); kvec=kfun(xvec); c =sum(kvec); c # 1.051
fvec=kvec/c; sum(fvec) # 1
print(rbind(xvec,fvec)[,1:9],digits=3)
# xvec 1.000 3.000 5.000 7.0000 9.0000 11.0000 13.00000 1.50e+01 1.70e+01
# fvec 0.175 0.468 0.248 0.0816 0.0214  0.0049  0.00103 2.02e-04 3.78e-05
sum(xvec*kvec)/sum(kvec)  # 3.6527
# Check that 41 is large enough:
xvec=seq(1,101,2); kvec=kfun(xvec); sum(xvec*kvec)/sum(kvec)
    # 3.6527 (same)
# Sample from the distribution
xvec=seq(1,41,2); kvec=kfun(xvec); J=100000; set.seed(332);
samp = sample(x=xvec,size=J,replace=TRUE,prob=fvec)
est =mean(samp); std=sd(samp); ci=est+c(-1,1)*qnorm(0.975)*std/sqrt(J)
c(est,ci) # 3.6494 3.6374 3.6615
```

5.10 The inversion technique

Suppose we wish to sample x, a value of a *continuous* random variable X with cdf $F_X(x)$. One way to do this is using the *inversion technique*, defined as follows, with the underlying theorem and proof shown below.

First derive the quantile function of X, denoted $F_X^{-1}(p)$ $(0 < p < 1)$. (This can be done by setting $F_X(x)$ to p and solving for x.)

Next, generate a random number u from the standard uniform distribution. (It will be assumed that this can be done easily, e.g. using runif() in R.)

Then return $x = F_X^{-1}(u)$ as a value sampled from the distribution of X.

Theorem 5.1: Suppose that X is a continuous random variable with cdf $F_X(x)$ and quantile function $F_X^{-1}(p)$. Let $U \sim U(0,1)$, independently of X, and define $R = F_X^{-1}(U)$. Then R has the same distribution as X.

Proof of Theorem 5.1: Observe that U has cdf $F_U(u) = u, 0 < u < 1$. This implies that R has cdf

$$F_R(r) = P(R \leq r) = P(F_X(F_X^{-1}(U)) \leq F_X(r)) = P(U \leq F_X(r)) = F_X(r) .$$

Thus, R has the same cdf as X and therefore the same distribution.

Note: A complication with the inversion technique may arise if there is difficulty deriving the quantile function $F_X^{-1}(p)$. In that case, since the task is fundamentally to solve $F_X(x) = u$ for x, it may be useful to employ the Newton-Raphson algorithm to the problem of solving the equation $g(x) = 0$, where $g(x) = F_X(x) - u$.

Exercise 5.10 Practice at the inversion technique

(a) Using $u = 0.371$ as a value from the standard uniform distribution, obtain a value from the standard exponential distribution. Then generate a large random sample $u_1, ..., u_J \sim iid\ U(0,1)$ (of size $J = 1,000$ say) and use this to create a random sample of the same size from the standard exponential distribution. Check your results by calculating an estimate of the mean of that distribution and also a 95% CI for that mean. Compare your results with the true value of that mean, namely 1.

(b) Using $u = 0.371$ as a value from the standard uniform distribution, obtain a value from the gamma distribution with mean and variance both equal to 2. Then generate a large random sample $u_1,...,u_K \sim iid\ U(0,1)$ (of size $J = 1,000$, say) and use this to create a random sample of the same size from the said gamma distribution. Check your results by calculating an estimate of the mean of that distribution and also a 95% CI for that mean. Compare your results with the true value, namely 2.

Solution to Problem 5.10

(a) Let $X \sim G(1,1)$ with density function $f(x) = e^{-x}$, $x > 0$, and cdf

$$F(x) = \int_0^x e^{-t}dt = 1 - e^{-x}, x > 0.$$ The quantile function here is the solution

of $1 - e^{-x} = p$, namely $F^{-1}(p) = -\log(1-p)$.

So a value from the standard exponential distribution is easily computed as $x = F^{-1}(u) = -\log(1-0.371) = 0.463624$.

Taking $J = 1,000$, we now generate $u_1,...,u_J \sim iid\ U(0,1)$ in R using the runif() function, and then calculate $x_j = -\log(1-u_j)$ for each $j = 1,...,J$.

This results in the required sample $x_1,...,x_J \sim iid\ G(1,1)$. Using this sample, the MC estimate of $\mu = EX$ is 0.9967, and a 95% CI for μ is (0.9322, 1.0613). We see that the CI contains the true value being estimated (i.e. 1).

(b) Here, $X \sim G(2,1)$ with mean 2/1, variance $2/1^2 = 2$, pdf $f(x) = xe^{-x}$ and cdf

$$F(x) = \int_0^x te^{-t}dt = \left[t(-e^{-t})\Big|_0^x\right] - \int_0^x 1(-e^{-t})dt$$

$$= -xe^{-x} + 0 + \left[-e^{-t}\Big|_0^x\right] = -xe^{-x} - e^{-x} + 1 = 1 - (x+1)e^{-x}.$$

We see that the quantile function of X, $F^{-1}(p)$, does not have a closed form expression, since it is the root of the function
$$g(x) = F(x) - p = 1 - (x+1)e^{-x} - p$$
(i.e. the solution of $g(x) = 0$).

However, for any p we can obtain that root using the Newton-Raphson algorithm by iterating

$$x_{j+1} = x_j - \frac{g(x_j)}{g'(x_j)} \quad \text{where } g'(x) = F'(x) - 0 = f(x) = xe^{-x}$$

$$= x_j - \left(\frac{1 - (x_j + 1)e^{-x_j} - p}{x_j e^{-x_j}} \right).$$

With $p = u = 0.371$ and starting arbitrarily at $x_0 = 1$, we get the sequence:
$$1.0000, 1.2902, 1.2939, 1.2939, 1.2939, 1.2939, 1.2939.....$$

So we return 1.2939 as a value from the $G(2,1)$ distribution.

As a check, we use the pgamma() function in R to confirm that $F_X(1.2939) = 0.371$ as follows:

```
pgamma(1.2939,2,1) # 0.37101
```

Taking $K = 1,000$, we now generate $u_1,...,u_K \sim iid\ U(0,1)$ in R using the runif() function, and then for $k = 1,...,K$ we solve
$$1 - (x_k + 1)e^{-x_k} = u_k \text{ for } x_k$$
using the NR algorithm *each time*. This procedure results in the sample,
$$x_1,...,x_K \sim iid\ G(2,1).$$

Using this sample, an estimate of $\mu = EX$ is 1.9631, and a 95% CI for μ is (1.8815, 2.0446). We see that the CI contains the true value, 2.

R Code for Problem 5.10

```
options(digits=5)

# (a)
-log(1-0.371) #  0.463624
J=1000; set.seed(221); uv=runif(J,0,1)
xv=-log(1-uv)   # Generate a random sample of size 1000 from the G(1,1) dsn
est=mean(xv); std=sd(xv); ci=est+c(-1,1)*qnorm(0.975)*std/sqrt(J)
c(est,ci) # 0.99673 0.93216 1.06130
```

```
# (b)
u=0.371; x=1; xv=x; for(j in 1:7) { x=x-(1-(x+1)*exp(-x)-u)/(x*exp(-x)); xv=c(xv,x) }
xv # 1.0000 1.2902 1.2939 1.2939 1.2939 1.2939 1.2939 1.2939
pgamma(x,2,1) # 0.371    Just checking that F(1.293860) = 0.371
pgamma(1.2939,2,1) # 0.37101


K=1000; xvec=rep(NA,K); set.seed(332); for(k in 1:K){
   u=runif(1); x=1; for(j in 1:10)  x=x-(1-(x+1)*exp(-x)-u)/(x*exp(-x))
   xvec[k]=x    } # Generate a random sample of size 1000 from the G(2,1) dsn
est=mean(xvec); std=sd(xvec)
ci=est+c(-1,1)*qnorm(0.975)*std/sqrt(K)
c(est,ci) # 1.9631 1.8815 2.0446
```

5.11 Random number generation via compositions

Sometimes the most convenient way to sample from a distribution is to express it as a function (or composition) of two or more random variables which are easy to sample from. For example, to obtain two independent values from the standard normal distribution we may use the well-known *Box-Muller algorithm*, as follows.

Sample $u_1, u_2 \sim iid\ U(0,1)$ and let:
$$z_1 = \sqrt{-2\log u_1}\ \cos(2\pi u_2)$$
$$z_2 = \sqrt{-2\log u_1}\ \sin(2\pi u_2).$$

It can be shown that $z_1, z_2 \sim iid\ N(0,1)$. If we only need one value from the standard normal distribution then we may arbitrarily discard z_2 and return only z_1.

Exercise 5.11 Sampling from the double exponential distribution

Suppose we wish to sample a value $x \sim f(x)$, where
$$f(x) = (1/2)e^{-|x|}, x \in \Re.$$

Describe how to obtain x as a composition of two other values than can be easily sampled.

Solution to Exercise 5.11

Let R and Y be independent random variables such that $R \sim Bern(0.5)$ and $Y \sim G(1,1)$. Then $U = (2R-1)Y$ has the same distribution as X.

This is because R is equally likely to be 0 as it is to be 1, and so $2R-1$ is equally likely to be -1 as it is to be $+1$. So there is a 50% chance that U will be exponential ($G(1,1)$) and a 50% chance that U will be *negative* exponential. So, obviously U has exactly the same distribution as X. For a formal proof, see the Note below.

We see that a method for obtaining a value $x \sim f(x)$ is to independently sample $r \sim Bern(0.5)$ and $y \sim G(1,1)$, and then calculate $x = (2r-1)y$.

Note: The cdf of $U = (2R-1)Y$ is

$$\begin{aligned}
F(u) &= P(U \leq u) \\
&= P((2R-1)Y \leq u) \\
&= EP((2R-1)Y \leq u \mid R) \\
&= P(R=0)P((2R-1)Y \leq u \mid R=0) \\
&\quad +P(R=1)P((2R-1)Y \leq u \mid R=1) \\
&= \frac{1}{2}P(-Y \leq u \mid R=0) + \frac{1}{2}P(+Y \leq u \mid R=1) \\
&= \frac{1}{2}P(Y \geq -u) + \frac{1}{2}P(Y \leq u) \\
&= \begin{cases} (1/2)e^{-(-u)} + (1/2)(0), & u < 0 \\ (1/2)(1) + (1/2)(1-e^{-u}), & u \geq 0 \end{cases} \\
&= \begin{cases} (1/2)e^{u}, & u < 0 \\ 1-(1/2)e^{-u}, & u \geq 0. \end{cases}
\end{aligned}$$

So U has pdf $f(u) = F'(u) = \begin{cases} (1/2)e^{u}, & u < 0 \\ 0-(1/2)e^{-u}(-1), & u \geq 0 \end{cases}$.

That is, $f(u) = \frac{1}{2}e^{-|u|}, -\infty < u < \infty$, which is the same the pdf of X.

Exercise 5.12 Sampling from a triangular distribution

Suppose we want to sample $x \sim f(x)$ where $f(x) = \begin{cases} x, & 0 < x < 1 \\ 2-x, & 1 < x < 2 \end{cases}$.

Describe how two random variables can be combined to obtain x.

Solution to Exercise 5.12

Sample the two random variables $r \sim Bern(0.5)$ and $y \sim Beta(2,1)$. Then calculate $x = ry + (1-r)(2-y)$. This way, there is a 50% chance that x will equal y, whose pdf is $f(y) = 2y, 0 < y < 1$, and a 50% chance that x will equal $z = 2 - y$, whose pdf is $f(z) = 2(2-z), 1 < z < 2$.

A second solution is as follows. Sample $u_1, u_2 \sim iid \ U(0,1)$ and calculate $x = u_1 + u_2$. It can easily be shown that a value of x formed in this way has the triangular pdf in question.

5.12 Rejection sampling

Some distributions are difficult to sample from using any of the already mentioned methods. For example, when applying the inversion technique, solving the equation $F(x) = u$ may be problematic even with the aid of the Newton-Raphson algorithm (e.g. due to instability unless starting at very close to the solution).

In such cases, one convenient and easy way to obtain a value from the distribution of interest may be via *rejection sampling* (also known as the *rejection method* or the *acceptance-rejection* method). This method works as follows.

Suppose we want to generate a random number from a target distribution with density $f(x)$. This target distribution may be continuous or discrete.

We must first decide on a suitable *envelope distribution* with *envelope density* $h(x)$. (These are also called the *majorising distribution* and *majorising density*.) Ideally, the chosen density $h(x)$ is similar in shape to $f(x)$ and relatively easy to sample from.

We next define the following quantities:

$$c = \max_{x} \left(\frac{f(x)}{h(x)} \right)$$

$$p(x) = \frac{f(x)}{ch(x)}.$$

The idea here is that $f(x)$ lies entirely beneath $ch(x)$ except that it touches $ch(x)$ at maybe only one point. Then $p(x)$, which is called the *acceptance probability*, appropriately lies between 0 and 1 (inclusive). Figure 5.10 illustrates this setup. The rejection algorithm is as follows:

1. Sample a *proposed value* (or *candidate*) $x' \sim h(x)$.

2. Calculate the *acceptance probability* $p = p(x') = \dfrac{f(x')}{ch(x')}$.

3. Generate a standard uniform value $u \sim U(0,1)$.

4. Decide whether to accept or reject the candidate, as follows:
   If $u < p$ then *accept* x', meaning return $x = x'$ and STOP.
   If $u > p$ then *reject* x', meaning go to Step 1 and REPEAT.

Steps 1 to 4 are repeated as many times as necessary until an acceptance occurs, resulting in $x = x'$. The finally accepted value x is an observation from $f(x)$. Repeating the entire procedure above another $J - 1$ times independently will result in a random sample of size J from $f(x)$.

Figure 5.10 illustrates, with:

$f(x)$ = density of the *Beta*(4,8) distribution
$h(x)$ = density of the *Beta*(2,2) distribution

$$c = \max_{x} \left(\frac{f(x)}{h(x)} \right) = 2.45$$

$x' = 0.4$ (example of a candidate)

$$p = p(x') = \frac{f(x')}{ch(x')} = \frac{2.365}{3.524} = 0.671.$$

In this case, if we sample $u = 0.419$ (for example), then we accept x' and return $x = 0.4$. If, however, we sample $u = 0.705$ (say), then we reject x' and propose another x', etc.

Figure 5.10 Illustration of the rejection sampling algorithm

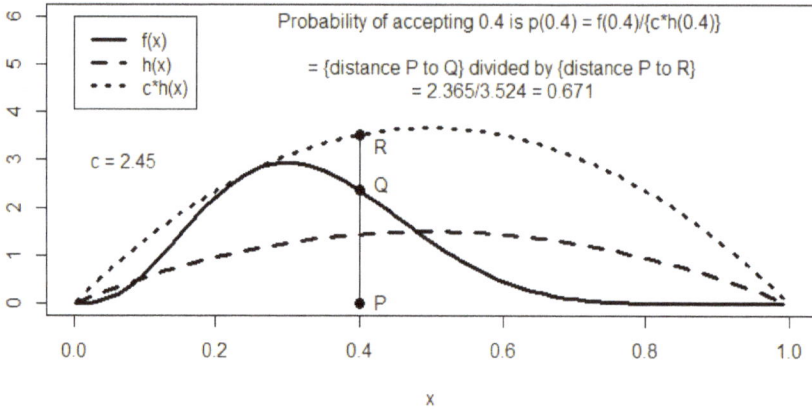

Note 1: The rejection sampling algorithm as defined here also works with $f(x)$ and $h(x)$ in the equations replaced by any *kernels* of the target and envelope distributions, respectively.

Note 2: The *overall acceptance rate* is the unconditional probability of acceptance and equals the area under $f(x)$ divided by the area under $ch(x)$, which is obviously $1/c$ $(= 0.409$ in our example).

The *wastage* may be defined as the overall probability of rejection, namely $1-1/c$, and this is simply the area between $f(x)$ and $ch(x)$ $(= 0.591$ in our example).

Note 3: If we consider the experiment of proposing values repeatedly until the next acceptance, then the number of candidates follows a geometric distribution with parameter $1/c$, and so the expected number of candidates (until acceptance) is $1/(1/c) = c$.

Note 4: There are two basic principles which must be considered in rejection sampling:

(i) The envelope density $h(x)$ should be similar to the target density $f(x)$ since this will minimise wastage, i.e. minimise the average number of proposals per acceptance, c, and hence optimise the computer time required.

(ii) The envelope distribution should be easy to sample from.

Note 5: The idea of rejection sampling can be used to give an intuitively appealing account of how Bayes' theorem works. In this regard, see Smith and Gelfand (1992).

Note 6: How rejection sampling works can most easily be explained by considering the case where $f(x)$ defines a simple discrete distribution. This is the subject of the next exercise.

R Code for Section 5.12

```
X11(w=8,h=4.5); par(mfrow=c(1,1))

plot(c(0,1), c(0,6),type="n",xlab="x",ylab="")
xv=seq(0.001,0.999,0.01); hxv=dbeta(xv,2,2); lines(xv,hxv,lty=2,lwd=3)

kfun=function(x){ dbeta(x,4,8) }
    # We could specify any positive function here        (*)
k0=integrate(f=kfun,lower=0,upper=1)$value
    # This calculates the normalising constant
fxv=kfun(xv)/k0;   # This ensures f(x) as defined at (*) is a proper density

lines(xv,fxv,lty=1,lwd=3)
c=max(fxv/hxv); c #  2.4472
lines(xv,c*hxv,lty=3,lwd=3)
legend(0,6,c("f(x)","h(x)","c*h(x)"),lty=c(1,2,3),lwd=c(3,3,3))
text(0.07,3,"c = 2.45")

xval=0.4; lines(c(xval,xval),c(0, c*dbeta(xval,2,2)),lty=1,lwd=1)
points(rep(xval,3),  c(0,kfun(xval)/k0 ,c*dbeta(xval,2,2)) ,
    pch=rep(16,3), cex=rep(1.2,3))
text(0.43,0.05,"P"); text(0.43,2.5,"Q"); text(0.43,3.3,"R");
c(0,kfun(xval)/k0 ,c*dbeta(xval,2,2))
    # 0.0000 2.3649 3.5239 2.3649/3.5239 # 0.6711
text(0.6,5.2,"Probability of accepting 0.4 is p(0.4) = f(0.4)/{c*h(0.4)} \n
= {distance P to Q} divided by {distance P to R}\n= 2.365/3.524 = 0.671")
c(0,kfun(xval)/k0 ,c*dbeta(xval,2,2)) # 0.0000 2.3649 3.5239
```

Exercise 5.13 Illustration of rejection sampling

Consider the $Bin(2,1/2)$ distribution with pdf
$$f(x) = \begin{cases} 1/4, & x = 0, 2 \\ 1/2, & x = 1 \end{cases},$$
and suppose we want to sample from this using rejection method envelope $g(x) = 1/3, \, x = 0,1,2$, i.e. the density of the discrete uniform distribution over the integers 0, 1 and 2. Show that the rejection sampling algorithm returns 0, 1 and 2 with the correct probabilities.

Solution to Exercise 5.13

Here: $c = \max_{x} \left(\dfrac{f(x)}{g(x)} \right) = \dfrac{1/2}{1/3} = \dfrac{3}{2}, \quad p(x) = \dfrac{f(x)}{cg(x)} = \begin{cases} 1/2, & x = 0, 2 \\ 1, & x = 1 \end{cases}.$

Now, suppose that we propose a very large number of proposed values from $g(x)$. Then:
- about 1/3 of these will be 0, of which about 1/2 will be accepted
- about 1/3 of these will be 1, of which (fully) *all* will be accepted
- about 1/3 of these will be 2, of which about 1/2 will be accepted.

We see that about 2/3 of all the proposed values will be accepted, and of these about 25% will be 0, 50% will be 1, and 25% will be 2. About 1/3 of the candidates will be rejected, about half of these being 0 and half being 2. The overall acceptance rate is $1/c = 1/(3/2) = 2/3$, and the wastage is $1 - 1/c = 1/3$. On average, $c = 1.5$ candidates will have to be proposed until an acceptance. Thus, generation of 1,000 $Bin(2,1/2)$ values (say) will require about 1,500 candidates.

5.13 Methods based on the rejection algorithm

The rejection method may be used in conjunction with many other methods. For example, the Box-Muller algorithm (mentioned earlier) is a basis for the *Marsaglia polar method* for sampling from a normal distribution. This method involves generating
$$u_1, u_2 \sim iid \ U(0,1)$$
repeatedly until
$$s \equiv (2u_1 - 1)^2 + (2u_2 - 1)^2 < 1$$
and then returning $z_i = (2u_i - 1)\sqrt{-2(\log s)/s}$, $i = 1,2$.

The result will (eventually) be the required sample

$z_1, z_2 \sim iid\ N(0,1)$.

This algorithm includes a condition for rejecting the sample values u_1, u_2 and involves iterating until these values are accepted (as a pair). The procedure may be less efficient than the Box-Muller algorithm (which does not involve rejection sampling and never requires more than two standard uniform variates) but avoids the computation of sines and cosines.

5.14 Monte Carlo methods in Bayesian inference

Most of the ideas above in this chapter are directly applicable to Bayesian inference. Suppose we have derived a posterior distribution or density $f(\theta\,|\,x)$ but it is complicated and difficult to work with directly. Then we can try to generate a random sample from that posterior with a view to estimating all the required inferential quantities (e.g. point and interval estimates) via the method of Monte Carlo.

First, denote the Monte Carlo sample as $\theta_1, ..., \theta_J \sim iid\ f(\theta\,|\,x)$. Then, the MC estimate of the posterior mean of θ, namely

$$\hat{\theta} = E(\theta\,|\,x) = \int \theta f(\theta\,|\,x)d\theta,$$

is

$$\bar{\theta} = \frac{1}{J}\sum_{j=1}^{J}\theta_j \ \ \text{(the MC sample mean)},$$

and a $1-\alpha$ CI for $\hat{\theta}$ is

$$\left(\bar{\theta} \pm z_{\alpha/2}\frac{s_\theta}{\sqrt{J}}\right),$$

where

$$s_\theta^2 = \frac{1}{J-1}\sum_{j=1}^{J}(\theta_j - \bar{\theta})^2.$$

Also, a MC estimate of the $1-\alpha$ CPDR for θ is $(\hat{q}_{\alpha/2}, \hat{q}_{1-\alpha/2})$, where $\hat{q}_p$ is the empirical p-quantile of $\theta_1, ..., \theta_J$, and the MC estimate of the posterior median is $\hat{q}_{1/2}$, etc.

Further, when the posterior density $f(\theta \mid x)$ does not have a closed form expression (as is often the case), it can be estimated by smoothing a probability histogram of $\theta_1,...,\theta_J$.

Once an estimate of the posterior density has been obtained, the mode of that estimate defines the MC estimate of the posterior mode.

Suppose we are interested in some posterior probability
$$p = P(\theta \in A \mid y)$$
(where A is a subset of the parameter space).

Then, the MC estimate of p is
$$\hat{p} = \frac{1}{J}\sum_{j=1}^{J} I(\theta_j \in A),$$
i.e. the proportion of the θ_j values which lie in A, and a $1-\alpha$ CI for p is
$$\left(\hat{p} \pm z_{\alpha/2}\sqrt{\hat{p}(1-\hat{p})/J}\right).$$

Suppose we are interested in a function of the parameter, $\psi = g(\theta)$. Then regardless of how complicated g is, we can perform MC inference on ψ easily. Simply calculate $\psi_j = g(\theta_j)$ for each $j = 1,...,J$. This results in a random sample from the posterior distribution of ψ, namely the values
$$\psi_1,...,\psi_J \sim iid \; f(\psi \mid x).$$

One may then apply any of the ideas above, just as before. For example, the posterior mean of ψ, namely
$$\hat{\psi} = E(\psi \mid x) = \int \psi f(\psi \mid x)d\psi = \int g(\theta)f(\theta \mid x)d\theta,$$
can be estimated by its MC estimate,
$$\overline{\psi} = \frac{1}{J}\sum_{j=1}^{J}\psi_j,$$
and a $1-\alpha$ CI for $\hat{\psi}$ is
$$\left(\overline{\psi} \pm z_{\alpha/2}\frac{s_\psi}{\sqrt{J}}\right),$$
where
$$s_\psi^2 = \frac{1}{J-1}\sum_{j=1}^{J}(\psi_j - \overline{\psi})^2.$$

Exercise 5.14 MC inference under the normal-normal-gamma model

Recall the Bayesian model:
$$(y_1,\ldots,y_n \mid \mu,\lambda) \sim iid \; N(\mu, 1/\lambda)$$
$$f(\mu,\lambda) \propto 1/\lambda, \; \mu \in \Re, \lambda > 0.$$

Suppose we observe the data vector $y = (y_1,\ldots,y_n) = (2.1, 3.2, 5.2, 1.7)$.

(a) Generate $J = 1{,}000$ values from the posterior distribution of μ. Use this sample to perform MC inference on μ. Illustrate your inferences with a suitable graph.

(b) Generate $J = 1{,}000$ values from the posterior distribution of λ. Use this sample to perform MC inference on λ. Illustrate your inferences with a suitable graph.

(c) Use MC methods to estimate the *signal to noise ratio* (SNR), defined as $\gamma = \mu/\sigma = \mu\sqrt{\lambda}$. Illustrate your inferences with a suitable graph.

Solution to Exercise 5.14

(a) Recall that the marginal posterior distribution of μ is given by
$$\left(\frac{\mu - \bar{y}}{s/\sqrt{n}} \, \middle| \, y \right) \sim t(n-1).$$

So we generate $w_1,\ldots,w_J \sim iid \; t(n-1)$ and then calculate
$$\mu_j = \bar{y} + \frac{s}{\sqrt{n}} w_j, \quad j = 1,\ldots,J.$$

We then use the sample $\mu_1,\ldots,\mu_J \sim iid \; f(\mu \mid y)$ for MC inference on μ. Thereby, we estimate μ's posterior mean $\hat{\mu} = E(\mu \mid y)$ by $\bar{\mu} = 3.077$ with $(3.001, 3.153)$ as the 95% MC CI for $\hat{\mu}$. The MC estimate of μ's 95% CPDR is $(0.685, 5.507)$.

We now compare the above estimates with the true values:
$$\hat{\mu} = \bar{y} = 3.050$$
$$95\% \text{ CPDR for } \mu = \left(\bar{y} \pm t_{0.025}(n-1)\frac{s}{\sqrt{n}} \right) = (0.556, 5.544).$$

We observe that the true posterior mean is contained in the 95% MC CI for that mean. Figure 5.11 provides a comparison of the above Monte Carlo and 'exact' inferences.

Note 1: The formula for the exact posterior density is

$$f(\mu \mid y) = f(w \mid y) \left| \frac{dw}{d\mu} \right| = f_{t(n-1)} \left(\frac{\mu - \bar{y}}{s / \sqrt{n}} \right) \times \left| \frac{1}{s / \sqrt{n}} \right|$$

$$= \frac{\Gamma\left(\frac{(n-1)+1}{2} \right)}{\Gamma\left(\frac{n-1}{2} \right) \sqrt{(n-1)\pi}} \left(1 + \frac{\left(\frac{\mu - \bar{y}}{s / \sqrt{n}} \right)^2}{n-1} \right)^{-\left(\frac{(n-1)+1}{2} \right)} \times \frac{\sqrt{n}}{s}, \quad \mu \in \Re.$$

Note 2: The MC sample $\mu_1, ..., \mu_J \sim iid\ f(\mu \mid y)$ could also be obtained using the following results:

$$(\lambda \mid y) \sim Gamma\left(\left(\frac{n-1}{2} \right), \left(\frac{n-1}{2} \right) s^2 \right)$$

$$(\mu \mid y, \lambda) \sim N\left(\bar{y}, \frac{1}{n\lambda} \right).$$

Thus, using the method of composition and the identity
$$f(\mu, \lambda \mid y) = f(\lambda \mid y) f(\mu \mid y, \lambda),$$
we first sample

$$\lambda_1, ..., \lambda_J \sim Gamma\left(\left(\frac{n-1}{2} \right), \left(\frac{n-1}{2} \right) s^2 \right),$$

and then sample

$$\mu_j \sim N\left(\bar{y}, \frac{1}{n\lambda_j} \right) \text{ for each } j = 1, ..., J.$$

The result of this procedure is
$$(\mu_1, \lambda_1), ..., (\mu_J, \lambda_J) \sim iid\ f(\mu, \lambda \mid y),$$
and thereby
$$\mu_1, ..., \mu_J \sim iid\ f(\mu \mid y),$$
as before, after discarding all of the λ_j values.

Figure 5.11 Monte Carlo inference on the normal mean

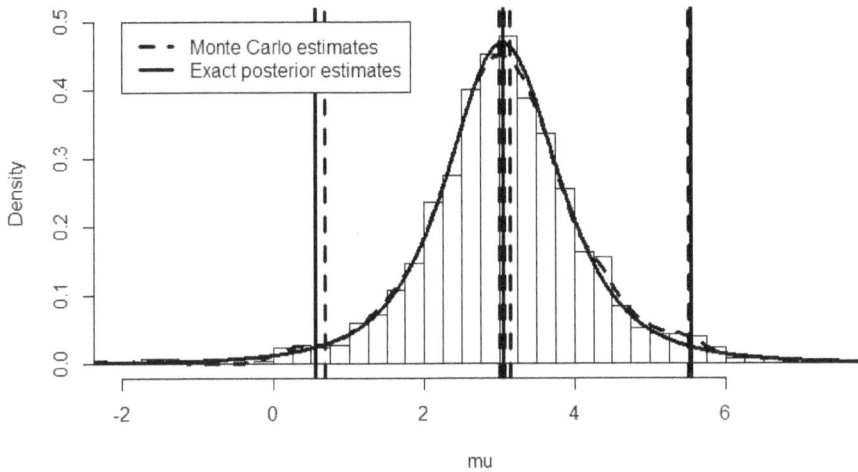

(b) One way to obtain a MC sample from the marginal posterior distribution of λ is as indicated in Note 2 of part (a). Alternatively, we can make use of the result

$$(\lambda \,|\, y, \mu) \sim Gamma\left(\frac{n}{2}, \frac{n}{2} s_\mu^2\right), \text{ where } s_\mu^2 = \frac{1}{n}\sum_{i=1}^{n}(y_i - \mu)^2.$$

So, again by the method of composition, but this time using the identity
$$f(\mu, \lambda \,|\, y) = f(\mu \,|\, y) f(\lambda \,|\, y, \mu),$$
we make use of the sample already generated in (a) and sample

$$\lambda_j \sim Gamma\left(\frac{n}{2}, \frac{n}{2} s_{\mu_j}^2\right)$$

for each $j = 1, ..., J$. The result is $(\mu_1, \lambda_1), ..., (\mu_J, \lambda_J) \sim iid \, f(\mu, \lambda \,|\, y)$, and thereby $\lambda_1, ..., \lambda_J \sim iid \, f(\lambda \,|\, y)$ (after discarding all of the μ_j values).

Implementing this procedure (i.e. making use of the simulated values in (a)) we obtain the required sample, $\lambda_1, ..., \lambda_J \sim iid \, f(\lambda \,|\, y)$, and use it for MC inference. Thereby we estimate λ's posterior mean $\hat{\lambda} = E(\lambda \,|\, y)$ by $\bar{\lambda} = 0.3998$ with (0.3804, 0.4192) as the 95% MC CI for $\hat{\lambda}$. The MC estimate of λ's 95% CPDR is (0.0347, 1.2828).

We now compare the above estimates with the true values:

$$\hat{\lambda} = \frac{1}{s^2} = 0.4071$$

$$95\% \text{ CPDR} = \left(F^{-1}_{G\left(\frac{n-1}{2}, \frac{n-1}{2}s^2\right)}(0.025), F^{-1}_{G\left(\frac{n-1}{2}, \frac{n-1}{2}s^2\right)}(0.975) \right)$$

$$= (0.0293, 1.2684).$$

We see that the true posterior mean is contained in the 95% MC CI for that mean. Figure 5.12 illustrates these Monte Carlo and 'exact' inferences.

Figure 5.12 Monte Carlo inference on the precision parameter

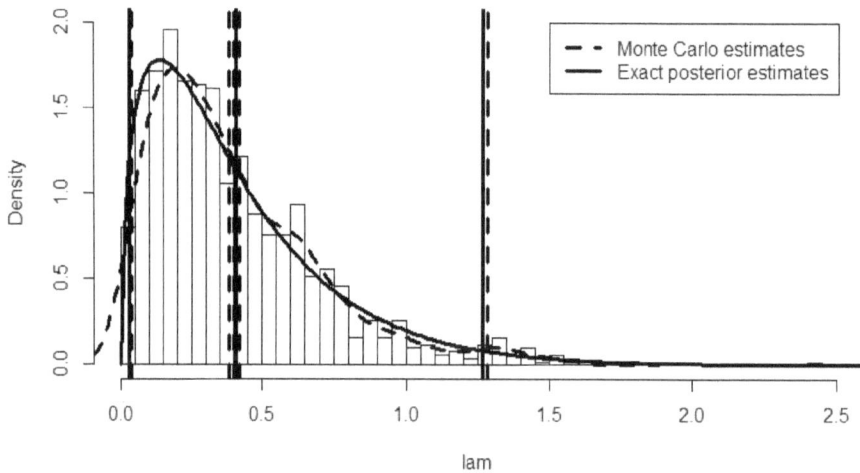

(c) Using the values sampled in (a) and (b), we now calculate $\gamma_j = \mu_j \sqrt{\lambda_j}$ for each $j = 1,..,J$, and hence obtain a MC sample $\gamma_1,...,\gamma_J \sim iid\ f(\gamma\,|\,y)$, which can then be used to perform MC inference on γ. (NB: The symbols 'γ' and 'γ' are typographically equivalent.) Implementing this strategy, we estimate γ's posterior mean by 1.800, with (1.745 1.854) as a 95% CI for that mean, and we estimate γ's 95% CPDR as (0.228, 3.543).

Figure 5.13 illustrates these Monte Carlo estimates. Also shown are:
- the exact posterior mean of γ, which is $\hat{\gamma} = E(\gamma\,|\,y) = 1.793$
- the exact 95% CPDR for γ, which is (0.0733, 3.5952)
- the exact posterior density of γ
- the MLE of γ, which is $\tilde{\gamma} = \bar{y}/s = 3.05/1.567 = 1.946$.

See the Note and R Code below for details of these calculations.

Figure 5.13 Monte Carlo inference on the signal to noise ratio

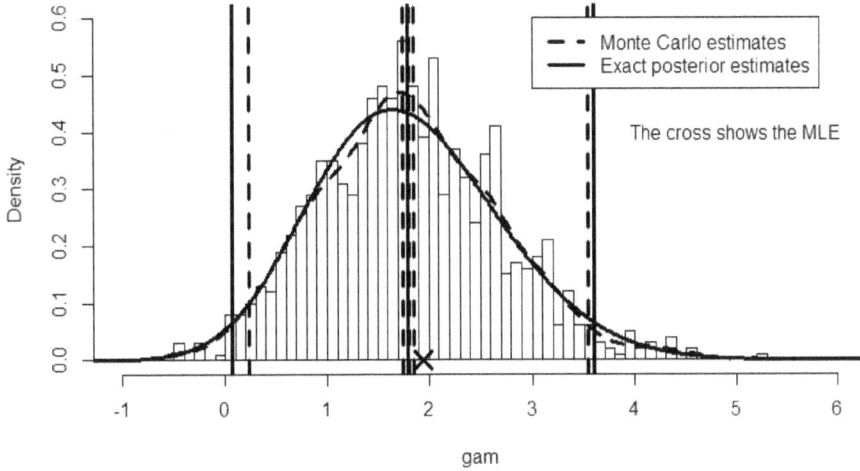

Note: The conditional posterior distribution of $\gamma = \mu\sqrt{\lambda}$ given λ is

$$(\gamma \mid y, \lambda) \sim N((\sqrt{\lambda})\bar{y}, (\sqrt{\lambda})^2 1 / (n\lambda)) \sim N(\bar{y}\sqrt{\lambda}, 1/n).$$

This follows from the uninformative normal-normal model, i.e. from the fact that

$$(\mu \mid y, \lambda) \sim N(\bar{y}, 1 / (n\lambda)).$$

So the posterior density of γ may be obtained numerically according to

$$f(\gamma \mid y) = E_\lambda\{f(\gamma \mid y, \lambda) \mid y\} = \int_0^\infty f(\gamma \mid y, \lambda) f(\lambda \mid y) d\lambda,$$

where:

$$f(\gamma \mid y, \lambda) = f_{N(\bar{y}\sqrt{\lambda}, 1/n)}(\gamma) = \frac{n}{\sqrt{2\pi}} e^{-\frac{n}{2}(\gamma - \bar{y}\sqrt{\lambda})^2}, \gamma \in \Re$$

$$f(\lambda \mid y) = f_{G\left(\frac{n-1}{2}, \frac{n-1}{2}s^2\right)}(\lambda) = \frac{\left(\frac{n-1}{2}s^2\right)^{\frac{n-1}{2}} \lambda^{\frac{n-1}{2}-1} e^{-\lambda\frac{n-1}{2}s^2}}{\Gamma\left(\frac{n-1}{2}\right)}, \lambda > 0.$$

Also (as shown in a previous exercise), the posterior mean of γ is exactly

$$\hat{\gamma} = E(\mu\sqrt{\lambda} \mid y) = E\{E(\mu\sqrt{\lambda} \mid y, \lambda) \mid y\}$$

$$= \frac{\bar{y}}{s} \times \frac{\Gamma\left(\dfrac{n-1}{2} + \dfrac{1}{2}\right)}{\left(\dfrac{n-1}{2}\right)^{1/2} \Gamma\left(\dfrac{n-1}{2}\right)} = 1.793$$

(after some algebra).

The exact 95% CPDR for γ may be obtained by using the optim() function to minimise

$$g(L,U) = \left(F_\gamma(U \mid y) - F_\gamma(L \mid y) - 0.95\right)^2 + \left(f_\gamma(U \mid y) - f_\gamma(L \mid y)\right)^2$$

$$= \left(\int_L^U \left(\int_0^\infty f(\gamma \mid y, \lambda) f(\lambda \mid y) d\lambda\right) d\gamma - 0.95\right)^2$$

$$+ \left(\int_0^\infty f_\gamma(U \mid y, \lambda) f(\lambda \mid y) d\lambda - \int_0^\infty f_\gamma(L \mid y, \lambda) f(\lambda \mid y) d\lambda\right)^2,$$

with the result being $(L, U) = (0.0733, 3.5952)$.

R Code for Exercise 5.14

```
# (a)
y=c(2.1, 3.2, 5.2, 1.7); n=length(y); ybar=mean(y); s=sd(y); s # 1.567
J=1000; set.seed(144); options(digits=4)
wv=rt(J,n-1); muv=ybar+s*wv/sqrt(n)
mubar=mean(muv); muci=mubar + c(-1,1)*qnorm(0.975)*sd(muv)/sqrt(J)
mucpdr=quantile(muv,c(0.025,0.975))
c(mubar,muci,mucpdr) # 3.0770 3.0012 3.1528 0.6848 5.5069
muhat=ybar; mucpdrtrue= ybar+(s/sqrt(n))*qt(c(0.025,0.975),n-1)
c(muhat,mucpdrtrue) # 3.050 0.556 5.544

X11(w=8,h=5); par(mfrow=c(1,1))

hist(muv,prob=T,xlab="mu",xlim=c(-2,7.5), ylim=c(0,0.5),main="",
        breaks=seq(-20,20,0.25))
muvec=seq(-20,20,0.01);
postvec=dt( (muvec-ybar)/(s/sqrt(n)) , n-1 ) / (s/sqrt(n))
```

```
lines(muvec,postvec, lty=1,lwd=3)
lines(density(muv),lty=2,lwd=3)
abline(v=c(mubar,muci,mucpdr),lty=2,lwd=3)
abline(v=c(ybar, mucpdrtrue)  , lty=1,lwd=3)
legend(-2,0.5,c("Monte Carlo estimates","Exact posterior estimates"),
        lty=c(2,1),lwd=c(3,3),bg="white")


# (b)
lamv=rep(NA,J); set.seed(332)
for(j in 1:J) lamv[j] = rgamma(1,n/2,(n/2)*mean((y-muv[j])^2))


lambar=mean(lamv); lamci=lambar + c(-1,1)*qnorm(0.975)*sd(lamv)/sqrt(J)
lamcpdr=quantile(lamv,c(0.025,0.975))
c(lambar, lamci, lamcpdr) # 0.39980 0.38040 0.41920 0.03465 1.28283
lamhat=1/s^2; lamcpdrtrue= qgamma(c(0.025,0.975),(n-1)/2,((n-1)/2)*s^2)
c(lamhat, lamcpdrtrue) # 0.40706 0.02928 1.26844


hist(lamv,prob=T,xlab="lam",xlim=c(0,2.5), ylim=c(0,2),main="",
        breaks=seq(0,3,0.05))
lamvec=seq(0,3,0.01) ; lampostvec= dgamma(lamvec,(n-1)/2,((n-1)/2)*s^2)
lines(lamvec, lampostvec, lty=1,lwd=3)
lines(density(lamv),lty=2,lwd=3)
abline(v=c(lambar, lamci, lamcpdr),lty=2,lwd=3)
abline(v=c(1/s^2,  lamcpdrtrue),  lty=1,lwd=3)
legend(1.5,2,c("Monte Carlo estimates","Exact posterior estimates"),
        lty=c(2,1),lwd=c(3,3),bg="white")


# (c)
gamv=muv*sqrt(lamv)


gambar=mean(gamv); gamci=gambar + c(-1,1)*qnorm(0.975)*sd(gamv)/sqrt(J)
gamcpdr=quantile(gamv,c(0.025,0.975))
c(gambar, gamci, gamcpdr) # 1.7997 1.7453 1.8540 0.2284 3.5433
mle=ybar/s; mle # 1.946


gamhat=(ybar/s)*gamma(0.5+(n-1)/2)/(sqrt((n-1)/2)*gamma((n-1)/2))
print(c(ybar,s,gamhat),digits=8) # 3.0500000 1.5673757 1.7928178
intfun=function(lam,gam, ybar=3.05,s=1.5673757,n=4){
  dnorm(gam,ybar*sqrt(lam),1/sqrt(n))*dgamma(lam,(n-1)/2,s^2*(n-1)/2)  }
```

```
integrate(function(gam) {
        sapply(gam, function(gam) {
        integrate(function(lam) {
        sapply(lam, function(lam) intfun(lam,gam) )
        }, 0, Inf)$value  }) }, -Inf, Inf)
   # 1 with absolute error < 4.7e-07   OK (Just checking)

integrate(function(gam) {
        sapply(gam, function(gam) {
        integrate(function(lam) {
        sapply(lam, function(lam) gam*intfun(lam,gam) )
        }, 0, Inf)$value  }) }, -Inf, Inf)
   # 1.793 with absolute error < 4.7e-06   OK (Agrees with exact calculation)

gamvec=seq(-5,10,0.01); fgamvec=gamvec

for(i in 1:length(gamvec)){
        fgamvec[i]=integrate( f=intfun, lower=0, upper=Inf,
                gam=gamvec[i])$value   }
 plot(gamvec,fgamvec) # OK

L=-0.1; U=4.2  # Testing....
integrate(function(gam) {
        sapply(gam, function(gam) {
        integrate(function(lam) {
        sapply(lam, function(lam) intfun(lam,gam) )
        }, 0, Inf)$value  }) }, L,U)
   # 0.9823 with absolute error < 4.3e-08  OK

integrate( f=intfun, lower=0, upper=Inf, gam=U)$value -
  integrate( f=intfun, lower=0, upper=Inf, gam=L)$value   # -0.02074  OK

gfun=function(v){  L=v[1]; U=v[2]
( integrate(function(gam) {
        sapply(gam, function(gam) {
        integrate(function(lam) {
        sapply(lam, function(lam) intfun(lam,gam) )
        }, 0, Inf)$value  }) }, L,U)$value  - 0.95  )^2 +
( integrate( f=intfun, lower=0, upper=Inf, gam=U)$value -
    integrate( f=intfun, lower=0, upper=Inf, gam=L)$value  )^2  }

gfun(v=c(-0.1,4.2)) # 0.001473  OK
gfun(v=c(1,3)) #   0.08562 OK
```

```
res0=optim(par=c(0,4),fn=gfun)$par
res0 # 0.07334 3.59516
res1=optim(par=res0,fn=gfun)$par
res1 # 0.07332 3.59518
res2=optim(par=res1,fn=gfun)$par
res2 # 0.07332 3.59518  OK


L=res2[1]; U=res2[2]  # Now check...

integrate(function(gam) {
        sapply(gam, function(gam) {
        integrate(function(lam) {
        sapply(lam, function(lam) intfun(lam,gam) )
        }, 0, Inf)$value }) }, L,U)
  # 0.95 with absolute error < 3.2e-07
integrate( f=intfun, lower=0, upper=Inf, gam=L)$value  # 0.06598
integrate( f=intfun, lower=0, upper=Inf, gam=U)$value # 0.06598  All OK

hist(gamv,prob=T,xlab="gam",xlim=c(-1,6), ylim=c(0,0.6),main="",
        breaks=seq(-2,7,0.1))
lines(density(gamv),lty=2,lwd=3)
abline(v=c(gambar, gamci, gamcpdr),lty=2,lwd=3)
points(mle,0,pch=4,lwd=3,cex=2)
lines(gamvec,fgamvec,lty=1,lwd=3)
abline(v=c(gamhat,L,U),lty=1,lwd=3)
legend(3,0.6,c("Monte Carlo estimates","Exact posterior estimates"),
        lty=c(2,1),lwd=c(3,3),bg="white")
text(5,0.4,"The cross shows the MLE")
```

5.15 MC predictive inference via the method of composition

Suppose that in the context of a Bayesian model defined by $f(y|\theta)$ and $f(\theta)$, we wish to predict a value x whose distribution is specified by $f(x|y,\theta)$. Recall that the posterior predictive density is

$$f(x|y) = \int f(x|y,\theta)f(\theta|y)d\theta.$$

If this density is complicated, we may choose to perform MC predictive inference on x using a sample $x_1,...,x_J \sim iid\ f(x|y)$. The question then arises as to how such a sample may be obtained.

One answer is to sample from $f(x \mid y)$ *directly*. But that may be difficult since $f(x \mid y)$ is complicated. Another answer is to apply the method of composition through the equation

$$f(x, \theta \mid y) = f(x \mid y, \theta) f(\theta \mid y).$$

This means that we should first sample $\theta' \sim f(\theta \mid y)$ and then sample $x' \sim f(x \mid y, \theta')$, the result being $(x', \theta') \sim f(x, \theta \mid y)$. If we then discard θ', the result is the required $x' \sim f(x \mid y)$. Implementing this process a total of J times results in the required sample, $x_1, ..., x_J \sim iid\ f(x \mid y)$.

Exercise 5.15 Monte Carlo prediction in the binomial-beta model

The probability of heads coming up on a bent coin follows a standard uniform distribution *a priori*. We toss the coin 50 times and get 28 heads. Estimate using Monte Carlo the probability that heads will come up on at least six of the next 10 tosses of the same bent coin.

Solution to Exercise 5.15

Recall that the binomial-beta model:

$$(y \mid \theta) \sim Bin(n, \theta)$$
$$\theta \sim Beta(\alpha, \beta),$$

for which the posterior distribution is given by

$$(\theta \mid y) \sim Beta(\alpha + y, \beta + n - y).$$

Earlier we showed that if the future data x has distribution defined by

$$(x \mid y, \theta) \sim Bin(m, \theta),$$

then posterior predictive distribution is given by

$$f(x \mid y) = \binom{m}{x} \frac{B(y + x + \alpha, n - y + m - x + \beta)}{B(y + \alpha, n - y + \beta)}, \quad x = 0, ..., m.$$

Rather than trying to sample from this distribution directly, we may do the following:

Sample $\theta' \sim Beta(\alpha + y, \beta + n - y)$

Sample $x' \sim Bin(m, \theta')$.

Discarding θ', we obtain the required sample value, $x' \sim f(x \mid y)$.

In the situation here: $\alpha = \beta = 1$, $n = 50$, $y = 32$, $m = 10$.

Implementing the above sampling strategy $J = 10,000$ times with these specifications, we obtain a large MC sample, $x_1,...,x_J \sim f(x \mid y)$.

It is found that 7,084 of the sample values are at least 6. So we estimate $p = P(X \geq 6 \mid y)$ by $\hat{p} = 0.7084$. A 95% CI for p is then
$$(\hat{p} \pm 1.96\sqrt{\hat{p}(1-\hat{p})/J}) = (0.6995, 0.7173).$$

For interest, we also work out the probability exactly as
$$p = \sum_{x=6}^{10} f(x \mid y) = 0.7030 \quad \text{(correct to 4 decimals)}$$
and note that this value lies in the 95% CI obtained using MC methods.

R Code for Exercise 5.15

```
options(digits=5)
n=50; y=32;   alp=1;bet=1;    a=alp+y; b=bet+n-y;    m=10;  J=10000
set.seed(443); tv=rbeta(J,a,b); xv=rbinom(J,m,tv)
phat=length(xv[xv>=6])/J;
ci=phat+c(-1,1)*qnorm(0.975)*sqrt(phat*(1-phat)/J)
c(phat,ci) # 0.70840 0.69949 0.71731

xvec=0:m; fxgiveny=
  choose(m,xvec)*beta(y+xvec+alp,n-y+m-xvec+bet)/beta(y+alp,n-y+bet)
sum(fxgiveny)  #1   Just checking
sum(fxgiveny[xvec>=6]) # 0.70296
```

5.16 Rao-Blackwell methods for estimation and prediction

Consider a Bayesian model with two parameters given by a specification of $f(y \mid \theta, \psi)$ and $f(\theta, \psi)$, and suppose that we obtain a sample from the joint posterior distribution of the two parameters, say
$$(\theta_1, \psi_1),...,(\theta_J, \psi_J) \sim iid \; f(\theta, \psi \mid y).$$

As we have seen, an unbiased Monte Carlo estimate of θ's posterior mean, $\hat{\theta} = E(\theta \mid y)$, is $\bar{\theta} = (1/J)\sum_{j=1}^{J} \theta_j$, with an associated MC $1 - \alpha$ CI for $\hat{\theta}$ given by $(\bar{\theta} \pm z_{\alpha/2} s_\theta / \sqrt{J})$, where s_θ is the sample standard deviation of $\theta_1,...,\theta_J$.

Now observe that

$$\hat{\theta} = E\{E(\theta \mid y, \psi) \mid y\} = \int E(\theta \mid y, \psi) f(\psi \mid y) d\psi \,.$$

This implies that another unbiased Monte Carlo estimate of $\hat{\theta}$ is

$$\bar{e} = \frac{1}{J} \sum_{j=1}^{J} e_j \,,$$

where

$$e_j = E(\theta \mid y, \psi_j),$$

and another $1 - \alpha$ CI for $\hat{\theta}$ is

$$(\bar{e} \pm z_{\alpha/2} s_e / \sqrt{J}) \,,$$

where s_e is the sample standard deviation of $e_1, ..., e_J$.

If possible, this second method of Monte Carlo inference is preferable to the first because it typically leads to a shorter CI. We call this second method *Rao-Blackwell* (RB) *estimation*. The first (original) method may be called *direct Monte Carlo estimation* or *histogram estimation*.

The same idea extends to estimation of the entire marginal posterior density of θ, because this can be written

$$f(\theta \mid y) = \int f(\theta \mid y, \psi) f(\psi \mid y) d\psi = E_\psi \{ f(\theta \mid y, \psi) \mid y \}.$$

Thus, the Rao-Blackwell estimate of $f(\theta \mid y)$ is

$$\tilde{f}(\theta \mid y) = \frac{1}{J} \sum_{j=1}^{J} f(\theta \mid y, \psi_j),$$

as distinct from the ordinary histogram estimate obtained by smoothing a probability histogram of $\theta_1, ..., \theta_J$.

The idea further extends to predictive inference, where we are interested in a future quantity x defined by a specification of $f(x \mid y, \theta, \psi)$.

The direct MC estimate of the predictive mean, namely

$$\hat{x} = E(x \mid y),$$

is

$$\bar{x} = (1 / J) \sum_{j=1}^{J} x_j \,,$$

where

$$x_1, ..., x_J \sim iid \; f(x \mid y)$$

(e.g. as obtained via the method of composition).

A superior estimate is the Rao-Blackwell estimate given by

$$\overline{E} = \frac{1}{J}\sum_{j=1}^{J} E_j,$$

where there is now a choice from the following:

$$E_j = E(x\,|\,y,\theta_j,\psi_j)$$

$$\text{or} \quad E_j = E(x\,|\,y,\psi_j)$$

$$\text{or} \quad E_j = E(x\,|\,y,\theta_j).$$

This estimator ($\overline{E}$) is based on the identities

$$\hat{x} = E\{E(x\,|\,y,\theta,\psi)\,|\,y\} = E\{E(x\,|\,y,\psi)\,|\,y\} = E\{E(x\,|\,y,\theta)\,|\,y\}.$$

Note: The first of the three choices for E_j is typically the easiest to calculate but also leads to the least improvement over the ordinary 'histogram' predictor, $\overline{x} = (1/J)\sum_{j=1}^{J} x_j$.

Likewise, the Rao-Blackwell estimate of the entire posterior predictive density $f(x\,|\,y)$ is

$$\tilde{f}(x\,|\,y) = \frac{1}{J}\sum_{j=1}^{J} f_j(x),$$

where there is a choice from the following:

$$f_j(x) = f(x\,|\,y,\theta_j,\psi_j)$$

$$\text{or} \quad f_j(x) = f(x\,|\,y,\psi_j)$$

$$\text{or} \quad f_j(x) = f(x\,|\,y,\theta_j).$$

Exercise 5.16 Practice at Rao-Blackwell estimation in the normal-normal-gamma model

Recall the Bayesian model:

$$(y_1,\ldots,y_n\,|\,\mu,\lambda) \sim iid\ N(\mu,1/\lambda)$$

$$f(\mu,\lambda) \propto 1/\lambda,\ \mu \in \Re, \lambda > 0.$$

Suppose that we observe the vector $y = (y_1,\ldots,y_n) = (2.1, 3.2, 5.2, 1.7)$.

Generate $J = 100$ values from the joint posterior distribution of μ and λ and use these values as follows. Calculate the direct Monte Carlo estimate and the Rao-Blackwell estimate of λ's marginal posterior mean.

In each case, report the associated 95% CI for that mean. Compare your results with the true value of that mean. Produce a probability histogram of the simulated λ-values. Overlay a smooth of this histogram and the Rao-Blackwell estimate of λ's marginal posterior density. Also overlay the exact density.

Solution to Exercise 5.16

Recall from Equation (3.3) in Exercise 3.11 that:

$$(\lambda \mid y) \sim Gamma\left(\left(\frac{n-1}{2}\right), \left(\frac{n-1}{2}\right)s^2\right)$$

$$(\mu \mid y, \lambda) \sim N\left(\bar{y}, \frac{1}{n\lambda}\right).$$

So we first sample

$$\lambda' \sim Gamma\left(\left(\frac{n-1}{2}\right), \left(\frac{n-1}{2}\right)s^2\right),$$

and then we sample

$$\mu' \sim N\left(\bar{y}, \frac{1}{n\lambda'}\right).$$

The result is

$$(\mu', \lambda') \sim f(\mu, \lambda \mid y).$$

Repeating many times, we get

$$(\mu_1, \lambda_1), ..., (\mu_J, \lambda_J) \sim iid \ f(\mu, \lambda \mid y).$$

The histogram estimate of $\hat{\lambda} = E(\lambda \mid y)$ works out as $\bar{\lambda} = 0.4142$, with 95% CI (0.4076, 0.4209).

Next let $e_j = E(\lambda \mid y, \mu_j)$.

Then the Rao-Blackwell estimate of $\hat{\lambda}$ is $\bar{e} = 0.4073$, with associated 95% CI (0.4047, 0.4100).

It will be observed that this second CI is narrower than the first (having width 0.0053 compared with 0.0133). It will also be observed that both CIs contain the true value, $\hat{\lambda} = 1/s^2 = 0.4071$.

Figure 5.14 shows:

- a probability histogram of $\lambda_1, ..., \lambda_J$

- a smooth of that probability histogram

- the true marginal posterior density, namely
  $$f(\lambda \mid y) = f_{Gamma\left((n-1)/2, s^2(n-1)/2\right)}(\lambda)$$

- the Rao-Blackwell estimate of $f(\lambda \mid y)$ as given by
  $$\tilde{f}(\lambda \mid y) = \frac{1}{J} \sum f_{Gamma\left(n/2, s_j^2 n/2\right)}(\lambda) \quad \text{where} \quad s_j^2 = \frac{1}{n} \sum_{i=1}^{n} (y_i - \mu_j)^2 .$$

Note: The Rao-Blackwell estimate here is based on the result
$$(\lambda \mid y, \mu) \sim Gamma\left(\frac{n}{2}, \frac{n}{2}\left[\frac{1}{n}\sum_{i=1}^{n}(y_i - \mu)^2\right]\right).$$

It will be observed that the Rao-Blackwell estimate of λ's posterior density is fairly close. The histogram estimate is much less accurate and incorrectly suggests that λ has some probability of being negative.

Figure 5.14 Illustration of Rao-Blackwell estimation

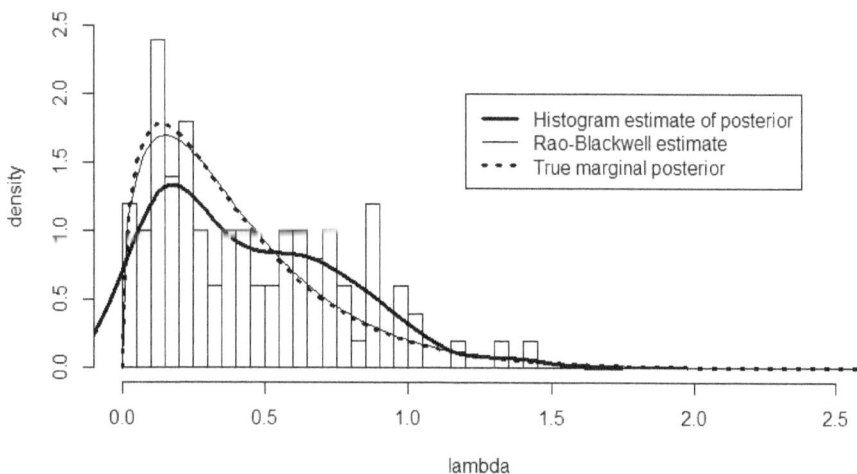

R Code for Exercise 5.16

```
options(digits=4)

# (a)
y=c(2.1, 3.2, 5.2, 1.7); n=length(y); ybar=mean(y); s=sd(y); s2=s^2
J=100; set.seed(254); lamv=rgamma(J,(n-1)/2,s2*(n-1)/2);
muv=rnorm(J,ybar,1/sqrt(n*lamv)); est0=1/s^2
est1=mean(lamv); std1=sd(lamv); ci1=est1 + c(-1,1)*qnorm(0.975)*std1/sqrt(J)
ev=rep(NA,J); for(j in 1:J){ muval=muv[j]; ev[j]=1/mean((y-muval)^2) }
est2=mean(ev); std2=sd(ev); ci2=est2 + c(-1,1)*qnorm(0.975)*std2/sqrt(J)
rbind( c(est0,NA,NA,NA),  c(est1,ci1,ci1[2]-ci1[1]),  c(est2,ci2,ci2[2]-ci2[1]) )
# [1,] 0.4071   NA    NA    NA
# [2,] 0.4396 0.3767 0.5026 0.12589
# [3,] 0.4150 0.3892 0.4408 0.05166


# (b)
X11(w=8,h=5); par(mfrow=c(1,1))
hist(lamv,xlab="lambda",ylab="density",prob=T,xlim=c(0,2.5),
      ylim=c(0,2.5),main="",breaks=seq(0,4,0.05))
lines(density(lamv),lty=1,lwd=3)
lamvec=seq(0,3,0.01); RBvec=lamvec; smu2v=1/ev
for(k in 1:length(lamvec)){      lamval=lamvec[k]
      RBvec[k]=mean(dgamma(lamval,n/2,(n/2)*smu2v))  }
lines(lamvec,RBvec,lty=1,lwd=1)
lines(seq(0,3,0.005),dgamma(seq(0,3,0.005),(n-1)/2,s2*(n-1)/2), lty=3,lwd=3)
legend(1.2,2,c("Histogram estimate of posterior","Rao-Blackwell estimate",
      "True marginal posterior"), lty=c(1,1,3),lwd=c(3,1,3))
```

5.17 MC estimation of posterior predictive p-values

Recall the theory of posterior predictive p-values whereby, in the context of a Bayesian model specified by $f(y|\theta)$ and $f(\theta)$, we test H_0 versus H_1 by choosing a suitable test statistic $T(y,\theta)$.

The posterior predictive p-value is then
$$p = P(T(x,\theta) \geq T(y,\theta)|y)$$
(or something similar, e.g. with $\geq$ replaced by $\leq$), calculated under the implicit assumption that H_0 is true.

If the calculation of p is problematic, a suitable Monte Carlo strategy is as follows:

1. Generate a random sample from the posterior,
$$\theta_1,...,\theta_J \sim iid\ f(\theta\,|\,y)\,.$$

2. Generate $x_j \sim\!\perp f(y\,|\,\theta_j)\,,j=1,...,J$
(so that $x_1,...,x_J \sim iid\ f(x\,|\,y)$).

3. For each $j=1,...,J$ calculate $T_j = T(x_j,\theta_j)$ and $I_j = I(T_j \geq T)$, where $T = T(y,\theta)\,.$

4. Estimate p by $\hat{p} = \dfrac{1}{J}\displaystyle\sum_{j=1}^{J} I_j$ with associated $1-\alpha$ CI
$$\left(\hat{p} \pm z_{\alpha/2}\sqrt{\frac{\hat{p}(1-\hat{p})}{J}}\right).$$

Exercise 5.17 Testing for independence in a sequence of Bernoulli trials

A bent coin has some chance of coming up heads whenever it is tossed. Our uncertainty about that chance may be represented by the standard uniform distribution.

The bent coin is tossed 10 times. Heads come up on the first seven tosses and tails come up on the last three tosses.

Using Bayesian methods, test that the 10 tosses were *independent*.

Solution to Exercise 5.17

The observed number of runs (of heads or tails in a row) is 2, which seems rather small.

Let y_i be the indicator for heads on the ith toss, $(i = 1,...,n)$ $(n = 10)$, and let θ be the unknown probability of heads coming up on any single toss.

Also let x_i be the indicator for heads coming up on the ith of the *next n* tosses of the same coin, tossed independently each time.

Further, let $y = (y_1,..., y_n)$ and $x = (x_1,..., x_n)$, and choose the test statistic as

$$T(y,\theta) = R(y),$$

defined as the number of runs in the vector y.

Then an appropriate posterior predictive p-value is

$$p = P(R(x) \le R(y) \mid y),$$

where $y = (1,1,1,1,1,1,1,0,0,0)$ and $R(y) = 2$.

Under the Bayesian model:

$$(y_1,..., y_n \mid \theta) \sim iid\ Bern(\theta)$$
$$\theta \sim U(0,1),$$

the posterior is given by

$$(\theta \mid y) \sim Beta(y_T + 1, n - y_T + 1),$$

where $y_T = y_1 + ... + y_n = 7$.

With $J = 10,000$, we now generate

$$\theta_1,..., \theta_J \sim iid\ Beta(8,4).$$

After that, we do the following for each $j = 1,..., J$:

1. Sample $x_1^j,..., x_n^j \sim iid\ Bern(\theta_j)$ and form the vector

$$x^j = (x_1^j,..., x_n^j).$$

2. Calculate $R_j = R(x^j)$ (i.e. calculate the number of runs in $(x_1^j,..., x_n^j)$).

3. Obtain $I_j = I(R_j \le R)$, where $R = R(y) = 2$.

Thereby we estimate p by

$$\hat{p} = \frac{1}{J}\sum_{j=1}^{J} I_j = 0.0995,$$

with 95% CI

$$\left(\hat{p} \pm 1.96\sqrt{\frac{\hat{p}(1-\hat{p})}{J}} \right) = (0.0936, 0.1054).$$

So the posterior predictive p-value is about 10 percent, which may be considered as statistically non-significant. That is, there is insufficient evidence (at the 5% level of significance, say) to conclude that the 10 tosses of the coin were somehow dependent.

Note 1: Using a suitable formula from runs theory, the exact value of p could be obtained as

$$p = \int_0^1 P(R(x) \leq 2 \mid \theta) f_{Beta(8,4)}(\theta) d\theta$$

$$= \int_0^1 \left\{ \sum_{x_T=0}^n P(R(x) \leq 2 \mid \theta, x_T) f(x_T \mid \theta) \right\} f_{Beta(8,4)}(\theta) d\theta \, ,$$

where:

- $P(R(x) \leq 2 \mid \theta)$ is the exact probability that 2 or fewer runs will result on 10 Bernoulli trials if each has probability of success θ

- $P(R(x) \leq 2 \mid \theta, x_T)$ is the probability of 2 or fewer runs will result when x_T 1s and $n - x_T$ 0s are placed in a row

- $f(x_T \mid \theta) = \binom{n}{x_T} \theta^{x_T} (1-\theta)^{n-x_T}$ is the binomial density with

    parameters n and θ, evaluated at x_T.

Note 2: It is of interest to recalculate p using data which seems even more 'extreme', for example,
$$y = (1,1,1,1,1, \ 1,1,1,1,1, \ 1,1,1,1,0, \ 0,0,0,0,0) \, .$$

For this data, $R(y) = 2$ again but with $n = 20$ and $y = 14$. In this case,
$$(\theta \mid y) \sim Beta(y_T + 1, n - y_T + 1) \sim Beta(15, 7),$$
and we obtain the estimate $\hat{p} = 0.0088$ with 95% CI (0.0070 0.0106).

Thus there is, as was to be expected, much stronger statistical evidence to reject the null hypothesis of independence.

R Code for Exercise 5.17

```
R=function(v){m=length(v); sum(abs(v[-1]-v[-m]))+1}
        # Calculates the runs in vector v
R(c(1,1,1,0,1)) # 3      testing …
R(c(1,1)) # 1
R(c(1,0,1,0,1)) # 5
R(c(0,0,1,1,1)) # 2
R(c(1,0,0,1,1,0,0,1,1,1,1,0)) # 6      …. all OK

n=10; J=10000; lv=rep(0,J); set.seed(214); tv=rbeta(J,8,4)
for(j in 1:J){ xj=rbinom(n,1,tv[j]); if(R(xj)<=2) lv[j]=1 }
p=mean(lv); ci=p+c(-1,1)*qnorm(0.975)*sqrt(p*(1-p)/J)
c(p,ci) # 0.09950 0.09363 0.10537

n=20; J=10000; lv=rep(0,J); set.seed(214); tv=rbeta(J,15,7)
for(j in 1:J){ xj=rbinom(n,1,tv[j]); if(R(xj)<=2) lv[j]=1 }
p=mean(lv); ci=p+c(-1,1)*qnorm(0.975)*sqrt(p*(1-p)/J)
c(p,ci) # 0.008800 0.006969 0.010631
```

CHAPTER 6

MCMC Methods Part I

6.1 Introduction

Monte Carlo methods were introduced in the last chapter. These included basic techniques for generating a random sample and methods for using such a sample to estimate quantities such as difficult integrals. This chapter will focus on advanced techniques for generating a random sample, in particular the class of techniques known as *Markov chain Monte Carlo* (MCMC) methods. Applying an MCMC method involves designing a suitable Markov chain, generating a large sample from that chain for a burn-in period until stochastic convergence, and making appropriate use of the values following that burn-in period.

Like other iterative techniques such as the Newton-Raphson and Expectation-Maximisation algorithms, MCMC methods require an arbitrary starting point (or vector) and then involve iterating repeatedly until convergence. But MCMC methods are distinguished from these other methods by the fact that the update at each iteration is not deterministic but stochastic, with the probability distributions involved dependent on results from the previous iteration.

Typically, MCMC methods are used to sample from *multivariate* probability distributions rather than *univariate* ones. This is because a univariate distribution can usually be sampled from using simpler methods. Nevertheless, we will begin our discussion of MCMC methods with a description of the *Metropolis algorithm* for sampling from univariate distributions, because that algorithm constitutes a basic building block for the more advanced methods.

6.2 The Metropolis algorithm

Suppose that we wish to sample from a univariate distribution with pdf $f(x)$ for which rejection sampling and the other techniques described previously are problematic (say). Then another way to proceed is via the *Metropolis algorithm*. This is an example of *Markov chain Monte Carlo* (MCMC) *methods*. The Metropolis algorithm may be described as follows.

As with the Newton-Raphson algorithm, we begin by specifying an initial value of x, call it x_0. We then also need to specify a suitable *driver distribution* which is easy to sample from, defined by a pdf,
$$g(t \mid x).$$

For now, we will assume the driver to be *symmetric*, in the sense that
$$g(t \mid x) = g(x \mid t),$$
or more precisely,
$$g(t = a \mid \theta = b) = g(t = b \mid \theta = a) \quad \forall \, a, b \in \mathfrak{R}.$$

Note: The driver distribution may also be non-symmetric, but this case will be discussed later.

We then do the following iteratively for each $j = 1,2,3,...,K$ (where K is 'large'):

(a) Generate a *candidate value* of x by sampling $x'_j \sim g(t \mid x_{j-1})$. We call x'_j the *proposed value* and $g(t \mid x_{j-1})$ the *proposal density*.

(b) Calculate the *acceptance probability* as $p = \dfrac{f(x'_j)}{f(x_{j-1})}$.

Note: If $p > 1$ then we take $p = 1$. Also, if x'_j is outside the range of possible values for the random variable x, then $f(x'_j) = 0$ and so $p = 0$.

(c) *Accept* the proposed value x'_j with probability p.
  To determine if x'_j is accepted, generate $u \sim U(0,1)$
  (independently). If $u < p$ then *accept* x'_j, and otherwise *reject* x'_j.

(d) If x'_j has been accepted then let $x_j = x'_j$, and otherwise let
  $x_j = x_{j-1}$ (i.e. *repeat* the last value x_{j-1} in the case of a rejection).

This procedure results in the realisation of a Markov chain, $x_0, x_1, x_2, ..., x_K$. The last value of this chain, x_K, may be taken as an observation from $f(x)$, at least approximately. The approximation will be extremely good if K is sufficiently large.

If we want a random sample of size J from $f(x)$, then the whole procedure can be repeated another $J-1$ times, each time using either the same starting value x_0 or a different one.

If K is sufficiently large, stochastic convergence will be achieved within K iterations, regardless of the point(s) from which the algorithm is started. Relabelling the last value, x_K, in the jth chain as x_j ($j=1,...,J$) leads the required sample, namely $x_1,...,x_J \sim iid\ f(x)$.

Generating a chain of length K a large number times J may be considered wasteful of computer resources. So typically only one long chain is generated, of length $K = B + J$, where B is sufficiently large for stochastic convergence to be achieved from the single starting value, x_0, and J is again the required sample size. Discarding the results of the first B iterations (called the *burn-in*, including also x_0) and relabelling the last J values of the chain appropriately, the result will be the sample $x_1,...,x_J \sim f(x)$.

A problem with this second method of generating the sample values is that they will be *autocorrelated* to some extent i.e. not a truly random (iid) sample from the distribution $f(x)$. We will later discuss this issue and how to deal with the problems that may arise from it. For the moment, we stress that $x_1,...,x_J$ will be *approximately* a random sample from $f(x)$. Moreover, if J is sufficiently large, then these values will be *effectively* independent. This means that a probability histogram of these values will in fact converge to $f(x)$ as J tends to infinity.

Exercise 6.1 A simple application of the Metropolis algorithm

Illustrate the Metropolis algorithm by generating a sample of size 400 from the distribution defined by the density
$$f(x) = 6x^5, 0 < x < 1.$$

Note: This is just the *Beta*(6,1) density and could be sampled from easily in many other ways.

Solution to Exercise 6.1

Let us specify the driver distribution as the uniform distribution from $x-c$ to $x+c$, where c is a tuning parameter whose value is to be determined (as discussed further below). Thus the driver density is

$$g(t \mid x) = \frac{1}{2c}, \, x-c < t < x+c,$$

or equivalently

$$g(t \mid x) = \frac{1}{2c} I(|t-x| < c).$$

Note: This driver is symmetric, since
$$g(t = a \mid x = b) = g(t = b \mid x = a) \quad \forall \, a,b \in \mathfrak{R}.$$

The jth iteration of the algorithm involves first sampling a *candidate value* (or *proposed value*) from the driver distribution centred at the last value, namely

$$x'_j \sim U(x_{j-1} - c, x_{j-1} + c),$$

and then accepting this candidate value with probability

$$p = \frac{f(x = x'_j)}{f(x = x_{j-1})} = \frac{\cancel{6}x'^5_j}{\cancel{6}x^5_{j-1}} = \left(\frac{x'_j}{x_{j-1}} \right)^5, \tag{6.1}$$

where p is taken to be:

0 in the case where $x'_j < 0$ or $x'_j > 1$

1 in the case where $x_{j-1} < x'_j \leq 1$.

Note: The cancellation of 6s in (6.1) illustrates an attractive feature of the Metropolis algorithm generally: only the *kernel* of the sampling density is needed. Here, the kernel of the sampling density $f(x) = 6x^5$ is $k(x) = x^5$. This fact can be very useful in more complicated situations where only the kernel of the sampling density is known.

Starting from $x_0 = 0.1$ and with $c = 0.15$ (arbitrarily), we obtain a Markov chain of length $K = 500$, with values as illustrated in Figure 6.1.

Some of the values of this chain are as follows:

$$x_0, ..., x_{10}, ...,$$

$$x_{301}, ..., x_{310}, ...,$$

$$x_{491}, ..., x_{500} =$$

0.1000, 0.1000, 0.1000, 0.1000, 0.1000, <u>0.1861</u>, 0.2650, 0.2650, 0.4065, 0.4388, 0.4388, ...,

0.9261, 0.9987, 0.9987, 0.9987, 0.9987, 0.9725, 0.8889, 0.8889, 0.9672, 0.9315, ...,

0.8058, 0.6811, 0.6073, 0.4587, 0.4353, 0.3462, 0.3462, 0.4177, 0.4177, 0.4656.

Note: There were four rejections until the first acceptance, at iteration 5, where $x_5 = x_5' = 0.1861$, as underlined above.

Figure 6.2 shows a probability histogram of the last $J = 400$ values, together with the exact density of x. It would appear that stochastic equilibrium has been achieved by about iteration 50. So we may, very conservatively, discard the first $B = 100$ iterations as the burn-in.

The *acceptance rate* (AR) for this Markov chain is found to be 64%, meaning that 320 of the 500 candidate values x_j' were accepted and 36% (or 180) were rejected.

Figure 6.1 Trace of sample values with tuning constant c = 0.15

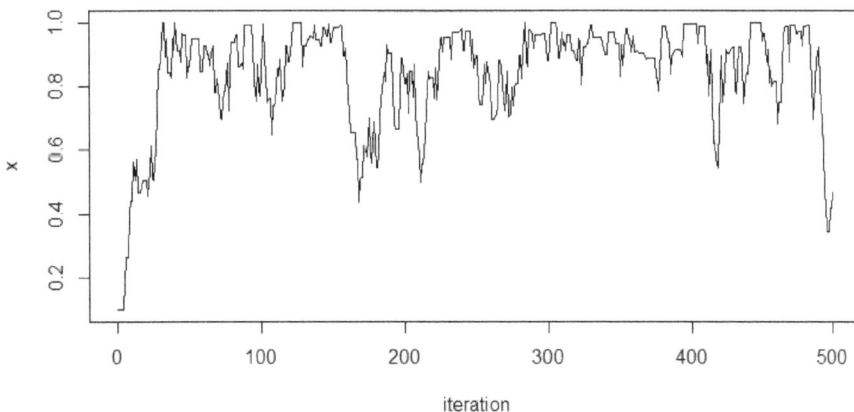

iteration

Figure 6.2 Probability histogram with tuning constant c = 0.15

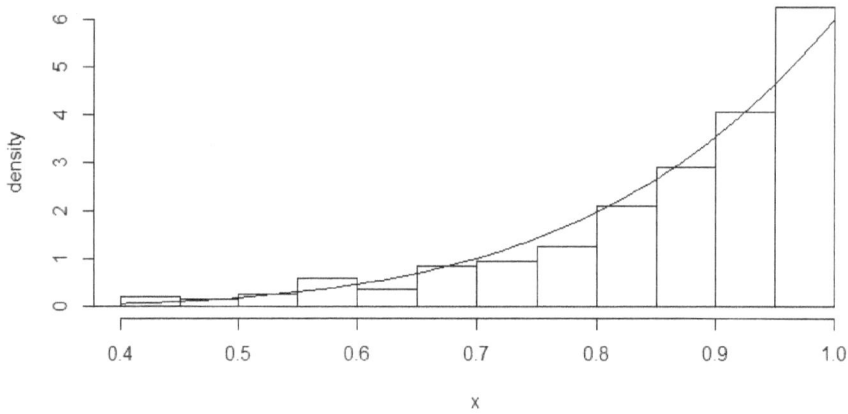

Changing the tuning parameter

What happens if we make the tuning parameter $c = 0.15$ *larger*? Figures 6.3 and 6.4 are a repeat of Figures 6.1 and 6.2, respectively, but using simulated values from a run of the Metropolis algorithm with $c = 0.65$.

In this case the acceptance rate is only 20.8% and the histogram is a poorer estimate of the true density (to which it would however converge as $J \rightarrow \infty$). We say that the algorithm is now displaying *poor mixing* compared to results in the first run of 500 where $c = 0.15$.

What happens if we make $c = 0.15$ *smaller*? Figures 6.5 and 6.6 are a repeat of Figures 6.1 and 6.2, respectively, but using simulated values from a run of the Metropolis algorithm with $c = 0.05$.

In this case the acceptance rate is higher at 83%, there is greater autocorrelation, and the histogram is again a poorer estimate of the true density (to which it would however still converge as $J \rightarrow \infty$). We again say that the algorithm is *mixing poorly*.

It is important to stress that even if the algorithm is mixing poorly (whether this be due to the tuning constant being too large or too small), it will eventually (with a sufficiently large value of J) yield a sample that is useful for inference to the desired degree of precision. Tweaking the tuning constant is merely a device for optimising computational efficiency.

Figure 6.3 Trace of sample values with tuning constant $c = 0.65$

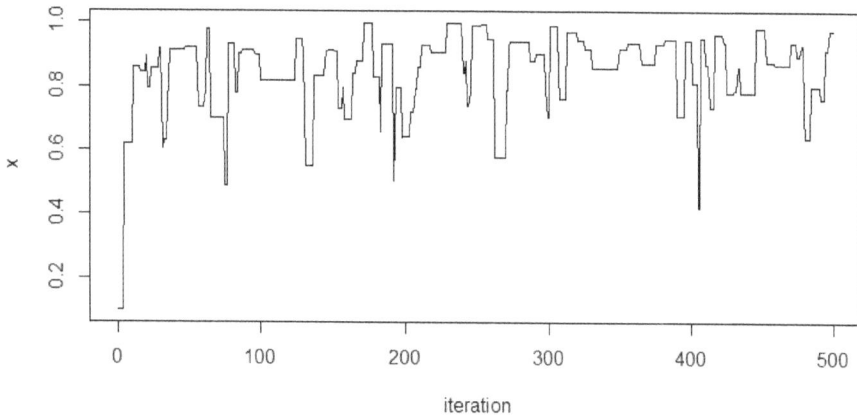

Figure 6.4 Probability histogram with tuning constant $c = 0.65$

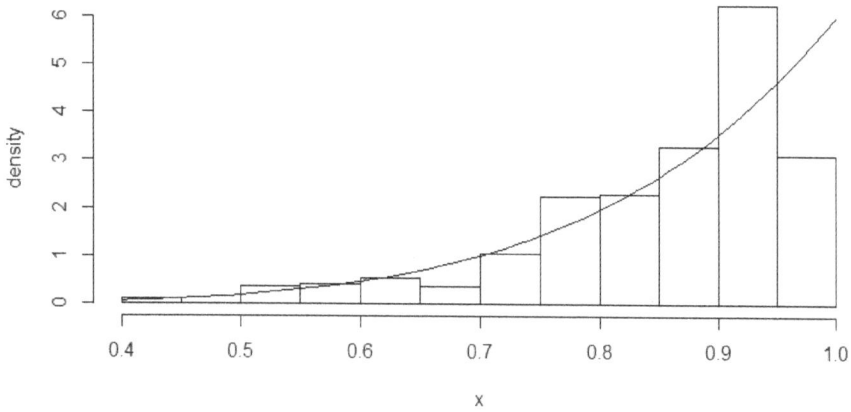

Figure 6.5 Trace of sample values with tuning constant $c = 0.05$

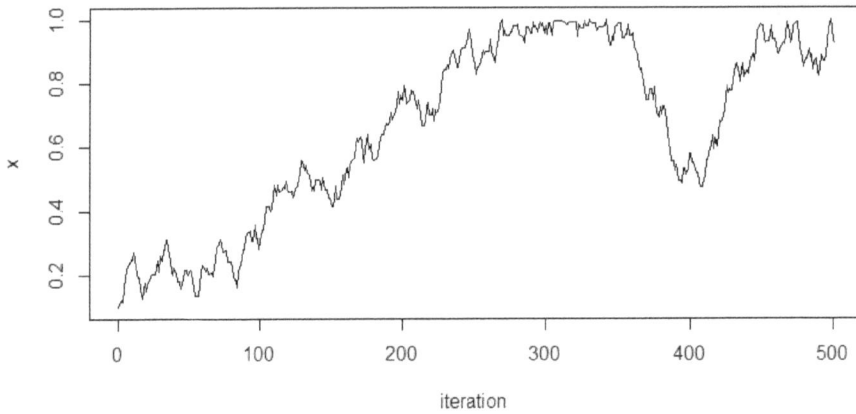

Figure 6.6 Probability histogram with tuning constant $c = 0.05$

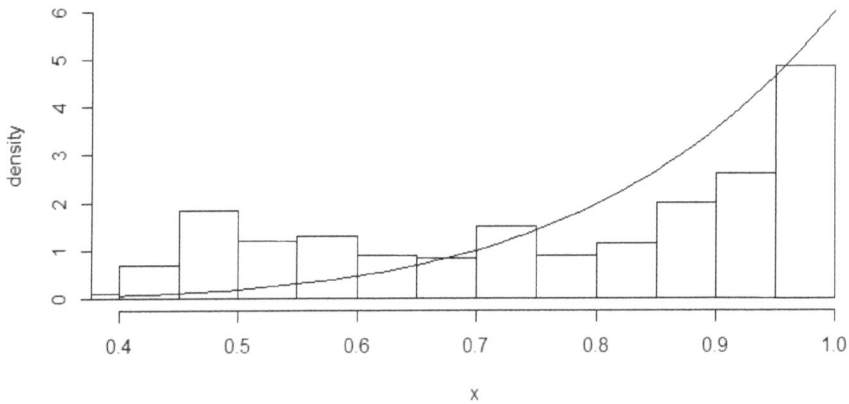

R Code for Exercise 6.1

```
MET <- function(K,x,c){
# This function performs the Metropolis algorithm for a simple model.
# Inputs:      K = total number of iterations
#              x = initial value of x
#              c = tuning parameter.
# Outputs:     $vec = vector of (K+1) values of x
#              $ar = acceptance rate
```

```
vec <- x; ct <- 0
for(j in 1:K){
        prop <- runif(1,x-c,x+c)
        p <- 0
        if((prop>0) && (prop<1))   p <- (prop/x)^5
        u <- runif(1)
        if(u < p){
                x <- prop
                ct <- ct + 1
                }
        vec <- c(vec,x)
        }
ar <- ct/K
list(vec=vec,ar=ar)
}


K <- 500;  X11(w=8,h=4.5); par(mfrow=c(1,1))

set.seed(316); res <- MET(K=K,x=0.1,c=0.15)
plot(0:K,res$vec,type="l",xlab="iteration",ylab="x", main="")

hist(res$vec[-(1:101)],prob=T,xlim=c(0.4,1),ylim=c(0,6),
        xlab="x",ylab="density",main="")
lines(seq(0.4,1,0.01),6*seq(0.4,1,0.01)^5);   res$ar  # 0.64

print(res$vec[1+c(0,1:10,301:310,491:500)], digits=4)
# [1] 0.1000 0.1000 0.1000 0.1000 0.1000 0.1861 0.2650 0.2650 0.4065 0.4388
# [11] 0.4388 0.9261 0.9987 0.9987 0.9987 0.9987 0.9725 0.8889 0.8889 0.9672
# [21] 0.9315 0.8058 0.6811 0.6073 0.4587 0.4353 0.3462 0.3462 0.4177 0.4177
# [31] 0.4656

set.seed(322); res <- MET(K=K,x=0.1,c=0.65)
plot(0:K,res$vec,type="l",xlab="iteration",ylab="x", main=" ")

hist(res$vec[-(1:101)],prob=T,xlim=c(0.4,1),ylim=c(0,6),xlab="x",
        ylab="density", main=" ")
lines(seq(0.4,1,0.01),6*seq(0.4,1,0.01)^5);   res$ar  # 0.208

set.seed(302); res <- MET(K=K,x=0.1,c=0.05)
plot(0:K,res$vec,type="l",xlab="iteration",ylab="x",main=" ")

hist(res$vec[-(1:101)],prob=T,xlim=c(0.4,1),ylim=c(0,6),xlab="x",
        ylab="density", main=" ")
lines(seq(0.4,1,0.01),6*seq(0.4,1,0.01)^5);   res$ar  # 0.83
```

Exercise 6.2 Sampling from a normal distribution via the Metropolis algorithm

Use the Metropolis algorithm and a uniform driver to sample 10,000 values from the standard normal distribution.

Check your result by comparing the sample mean and sample standard deviation of your sample to the true theoretical values, 0 and 1.

Calculate a Monte Carlo 95% confidence interval for the normal mean, 0.

Solution to Exercise 6.2

Since $f(x) \propto e^{-\frac{1}{2}x^2}$, the acceptance probability at iteration j is given by

$$p = \frac{f(x = x'_j)}{f(x = x'_{j-1})} = \frac{\frac{1}{\sqrt{2\pi}} e^{-\frac{1}{2}(x'_j)^2}}{\frac{1}{\sqrt{2\pi}} e^{-\frac{1}{2}(x'_{j-1})^2}} = \exp\left\{\frac{\left(x_{j-1}\right)^2 - \left(x'_j\right)^2}{2}\right\}.$$

Using the same uniform driver as in Exercise 6.1, $x_0 = 5$ and $c = 2.5$ (where this tuning constant was chosen after some experimentation), we obtain a Markov chain of length $K = 10,500$, as shown in Figure 6.7.

Figure 6.8 shows a histogram of the last $J = 10,000$ values, together with the standard normal density overlaid.

We have very conservatively discarded the first $B = 500$ iterations as the burn-in. The acceptance rate for this Markov chain is 56.1%.

The average of the J sampled values is 0.0355 (close to 0) and their sample standard deviation is 1.0047 (close to 1). These values lead to a 95% CI for the normal mean equal to (0.0158, 0.0552). We note that this CI does *not* contain the true value, 0, as one might expect. The underlying issue behind this fact will be discussed generally in the next section.

Figure 6.7 Trace of sample values

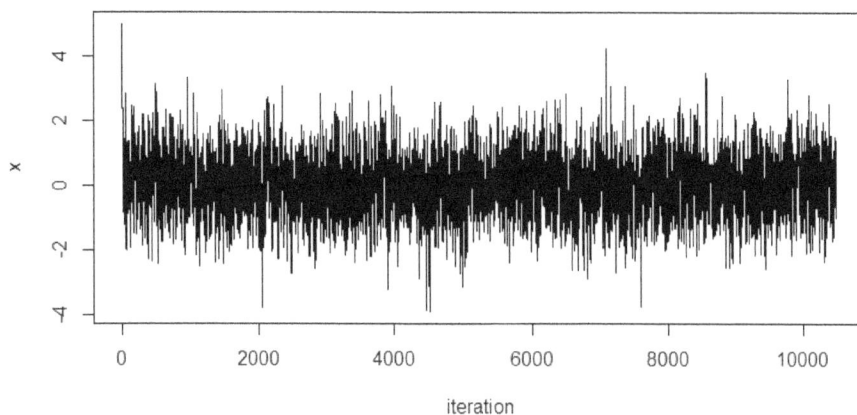

Figure 6.8 Probability histogram

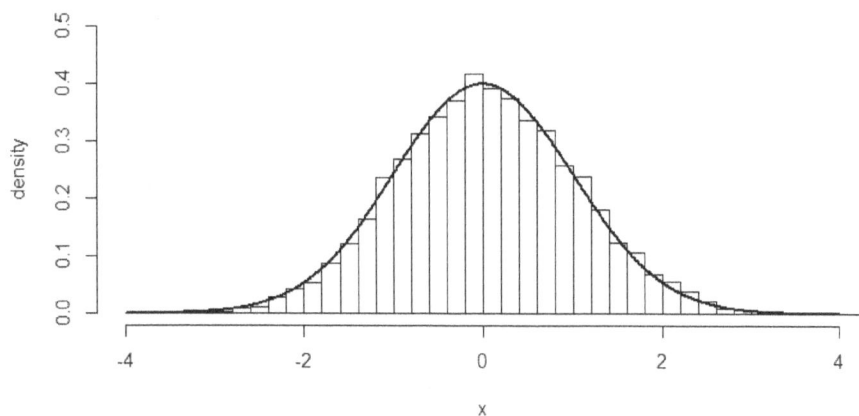

R Code for Exercise 6.2

```
MET <- function(K,x,c){
# This function performs the Metropolis algorithm to sample from the
# standard normal dsn.
# Inputs:       K = total number of iterations
#               x = initial value of x
#               c = tuning parameter.
# Outputs:      $vec = vector of (K+1) values of x
#               $ar = acceptance rate.
```

```
vec = x; ct = 0
for(j in 1:K){    prop = runif(1,x-c,x+c)
                  p = exp(-0.5*(prop^2-x^2));        u = runif(1)
                  if(u <= p){ x = prop;      ct = ct + 1        }
                  vec <- c(vec,x)   }
ar = ct/K;   list(vec=vec,ar=ar)   }


B=500; J = 10000; K = B + J
set.seed(117); res <- MET(K=K,x=5,c=2.5);   res$ar   # 0.548381
X11(w=8,h=4.5); par(mfrow=c(1,1))

plot(0:K,res$vec,type="l",xlab="iteration",ylab="x",main=" ")

hist(res$vec[-(1:(B+1))],prob=T,xlim=c(-4,4),ylim=c(0,0.5),xlab="x",
        ylab="density",nclass=50, main=" ")
lines(seq(-4,4,0.01),dnorm(seq(-4,4,0.01)),lwd=2)
est=mean(res$vec[-(1:(B+1))]); std=sd(res$vec[-(1:(B+1))])
ci=est+c(-1,1)*qnorm(0.975)*std/sqrt(10000)
c(est,std,ci) # 0.03550254 1.00470749 0.01581064 0.05519445
```

6.3 The batch means method

As stated earlier, the output from the Metropolis algorithm leads to a sample, $x_1, ..., x_J$, from the target density, $f(x)$, which exhibits some degree of positive *autocorrelation*.

This does not present a major problem when one is interested in calculating only *point* estimates. For example, if we wish to estimate the distribution mean $EX = \int xf(x)dx$, each sample value x_j has expected value EX, and this is true regardless of how severely the simulated values are correlated (assuming that all the simulated values are collected *after* stochastic convergence). Therefore, the expected value of the Monte Carlo mean is also exactly EX (or very nearly so).

However, when one uses a severely and positively autocorrelated Monte Carlo sample to calculate the standard $1 - \alpha$ confidence interval for a quantity such as EX, the true coverage probability of that interval may be far *less* than the intended nominal value of $1 - \alpha$.

One way of dealing with this problem is to generate J independent chains and take the last value in each chain. Note that this was our original formulation of the Metropolis algorithm (i.e. for sampling a single value).

274

Another option is to generate a single long chain, of length $K = B + 10J$ (say) and thin it out by recording only every 10th value in the chain after burn-in. Even so, there will still be some autocorrelation remaining in the J resulting values. The autocorrelation could be reduced further by changing 10 to 100, say; but this would be at the cost of a 10-fold increase in computer time needed.

A more efficient solution to the autocorrelation problem is the *batch means method*. We will now describe how this works for when we wish to construct a $1 - \alpha$ CI for $EX = \int x f(x) dx$ based on an autocorrelated sample $x_1, ..., x_J \sim iid\ f(x)$.

The *batch means CI* will be different from the *ordinary CI*, namely $(\bar{x} \pm 1.96 s_x / \sqrt{J})$, where $\bar{x}$ and s_x are the sample mean and sample standard deviation of $x_1, ..., x_J$. The batch means CI is obtained as follows.

First, break up the J sample values into m batches of size n each, so that:

Batch 1 contains values $1, ..., n$ (the first n values)

Batch 2 contains values $n + 1, ..., 2n$ (the next n values)

....

Batch m contains values $(m - 1)n + 1, ..., J$ (the last n values).

Next: Let y_k be the mean of the n x_j-values in the kth batch ($k = 1, ..., m$).

Let s_y^2 be the sample variance of $y_1, ..., y_m$.

Note: Thus $s_y^2 = \dfrac{1}{m-1} \sum_{k=1}^{m} (y_k - \bar{y})^2$, where $\bar{y} = \dfrac{1}{m} \sum_{k=1}^{m} y_k = \bar{x}$ is the mean of the batch means and identical to the mean of all J x_j-values.

Finally, compute the $1 - \alpha$ batch means CI for EX as $(\bar{x} \pm 1.96 s_y / \sqrt{m})$.

Discussion

The rationale for the batch means method is as follows. If the batch size n is sufficiently large then, by the central limit theorem,

$$y_1,..., y_m \sim iid\ N(\mu, \sigma^2 / n),$$

where $\mu = E(x_j)$ and $\sigma^2 = Var(x_j)$.

Consequently,

$$\bar{y} \sim N\left(\mu, \frac{\sigma^2 / n}{m}\right) \sim N\left(\mu, \frac{\sigma^2}{J}\right),$$

since $J = mn$.

Therefore a $1 - \alpha$ CI for μ is

$$(\bar{y} \pm z_{\alpha/2} r / \sqrt{J}),$$

where r is an estimate of σ.

Now, an unbiased estimator of σ^2 / n is s_y^2.

So an unbiased estimator of σ^2 is ns_y^2.

It follows that a $1 - \alpha$ CI for μ is

$$(\bar{x} \pm z_{\alpha/2} \sqrt{ns_y^2} / \sqrt{J}) = (\bar{x} \pm z_{\alpha/2} s_y / \sqrt{m}).$$

Exercise 6.3 Testing the batch means method

We wish to perform Monte Carlo estimation of the expected value of X whose pdf is given by $f(x) \propto x^2, 0 < x < 2$.

Note: Here, $X \sim 2Beta(3,1)$ and so $EX = 2 \times 3 / (3+1) = 1.5$.

(a) Use the Metropolis algorithm to generate a sample of size $J = 1,000$ from X's distribution after a burn-in of 100.

Then use this sample to estimate EX, together with a 95% confidence interval for EX. For this CI use the formula $(\bar{x} \pm 1.96 s / \sqrt{J})$, where s^2 is the sample variance of the J sampled X-values. Also draw a histogram of the J X-values overlaid with the exact pdf of X.

(b) Use the output from the Metropolis algorithm in (a) to construct another 95% CI for EX, one using the *batch means method*, as follows:

> Divide the $J = 1,000$ iterations into $m = 20$ consecutive batches, each having $n = 50$ values of X.

> Let y_k be the average of the n X-values in the kth batch $(k = 1,...,m)$.

> Let s_y^2 be the sample variance of the m batch means $y_1,..., y_m$.

> Let the confidence interval for EX be $(\bar{x} \pm 1.96 s_y / \sqrt{m})$.

(c) Conduct a Monte Carlo experiment to assess the quality of the two CIs for EX in (a) and (b).

Do this by implementing the following three-step procedure a total of $R = 100$ times:

> **(i)** Run the Metropolis algorithm in (a) so as to generate $J = 1,000$ observations from $f(x)$.

> **(ii)** Calculate the CI in (a) and count 1 if 1.5 is in it.

> **(iii)** Calculate the CI in (b) and count 1 if 1.5 is in it.

Now divide the total count from (ii) by R to get an unbiased point estimate of the probability that the ordinary CI for EX in (a) contains EX.

Similarly, divide the two total count from (iii) by R to get an unbiased point estimate of the probability that the batch means CI for EX in (b) contains EX.

Also produce 95% CIs for the two probabilities just mentioned.

(d) Repeat the experiment in (c) but with the following in place of (i):

> Generate $J = 1,000$ observations from X's distribution using the rbeta() function.

Solution to Exercise 6.3

(a) Let us specify a uniform driver centred at the last value and with half-width h. We now iterate as follows after choosing a suitable starting value of x:

Sample $x' \sim U(x-h, x+h)$.

If x' is outside the interval $(0,2)$ then automatically reject x'.

Otherwise accept x' with probability $\min(1, p)$, where
$$p = x'^2 / x^2.$$

Starting from $x = 1$ with $h = 0.7$, we get an acceptance rate of 55% and simulated values as depicted in Figures 6.9 and 6.10.

Taking the last 1,000 values of x as a random sample from $f(x)$ we estimate EX as 1.539, with ordinary 95% CI (1.467, 1.611). We note that this CI does *not* contain the true value, 1.5.

Figure 6.9 Trace of sample values

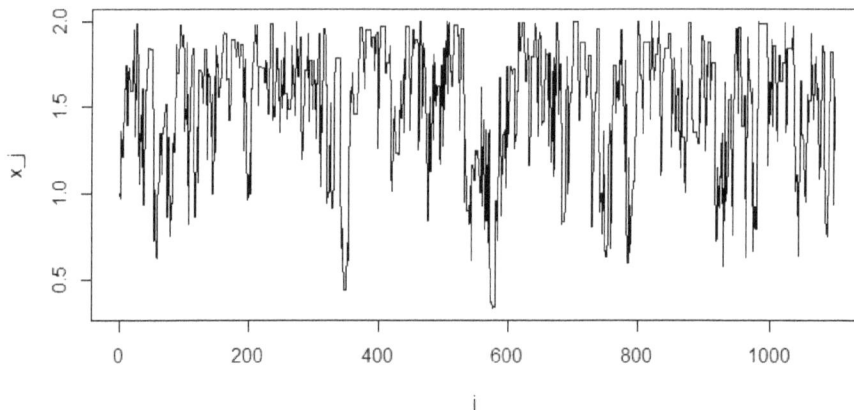

Figure 6.10 Histogram of sample values

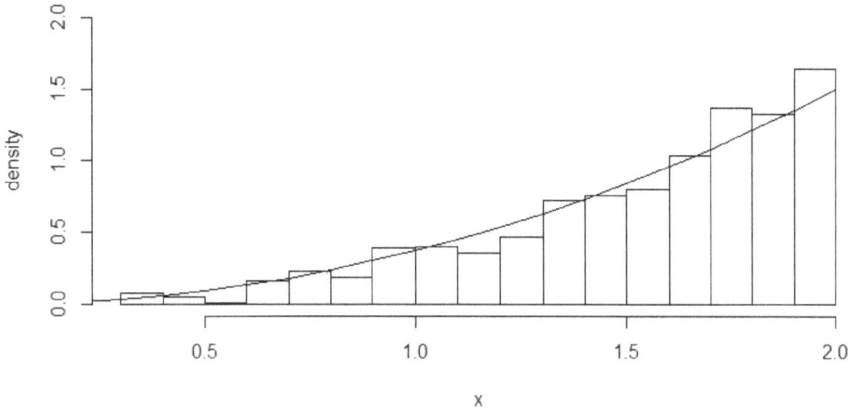

(b) Applying the batch means method with $m = 20$ and $n = 50$, we estimate *EX* as 1.539 again, but with 95% CI (1.467, 1.611). Note that this CI is *wider* than the CI in (a) and *does* contain the true value, 1.5.

(c) After conducting the experiment we estimate p_1, the true probability content of the *ordinary* 95% CI in (a), as 52.0%, with 95% CI 42.2% to 61.8%.

We also estimate p_2, the true probability content of the *batch means* 95% CI in (b) (with $m = 20$ and $n = 50$), as 90.0%, with 95% CI 84.1% to 95.9%.

We see that in this example the batch means method has performed far better than the ordinary method for constructing 95% CIs for *EX* from the output of a Metropolis algorithm.

(d) Generating each value of X as twice a random number from the *Beta*(3,1) distribution, we estimate p_1 by 92.0%, with 95% CI 86.7% to 97.3%. We also estimate p_2 by 90.0%, with 95% CI 84.1% to 95.9%.

We see that the two CIs have performed about equally well when calculated using a *truly* random sample from X's distribution. In such situations, the batch means CI is in fact slightly inferior and the ordinary CI should be used.

279

R Code for Exercise 6.3

```
# (a)
MET <- function(Jp,x,h){
# This function implements a simple Metropolis algorithm.
# Inputs:        Jp = total number of iterations
#                x = starting value of x
#                h = halfwidth of uniform driver.
# Outputs:       $xv  = vector of x-values of length (Jp + 1)
#                $ar = acceptance rate.
xv <- x; ct <- 0
for(j in 1:Jp){    xprop <- runif(1,x-h,x+h)
                   if( (xprop>0) && (xprop<2) ){
                          p <- xprop^2 / x^2;  u <- runif(1)
                          if(u < p){ x <- xprop; ct <- ct + 1 } }
                   xv <- c(xv,x)   }
list(xv=xv,ar=ct/Jp)   }


Jp <- 1100; set.seed(151); res <- MET(Jp=Jp,x=1,h=0.7); res$ar   # 0.5454545

X11(w=8, h=4.5); par(mfrow=c(1,1));

plot(0:Jp,res$x,type="l",xlab="j",ylab="x_j")
xv <- res$xv[-c(1:101)]; J= length(xv)

hist(xv,xlab="x",prob=T,ylim=c(0,2),nclass=20,ylab="density", main="")
xvec <- seq(0,2,0.1); fvec <- (3/8)*xvec^2; lines(xvec,fvec)

EXhat <- mean(xv); sdhat <- sqrt(var(xv)); sdhat   # 0.3755086
EXci <- EXhat + c(-1,1)*qnorm(0.975)*sdhat/sqrt(J)
c(EXhat,EXci)   # 1.538984 1.515710 1.562258


# (b)
m <- 20; n <- 50; yv <- rep(NA,m)
for(k in 1:m){              xvsub <- xv[ ((k-1)*n+1):(k*n)  ]
                            yv[k] <- mean(xvsub)   }
sdhat2 <- sqrt(n*var(yv)); sdhat2   # 1.15783
EXci <- EXhat + c(-1,1)*qnorm(0.975)*sdhat2/sqrt(J)
c(EXhat,EXci)   # 1.538984 1.467222 1.610746


# (c)
R<- 100; m <- 20; n <- 50; J <- 1000; burn <- 100; EX <- 1.5; ct1 <- 0; ct2 <- 0;
yv <- rep(NA,m); set.seed(214)
```

```
for(r in 1:R){
        xv <- MET(Jp=burn+J,x=1,h=0.7)$xv[-c(1:101)]
        # xv <- rbeta(J,3,1)*2    # for use in (d) (see below)
        for(k in 1:m){              xvsub <- xv[ ((k-1)*n+1):(k*n)  ]
                                    yv[k] <- mean(xvsub)  }
        EXhat <- mean(xv); sdhat1 <- sqrt(var(xv)); sdhat2 <- sqrt(n*var(yv))
        ci1 <- EXhat + c(-1,1)*qnorm(0.975)*sdhat1/sqrt(J)
        ci2 <- EXhat + c(-1,1)*qnorm(0.975)*sdhat2/sqrt(J)
        if( (EX >= ci1[1]) && (EX <= ci1[2])) ct1 <- ct1 + 1
        if( (EX >= ci2[1]) && (EX <= ci2[2])) ct2 <- ct2 + 1  }
date() # took 2 secs

p1 <- ct1/R; p2 <- ct2/R
p1ci <- p1 + c(-1,1)*qnorm(0.975)*sqrt(p1*(1-p1)/R)
p2ci <- p2 + c(-1,1)*qnorm(0.975)*sqrt(p2*(1-p2)/R)
c(p1,p1ci)  # 0.5200000 0.4220802 0.6179198
c(p2,p2ci)  # 0.9000000 0.8412011 0.9587989

# (d)
# Repeat code in (c) but with the line
# "xv  <- MET(Jp=burn+J,x=1,h=0.7)$xv[-c(1:101)]"
# replaced by the line "xv  <- rbeta(J,3,1)*2".

# The results should be:
#       c(p1,p1ci)  # 0.9200000 0.8668275 0.9731725
#       c(p2,p2ci)  # 0.9000000 0.8412011 0.9587989
```

Exercise 6.4 Bayesian inference via the Metropolis algorithm

The prior on a normal mean μ is uniform from zero to infinity. Values are sampled repeatedly from the $N(\mu,1)$ distribution until $n = 4$ positive values have been observed, resulting in the data: 0.1, 0.2, 1.9, 0.8.

Find the posterior mean of μ in the following ways:

(a) exactly, using numerical integration in R

(b) approximately, using a Monte Carlo method that does *not* involve Markov chains

(c) approximately, using the Metropolis algorithm with a normal driver.

Solution to Exercise 6.4

(a) The posterior density of μ is

$$f(\mu \mid y) \propto f(\mu)f(y \mid \mu) \propto 1 \times \prod_{i=1}^{n} \frac{e^{-\frac{1}{2}(y_i - \mu)^2}}{1 - \Phi(-\mu)},$$

since $P(y > 0 \mid \mu) = 1 - P\left(z < \dfrac{0 - \mu}{1}\right) = 1 - \Phi(-\mu)$.

Thus $f(\mu \mid y) \propto \left(1 - \Phi(-\mu)\right)^{-n} \exp\left(-\dfrac{1}{2} \sum_{i=1}^{n} (y_i - \mu)^2\right)$

$$= \left(1 - \Phi(-\mu)\right)^{-n} \exp\left(-\dfrac{1}{2}\left[(n-1)s^2 + n(\bar{y} - \mu)^2\right]\right)$$

$$\propto \left(1 - \Phi(-\mu)\right)^{-n} \exp\left(-\dfrac{1}{2} n(\mu - \bar{y})^2\right)$$

$$\equiv k(\mu), \quad \mu > 0 \quad \text{(this is the \textit{kernel} of the posterior density).}$$

Thus $\hat{\mu} = E(\mu \mid y) = \dfrac{\displaystyle\int_0^\infty \mu^1 k(\mu)d\mu}{\displaystyle\int_0^\infty \mu^0 k(\mu)d\mu} = \dfrac{I_1}{I_0}$,

where $I_q = \displaystyle\int_0^\infty \mu^q k(\mu)d\mu$, $q = 0,1$.

Using integrate() in R we obtain $I_0 = 4.328041$, $I_1 = 2.328058$ and hence $\hat{\mu} = 0.5379$.

(b) Observe that $\hat{\mu} = \dfrac{\displaystyle\int_0^\infty \mu^1 \left(1 - \Phi(-\mu)\right)^{-n} h(\mu)d\mu}{\displaystyle\int_0^\infty \mu^0 \left(1 - \Phi(-\mu)\right)^{-n} h(\mu)d\mu}$,

where $h(\mu) = \dfrac{\dfrac{\sqrt{n}}{\sqrt{2\pi}} \exp\left(-\dfrac{n}{2}(\mu - \bar{y})^2\right)}{1 - \Phi\left(\dfrac{0 - \bar{y}}{1/\sqrt{n}}\right)}$.

Note: $h(\mu)$ is the density of the $N(\bar{y}, 1/n)$ distribution restricted to the positive real line.

Thus $\hat{\mu} = \dfrac{E_1}{E_0}$, where: $E_q = E\left\{\mu^q \left(1 - \Phi(-\mu)\right)^{-n}\right\}$, $q = 0, 1$

$$\mu \sim h(\mu) \sim N(\bar{y}, 1/n)I(\mu > 0).$$

Note: At this point we 'forget' about the posterior distribution of μ.

We see that a non-Markov chain Monte Carlo estimate of $\hat{\mu}$ is

$$\tilde{\mu} = \frac{\tilde{E}_1}{\tilde{E}_0},$$

where: $\tilde{E}_q = \dfrac{1}{J}\sum_{j=1}^{J} \mu_j^q \left(1 - \Phi(-\mu_j)\right)^{-n}$

$$\mu_1, ..., \mu_J \sim iid\ h(\mu).$$

Note: To obtain the required sample here, we repeatedly sample $\mu \sim N(\bar{y}, 1/n)$ until J positive values have been achieved.

Implementing this strategy in R using the rnorm() function with a Monte Carlo sample size of J = 100,000, we obtain $\tilde{E}_0$ = 3.7059926, $\tilde{E}_1$ = 1.9900593 and hence $\tilde{\mu}$ = 0.5370.

(c) Using the Metropolis algorithm and a normal driver distribution with standard deviation 0.5, we obtain a Markov chain of size 10,000 following a burn-in of size 100. The acceptance rate is found to be 59%.

Then taking every 10th value results in a very nearly uncorrelated sample of size 1,000 from the posterior distribution of μ. Using these 1,000 values, leads to the estimate $\hat{\mu}$ by 0.5297, with associated 95% CI equal to (0.5047, 0.5547).

We note that the true exact value calculated in (a), 0.5379, is contained in this CI.

R Code for Exercise 6.4

```
# (a)
y=c(0.1, 0.2, 1.9, 0.8); n = length(y); ybar=mean(y); c(n,ybar) # 4.00 0.75
kfun=function(mu){    exp(-0.5*n*(mu-ybar)^2) / (1-pnorm(-mu))^n    }
topfun=function(mu){ mu * kfun(mu) }
par(mfrow=c(2,1)); muvec=seq(0,5,0.1)
plot(muvec,kfun(muvec),type="l"); abline(h=0,lty=3)    # OK
plot(muvec,topfun(muvec),type="l"); abline(h=0,lty=3)   # OK
top=integrate(f=topfun,lower=0,upper=5)$value
bot=integrate(f=kfun,lower=0,upper=5)$value
c(bot,top,top/bot) # 4.328041 2.328058 0.537901

# (b)
J=110000; set.seed(551); samp=rnorm(J,ybar,1/sqrt(n))
samppos=samp[samp>0]; length(samppos) # 102763
samppos=samppos[1:100000]
numer=mean(samppos*(1-pnorm(-samppos))^(-n) )
denom=mean( (1-pnorm(-samppos))^(-n) )
c(numer,denom,numer/denom) # 1.9900593 3.7059926 0.5369842

# (c)
MET <- function(K,mu,del,y){
# This function implements a simple Metropolis algorithm.
# Inputs:       K = total number of iterations
#               mu = starting value of mu
#               del = standard deviation of normal driver
#               y = data vector
# Outputs:      $muv  = vector of mu-values of length (K + 1)
#               $ar = acceptance rate
muv = mu; ct = 0; n=length(y); ybar=mean(y)
kfun=function(mu,ybar,n){    exp(-0.5*n*(mu-ybar)^2) / (1-pnorm(-mu))^n    }
for(j in 1:K){    muprop = rnorm(1,mu,del)
                if( muprop>0 ){
        p=kfun(mu=muprop,ybar=ybar,n=n)/kfun(mu=mu,ybar=ybar,n=n)
                        u=runif(1); if(u < p){ mu = muprop; ct = ct + 1 }    }
                muv = c(muv,mu)   }
list(muv=muv,ar=ct/K)   }

K=10100; set.seed(352); res= MET(K=K,mu=1,del=0.5,y=y)
res$ar # 0.590297
mean(res$muv)  # 0.5303868  = preliminary estimate

plot(0:K,res$muv,type="l")
```

```
vec1=res$muv[-(1:101)]
print(acf(vec1)$acf[1:10],digits=2) # Evidence of strong autocorrelation
        # 1.00 0.78 0.61 0.48 0.39 0.30 0.24 0.19 0.14 0.11

v=vec1[seq(10,10000,10)]  # Take every 10th value only
print(acf(v)$acf[1:10],digits=2) #  No apparent residual autocorrelation
# 1.0000  0.0534  0.0014  0.0331 -0.0089 -0.0041 0.0034  0.0087 0.0102  0.0133

J=length(v); J # 1000
est=mean(v); std=sd(v); ci=est+c(-1,1)*qnorm(0.975)*std/sqrt(J)
c(est,std,ci) #  0.5296887 0.4039238 0.5046537 0.5547237
```

6.4 Computational issues

Numerical issues may arise when attempting to calculate the acceptance probability

$$p = f(x'_j) / f(x_{j-1})$$

due to $f(x'_j)$ or $f(x_{j-1})$ being too large or too small for R to handle.

One relevant fact here is that in R on most computers (at present), 5e-324 (meaning 5×10^{-324}) is the smallest representable non-zero number. This problem can often be resolved by calculating p as

$$p = \exp(q)$$

after first computing

$$q = \log f(x'_j) - \log f(x_{j-1}),$$

but even this formulation may not be sufficient in every situation.

It may sometimes also be necessary to replace the calculation of a function, say $h(r)$, by

$$h(\max(r, 5e - 324))$$

if that function requires a non-zero argument r which is likely to be reported by R as 0 (because the exact value of r is likely to be between 0 and $5e - 324$).

Further, and by the same token, if

$$0 < h(\max(r, 5e - 324)) < 5e - 324$$

then R will report a value of 0. In that case, if a non-zero value of h is absolutely required (for some subsequent calculation) then the code for $h(r)$ should be replaced by code which returns

$$\max(h(\max(r, 5e - 324)), 5e - 324).$$

6.5 Non-symmetric drivers and the general Metropolis algorithm

In some cases, applying the Metropolis algorithm as described above may lead to poor mixing, even after experimentation to decide on the most suitable value of the tuning constant.

For example, if the random variable of interest is strictly positive with a pdf $f(x)$ which is positively skewed and highly concentrated just above 0 (for example, if $f(x) \to \infty$ as $x \downarrow 0$), proposing a value *symmetrically* distributed around the last value may lead to many candidate values which are *negative* and therefore *automatically rejected*.

In such cases, the support of X may not be properly represented, and it may be preferable to choose a different type of driver distribution, one which adapts 'cleverly' to the current state of the Markov chain.

This can be achieved using the *general Metropolis algorithm* which allows for non-symmetric driver distributions. As before, let $g(t \mid x)$ denote a driver density, where t denotes the proposed value and x is the last value in the chain. Then at iteration j, after generating a proposed value from the driver distribution,
$$x'_j \sim g(t \mid x = x_{j-1}),$$
the *acceptance probability* is
$$p = \frac{f(x'_j)}{f(x_{j-1})} \times \frac{g(x_{j-1} \mid x'_j)}{g(x'_j \mid x_{j-1})}.$$

Note 1: Previously, when $g(t \mid x)$ was assumed to be symmetric,
$$\frac{g(x_{j-1} \mid x'_j)}{g(x'_j \mid x_{j-1})} = 1.$$

Note 2: To calculate p, the best strategy is to let
$$p = \exp(q)$$
after first computing
$$q = \log f(x'_j) - \log f(x_{j-1})$$
$$+ \log g(x_{j-1} \mid x'_j) - \log g(x'_j \mid x_{j-1}).$$

Exercise 6.5 A Metropolis algorithm with a non-symmetric driver

Generate a random sample of size 10,000 from the distribution defined by the pdf

$$f(x) = \begin{cases} \dfrac{1}{4}x^{-1/2}, & 0 < x < 1 \\[2mm] \dfrac{1}{2}e^{-(x-1)}, & x > 1 \end{cases}$$

using the Metropolis algorithm and a non-symmetric driver with density of the form

$$g(t \mid x) = f_{G(x\delta,\delta)}(t) = \frac{\delta^{x\delta} t^{x\delta-1} e^{-x\delta}}{\Gamma(x\delta)}, t > 0,$$

or equivalently, a driver defined by

$$(t \mid x) \sim G(x\delta, \delta).$$

Check your results by plotting a probability histogram of the sample values and overlaying the target density, $f(x)$. Also discuss why this driver is suitable in this situation.

Solution to Exercise 6.5

At each iteration j the proposed value is generated by sampling

$$x'_j \sim G(x_{j-1}\delta, \delta).$$

The rationale for this choice of driver is that the proposed value is certainly *positive*, it has:

mean $\quad x_{j-1}\delta / \delta = x_{j-1}$

variance $\quad x_{j-1}\delta / \delta^2 = x_{j-1} / \delta$.

Thus the candidate x'_j is *guaranteed* to be in the appropriate range ($\Re^+$), and it is *centred* at the last value (x_{j-1}).

Also, its variance around that last value is *proportional* to it (by a factor of $1/\delta$). This ensures that values near zero are appropriately 'explored' by the Markov chain.

With this driver, the acceptance probability at iteration j is

$p = \exp(q)$,

where:

$q = \log f(x_j') - \log f(x_{j-1}) + \log g(x_{j-1} | x_j') - \log g(x_j' | x_{j-1})$.

$\log f(x) = I(0 < x < 1)\{-0.5\log x - \log 4\} + I(x > 1)\{1 - x - \log 2\}$

$\log g(t | x) = x\delta \log \delta + (x\delta - 1)\log t - x\delta - \log \Gamma(x\delta)$.

Even with this use of the logarithmic function, computational issues arose in R on account of limitations with the functions rgamma() and lgamma(). These limitations are acknowledged in the help files for these functions in R.

To give an example:

```
set.seed(321)
v = rgamma(10000,0.001,0.001)
             # Large sample from the G(0.001,0.001) distribution.
mean(v) # 0.5827886
             # This is clearly wrong since the mean is 0.001/0.001 = 1.
length(v[v==0]) # 4777
             # Almost HALF of the values are EXACTLY zero.
```

The R code was appropriately modified so that whenever very small but non-zero values were reported as zero by R (and problems ensued or potentially ensued because of this) those values were changed in the code to 5e-324 (the smallest representable non-zero number in R).

With the above specification and fixes, the Metropolis algorithm was run for 10,000 iterations following a burn-in of size 100 and starting at 1. The value of δ used was 1.3 and this resulted in an acceptance rate of 53% as well as good mixing. Figure 6.11 shows the resulting trace of all 10,101 values of x, and Figure 6.12 shows the required probability histogram of the last 10,000 values, together with the exact density $f(x)$ overlaid.

Note: Applying a gamma driver here (in an attempt to improve the 'vanilla' version of the Metropolis algorithm) created problems, due to numerical issues in R associated with the gamma distribution. With some modifications, we were in the end able to make things work. Another choice of nonsymmetric driver distribution is the lognormal, and we leave it as an additional exercise to examine this option in detail.

Figure 6.11 Trace of simulated values

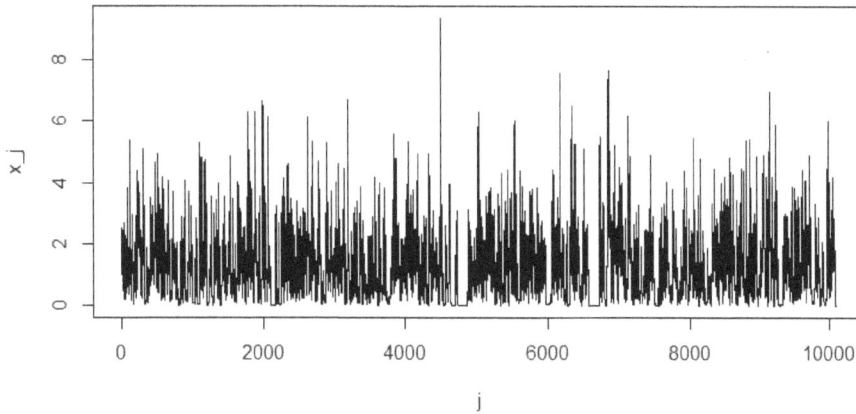

Figure 6.12 Histogram and true density

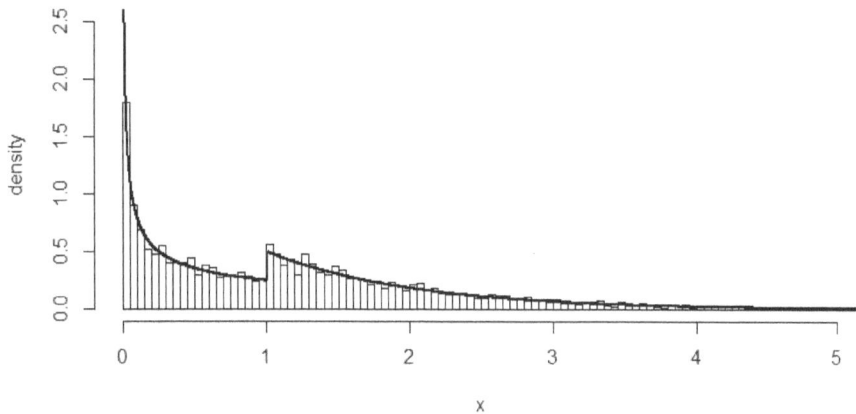

R Code for Exercise 6.5

```
set.seed(321);    v = rgamma(10000,0.001,0.001)
        # Large sample from the G(0.001,0.001) distribution.
mean(v) # 0.5827886   This is clearly wrong since the mean is 0.001/0.001 = 1.
length(v[v==0]) # 4777   Almost HALF of the values are EXACTLY zero.
logffun=function(x){ res=-0.5*log(x)-log(4); if(x>1) res=1-x-log(2); res }
loggfun=function(t,x,del){
    x*del*log(del)+(x*del-1)*log(t)-t*del-lgamma(max( x*del, 5e-324 )) }
```

```
MET <- function(K,x,del){  # This function implements a simple Metropolis alg.
# Inputs: K = total number of iterations, x = starting value of x,
# del = tuning constant in driver
# Outputs: $xv = vector of x-values of length (K + 1),  $ar = acceptance rate
xv = x;   ct = 0
for(j in 1:K){   xp = max(  rgamma(1,x*del,del),   5e-324  )
                 logp = logffun(x=xp) - logffun(x=x) +
                        loggfun(t=x,x=xp,del=del) - loggfun(t=xp,x=x,del=del)
                 p = exp(logp);   u = runif(1);   if(u < p){ x = xp; ct = ct + 1 }
                 xv = c(xv,x)   }
list(xv=xv,ar=ct/K)   }


X11(w=8,h=4.5); par(mfrow=c(1,1)); K = 10100;
set.seed(319); res = MET(K=K,x=1,del=1.3); res$ar   # 0.5324752
plot(0:K,res$xv,type="l",xlab="j",ylab="x_j")
xv <- res$xv[-c(1:101)]


hist(xv,xlab="x",prob=T,ylim=c(0,2.5),xlim=c(0,5), ylab="density", main="",
   breaks=seq(0,20,0.05)   )
xvec=seq(0,10,0.001); fvec=xvec;
for(i in 1:length(xvec)) fvec[i]=exp(logffun(xvec[i]))
lines(xvec,fvec,lwd=2)


summary(res$xv)
#  Min.  1st Qu.  Median   Mean 3rd Qu.    Max.
#  0.004243 0.309400 1.034000 1.218000 1.738000 9.356000  (OK, as Min > 0)
```

6.6 The Metropolis-Hastings algorithm

We have introduced Markov chain Monte Carlo methods with a detailed discussion of the Metropolis algorithm. As already noted, this algorithm is limited and rarely used on its own because it can only be used to sample from *univariate* distributions. Typically, other methods will be better suited to the task of sampling from a univariate distribution.

We now turn to the *Metropolis-Hastings* (MH) *algorithm*, a generalisation of the Metropolis algorithm that can be used to sample from a very wide range of *multivariate* distributions. This algorithm is very useful and has been applied in many difficult statistical modelling settings.

First let us again review the Metropolis algorithm for sampling from a univariate density, $f(x)$. This involves choosing an arbitrary starting value of x, a suitable driver density $g(t \mid x)$ and then repeatedly proposing a value $x' \sim g(t \mid x)$, each time accepting this value with probability

$$p = \frac{f(x')}{f(x)} \times \frac{g(x \mid x')}{g(x' \mid x)}$$

(or $p = \dfrac{f(x')}{f(x)}$ in the case of a symmetric driver).

Each proposal and then either acceptance or rejection constitutes one iteration of the algorithm and may be referred to as a *Metropolis step*.

Performing K iterations, each consisting of a single Metropolis step, results in a Markov chain of values which may be denoted $x^{(0)}, x^{(1)}, ..., x^{(K)}$.

Assuming that stochastic equilibrium has been attained within B iterations (B standing for *burn-in*) the last $J = K - B$ values may be renumbered so as to yield the required sample, $x^{(1)}, ..., x^{(J)} \sim iid\ f(x)$.

The *Metropolis-Hastings* (MH) *algorithm* is a generalisation of this procedure to the case where x is a vector of length M (say).

The bivariate MH algorithm

For simplicity we will first focus on the bivariate case ($M = 2$). Thus, suppose we wish to generate a random sample from the distribution of a random vector $X = (X_1, X_2)$ with pdf $f(x)$, where $x = (x_1, x_2)$ denotes a value of X.

First, choose an initial value of $x = (x_1, x_2)$.

Then choose two suitable driver distributions or densities:

$$g_1(t \mid x_1, x_2)$$
$$g_2(t \mid x_1, x_2).$$

Next perform the following two Metropolis steps:

1. Propose a candidate value of x_1 by sampling
$$x_1' \sim g_1(t \mid x_1, x_2),$$
and accept this value with probability
$$p_1 = \frac{f(x_1' \mid x_2)}{f(x_1 \mid x_2)} \times \frac{g_1(x_1 \mid x_1', x_2)}{g_1(x_1' \mid x_1, x_2)}.$$
(In the case of an acceptance, let $x_1 = x_1'$,
and otherwise leave x_1 unchanged.)

2. Propose a candidate value of x_2 by sampling
$$x_2' \sim g_2(t \mid x_1, x_2),$$
and accept this value with probability
$$p_2 = \frac{f(x_2' \mid x_1)}{f(x_2 \mid x_1)} \times \frac{g_2(x_2 \mid x_1, x_2')}{g_2(x_2' \mid x_1, x_2)}.$$
(In the case of an acceptance, let $x_2 = x_2'$,
and otherwise leave x_2 unchanged.)

This completes the first iteration of the MH algorithm.

The initial value of $x = (x_1, x_2)$ may be denoted
$$x^{(0)} = (x_1^{(0)}, x_2^{(0)}),$$
and the current value of the Markov chain may be denoted
$$x^{(1)} = (x_1^{(1)}, x_2^{(1)}).$$

Performing another iteration of the MH algorithm as above (starting from $x = x^{(1)}$) leads to the next value,
$$x^{(2)} = (x_1^{(2)}, x_2^{(2)}),$$
and so on.

Continuing in this fashion results in a Markov chain of vectors,
$$x^{(0)}, x^{(1)}, ..., x^{(K)}.$$

Assuming that stochastic equilibrium has been attained within B iterations, the last $J = K - B$ vectors may be renumbered consecutively to yield the required sample,
$$x^{(1)}, ..., x^{(J)} \sim iid\ f(x),$$
where $x^{(j)} = (x_1^{(j)}, x_2^{(j)})$.

Note 1: This multivariate sample can then be used to perform *marginal* inferences. For example, by discarding all the $x_2^{(j)}$ values, we obtain a sample from the marginal posterior distribution of x_1, namely

$$x_1^{(1)}...,x_1^{(J)} \sim iid \ f(x_1).$$

This technique would be useful if obtaining a sample from $f(x_1)$ *directly* were for any reason problematic. For example, the marginal density

$$f(x_1) = \int f(x_1,x_2)dx_2$$

might be difficult to derive explicitly or sample from.

Note 2: Observe that

$$\frac{f(x_1'|x_2)}{f(x_1|x_2)} = \frac{f(x_1',x_2)/f(x_2)}{f(x_1,x_2)/f(x_2)}, \text{ etc.}$$

Thus the two acceptance probabilities could also be written as:

$$p_1 = \frac{f(x_1',x_2)}{f(x_1,x_2)} \frac{g_1(x_1|x_1',x_2)}{g_1(x_1'|x_1,x_2)}$$

$$p_2 = \frac{f(x_2',x_1)}{f(x_2,x_1)} \times \frac{g_2(x_2|x_1,x_2')}{g_2(x_2'|x_1,x_2)}.$$

The trivariate MH algorithm

The Metropolis-Hastings algorithm for sampling from the *trivariate* distribution ($M = 3$) of a vector random variable $X = (X_1,X_2,X_3)$ involves choosing an initial value of the vector

$$x = (x_1,x_2,x_3),$$

specifying three driver densities:

$$g_1(t|x_1,x_2,x_3)$$
$$g_2(t|x_1,x_2,x_3)$$
$$g_3(t|x_1,x_2,x_3),$$

and repeatedly iterating three Metropolis steps as follows:

1. Propose a candidate value of x_1 by sampling

$$x_1' \sim g_1(t \mid x_1, x_2, x_3),$$

and accept this value with probability

$$p_1 = \frac{f(x_1' \mid x_2, x_3)}{f(x_1 \mid x_2, x_3)} \times \frac{g_1(x_1 \mid x_1', x_2, x_3)}{g_1(x_1' \mid x_1, x_2, x_3)}$$

2. Propose a candidate value of x_2 by sampling

$$x_2' \sim g_2(t \mid x_1, x_2, x_3),$$

and accept this value with probability

$$p_2 = \frac{f(x_2' \mid x_1, x_3)}{f(x_2 \mid x_1, x_3)} \times \frac{g_2(x_2 \mid x_1, x_2', x_3)}{g_2(x_2' \mid x_1, x_2, x_3)}.$$

3. Propose a candidate value of x_3 by sampling

$$x_3' \sim g_3(t \mid x_1, x_2, x_3),$$

and accept this value with probability

$$p_3 = \frac{f(x_3' \mid x_1, x_2)}{f(x_3 \mid x_1, x_2)} \times \frac{g_3(x_3 \mid x_1, x_2, x_3')}{g_3(x_3' \mid x_1, x_2, x_3)}.$$

As before, continuing in this fashion until stochastic equilibrium has been achieved, and then for another J iterations, leads to the random sample $x^{(1)}, ..., x^{(J)} \sim iid\ f(x)$, where now $x^{(j)} = (x_1^{(j)}, x_2^{(j)}, x_3^{(j)})$.

Note: As before, the $x_1^{(j)}$ values on their own then constitute a sample from the marginal distribution of x_1, whose density is now

$$f(x_1) = \iint f(x_1, x_2, x_3) dx_2 dx_3,$$

and the three acceptance probabilities can also be expressed as

$$p_1 = \frac{f(x_1', x_2, x_3)}{f(x_1, x_2, x_3)} \frac{g_1(x_1 \mid x_1', x_2, x_3)}{g_1(x_1' \mid x_1, x_2, x_3)}, \quad \text{etc.}$$

The general MH algorithm

These ideas extend naturally and in an obvious fashion to higher values of M. Thus, for sampling from an M-variate distribution with density $f(x_1, ..., x_M)$, the MH algorithm involves choosing a starting value

$$x = (x_1, ..., x_M),$$

specifying M drivers,

$$g_m(t \mid x_1,...,x_M) \quad (m=1,...,M),$$

and repeatedly iterating M steps as follows:

1. Propose a candidate value of x_1 by sampling

    $$x_1' \sim g_1(t \mid x_1,...,x_M),$$

    and accept this value with probability

    $$p_1 = \frac{f(x_1' \mid x_2,...,x_M)}{f(x_1 \mid x_2,...,x_M)} \times \frac{g_1(x_1 \mid x_1',...,x_M)}{g_1(x_1' \mid x_1,...,x_M)}$$

2. Propose a candidate value of x_2 by sampling

    $$x_2' \sim g_2(t \mid x_1,...,x_M),$$

    and accept this value with probability

    $$p_2 = \frac{f(x_2' \mid x_1,x_3,...,x_M)}{f(x_2 \mid x_1,x_3,...,x_M)} \times \frac{g_2(x_2 \mid x_1,x_2',x_3,...,x_M)}{g_2(x_2' \mid x_1,x_2,x_3,...,x_M)}$$

    . . . .

M. Propose a candidate value of x_M by sampling

    $$x_M' \sim g_M(t \mid x_1,...,x_M),$$

    and accept this value with probability

    $$p_M = \frac{f(x_M' \mid x_1,...,x_{M-1})}{f(x_M \mid x_1,...,x_{M-1})} \times \frac{g_M(x_M \mid x_1,...,x_{M-1},x_M')}{g_M(x_M' \mid x_1,...,x_{M-1},x_M)}.$$

As before, continuing in this fashion until stochastic equilibrium and then for J more iterations leads to the sample $x^{(1)},...,x^{(J)} \sim iid\ f(x)$, where now $x^{(J)} = (x_1^{(J)},...,x_M^{(J)})$.

Note: Again, the $x_1^{(J)}$ values on their own then constitute a sample from the marginal distribution of x_1, whose density is now

$$f(x_1) = \int \cdots \int f(x_1,...,x_M)dx_2...dx_M),$$

and the M acceptance probabilities can also be expressed as

$$p_1 = \frac{f(x_1',...,x_M)}{f(x_1,...,x_M)} \frac{g_1(x_1 \mid x_1',x_2,...,x_M)}{g_1(x_1' \mid x_1,x_2,...,x_M)}, \text{ etc.}$$

Exercise 6.6 MH algorithm applied to a bent coin which is tossed an unknown number of times

Suppose that five heads have come up on an unknown number of tosses of a bent coin.

Before the experiment, we believed the coin was going to be tossed a number of times equal to 1, 2, 3, ..., or 9, with all possibilities equally likely. As regards the probability of heads coming up on a single toss, we deemed no value more or less likely than any other value. We also considered the probability of heads as unrelated to the number of tosses.

Find the marginal posterior distribution and mean of the number of tosses and of the probability of heads, respectively. Also find the number of heads we can expect to come up if the coin is tossed again the same number of times.

Do all this via Monte Carlo by designing and implementing a suitable MH algorithm.

Note: This problem was solved analytically in Exercise 3.10.

Solution to Exercise 6.6

As in Exercise 3.10, the relevant Bayesian model is:

$$(y \mid \theta, n) \sim Binomial(n, \theta)$$
$$(\theta \mid n) \sim U(0,1)$$
$$f(n) = 1/k, \quad n = 1,...,k, \quad k = 9,$$

and the joint posterior density of the two parameters n and θ is

$$f(n, \theta \mid y) \propto f(n) f(\theta \mid n) f(y \mid n, \theta)$$

$$\propto \frac{n! \theta^y (1-\theta)^{n-y}}{(n-y)!}$$

$$\equiv h(n, \theta), \quad 0 < \theta < 1, \quad n = y, y+1, ..., k.$$

Let us now specify the driver for n as discrete uniform over the integers from $n-r$ to $n+r$, where r is a tuning parameter.

Also let the driver for θ be uniform from $\theta - c$ to $\theta + c$, where c is another tuning parameter.

Note: These drivers may also be expressed by writing the distributions explicitly as:

$$n' \sim DU(n-r, n-r+1, ..., n+r)$$
$$\theta' \sim U(\theta - c, \theta + c),$$

or by writing the driver densities explicitly as:

$$g_1(t \mid n, \theta) = \frac{1}{2r+1}, \quad t = n-r, n-r+1, ..., n+r$$

$$g_2(t \mid n, \theta) = \frac{1}{2c}, \quad \theta - c < t < \theta + c.$$

Noting that both drivers are symmetric, a suitable MH algorithm may be defined by the following two steps at each iteration:

1. Propose a value
    $$n' \sim DU(n-r, ..., n+r),$$
    and accept this value with probability
    $$p_1 = \frac{f(n', \theta \mid y)}{f(n, \theta \mid y)} = \frac{h(n', \theta)}{h(n, \theta)} = \frac{n'! \theta^y (1-\theta)^{n'-y} / (n'-y)!}{n! \theta^y (1-\theta)^{n-y} / (n-y)!}$$
    $$= \frac{n'! (1-\theta)^{n'} / (n'-y)!}{n! (1-\theta)^{n} / (n-y)!}.$$

2. Propose a value
    $$\theta' \sim U(\theta - c, \theta + c),$$
    and accept this value with probability
    $$p_2 = \frac{f(n, \theta' \mid y)}{f(n, \theta \mid y)} = \frac{h(n, \theta')}{h(n, \theta)} = \frac{n! \theta'^y (1-\theta')^{n-y} / (n-y)!}{n! \theta^y (1-\theta)^{n-y} / (n-y)!}$$
    $$= \frac{\theta'^y (1-\theta')^{n-y}}{\theta^y (1-\theta)^{n-y}}.$$

Note: The proposed value n' should automatically be rejected if it is outside the set $\{5, ..., 9\}$ (because then $f(n', \theta \mid y) = 0$), and otherwise automatically accepted if $p_1 > 1$. If $n' = n$ then $p_1 = 1$, again leading to automatic acceptance.

Likewise, the proposed value θ' should be automatically rejected if it is outside the interval $(0,1)$, and otherwise automatically accepted if $p_2 > 1$.

Setting $c = 0.3$ and $r = 1$ (after some experimentation) and starting from $n = 7$ and $\theta = 0.5$, the MH algorithm converged very quickly, with acceptance rates of 73% for n and 58% for θ over a total of 10,100 iterations.

The first 100 iterations were thrown away as the burn-in, and then every 20th value (only) was recorded so as to thereby yield an approximately random sample of size $J = 500$ from the joint posterior distribution of n and θ, namely $(n_1, \theta_1), ..., (n_J, \theta_J) \sim iid\ f(n, \theta \mid y)$.

Figures 6.13 and 6.14 (pages 299 and 300) show the traces for all 10,101 values of n and θ, respectively, and Figures 6.15 and 6.16 (pages 300 and 301) show the traces for the final 500 values of n and θ, respectively.

Figure 6.17 (page 301) shows the corresponding sample ACFs (autocorrelation functions), labelled nv0 and thv0 for the last 10,000 values of n and θ, respectively, and labelled nv and thv for the final 500 values of n and θ. The thinning process has dramatically reduced the high serial correlation.

The final bivariate sample of size $J = 500$ was used for Monte Carlo inference in the usual way, with the following results.

The MC estimate of $\hat{n} = E(n \mid y)$ $(= 6.744)$ was $\bar{n} = 6.708$, with 95% CI (6.587, 6.829).

The Monte Carlo estimate of $\hat{\theta} = E(\theta \mid y)$ $(= 0.7040)$ was $\bar{\theta} = 0.7097$, with 95% CI (0.6943, 0.7252). Also, the 95% CPDR estimate for θ was (0.3547, 0.9886).

Figure 6.18 (page 302) is a probability histogram of the almost random sample $n_1, ..., n_J \sim iid\ f(n \mid y)$, and Figure 6.19 (page 302) is a probability histogram of the almost random sample $\theta_1, ..., \theta_J \sim iid\ f(\theta \mid y)$.

Each histogram is overlaid with a nonparametric density estimate based on the histogram, as well as with the true marginal posterior density.

Each histogram also includes vertical lines showing the true distribution mean, the MC estimate of that mean, and the 95% CI for that mean.

Figure 6.19 also displays the 95% CPDR estimate for θ.

Note 1: The histogram of n-values in Figure 6.18 (page 302) is itself an estimate of $f(n\,|\,y)$. The short vertical lines in the histogram indicate the MC 95% CIs for $f(n\,|\,y)$.

For example, the height of the bar above 6 is the proportion of sample values $n_1,...,n_J$ equal to 6, which is $117/500 = 0.234$, and the short vertical bar above 6 is the MC 95% CI for $P(n = 6\,|\,y)$, which is $(0.234 \pm 1.96\sqrt{0.234(1-0.234)/500}) = (0.1969, 0.2711)$.

Note 2: The histogram of θ-values in Figure 6.19 (page 302) in fact shows two posterior density estimates. The first and simplest estimate tapers towards zero as θ approaches 1. The second estimate was obtained using a special mathematical device that was applied so as to 'force' the density estimate to be relatively high near 1. For values of θ less than about 0.8, the two density estimates are virtually identical. Details of said mathematical device can be found in the R code below.

Figure 6.13 Trace of 10,101 n-values

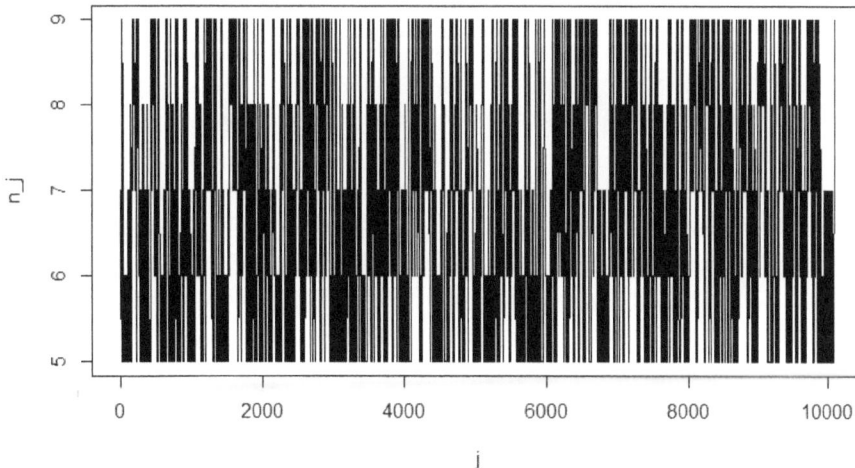

Figure 6.14 Trace of 10,101 θ-values

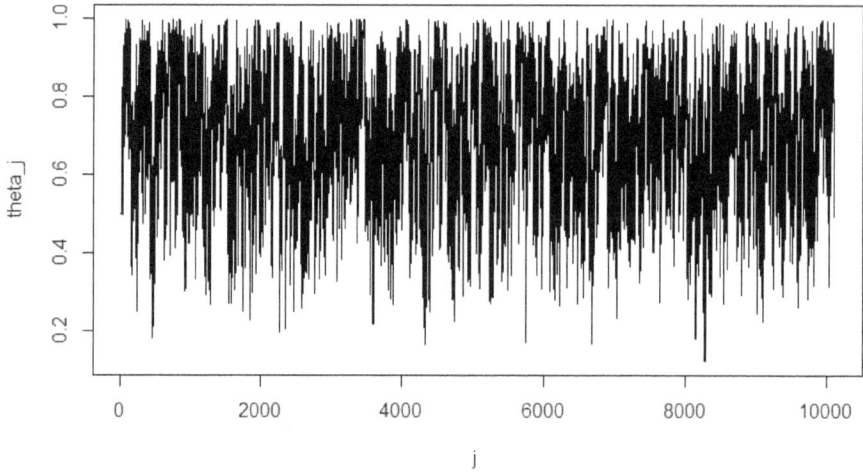

Figure 6.15 Trace of 500 n-values

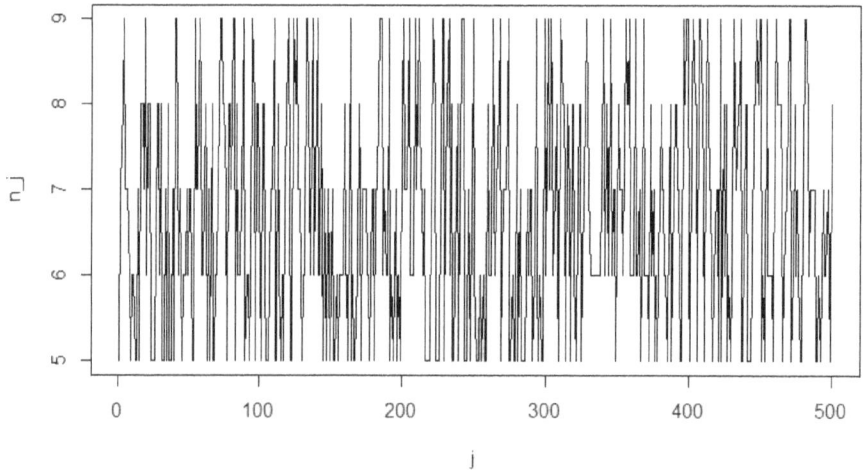

Figure 6.16 Trace of 500 θ-values

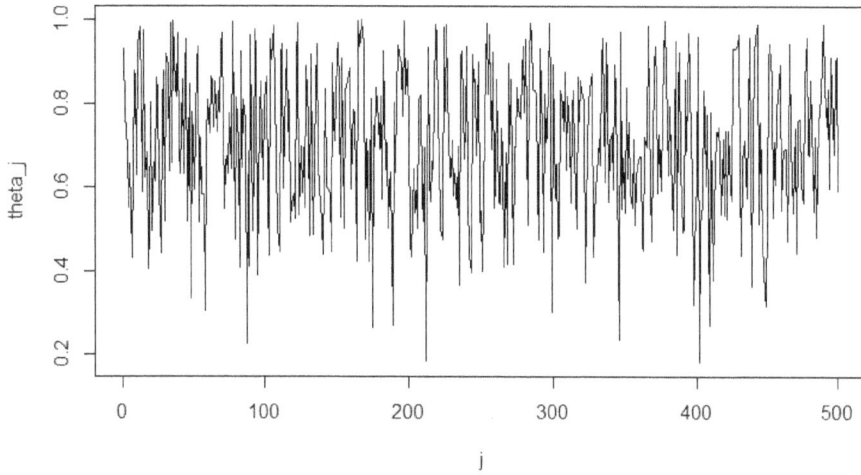

Figure 6.17 Sample autocorrelation functions

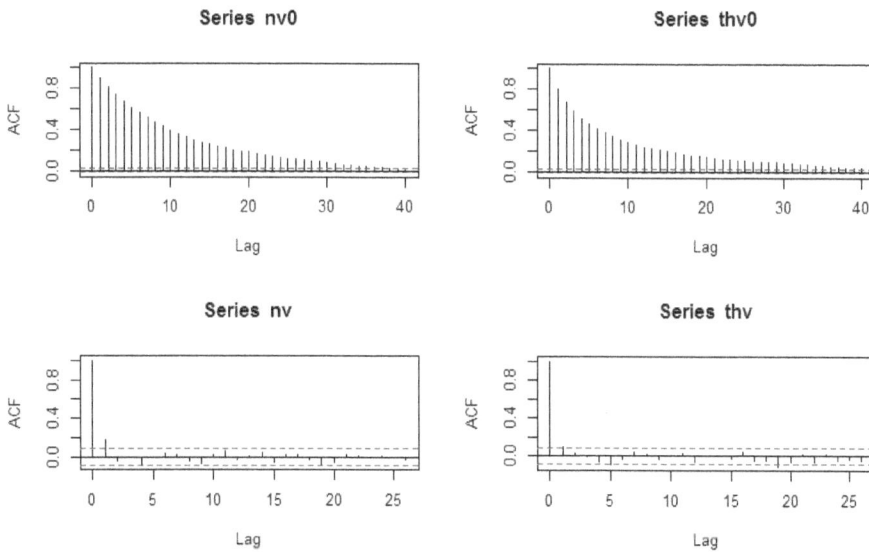

Figure 6.18 Probability histogram of 500 *n*-values

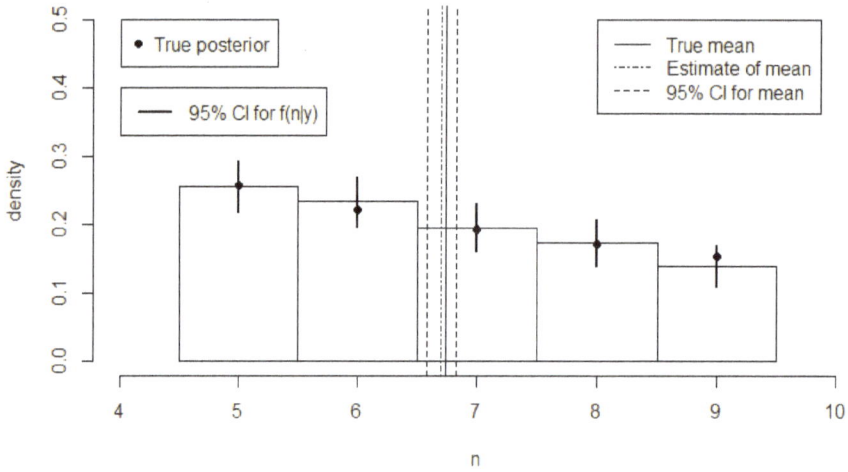

Figure 6.19 Probability histogram of 500 θ-values

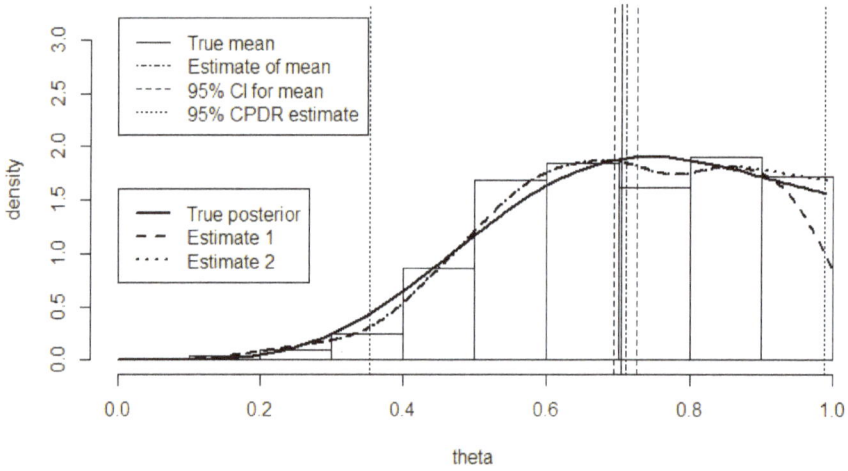

R Code for Exercise 6.6

```
# NB: Some of this R Code was copied from a previous exercise


y <- 5; k <- 9; options(digits=4)
nvec <- y:k;   avec <- 1/(nvec+1);  sumavec <- sum(avec); sumavec  # 0.6456
fny <- avec/sumavec;   rbind(nvec,avec,fny)
# nvec 5.0000 6.0000 7.0000 8.0000 9.0000
# avec 0.1667 0.1429 0.1250 0.1111 0.1000
#  fny 0.2581 0.2213 0.1936 0.1721 0.1549


nhat <- sum(nvec*fny); nhat   # 6.744
thhat <- sum( fny * (y+1)/(nvec+2) ); thhat  # 0.704
xhat <- sum( fny * nvec * (y+1)/(nvec+2) ); xhat  # 4.592
thvec <- seq(0,0.99,0.01); fthyvec <- thvec
for(i in 1:length(thvec)) fthyvec[i] <- sum( fny * dbeta(thvec[i],y+1,nvec-y+1) )


X11(w=8,h=6); par(mfrow=c(2,1))
plot(nvec,fny,type="n",xlab="n",ylab="f(n|y)",ylim=c(0,0.4))
points(nvec,fny,pch=16,cex=1);  abline(v=nhat)
plot(thvec,fthyvec,type="n",xlab="theta",ylab="f(theta|y) ",ylim=c(0,2.5))
lines(thvec,fthyvec,lwd=3);   abline(v=thhat)


# Code for Metropolis-Hastings algorithm -------------------------------------
MH = function(Jp,n,th,c,r,y,k){
# This function performs the Metropolis-Hastings algorithm for a simple model.
# Inputs:      Jp = total number of iterations
#              n, th = intial values of n and theta
#              r, c = tuning parameters for n and theta
#              y, k = number of successes, maximum value of n
# Outputs:     $nvec = vector of (Jp+1) values of n
#              $thvec = vector of (Jp+1) values of theta
#              $nar, $thar = acceptance rates for n and theta.
nvec = n; thvec = th; nct = 0; thct = 0
logfun = function(n,th,y){  # Calculates the log of the joint posterior kernel
        lgamma(n+1) + y*log(th) + (n-y)*log(1-th) - lgamma(n-y+1)    }
for(j in 1:Jp){
        nprop = sample((n-r):(n+r),1)
        if(nprop >= y)  if(nprop <= k){
                if(nprop == n)   nct = nct + 1
                if(nprop != n){
                   logp1 = logfun(n=nprop,th=th,y=y) - logfun(n=n,th=th,y=y)
                   p1 = exp(logp1);   u <- runif(1)
                   if(u < p1){ n = nprop; nct = nct + 1}
```

```
                      }
                }
        thprop = runif(1,th-c,th+c)
        if(thprop > 0) if(thprop < 1){
                logp2 = logfun(n=n,th=thprop,y=y) - logfun(n=n,th=th,y=y)
                p2 = exp(logp2);   u = runif(1)
                if(u < p2){ th = thprop; thct = thct + 1}
                }
        nvec = c(nvec,n); thvec = c(thvec,th)
        }
nar = nct/Jp; thar = thct/Jp;   list(nvec=nvec,thvec=thvec,nar=nar,thar=thar)   }
# END

X11(w=8,h=5); par(mfrow=c(1,1))
Jp = 10100; set.seed(135);   res = MH(Jp=Jp,n=7,th=0.5,c=0.3,r=1,y=5,k=9)
c(res$nar,res$thar)  # 0.7344 0.5847

plot(0:Jp,res$nvec,type="l", xlab="j",ylab="n_j")
plot(0:Jp,res$thvec,type="l", xlab="j",ylab="theta_j")

burn = 100; nv0 = res$nvec[-(1:(burn+1))]; thv0 = res$thvec[-(1:(burn+1))]
nv=nv0[seq(20,10000,20)]; thv=thv0[seq(20,10000,20)]; J=500

plot(1:J,nv,type="l", xlab="j",ylab="n_j")
plot(1:J,thv,type="l", xlab="j",ylab="theta_j")

par(mfrow=c(2,2));acf(nv0); acf(thv0); acf(nv); acf(thv)

nbar = mean(nv); nci = nbar + c(-1,1)*qnorm(0.975)*sd(nv)/sqrt(J)
c(nbar,nci)  # 6.708 6.587 6.829
thbar = mean(thv); thci = thbar + c(-1,1)*qnorm(0.975)*sd(thv)/sqrt(J)
thcpdr = quantile(thv,c(0.025,0.975))
c(thbar,thci,thcpdr)  # 0.7097 0.6943 0.7252 0.3547 0.9886

nvals=5:9; fvals=summary(as.factor(nv)); pvals=fvals/J
Lvals=pvals-qnorm(0.975)*sqrt(pvals*(1-pvals)/J)
Uvals=pvals+qnorm(0.975)*sqrt(pvals*(1-pvals)/J)

rbind(nvals,fvals,pvals,Lvals,Uvals)
# nvals  5.0000   6.0000  7.0000  8.0000  9.0000
# fvals 128.0000 117.0000 98.0000 87.0000 70.0000
# pvals  0.2560   0.2340  0.1960  0.1740  0.1400
# Lvals  0.2177   0.1969  0.1612  0.1408  0.1096
# Uvals  0.2943   0.2711  0.2308  0.2072  0.1704
```

```
par(mfrow=c(1,1))
hist(nv,prob=T,xlim=c(4,10),ylim=c(0,0.5),xlab="n",breaks=seq(4.5,9.5,1),
        main="", ylab="density")
points(nvec,fny,pch=16);   abline(v=nhat)
for(i in 1:length(nvals)) lines(rep(nvals[i],2),c(Lvals[i],Uvals[i]),lwd=2)
abline(v=nbar,lty=4); abline(v=nci,lty=2)
legend(8,0.5,c("True mean","Estimate of mean","95% CI for mean"),lty=c(1,4,2))
legend(4,0.5,c("True posterior"),pch=16,cex=1)
legend(4,0.4,c("95% CI for f(n|y)"),lty=1,lwd=2)

hist(thv,prob=T,xlim=c(0,1),ylim=c(0,3.2),xlab="theta",
        main="", ylab="density")
lines(thvec,fthyvec,lwd=2);   abline(v=thhat)
thdensity <- density( c(thv,1+abs(1-thv)),   from=0, to=1,width=0.2)
lines(density(thv,from=0,to=1,width=0.2),lty=2,lwd=2)
   # Note: This is the simplest way to estimate the density
lines(thdensity$x,thdensity$y*2,lty=3,lwd=2)
   # Note: This density estimate is forced to be higher at theta=1
abline(v=thbar,lty=4); abline(v=thci,lty=2); abline(v=thcpdr,lty=3)
legend(0,3.2,c("True mean","Estimate of mean","95% CI for mean",
                "95% CPDR estimate"),lty=c(1,4,2,3))
legend(0,1.6,c("True posterior","Estimate 1","Estimate 2"),lty=c(1,2,3),lwd=2)
```

6.7 Independence drivers and block sampling

The Metropolis-Hastings algorithm is very flexible and allows for a lot of choice in the way it is designed. In any particular application, many different MH algorithms will work, but some may perform better than others, meaning they will result in better mixing and faster convergence towards stochastic equilibrium. This will have a lot to do with how the random variables involved are set up and parameterised, what driver distributions are specified, and which tuning parameters are then chosen for completely defining those driver distributions.

For example, the driver distribution for a component x_m of the vector $x = (x_1,...,x_M)$ may be chosen so that it depends only on the last value of itself. In that case, $g_m(t \mid x_1, x_2,..., x_M)$ can also be written $g_m(t \mid x_m)$.

In fact, this is the norm in practice, and it was the case for both drivers in the last exercise.

It is also permissible to choose the mth driver so that it doesn't depend on *any* of the current values of the Markov chain, including itself. In that case, the driver $g_m(t \mid x_1, x_2, ..., x_M)$ may be written $g_m(t)$ and be referred to as an *independence driver*.

Also, one may 'bundle' any of the M random variables into *blocks* and thereby reduce the number of actual Metropolis steps per iteration. For example, instead of doing a Metropolis step for each of x_3 and x_4 at each iteration, one may do a single Metropolis step as follows:

> Create a candidate value of (x_3, x_4) by sampling
> $$(x_3', x_4') \sim g_{34}(t, u \mid x_3, x_4) \text{ (say)},$$
> and then accept this candidate (x_3', x_4') with probability
> $$p_{34} = \frac{f(x_1, x_2, x_3', x_4', x_5, ..., x_M)}{f(x_1, x_2, x_3, x_4, x_5, ..., x_M)}$$
> $$\times \frac{g_{34}(x_3, x_4 \mid x_1, x_2, x_3', x_4', x_5, ..., x_M)}{g_{34}(x_3', x_4' \mid x_1, x_2, x_3, x_4, x_5, ..., x_M)}.$$

This idea can be used to improve mixing and speed up the rate of convergence but may require more work sampling from the bivariate driver and determining the optimal tuning constant. Note that to sample (x_3', x_4'), it may be possible to do this in two steps via the method of composition according to
$$g_{34}(t, u \mid x_3, x_4) = g_3(t \mid x_3, x_4) g_{4|3}(u \mid x_3, x_4, t).$$

6.8 Gibbs steps and the Gibbs sampler

One important possibility is to give the driver for x_m exactly the same distribution as the conditional distribution of x_m given all the other values.

In that case, the proposal density is
$$g_m(t \mid x_1, ..., x_M) = f(x_m = t \mid x_1, ..., x_{m-1}, x_{m+1}, ..., x_M).$$

With this choice, the acceptance probability equals
$$p_m = \frac{f(x_m' \mid x_1, ..., x_{m-1}, x_{m+1}, ..., x_M)}{f(x_m \mid x_1, ..., x_{m-1}, x_{m+1}, ..., x_M)} \times \frac{f(x_m \mid x_1, ..., x_{m-1}, x_{m+1}, ..., x_M)}{f(x_m' \mid x_1, ..., x_{m-1}, x_{m+1}, ..., x_M)}$$
$$= 1 \quad \text{(that is, 100\%)}.$$

This means that the candidate value x'_m is *definitely* accepted at every iteration. In that case we call the mth step of the Metropolis-Hastings algorithm a *Gibbs step.*

If all the Metropolis steps are Gibbs steps then the algorithm may also be called a *Gibbs sampler.*

Note: In the case $M = 1$, the Gibbs sampler equates to sampling directly from the distribution of interest, with no stochastic dependence between values of the resulting chain.

Thus a Gibbs sampler for sampling from a multivariate distribution
$$f(x) = f(x_1, ..., x_M)$$
may be defined as iteratively sampling from the full conditional densities:
$$f(x_1 \mid x_2, x_3, ..., x_M)$$
$$f(x_2 \mid x_1, x_3, ..., x_M)$$

$$.............................$$

$$f(x_M \mid x_1, x_2, ..., x_{M-1}),$$
where each of these is proportional to $f(x_1, ..., x_M)$, for example, where

$$f(x_1 \mid x_2, x_3, ..., x_M) = \frac{f(x_1, x_2, x_3, ..., x_M)}{f(x_2, x_3, ..., x_M)}$$
$$\overset{x_1}{\propto} f(x_1, x_2, x_3, ..., x_M).$$

Note: We could also write the mth conditional density as
$$f(x_m \mid x_{-m}),$$
where
$$x_{-m} = (x_1, ..., x_{m-1}, x_{m+1}, ..., x_M)$$
denotes the vector x with the mth component removed.

In any case, the mth distribution can be obtained by examining the joint density of all the variables seeing that joint density as a density function of only x_m.

An advantage of the Gibbs sampler is that it produces 'good mixing', on account of no 'wastage' due to rejections. A disadvantage is that sampling from all the required exact conditional distributions may not be easy or even possible.

The Metropolis-Hastings algorithm is a very versatile tool that will work in almost every situation with the least amount of mathematical effort. The Gibbs sampler performs better but is practically feasible only in some special cases.

A general recommendation in any given situation is to begin by specifying a 'pure' Metropolis-Hastings algorithm, and then to examine each of its M Metropolis steps with a view to converting it into a Gibbs step if that is not too much effort. If the resulting Metropolis-Hastings algorithm consists of at least one Gibbs step and at least one Metropolis step, it may also be referred to as a *Metropolis-Hastings within Gibbs sampler*.

Example

As an example of converting a Metropolis step into a Gibbs step, recall the joint posterior density in Exercise 6.5:

$$f(n,\theta \mid y) \propto \frac{n!\theta^y(1-\theta)^{n-y}}{(n-y)!}, \quad 0<\theta<1, \ n=y,y+1,...,k \,.$$

This density was used as a basis for the following Metropolis step for θ at each iteration:

2. Propose a value $\theta' \sim U(\theta-c,\theta+c)$, and accept this value with
   probability $p_2 = \dfrac{\theta'^y(1-\theta')^{n-y}}{\theta^y(1-\theta)^{n-y}}$.

Instead of this Metropolis step at each iteration, it would be better and also easier to apply a Gibbs step which involves sampling the next value of θ directly from the $Beta(y+1,n-y+1)$ distribution.

Equivalently, one could write that Gibbs step as:

2. Draw $\theta \sim Beta(y+1,n-y+1)$.

Now consider the Metropolis step for n in Exercise 6.5:

1. Propose a value $n' \sim DU(n-r,...,b+r)$, and accept this value
   with probability $p_1 = \dfrac{f(n',\theta \mid y)}{f(n,\theta \mid y)} = \dfrac{n'!(1-\theta)^{n'}/(n'-y)!}{n!(1-\theta)^n/(n-y)!}$.

Unfortunately, the kernel of $f(n,\theta \mid y)$ when seen as a function of n alone (i.e. $n!(1-\theta)^n / (n-y)!$) does not suggest a well-known distribution. However, with a little effort, it is still possible to convert the Metropolis step for n into a Gibbs step, as follows:

1. Calculate $q(n) = n!(1-\theta)^n / (n-y)!$ for each $n = 5,...,9$.
   Calculate $q_T = q(5) + ... + q(9)$.
   Hence obtain $f(n \mid y,\theta) = q(n) / q_T$.
   Draw $n \sim f(n \mid y,\theta)$ (now easy).

Exercise 6.7 Sampling from a normal-normal-gamma model via MCMC

Consider the *general* normal-normal-gamma model given by:
$$(y_1,...,y_n \mid \mu,\lambda) \sim iid\ N(\mu,\lambda)$$
$$(\mu \mid \lambda) \sim N(\mu_0,\sigma_0^2)$$
$$\lambda \sim G(\alpha,\beta).$$

Suppose that $\mu_0 = 10$, $\sigma_0 = 2$, $\alpha = 3$, $\beta = 6$ and $n = 40$.

(a) Generate $y = (y_1,...,y_n)$ from the model using these constants.

(b) Design a suitable Metropolis-Hastings algorithm in this setting. Then apply it y in (a) so as to generate a random sample of size $J = 5,000$ from the bivariate posterior distribution of μ and λ. Illustrate the sample with appropriate trace plots and probability histograms.

(c) Repeat (b) but with a Gibbs sampler in place of the MH algorithm.

Solution to Exercise 6.7

(a) Using the specified values, we generated the parameters
$$\lambda = 0.1292 \text{ and } \mu = 11.95$$
from their independent prior distributions.

We then generated $n = 40$ values from the $N(\mu,\sigma^2)$ distribution with $\sigma = 1/\sqrt{\lambda} = 2.782$. The sample mean and standard deviation of these values were 12.28 and 2.592. A histogram of the sample values is shown in Figure 6.20. Overlaid is the $N(\mu,\sigma^2)$ density.

Figure 6.20 Probability histogram of 40 y-values

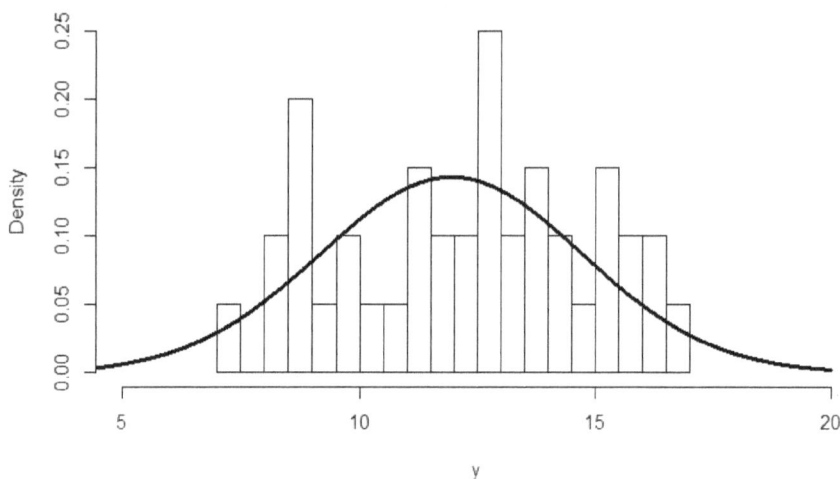

(b) The joint posterior density of μ and λ is

$$f(\mu,\lambda\,|\,y) \propto f(\lambda)f(\mu\,|\,\lambda)f(y\,|\,\lambda,\mu)$$

$$\propto \lambda^{\alpha-1}e^{-\beta\lambda} \times e^{-\frac{1}{2\sigma_0^2}(\mu-\mu_0)^2} \times \prod_{i=1}^{n} \lambda^{\frac{1}{2}}e^{-\frac{\lambda}{2}(y_i-\mu)^2}$$

$$\propto \lambda^{\alpha+\frac{n}{2}-1} \exp\left\{-\lambda\beta - \frac{1}{2\sigma_0^2}(\mu-\mu_0)^2 - \frac{\lambda}{2}\sum_{i=1}^{n}(y_i-\mu)^2\right\}$$

$$\equiv k(\mu,\lambda) .$$

A suitable MH algorithm is then defined by the following two steps:

    1.   Draw a value $\mu' \sim U(\mu - c, \mu + c)$

          and accept it with probability $p_1 = \dfrac{k(\mu',\lambda)}{k(\mu,\lambda)}$.

    2.   Draw a value $\lambda' \sim U(\lambda - r, \lambda + r)$

          and accept it with probability $p_2 = \dfrac{k(\mu,\lambda')}{k(\mu,\lambda)}$.

Note: The best way to calculate the acceptance probabilities is as:

        $p_1 = \exp(q_1)$ and $p_2 = \exp(q_2)$,

after first deriving $q_1 = l(\mu',\lambda) - l(\mu,\lambda)$ and $q_2 = l(\mu,\lambda') - l(\mu,\lambda)$, where $l(\mu,\lambda) = \log k(\mu,\lambda)$

$$= \left(\alpha + \frac{n}{2} - 1\right)\log\lambda - \lambda\beta - \frac{1}{2\sigma_0^2}(\mu - \mu_0)^2 - \frac{\lambda}{2}\sum_{i=1}^{n}(y_i - \mu)^2.$$

The MH algorithm was started at $\mu = 0$ and $\lambda = 1$ with tuning constants $c = 0.1$ and $r = 0.01$, and run for a total of 6,000 iterations. The resulting traces are shown in Figures 6.21 and 6.22.

The acceptance rates for μ and λ were 92% and 92%. These rates were judged to be unduly high because they led to very strong serial correlation in the simulated values (i.e. poor mixing).

So the algorithm was run again from the same starting values but with $c = 0.9$ and $r = 0.08$ (both larger). This resulted in Figures 6.23 and 6.24 (pages 312 and 313), with much better mixing, faster convergence, and the better acceptance rates of 59% and 58%.

The last 5,000 pairs of values from this second run of the algorithm were then collected and used to produce the two histograms in Figures 6.25 and 6.26 (pages 313 and 314). Each histogram is overlaid by a density estimate of the corresponding posterior and shows a dot indicating the true value of the parameter (which was initially sampled from its prior).

Figure 6.21 Trace for μ

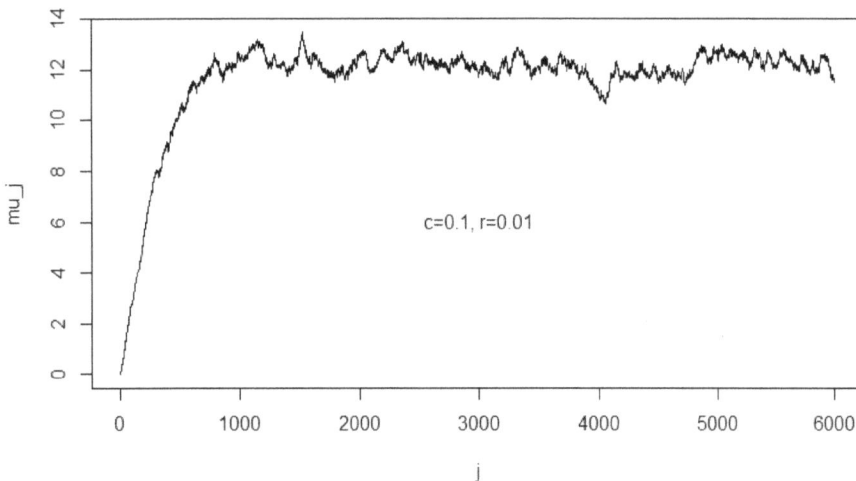

Figure 6.22 Trace for λ

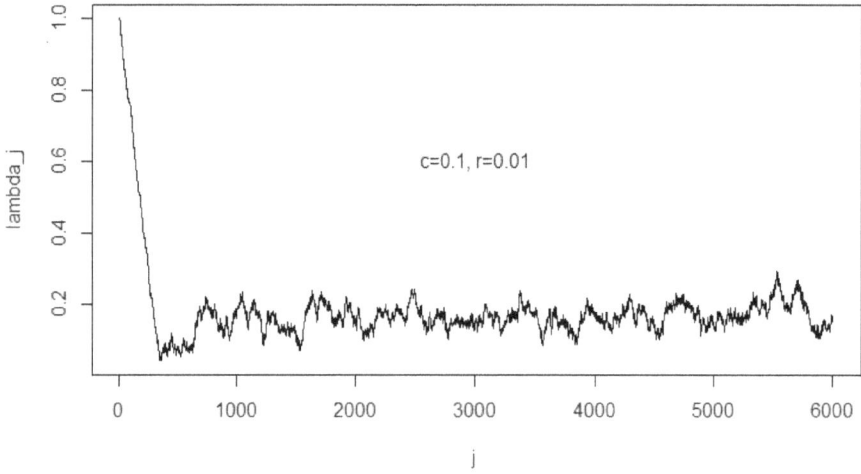

c=0.1, r=0.01

Figure 6.23 Improved trace for μ

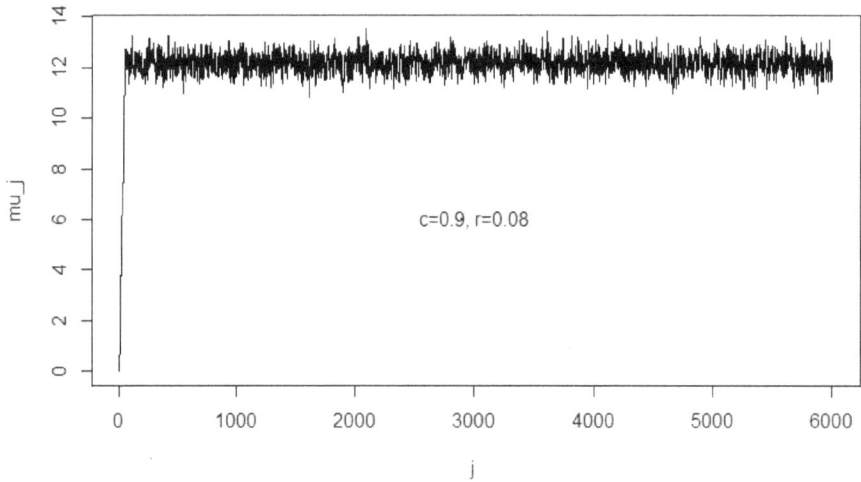

c=0.9, r=0.08

Figure 6.24 Improved trace for λ

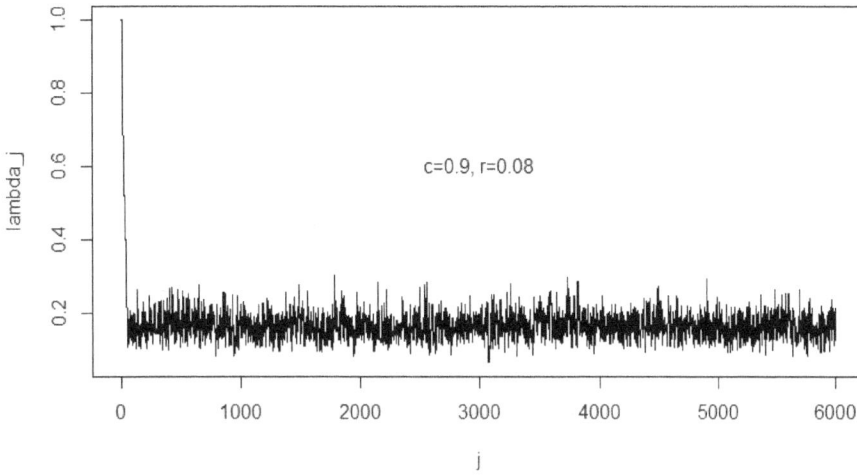

c=0.9, r=0.08

Figure 6.25 Histogram for μ

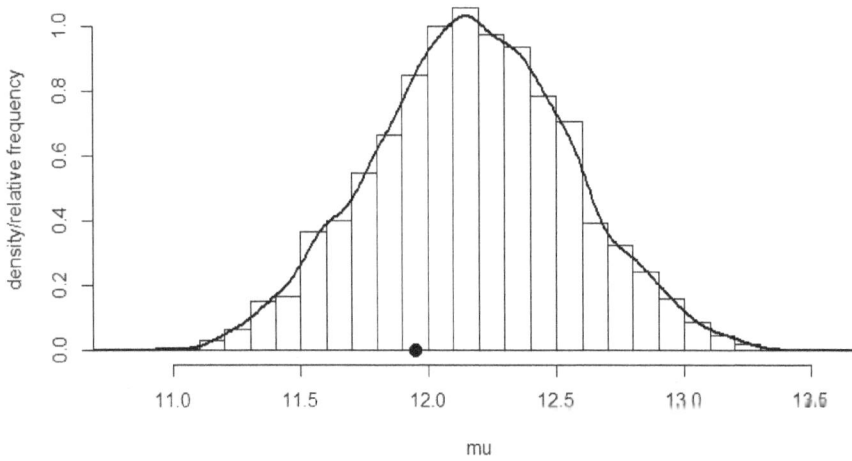

Figure 6.26 Histogram for λ

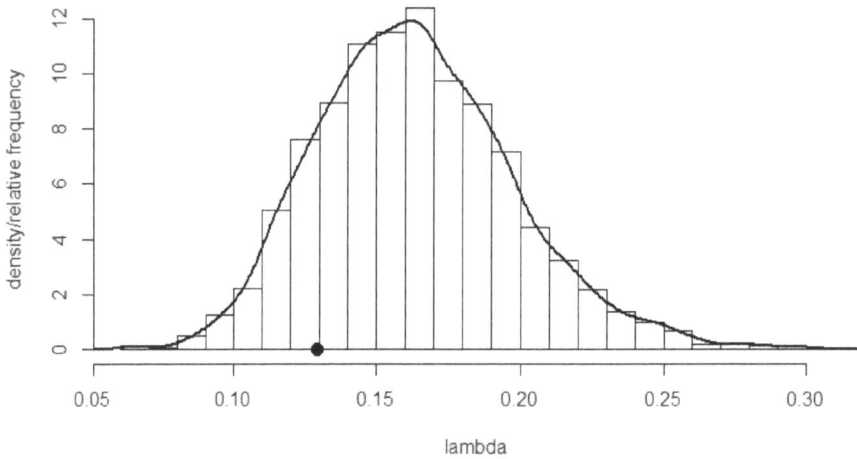

lambda

(c) Examining the kernel of the joint posterior in (b) and studying previous exercises (involving the *normal-normal model* and the *normal-gamma model*) we easily identify the two conditional distributions which define the Gibbs sampler. These are defined as follows:

1. Sample $\mu \sim f(\mu \mid y, \lambda) \sim N(\mu_*, \sigma_*^2)$, where: $\mu_* = (1-k)\mu_0 + k\bar{y}$,

$$\sigma_*^2 = k\frac{\sigma^2}{n} = \frac{k}{n\lambda}, \quad k = \frac{n}{n+\sigma^2/\sigma_0^2} = \frac{n}{n+(1/(\lambda\sigma_0^2))}, \quad \sigma^2 \equiv 1/\lambda.$$

2. Sample $\lambda \sim f(\lambda \mid y, \mu) \sim G\left(\alpha + \frac{n}{2}, \beta + \frac{1}{2}\left((n-1)s^2 + n(\mu - \bar{y})^2\right)\right)$.

This Gibbs sampler was started at $\mu = 0$ and $\lambda = 1$ and run for a total of 6,000 iterations. The resulting traces are shown in Figures 6.27 and 6.28.

The last 5,000 pairs of values were then collected and used to produce the histograms in Figures 6.29 and 6.30 (page 316). Each histogram is overlaid by a density estimate of the corresponding posterior and shows a dot indicating the true value of the parameter.

We see that the Gibbs sampler has produced very similar output to that in (b) as obtained using the Metropolis-Hastings algorithm, but with less effort (e.g. no need to worry about tuning constants) and with arguably better results.

By this we mean that the output from the Gibbs sampler exhibits far less serial correlation. This is evidenced clearly in Figure 6.31 (page 317), which shows the sample autocorrelation functions of the simulated values of μ and λ in (b) (top two subplots) and in (c) (bottom two subplots).

Figure 6.27 Trace for μ from Gibbs sampler

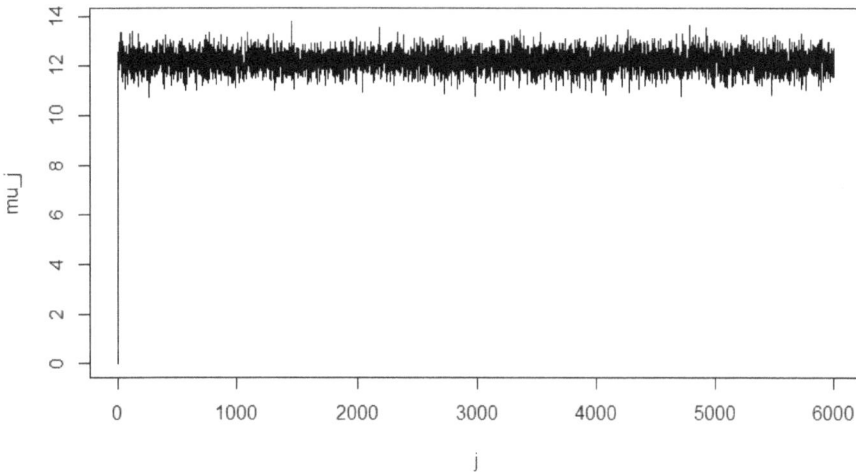

Figure 6.28 Trace for λ from Gibbs sampler

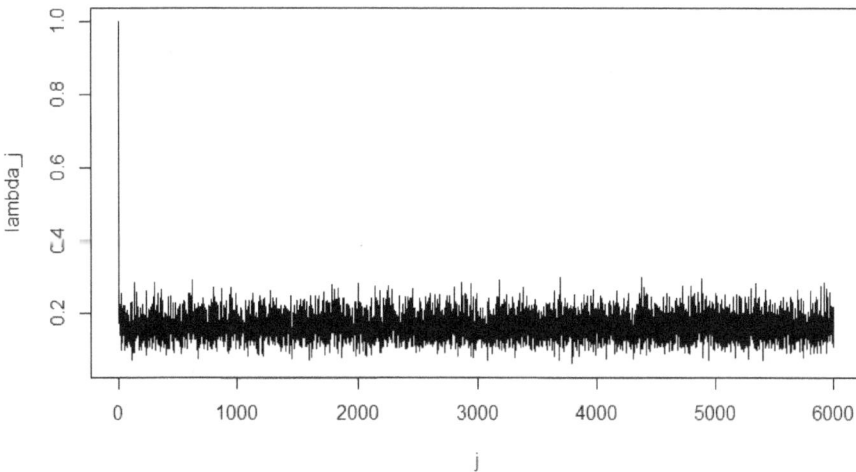

Figure 6.29 Histogram for μ from Gibbs sampler

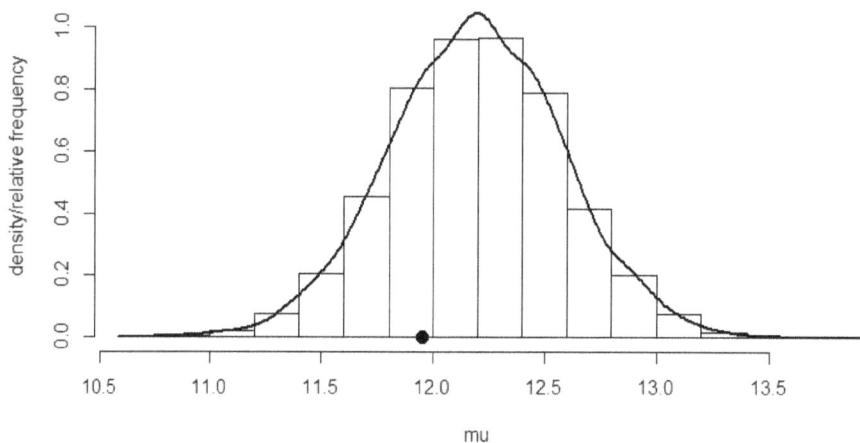

Figure 6.30 Histogram for λ from Gibbs sampler

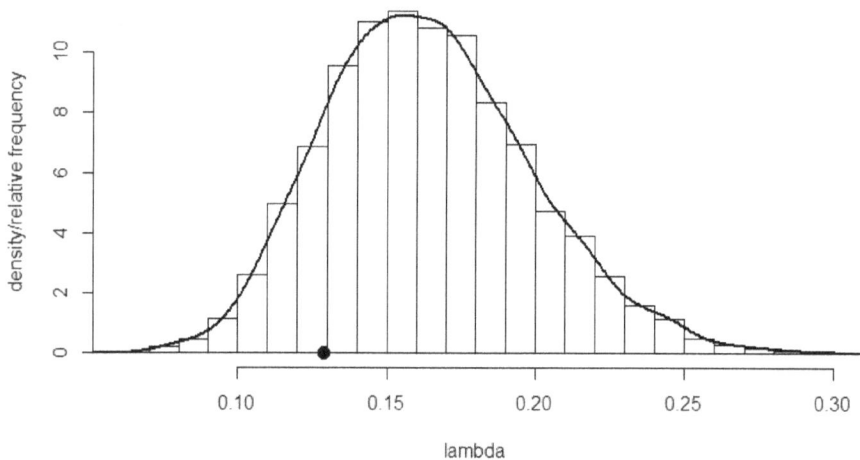

Figure 6.31 Sample autocorrelations

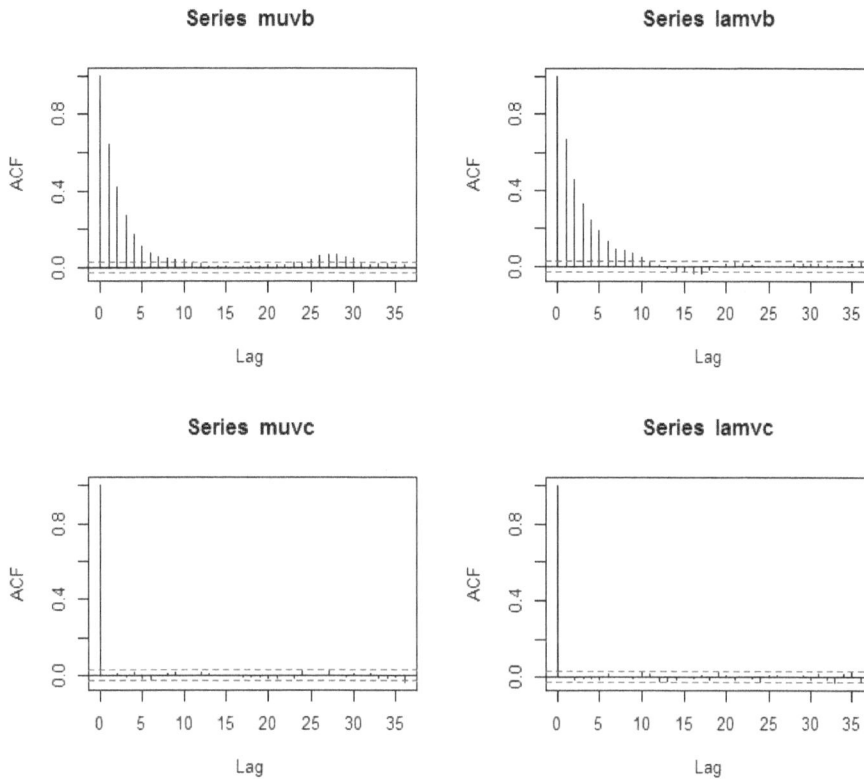

R Code for Exercise 6.7

```
# (a)
mu0=10; sig0=2; alp=3; bet=6; n=40; options(digits=4)
set.seed(226); lam=rgamma(1,alp,bet); mu=rnorm(1,mu0,sig0);
sig=1/sqrt(lam); y=rnorm(n,mu,sig)
c(lam, sig, sig^2, mu, mean(y), sd(y))
        # 0.1292  2.7822  7.7405 11.9511 12.2768  2.5919

X11(w=8,h=5); par(mfrow=c(1,1))

hist(y,prob=T,xlim=c(5,20),ylim=c(0,0.25),breaks=seq(7,17,0.5), main=" ")
yv=seq(0,20,0.01); lines(yv, dnorm(yv,mu,sig),lwd=3)
```

```
# (b)
MH <- function(Jp, mu, lam,  y,  c, r, alp=0, bet=0,  mu0=0, sig0=10000 ){
# This function implements a Metropolis-Hastings algorithm for the general
#  normal-normal-gamma model.
# Inputs:       Jp = total number of iterations
#                mu, lam = starting values of mu and lambda
#                y = vector of n observations
#                c, r = tuning parameters for mu and lambda
#                alp, bet = parameters of lambda's gamma prior (mean = alp/bet)
#                mu0, sig0 = mean and standard deviation of mu's normal prior
# Outputs:      $muv, $lamv = (Jp+1)-vectors of values of mu and lambda
#                $muar, $lamar = acceptance rates for mu  and lambda.
muv <- mu; lamv <- lam; ybar <- mean(y); n <- length(y); muct <- 0; lamct <- 0
logpost <- function(n,y,mu,lam,alp,bet,mu0,sig0){
   (alp + n/2-1)*log(lam) - bet*lam -
      0.5*lam*sum((y-mu)^2) -0.5*(mu-mu0)^2/sig0^2 }
for(j in 1:Jp){
       mup <- runif(1,mu-c,mu+c)    # propose a value of mu
       q1 <-
  logpost(n=n,y=y,mu=mup,  lam=lam,alp=alp,bet=bet,mu0=mu0,sig0=sig0)-
  logpost(n=n,y=y,mu=mu    ,lam=lam,alp=alp,bet=bet, mu0=mu0,sig0=sig0)
       p1 <- exp(q1)        # acceptance probability
       u <- runif(1); if(u < p1){ mu <- mup;  muct <- muct + 1  }
       lamp <- runif(1,lam-r,lam+r)    # propose a value of lambda
       if(lamp > 0){  # automatically reject if lamp < 0
         q2 <-
  logpost(n=n,y=y,mu=mu,lam=lamp,alp=alp,bet=bet, mu0=mu0,sig0=sig0)-
  logpost(n=n,y=y,mu=mu,lam=lam    ,alp=alp,bet=bet, mu0=mu0,sig0=sig0)
          p2 <- exp(q2)    # acceptance probability
          u <- runif(1); if(u < p2){ lam <- lamp;  lamct <- lamct + 1  }
         }
       muv <- c(muv,mu);  lamv <- c(lamv,lam)
       }
list(muv=muv,lamv=lamv,muar=muct/Jp,lamar=lamct/Jp)
}

Jp <- 6000; set.seed(331)
res <- MH(Jp=Jp,  mu=0,lam=1,  y=y,  c=0.1,r=0.01,  alp=3,bet=6,
        mu0=10,sig0=2)
c(res$muar,res$lamar)  # 0.9193 0.9165

plot(0:Jp,res$muv,type="l",xlab="j",ylab="mu_j"); text(3000,6,"c=0.1, r=0.01")
plot(0:Jp,res$lamv,type="l",xlab="j",ylab="lambda_j");
text(3000, 0.6,"c=0.1, r=0.01")
```

```
res <- MH(Jp=Jp,  mu=0,lam=1,  y=y,  c=0.9,r=0.08,  alp=3,bet=6,
       mu0=10,sig0=2)
c(res$muar,res$lamar)  # 0.5890 0.5757

plot(0:Jp,res$muv,type="l",xlab="j",ylab="mu_j"); text(3000,6,"c=0.9, r=0.08")

plot(0:Jp,res$lamv,type="l",xlab="j",ylab="lambda_j");
text(3000,0.6,"c=0.9, r=0.08")

burn <- 1000;  muv <- res$muv[-(1:(burn+1))];  lamv <- res$lamv[-(1:(burn+1))]
hist(muv,prob=T,xlab="mu",nclass=20,main="",
       ylab="density/relative frequency"); lines(density(muv),lwd=2);
points(mu,0,pch=16,cex=1.5)

hist(lamv,prob=T,xlab="lambda",nclass=20,main="",
       ylab="density/relative frequency"); lines(density(lamv),lwd=2)
points(lam,0,pch=16,cex=1.5)

# acf(muv)$acf[1:5] # 1.0000 0.6452 0.4175 0.2744 0.1770
# acf(lamv)$acf[1:5] # 1.0000 0.6641 0.4535 0.3300 0.2419
muvb= muv;  lamvb=lamv  # For use later

# (c)
GS = function(Jp, mu, lam,  y,  alp=0, bet=0,  mu0=0, sig0=10000 ){
# This function implements a Gibbs Sampler for the general  normal-normal-
gamma model.
# Inputs:        Jp = total number of iterations
#                mu, lam = starting values of mu and lambda
#                y = vector of n observations
#                alp, bet = parameters of lambda's gamma prior (mean = alp/bet)
#                mu0, sig0 = mean and standard deviation of mu's normal prior
# Outputs:       $muv, $lamv = (Jp+1)-vectors of values of mu and lambda
muv = mu; lamv = lam; n = length(y); ybar = mean(y); s2 = var(y); sig02 = sig0^2
for(j in 1:Jp){
       sig2=1/lam; k=n/(n+sig2/sig02); sig2star=k*sig2/n;
       mustar=(1-k)*mu0+k*ybar
       mu = rnorm(1,mustar,sqrt(sig2star))
       lam=rgamma( 1,  alp+0.5*n,   bet+0.5*((n-1)*s2+n*(mu-ybar)^2)  )
       muv = c(muv,mu); lamv = c(lamv,lam)   }
list(muv=muv,lamv=lamv)
}
```

```
Jp = 6000; set.seed(331)
res = GS(Jp=Jp,  mu=0,lam=1,  y=y,  alp=3,bet=6,   mu0=10,sig0=2)

plot(0:Jp,res$muv,type="l",xlab="j",ylab="mu_j");

plot(0:Jp,res$lamv,type="l",xlab="j",ylab="lambda_j");

burn <- 1000; muv <- res$muv[-(1:(burn+1))]; lamv <- res$lamv[-(1:(burn+1))]

hist(muv,prob=T,xlab="mu",nclass=20,main="",ylim=c(0,1.1),
        ylab="density/relative frequency"); lines(density(muv),lwd=2);
points(mu,0,pch=16,cex=1.5)

hist(lamv,prob=T,xlab="lambda",nclass=20,main="",
        ylab="density/relative frequency"); lines(density(lamv),lwd=2)
points(lam,0,pch=16,cex=1.5)

muvc=muv; lamvc=lamv

X11(w=8,h=7); par(mfrow=c(2,2))

acf(muvb)$acf[1:5] # 1.0000 0.6452 0.4175 0.2744 0.1770
acf(lamvb)$acf[1:5] # 1.0000 0.6641 0.4535 0.3300 0.2419

acf(muvc)$acf[1:5] # 1.0000000 -0.0004031  0.0079520 -0.0073517  0.0135979
acf(lamvc)$acf[1:5] # 1.000000  0.002873 -0.011504 -0.006671 -0.001769
```

CHAPTER 7

MCMC Methods Part 2

7.1 Introduction

In the last chapter we introduced a set of very powerful tools for generating samples required for Bayesian Monte Carlo inference, namely Markov chain Monte Carlo (MCMC) methods. The topics we covered included the Metropolis algorithm, the Metropolis Hastings algorithm and the Gibbs sampler.

We now present one more topic, stochastic data augmentation, and provide some further exercises in MCMC. These exercises will illustrate how many statistical problem can be cast in the Bayesian framework and how easily inference can then proceed relative to the classical framework.

The examples below include simple linear regression, logistic regression (an example of generalised linear modelling and survival analysis), autocorrelated Bernoulli data, and inference on the unknown bounds of a uniform distribution.

7.2 Data augmentation

Data augmentation (DA) is a method for using unobserved data or latent variables so as to simplify and facilitate an iterative optimisation or sampling algorithm. There are two basic types of DA: *deterministic DA* and *stochastic DA*. An example of the former is the EM algorithm as described earlier. Stochastic DA is illustrated in the following example.

Example of stochastic data augmentation

Suppose we wish to sample from a univariate distribution defined by a density $f(x)$ but that this is difficult to do directly. But then, also suppose that we can factor this density as

$$f(x) \propto g(x)h(x),$$

where:

$$g(x) = \int q(u \mid x)du$$

$q(u \mid x)$ is the kernel of conditional density for a latent random variable u given x which is easy to sample from

$q(u \mid x)h(x)$ defines the kernel of a conditional density for x given u which is easy to sample from; call this kernel $k(x \mid u)$.

In such a situation we may define the joint distribution of u and x by the density

$$f(u,x) \propto q(u \mid x)h(x).$$

Then, since both of the conditional distributions (of u given x, and of x given u) are easy to sample from, we may define a suitable Gibbs sampler by the following two steps:

(i) Sample $u' \sim q(u \mid x)$
(ii) Sample $x' \sim k(x \mid u')$.

Running this Gibbs sampler will eventually result in a random sample

$$(u_1, x_1), ..., (u_J, x_J) \sim iid \ f(u,x).$$

Discarding the simulated latent variables $u_1, ..., u_J$ then yields the desired sample,

$$x_1, ..., x_J \sim iid \ f(x).$$

This idea can be extended in a straightforward fashion to sampling from a multivariate distribution, i.e. where x is a vector. In such cases, it may be necessary to define several latent variables in the fashion described above.

Exercise 7.1 Sampling with the aid of stochastic data augmentation

We wish to find the mean of a random variable with density

$$f(x) \propto \frac{e^{-x}}{x+1}, \ x > 0.$$

(a) Calculate the exact value of EX using numerical integration techniques.

(b) Estimate EX using a Monte Carlo sample obtained via *rejection sampling*.

(c) Estimate *EX* using a Monte Carlo sample obtained via the *Metropolis algorithm*.

(d) Estimate *EX* using a Monte Carlo sample obtained via a *Gibbs sampler* designed using the principles of *data augmentation*.

Note 1: We have already seen the above density $f(x)$ in the context of a previous exercise.

Note 2: The intent of this exercise is threefold:

(i) to illustrate stochastic data augmentation
(ii) to provide additional practice at several Monte Carlo techniques
(iii) to introduce an idea that will be useful later when attempting finite population inference under biased sampling without replacement.

Solution to Exercise 7.1

(a) Let the kernel be $k(x) = \dfrac{e^{-x}}{x+1}$.

Then, using the integrate() function in R, we obtain

$$\int_0^\infty k(x)dx = 0.59635 \text{ and } \int_0^\infty xk(x)dx = 0.40365.$$

So $EX = 0.40365/0.59635 = 0.6769$.

(b) A suitable envelope is the standard exponential density
$$h(x) = e^{-x}, x > 0,$$
for which the acceptance probability is
$$p(x) - \frac{k(x)}{ch(x)},$$
where
$$c = \max\frac{k(x)}{h(x)} = \max\frac{e^{-x}/(x+1)}{e^{-x}} = 1.$$

Thus $p(x) = \dfrac{1}{x+1}$.

Applying this algorithm we obtained a random sample of size $J = 1,000$ using a total of 1,651 draws from the envelope. (Thus the acceptance rate was $1,000/1651 = 61\%$.) Using this Monte Carlo sample, we estimated EX as 0.6875 with 95% CI (0.6402, 0.7349).

Figure 7.1 shows a trace plot of the simulated values and (just for interest) the associated sample ACF of these values (showing the complete absence of autocorrelation), respectively.

Figure 7.1 Trace plot and sample ACF

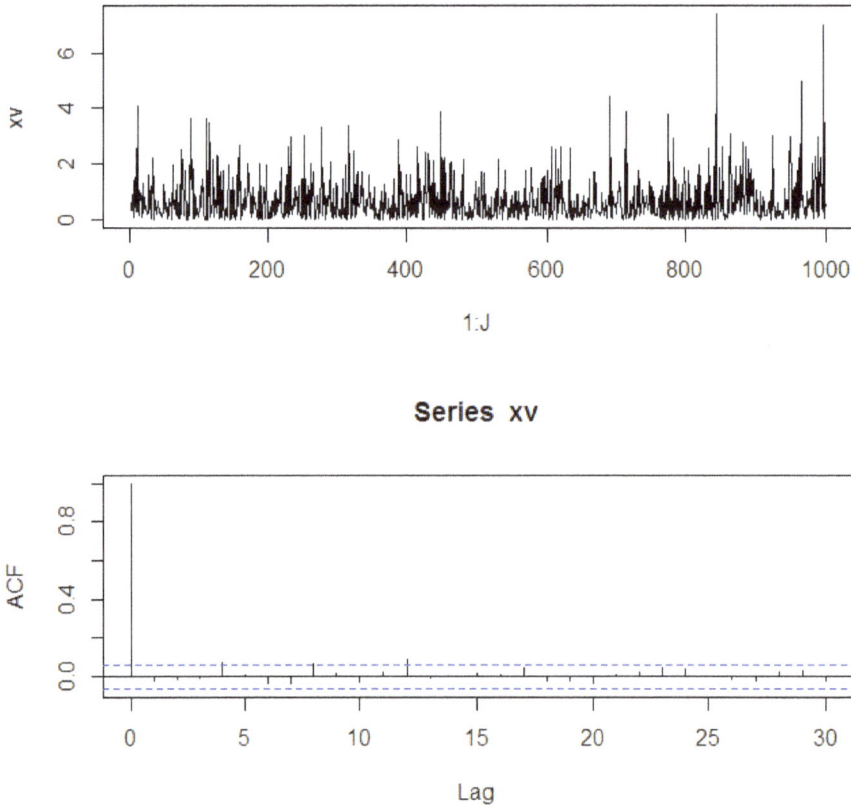

Series xv

(c) Using a normal driver distribution centred at the last value and with standard deviation 0.6 we ran a Metropolis algorithm for 40,500 iterations, starting at $x = 1$. We kept every 40th sampled value after first discarding the first 500 iterations as the burn-in. Using the resulting Monte Carlo sample of size 1,000, we estimated EX as 0.7049 with 95% CI (0.6561, 0.7537). The overall acceptance rate of the algorithm was 58%. Figure 7.2 shows a trace plot of all 40,500 simulated values, the sample ACF of those values (showing a very strong autocorrelation), a trace plot of the 1,000 values used for inference, and the sample ACF of those values (showing very little autocorrelation).

Figure 7.2 Trace plots and sample ACFs

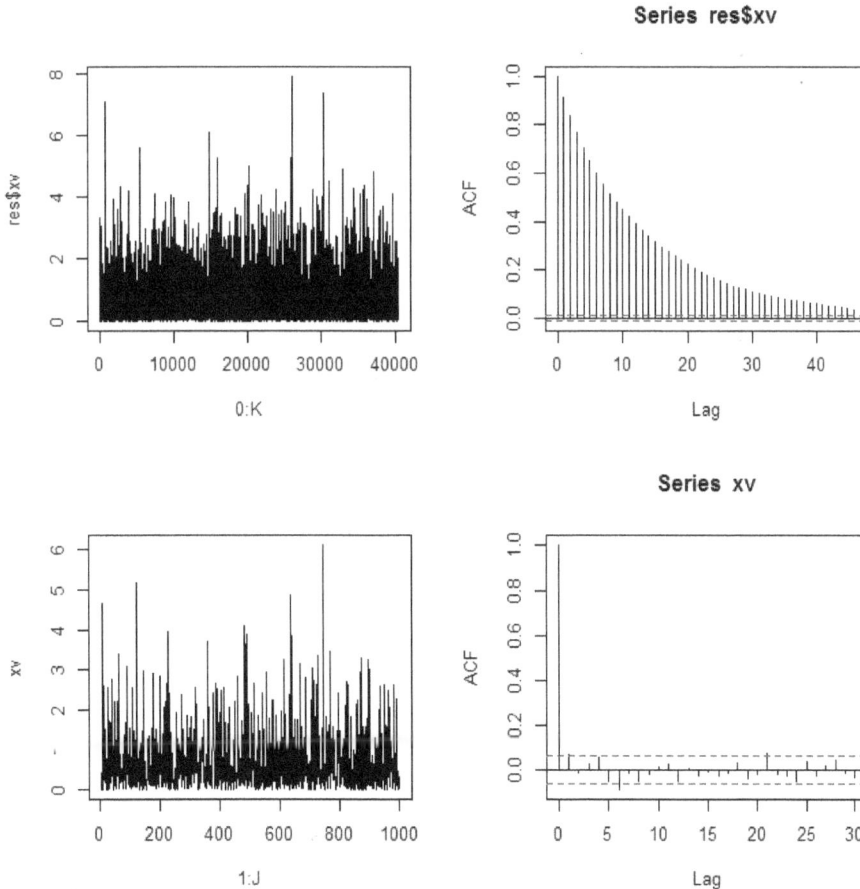

(d) Observe that $\dfrac{1}{x+1} = \displaystyle\int_0^\infty e^{-(x+1)w}\,dw$.

Therefore $f(x) = \dfrac{1}{x+1} e^{-x} \propto \displaystyle\int_0^\infty e^{-(x+1)w} e^{-x}\,dw$.

Hence we may define an artificial latent variable w such that the joint density of w and x is

$$f(w,x) \propto e^{-(x+1)w} e^{-x},\ w > 0,\ x > 0 .$$

We see that:

$$f(w\mid x) \propto f(w,x) \overset{w}{\propto} e^{-(x+1)w},\ w > 0$$

$$f(x\mid w) \propto f(w,x) \overset{x}{\propto} e^{-(w+1)x},\ x > 0 .$$

So, a Gibbs sampler for sampling from $f(w,x)$ is defined by the two densities:

$$f(w\mid x) = (x+1)e^{-(x+1)w},\ w > 0$$

$$f(x\mid w) = (w+1)e^{-(w+1)x},\ x > 0 ,$$

or equivalently by the two steps:

Sample $w \sim Gamma(1, x+1)$

Sample $x \sim Gamma(1, w+1)$.

Starting at $x = 1$, we ran this Gibbs sampler for 5,100 iterations. We then kept every 5th sampled value after first discarding the first 100 iterations as the burn-in. Using the resulting Monte Carlo sample of size 1,000 we estimated EX as 0.7172 with 95% CI (0.6671, 0.7673).

Figure 7.3 shows a trace plot of all 5,100 simulated values, their sample ACF (showing a slight autocorrelation), a trace plot of the 1,000 values used for inference, and the sample ACF of these 1,000 values (showing very little autocorrelation).

Note that similar plots could also be produced for the simulated latent variable, w. Also note how data augmentation and a Gibbs sampler have resulted in a usable Monte Carlo sample more easily and effectively than the Metropolis algorithm.

Figure 7.3 Trace plots and sample ACFs

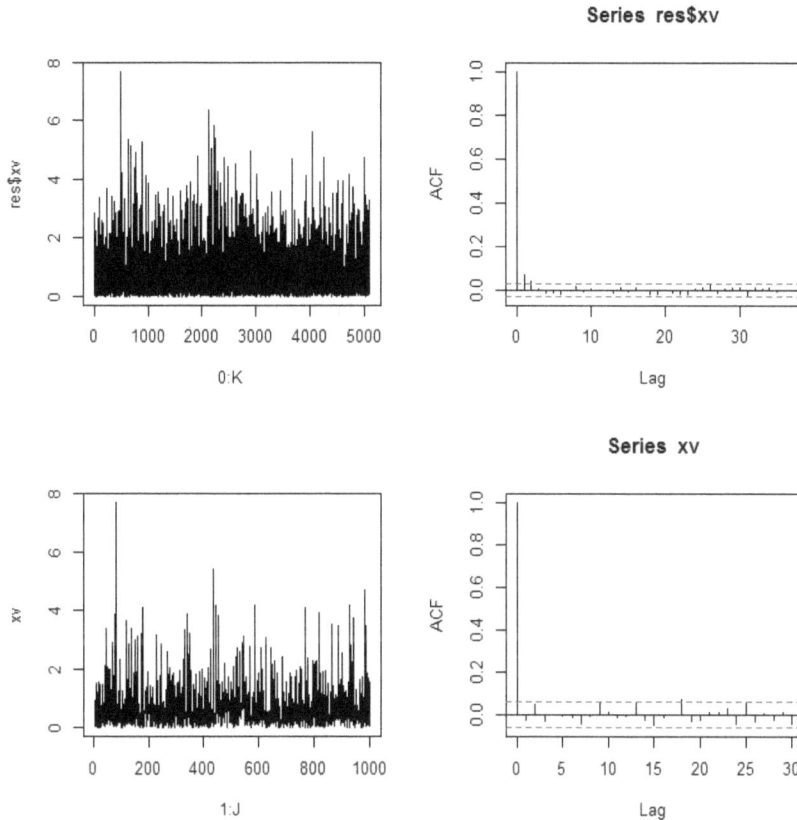

Series res$xv

Series xv

R Code for Exercise 7.1

```
# (a)
options(digits=5); kfun=function(x){ exp(-x)/(x+1) }
c=integrate(f=kfun,lower=0,upper=Inf)$value; c # 0.59635
xkfun =function(x){ x*exp(-x)/(x+1) }
top=integrate(f=xkfun,lower=0,upper=Inf)$value; top # 0.40365
EX=top/c; EX # 0.67688

# (b)
J=1000; xv=rep(NA,J); ct=0; set.seed(331)
for(j in 1:J){ acc=F; while(acc==F){ ct=ct+1
        x=rgamma(1,1,1); p=1/(x+1); u=runif(1); if(u<p){ acc=T; xv[j]=x } } }
xbar=mean(xv); ci=xbar + c(-1,1)*qnorm(0.975)*sd(xv)/sqrt(J)
c(ct,xbar,ci) # 1651.00000   0.68754   0.64016   0.73492
par(mfrow=c(2,1)); plot(1:J,xv,type="l")
acf(xv)$acf[1:5] # 1.0000000 -0.0205516 -0.0100987 -0.0040018 0.0732520
```

327

```
# (c)
MET <- function(K,x,c){
# This function applies the Metropolis algorithm to sampling from
# f(x)~exp(-x)/(x+1),x>0.
# Inputs:        K = total number of iterations
#                x = intial value of x,      c = standard deviation of normal driver
# Ouputs:     $xv = vector of (K+1) values of x,       $ar = acceptance rate
xv = x; ct = 0
for(j in 1:K){
        xp = rnorm(1,x,c)
        if(xp>0) {
                q =  (-xp-log(xp+1)) - (-x-log(x+1)); p = exp(q);   u = runif(1)
                if(u < p){ x = xp; ct = ct + 1 }
                }
        xv <- c(xv,x)     }
ar = ct/K;   list(xv=xv,ar=ar)   }
```

```
K=40500; set.seed(298); res <- MET(K=K,x=1,c=0.6);   res$ar  # 0.53896
par(mfrow=c(2,2)); plot(0:K, res$xv,type="l")
acf(res$xv)$acf[1:5] # 1.00000 0.91458 0.83710 0.76808 0.70716
xv=res$xv[-(1:501)][seq(40,40000,40)];  plot(1:J,xv,type="l")
acf(xv)$acf[1:5] # 1.0000000 0.0727149 -0.0088327 0.0265807 0.0592275
xbar=mean(xv); ci=xbar + c(-1,1)*qnorm(0.975)*sd(xv)/sqrt(J)
c(xbar,ci) # 0.70491 0.65614 0.75368
```

```
# (d)
GIBBS <- function(K,x){
# This generates a sample using the Gibbs sampler and data augmentation.
# Inputs:  K = total number of iterations,   x = initial value of x
# Ouputs: $xv = vector of (K+1) values of x, $wv = vector of (K+1) values of w
xv = x; wv=NA; for(j in 1:K){
        w=rgamma(1,1,x+1); x=rgamma(1,1,w+1); xv=c(xv,x); wv=c(wv,w) }
list(xv=xv,wv=wv)   }
```

```
K=5100; set.seed(319); res <- GIBBS(K=K,x=1)
par(mfrow=c(2,2)); plot(0:K, res$xv,type="l")
acf(res$xv)$acf[1:5] # 1.0000000 0.0692628 0.0407747 0.0053119 -0.0133717
xv=res$xv[-(1:101)][seq(5,5000,5)];  plot(1:J,xv,type="l")
acf(xv)$acf[1:5]
   # 1.0000e+00 -2.4435e-02  4.5681e-02 -3.1778e-02  2.7116e-05
xbar=mean(xv); ci=xbar + c(-1,1)*qnorm(0.975)*sd(xv)/sqrt(J)
c(xbar,ci) # 0.71720 0.66711 0.76729
```

Exercise 7.2 Comparison of classical and Bayesian simple linear regression (and practice at various statistical techniques)

Consider the following simple linear regression model:
$$Y_i \sim\!\perp N(\mu_i, \sigma^2), \; i = 1,...,n,$$
where
$$\mu_i = a + bx_i$$
(linear predictor for a value with covariate x_i).

(a) Generate a data vector $y = (y_1, ..., y_n)$ from the model, using:
$$n = 10, \; a = 5, \; b = 2, \; \sigma = 2,$$
and with covariates
$$x_i = i$$
for all $i = 1,...,n$.

(b) Conduct a *classical* analysis of the data in (a). Report the MLEs and 95% CIs for a and b. Also create a single graph which shows:

- the data values

- the true regression line $E(Y \mid x) = a + bx$

- the fitted regression line $\hat{E}(Y \mid x) = \hat{a} + \hat{b}x$

- two lines showing the 95% CI for the regression line

- two lines showing the 95% prediction interval at each value of x.

(c) Perform a Bayesian analogue of the inference in (b) using the Metropolis-Hastings algorithm and a Monte Carlo sample of size $J = 2{,}000$.

Use a suitable joint uninformative and improper prior for the three parameters in the model.

(d) Create a single graph showing all the information in the two graphs in (b) and (c).

Note: The Bayesian analysis in (c) could also be performed via the Gibbs sampler.

Solution to Exercise 7.2

(a) The simulated data are shown in Table 7.1. Note that $x_i = i$.

Table 7.1 Simulated data

i	1	2	3	4	5
y_i	5.879	8.54	14.12	13.14	15.26

i	6	7	8	9	10
y_i	20.43	19.92	18.47	21.63	24.11

(b) The MLE of b is $\hat{b} = \dfrac{\sum_{i=1}^{n}(x_i - \overline{x})(y_i - \overline{y})}{\sum_{i=1}^{n}(x_i - \overline{x})^2} = 1.836,$

and the MLE of a is then $\hat{a} = \overline{y} - \hat{b}\overline{x} = 6.051.$

An unbiased estimate of $\sigma^2 (= 1/\lambda = 4)$ is

$$s^2 = \frac{1}{n-2}\sum_{i=1}^{n}(y_i - \{\hat{a} + \hat{b}x_i\})^2 = 3.816.$$

Let: $X = \begin{pmatrix} 1 & 1 \\ 1 & 2 \\ \vdots & \vdots \\ 1 & n \end{pmatrix}$

$M = \begin{pmatrix} m_{11} & m_{12} \\ m_{21} & m_{22} \end{pmatrix} = (X'X)^{-1}.$

A 95% CI for a is then

$$\left(\hat{a} \pm t_{0.025}(8)s\sqrt{m_{11}}\right) = (1.340, 2.332),$$

and a 95% CI for b is

$$\left(\hat{b} \pm t_{0.025}(8)s\sqrt{m_{22}}\right) = (2.973, 9.128).$$

Also, a 95% CI for $E(Y \mid x) = a + bx$ is

$$\left((\hat{a} + \hat{b}x) \pm t_{0.025}(8)s \sqrt{(1 \quad x) M \binom{1}{x}} \right),$$

and a 95% prediction interval for a new observation Y with covariate x is

$$\left((\hat{a} + \hat{b}x) \pm t_{0.025}(8)s \sqrt{1 + (1 \quad x) M \binom{1}{x}} \right).$$

The required graph is shown in Figure 7.4.

Figure 7.4 Classical inference

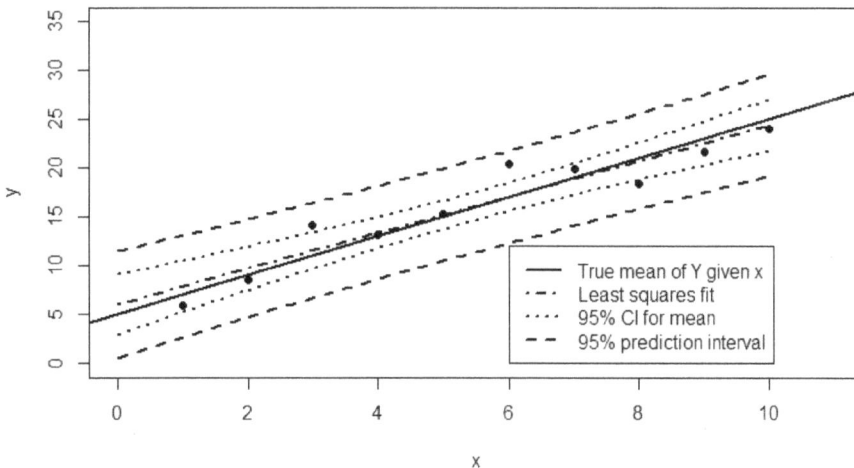

(c) A suitable Bayesian model is given by:

$$(Y_i \mid a, b, \lambda) \sim \perp N(a + bx_i, 1/\lambda), \, i = 1, \dots, n$$

$$f(a, b, \lambda) \propto 1/\lambda, \quad a, b \in \Re, \quad \lambda > 0 \quad \text{(where } \lambda = 1/\sigma^2 \text{)}.$$

Let us now solve this Bayesian model so as to estimate the posterior means and 95% CPDRs for a and b. The joint posterior density of the three model parameters is

$$f(a, b, \lambda \mid y) \propto \frac{1}{\lambda} \times \prod_{i=1}^{n} \sqrt{\lambda} \exp\left\{ -\frac{\lambda}{2} (y_i - \mu_i)^2 \right\}$$

(where $\mu_i = a + bx_i$ as already defined).

Hence the joint log-posterior density (up to an additive constant) is

$$l(a,b,\lambda) = \left(\frac{n}{2}-1\right)\log\lambda - \frac{\lambda}{2}\sum_{i=1}^{n}(y_i - \mu_i)^2 \ .$$

Applying the MH algorithm for 2,500 iterations, we obtain traces for the three parameters as shown in Figure 7.5. The horizontal lines show the true values of the three parameters. The fourth subplot (bottom right) is a histogram of the last 2,000 values of b simulated.

Figure 7.5 Results of a MH algorithm

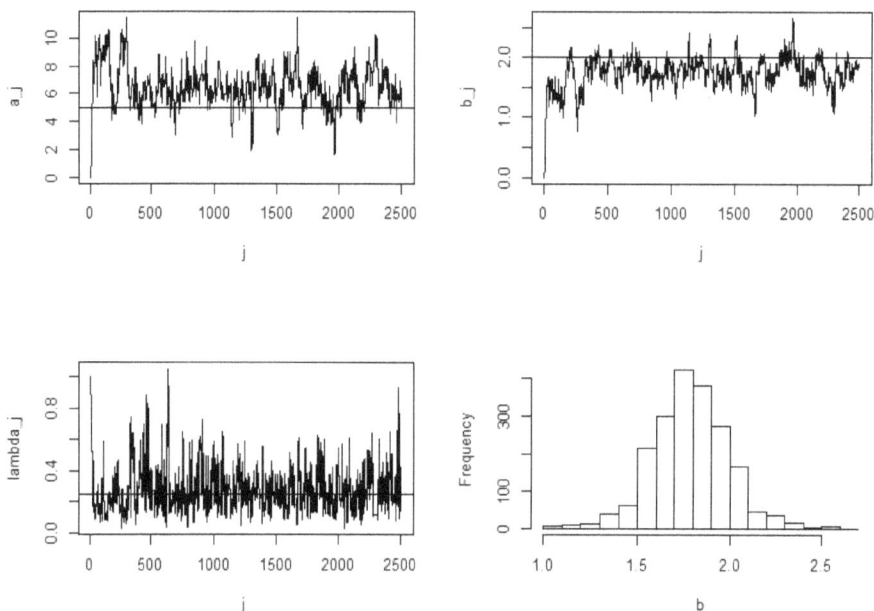

Using output from the last 2,000 iterations only, we estimate the posterior mean and 95% CPDR for a (= 5) as 6.3445 and (3.578, 8.808), and the same for b (= 2) are about 1.7881 and (1.392, 2.234).

Figure 7.6 shows the Bayesian analogue of Figure 7.5 in part (b).

Figure 7.6 Bayesian inference

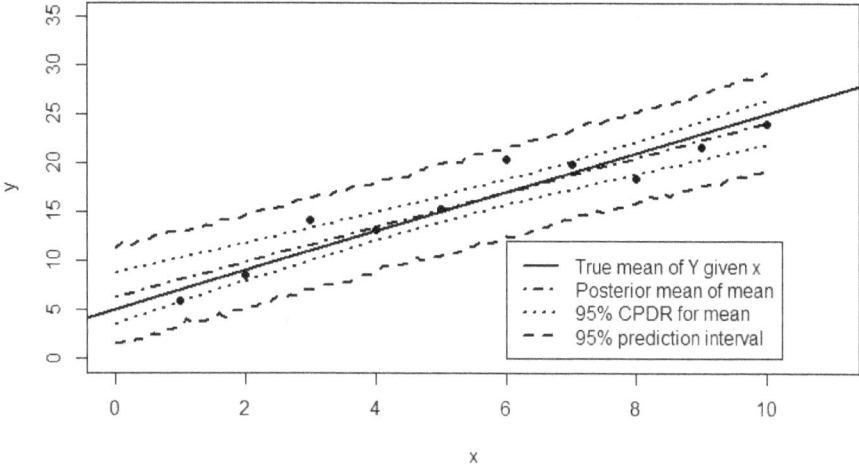

(d) The required graph is shown in Figure 7.7.

Figure 7.7 Comparison of inferences

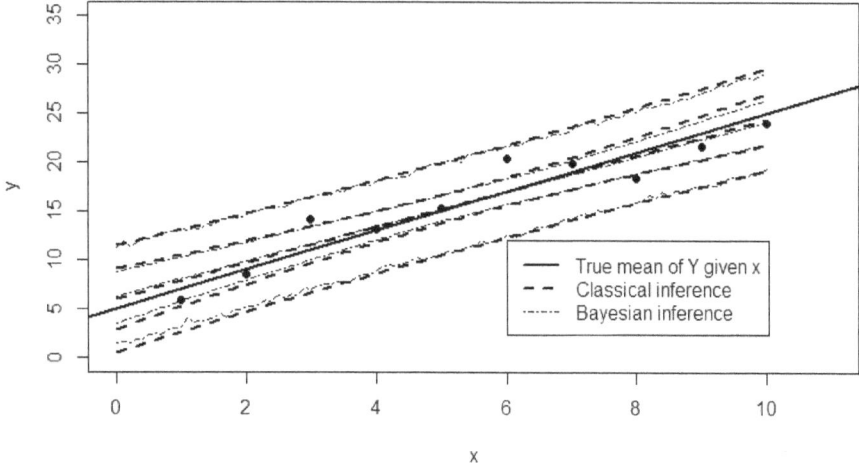

R Code for Exercise 7.2

```
# (a) ************************************************
options(digits=4)
n <- 10; a <- 5; b <- 2; lam <- 0.25; sig <- 1/sqrt(lam);   c(sig,sig^2)  # 2 4
xdat <- 1:n;    set.seed(123); ydat <- rnorm(n,a+b*xdat,sig)
rbind(xdat,ydat)
# xdat 1.000 2.00 3.00 4.00 5.00 6.00 7.00 8.00 9.00 10.00
# ydat 5.879 8.54 14.12 13.14 15.26 20.43 19.92 18.47 21.63 24.11


# (b)  ************************************************************
fit <- lm(ydat ~ xdat);   summary(fit)
#           Estimate Std. Error t value Pr(>|t|)
# (Intercept)   6.051     1.335    4.53  0.0019 **
# xdat          1.836     0.215    8.54  2.7e-05 ***


ahat <- coef(fit)[[1]]; bhat <- coef(fit)[[2]]
sse <- sum((ydat-(ahat+bhat*xdat))^2)
sig2hat <- sse/(n-2); lamhat <- 1/sig2hat
c(sse,sig2hat,lamhat)  # 30.532  3.816  0.262

df <- length(ydat)-length(fit$coef)
aCI <- ahat + c(-1,1)*qt(0.975,df)*sqrt(sig2hat*summary(fit)$cov.unscaled[1,1])
aCI       # 2.973 9.128
bCI <- bhat + c(-1,1)*qt(0.975,df)*sqrt(sig2hat*summary(fit)$cov.unscaled[2,2])
bCI       # 1.340 2.332

xxv <- seq(0,n,0.1); nn <- length(xxv)
Xmat <- cbind(1,xxv)
muhat <- Xmat %*% fit$coef
muhatvar <- sig2hat * diag(Xmat %*% summary(fit)$cov.unscaled %*% t(Xmat))
df <- length(ydat)-length(fit$coef)
muhatlb <- muhat - qt(0.975,df) * sqrt(muhatvar)
muhatub <- muhat + qt(0.975,df) * sqrt(muhatvar)

predlb <- muhat - qt(0.975,df) * sqrt(sig2hat+muhatvar)
predub <- muhat + qt(0.975,df) * sqrt(sig2hat+muhatvar)

X11(w=8,h=5); par(mfrow=c(1,1))  # Figure
plot(xdat, ydat, pch=16, xlim=c(0,11),ylim=c(0,35),xlab="x",ylab="y"  )
abline(c(a,b),lwd=2);
lines(c(0,n),c(fit$coef[1], fit$coef[1]+ fit$coef[2]*n),lty=4, lwd=2)
lines(xxv,muhatlb,lty=3,lwd=2)
```

```
lines(xxv,muhatub,lty=3,lwd=2)
lines(xxv, predlb,lty=2,lwd=2)
lines(xxv, predub,lty=2,lwd=2)
legend(6,12,c("True mean of Y given x","Least squares fit","95% CI for mean",
            "95% prediction interval"),lty=c(1,4,3,2),lwd=rep(2,4))


# (c) ********************************************************
MH.SLR <-  function(Jp,  x, y,   a, b, lam,    asd, bsd, lamsd){
# This function implements a Metropolis Hastings algorithm for a
# simple linear regression model with uninformative priors.
# Inputs:        Jp = total number of iterations
#                x = vector of covariates
#                y = vector of observations
#                a,b,lam = starting values of a,b,lambda
#                asd,bsd,lamsd = st. dev.s of drivers for a,b,lambda.
# Outputs:       $av,$bv,$lamv = (Jp+1)-vectors of values of a,b,lambda
#                $aar,$bar,$lamar = acceptance rates for a,b,lambda.
av <- a; bv <- b; lamv <- lam; ybar <- mean(y); n <- length(y)
act <- 0; bct <- 0; lamct <- 0
logpost <- function(n, x, y, a, b, lam){ #  logposterior
        (n/2 - 1) * log(lam) - 0.5 * lam * sum((y - a - b * x)^2)   }
for(j in 1:Jp) {
        ap <- rnorm(1, a, asd)  # propose a value of a
        k <-     logpost(n=n, x=x, y=y, a=ap,     b=b, lam=lam) -
                 logpost(n=n, x=x, y=y, a=a,      b=b, lam=lam)
        p <- exp(k)   # acceptance probability
        u <- runif(1);   if(u < p) {  a <- ap;  act <- act + 1  }
        bp <- rnorm(1, b, bsd)  # propose a value of b
        k <-     logpost(n=n, x=x, y=y, a=a, b=bp, lam=lam) -
                 logpost(n=n, x=x, y=y, a=a, b=b, lam=lam)
        p <- exp(k)  # acceptance probability
        u <- runif(1);   if(u < p) {  b <- bp;  bct <- bct + 1  }
        lamp <- rnorm(1, lam, lamsd)  # propose a value of lambda
        if(lamp > 0) { # automatically reject if lamp < 0
          k <-  logpost(n=n, x=x, y=y, a=a, b=b, lam=lamp) -
                    logpost(n=n, x=x, y=y, a=a, b=b, lam=lam)
          p <- exp(k) # acceptance probability
          u <- runif(1);   if(u < p) {  lam <- lamp;   lamct <- lamct + 1  }
          }
        av <- c(av, a);   bv <- c(bv, b);   lamv <- c(lamv, lam)
        }
list(av = av, bv = bv, lamv = lamv, aar = act/Jp, bar = bct/Jp, lamar = lamct/Jp)
}
```

```
Jp <- 2500; set.seed(441)
mh <- MH.SLR(Jp=Jp,   x=xdat,y=ydat, a=0,b=0,lam=1,
                 asd=1.2,bsd=0.2,lamsd=0.2)
c(mh$aar,mh$bar,mh$lamar)    # 0.5228 0.5008 0.5132

X11(w=8,h=6); par(mfrow=c(2,2)) # Figure
plot(0:Jp,mh$av,xlab="j",ylab="a_j",type="l"); abline(h=a)
plot(0:Jp,mh$bv,xlab="j",ylab="b_j", type="l"); abline(h=b)
plot(0:Jp,mh$lamv,xlab="j",ylab="lambda_j", type="l"); abline(h=lam)
hist(mh$bv[-(1:501)],main="",xlab="b")

burn <- 500; J <- Jp - burn;  J # 2000
av <- mh$av[-c(1:(burn+1))];     abar <-mean(av)
bv <- mh$bv[-c(1:(burn+1))]; bbar <- mean(bv)
lamv <- mh$lamv[-c(1:(burn+1))]; lambar <- mean(lamv)

sig2bar <- mean(1/lamv)
c(abar,bbar,lambar,sig2bar)  # 6.3445 1.7881 0.2758 4.7505

quantile(av,c(0.025,0.975))  # 3.578 8.808
quantile(bv,c(0.025,0.975)) # 1.392 2.234

cpdrLBs <- xxv; cpdrUBs <- xxv;   predLBs <- xxv; predUBs <- xxv; set.seed(171)
for(i in 1:nn){
        mus <- av + bv*xxv[i]
        cpdrLBs[i] <- quantile(mus,0.025)
        cpdrUBs[i] <- quantile(mus,0.975)
        sim <- rnorm(J,mus,1/sqrt(lamv))
        predLBs[i] <- quantile(sim,0.025)
        predUBs[i] <- quantile(sim,0.975)
        }

X11(w=8,h=5); par(mfrow=c(1,1))  # Figure
plot(xdat,ydat,pch=16,xlim=c(0,11),ylim=c(0,35),xlab="x",ylab="y"  )
abline(c(a,b),lwd=2); lines(c(0,n),c(abar, abar + bbar *n),lty=4, lwd=2);
lines(xxv,cpdrLBs,lty=3,lwd=2)
lines(xxv,cpdrUBs,lty=3, lwd=2)
lines(xxv,predLBs,lty=2, lwd=2)
lines(xxv,predUBs,lty=2, lwd=2)
legend(6,12,c("True mean of Y given x","Posterior mean of mean",
  "95% CPDR for mean","95% prediction interval"),lty=c(1,4,3,2),lwd=rep(2,4))
```

```
# (d)  ***************************************************

X11(w=8,h=5); par(mfrow=c(1,1))  # Figure
plot(xdat,ydat,pch=16, xlim=c(0,11),ylim=c(0,35),xlab="x",ylab="y" )
abline(c(a,b),lwd=2)  # True regression line
# Classical lines
lines(c(0,n),c(fit$coef[1], fit$coef[1]+ fit$coef[2]*n),lty=2, lwd=2)
lines(xxv,muhatlb,lty=2, lwd=2);    lines(xxv,muhatub,lty=2, lwd=2)
lines(xxv, predlb,lty=2, lwd=2);    lines(xxv, predub,lty=2, lwd=2)
# Bayesian lines
lines(c(0,n),c(abar,abar+n*bbar),lty=4, lwd=1)
lines(xxv,cpdrLBs,lty=4, lwd=1);    lines(xxv,cpdrUBs,lty=4, lwd=1)
lines(xxv,predLBs,lty=4, lwd=1);    lines(xxv,predUBs,lty=4, lwd=1)

legend(6,12,c("True mean of Y given x",
        "Classical inference","Bayesian inference"),lty=c(1,2,4), lwd=c(2,2,1))
```

Exercise 7.3 Comparison of classical and Bayesian logistic regression (an example of GLMs) (and practice at various statistical techniques)

Table 7.2 shows data on the number of rats who died in each of $n = 10$ experiments within one month of being administered a particular dose of radiation. For example in Experiment 3, a total of 40 rats were exposed to radiation for 3.6 hours, and 23 of them died within one month. Thus an estimate of the probability of a rat dying within one month if it is exposed to 3.6 hours of radiation is 23/40 = 57.5%.

Table 7.2 Rat mortality data

i	n_i	x_i	y_i	y_i/n_i	$=$	$\hat{p}_i$
1	10	0.1	1	1/10	=	0.1
2	30	1.4	0	0/30	=	0
3	40	3.6	23	23/40	=	0.575
4	20	3.8	12	12/20	=	0.6
5	15	5.2	8	8/15	=	0.5333
6	46	6.1	32	32/46	=	0.696
7	12	8.7	10	10/12	=	0.833
8	37	9.1	35	35/37	=	0.946
9	23	9.1	19	19/23	=	0.826
10	8	13.6	8	8/8	=	1

Consider the following logistic regression model for these data:
$$Y_i \sim\!\!\perp Bin(n_i, p_i), \quad i = 1,...,n,$$
where:

$$p_i = \frac{1}{1+\exp(-z_i)} \qquad \text{(probability of a 'success' for experiment } i)$$

$$z_i = a + bx_i \qquad \text{(linear predictor).}$$

(a) Find the ML estimates of a and b using the glm() function in R. For each parameter also calculate a suitable 95% CI.

(b) Find the ML estimates and associated 95% CIs in R using your own code for the Newton-Raphson algorithm and without using the glm() function.

(c) Find the ML estimates using a modification of the Newton-Raphson algorithm which does not require the inversion of matrices.

(d) Suppose that a and b are assigned independent flat priors over the whole real line. Thus consider the Bayesian model:

$$(Y_i \mid a, b) \sim\!\!\perp Bin(n_i, p_i), \quad i = 1,...,n$$

$$p_i = \frac{1}{1+\exp(-z_i)} \qquad \text{(probability of death for experiment } i)$$

$$z_i = a + bx_i \qquad \text{(linear predictor)}$$

$$f(a,b) \propto 1, \quad a,b \in \Re.$$

Use the MH algorithm to get a sample of $J = 10,000$ observations from $f(a,b \mid y)$, where $y = (y_1, ..., y_n)$.

Hence estimate the posterior means of a and b, together with 95% MC CIs for these estimates, and also estimate the 95% CPDRs.

Show graphs of the traces and histograms. Overlay the MC estimates and MLEs over the traces, together with 95% CPDRs and CIs, respectively. Also, overlay kernel density estimates over the histograms.

(e) Use the sample in (d) to estimate $p(x)$, the probability of a rat dying if it is exposed to x hours of radiation, for each $x = 0,1,2,...,15$.

Graph these results with a line in a figure which also shows the 10 $\hat{p}_i$ values.

Also include:
- the MC 95% CI for each estimate of $p(x)$ (i.e. for each $E\{p(x)\,|\,y\}$)
- the MC 95% CPDR for each $p(x)$
- the MLE of each $p(x)$ using standard GLM procedures,
  together with associated large-sample 95% CIs.

(f) Suppose that 20 more rats are about to be exposed to exactly five hours of radiation. Use the sample in (d) to estimate how many of these 20 rats will die, together with a 95% CI for your estimate. Also construct an approximate 95% prediction region for the number of rats that will die and report the estimated actual probability content of this region.

(g) Use the sample in (d) to estimate $LD50$, the lethal dose of radiation at which 50% of rats die, together with a 95% CPDR. Also compute an estimate and 95% CI for $LD50$ using standard GLM techniques.

(h) Consider the Bayesian model and data in (d). Modify the model suitably so as to constrain the probability of death at a dose of zero to be exactly zero. Estimate the parameters in the new model and draw a graph similar to the one in (e) which shows the posterior probability of death for each dose x from zero to 15, together with the associated 95% CPDRs.

Solution to Exercise 7.3

(a) Using the glm() function in R, we find that the MLE and 95% CI for a are –2.156 and (–2.9998, –1.3113). Also, the MLE and 95% CI for b are 0.5028 and (0.3456, 0.6601).

(b) Since the priors on a and b are flat, finding the maximum likelihood estimate of (a,b) is the same as finding the posterior mode of (a,b). Now, the posterior density of a and b is

$$f(a,b\,|\,y) \propto \prod_{i=1}^{n} p_i^{y_i}(1-p_i)^{n_i-y_i}\,.$$

So the log-posterior is

$$l(a,b) = \log f(a,b\,|\,y) = \sum_{i=1}^{n} q_i\,,$$

where $q_i = y_i \log p_i + (n_i - y_i)\log(1-p_i)$
$$= y_i z_i - n_i \log(1+\exp(z_i)) \quad \text{(after some algebra)}.$$

Let: $d_{1i} = \dfrac{dq_i}{da} = y_i - n_i p_i$, $\qquad d_{2i} = \dfrac{dq_i}{db} = (y_i - n_i p_i)x_i$

$$d_{11i} = \frac{d^2 q_i}{da^2} = -n_i p_i (1 - p_i), \qquad d_{12i} = \frac{d^2 q_i}{da\,db} = -n_i p_i (1 - p_i)x_i$$

$$d_{22i} = \frac{d^2 q_i}{db^2} = -n_i p_i (1 - p_i)x_i^2, \qquad d_1 = \sum_{i=1}^{n} d_{1i}, \; d_2 = \sum_{i=1}^{n} d_{2i}$$

$$d_{11} = \sum_{i=1}^{n} d_{11i}, \qquad d_{22} = \sum_{i=1}^{n} d_{22i}, \qquad d_{12} = \sum_{i=1}^{n} d_{12i}$$

$$v = \begin{pmatrix} a \\ b \end{pmatrix}, \quad D = D(v) = \begin{pmatrix} d_1 \\ d_2 \end{pmatrix}, \quad M = M(v) = \begin{pmatrix} d_{11} & d_{12} \\ d_{12} & d_{22} \end{pmatrix}.$$

Then the NR algorithm is defined by

$$v_t = v_{t-1} - M(v_{t-1})^{-1} D(v_{t-1}), \quad t = 1,2,3,\dots.$$

Starting from the origin, the iterates of a and b are as shown in Table 7.3.

Table 7.3 Results of a Newton-Raphson algorithm

t	0	1	2	3	4	5
a_t	0	−1.474	−2.013	−2.148	−2.156	−2.156
b_t	0	0.3369	0.4670	0.5008	0.5028	0.5028

Thus the MLEs of a and b are $\hat{a} = -2.156$ and $\hat{b} = 0.5028$. This agrees perfectly with the results in (a).

A 95% CI for a is $\left(\hat{a} \pm t_{0.025}(8)s_a\right)$ and a 95% CI for b is $\left(\hat{b} \pm t_{0.025}(8)s_b\right)$, where:

$t_{0.025}(8) = 2.306$

s_a^2 is the top left element of V

s_b^2 is the bottom right element of V

$V = \phi(X'WX)^{-1}$ (a 2 by 2 matrix)

$$\phi = 1, \; X = \begin{pmatrix} 1 & x_1 \\ 1 & x_2 \\ \vdots & \vdots \\ 1 & x_n \end{pmatrix}, \quad W = diag(w_1,\dots,w_n), \; w_i = \frac{\omega_i}{V(\mu_i)g'(\mu_i)^2}$$

$$\omega_i = n_i, \mu_i = \hat{p}_i = \frac{1}{1+\exp(-\hat{z}_i)} \quad \text{(MLE of the probability at } x = x_i)$$

$$\hat{z}_i = \hat{a} + \hat{b}x_i \quad \text{(MLE of linear predictor at } x = x_i)$$

$$V(\mu) = \mu(1-\mu), \quad g(\mu) = \log\left(\frac{\mu}{1-\mu}\right) \quad \text{(logit link function)}$$

$$g'(\mu) = \frac{1}{\mu^2(1-\mu)^2}.$$

We find that $w_i = n_i\hat{p}_i(1-\hat{p}_i)$. Numerically, we find that 95% CIs for a and b are (–3.000, –1.311) and (0.3456, 0.6601), respectively. These results agree with those in (a).

(c) At each iteration $t = 1,2,3,4,...$, we:

1. Fix b and perform a NR step towards maximising wrt a:
$$a_{t+1} = a_t - d_1(a_t)/d_{11}(a_t)$$

2. Fix a and perform a NR step towards maximising wrt b:
$$b_{t+1} = b_t - d_2(b_t)/d_{22}(b_t).$$

Starting from the origin $(a, b) = (0,0)$ we obtain the results in Table 7.4.

Table 7.4 Results of a search algorithm

t	0	1	2	3	4
a_t	0	0.4564	–0.45034	–0.06132	–0.7294
b_t	0	0.1401	0.09223	0.20571	0.1690

t	20	21	99	100
a_t	–1.8585	–1.8619	–2.1555	–2.1555
b_t	0.4424	0.4532	0.5028	0.5028

We see that this modified and simpler algorithm converges more slowly than plain NR. Also, it is less stable, as it fails to converge if started from $(a, b) = (0.3, 0.3)$, unlike plain NR. Both algorithms fail to converge if started from (0.5, 0.5). (See the R code below for details.)

(d) We apply the Metropolis Hastings algorithm with a burn-in of 500 and starting from the origin to get a sample of size of $J = 10,000$ from $f(a,b\,|\,y)$. The acceptance rates were 37% for a and 55% for b. The Markov chain was not thinned for subsequent inference, meaning that the CIs obtained below are perhaps narrower than they should be.

The MC estimate of $E(a\,|\,y)$ is -2.207 (similar to the MLE, -2.156), with 95% CI $(-2.214, -2.199)$ and 95% CPDR $(-2.963, -1.521)$.

The MC estimate of $E(b\,|\,y)$ is 0.5145 (similar to the MLE, 0.5028), with 95% CI $(0.5132, 0.5158)$ and 95% CPDR $(0.3895, 0.6605)$.

Traces and histograms of the sampled values of a and b are shown in Figure 7.8.

Figure 7.8 Results of MH algorithm

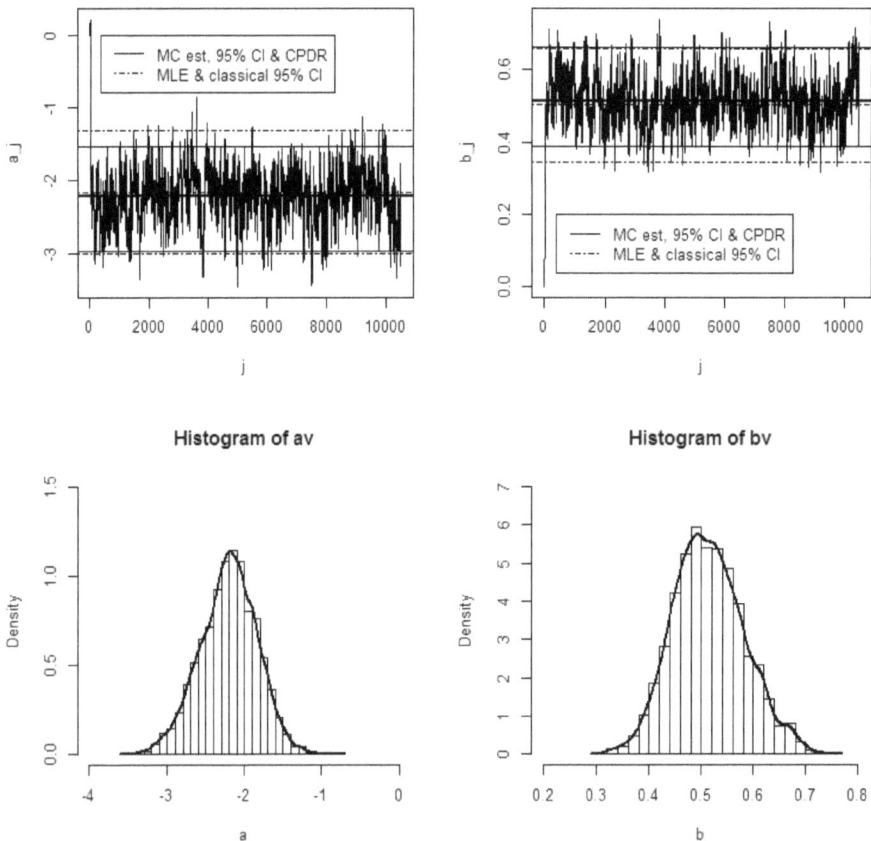

(e) The required results are shown in Figure 7.9.

Note: Figure 7.9 shows that the probability of a rat dying when given no radiation is about 10%. We should interpret this result and the graph near $x = 0$ with caution. Ideally, we would conduct another experiment with only small values of x and a second logistic regression, perhaps using the log of x as the explanatory variable. On the other hand, maybe the 10% figure is reasonable because rats could die within one month for reasons other than radiation. Alternatively, we could modify our model so as to force $p(0) = 0$ (see (h) below).

Figure 7.9 Mortality rate estimates

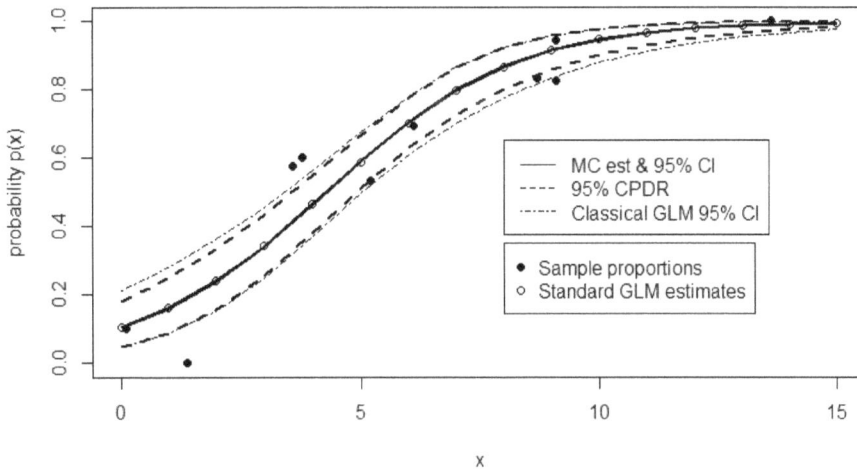

(f) Let d be the number of rats which will die if exposed to radiation for five hours. Then

$$(d \,|\, y, a, b) \sim Bin(20, p(a,b)),$$

where

$$p(a,b) = 1/(1 + \exp(-a - 5b)).$$

We can now apply the method of composition whereby

$$f(d,a,b\,|\,y) = f(d\,|\,y,a,b)f(a,b\,|\,y).$$

Thus for each sampled (a,b) we calculate $p(a,b)$ and sample from the binomial distribution of d above. The frequencies of the resulting 10,000 values of d are shown in Table 7.5

Table 7.5 Simulated frequencies of rats dying

d	3	4	5	6	7	8
frequency	1	3	20	75	217	472

d	9	10	11	12	13	14
frequency	845	1188	1562	1733*	1546	1123

d	15	16	17	18	19
frequency	709	332	131	37	6

Using the 10,000 values of d, our estimate of d is 11.81 (the average of the 10,000 values), with (11.76, 11.85) as the 95% MC CI for d's posterior mean. We feel about 95.1% confident that the number of rats which die will be between 8 and 16, inclusive (since 95.1% of the simulated d values are in this range). Also, it is most likely that 12 of the 20 rats will die, because the MC estimate of $Mode(d \mid y)$ is 12 (since $d = 12$ above has the highest frequency, namely 1,733, as marked by an asterisk).

(g) First observe that the *LD50* is the value of x such that $p(x) = 0.5$.

Solving $1/(1 + \exp(-a - bx)) = 0.5$, we get $x = LD50 = -a/b$.

Using the sample of 10,000 in part (f), we estimate the posterior mean of *LD50* is 4.279, with 95% MC CI (4.273, 4.286). The MC 95% CPDR for *LD50* is (3.584, 4.916). Thus we can be 95% confident that the dose required to kill half of a large number of rats is between 3.6 and 4.9.

Using standard GLM procedures and the delta method we estimate *LD50* as 4.287 (the MLE) with 95% CI (3.532, 5.042). Thus we can be 95% confident that the dose required to kill half of a large number of rats is between 3.5 and 5.0. We see that Bayesian and classical methods have resulted in inferences which are very similar.

(h) An alternative to the logistic model in (d), one with zero probability of death at zero dosage of radiation, is as follows:

$$(Y_i \mid a, b) \sim \perp Bin(n_i, p_i), \quad i = 1, \dots, n$$

$$p_i = 1 - \exp(-z_i), \quad z_i = ax_i + bx_i^2$$

$$f(a, b) \propto 1, \quad a, b > 0.$$

Running a suitable modification of the MH algorithm in (d), we estimate a and b as 0.11 and 0.017, with respective 95% CPDRs (0.04, 0.20) and (0.004, 0.030). The required graph is shown in Figure 7.10.

Figure 7.10 Modified mortality rate estimates

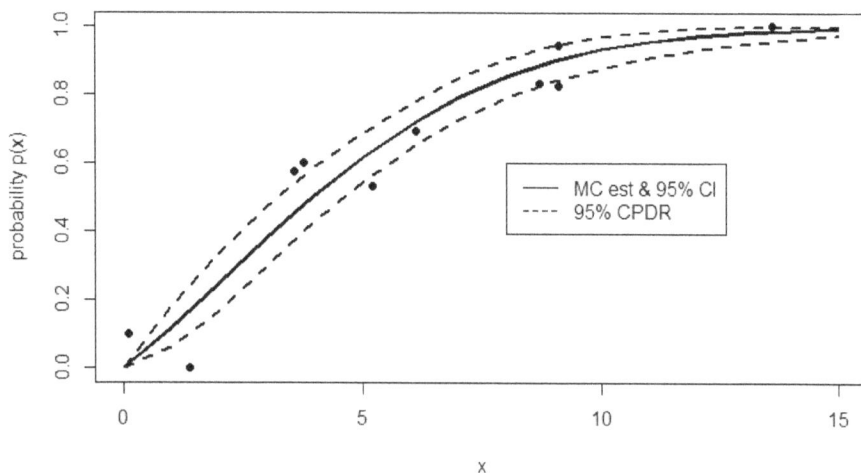

R Code for Exercise 7.3

```
# (a) *******************************************************
nvec <- c(10,30,40,20,15,46,12,37,23,8)
xvec <- c(0.1,1.4,3.6,3.8,5.2,6.1,8.7,9.1,9.1,13.6)
yvec <- c(1,0,23,12,8,32,10,35,19,8)
pvec <- yvec/nvec
options(digits=4)
cbind(xvec,nvec,yvec,pvec)
#     xvec nvec yvec   pvec
# [1,] 0.1  10    1 0.1000
# [2,] 1.4  30    0 0.0000
# [3,] 3.6  40   23 0.5750
# [4,] 3.8  20   12 0.6000
# [5,] 5.2  15    8 0.5333
# [6,] 6.1  46   32 0.6957
# [7,] 8.7  12   10 0.8333
# [8,] 9.1  37   35 0.9459
# [9,] 9.1  23   19 0.8261
# [10,] 13.6  8    8 1.0000
```

```
fit <- glm(pvec~xvec,family=binomial(link=logit),weights=nvec)
fit$coef      # -2.1555    0.5028
summary(fit)$cov.unscaled
#          (Intercept)    xvec
# (Intercept)    0.13404 -0.022442
# xvec          -0.02244 0.004651

alpse <- sqrt(summary(fit)$cov.unscaled[1,1])
fitalpci <- fit$coef[1] + c(-1,1)*qt(0.975,8)*alpse
c(alpse,fitalpci)   # 0.3661 -2.9998 -1.3113

betse <- sqrt(summary(fit)$cov.unscaled[2,2])
fitbetci <- fit$coef[2] + c(-1,1)* qt(0.975,8)*betse
c(betse,fitbetci)   # 0.0682 0.3456 0.6601

# (b) ****************************************************

NR.LOGISTIC <- function(m,alp,bet,xv,nv,yv){
# Performs logistic regression via the Newton-Raphson algorithm.
# Inputs:      m = number of iterations
#              alp, bet = starting values of alpha and beta
#              xv, nv, yv = vectors of covariates, sample sizes and
#                           numbers of successes, respectively.
# Outputs:     $alpv = vector of (m+1) alpha values
#              $betv = vector of (m+1) beta values

alpv <- alp; betv <- bet; ve <- c(alp,bet)
for(t in 1:m){
        pv <- 1/(1+exp(-alp-bet*xv))
        d1 <- sum(yv - nv*pv);  d2 <- sum((yv - nv*pv)*xv)
        d11 <- -sum(nv*pv*(1-pv));     d12 <- -sum(nv*pv*(1-pv)*xv)
        d22 <- -sum(nv*pv*(1-pv)*xv^2)
        D <- c(d1,d2)
        M <- matrix(c(d11,d12,d12,d22),nrow=2)
        ve <- ve - solve(M) %*% D
        alp <- ve[1]; bet <- ve[2]
        alpv <- c(alpv,alp); betv <- c(betv,bet)
        }
list(alpv=alpv,betv=betv)
}
```

```
options(digits=4)
nrres <- NR.LOGISTIC(m=20,alp=0,bet=0,xv=xvec,nv=nvec,yv=yvec)
nrres
#   $alpv:      [1]  0.000 -1.474 -2.013 -2.148 -2.156 -2.156 ....
#   $betv:      [1]  0.0000 0.3369 0.4670 0.5008 0.5028 0.5028 ....

NR.LOGISTIC(m=20,alp=0.3,bet=0.3,xv=xvec,nv=nvec,yv=yvec)
#   $alpv:      [1]  0.000 -1.474 -2.013 -2.148 -2.156 -2.156 ....
#   $betv:      [1]  0.0000 0.3369 0.4670 0.5008 0.5028 0.5028 ....

NR.LOGISTIC(m=20,alp=0.5,bet=0.5,xv=xvec,nv=nvec,yv=yvec)
# Error in solve.default(M) :
#   system is computationally singular: reciprocal condition
# number = 9.01649e-18

alpmle <- nrres$alp[21]; betmle <- nrres$bet[21]
X <- cbind(1,xvec)
zmle <- alpmle + betmle*xvec   # linear predictor
pmle <- 1/(1 + exp(-zmle))
wtvec <- nvec*pmle*(1-pmle)
W <- diag(wtvec)
varmat <- solve(t(X) %*% W %*% X)
varmat
#       0.13404         -0.022442
#       -0.02244        0.004651

qt(0.975,8)      # 2.306
alpmle + c(-1,1)*qt(0.975,8)*sqrt(varmat[1,1]) # -3.000 -1.311
betmle + c(-1,1)*qt(0.975,8)*sqrt(varmat[2,2]) #  0.3456 0.6601

# (c) **************************************************

NRMOD.LOGISTIC <- function(m,alp,bet,xv,nv,yv){
# Performs logistic regression via a modification of the Newton-Raphson
# algorithm.
# Inputs:       m = number of iterations
#               alp, bet = starting values of alpha and beta
#               xv, nv, yv = vectors of covariates, sample sizes and
#                       numbers of successes, respectively.
# Outputs:      $alpv = vector of (m+1) alpha values
#               $betv = vector of (m+1) beta values
alpv <- alp; betv <- bet; ve <- c(alp,bet)
```

```
for(t in 1:m){
        pv <- 1/(1+exp(-alp-bet*xv))
        d1 <- sum(yv - nv*pv)
        d2 <- sum((yv - nv*pv)*xv)
        d11 <- -sum(nv*pv*(1-pv))
        d22 <- -sum(nv*pv*(1-pv)*xv^2)
        alp <- alp - d1/d11
        bet <- bet - d2/d22
        alpv <- c(alpv,alp); betv <- c(betv,bet)
        }
list(alpv=alpv,betv=betv)
}


resnr <- NRMOD.LOGISTIC(m=100,alp=0,bet=0,xv=xvec,nv=nvec,yv=yvec)
inc=c(1,2,3,4,5,21,22,100,101); rbind(inc-1,resnr$alpv[inc], resnr$betv[inc])
# [1,]   0 1.0000  2.00000  3.00000  4.0000 20.0000 21.0000 99.0000 100.0000
# [2,]   0 0.4564 -0.45034 -0.06132 -0.7294 -1.8585 -1.8619 -2.1555  -2.1555
# [3,]   0 0.1401  0.09223  0.20571  0.1690  0.4424  0.4532  0.5028   0.5028


resnr <- NRMOD.LOGISTIC(m=100,alp=0.3,bet=0.3,xv=xvec,nv=nvec,yv=yvec)
rbind(inc-1,resnr$alpv[inc], resnr$betv[inc])
# [1,]  0.0 1.00000 2.0000  3.00 4.000e+00  20  21  99  100
# [2,]  0.3 -1.72625 2.1776 -31.10 4.023e+15 NaN NaN NaN  NaN
# [3,]  0.3 -0.01407 0.6942 -21.36 2.861e+18 NaN NaN NaN  NaN


resnr <- NRMOD.LOGISTIC(m=100,alp=0.5,bet=0.5,xv=xvec,nv=nvec,yv=yvec)
rbind(inc-1,resnr$alpv[inc], resnr$betv[inc])
# [1,]  0.0 1.000   2.0   3   4  20  21  99  100
# [2,]  0.5 -4.532 828.1 -Inf NaN NaN NaN NaN NaN
# [3,]  0.5 -1.090 3101.9 -Inf NaN NaN NaN NaN NaN


# (d) ************************************************


xvdata <- c(0.1,1.4,3.6,3.8,5.2,6.1,8.7,9.1,9.1,13.6)
yvdata <- c(1,0,23,12,8,32,10,35,19,8)
nvdata <- c(10,30,40,20,15,46,12,37,23,8)
pvdata <- yvdata/nvdata


MHLR <- function(burn,J,a0,b0,xv,yv,nv,sa,sb){
# Performs the Metropolis-Hastings algorithm for a logistic regression model.
# Inputs:       burn = number of iterations for burn-in
#               J = required number of Monte Carlo simulations
#               a0 = starting value of alpha
#               b0 = starting value of beta
```

```
#                   xv = vector of xi values (length n)
#                   yv = vector of yi observations
#                   nv = vector of ni values
#                   sa, sb = standard deviations of the two normal driver fns.
# Outputs:          $av = vector of (burn+J+1) values of alpha (incl. starting value)
#                   $bv = vector of (burn+J+1) values of beta (incl. starting value)
#                   $ara = acceptance rate for alpha (over last J iterations)
#                   $arb = acceptance rate for beta.

logfun <- function(a,b,xv,yv,nv){
        phatv <- 1/(1+exp(-a-b*xv))
        sum( yv*log(phatv) + (nv-yv)*log(1-phatv) )
        }


n <- length(yv);   a <- a0;  b <- b0
its <- burn + J              # total number of iterations
av <- c(a, rep(NA,its));
bv <- c(b, rep(NA,its))   # vectors of simulated a & b values
arav <- c(NA, rep(0,its));   arbv <- c(NA, rep(0,its))
                # acceptance rate vectors for a and b

for(j in 1:its){
        a2 <- rnorm(1,a,sa)
        logpr <- logfun(a=a2,b=b,xv=xv,yv=yv,nv=nv)-
                logfun(a=a,b=b, xv=xv,yv=yv,nv=nv)
        pr <- exp(logpr); u <- runif(1)
        if(u<pr){  a <- a2; arav[j+1] <- 1 }

        b2 <- rnorm(1,b,sb)
        logpr <- logfun(a=a,b=b2, xv=xv,yv=yv,nv=nv)-
                logfun(a=a,b=b, xv=xv,yv=yv,nv=nv)
        pr <- exp(logpr); u <- runif(1)
        if(u<pr){ b <- b2; arbv[j+1] <- 1 }

        av[j+1] <- a;     bv[j+1] <- b
        }

ara <- sum(arav[(burn+2):(its+1)])/J
arb <- sum(arbv[(burn+2):(its+1)])/J  # acceptance rates for a & b

list(av=av,bv=bv,ara=ara,arb=arb)
}
```

```
burn <- 500; K <- 10000; its <- burn + K; set.seed(221); date() #
res <- MHLR(burn=burn,J=K,a0=0,b0=0,xv=xvdata,
    yv=yvdata,nv=nvdata,sa=0.5,sb=0.05); date() # 10000 Took 1 second
c(res$ara,res$arb)          # 0.3650 0.5544
par(mfrow=c(2,1)); plot(res$av,type="l"); plot(res$bv,type="l")   # OK

options(digits=4); J = K; thin=1
    # thin=1 means no thinning (for experimentation)
av <- res$av[-(1:(burn+1))][seq(thin,K,thin)]; length(av) # 10000
acf(av)$acf[1:5] # 1.0000 0.9283 0.8756 0.8324 0.7945
                # (very high autocorrelation)
ahat <- mean(av); aci <- ahat + c(-1,1) * qnorm(1-0.05/2)*sqrt(var(av)/J)
acpdr <- quantile(av,c(0.025,0.975))
c(ahat,aci,acpdr)        #    -2.207 -2.214 -2.199 -2.963 -1.521

bv <- res$bv[-(1:(burn+1))][seq(thin,K,thin)]; length(bv) # 10000
acf(bv)$acf[1:5] # 1.0000 0.9363 0.8892 0.8481 0.8109
bhat <- mean(bv); bci <- bhat + c(-1,1) * qnorm(1-0.05/2)*sqrt(var(bv)/J)
bcpdr <- quantile(bv,c(0.025,0.975))
c(bhat,bci,bcpdr)   # 0.5145 0.5132 0.5158 0.3895 0.6605

dena <- density(av);      denb <- density(bv)
fit <- glm(pvdata~xvdata,family=binomial(link=logit),weights=nvdata)
fit$coef #              -2.1555     0.5028

ase <- sqrt(summary(fit)$cov.unscaled[1,1])
fitaci <- fit$coef[1] + c(-1,1)*qt(0.975,8)*ase
c(ase,fitaci)   # 0.3661 -2.9998 -1.3113

bse <- sqrt(summary(fit)$cov.unscaled[2,2])
fitbci <- fit$coef[2] + c(-1,1)* qt(0.975,8)*bse
c(bse,fitbci)   # 0.0682 0.3456 0.6601

X11(w=8,h=8); par(mfrow=c(2,2))
plot(0:its,res$av,type="l",xlab="j",ylab="a_j")
        abline(h=c(ahat,aci,acpdr))
        abline(h=c(fit$coef[1],fitaci),lty=4)
    legend(400,0,c("MC est, 95% CI & CPDR",
                                "MLE & classical 95% CI"),lty=c(1,4))
plot(0:its,res$bv,type="l", xlab="j",ylab="b_j")
        abline(h=c(bhat,bci,bcpdr))
        abline(h=c(fit$coef[2],fitbci),lty=4)
    legend(400,0.2,c("MC est, 95% CI & CPDR",
                                "MLE & classical 95% CI"),lty=c(1,4))
```

```
hist(av,prob=T, xlim=c(-4,0),ylim=c(0,1.5),nclass=20,xlab="a")
lines(dena$x,dena$y,lwd=2)
hist(bv,prob=T, xlim=c(0.2,0.8),ylim=c(0,7),nclass=20,xlab="b")
lines(denb$x,denb$y,lwd=2)


# (e) ***********************************************

xxv <- seq(0,15,1); len <- length(xxv)
ppv <- xxv; ppci1 <- xxv; ppci2 <- xxv; ppcpdr1 <- xxv; ppcpdr2 <- xxv

for(i in 1:len){
        xx <- xxv[i]
        ppsim <- 1/(1+exp(-av-bv*xx))
        pp <- mean(ppsim)
        ppci <- pp + c(-1,1)*qnorm(0.975)*sqrt(var(ppsim)/J)
        ppcpdr <- quantile(ppsim,c(0.025,0.975))
        ppv[i] <- pp       # MC estimate of E(p|xx) and so indirectly of p at x=xx
        ppci1[i] <- ppci[1]; ppci2[i] <- ppci[2]
        ppcpdr1[i] <- ppcpdr[1]; ppcpdr2[i] <- ppcpdr[2]
        }

Xmat <- cbind(1,xxv)
etahat <- Xmat %*% fit$coef        # NB: fit was created in (a)
pihat <- 1/(1+exp(-etahat))
etahatvar<- diag ( Xmat %*% summary(fit)$cov.unscaled %*% t(Xmat) )
df <- length(yvdata)-length(fit$coef)       # 10-2=8
etahatub <- etahat +  qt(0.975,df) * sqrt(etahatvar)
etahatlb <- etahat -  qt(0.975,df) * sqrt(etahatvar)
pihatub <- 1/(1+exp(-etahatub))
pihatlb <- 1/(1+exp(-etahatlb))

X11(w=8,h=5); par(mfrow=c(1,1))
plot(c(0,15),c(0,1),type="n",xlab="x",ylab="probability p(x)")
points(xvdata,pvdata,pch=16);    lines(xxv,ppv)
lines(xxv,ppci1,lwd=2); lines(xxv,ppci2,lwd=2)
lines(xxv,ppcpdr1,lty=2,lwd=2); lines(xxv,ppcpdr2,lty=2,lwd=2)
points(xxv,pihat);       lines(xxv,pihatlb,lty=4);       lines(xxv,pihatub,lty=4)
legend(8,0.65,  c("MC est & 95% CI","95% CPDR","Classical GLM 95% CI"),
        lty=c(1,2,4))
legend(8,0.35,c("Sample proportions","Standard GLM estimates"),pch=c(16,1))
# pphatv <- 1/(1+exp(-ahat-bhat*xxv))
# lines(xxv,pphatv,lty=3)    # This alternative estimate is practically
                             # indistinguishable from ppv and so is not plotted
```

```
# (f) *****************************************************

p5v <- 1/(1+exp(-av-bv*5)); set.seed(331); dv <- rbinom(J,20,p5v)
hist(dv,prob=T,breaks=seq(-0.5,20.5,1))
summary(as.factor(dv))
#3  4  5  6  7  8  9  10 11 12 13 14 15 16 17 18 19
#1  3  20 75 217 472 845 1188 1562 1733 1546 1123 709 332 131 37  6

dhat <- mean(dv); dci <- dhat + c(-1,1)*qnorm(0.975)*sqrt(var(dv)/J)
dcpdr <- quantile(dv,c(0.025,0.975))
c(dhat,dci,dcpdr)    # 11.81 11.76 11.85 7.00 16.00

dv2 <- dv[dv>=7]; dv3 <- dv2[dv2<=16]; length(dv3)/J   # 0.9727
dv2 <- dv[dv>=8]; dv3 <- dv2[dv2<=16]; length(dv3)/J   # 0.951 OK (>= 95%)
dv2 <- dv[dv>=7]; dv3 <- dv2[dv2<=15]; length(dv3)/J   # 0.9395  (too small)

dhat2 <- mean(p5v)    # alternative method
qbinom(c(0.025,0.975),20,dhat2)   # 7 16

# (g) *****************************************************

Lv <- -av/bv; Lhat <- mean(Lv); Lci <- Lhat + c(-1,1)*qnorm(0.975)*sqrt(var(Lv)/J)
Lcpdr <- quantile(Lv,c(0.025,0.975))
c(Lhat,Lci,Lcpdr)  # 4.279 4.273 4.286 3.584 4.916
cf <- coef(fit); Lmle <-  -cf[1]/cf[2]; deriv <-  c( -1/cf[2] , cf[1]/cf[2]^2 )
Lvar <- t(deriv) %*% summary(fit)$cov.unscaled %*% deriv
Lci2 <- Lmle + c(-1,1)*qt(0.975,8) * sqrt(Lvar)
c(Lmle,Lci2) # 4.287     3.532     5.042

# (h) *****************************************************

xvdata <- c(0.1,1.4,3.6,3.8,5.2,6.1,8.7,9.1,9.1,13.6)
yvdata <- c(1,0,23,12,8,32,10,35,19,8)
nvdata <- c(10,30,40,20,15,46,12,37,23,8)
pvdata <- yvdata/nvdata
```

```
MHLRZC <- function(burn,J,a0,b0,xv,yv,nv,sa,sb){

# Performs the Metropolis-Hastings algorithm for a logistic regression model
# modified to have a zero constraint.

# Inputs:       burn = number of iterations for burn-in
#               J = required number of Monte Carlo simulations
#               a0 = starting value of alpha
#               b0 = starting value of beta
#               xv = vector of xi values (length n)
#               yv = vector of yi observations
#               nv = vector of ni values
#               sa, sb = standard deviations of the two normal driver fns.

# Outputs:      $av = vector of (burn+J+1) values of alpha (incl. starting value)
#               $bv = vector of (burn+J+1) values of beta (incl. starting value)
#               $ara = acceptance rate for alpha (over last J iterations)
#               $arb = acceptance rate for beta.

logfun <- function(a,b,xv,yv,nv){
        phatv <-  1 - exp( -a*xv -  b*xv^2  )        # The main change is here
        sum( yv*log(phatv) + (nv-yv)*log(1-phatv) )  }
n <- length(yv);   a <- a0;  b <- b0
its <- burn + J              # total number of iterations
av <- c(a, rep(NA,its));    bv <- c(b, rep(NA,its))   # vectors of simulated a & b
values
arav <- c(NA, rep(0,its));   arbv <- c(NA, rep(0,its))
                # acceptance rate vectors for a and b
for(j in 1:its){
        a2 <- rnorm(1,a,sa)
        if(a2 > 0){
           logpr <- logfun(a=a2,b=b,xv=xv,yv=yv,nv=nv)-
                   logfun(a=a,b=b, xv=xv,yv=yv,nv=nv)
           pr <- exp(logpr); u <- runif(1)
           if(u<pr){  a <- a2; arav[j+1] <- 1 }
           }
        b2 <- rnorm(1,b,sb)
        if(b2 > 0){
           logpr <- logfun(a=a,b=b2, xv=xv,yv=yv,nv=nv)-
                   logfun(a=a,b=b, xv=xv,yv=yv,nv=nv)
           pr <- exp(logpr); u <- runif(1)
           if(u<pr){  b <- b2; arbv[j+1] <- 1 }
           }
        av[j+1] <- a;      bv[j+1] <- b      }
```

```
ara <- sum(arav[(burn+2):(its+1)])/J
arb <- sum(arbv[(burn+2):(its+1)])/J   # acceptance rates for a & b
list(av=av,bv=bv,ara=ara,arb=arb)
}
burn <- 500; J <- 10000; its <- burn + J; set.seed(111)
res <- MHLRZC(burn=burn,J=J,a0=0.1,b0=0.01,
        xv=xvdata,yv=yvdata,nv=nvdata,sa=0.03,sb=0.005)
 c(res$ara,res$arb)          # 0.5686 0.5637   OK
par(mfrow=c(2,1)); plot(res$av,type="l"); plot(res$bv,type="l")   # OK
options(digits=4)
av <- res$av[-(1:(burn+1))]; ahat <- mean(av)
aci <- ahat + c(-1,1) * qnorm(1-0.05/2)*sqrt(var(av)/J)
acpdr <- quantile(av,c(0.025,0.975))
c(ahat,aci,acpdr)         #    0.10921 0.10842 0.11000 0.03622 0.19256

bv <- res$bv[-(1:(burn+1))]; bhat <- mean(bv)
bci <- bhat + c(-1,1) * qnorm(1-0.05/2)*sqrt(var(bv)/J)
bcpdr <- quantile(bv,c(0.025,0.975))
c(bhat,bci,bcpdr)   # 0.016683 0.016552 0.016814 0.003641 0.029898

xxv <- seq(0,15,1); len <- length(xxv)
ppv <- xxv; ppci1 <- xxv; ppci2 <- xxv; ppcpdr1 <- xxv; ppcpdr2 <- xxv

for(i in 1:len){
        xx <- xxv[i]
        ppsim <- 1-exp(-av*xx-bv*xx^2)
        pp <- mean(ppsim)
        ppci <- pp + c(-1,1)*qnorm(0.975)*sqrt(var(ppsim)/J)
        ppcpdr <- quantile(ppsim,c(0.025,0.975))
        ppv[i] <- pp       # MC estimate of E(p|xx) and so indirectly of p at x=xx
        ppci1[i] <- ppci[1]; ppci2[i] <- ppci[2]
        ppcpdr1[i] <- ppcpdr[1]; ppcpdr2[i] <- ppcpdr[2]
        }

X11(w=8,h=5); par(mfrow=c(1,1))

plot(c(0,15),c(0,1),type="n",xlab="x",ylab="probability p(x)")
points(xvdata,pvdata,pch=16);    lines(xxv,ppv)
lines(xxv,ppci1,lwd=2); lines(xxv,ppci2,lwd=2)
lines(xxv,ppcpdr1,lty=2,lwd=2); lines(xxv,ppcpdr2,lty=2,lwd=2)

legend(8,0.6,  c("MC est & 95% CI","95% CPDR"),lty=c(1,2))
```

Exercise 7.4 Autocorrelated Bernoulli data (and practice at various statistical techniques)

Consider the following Bayesian model for a sequence of identically distributed but possibly dependent and serially autocorrelated Bernoulli random variables y_i:

$$(y_i \mid a,b,y_{i-1},y_{i-2},y_{i-3},...) \sim Bernoulli(p_i), \quad i=0,\pm 1,\pm 2,...$$

$$p_i = \frac{1}{1+\exp\{-(a+by_{i-1})\}}$$

$$f(a,b) \propto 1, \quad a,b \in \Re.$$

Suppose that the data is $y=(y_1,...,y_n)=(1,1,1,1,1, \; 1,1,0,0,0)$.

Use the Metropolis-Hastings algorithm to generate a random sample of $J=10,000$ values from the joint posterior distribution of a and b. Use this sample to estimate the posterior means and 95% CPDRs for a and b. Also estimate $P(b<0 \mid y)$.

Solution to Exercise 7.4

The first thing we need to do is work out the probability that $Y_1 = 1$ conditional on a and b but *not* conditional on y_0 (since y_0 is not known). With an implicit conditioning on a and b, observe by the law of total probability that

$$P(Y_1=1)=P(Y_0=0)P(Y_1=1 \mid Y_0=0)+P(Y_0=1)P(Y_1=1 \mid Y_0=1)$$
$$= \{1-P(Y_1=1)\}P(Y_1=1 \mid Y_0=0)+P(Y_1=1)P(Y_1=1 \mid Y_0=1).$$

Solving for $P(Y_1=1)$, we get

$$q_1 \equiv P(Y_1=1 \mid a,b) = \frac{1+\exp(a+b)}{2+\exp(a+b)+\exp(-a)}.$$

Hence, with $p_i = P(Y_i=1 \mid a,b,y_{i-1}) = \dfrac{1}{1+\exp(-a-by_{i-1})}$

(as already defined), the joint posterior pdf of a and b is

$$f(a,b \mid y) \propto f(a,b)f(y \mid a,b)$$

$$\overset{a,b}{\propto} 1 \times f(y_1 \mid a,b)\prod_{i=2}^{n} f(y_i \mid a,b,y_{i-1}) = q_1^{y_1}(1-q_1)^{1-y_1}\prod_{i=2}^{n} p_i^{y_i}(1-p_i)^{1-y_i}.$$

So the log of the posterior density is given by
$$l(a,b) \equiv \log f(a,b \mid y)$$
$$= c + y_1 \log q_1 + (1 - y_1) \log(1 - q_1)$$
$$+ \sum_{i=2}^{n} \{ y_i \log p_i + (1 - y_i) \log(1 - p_i) \}.$$

Using normal drivers for both a and b, we implement a Metropolis-Hastings algorithm and thereby, following a burn-in of size $B = 1,000$, obtain an approximately random MC sample of size $J = 10,000$, which we will denote by
$$(a_j, b_j) \sim iid \; f(a,b \mid y), \quad j = 1, ..., J.$$

From this MC sample we estimate a by –2.337 with 95% CPDR (–6.3980, 0.8313), and b by 5.411 with 95% CPDR (0.9098, 11.8691). We also estimate $P(b < 0 \mid y)$ by 0.081.

The traces of a and b over all 11,000 iterations, and histograms of the last 10,000 values of a and b, respectively, are shown in Figure 7.11, together with posterior density estimates.

Note: In an earlier exercise we considered a posterior predictive p-value for the null hypothesis that the sequence in the present exercise consists of values that are iid.

That p-value was estimated as 0.0995 with 95% CI (0.0936, 0.1054). The estimate 0.081 of $P(b < 0 \mid y)$ in the present exercise may be interpreted in a similar way to the p-value 0.0995.

In this case the appropriate p-value is *one-sided*.

If we wish to do a *two-sided* test, in the present context, $b = 0$ versus $b \neq 0$, then the p-value may be calculated as *twice* the *minimum* of $P(b < 0 \mid y)$ and $P(b > 0 \mid y)$.

Clearly, if the posterior distribution of b is well above or well below zero, then the resulting two-sided p-value will appropriately be very close to zero.

Figure 7.11 Traces and histograms for *a* and *b*

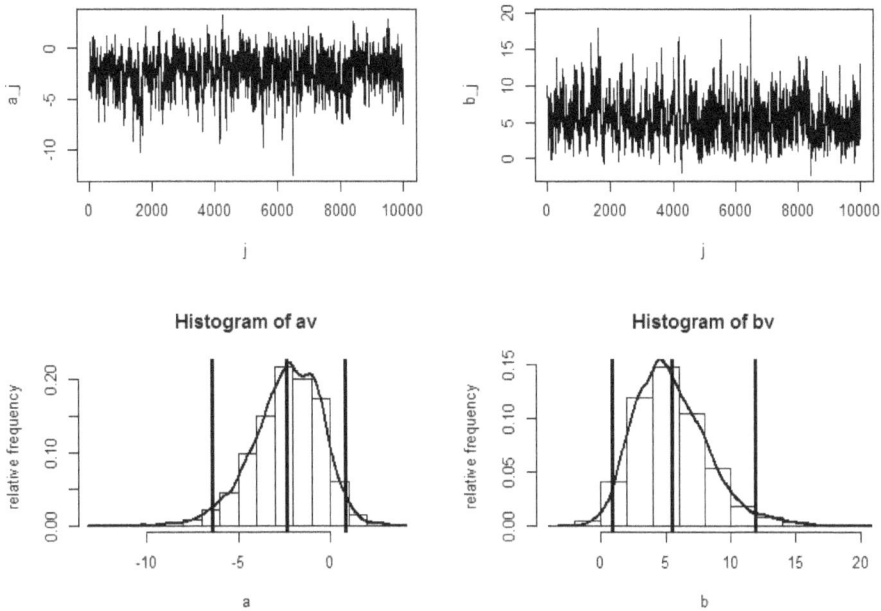

R Code for Exercise 7.4

```
yv <- c(1,1,1,1,1,  1,1,0,0,0);  n <- length(yv); ybar <- mean(yv); ydot <- sum(yv)

MHBD <- function(K,a,b,yv,sa,sb){
# Performs a Metropolis-Hastings algorithm for a binary dependence model.
# Inputs:       K = total number of iterations
#               a,b = starting values of a and b
#               yv = vector of 0-or-1 values (y1,...,yn)
#               sa, sb = standard deviations of the two normal driver fns.
# Outputs:      $av = vector of (K+1) values of a (incl. starting value)
#               $bv = vector of (K+1) values of b (incl. starting value)
#               $ara, $arb = acceptance rates for a and b.
n <- length(yv);  av <- a; bv <- b; cta <- 0; ctb <- 0
logfun <- function(a,b,yv,n){
        p1 = (1 + exp(a+b)) / (2 + exp(a+b) + exp(-a))      # p1
        p2ton <- 1/(1 + exp(-a-b*yv[-n]))                    # p2,...,pn
        pv <- c(p1,p2ton)                                    # p1,...,pn
        sum( yv*log(pv) + (1-yv)*log(1-pv) )       }
```

```
for(j in 1:K){
        a2 <- rnorm(1,a,sa) # proposed value of a
        logpr <- logfun(a=a2,b=b,yv=yv,n=n)-logfun(a=a,b=b,yv=yv,n=n)
        pr <- exp(logpr); u <- runif(1)
        if(u<pr){ a <- a2; cta <- cta + 1 }
        if(sb > 0){
          b2 <- rnorm(1,b,sb) # proposed value of b
          logpr <- logfun(a=a,b=b2,yv=yv,n=n)-logfun(a=a,b=b,yv=yv,n=n)
          pr <- exp(logpr); u <- runif(1)
          if(u<pr){ b <- b2; ctb <- ctb + 1 }
          }
        av <- c(av,a); bv <- c(bv,b)
        }
list(av=av,bv=bv,ara=cta/K,arb=ctb/K)
}

options(digits=4); set.seed(143); date() #
res <- MHBD(K=11000,a=0,b=0,yv=yv,sa=1.5,sb=2.2); date() #  Took 2 secs
c(res$ara,res$arb)  # 0.5575 0.5753 (acceptance rates for a and b) OK

X11(w=8,h=6); par(mfrow=c(2,1));  plot(res$av); plot(res$bv)  # OK

av <- res$av[1002:11001]; bv <- res$bv[1002:11001]; J=1000

abar <- mean(av); bbar <- mean(bv);
acpdr <- quantile(av,c(0.025,0.975));
bcpdr <- quantile(bv,c(0.025,0.975))

rbind(c(abar,acpdr),c(bbar,bcpdr))
# [1,] -2.337 -6.3980  0.8313
# [2,]  5.411  0.9098 11.8691

pr <- length(bv[bv<0])/J; pr # 0.081

X11(w=8,h=6); par(mfrow=c(2,2));

plot(av,type="l",xlab="j",ylab="a_j",cex=1.2)
plot(bv,type="l",xlab="j",ylab="b_j",cex=1.2)

hist(av,prob=T,xlab="a",ylab="relative frequency",cex=1.2);
        abline(v=c(abar,acpdr), lty=1,lwd=3); lines(density(av),lwd=2)
hist(bv,prob=T,xlab="b",ylab="relative frequency",cex=1.2);
        abline(v=c(bbar,bcpdr), lty=1,lwd=3); lines(density(bv),lwd=2)
```

358

Exercise 7.5 Inference on the bounds of a uniform distribution

Consider the following Bayesian model:

$$(y_1,...,y_n \mid a,b) \sim iid \; U(a,b)$$
$$(a \mid b) \sim U(0,b)$$
$$b \sim U(0,1).$$

Generate a random sample of size $n = 20$ from the model with $a = 0.6$ and $b = 0.8$. Then apply MCMC methods to generate a random sample from the joint posterior of a and b. Then use this sample to perform Monte Carlo inference on $m = E(y_i \mid a,b) = (a+b)/2$.

Solution to Exercise 7.5

Rounding to four decimals, the generated sample values are as shown in Table 7.6.

Table 7.6 Sample values

i	1	2	3	4	5
y_i	0.7846	0.7572	0.6381	0.7626	0.6105

i	6	7	8	9	10
y_i	0.6990	0.7728	0.7113	0.7314	0.7435

i	11	12	13	14	15
y_i	0.6324	0.7072	0.7493	0.7979	0.6182

i	16	17	18	19	20
y_i	0.7652	0.7883	0.7194	0.6211	0.6054

Note: The range of this data is from 0.6054 to 0.7979. This tells us immediately that $0 \leq a \leq 0.6054$ and $0.7979 \leq b \leq 1$.

Now, the joint posterior density of a and b is

$$f(a,b \mid y) \propto f(a,b,y) = f(b)f(a \mid b)f(y \mid a,b)$$

$$\propto \frac{I(0 < b < 1)}{1} \times \frac{I(0 < a < b)}{b} \times \prod_{i=1}^{n} \frac{I(a < y_i < b)}{b-a}$$

$$= \frac{1}{b(b-a)^n}, \quad 0 < a < b < 1, \quad a < \min y_i < \max y_i < b.$$

So the two conditional posterior distributions are defined by:

$$f(a \mid y,b) \propto \frac{1}{(b-a)^n}, \quad 0 < a < \min(y_i)$$

$$f(b \mid y,a) \propto \frac{1}{b(b-a)^n}, \quad \max(y_i) < b < 1.$$

Neither of these conditionals defines a well-known distribution. So we will apply a 'pure' Metropolis-Hastings algorithm (rather than a Gibbs sampler).

With a' and b' denoting the proposed values of a and b, the acceptance probabilities at the two steps are:

$$p_a = \frac{f(a' \mid y,b)}{f(a \mid y,b)} = \frac{1/(b-a')^2}{1/(b-a)^2} = \left(\frac{b-a}{b-a'}\right)^n$$

$$p_b = \frac{f(b' \mid y,a)}{f(b \mid y,a)} = \frac{1/(b'(b'-a)^2)}{1/(b(b-a)^2)} = \frac{b}{b'}\left(\frac{b-a}{b'-a}\right)^n.$$

The following drivers were chosen:

$$a' \sim N(a,r^2)$$
$$b' \sim N(b,t^2).$$

Starting at $a = 0.1$ and $b = 0.9$, and using the tuning constants $r = 0.008$ and $t = 0.01$, the algorithm was run for 2,500 iterations. The resulting trace plots are shown in Figure 7.12.

We see that stochastic convergence was achieved within 500 iterations. The acceptance rates over the last 2,000 iterations were 62% and 58% for a and b, respectively.

Figure 7.12 Traces for *a* and *b*

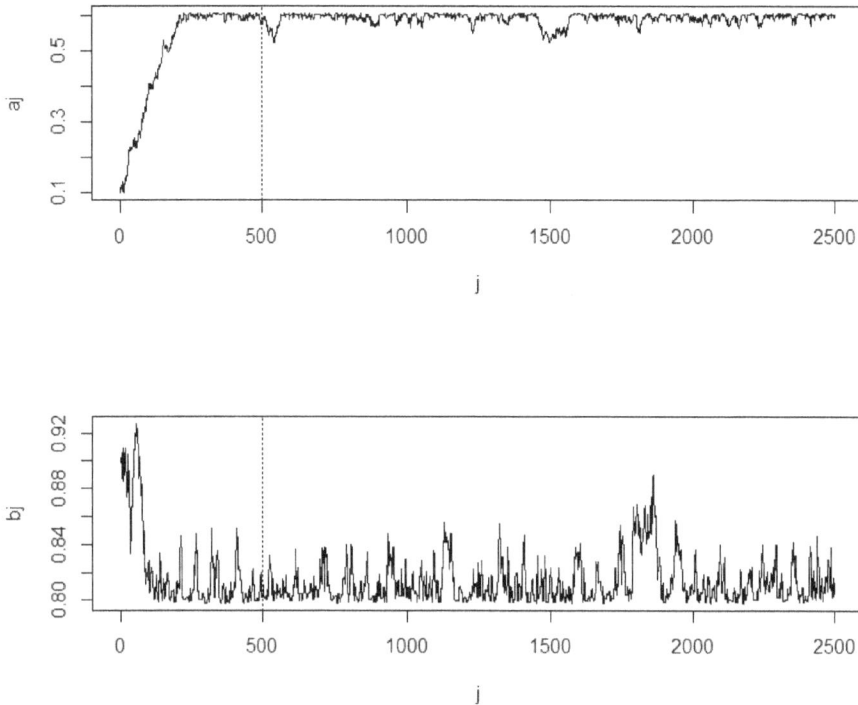

The algorithm was then run for a further 50,000 iterations, starting at the last values in the previous run ($a = 0.5979$ and $b = 0.8123$). The acceptance rates were now 61% and 54%, and this second run took 14 seconds of computer time.

Then every 50th value was recorded so as to yield a final random sample of size $J = 1,000$ from the joint posterior distribution of a and b, i.e.

$$(a_1, b_1), \dots, (a_J, b_J) \sim iid \ f(a, b \mid y).$$

As a check, the sample ACF of each sample of size 1,000 was calculated. Figure 7.13 shows the ACF estimates for a and b, and these provide no evidence for residual autocorrelation in either series.

Figure 7.13 Sample ACFs for *a* and *b*

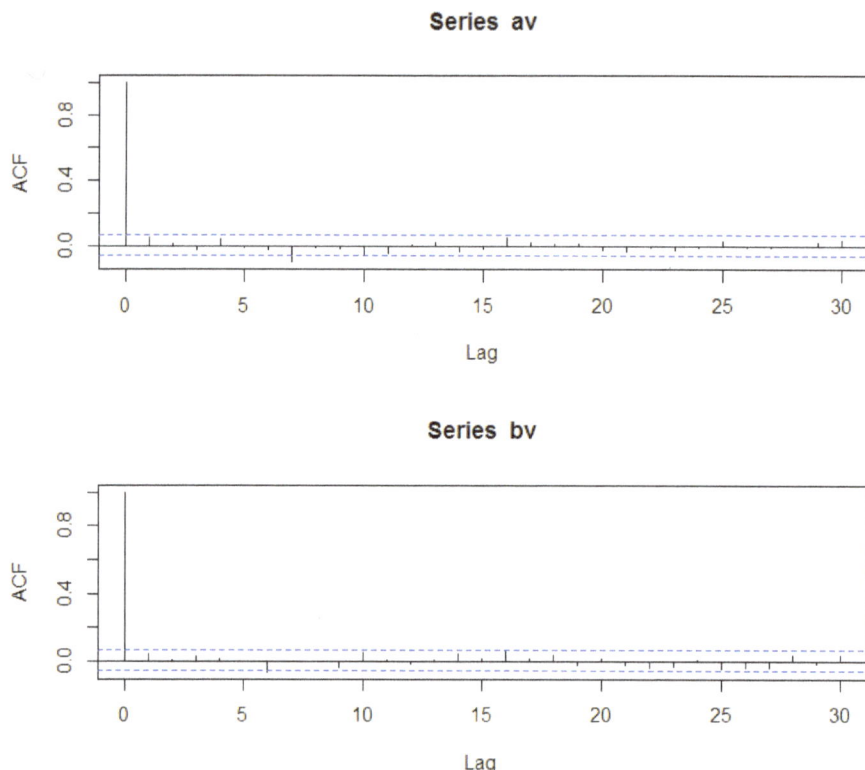

Series av

Series bv

A random sample from the posterior distribution of the mean

$$m = E(y_i \mid a,b) = (a+b)/2$$

was then formed by calculating

$$m_j = (a_j + b_j)/2.$$

We thereby obtained the random sample

$$m_1,...,m_J \sim iid \ f(m \mid y).$$

This Monte Carlo sample was used to estimate $\hat{m} = E(m \mid y)$ by 0.7013, with 95% CI (0.7008, 0.7019). The estimated 95% CPDR for *m* was (0.6837, 0.7173).

Figure 7.14 is a histogram of the 1,000 values of *m*, overlaid by a density estimate of $f(m \mid y)$, with the vertical lines showing the point and interval estimates reported above.

Figure 7.14 Inference on $m = (a + b)/2$

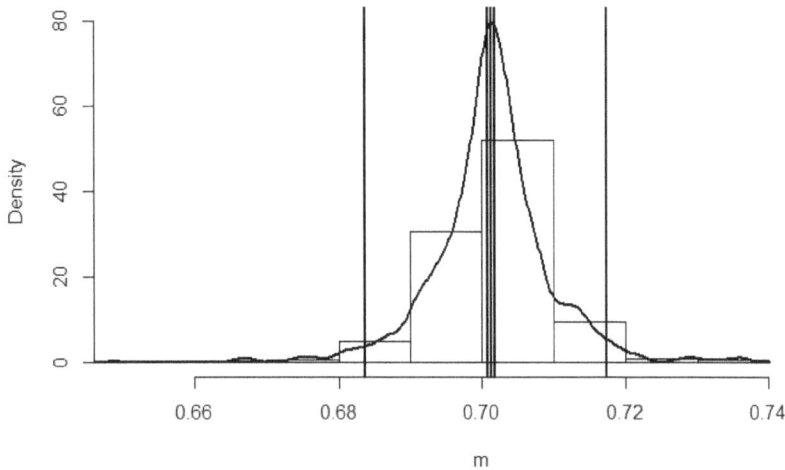

m

R Code for Problem 7.5

```
options(digits=4)

MH = function(B,J=1000,y,a,b,r,t){
# This function performs a Metropolis-Hastings algorithm for a model involving
3 uniforms.
# Inputs:        B = burn-in length
#                J = desired Monte Carlo size
#                y = (y1,...,yn) = data (yi ~ iid U(a,b))
#                a = starting value of a (a ~ U(0,b))
#                b = starting value of b (b ~ U(0,1))
#                r,t = tuning constants for a & b, respectively
# Outputs:       $av = (1+B+J) vector of a-values
#                $bv = (1+b+J) vector of b-values
#                $ar = acceptance rate for a (over last J iterations)
#                $br = acceptance rate for b (over last J iterations)
av = a; bv = b; an=0; bn=0; miny=min(y); maxy=max(y); n=length(y);
for(j in 1:(B+J)){
        ap = rnorm(1,a,r)
        if((0<ap)&&(ap<miny)){
                p = ((b-a)/(b-ap))^n; u = runif(1)
                if(u<p){ a=ap; if(j>B) an=an+1 }  }
        bp = rnorm(1,b,t)
        if((maxy<bp)&&(bp<1)){
                q = (b/bp)*((b-a)/(bp-a))^n; v = runif(1)
```

363

```
                if(v<q){ b=bp; if(j>B) bn=bn+1 }  }
        av=c(av,a); bv=c(bv,b)
        }
ar = an/J; br=bn/J; list(av=av,bv=bv,ar=ar,br=br)     }


set.seed(337); ydata = runif(20,0.6,0.8); round(ydata,4)
# [1] 0.7846 0.7572 0.6381 0.7626 0.6105 0.6990 0.7728 0.7113 0.7314
# [10] 0.7435 0.6324 0.7072 0.7493 0.7979 0.6182 0.7652 0.7883 0.7194
# [19] 0.6211 0.6054
summary(ydata)
#   Min. 1st Qu. Median   Mean 3rd Qu.   Max.
#   0.605  0.637  0.725  0.711  0.763  0.798


B = 500; J = 2000; set.seed(232)
mh = MH(B=B,J=J,y=ydata, a=0.1,b=0.9,r=0.008,t=0.01)
c(mh$ar,mh$br) # 0.616 0.576
X11(w=8,h=7); par(mfrow=c(2,1))
plot(0:(B+J),mh$av,type="l",main="",xlab="j",ylab="aj")
        abline(v=B,lty=3)
plot(0:(B+J),mh$bv,type="l", main="",xlab="j",ylab="bj")
        abline(v=B,lty=3)
alast= mh$av[length(mh$av)]; blast= mh$bv[length(mh$bv)]
c(alast,blast) # 0.5979 0.8123


B=0; J = 50000; set.seed(230); date()
mh = MH(B=B,J=J,y=ydata, a=alast,b=blast,r=0.008,t=0.01)
date() # Takes about 14 seconds
c(mh$ar,mh$br) # 0.6141 0.5434
av=mh$av[-1][seq(05,50000,50)]; J = length(av); J #  1000
bv=mh$bv[-1][seq(50,50000,50)];
acf(av)$acf[1:5] # 1  0.04828  0.01193 -0.02745  0.03983   OK
acf(bv)$acf[1:5] # 1 0.038617 0.007026 0.030259 0.011678   OK
mv=0.5*(av+bv)
# acf(mv)$acf[1:5] # 1  -0.001121 -0.020770  0.001872 -0.008731 OK


X11(w=8,h=5); par(mfrow=c(1,1))
hist(mv,prob=T,xlab="m",main="",
        xlim=c(0.65,0.75), ylim=c(0,80))
lines(density(mv),lwd=2)
est=mean(mv); ci=est+c(-1,1)*qnorm(0.975)*sd(mv)/sqrt(J)
cpdr=quantile(mv,c(0.025,0.975))
print(c(est,ci,cpdr),digits=4)  # 0.7013 0.7008 0.7019 0.6837 0.7173
abline(v= c(est,ci,cpdr),lwd=2)
```

CHAPTER 8

Inference via **WinBUGS**

8.1 Introduction to **BUGS**

We have illustrated the usefulness of MCMC methods by applying them to a variety of statistical contexts. In each case, specialised R code was used to implement the chosen method. Writing such code is typically time consuming and requires a great deal of attention to details such as choosing suitable tuning constants in the Metropolis-Hastings algorithm.

A software package which can greatly assist with the application of MCMC methods is WinBUGS. This stands for:

Bayesian Inference **U**sing **G**ibbs **S**ampling for Microsoft **Win**dows.

The BUGS Project was started in 1989 by a team of statisticians in the UK (at the Medical Research Council Biostatistics Unit, Cambridge, and Imperial College School of Medicine, London) and developed until the latest version WinBUGS 1.4.3 was released in 2007.

WinBUGS 1.4.3 is a stable version of BUGS which is suitable for routine use, even today.

Since 2007, development of BUGS has focused on OpenBUGS, an open source version of the package. In what follows we will only refer to WinBUGS 1.4.3. This is freely available from the official website:

http://www.mrc-bsu.cam.ac.uk/software/bugs/

Figure 8.1 shows this website (as it appeared on 18 February 2015).

Figure 8.2 shows the Wikipedia article on WinBUGS (on the same day):

http://en.wikipedia.org/wiki/WinBUGS

The preferred reference for citing WinBUGS in scientific papers is:

Lunn, D.J., Thomas, A., Best, N., and Spiegelhalter, D. (2000). WinBUGS – A Bayesian modelling framework: Concepts, structure, and extensibility. *Statistics and Computing*, **10**: 325–337.

Figure 8.1 Official website for WinBUGS

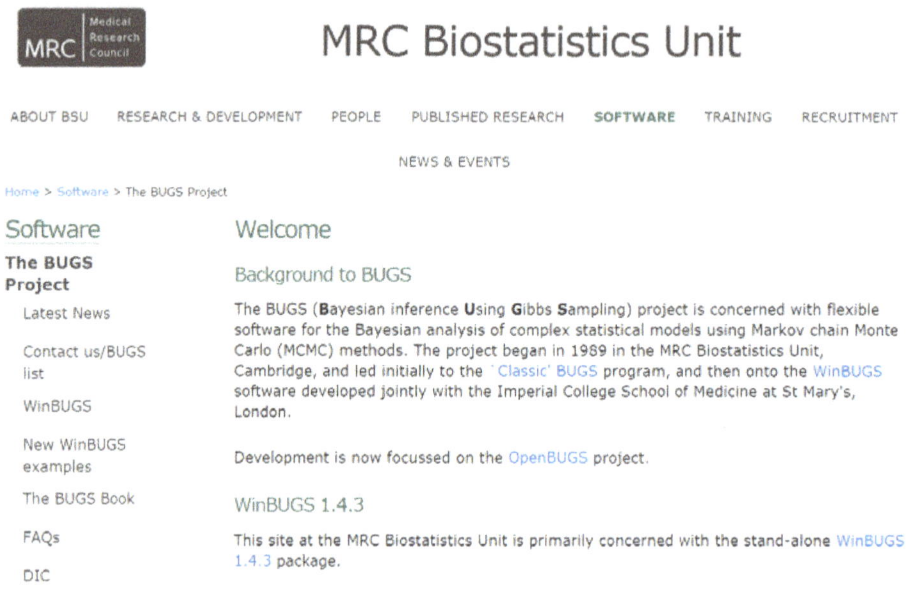

Figure 8.2 Wikipedia article on WinBUGS

8.2 A first tutorial in BUGS

Consider the following Bayesian model:

$$y_1,...,y_n \mid \mu,\tau \sim iid\ Normal(\mu,\sigma^2) \quad (\tau = 1/\sigma^2)$$

$$\mu \mid \tau \sim Normal(\mu_0,\sigma_0^2)$$

$$\tau \sim Gamma(\alpha,\beta) \quad (E\tau = \alpha/\beta)$$

where $\mu_0 = 0$, $\sigma_0^2 = 10,000$ and $\alpha = \beta = 0.001$.

Suppose the data is $y = (y_1,...,y_n) = (2.4, 1.2, 5.3, 1.1, 3.9, 2.0)$, and we wish to find the posterior mean and 95% posterior interval for each of μ and $\gamma = \mu\sqrt{\tau}$ (the signal to noise ratio).

To perform this in WinBUGS 1.4.3, open a new window (select 'File' and then 'New' in the BUGS toolbar), and type the following BUGS code:

```
model
{
for(i in 1:n){
        y[i] ~ dnorm(mu, tau)
        }
mu ~ dnorm(0,0.0001)
tau ~ dgamma(0.001, 0.001)
gam <- mu*sqrt(tau)
}

list( n=6, y=c(2.4,1.2,5.3,1.1,3.9,2.0) )

list(tau=1)
```

Alternatively, copy this text from a Word document into a Notepad file, and then copy the text from the Notepad file into the WinBUGS window.

Note: Do not copy text from Word to WinBUGS *directly* or you may get an error message.

The WinBUGS window should then look as depicted in Figure 8.3.

Figure 8.3 WinBUGS window with code

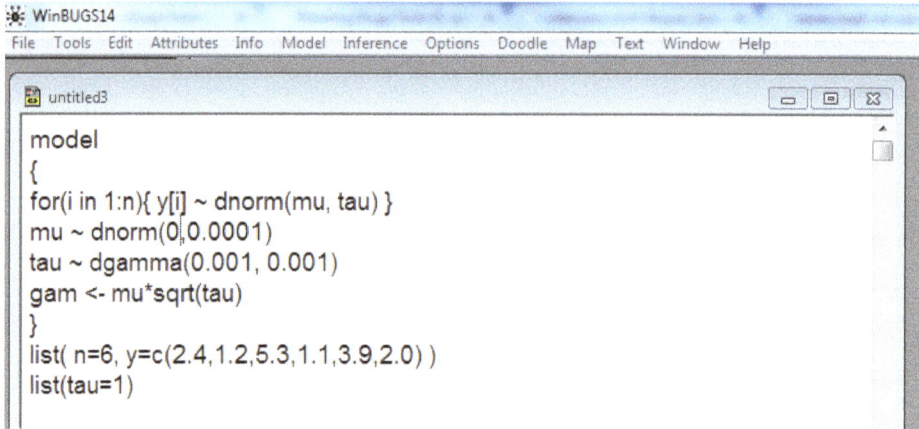

Next, select 'Model' (in the WinBUGS toolbar) and then 'Specification'.

Then highlight the word 'model' (in the BUGS code above) and click on 'check model' in the 'Specification Tool'.

Then highlight the first word 'list', click on 'load data' and click on 'compile'.

Then highlight the second word 'list', click on 'load inits' and click on 'gen inits'.

Next, select 'Inference' and then 'Samples'. Then, in the 'Sample Monitor Tool' which appears, type 'mu' in the 'node' box, click 'set', type 'gam' in the 'node' box and click 'set' again.

Then click 'Model' and 'Update'.

In the 'Update Tool' which appears, change '1000' to '1500' and click 'update'. This will implement 1,500 iterations of an MCMC algorithm.

Next type '*' (an asterisk) in the 'node' box, change '1' to '501' in the 'beg' box (meaning beginning) and click 'stats' (statistics).

This should produce something similar to what is shown in Figure 8.4 and Table 8.1.

Figure 8.4 Tools and node statistics in **WinBUGS**

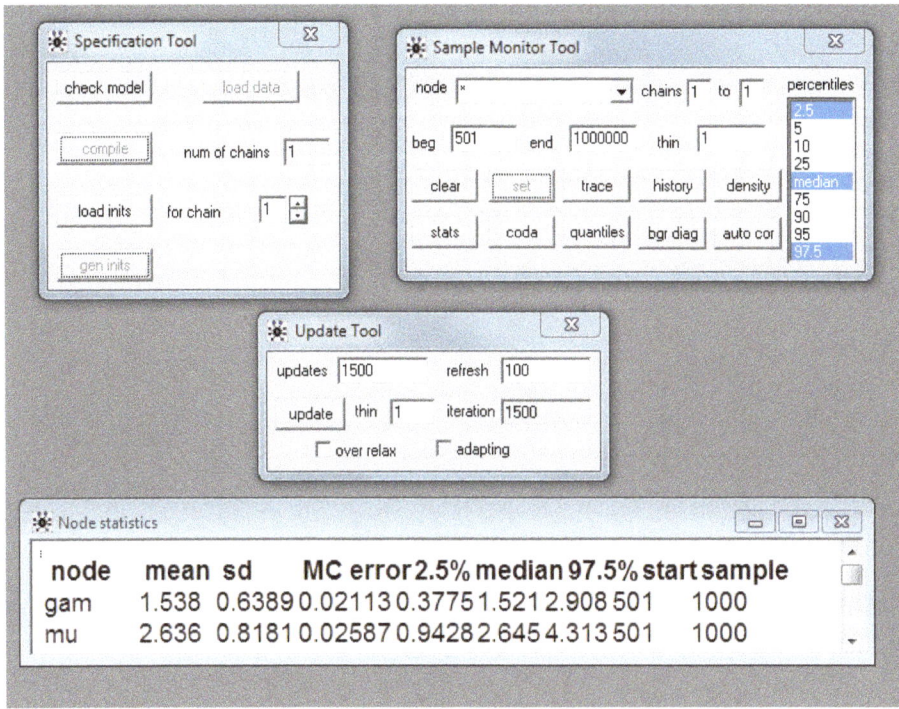

Table 8.1 Node statistics in **WinBUGS** (as in Figure 8.4)

node	mean	sd	MC error	2.5%	median	97.5%	start	sample
gam	1.538	0.6389	0.02113	0.3775	1.521	2.908	501	1000
mu	2.636	0.8181	0.02587	0.9428	2.645	4.313	501	1000

From these results, we see that the posterior mean and 95% posterior interval for μ are about 2.64 and (0.94, 4.31), and the same quantities for γ are about 1.54 and (0.38, 2.91).

To obtain more precise inference we could repeat the above procedure with a larger Monte Carlo sample size (e.g. 10,000 rather than 1,000).

Note: If $\sigma_0 = \infty$ and $\alpha = \beta = 0$, the posterior mean and 95% CPDR for μ are exactly

$$\bar{y} = 2.65$$

(i.e. the sample mean) and

$$(\bar{y} \pm t_{0.025}(n-1)/\sqrt{n}) = (0.92, 4.38)$$

(where s is the sample standard deviation).

The posterior mean and CPDR for γ do not have such simple formulae.

To see line plots of the simulated values, click on 'history' (in the 'Sample Monitor Tool'), and to view smoothed histograms of them, click 'density'. Figure 8.5 illustrates.

Figure 8.5 Line plots and smoothed histograms in WinBUGS

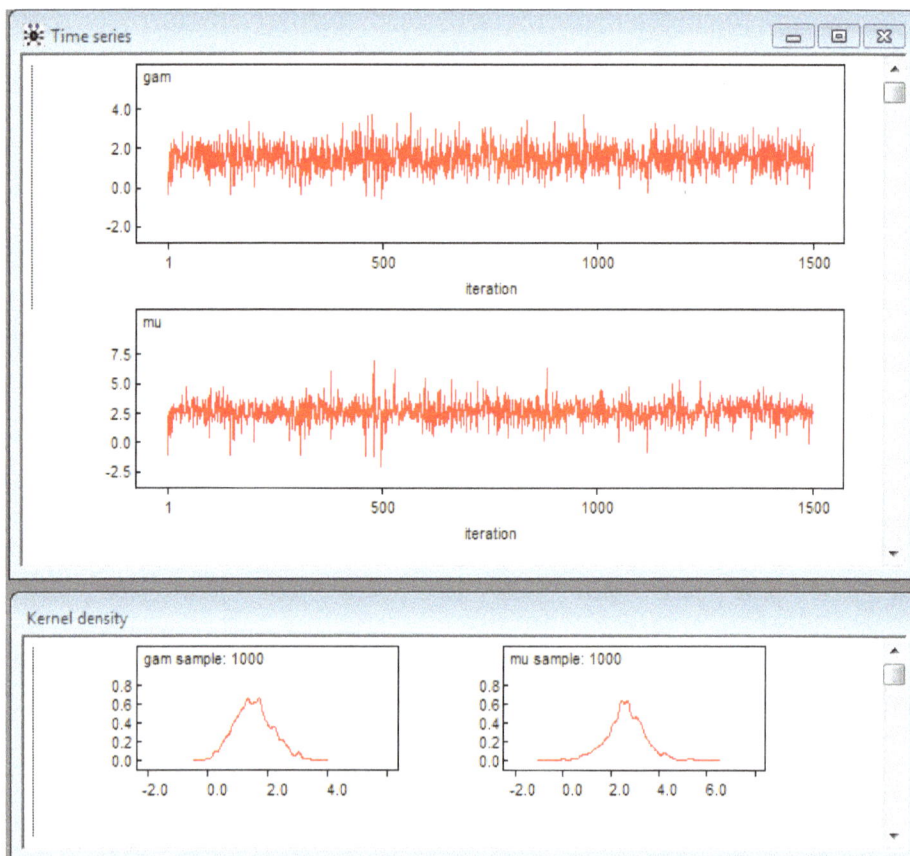

To transfer the simulated values from WinBUGS into R (for further analysis) click on 'coda'. Two boxes will appear, one called 'CODA index' with the following:

gam	1	1000
mu	1001	2000

The other box, called 'CODA for chain 1', should have two columns and 2,000 rows and look as follows:

501	1.298
502	1.307
503	1.478
....	
1498	0.8303
1499	1.993
1500	2.326
501	1.812
502	1.999
503	2.8
....	
1498	1.628
1499	2.161
1500	2.748

Next, copy the contents of 'CODA for chain 1' into a Notepad file called 'out.txt' (say). Save that file somewhere, e.g. onto the desktop.

Then begin a session in R and proceed as follows:

```
out <- read.table(file=file.choose())   # Navigate to and choose 'out.txt'

dim(out) # 2000   2

gamv <- out[1:1000,2]; muv <- out[1001:2000,2]

par(mfrow=c(2,1)); hist(muv, breaks=20); hist(gamv, breaks=20)
```

This should result in the graphs shown in Figure 8.6.

Figure 8.6 Histograms in R using output from WinBUGS

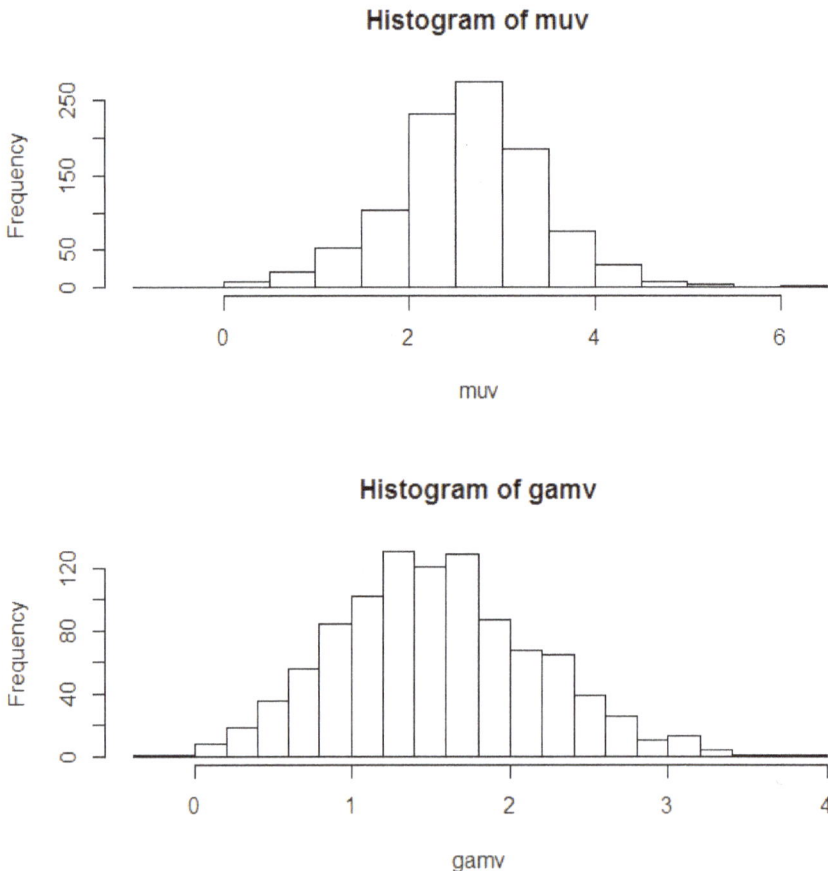

Histogram of muv

Histogram of gamv

One can then use the MCMC output in many other ways, e.g. to simulate from a posterior predictive distribution via the method of composition.

As an alternative, it is possible to run WinBUGS directly from R after installing the appropriate packages. (This will be done in a future exercise). But this method is really only for production runs and is not recommended during the experimentation stage of an analysis.

For more information on BUGS, click on 'Help' and 'User manual' in the toolbar. Also see 'Examples Vol I' and 'Examples Vol II' for several dozen worked examples in BUGS. The examples are very user-friendly. They contain data, code and everything one needs to reproduce the results shown. Figure 8.7 shows various excerpts from these files.

Figure 8.7 Exerpts from the WinBUGS 1.4.3 User Manual
(several pages)

WinBUGS User Manual

Version 1.4, January 2003

Upgraded to:
Version 1.4.3 (please see here for details)
August 6th. 2007

Beware: MCMC sampling can be dangerous!

Contents

Introduction ⇒⇐	Compound Documents ⇒⇐
Model Specification ⇒⇐	DoodleBUGS: The Doodle Editor ⇒⇐
The Model Menu ⇒⇐	The Inference Menu ⇒⇐
The Info Menu ⇒⇐	The Options Menu ⇒⇐
Batch-mode: Scripts	Tricks: Advanced Use of the BUGS Language ⇒⇐
WinBUGS Graphics ⇒⇐	Tips and Troubleshooting ⇒⇐
Tutorial ⇒⇐	Changing MCMC Defaults (advanced users only) ⇒⇐
Distributions ⇒⇐	References

Distributions   ☐ ☐ ☒

Distributions

BUGS

Contents

Discrete Univariate   [top | home]
Bernoulli
`r ~ dbern(p)`

$$p^r(1-p)^{1-r}; \quad r = 0,1$$

Binomial
`r ~ dbin(p, n)`

$$\frac{n!}{r!(n-r)!}p^r(1-p)^{n-r}; \quad r = 0,...,n$$

Categorical
`r ~ dcat(p[])`

$$p[r]; \quad r = 1,2,...,\dim(p); \quad \sum_i p[i] = 1$$

Negative Binomial
`x ~ dnegbin(p, r)`

$$\frac{(x+r-1)!}{x!(r-1)!}p^r(1-p)^x; \quad x = 0,1,2,...$$

Poisson
`r ~ dpois(lambda)`

$$e^{-\lambda}\frac{\lambda^r}{r!}; \quad r = 0,1,...$$

Continuous Univariate   [top | home]
Beta
`p ~ dbeta(a, b)`

$$p^{a-1}(1-p)^{b-1}\frac{\Gamma(a+b)}{\Gamma(a)\Gamma(b)}; \quad 0 < p < 1$$

Chi-squared
`x ~ dchisqr(k)`

$$\frac{2^{-k/2}x^{k/2-1}e^{-x/2}}{\Gamma(\frac{k}{2})}; \quad x > 0$$

Examples Volume I ▭ ▭

BUGS **Examples Volume 1**

Rats: Normal hierarchical model

Pump: conjugate gamma-Poisson hierarchical model

Dogs: log linear binary model

Seeds: random effects logistic regression

Surgical: institutional ranking

Salm: extra-Poisson variation in dose-response study

Equiv: bioequivalence in a cross-over trial

Dyes: variance components model

Stacks: robust and ridge regression

Epil: repeated measures on Poisson counts

Blocker: random effects meta-analysis of clinical trials

Oxford: smooth fit to log-odds ratios in case control studies

LSAT: latent variable models for item-response data

Bones: latent trait model for multiple ordered catagorical responses

Inhalers: random effects model for ordinal responses from a cross-over trial

Mice: Weibull regression in censored survival analysis

Kidney: Weibull regression with random effects

Leuk: survival analysis using Cox regression

Cox regression with frailties

References:
Sorry - an on-line version of the references is currently unavailable.
Please refer to the existing Examples documentation available from
http://www.mrc-bsu.cam.ac.uk/bugs.

BUGS Rats: a normal hierarchical model

This example is taken from section 6 of Gelfand et al (1990), and concerns 30 young rats whose weights were measured weekly for five weeks. Part of the data is shown below, where Y_{ij} is the weight of the ith rat measured at age x_j.

| | Weights Y_{ij} of rat i on day x_j | | | | |
	$x_j = 8$	15	22	29	36
Rat 1	151	199	246	283	320
Rat 2	145	199	249	293	354
....					
Rat 30	153	200	244	286	324

A plot of the 30 growth curves suggests some evidence of downward curvature.

The model is essentially a random effects linear growth curve

$$Y_{ij} \sim Normal(\alpha_i + \beta_i(x_j - x_{bar}), \tau_c)$$

$$\alpha_i \sim Normal(\alpha_c, \tau_\alpha)$$

$$\beta_i \sim Normal(\beta_c, \tau_\beta)$$

where $x_{bar} = 22$, and τ represents the *precision* (1/variance) of a normal distribution. We note the absence of a parameter representing correlation between α_i and β_i unlike in Gelfand et al 1990. However, see the Birats example in Volume 2 which does explicitly model the covariance between α_i and β_i. For now, we standardise the x_j's around their mean to reduce dependence between α_i and β_i in their likelihood: in fact for the full balanced data, complete independence is achieved. (Note that, in general, prior independence does not force the posterior distributions to be independent).

α_c, τ_α, β_c, τ_β, τ_c are given independent "noninformative" priors, with two alternatives considered for τ_α and τ_β: prior 1 is uniform on the scale of the standard deviations $\sigma_\alpha = 1/sqrt(\tau_\alpha)$ and $\sigma_\beta = 1/sqrt(\tau_\beta)$, and prior 2 is a gamma(0.001, 0.001) on the precisions τ_α and τ_β. Interest particularly focuses on the intercept at zero time (birth), denoted $\alpha_0 = \alpha_c - \beta_c x_{bar}$.

Graphical model for rats example (using prior 1):

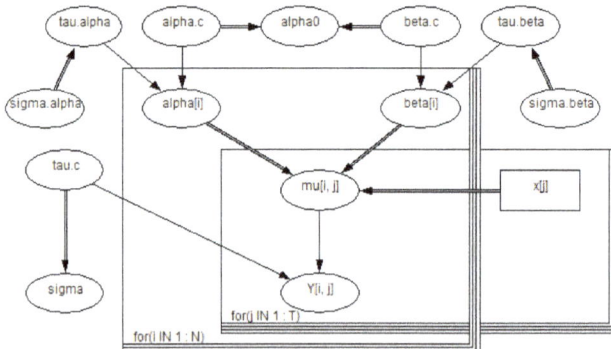

Note: The last graphic shown is called a Doodle. WinBUGS has a facility whereby the user can create such a diagram and have the code generated automatically.

BUGS *language for rats example:*

```
model
{
    for( i in 1 : N ) {
        for( j in 1 : T ) {
            Y[i , j] ~ dnorm(mu[i , j],tau.c)
            mu[i , j] <- alpha[i] + beta[i] * (x[j] - xbar)
        }
        alpha[i] ~ dnorm(alpha.c,tau.alpha)
        beta[i] ~ dnorm(beta.c,tau.beta)
    }
    tau.c ~ dgamma(0.001,0.001)
    sigma <- 1 / sqrt(tau.c)
    alpha.c ~ dnorm(0.0,1.0E-6)
    # Choice of prior of random effects variances
    # Prior 1: uniform on SD
    sigma.alpha~ dunif(0,100)
    sigma.beta~ dunif(0,100)
    tau.alpha<-1/(sigma.alpha*sigma.alpha)
    tau.beta<-1/(sigma.beta*sigma.beta)

    #Prior 2: (not recommended)
    #tau.alpha ~ dgamma(0.001,0.001)
    #tau.beta ~ dgamma(0.001,0.001)

    beta.c ~ dnorm(0.0,1.0E-6)

    alpha0 <- alpha.c - xbar * beta.c
}
```

Data ⇒list(x = c(8.0, 15.0, 22.0, 29.0, 36.0), xbar = 22, N = 30, T = 5,
Y = structure(
.Data =   c(151, 199, 246, 283, 320,
145, 199, 249, 293, 354,
147, 214, 263, 312, 328,
155, 200, 237, 272, 297,
135, 188, 230, 280, 323,
159, 210, 252, 298, 331,
141, 189, 231, 275, 305,
159, 201, 248, 297, 338,
177, 236, 285, 350, 376,
134, 182, 220, 260, 296,
160, 208, 261, 313, 352,
143, 188, 220, 273, 314,
154, 200, 244, 289, 325,
171, 221, 270, 326, 358,
163, 216, 242, 281, 312,
160, 207, 248, 288, 324,
142, 187, 234, 280, 316,
156, 203, 243, 283, 317,
157, 212, 259, 307, 336,
152, 203, 246, 286, 321,
154, 205, 253, 298, 334,
139, 190, 225, 267, 302,
146, 191, 229, 272, 302,
157, 211, 250, 285, 323,
132, 185, 237, 286, 331,
160, 207, 257, 303, 345,
169, 216, 261, 295, 333,
157, 205, 248, 289, 316,
137, 180, 219, 258, 291,
153, 200, 244, 286, 324),
.Dim = c(30,5)))⇐

Inits1 ⇒list(alpha = c(250, 250, 250, 250, 250, 250, 250, 250, 250, 250, 250, 250, 250, 250, 250,
250, 250, 250, 250, 250, 250, 250, 250, 250, 250, 250, 250, 250, 250, 250),
beta = c(6, 6, 6, 6, 6, 6, 6, 6, 6, 6, 6, 6, 6, 6, 6,
6, 6, 6, 6, 6, 6, 6, 6, 6, 6, 6, 6, 6, 6, 6),
alpha.c = 150, beta.c = 10,
tau.c = 1, sigma.alpha = 1, sigma.beta = 1)⇐

Results

A 1000 update burn in followed by a further 10000 updates gave the parameter estimates:

node	mean	sd	MC error	2.5%	median	97.5%	start	sample
alpha0	106.6	3.65	0.04151	99.43	106.5	113.9	1001	10000
beta.c	6.185	0.1102	0.001294	5.967	6.185	6.404	1001	10000
sigma	6.074	0.4673	0.007724	5.247	6.044	7.068	1001	10000

Predictions:

node	mean	sd	MC error	2.5%	median	97.5%	start	sample
Y[26,2]	204.5	8.74	0.1159	187.0	204.4	221.7	1001	10000
Y[26,3]	250.0	10.27	0.1642	229.7	249.9	270.1	1001	10000
Y[26,4]	295.4	12.64	0.2092	270.3	295.3	320.3	1001	10000
Y[26,5]	340.6	15.32	0.284	310.2	340.5	370.5	1001	10000
beta.c	6.575	0.1507	0.003708	6.281	6.573	6.875	1001	10000

Trace plots and density estimates:

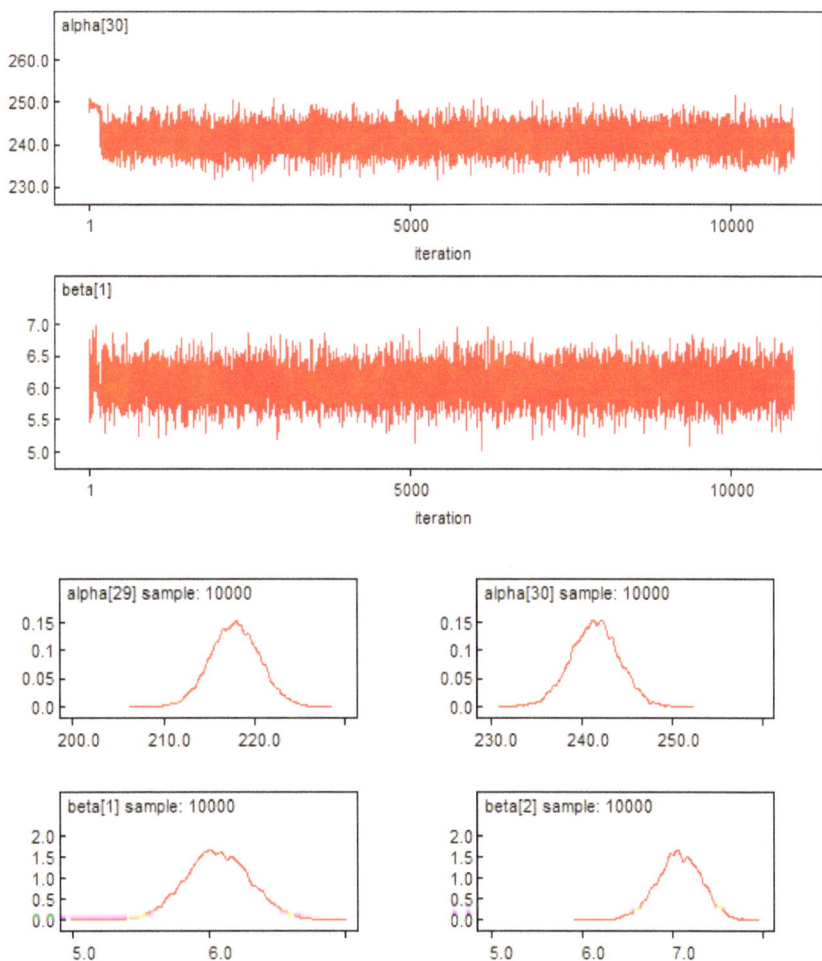

(End of Figure 8.7)

Exercise 8.1 Simple linear regression via WinBUGS

Use WinBUGS to perform a simple linear regression on the data in Table 8.2 (which is the same as Table 7.1 in Exercise 7.2).

Table 8.2 Regression data

$x_i (= i)$	1	2	3	4	5
y_i	5.879	8.54	14.12	13.14	15.26

i	6	7	8	9	10
y_i	20.43	19.92	18.47	21.63	24.11

Solution to Exercise 8.1

Using the following WinBUGS code, we obtain the results in Table 8.3:

```
model{
for(i in 1:n){
  mu[i] <- a + b*x[i]
  y[i] ~ dnorm(mu[i],lam)
  }
a ~ dnorm(0.0,0.001)
b ~ dnorm(0.0,0.001)
lam ~ dgamma(0.001,0.001)
}

# data
list(n = 10, x = c(1,2,3,4,5,6,7,8,9,10), y=c(5.879,8.54,14.12,
13.14,15.26,20.43,19.92,18.47,21.63,24.11))

# inits
list(a=0,b=0,lam=1)
```

Table 8.3 Results of regression performed using WinBUGS

node	mean	sd	MC error	2.5%	median	97.5%	start	sample
a	6.039	1.532	0.01646	2.955	6.051	9.107	1001	10000
b	1.836	0.247	0.00266	1.342	1.834	2.334	1001	10000
lam	0.2625	0.1313	0.001602	0.07259	0.2404	0.5788	1001	10000

Using the results in Table 8.3, we estimate a by 6.039 with 95% CPDR (2.955, 9.107), and we estimate b by 1.836 with 95% CPDR (1.342, 2.334).

It may be noted that these results are very similar to those obtained via classical techniques in an earlier exercise: 6.051 and (2.973, 9.128) for a, and 1.836 and (1.340, 2.332) for b.

Figure 8.8 shows trace plots and density estimates produced as part of the WinBUGS output.

Figure 8.8 Graphical output from **WinBUGS** regression

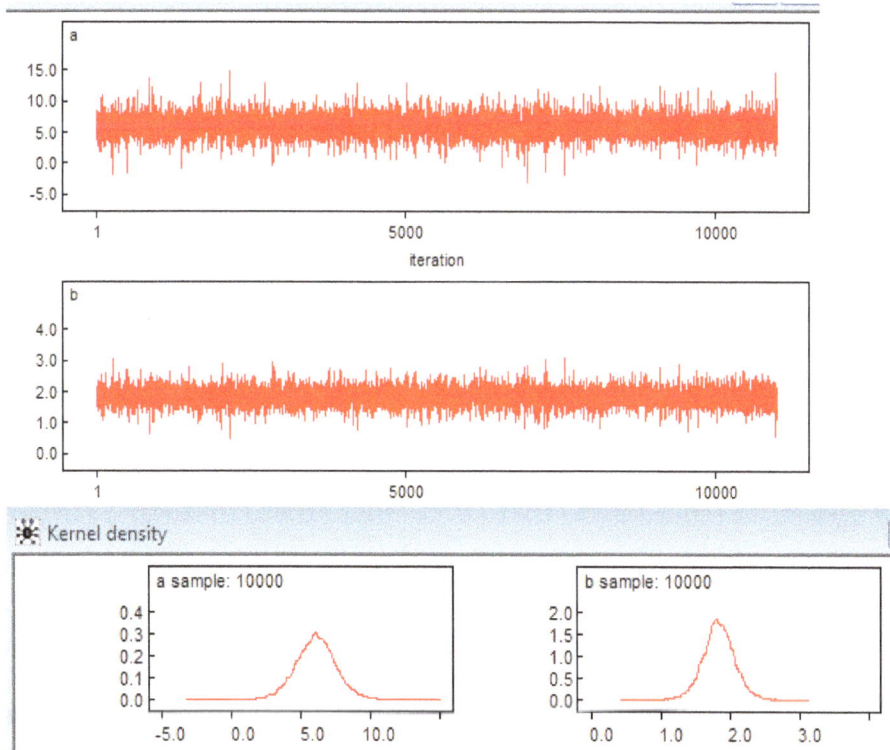

Exercise 8.2 Logistic regression via WinBUGS

Consider the data in Table 8.4, which is the same as in Table 7.2 of Exercise 7.3 (where, for example, in Experiment 3 a total of 40 rats were exposed to radiation for 3.6 hours, and 23 of them died within one month).

Table 8.4 Rat mortality data

i	n_i	x_i	y_i	y_i/n_i	$=$	$\hat{p}_i$
1	10	0.1	1	1/10	$=$	0.1
2	30	1.4	0	0/30	$=$	0
3	40	3.6	23	23/40	$=$	0.575
4	20	3.8	12	12/20	$=$	0.6
5	15	5.2	8	8/15	$=$	0.5333
6	46	6.1	32	32/46	$=$	0.696
7	12	8.7	10	10/12	$=$	0.833
8	37	9.1	35	35/37	$=$	0.946
9	23	9.1	19	19/23	$=$	0.826
10	8	13.6	8	8/8	$=$	1

Use WinBUGS to estimate the parameters in the following logistic regression model for these data:
$$Y_i \sim\perp Bin(n_i, p_i), \quad i = 1,...,n,$$
where:

$$p_i = \frac{1}{1+\exp(-z_i)} \qquad \text{(probability of a 'success' for experiment } i\text{)}$$

$$z_i = a + bx_i \qquad \text{(linear predictor)}.$$

In your results, also include inference on *LD50*, the dose at which 50% of rats will die ($= -a/b$), and on d, defined as the number of rats that will die out of 20 that are exposed to five hours of radiation.

Solution to Exercise 8.2

Applying the following WinBUGS code, we obtain the results in Table 8.5:

```
model
{
for(i in 1:N){
  z[i] <- a + b*x[i]
  logit(p[i])<- z[i]
  y[i] ~ dbin(p[i],n[i])
  }
a ~ dnorm(0.0,0.001)
b ~ dnorm(0.0,0.001)
logit(p5) <- a+5*b
d ~ dbin(p5,20)
LD50 <- -a/b
}

# data
list(N=10,n=c(10,30,40,20,15,46,12,37,23,8),
x=c(0.1,1.4,3.6,3.8,5.2,6.1,8.7,9.1,9.1,13.6),
y=c(1,0,23,12,8,32,10,35,19,8))

# inits
list(a=0,b=0)
```

Table 8.5 Results of logistic regression performed using WinBUGS

node	mean	sd	MC error	2.5%	median	97.5%	start	sample
LD50	4.273	0.3373	0.00464	3.587	4.285	4.899	1001	10000
a	-2.177	0.3726	0.01041	-2.922	-2.168	-1.478	1001	10000
b	0.5082	0.06962	0.001964	0.3794	0.5059	0.6501	1001	10000
d	11.79	2.344	0.02447	7.0	12.0	16.0	1001	10000
p5	0.5895	0.03946	3.174E4	0.5125	0.5896	0.6664	1001	10000

Thus, we estimate a by -2.177 with 95% CPDR $(-2.922, -1.478)$, etc.

These results are very similar to those obtained via classical techniques in Exercise 7.3, namely -2.156 and $(-3.000, -1.311)$ for a, etc.

Figure 8.9 shows some traces and density estimates produced as part of the WinBUGS output. Here, 'p5' represents the probability of a rat dying within one month if exposed to five hours of radiation. We chose to monitor this node so as to estimate its posterior density

Figure 8.9 Graphical output from WinBUGS logistic regression

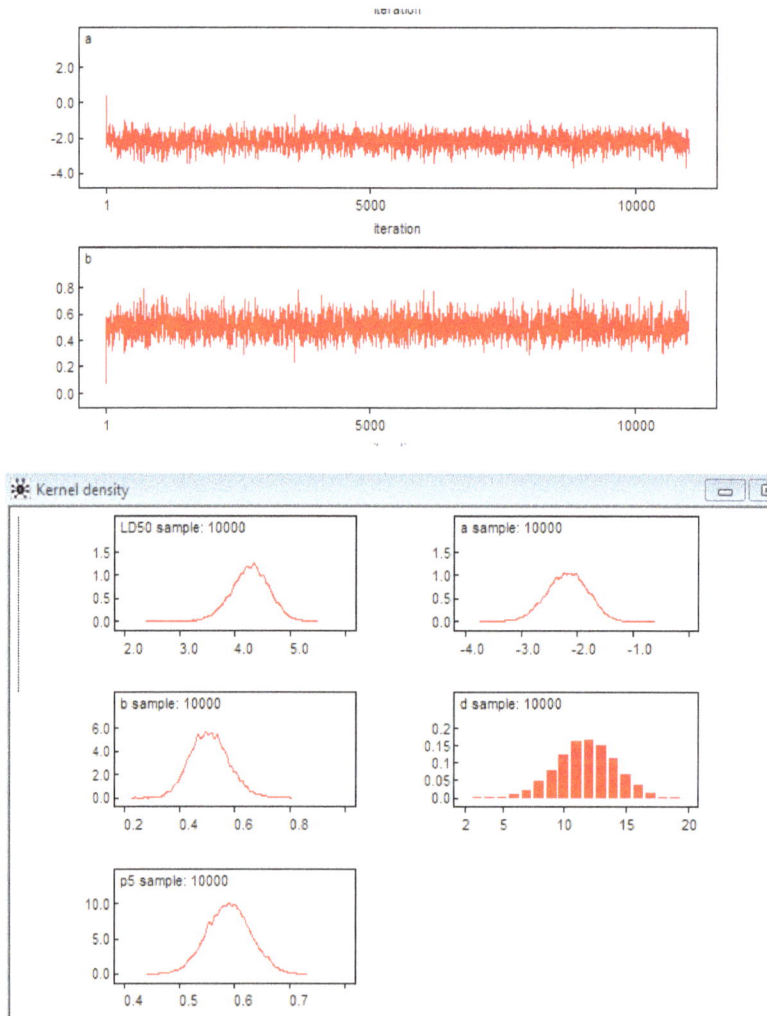

Exercise 8.3 Inference on a uniform distribution via WinBUGS

Consider the following Bayesian model:
$$(y_1,...,y_n \mid a,b) \sim iid\ U(a,b)$$
$$(a \mid b) \sim U(0,b)$$
$$b \sim U(0,1).$$

Suppose that $n = 20$ data values from this model with $a = 0.6$ and $b = 0.8$ are as shown in Table 8.6 (which is the same as Table 7.6 in Exercise 7.5).

Table 8.6 Sample values from a uniform distribution

i	1	2	3	4	5
y_i	0.7846	0.7572	0.6381	0.7626	0.6105

i	6	7	8	9	10
y_i	0.6990	0.7728	0.7113	0.7314	0.7435

i	11	12	13	14	15
y_i	0.6324	0.7072	0.7493	0.7979	0.6182

i	16	17	18	19	20
y_i	0.7652	0.7883	0.7194	0.6211	0.6054

Use WinBUGS to generate a random sample from the joint posterior distribution of the parameters a and b. Then use this sample to estimate the mean of the uniform distribution, namely
$$m = E(y_i \mid a,b) = (a+b)/2.$$

Solution to Exercise 8.3

Applying the following WinBUGS code we obtain the results in Table 8.7:

```
model
{
for(i in 1:n){ y[i] ~ dunif(a,b) }
b ~ dunif(0,1)
a ~ dunif(0,b)
m <- (a+b)/2
}

list( n=20, y=c(  0.7846, 0.7572, 0.6381, 0.7626, 0.6105,
                  0.6990, 0.7728, 0.7113, 0.7314, 0.7435,
                  0.6324, 0.7072, 0.7493, 0.7979, 0.6182,
                  0.7652, 0.7883, 0.7194, 0.6211, 0.6054) )

list(a=0.1, b=0.9)
```

Table 8.7 Results of WinBUGS analysis for a uniform distribution

node	mean	sd	MC error	2.5%	median	97.5%	start	sample
a	0.594	0.01184	1.996E-4	0.5623	0.5977	0.6051	1001	10000
b	0.8091	0.01187	2.004E-4	0.7982	0.8054	0.841	1001	10000
m	0.7016	0.008201	1.388E-4	0.6844	0.7015	0.7187	1001	10000

Using the results in Table 8.7, we estimate m by 0.7016, with 95% CI (0.7013, 0.7019) for m's posterior mean.

We also estimate the 95% CPDR for m as (0.6844, 0.7187).

Note 1: The CI here was obtained in R using the following code:

```
0.7016 +c(-1,1)*qnorm(0.975)*0.0001388
```

Another CI is (0.7014, 0.7018), obtained using the code:

```
0.7016 +c(-1,1)*qnorm(0.975)*0.008201/sqrt(10000)
```

But this second CI is 'inferior' to (0.7013, 0.7019) because it ignores the autocorrelation in the simulated values. The fact that the second CI is shorter corresponds to the fact that its true coverage probability is less than the nominal and desired 95%.

Note 2: These inferences (above Note 1) are similar to those obtained in the solution to Exercise 7.5 using custom-written R code: 0.7013 with 95% CI (0.7008, 0.7019) and 95% CPDR estimate (0.6837, 0.7173).

Note 3: The CI in Note 2 is wider than the CI (0.7013, 0.7019) because it is based on a smaller Monte Carlo sample size (of 1,000 rather than 10,000). If we use only iterations 1,001 to 2,000 from the WinBUGS output, we get

```
m    0.7016 0.008287  3.573E-4  0.6833 0.7016 0.7194  1001  1000
```

in place of the corresponding row of Table 8.7. Then, the 95% CI for *m*'s posterior mean becomes (0.7009, 0.7023), obtained via

```
0.7016 +c(-1,1)*qnorm(0.975)*0.0003573
```

This CI has a width of 0.0014, which is greater than 0.0006, the width of (0.7013, 0.7019), and closer to 0.0011, the width of the CI in Note 2.

Figure 8.10 shows some traces and density estimates produced as part of the WinBUGS output.

Figure 8.10 Graphs from WinBUGS analysis for a uniform distribution

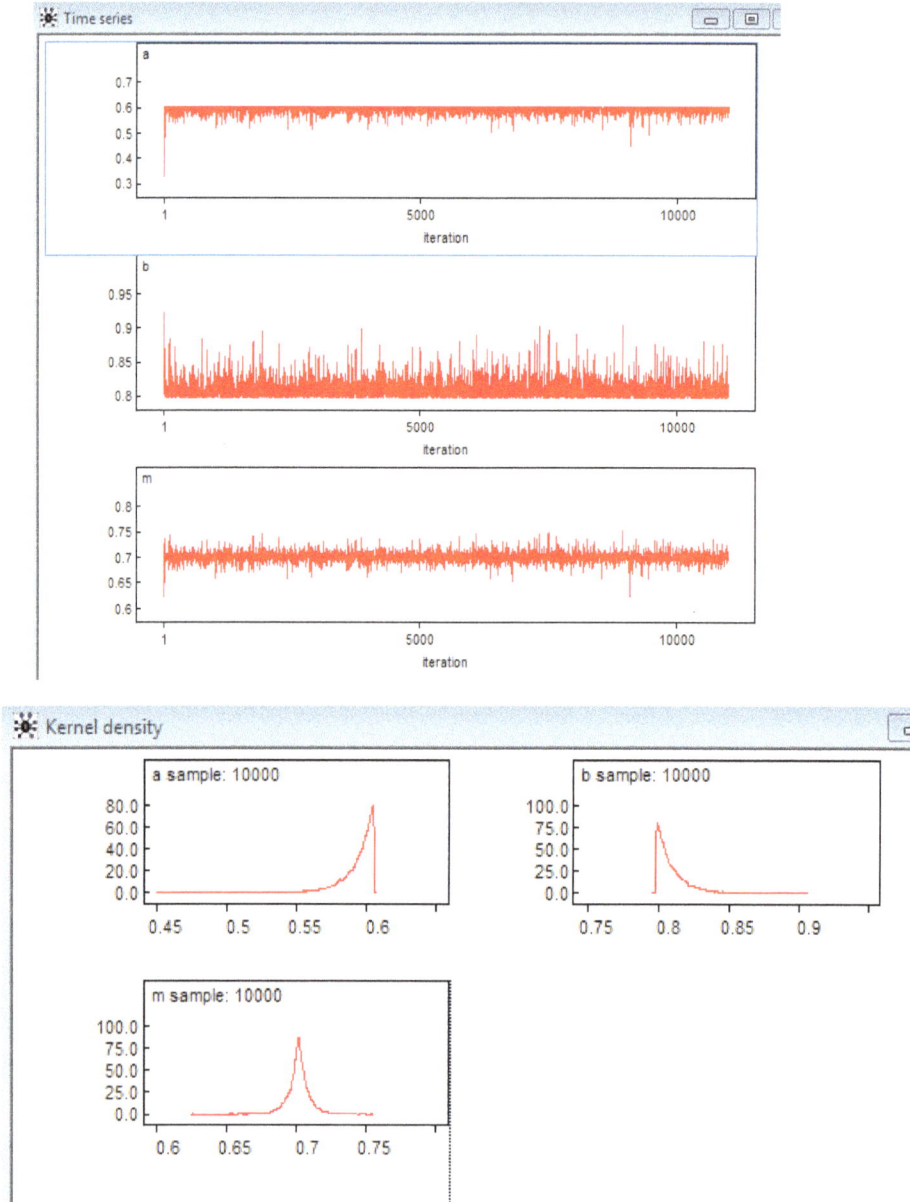

8.3 Tutorial on calling **BUGS** in R

The following is a short tutorial on how WinBUGS can be called within an R session. Some of the details may need to be changed depending on the configuration of files and directories in the computer being used.

First, assume that R (v3.01) is installed in **C:/R-3.0.1**

Also assume that WinBUGS (v4.1.3) is installed in **C:/WinBUGS14**

Open R and type

> install.packages("R2WinBUGS")

Note: You must have a connection to the internet for this to work. This command is required only once for each installed version of R.

Next, select a CRAN mirror when prompted. 'Melbourne' should work.

You should then see something like the following:

---
package 'coda' successfully unpacked and MD5 sums checked

package 'R2WinBUGS' successfully unpacked and MD5 sums checked, etc.

Then type

> library("R2WinBUGS")

Note: This loads the necessary functions and must be done at the beginning of each R session in which WinBUGS is to be called.

You should now see something like:

Loading required package: coda

Loading required package: lattice

Loading required package: boot, etc.

Next, create a file called **C:/R-3.0.1/BugsCode1.txt**
which contains the following code for a simple Bayesian model:

```
model

{

for(i in 1:n){ y[i] ~ dnorm(mu, tau) }

mu ~ dnorm(0,0.0001)

tau ~ dgamma(0.001, 0.001)

gam <- mu*sqrt(tau)

}
```

Next create a working directory, say **C:/R-3.0.1/BugsOut/**
and proceed in R as follows:

```
y <- c(2.4,1.2,5.3,1.1,3.9,2.0)

n <- length(y)

data <- list("n","y")

inits <- function(){   list(mu=0, tau=1.0)  }

parameters <- c("mu", "gam")

sim  <- bugs(data, inits, parameters,

        model.file= "C:/R-3.0.1/BugsCode1.txt",

        n.chains = 1, n.iter = 1500, n.burnin=500, DIC = FALSE,

    bugs.directory = "C:/WinBUGS14/",

    working.directory = "C:/R-3.0.1/BugsOut/")
```

This sets things up, starts WinBUGS, runs the BUGS code, closes
WinBUGS, and creates a number of files in the working directory, similar
to the ones shown in Figure 8.11.

Figure 8.11 Files created by running WinBUGS in R

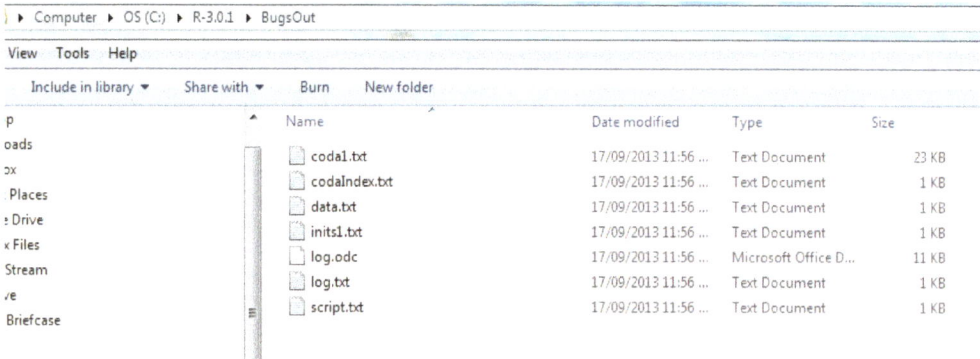

These files contain information which can then be accessed within R, for example as follows:

```
print(sim,digits=4)

# Inference for Bugs model at "C:/R-3.0.1/BugsCode1.txt", fit using WinBUGS,

#  1 chains, each with 1500 iterations (first 500 discarded)

#  n.sims = 1000 iterations saved

#     mean   sd   2.5%   25%  50%   75%  97.5%

# mu  2.6358 0.8185 0.9424 2.1760 2.645 3.1175 4.2984

# gam 1.5380 0.6392 0.3774 1.0935 1.521 1.9360 2.9061

par(mfrow=c(2,1))

hist(sim$sims.list$mu, breaks=20)

hist(sim$sims.list$gam, breaks=20)
```

After typing these commands, you should see two histograms similar to the ones shown in Figure 8.12. For more information on the bugs() function, simply type

```
help(bugs)
```

Figure 8.12 Histograms obtained in R after calling WinBUGS

Histogram of sim$sims.list$mu

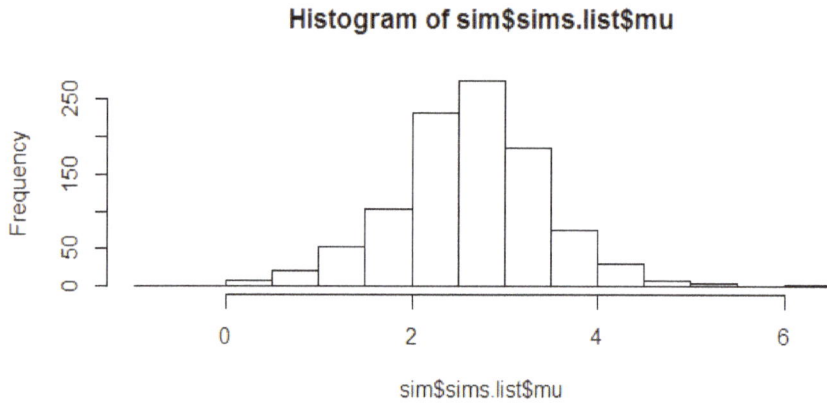

sim$sims.list$mu

Histogram of sim$sims.list$gam

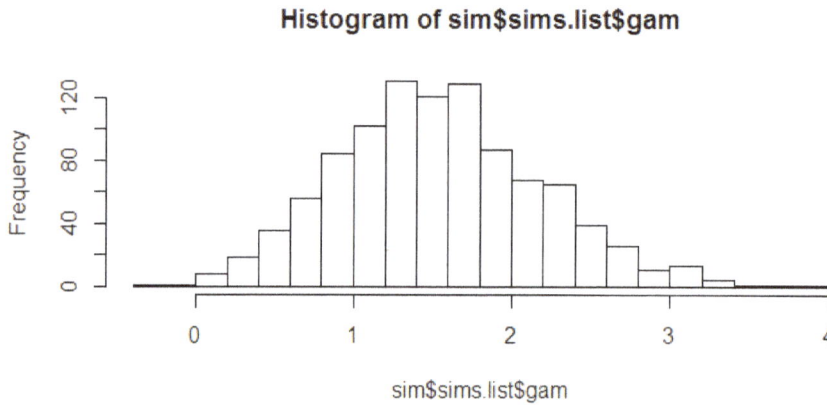

sim$sims.list$gam

Note: If your WinBUGS code has an error, the procedure will crash, with little to tell you what went wrong. In that case, first iron out any 'bugs' directly in WinBUGS, and only then run your WinBUGS code in R, as above.

Exercise 8.4 ARIMA modeling and forecasting with WinBUGS in R

Consider the well-known Total International Airline Passengers (TIAP) time series, as shown in Table 8.8. This series describes quarterly totals of international passengers for the period January 1949 to December 1960. (Here, Qtr1 refers to the period January–March, etc.)

Table 8.8 The TIAP time series

Year	Qtr1	Qtr2	Qtr3	Qtr4
1949	362	385	432	341
1950	382	409	498	387
1951	473	513	582	474
1952	544	582	681	557
1953	628	707	773	592
1954	627	725	854	661
1955	742	854	1023	789
1956	878	1005	1173	883
1957	972	1125	1336	988
1958	1020	1146	1400	1006
1959	1108	1288	1570	1174
1960	1227	1468	1736	1283

Using classical methods, fit a suitable ARIMA model to this time series.

Then forecast the time series forward for one up to twelve quarters.

Then repeat your analysis and forecasts using WinBUGS called from R.

Also create a single graph which compares both sets of forecasts.

Solution to Exercise 8.4

Figure 8.13 shows plots of the original times series x_t, its logarithm (showing stabilised variability), the difference of the logarithm (showing a removal of the trend), and y_t, the fourth seasonal difference of the first difference of the logarithm (showing that seasonality has been removed). The last two (bottom) plots are the sample ACF and sample PACF for y_t.

Figure 8.13 Plots for the TIAP time series

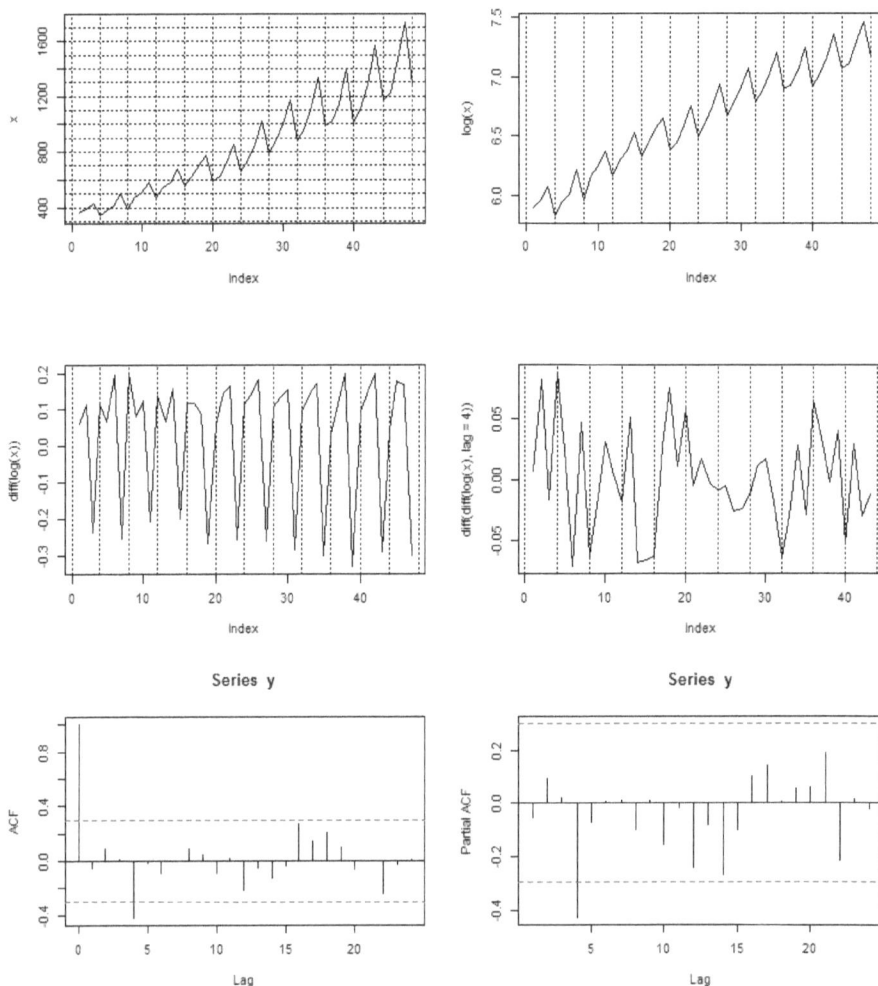

The last two plots in Figure 8.13 suggest SAR(1) or SMA(1) processes. Both fits pass standard diagnostic checks, the second being marginally better. Figure 8.14 shows some diagnostic plots for the SMA(1) fit (see the R Code below for further details).

Figure 8.14 Diagnostics for the SMA(1) fit to the TIAP time series

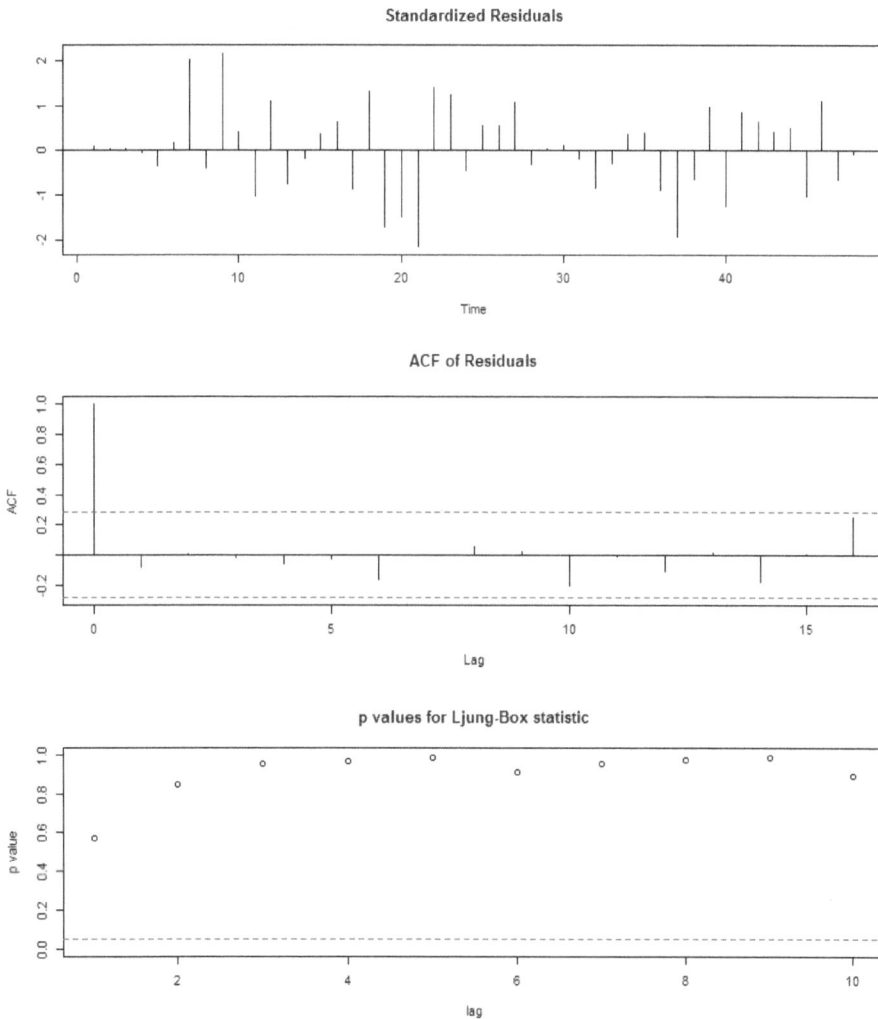

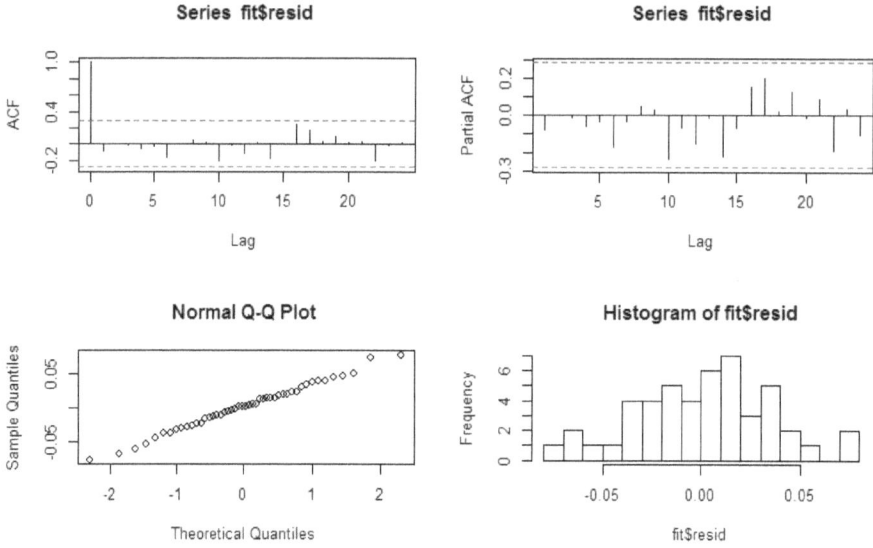

The chosen SMA(1) model for the TIAP time series x_t may be expressed by writing

$$y_t = \nabla_4 \nabla \log x_t,$$

where

$$y_t = w_t + \Theta_1 w_{t-4}, \quad w_t \sim iid\ N(0, \sigma^2).$$

The parameter estimates for this model are:

$$\hat{\Theta}_1 = -0.4927\ (SE = 0.1201)$$

$$\hat{\sigma}^2 = 0.0013.$$

Figure 8.15 shows the time series x_t plus predictions 12 quarters ahead based on the above fitted model. The dashed lines show the 95% prediction interval at each of the 12 future times points. (See the R code below for details regarding all calculations.)

Figure 8.15 Classical forecasts of the TIAP time series

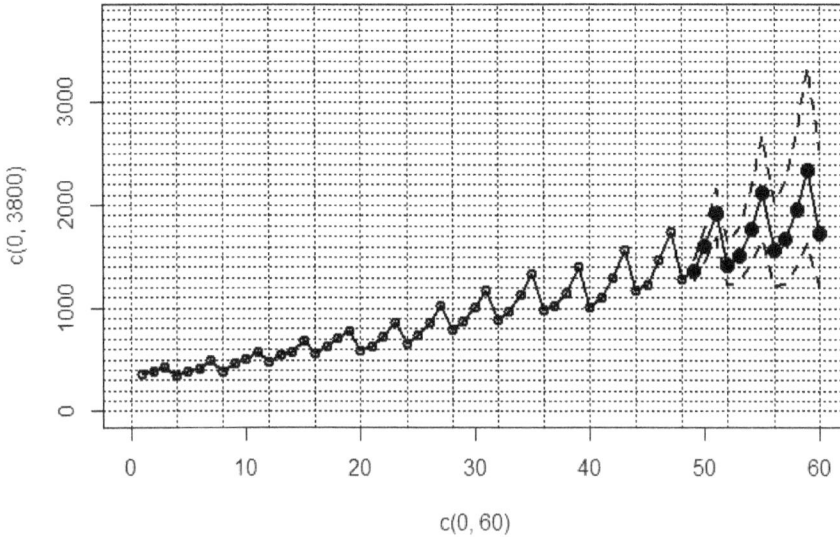

We now fit the same model to the time series but using MCMC via WinBUGS called from R. Some graphical output from the WinBUGS run is shown in Figure 8.16. (See the code below for details.)

Figure 8.17 shows the Bayesian analogue of the classical forecasts displayed in Figure 8.15.

To compare the classical and Bayesian analyses, we combine the two sets of forecasts into a single plot, as shown in Figure 8.18 (page 399). Figure 8.19 (page 399) is a detail in Figure 8.18.

Figure 8.16 Output from an analysis of the TIAP series using WinBUGS

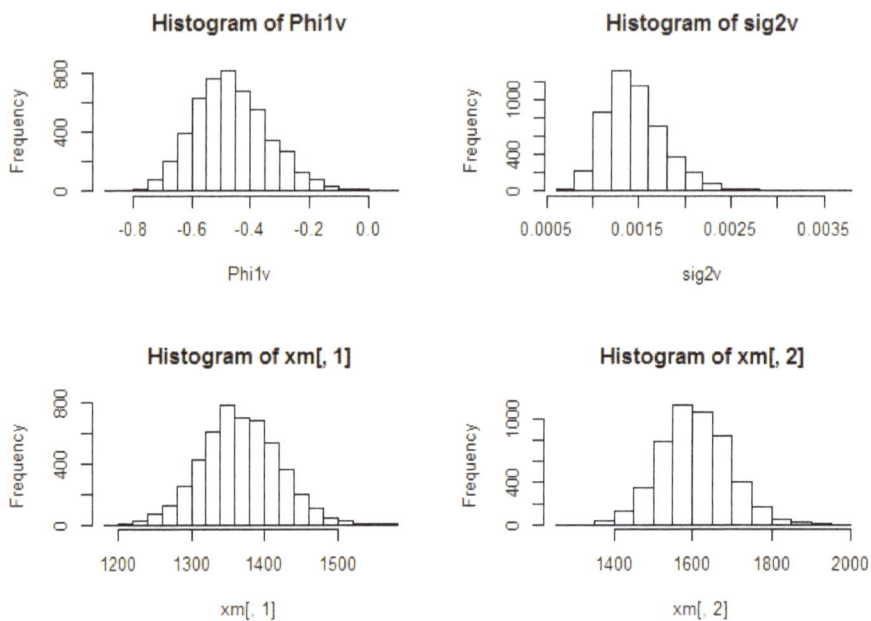

Figure 8.17 Bayesian forecasts of the TIAP time series

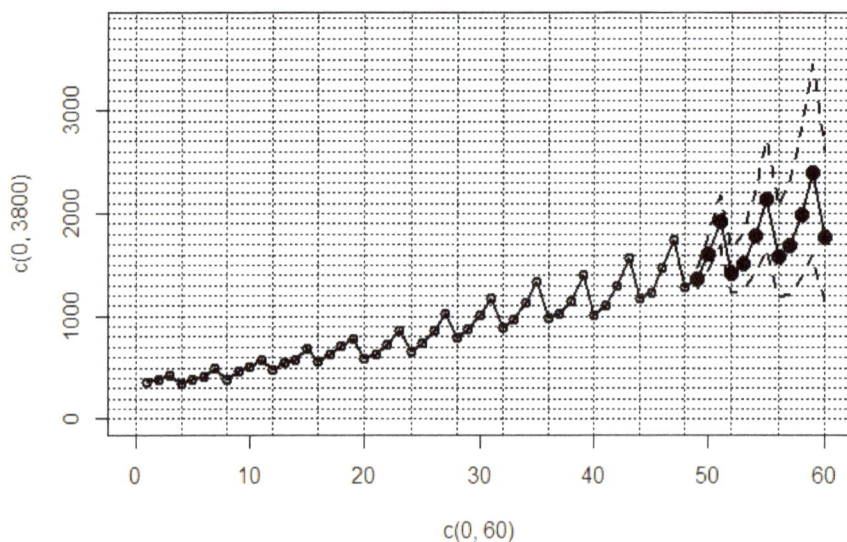

Figure 8.18 Comparison of forecasts for the TIAP time series

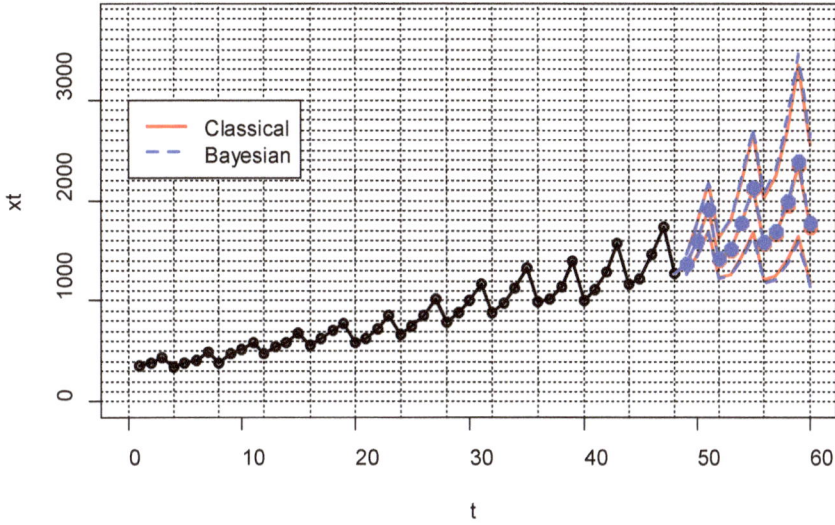

Figure 8.19 Detail in Figure 8.18

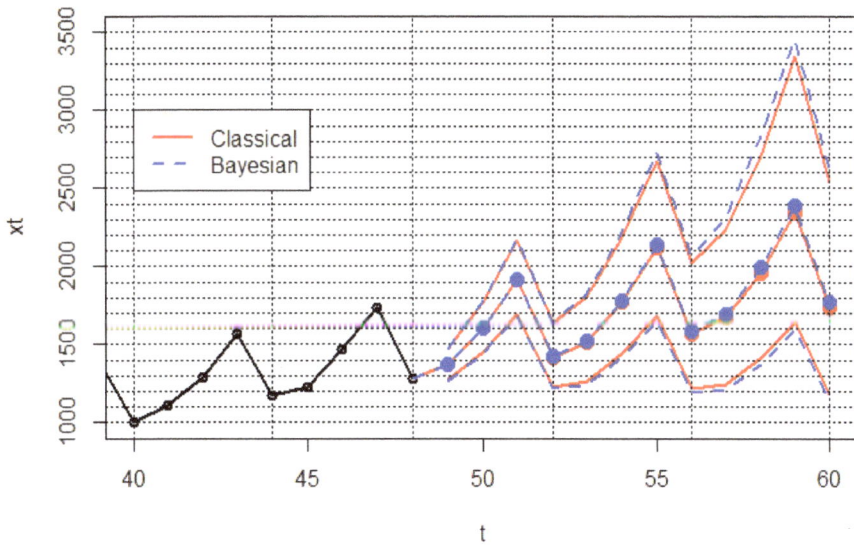

We see from Figures 8.18 and 8.19 that the two approaches to inference have yielded very similar results, at least as regards prediction.

The Bayesian approach has produced 95% prediction intervals which are slightly *wider* than those obtained via the classical approach.

It may be argued that such wider intervals are more appropriate, since the classical approach makes forecasts without taking into account any uncertainty in the parameter estimates.

By contrast, the Bayesian approach to forecasting does take into account that uncertainty.

To conclude, we report that the fitted model for the TIAP time series x_t is given by

$$y_t = \nabla_4 \nabla \log x_t \,,$$

with

$$y_t = \hat{w}_t + \hat{\Theta}_1 \hat{w}_{t-4} \,, \quad \hat{w}_t \sim iid\ N(0, \hat{\sigma}^2) \,,$$

where, via *classical* analysis:

$$\hat{\Theta}_1 = -0.4927\ (SE = 0.1201)$$

$$\hat{\sigma}^2 = 0.0013,$$

and where, via *Bayesian* analysis:

$$\hat{\Theta}_1 = -0.4661\ (SE = 0.1266)$$

$$\hat{\sigma}^2 = 0.0015.$$

R and WinBUGS Code for Exercise 8.4

```
# Classical analysis in R
# ============================================================

x <-
c(362, 385, 432, 341, 382, 409, 498, 387, 473, 513, 582, 474,
544, 582, 681, 557, 628, 707, 773, 592, 627, 725, 854, 661, 742,
854, 1023, 789, 878, 1005, 1173, 883, 972, 1125, 1336, 988, 1020,
1146, 1400, 1006, 1108, 1288, 1570, 1174, 1227, 1468, 1736, 1283 )
n <- length(x); n # 48

X11(w=8,h=9); par(mfrow=c(3,2))
plot(x,type="l");  abline(v=seq(0,48,4),h=seq(0,2000,100), lty=3)
plot(log(x),type="l"); abline(v=seq(0,48,4), lty=3)
plot(diff(log(x)),type="l"); abline(v=seq(0,48,4), lty=3)
plot(diff(diff(log(x),lag=4)),type="l"); abline(v=seq(0,48,4), lty=3)
y <- diff(diff(log(x),lag=4))
acf(y, lag=24)
pacf(y,lag=24)

fit1 <- arima( log(x),order=c(0,1,0), seasonal=list(order=c(1,1,0), period=4) )

tsdiag(fit1); fit1
#        sar1
#      -0.4990
# s.e.  0.1417
# sigma^2 estimated as 0.001310:  log lik. = 81.12,  aic = -158.24

fit2 <- arima( log(x),order=c(0,1,0), seasonal=list(order=c(0,1,1), period=4) )
tsdiag(fit2); fit2
#        sma1
#      -0.4927
# s.e.  0.1201
# sigma^2 estimated as 0.001306:  log lik. = 81.2,  aic = -158.4

# There's not much to distinguish the two fits.
# The second one is marginally better.
# Let's now display the diagnostics for that fit (again).
```

```
fit <- fit2;  tsdiag(fit)

# We see that the residuals from the fit are well-behaved,
# and their sample ACF is consistent with that of white noise.
# Let's also look at some other diagnostics. These turn out to be OK too.

X11(w=8,h=5); par(mfrow=c(2,2))
acf(fit$resid, lag=24)
pacf(fit$resid, lag=24)
qqnorm(fit$resid)
hist(fit$resid, nclass=12)

# Check whether to include a mean term
mean(y) # 0.0008141388
fit3 <- arima( y, order=c(0,0,0), seasonal=list(order=c(0,0,1), period=4),
        include.mean=T ); fit3
#       sma1   intercept
#     -0.4937  -0.0003  <--------- not significant
# s.e.  0.1204   0.0031
# So there's no need for an intercept term in the model.

# Let's now make some predictions.
logxpredict <- predict(fit, n.ahead=12)
xF <- exp(logxpredict$pred)
xL <- exp(logxpredict$pred - qnorm(0.975)* logxpredict$se)
xU <- exp(logxpredict$pred + qnorm(0.975)* logxpredict$se)

cbind(xF, xL, xU)
#       xF       xL       xU
# 49 1365.822 1272.412 1466.090
# 50 1602.240 1449.497 1771.079
# 51 1916.210 1694.939 2166.367
# 52 1418.253 1230.895 1634.130
# 53 1509.806 1264.357 1802.904
# 54 1771.148 1439.872 2178.641
# 55 2118.215 1677.977 2673.956
# 56 1567.764 1213.320 2025.751
# 57 1668.969 1244.652 2237.940
# 58 1957.861 1412.873 2713.066
# 59 2341.516 1640.034 3343.038
# 60 1733.037 1180.875 2543.381
```

```
X11(h=5); par(mfrow=c(1,1)); plot(c(0,60),c(0,3800), type="n")
lines(x, lwd=2);  points(x, lwd=2);
points((n+1):(n+12), xF, pch=16, cex=1.5);
lines(n:(n+12), c(x[n],xF), lty=1,lwd=2)
# points((n+1):(n+12), xL, pch=16);
lines((n+1):(n+12), xL, lty=2, lwd=2)
# points((n+1):(n+12), xU, pch=16);
lines((n+1):(n+12), xU, lty=2, lwd=2)
abline(v=seq(0,100,4),h=seq(0,4000,100), lty=3)  # OK....


# Bayesian reanalysis in R and WinBUGS
# ============================================================


# Assume that R (v3.0.1) is installed in C:/R-3.0.1
# and WinBUGS (v4.1.3) is installed in C:/WinBUGS14


install.packages("R2WinBUGS") # Not necessary if done previously

library("R2WinBUGS")  # Necessary every time R is started

# Make the following directory exists:  C:/R-3.0.1/BugsOut/
# Create a file called   C:/R-3.0.1/BugsCode2.txt   with the following:


# ---------------------------------------------------------------------
model
{
for(t in 1:n) { z[t] <- log(x[t])  }
for(t in 1:5){ y[t] <- 0; w[t] ~ dnorm(0,tau) }
for(t in 6:n){ y[t] <- z[t] - z[t-1] - z[t-4] + z[t-5] }
for(t in 6:N){  # N=n+12=60
        m[t] <- Phi1*w[t-4]
        y[t] ~ dnorm(m[t],tau)
        w[t] <- y[t] - m[t]
        }
tau ~ dgamma(0.001,0.001)
Phi1dum ~ dbeta(1,1);   Phi1 <- 2*Phi1dum-1
for(k in 1:12) {
        z[n+k] <- z[n+k-1] + z[n+k-4] - z[n+k-5] + y[n+k]
        x[n+k] <- exp(z[n+k])
        }
sig2 <- 1/tau
}
# ---------------------------------------------------------------------
```

```
# NB: We can't specify  Phi1 ~ dunif(-1,1). This causes an error.
# Update in March 2014: Phi1 ~ dunif(-1,1) works in WinBUGS 1.4.3.

x <- c(362, 385, 432, 341, 382, 409, 498, 387, 473, 513, 582, 474,
544, 582, 681, 557, 628, 707, 773, 592, 627, 725, 854, 661, 742,
854, 1023, 789, 878, 1005, 1173, 883, 972, 1125, 1336, 988, 1020,
1146, 1400, 1006, 1108, 1288, 1570, 1174, 1227, 1468, 1736, 1283,
NA,NA,NA,NA,   NA,NA,NA,NA,   NA,NA,NA,NA)

n <- 48;  N <- 60; data <- list("n","N","x")
inits <- function(){  list(tau=1, Phi1dum=0.5)  }
parameters <- c("sig2", "Phi1", "x")

sim  <- bugs(data, inits, parameters, n.thin=1,
         model.file= "C:/R-3.0.1/BugsCode2.txt",
         n.chains = 1, n.iter = 6000, n.burnin=1000, DIC = FALSE,
         bugs.directory = "C:/WinBUGS14/",
         working.directory =   "C:/R-3.0.1/BugsOut/")

# This starts WinBUGS, runs the BUGS code for 6000 iterations, closes
# WinBUGS, and creates a number of files in the working directory. These
# files contain information which can also be accessed within R, as follows.

print(sim,digits=4)

# Inference for Bugs model at "C:/R-3.0.1/BugsCode2.txt", fit using WinBUGS,
#  1 chains, each with 6000 iterations (first 1000 discarded)
#  n.sims = 5000 iterations saved

#       mean     sd    2.5%     25%     50%     75%    97.5%
# sig2   0.0015  0.0003  0.0009  0.0012  0.0014  0.0016  0.0022
# Phi1  -0.4661  0.1266 -0.6910 -0.5548 -0.4740 -0.3865 -0.1944
# x[49] 1367.1820  52.6189 1265.0000 1332.0000 1365.0000 1402.0000
# 1472.0000
# x[50] 1605.9746  86.2790 1443.0000 1547.0000 1603.0000 1662.0000
# 1781.0000
# x[51] 1918.2346 124.7788 1681.9750 1835.0000 1914.0000 2000.0000
# 2172.0250
# x[52] 1422.9222 107.4501 1220.9750 1350.0000 1420.0000 1491.0000
# 1641.0000
# x[53] 1517.8472 146.0119 1247.9750 1418.7499 1514.0000 1610.0000
# 1822.0000
# x[54] 1783.4306 201.9834 1415.0000 1645.0000 1777.0000 1908.2500
# 2217.0000
```

```
# x[55] 2133.7016 273.1291 1646.9750 1946.7500 2119.0000 2306.0000
# 2724.0000
# x[56] 1584.1955 223.5842 1187.9750 1431.0000 1576.0000 1720.2499
# 2066.0000
# x[57] 1693.4548 276.4929 1211.9750 1499.7499 1674.0000 1857.0000
# 2309.0750
# x[58] 1992.9153 364.3849 1370.9750 1742.7499 1968.0000 2204.0000
# 2837.0999
# x[59] 2388.4000 476.7169 1589.8999 2058.7500 2345.0000 2668.0000
# 3453.0250
# x[60] 1775.0647 381.9082 1137.0000 1511.0000 1735.0000 1992.0000
# 2628.1249


help(bugs)   # To get info on how to do the following...


Phi1v <- sim$sims.list$Phi1; sig2v <- sim$sims.list$sig2
xm <- sim$sims.list$x


par(mfrow=c(2,2))
hist(Phi1v, breaks=20); hist(sig2v, breaks=20)
hist(xm[,1], breaks=20); hist(xm[,2], breaks=20)


# Let's now make the forecasts of the series using the BUGS output.
xF2 <- xF; xL2 <- xL; xU2 <- xU; for(t in 1:12){
        xF2[t] <- mean(xm[,t])
        xL2[t] <- quantile(xm[,t], 0.025)
        xU2[t] <- quantile(xm[,t], 0.975)  } # Calc. estimates


par(mfrow=c(1,1)); plot(c(0,60),c(0,3800), type="n")
lines(x, lwd=2);  points(x, lwd=2)
points((n+1):(n+12), xF2, pch=16, cex=1.5);
lines(n:(n+12), c(x[n],xF2), lty=1,lwd=2)
lines((n+1):(n+12), xL2, lty=2, lwd=2)
lines((n+1):(n+12), xU2, lty=2, lwd=2)
abline(v=seq(0,100,4),h=seq(0,4000,100), lty=3)  #  OK.....
```

```
# Next we graph both sets of forecasts together in a single plot,
# and then produce a close-up in that single plot, as follows:

X11(h=5); par(mfrow=c(1,1));

plot(c(0,60),c(0,3800), type="n", xlab="t", ylab="xt")
lines(x, lwd=2);  points(x, lwd=2)
points((n+1):(n+12), xF, pch=16, cex=1.5, col="red");
lines(n:(n+12), c(x[n],xF), lty=1,lwd=2, col="red")
lines((n+1):(n+12), xL, lty=1, lwd=2, col="red")
lines((n+1):(n+12), xU, lty=1, lwd=2, col="red")
abline(v=seq(0,100,4),h=seq(0,4000,100), lty=3)
points((n+1):(n+12), xF2, pch=16, cex=1.5, col="blue" );
lines(n:(n+12), c(x[n],xF2), lty=2,lwd=2, col="blue ")
lines((n+1):(n+12), xL2, lty=2, lwd=2, col="blue ")
lines((n+1):(n+12), xU2, lty=2, lwd=2, col="blue ")
legend(0,3000,c("Classical","Bayesian"), lty=c(1,2),
         lwd=c(2,2), col=c("red", "blue"),   bg="white" )

par(mfrow=c(1,1))
plot(c(40,60),c(1000,3500), type="n", xlab="t", ylab="xt")
lines(x, lwd=2);  points(x, lwd=2)
points((n+1):(n+12), xF, pch=16, cex=1.5, col="red");
lines(n:(n+12), c(x[n],xF), lty=1,lwd=2, col="red")
lines((n+1):(n+12), xL, lty=1, lwd=2, col="red")
lines((n+1):(n+12), xU, lty=1, lwd=2, col="red")
abline(v=seq(0,100,4),h=seq(0,4000,100), lty=3)
points((n+1):(n+12), xF2, pch=16, cex=1.5, col="blue" );
lines(n:(n+12), c(x[n],xF2), lty=2,lwd=2, col="blue ")
lines((n+1):(n+12), xL2, lty=2, lwd=2, col="blue ")
lines((n+1):(n+12), xU2, lty=2, lwd=2, col="blue ")
legend(40,3000,c("Classical","Bayesian"), lty=c(1,2),
         lwd=c(2,2), col=c("red", "blue"),   bg="white" )
```

CHAPTER 9

Bayesian Finite Population Theory

9.1 Introduction

In this chapter we will focus on the topic of Bayesian methods for finite population inference in the sample survey context. We have previously touched on this topic when considering posterior predictive inference of 'future' values in the context of the normal-normal-gamma model. The topic will now be treated more generally and systematically.

There are many and various ways in which Bayesian finite population inference can be categorised, for example:

- situations with and without prior information being available
- sampling with and without replacement
- Monte Carlo based methods versus deterministic (or 'exact') methods
- situations with and without auxiliary information being available
- scenarios where a superpopulation variance is known and where it is unknown
- sampling with equal probabilities versus unequal probabilities
- sampling mechanisms that are ignorable versus nonignorable (i.e. biased)
- cases where the order of sampling is known versus where that order is unknown
- cases with full response versus where some sampled units fail to respond.

Each of these categories can in turn be broken down further. For example, Monte Carlo based techniques may or may not require Markov chain Monte Carlo methods for generating the sample required for inference. We see there is potentially a vast subject ground to cover.

We will begin with a description of some basic general concepts, notation and terminology in relation to finite population modelling in the Bayesian framework, with a focus on *simple random sampling without replacement* (SRSWOR). We then illustrate these ideas by way of a series of exercises which also feature some other concepts such as *simple random sampling with replacement* (SRSWR), nonignorable sampling schemes, and covariate data. Some of these ideas will be taken up again in later chapters.

We defer discussion of Bayesian finite population models involving *normal* (i.e. *Gaussian*) data to the next chapter (Chapter 10), where such models are the focus and treated in detail. In Chapter 11 we will discuss data transformations, inference on non-standard quantities of interest, and frequentist properties of Bayesian estimators in a finite population context, including the notions of model bias and design bias. Chapter 12 will focus on the issues of biased sampling and nonignorable nonresponse.

The exposition in Chapters 9 to 12 is largely theoretical but does include mention of several real world applications, including on-site sampling of recreation parks, oil discovery, and correcting for self-selection bias in volunteer surveys. Further discussion of the role that Bayesian methods and prior information play in survey sampling and finite population inference can be found in Rao (2011). This paper also lists many other papers and books on this and related topics, for example Ericson (1969) and Särndal, Swensson and Wretman (1992).

9.2 Finite population notation and terminology

Consider a finite population of N units labelled $i = 1,...,N$, and let y_i be the value of the ith unit for some observable variable of interest.

Define $y = (y_1,...,y_N)$ as the *population vector*.

Suppose that n units are selected from the finite population without replacement.

We refer to n as the *sample size* and to $m = N - n$ as the *nonsample size*.

Let $s = (s_1,...,s_n)$ be the vector of the ordered labels of the sampled units.

Also let $r = (r_1,...,r_m)$ be the vector of the ordered labels of the nonsampled units, i.e. those remaining.

Define $y_s = (y_{s_1},...,y_{s_n})$ to be the *sample vector*, and likewise define $y_r = (y_{r_1},...,y_{r_m})$ to be the *nonsample vector*.

Note 1: With the above definitions, it is always true that

$$s_1 < ... < s_n$$

and

$$r_1 < ... < r_m,$$

irrespective of the order in which the population units may actually be sampled. Also,

$$\{s_1, ..., s_n, r_1, ..., r_m\} = \{1, ..., N\}.$$

Note 2: For mathematical convenience, the population, sample and nonsample vectors may later sometimes be defined as the *column* vectors

$$y = (y_1, ..., y_N)' = \begin{pmatrix} y_1 \\ \vdots \\ y_N \end{pmatrix}, \ y_s = (y_{s_1}, ..., y_{s_n})' \text{ and } y_r = (y_{r_1}, ..., y_{r_m})',$$

respectively.

Also, the population vector may sometimes be written using upper case letters, as $Y = (Y_1, ..., Y_N)$ or $Y = (Y_1, ..., Y_N)'$. For the remainder of this chapter, these alternative notations will not be used.

Example: Suppose that we select $n = 3$ units from a finite population of size $N = 7$ and obtain units 4, 5 and 2 (in that order, or any other order).

Then the nonsample size is $m = N - n = 4$ and:

$$y = (y_1, ..., y_7)$$

$$s = (s_1, s_2, s_3) = (2, 4, 5), \qquad y_s = (y_2, y_4, y_5)$$

$$r = (r_1, r_2, r_3, r_4) = (1, 3, 6, 7), \qquad y_r = (y_1, y_3, y_6, y_7).$$

9.3 Bayesian finite population models

Consider a finite population vector y which may be thought of as having been generated from some probability distribution which depends on a parameter θ (possibly a vector).

Also suppose that a sample of size n is drawn from the finite population without replacement according to some probability distribution for s.

This scenario may be expressed in terms of a *Bayesian finite population model* with the following form:

$f(s \mid y, \theta)$ (the probability of obtaining sample s for given values of y and θ)

$f(y \mid \theta)$ (the model density of the finite population vector given θ)

$f(\theta)$ (the prior density of the parameter).

Suppose that we have data of the form $D = (s, y_s)$ and are interested in a quantity $Q = g(y, \theta)$, for some function g. Then the task is to determine the distribution of Q given D.

This distribution will be based on the joint distribution of the two unobserved quantities θ and y_r , given the two observed quantities, namely:

s (which tells us which units are sampled); and

y_s (the vector of the values of the sampled units).

Thus, inference on the quantity of interest $Q = g(y, \theta)$ is based on the density $f(Q \mid D)$, which in turn is based on the density

$$f(\theta, y_r \mid s, y_s) \propto f(\theta, y_s, y_r, s)$$
$$= f(\theta) f(y_s, y_r \mid \theta) f(s \mid y_s, y_r, \theta). \tag{9.1}$$

Note 1: The values of s and r here are fixed at their observed values defined by the data. Thus, given $D = (s, y_s)$, we may always express $Q = g(y, \theta)$ as $h((y_s, y_r), \theta)$ for some function h (which will in many cases be the same function as g), and there should be no ambiguity in the meaning of quantities such as $f(y_s, y_r \mid \theta)$ in (9.1).

Note 2: We have specified the sampling mechanism in terms of the quantity s which tells us *which* units are sampled but not the *order* in which they are sampled. In some cases it may be appropriate to replace $f(s \mid y, \theta)$ in the model by $f(L \mid y, \theta)$, where

$$L = (L_1, ..., L_n)$$

is the vector of the labels of the selected units *in the order that they are sampled*. L provides more information than s, which is a function of L.

Note 3: We have assumed that sampling is *without* replacement. If sampling is *with* replacement, it may be appropriate to replace $f(s \mid y, \theta)$ or $f(L \mid y, \theta)$ in the model by $f(I \mid y, \theta)$, where

$$I = (I_1, ..., I_N),$$

and where I_i is the number of times that population unit i is sampled.

In this case it may be necessary to modify the notation to account for the number of *distinct* units sampled, previously the fixed constant n, due to the possibility of multiple selections under sampling with replacement.

Example 1: Suppose that we sample units 4, 5 and 2, in that order. Then $L = (L_1, L_2, L_3) = (4, 5, 2)$ and $s = (2, 4, 5)$. Note that s is a function of L.

Example 2: Suppose we sample units 3, 5 and 3, in that order. Then $L = (L_1, L_2, L_3) = (3, 5, 3)$ and $I = (0, 0, 2, 0, 1)$. In this case, we write $s = (s_1, ..., s_d) = (s_1, s_2) = (3, 5)$ as the ordered vector of distinct labels for the units sampled. Here, d is the number of distinct units sampled (a random variable with realised value 2), in contrast to n, the total number of selections (a fixed constant equal to 3). Note that d is a function of s, which is a function of I, which in turn is a function of L.

9.4 Two types of sampling mechanism

There are basically two types of sampling mechanism in the context of the above model, data and quantity of interest. These two types correspond to two distinct cases, as follows:

(i) where $f(Q \mid D)$ *remains exactly the same* if the sampling density $f(s \mid y_s, y_r, \theta)$ is omitted from the calculation at equation (9.1); in this case we say that the sampling mechanism is ***ignorable*** (or ***unbiased***)

(ii) where $f(Q \mid D)$ *changes in some way* if the sampling density $f(s \mid y_s, y_r, \theta)$ is omitted from the calculation at equation (9.1); we then say the sampling mechanism is ***nonignorable*** (or ***biased***).

Perhaps the simplest example of an ignorable sampling mechanism is *simple random sampling without replacement* (SRSWOR), for which

$$f(s \mid y, \theta) = \binom{N}{n}^{-1}, \quad s \in S(s),$$

where

$$S(s) = \{(1, ..., n), (1, 2, ..., n-1, n+1), ..., (N-n+1, ..., N)\}.$$

is the sample space for s (the set of all possible combinations of n integers taken from N).

In this case, $f(s \mid y, \theta)$ does not depend on y or θ at all and so may also be written simply as $f(s)$. This then guarantees that

$$f(s \mid y_s, y_r, \theta) = f(s) = \binom{N}{n}^{-1}$$

at the single observed value of s, whatever that value may be.

Therefore, the joint density of the two unknowns is

$$f(\theta, y_r \mid s, y_s) \propto f(\theta, y_s, y_r, s)$$
$$= f(\theta) f(y_s, y_r \mid \theta) f(s \mid y_s, y_r, \theta)$$
$$\propto f(\theta) f(y_s, y_r \mid \theta) \times 1,$$

which is the same as (9.1) but with $f(s \mid y_s, y_r, \theta)$ omitted.

This result tells us that $f(Q \mid D)$ will be the same when the sampling mechanism density $f(s \mid y_s, y_r, \theta)$ is 'ignored' in the model, so to speak.

9.5 Two types of inference

There are basically two types of inference in the context of the above model, data and quantity of interest:

(a) where Q does not depend on y, in which case inference is on $Q = g(\theta)$ (a function of only θ) and may be called *analytic inference* or *infinite population inference* or *superpopulation inference*

(b) where Q does not depend on θ, in which case inference is on $Q = g(y)$ (a function of only y) and may be called *descriptive inference* or *finite population inference* or *predictive inference.*

9.6 Analytic inference

In the case of analytic inference, this is based solely on the *posterior density* of the model parameter θ, namely

$$f(\theta \mid D) = f(\theta \mid s, y_s)$$

$$\propto f(\theta, s, y_s)$$

$$= \int f(\theta, s, y_r, y_s) dy_r$$

$$= f(\theta) \int f(y_s, y_r \mid \theta) f(s \mid y_s, y_r, \theta) dy_r .$$

Now suppose further that the sampling mechanism is ignorable. In that case,

$$f(\theta \mid D) \propto f(\theta) \int f(y_s, y_r \mid \theta) \times 1 dy_r$$

since $f(s \mid y_s, y_r, \theta)$ may be ignored

$$= f(\theta) f(y_s \mid \theta) \int f(y_r \mid y_s, \theta) dy_r$$

since $f(y_s, y_r \mid \theta) = f(y_s \mid \theta) f(y_r \mid \theta, y_s)$

$$= f(\theta) f(y_s \mid \theta)$$

since $\int f(y_r \mid y_s, \theta) dy_r = 1$ for all θ.

Thus the posterior density of θ is obtained in exactly the same way as in previous chapters.

Note: As stressed earlier, it is to be understood that s in $f(y_s \mid \theta)$ here is fixed at its observed value. With this understanding, we will sometimes abbreviate $f(\theta \mid D) = f(\theta \mid s, y_s)$ as $f(\theta \mid y_s)$.

Example: If $s = (2,4,5)$, then y_s means specifically (y_2, y_4, y_5). Thus, in this context, y_s does *not* refer to the vector $(y_{s_1}, y_{s_2}, y_{s_3})$ with the subscripts s_1, s_2, s_3 as random variables.

9.7 Descriptive inference

In the case of descriptive inference, this is based solely on the *predictive density* of the nonsample vector y_r, namely

$$f(y_r \mid D) = f(y_r \mid s, y_s) \; \propto \; f(s, y_s, y_r)$$

$$= \int f(\theta, s, y_r, y_s) d\theta = \int f(\theta) f(y_s, y_r \mid \theta) f(s \mid y_s, y_r, \theta) d\theta .$$

Now suppose further that the sampling mechanism is ignorable. In that special case,

$$f(y_r \mid D) \propto \int f(\theta) f(y_s, y_r \mid \theta) \times 1 d\theta$$

$$\text{since } f(s \mid y_s, y_r, \theta) \text{ may be ignored}$$

$$= \int f(y_r \mid y_s, \theta) f(\theta) f(y_s \mid \theta) d\theta$$

$$\propto \int f(y_r \mid y_s, \theta) f(\theta \mid y_s) d\theta$$

$$\text{since } f(\theta \mid y_s) \propto f(\theta) f(y_s \mid \theta) .$$

So the predictive density of y_r is obtained in exactly the same way as in previous chapters.

Note: As before, it is to be understood that s and r in $f(y_r \mid y_s, \theta)$ and $f(\theta \mid y_s)$ are fixed at their observed values. With this understanding, we will sometimes write

$$f(y_r \mid D) = f(y_r \mid s, y_s) \qquad \text{as} \qquad f(y_r \mid y_s).$$

More generally, we will sometimes write

$$f(\theta, y_r \mid D) = f(\theta, y_r \mid s, y_s) \quad \text{as} \quad f(\theta, y_r \mid y_s),$$

and

$$f(Q \mid D) = f(Q \mid s, y_s) \qquad \text{as} \qquad f(Q \mid y_s).$$

Example: If $s = (2,4,5)$ and $N = 7$ then y_s means (y_2, y_4, y_5) and y_r means (y_1, y_3, y_6, y_7).

Exercise 9.1 A Bernoulli finite population model with *ignorable* *sampling*

A finite population of size $N = 4$ consists of values that are independently and identically distributed (iid) Bernoulli with parameter θ, where θ is a priori equally likely to be 1/4 or 1/2 (with no other possibilities).

We sample $n = 2$ units from the finite population according to SRSWOR.

Units 2 and 4 are sampled, and both have the value 1.

(a) Find the posterior distribution of θ.

(b) Find the predictive distribution of the finite population total, namely $y_T = y_1 + \ldots + y_N$.

Solution to Exercise 9.1

(a) The Bayesian model here may be written:

$$f(s \mid y, \theta) = \binom{N}{n}^{-1} = \binom{4}{2}^{-1} = \frac{1}{6},$$
$$s = (1,2),(1,3),(1,4),(2,3),(2,4),(3,4)$$
$$f(y \mid \theta) = \prod_{i=1}^{N} \theta^{y_i}(1-\theta)^{1-y_i}$$

(the model density of the finite population values)
$$f(\theta) = 1/2, \quad \theta = 1/4, 1/2 \quad \text{(the prior density of the parameter)}.$$

The observed sample data is
$$D = (s, y_s) = ((s_1, s_2),(y_{s_1}, y_{s_2})) = ((2,4),(y_2, y_4)) = ((2,4),(1,1)),$$
and the nonsample vector is $y_r = (y_{r_1}, y_{r_2}) = (y_1, y_3) \in \{0,1\}^2$.

The sampling mechanism is ignorable, and so
$$f(\theta \mid D) \propto f(\theta)f(y_s \mid \theta) \propto 1 \times \prod_{i \in s} \theta^{y_i}(1-\theta)^{1-y_i}$$
$$= \theta^2 \quad \text{since } n = 2 \text{ and } y_i = 1 \text{ for all } i \in s$$
$$= \begin{cases} (1/4)^2 = 1/16, & \theta = 1/4 \\ (1/2)^2 = 4/16, & \theta = 1/2. \end{cases}$$

It follows that $f(\theta \mid D) = \begin{cases} 1/5, & \theta = 1/4 \\ 4/5, & \theta = 1/2. \end{cases}$

(b) Next, observe that $f(y_r \mid D, \theta) = \begin{cases} (1-\theta)^2, & y_r = (0,0) \\ (1-\theta)\theta, & y_r = (0,1) \\ \theta(1-\theta), & y_r = (1,0) \\ \theta^2, & y_r = (1,1). \end{cases}$

This implies that

$$f(y_r \mid D) = \sum_\theta f(y_r \mid D, \theta) f(\theta \mid D)$$

$$= \begin{cases} \left(1-\dfrac{1}{4}\right)^2 \dfrac{1}{5} + \left(1-\dfrac{2}{4}\right)^2 \dfrac{4}{5} = \dfrac{25}{80}, & y_r = (0,0) \\[2mm] \left(1-\dfrac{1}{4}\right)\dfrac{1}{4}\dfrac{1}{5} + \left(1-\dfrac{2}{4}\right)\dfrac{2}{4}\dfrac{4}{5} = \dfrac{19}{80}, & y_r = (0,1) \\[2mm] \dfrac{1}{4}\left(1-\dfrac{1}{4}\right)\dfrac{1}{5} + \dfrac{2}{4}\left(1-\dfrac{2}{4}\right)\dfrac{4}{5} = \dfrac{19}{80}, & y_r = (1,0) \\[2mm] \left(\dfrac{1}{4}\right)^2 \dfrac{1}{5} + \left(\dfrac{2}{4}\right)^2 \dfrac{4}{5} = \dfrac{17}{80}, & y_r = (1,1). \end{cases}$$

The nonsample total is $y_{rT} = y_1 + y_3$, with three possible possible values:
$$0 + 0 = 0$$
$$0 + 1 = 1 + 0 = 1$$
$$1 + 1 = 2.$$

Therefore $f(y_{rT} \mid D) = \begin{cases} 25/80, & y_{rT} = 0 \\ 38/80, & y_{rT} = 1 \\ 17/80, & y_{rT} = 2. \end{cases}$

The finite population total is $y_T = y_{sT} + y_{rT}$, where $y_{sT} = y_2 + y_4 = 1+1$ $= 2$ is the sample total. It follows that the required predictive density of the finite population total is

$$f(y_T \mid D) = \begin{cases} 25/80, & y_T = 2+0 = 2 \\ 38/80, & y_T = 2+1 = 3 \\ 17/80, & y_T = 2+2 = 4. \end{cases}$$

Exercise 9.2 A Bernoulli finite population model with *nonignorable sampling*

A finite population of size $N = 4$ consists of values that are conditionally iid Bernoulli with parameter θ, where θ is a priori equally likely to be 1/4 or 1/2 (with no other possibilities).

We sample $n = 2$ units from the finite population without replacement in such a way that the probability of selecting a sample is *proportional to the sum of the values in that sample*.

Units 2 and 4 are sampled, and both have the value 1.

(a) Find the posterior distribution of θ.

(b) Find the predictive distribution of the finite population total, namely
$$y_T = y_1 + ... + y_N$$

(c) Find the *conditional* posterior distribution of θ given the nonsample vector, and then employ this distribution to check your answer to (a) using results in (b).

(d) Find the following probabilities of selection into the sample:

    (i) $P(i \in s \mid y, \theta)$ **(ii)** $P(i \in s \mid y)$

    (iii) $P(i \in s \mid \theta)$ **(iv)** $P(i \in s)$.

Solution to Exercise 9.2

(a) The Bayesian model here may be written:
$$f(s \mid y, \theta) \propto y_{sT}, \quad s = (1,2),(1,3),(1,4),(2,3),(2,4),(3,4)$$
$$f(y \mid \theta) = \prod_{i=1}^{N} \theta^{y_i}(1-\theta)^{1-y_i}$$
                     (the model density of the finite population values)
$$f(\theta) = 1/2, \ \theta = 1/4, 1/2 \quad \text{(the prior density of the parameter)}.$$

The observed sample data is
$$D = (s, y_s) = ((s_1, s_2), (y_{s_1}, y_{s_2})) = ((2,4),(y_2, y_4)) = ((2,4),(1,1)),$$
and the nonsample vector is
$$y_r = (y_{r_1}, y_{r_2}) = (y_1, y_3) \in \{0,1\}^2.$$

In this case the sampling mechanism is **nonignorable** and the first thing we should do is determine the exact form of the sampling density of $s = (s_1, s_2)$. Now,

$$f(s \mid y, \theta) = cy_{sT} = c(y_{s_1} + y_{s_2})$$

for some constant c such that

$$1 = \sum_s f(s \mid y, \theta)$$

$$= c\{(y_1 + y_2) + (y_1 + y_3) + (y_1 + y_4) + (y_2 + y_3) + (y_2 + y_4) + (y_3 + y_4)\}$$

$$= c\{3(y_1 + y_2 + y_3 + y_4)\} = 3cy_T.$$

We see that $c = 1/(3y_T)$, and so

$$f(s \mid y, \theta) = \frac{y_{s_1} + y_{s_2}}{3y_T}, \quad s = (s_1, s_2) = (1,2),(1,3),(1,4),(2,3),(2,4),(3,4).$$

Note 1: This formula shows explicitly how the sampling mechanism depends on the values in the finite population vector y. It also shows that, conditional on y, the sampling mechanism does not depend on the superpopulation parameter θ.

Note 2: This formula is only true when the finite population total y_T is *positive*, i.e. when at least one of $y_1, ..., y_N$ is *nonzero*. In the case where all population values are zero, we have that $y_{sT} = y_{s_1} + y_{s_2} = 0$ for all possible samples $s = (s_1, s_2)$, and consequently $f(s \mid y, \theta) \propto 0$, which must be understood to mean that that no sample actually gets drawn. The fact that a sample *has* been observed implies $f(s \mid y, \theta) > 0$ for at least *one* value of s, which implies that at least *one* population value is positive, which in turn implies that $y_T > 0$. This would be true even if all the sample values were zero; but as it happens, at least one of them is positive (in fact both are), which *in itself* implies that $y_T > 0$.

We may now work out the joint density of all quantities in the model:

$$f(\theta, y_s, y_r, s) = f(\theta) f(y_s \mid \theta) f(y_r \mid \theta) f(s \mid y_s, y_r, \theta)$$

$$= \frac{1}{2} \times \left(\prod_{i \in s} \theta^{y_i} (1 - \theta)^{1 - y_i} \right) \times \left(\prod_{i \in r} \theta^{y_i} (1 - \theta)^{1 - y_i} \right) \times \frac{y_{s_1} + y_{s_2}}{3y_T}$$

$$= \frac{1}{2} \times \theta^2 \times \theta^{y_1+y_3} (1-\theta)^{2-y_1-y_3} \times \frac{1+1}{3(y_1+1+y_3+1)}$$

$$\propto \frac{\theta^{2+y_1+y_3}(1-\theta)^{2-y_1-y_3}}{2+y_1+y_3} .$$

So the posterior density of θ is
$$f(\theta \mid D) = f(\theta \mid s, y_s)$$
$$\propto f(\theta, s, y_s)$$

$$= \sum_{y_r} f(\theta, s, y_s, y_r)$$

$$\propto \theta^2 (1-\theta)^2 \sum_{y_1=0}^{1} \sum_{y_3=0}^{1} \left(\frac{\theta}{1-\theta} \right)^{y_1+y_3} \frac{1}{2+y_1+y_3}$$

$$\propto \theta^2 (1-\theta)^2 \left\{ \left(\frac{\theta}{1-\theta} \right)^{0+0} \frac{1}{2+0+0} + \left(\frac{\theta}{1-\theta} \right)^{0+1} \frac{1}{2+0+1} \right.$$

$$\left. + \left(\frac{\theta}{1-\theta} \right)^{1+0} \frac{1}{2+1+0} + \left(\frac{\theta}{1-\theta} \right)^{1+1} \frac{1}{2+1+1} \right\}$$

$$= \theta^2 (1-\theta)^2 \left\{ \frac{1}{2} + \left(\frac{\theta}{1-\theta} \right) \frac{1}{3} + \left(\frac{\theta}{1-\theta} \right) \frac{1}{3} + \left(\frac{\theta}{1-\theta} \right)^2 \frac{1}{4} \right\}$$

$$= \frac{1}{12} \left\{ 6\theta^2 (1-\theta)^2 + 8\theta^3 (1-\theta) + 3\theta^4 \right\}$$

$$= \begin{cases} \frac{1}{12} \left[6\left(\frac{1}{4}\right)^2 \left(1-\frac{1}{4}\right)^2 + 8\left(\frac{1}{4}\right)^3 \left(1-\frac{1}{4}\right) + 3\left(\frac{1}{4}\right)^4 \right], & \theta = \frac{1}{4} \\[4mm] \frac{1}{12} \left[6\left(\frac{2}{4}\right)^2 \left(1-\frac{2}{4}\right)^2 + 8\left(\frac{2}{4}\right)^3 \left(1-\frac{2}{4}\right) + 3\left(\frac{2}{4}\right)^4 \right], & \theta = \frac{2}{4} \end{cases}$$

$$= \begin{cases} \frac{1}{12(256)} [6(9)+8(3)+3(1)], & \theta = \frac{1}{4} \\[4mm] \frac{1}{12(256)} [6(16)+8(16)+3(16)], & \theta = \frac{2}{4} \end{cases}$$

$$\propto \begin{cases} 6(9)+8(3)+3(1) = 81, & \theta = \frac{1}{4} \\[4mm] 6(16)+8(16)+3(16) = 272, & \theta = \frac{2}{4}. \end{cases}$$

Now $81 + 272 = 353$, and so

$$f(\theta \mid D) = \begin{cases} 81/353 = 0.22946, & \theta = 1/4 \\ 272/353 = 0.77054, & \theta = 1/2. \end{cases}$$

(b) The predictive density of the nonsample vector
$$y_r = (y_{r_1}, y_{r_2}) = (y_1, y_3)$$
is
$$f(y_r \mid D) = f(y_r \mid s, y_s) \propto f(y_r, s, y_s)$$

$$= \sum_\theta f(\theta, s, y_s, y_r)$$

$$\propto \frac{1}{2 + y_1 + y_3} \sum_{\theta = 1/4, 2/4} \theta^{2 + y_1 + y_3} (1-\theta)^{2 - y_1 - y_3}$$

$$= \frac{1}{2 + y_1 + y_3} \left\{ \left(\frac{1}{4}\right)^{2 + y_1 + y_3} \left(1 - \frac{1}{4}\right)^{2 - y_1 - y_3} + \left(\frac{2}{4}\right)^{2 + y_1 + y_3} \left(1 - \frac{2}{4}\right)^{2 - y_1 - y_3} \right\}$$

$$= \frac{1}{(2 + y_1 + y_3)256} \left\{ 3^{2 - y_1 - y_3} + 16 \right\} \propto \frac{16 + 3^{2 - y_1 - y_3}}{2 + y_1 + y_3}$$

$$\propto \begin{cases} \dfrac{16 + 3^{2-0-0}}{2+0+0} = \dfrac{25}{2} = \dfrac{150}{12}, & (y_1, y_3) = (0,0) \\[2mm] \dfrac{16 + 3^{2-0-1}}{2+0+1} = \dfrac{19}{3} = \dfrac{76}{12}, & (y_1, y_3) = (0,1) \\[2mm] \dfrac{16 + 3^{2-1-0}}{2+1+0} = \dfrac{19}{3} = \dfrac{76}{12}, & (y_1, y_3) = (1,0) \\[2mm] \dfrac{16 + 3^{2-1-1}}{2+1+1} = \dfrac{17}{4} = \dfrac{51}{12}, & (y_1, y_3) = (1,1) \end{cases}$$

$$\propto \begin{cases} 75, & (y_1, y_3) = (0,0) \\ 38, & (y_1, y_3) = (0,1) \\ 38, & (y_1, y_3) = (1,0) \\ 24, & (y_1, y_3) = (1,1). \end{cases}$$

Now, $150 + 76 + 76 + 51 = 353$, and so

$$f(y_r \mid D) = \begin{cases} 150/353 = 0.42493, & y_r = (0,0) \\ 76/353 = 0.21530, & y_r = (0,1) \\ 76/353 = 0.21530, & y_r = (1,0) \\ 51/353 = 0.14448, & y_r = (1,1). \end{cases}$$

So the predictive density of the nonsample total,

$$y_{rT} = y_{r_1} + y_{r_2} = y_1 + y_3,$$

is $\qquad f(y_{rT} \mid D) = \begin{cases} 150/353 = 0.42493, & y_{rT} = 0 \\ 152/353 = 0.43059, & y_{rT} = 1 \\ 51/353 = 0.14448, & y_{rT} = 2. \end{cases}$

So the predictive density of the finite population total,

$$y_T = y_{sT} + y_{rT} = (1+1) + y_{rT},$$

is $\qquad f(y_T \mid D) = \begin{cases} 150/353 = 0.42493, & y_T = 2 \\ 152/353 = 0.43059, & y_T = 3 \\ 51/353 = 0.14448, & y_T = 4. \end{cases}$

(c) The *conditional* posterior density of θ given y_r is

$$f(\theta \mid y_s, y_r, s) \overset{\theta}{\propto} f(\theta, y_s, y_r, s)$$
$$\propto \theta^{2+y_1+y_3}(1-\theta)^{2-y_1-y_3}.$$

We now need to consider all the possible values of y_r, one by one.

For $y_r = (0,0)$:

$$f(\theta \mid y_s, y_r, s) \propto \theta^{2+0+0}(1-\theta)^{2-0-0} = \begin{cases} \left(\dfrac{1}{4}\right)^2\left(1-\dfrac{1}{4}\right)^2 = \dfrac{9}{256}, & \theta = \dfrac{1}{4} \\[3mm] \left(\dfrac{2}{4}\right)^2\left(1-\dfrac{2}{4}\right)^2 = \dfrac{16}{256}, & \theta = \dfrac{2}{4} \end{cases}$$

$$\Rightarrow f(\theta \mid y_s, y_r, s) = \begin{cases} 9/25, & \theta = 1/4 \\ 16/25, & \theta = 1/2. \end{cases}$$

For $y_r = (0,1)$:

$$f(\theta \mid y_s, y_r, s) \propto \theta^{2+0+1}(1-\theta)^{2-0-1} = \begin{cases} \left(\dfrac{1}{4}\right)^3\left(1-\dfrac{1}{4}\right)^1 = \dfrac{3}{256}, & \theta = \dfrac{1}{4} \\[3mm] \left(\dfrac{2}{4}\right)^3\left(1-\dfrac{2}{4}\right)^1 = \dfrac{16}{256}, & \theta = \dfrac{2}{4} \end{cases}$$

$$\Rightarrow f(\theta \mid y_s, y_r, s) = \begin{cases} 3/19, & \theta = 1/4 \\ 16/19, & \theta = 1/2. \end{cases}$$

For $y_r = (1,0)$:

$$f(\theta \mid y_s, y_r, s) \propto \theta^{2+1+0}(1-\theta)^{2-1-0} = \begin{cases} \left(\dfrac{1}{4}\right)^3 \left(1-\dfrac{1}{4}\right)^1 = \dfrac{3}{256}, & \theta = \dfrac{1}{4} \\[3mm] \left(\dfrac{2}{4}\right)^3 \left(1-\dfrac{2}{4}\right)^1 = \dfrac{16}{256}, & \theta = \dfrac{2}{4} \end{cases}$$

$$\Rightarrow f(\theta \mid y_s, y_r, s) = \begin{cases} 3/19, & \theta = 1/4 \\ 16/19, & \theta = 1/2. \end{cases}$$

For $y_r = (1,1)$:

$$f(\theta \mid y_s, y_r, s) \propto \theta^{2+1+1}(1-\theta)^{2-1-1} = \begin{cases} \left(\dfrac{1}{4}\right)^4 = \dfrac{1}{256}, & \theta = \dfrac{1}{4} \\[3mm] \left(\dfrac{2}{4}\right)^4 = \dfrac{16}{256}, & \theta = \dfrac{2}{4} \end{cases}$$

$$\Rightarrow f(\theta \mid y_s, y_r, s) = \begin{cases} 1/17, & \theta = 1/4 \\ 16/17, & \theta = 1/2. \end{cases}$$

Now,

$$f(\theta \mid y_s, s) = \sum_{y_r} f(\theta, y_r \mid y_s, s) = \sum_{y_r} f(\theta \mid y_s, y_r, s) f(y_r \mid y_s, s).$$

So, using results in (b), we have that:

$$f(\theta = 1/4 \mid y_s, s) = \sum_{y_r} f\left(\theta = \frac{1}{4} \middle| y_s, y_r, s\right) f(y_r \mid y_s, s)$$

$$= \frac{9}{25} \times \frac{150}{353} + \frac{3}{19} \times \frac{76}{353} + \frac{3}{19} \times \frac{76}{353} + \frac{1}{17} \times \frac{51}{353} = 0.22946$$

$$f(\theta = 1/2 \mid y_s, s) = \sum_{y_r} f\left(\theta = \frac{1}{2} \middle| y_s, y_r, s\right) f(y_r \mid y_s, s)$$

$$= \frac{16}{25} \times \frac{150}{353} + \frac{16}{19} \times \frac{76}{353} + \frac{16}{19} \times \frac{76}{353} + \frac{16}{17} \times \frac{51}{353} = 0.77054.$$

These results are all in agreement with those obtained in (a) using a different approach.

(d) (i) The probability of selecting unit i into the sample given y and θ is the same for all i, in particular $i = 1$, and so may be written

$$P(1 \in s \mid y, \theta) = \sum_{s:1 \in s} f(s \mid y, \theta) = \frac{1}{3 y_T} \{ (y_1 + y_2) + (y_1 + y_3) + (y_1 + y_4) \}$$

$$= \frac{y_T + 2 y_1}{3 y_T} = \frac{1}{3} + \frac{2 y_1}{3 y_T},$$

assuming that $y_T > 0$; otherwise, $P(1 \in s \mid y, \theta) = 0$.

Thus, for each $i = 1, \ldots, 4$ we have that $P(i \in s \mid y, \theta) = \begin{cases} \dfrac{1}{3} + \dfrac{2 y_i}{3 y_T}, & y_T > 0 \\[2mm] 0, & y_T = 0. \end{cases}$

As a check, we may ask whether the sum of these inclusion probabilities equals $n = 2$.

The answer is yes, assuming that y is such that $y_T > 0$; in that case,

$$\sum_{i=1}^{N} P(i \in s \mid y, \theta) = \sum_{i=1}^{4} \left(\frac{1}{3} + \frac{2 y_i}{3 y_T} \right) = \frac{4}{3} + \frac{2(y_1 + y_2 + y_3 + y_4)}{3 y_T} = 2 = n.$$

(ii) Since $P(i \in s \mid y, \theta)$ does not depend on θ, we also have

$$P(i \in s \mid y) = \begin{cases} \dfrac{1}{3} + \dfrac{2 y_i}{3 y_T}, & y_T > 0 \\[2mm] 0, & y_T = 0. \end{cases}$$

(iii) The probability of selecting unit i into the sample given θ is the same for all i, in particular $i = 1$, and so may be written

$$P(i \in s \mid \theta) = P(1 \in s \mid \theta) = \sum_{y} P(1 \in s \mid \theta, y) f(y \mid \theta)$$

$$= 0 \times P(y = (0,0,0,0) \mid \theta) + \sum_{y : y_T > 0} \left(\frac{1}{3} + \frac{2 y_1}{3 y_T} \right) \prod_{i=1}^{4} \theta^{y_i} (1 - \theta)^{1 - y_i}$$

$$= \sum_{y : y_T > 0} \left(\frac{1}{3} + \frac{2 y_1}{3 y_T} \right) \theta^{y_T} (1 - \theta)^{4 - y_T} = \begin{cases} 0.34180, & \theta = 1/4 \\ 0.46875, & \theta = 1/2. \end{cases}$$

These numbers were obtained by writing and implementing a suitable function in R (see the R code below).

(iv) The unconditional probability that any particular population unit i will be selected into the sample is

$$P(i \in s) = \sum_\theta P(i \in s \mid \theta) f(\theta)$$

$$= 0.34180 \times \frac{1}{2} + 0.46875 \times \frac{1}{2} = 0.40527.$$

To check this result, we note that the *sum of inclusion probabilities* should in this case be identical to the **expected** *sample size*.

The first of these quantities is $\displaystyle\sum_{i=1}^4 P(i \in s) = 4 \times 0.40527 = 1.6211$.

The second of these quantities can be obtained by first noting that

$$P(y_T = 0 \mid \theta) = (1-\theta)^4 = \begin{cases} (3/4)^4 = 81/256, \theta = 3/4 \\ (2/4)^4 = 16/256, \theta = 2/4. \end{cases}$$

This implies that

$$P(y_T = 0) = \sum_\theta P(y_T = 0 \mid \theta) f(\theta) = \frac{81}{256} \times \frac{1}{2} + \frac{16}{256} \times \frac{1}{2} = \frac{97}{512}$$

$$= 0.18945.$$

The sample vector has size 2 if $y_T > 0$, and size 0 if $y_T = 0$. So its expected size is $0 \times 0.18945 + 2 \times (1 - 0.18945) = 1.6211$, which is the same as $\displaystyle\sum_{i=1}^4 P(i \in s)$ above.

R Code for Exercise 9.2

```
# (a) & (b)

options(digits=5)

kern=function(th,yr){ th^(2+sum(yr))*(1-th)^(2-sum(yr))/(2+sum(yr)) }

kernth0.25 = kern(th=0.25,yr=c(0,0))+ kern(th=0.25,yr=c(0,1))+
            kern(th=0.25,yr=c(1,0))+ kern(th=0.25,yr=c(1,1))
kernth0.5 = kern(th=0.5,yr=c(0,0))+ kern(th=0.5,yr=c(0,1))+
            kern(th=0.5,yr=c(1,0))+ kern(th=0.5,yr=c(1,1))
```

```
postth=c(kernth0.25, kernth0.5)/( kernth0.25 + kernth0.5)
postth # 0.22946 0.77054


kernyr00 = kern(th=0.25,yr=c(0,0))+ kern(th=0.5,yr=c(0,0))
kernyr01 = kern(th=0.25,yr=c(0,1))+ kern(th=0.5,yr=c(0,1))
kernyr10 = kern(th=0.25,yr=c(1,0))+ kern(th=0.5,yr=c(1,0))
kernyr11 = kern(th=0.25,yr=c(1,1))+ kern(th=0.5,yr=c(1,1))


postyr =c(kernyr00,kernyr01,kernyr10,kernyr11)/
        (kernyr00+kernyr01+kernyr10+kernyr11)
postyr #  0.42493 0.21530 0.21530 0.14448


# (c)

sum(c(9/25,3/19,3/19,1/17)*postyr) # 0.22946  Correct
sum(c(16/25,16/19,16/19,16/17)*postyr) # 0.77054  Correct


# (d)

probfun=function(y,th){ yT=sum(y);      res=0
        if(yT>0)  res = ((1/3) + (2/3)*y[1]/yT) * th^yT * (1-th)^(4-yT)
        res }

mat1=matrix(c(0,0,0, 0,0,1, 0,1,0,       1,0,0, 0,1,1, 1,0,1, 1,1,0, 1,1,1),
        byrow=T, nrow=8,ncol=3)

mat2=rbind(mat1,mat1); ymat=cbind(c(rep(0,8),rep(1,8)),  mat2)


ymat
# [1,]  0  0  0  0
# [2,]  0  0  0  1
# .............................
# [15,]  1  1  1  0
# [16,]  1  1  1  1


prob0.25=0; for(i in 1:16) prob0.25 = prob0.25 + probfun(y=ymat[i,],th=0.25)
prob0.5=0; for(i in 1:16) prob0.5 = prob0.5 + probfun(y=ymat[i,],th=0.5)

c(prob0.25,prob0.5) # 0.34180 0.46875
(prob0.25+prob0.5)/2  # 0.40527
4*(prob0.25+prob0.5)/2  # 1.6211
c(97/512, 2*(1-97/512) ) # 0.18945 1.62109
```

Exercise 9.3 A finite population Bayesian model with SRSWOR

We sample $n = 2$ units from a finite population of $N = 4$ via SRSWOR.

If $\theta = 0$ then the finite population vector y is equally likely to be each of the following:
$\qquad$ (0,0,0,0), (0,0,0,1), (0,0,1,1), (0,1,1,1).

If $\theta = 1$ then the finite population vector y is equally likely to be each of the following:
$\qquad$ (1,1,1,1), (1,1,1,0), (1,1,0,0), (1,0,0,0).

A priori, the parameter θ is equally likely to be 0 or 1 (e.g. according to the toss of a coin).

Suppose we sample units 2 and 3, with values 1 and 1, respectively.

(a) Find the posterior distribution of θ.

(b) Find the predictive distribution of the finite population mean, namely $\bar{y} = (y_1 + ... + y_N) / N$.

Solution to Exercise 9.3

The easiest way to do this exercise is to first identify eight equally likely possibilities to start with. These possibilities are:

1. $\theta = 0, y = (0,0,0,0)$ with $\bar{y} = 0$
2. $\theta = 0, y = (0,0,0,1)$ with $\bar{y} = 1/4$
3. $\theta = 0, y = (0,0,1,1)$ with $\bar{y} = 1/2$
→ 4. $\theta = 0, y = (0,1,1,1)$ with $\bar{y} = 3/4$

→ 5. $\theta = 1, y = (1,1,1,1)$ with $\bar{y} = 1$
→ 6. $\theta = 1, y = (1,1,1,0)$ with $\bar{y} = 3/4$
7. $\theta = 1, y = (1,1,0,0)$ with $\bar{y} = 1/2$
8. $\theta = 1, y = (1,0,0,0)$ with $\bar{y} = 1/4$.

After observing $y_s = (y_2, y_3) = (1,1)$, there are only three possibilities remaining (4, 5 and 6 in the list, each highlighted by an arrow).

(a) Two possibilities out of the 3 correspond to $\theta = 1$ (namely 5 and 6) and one to $\theta = 0$ (namely 4); consequently, $f(\theta \mid D) = \begin{cases} 1/3, & \theta = 0 \\ 2/3, & \theta = 1 \end{cases}$, or equivalently, $(\theta \mid D) \sim Bern(2/3)$.

(b) Two possibilities out of the 3 correspond to $\bar{y} = 3/4$ (namely 4 and 6) and one to $\bar{y} = 1$ (namely 5); therefore $f(\bar{y} \mid D) = \begin{cases} 2/3, & \bar{y} = 3/4 \\ 1/3, & \bar{y} = 1 \end{cases}$.

Alternative solution

The above results can also be obtained by working through in the style of the solutions to previous exercises, as follows. Before the data is observed, the Bayesian model may be written:

$$f(s \mid y, \theta) = \binom{N}{n}^{-1} = \binom{4}{2}^{-1} = \frac{1}{6},$$
$$s = (1,2), (1,3), (1,4), (2,3), (2,4), (3,4)$$
$$f(y \mid \theta) = \frac{1}{4}, \quad y = (\theta, \theta, \theta, \theta), (\theta, \theta, \theta, 1-\theta),$$
$$(\theta, \theta, 1-\theta, 1-\theta), (\theta, 1-\theta, 1-\theta, 1-\theta)$$
$$f(\theta) = 1/2, \theta = 0, 1 \qquad \text{(the prior density of the parameter)}.$$

The observed data is $D = (s, y_s) = ((2,3), (1,1))$. At this particular value of the data:

$$f(s \mid y, \theta) = \frac{1}{6}, s = (2,3) \qquad \text{(the value of } s \text{ actually observed)}$$

$$f(y \mid \theta) = \frac{1}{4}, \quad y = (0,1,1,1) \text{ and } \theta = 0,$$

$$y \in \{(1,1,1,1), (1,1,1,0)\} \text{ and } \theta = 1 \quad \text{(where we need}$$
$$\text{only consider values of } y \text{ consistent with the data)}$$
$$f(\theta) = 1/2, \ \theta = 0, 1 \qquad \text{(since both values of } \theta \text{ are still possible,}$$
$$\text{i.e. consistent with the observed data)}.$$

With the quantities $s = (2,3)$, $y_s = (y_2, y_3) = (1,1)$ and $y_r = (y_1, y_4)$ all fixed at these values, the joint density of all quantities in the model may be written

$$f(\theta, s, y) = f(\theta, s, y_s, y_r) = f(\theta)f(y_s, y_r \mid \theta)f(s \mid y_s, y_r, \theta)$$

$$= \frac{I(\theta \in \{0,1\})}{2} \times \frac{I(y = (0,1,1,1), \theta = 0) + I(y \in \{(1,1,1,1),(1,1,1,0)\}, \theta = 1)}{4} \times \frac{1}{6}$$

$$\overset{\theta, y_r}{\propto} I(y_r = (0,1), \theta = 0) + I(y_r \in \{(1,1),(1,0)\}, \theta = 1).$$

(a) It follows that

$$f(\theta \mid D) \propto f(\theta, s, y_s) = \sum_{y_r} f(\theta, s, y)$$

$$\propto \begin{cases} \sum I(y_r = (0,1)) = 1, & \theta = 0 \\ \sum I(y_r \in \{(1,1),(1,0)\}) = 2, & \theta = 1. \end{cases}$$

After normalising, we see that $f(\theta \mid D) = \begin{cases} 1/3, \ \theta = 0 \\ 2/3, \ \theta = 1 \end{cases}.$

(b) Also,

$$f(y_r \mid D) \propto f(y_r, s, y_s) = \sum_{\theta} f(\theta, s, y_s, y_r)$$

$$\propto \begin{cases} \sum_{\theta=0}^{1} \left[I(y_r = (0,1), \theta = 0) + I(y_r \in \{(1,1),(1,0)\}, \theta = 1) \right] = 1, & y_r = (0,1) \\ \sum_{\theta=0}^{1} \left[I(y_r = (0,1), \theta = 0) + I(y_r \in \{(1,1),(1,0)\}, \theta = 1) \right] = 1, & y_r = (1,1) \\ \sum_{\theta=0}^{1} \left[I(y_r = (0,1), \theta = 0) + I(y_r \in \{(1,1),(1,0)\}, \theta = 1) \right] = 1, & y_r = (1,0) \end{cases}$$

which implies that $f(y_r \mid D) = 1/3, \quad y_r = (0,1),(1,1),(1,0)$.

Consequently, $f(y \mid D) = 1/3, \quad y = (0,1,1,1),(1,1,1,1),(1,1,1,0)$.

Now, the values of y listed here as possible given the observed data have means 3/4, 1 and 3/4, respectively.

It follows that the predictive density of the population mean is

$$f(\bar{y} \mid D) = \begin{cases} 2/3, & \bar{y} = 3/4 \\ 1/3, & \bar{y} = 1 \end{cases} \quad \text{(as was obtained previously).}$$

Exercise 9.4 Length-biased with-replacement sampling from a Poisson finite population

A finite population of size 9 consists of values that are conditionally iid Poisson with a mean whose prior distribution is gamma with both parameters zero (considered uninformative).

We sample 3 times from the finite population according to a with-replacement sampling scheme, where on each draw *the probability of selecting a unit is proportional to its value*.

Unit **2** is selected *once* and its value is **1**.
Unit **4** is selected *twice* and its value is **3**.

Find the posterior distribution of the Poisson mean and the predictive distribution of the nonsample total.

Also find these distributions under the (false) assumption that the sampling is SRSWR.

Then create two plots which suitably compare the four distributions indicated above.

Note: The concepts here involve a *biased sampling mechanism* and are relevant to *on-site sampling*, where for example we wish to estimate the total number of times that visitors (or *potential* visitors) to a recreational park actually visit there in some specified time period.

If we go to the site at random times to survey visitors, we are more likely to interview people who come very often relative to those who come only rarely. This means that we may end up over-estimating the popularity of the park—unless we make a suitable correction (downwards) to account for the (upwardly) biased sampling mechanism. If a potential visitor to the site doesn't come at all, then there is *zero* chance of sampling them.

If we wish to consider only the population of persons who *actually* visit the site in a given period (i.e. to exclude the *potential* visitors who do not visit), we may need to consider a truncated model involving the Poisson random variable conditional on it being non-zero. For further details and a discussion of the modelling issues here, see Shaw (1988).

Solution to Exercise 9.4

Generally, we are considering a sample of size n obtained with replacement from a finite population with values, $y_1,...,y_N$ which are conditionally Poisson with some mean λ, where the prior distribution on λ is gamma with parameters η and τ (and mean η/τ).

Let I_i be the number of times population unit i is sampled and define $I = (I_1,...,I_N)$. Then let $d = \sum_{i=1}^{N} I(I_i > 0)$ be the distinct sample size (the number of distinct population units sampled), and let $m = N - d$ be the nonsample size (the number of units not sampled).

In this scenario, we define the sample vector as $y_s = (y_{s_1},...,y_{s_d})$, where $s = (s_1,...,s_d)$ is the vector of the labels of the d distinct units that are sampled, and we define the nonsample vector as $y_r = (y_{r_1},...,y_{r_m})$, where $r = (r_1,...,r_m)$ is the vector of the labels of the m units that are not sampled.

Note: Here, s is a function of I, and so the data in this situation could also be written as $D = (I, y_s)$.

Since we are interested in the nonsample values only by way of their total y_{rT}, a suitable Bayesian finite population model in this context is:

$$f(I \mid y, \lambda) = \frac{n!}{\prod_{i=1}^{N} I_i!} \prod_{i=1}^{N} \left(\frac{y_i}{y_T}\right)^{I_i},$$

$$I \in \{(a_1,...,a_N) : a_i \in \{0,1,...,n\} \forall i, \; a_1 +...+ a_N = n\}$$

$$f(y_s, y_{rT} \mid \lambda) = \left(\prod_{i \in s} \frac{e^{-\lambda}\lambda^{y_i}}{y_i!}\right) \times \frac{e^{-m\lambda}(m\lambda)^{y_{rT}}}{y_{rT}!}$$

$$\lambda \sim G(\eta,\tau).$$

In our specific situation, $N = 9$, $n = 3$, and the data is
$$D = (I, y_s) = ((0,1,0,2,0,0,0,0,0),(1,3)),$$
meaning that unit 2 is selected once and its values is 1, and unit 4 is selected twice and its value is 3. Thus $d = 2$ and $m = 7$. Also, $\eta = 0$ and $\tau = 0$.

On the basis of these specifications, we wish to make inferences about λ and the nonsample total,

$$y_{rT} = y_1 + y_3 + y_5 + y_6 + y_7 + y_8 + y_9 .$$

Note: The probability of sampling unit 2 once and unit 4 twice (as is assumed to have occurred) equals

$$\frac{y_2}{y_T} \frac{y_4}{y_T} \frac{y_4}{y_T} + \frac{y_4}{y_T} \frac{y_2}{y_T} \frac{y_4}{y_T} + \frac{y_4}{y_T} \frac{y_4}{y_T} \frac{y_2}{y_T} = \frac{3!}{1!2!}\left(\frac{y_2}{y_T}\right)^1 \left(\frac{y_4}{y_T}\right)^2 = \frac{3!}{\prod_{i=1}^9 I_i!} \prod_{i=1}^9 \left(\frac{y_i}{y_T}\right)^{I_i}$$

and so is consistent with

$$f(I \mid y,\theta) = \frac{n!}{\prod_{i=1}^N I_i} \prod_{i=1}^N \left(\frac{y_i}{y_T}\right)^{I_i}$$

as specified in the general model.

For this exercise we will first derive the predictive distribution of y_{rT} and then use this to obtain the posterior distribution of λ only afterwards. The predictive density of y_{rT} is

$$f(y_{rT} \mid D) \propto \int f(y_{rT}, y_s, I, \lambda) d\lambda$$

$$= \int f(\lambda) f(y_s \mid \lambda) f(y_{rT} \mid \lambda) f(I \mid y_s, y_{rT}, \lambda) d\lambda$$

$$\propto \int_0^\infty \lambda^{\eta-1} e^{-\tau\lambda} \times \left(\prod_{i\in s} e^{-\lambda} \lambda^{y_i}\right) \times \frac{e^{-m\lambda}(m\lambda)^{y_{rT}}}{y_{rT}!} \times \frac{1}{y_T^n} d\lambda$$

$$\text{(note that } \prod_{i=1}^N \left(\frac{1}{y_T}\right)^{I_i} = \left(\frac{1}{y_T}\right)^n \text{)}$$

$$= \frac{m^{y_{rT}}}{y_{rT}!} \frac{1}{y_T^n} \int_0^\infty \lambda^{\eta+y_{sT}+y_{rT}-1} e^{-\lambda(\tau+d+m)} d\lambda$$

$$= \frac{m^{y_{rT}}}{y_{rT}!} \frac{1}{y_T^n} \times \frac{\Gamma(\eta+y_{sT}+y_{rT})}{(\tau+d+m)^{\eta+y_{sT}+y_{rT}}} \times 1 .$$

Thus

$$f(y_{rT} \mid D) = \frac{k(y_{rT})}{c}, \quad y_{rT} = 0,1,2,\dots ,$$

where

$$k(y_{rT}) = \left(\frac{N-d}{N+\tau}\right)^{y_{rT}} \frac{\Gamma(\eta + y_{sT} + y_{rT})}{y_{rT}!(y_{sT} + y_{rT})^n}$$

and

$$c = \sum_{y_{rT}=0}^{\infty} k(y_{rT}).$$

Note: Here, $d + m = N$, and so

$$(\tau + d + m)^{\eta + y_{sT} + y_{rT}} = (N+\tau)^{\eta + y_{sT} + y_{rT}} \propto (N+\tau)^{y_{rT}}.$$

We may approximate $f(y_{rT} \mid D)$ by calculating $k(y_{rT})$ only for $y_{rT} = 0,1,2,\dots,M$ for some large integer M (in practice we used 100) for which $k(y_{rT})$ is sufficiently close to zero.

Using the predictive density of y_{rT}, we can now obtain the posterior density of λ as

$$f(\lambda \mid D) = \sum_{y_{rT}=0}^{\infty} f(\lambda \mid D, y_{rT}) f(y_{rT} \mid D),$$

where

$$(\lambda \mid D, y_{rT}) \sim G(\eta + y_{sT} + y_{rT}, \tau + N).$$

Note: This result is obvious but can also be obtained as follows:

$$f(\lambda \mid D, y_{rT}) = f(\lambda \mid s, y_s, y_{rT}) \propto f(\lambda, s, y_s, y_{rT}) \propto f(\lambda) f(y_s, y_{rT} \mid \lambda)$$

$$\propto \lambda^{\eta-1} e^{-\tau\lambda} \left(\prod_{i \in s} \frac{e^{-\lambda} \lambda^{y_i}}{y_i!} \right) \times \frac{e^{-m\lambda} (m\lambda)^{y_{rT}}}{y_{rT}!}$$

$$\propto \lambda^{\eta-1} e^{-\tau\lambda} \times e^{-d\lambda} \lambda^{y_{sT}} \times e^{-m\lambda} \lambda^{y_{rT}}$$

$$= \lambda^{\eta + y_{sT} + y_{rT} - 1} e^{-\lambda(\tau + N)} \quad \text{(since } d + m = N\text{)}.$$

We see that $f(\lambda \mid D)$ is an infinite mixture of gamma densities where the weight assigned to each one is the corresponding (marginal) predictive density of y_{rT}.

Note: An alternative way to derive $f(\lambda \mid D)$ is using the equation

$$f(\lambda \mid D) \propto \sum_{y_{rT}} f(y_{rT}, y_s, I, \lambda).$$

The case of SRSWR

In the case of SRSWR, the sampling density

$$f(I \mid y, \lambda) = \frac{n!}{\prod_{i=1}^{N} I_i} \prod_{i=1}^{N} \left(\frac{y_i}{y_T} \right)^{I_i}$$

changes to

$$f(I \mid y, \lambda) = \frac{n!}{\prod_{i=1}^{N} I_i} \prod_{i=1}^{N} \left(\frac{1}{N} \right)^{I_i} = \frac{n!}{N^n \prod_{i=1}^{N} I_i},$$

which we note does not depend on λ or y_{rT} and so can be 'ignored'.

The result is then almost the same as before, the only difference being that the term

$$\prod_{i=1}^{N} y_T^{I_i} = y_T^n = (y_{sT} + y_{rT})^n$$

in

$$k(y_{rT}) = \left(\frac{N-d}{N+\tau} \right)^{y_{rT}} \frac{\Gamma(\eta + y_{sT} + y_{rT})}{y_{rT}!(y_{sT} + y_{rT})^n}$$

is replaced by 1.

Thus under SRSWR we find that

$$f(y_{rT} \mid D) = \frac{K(y_{rT})}{C}, \quad y_{rT} = 0, 1, 2, \ldots,$$

where

$$K(y_{rT}) = \left(\frac{N-d}{N+\tau} \right)^{y_{rT}} \frac{\Gamma(\eta + y_{sT} + y_{rT})}{y_{rT}! \times 1}$$

and

$$C = \sum_{y_{rT}=0}^{\infty} K(y_{rT}).$$

As regards the posterior distribution of λ under SRSWR, this need no longer be expressed as an infinite mixture of gamma distributions but simply as

$$(\lambda \mid D) \sim G(\eta + y_{sT}, \tau + d).$$

Figure 9.1 shows the posterior density $f(\lambda \mid D)$ under the length-biased and SRSWR assumptions, respectively.

We see that the inference under the assumption of length-bias is the lower of the two. This is because it appropriately corrects for large finite population values being more likely to be selected. If we 'ignore' the fact that large values are more likely to be selected. then we will erroneously over-estimate the superpopulation mean, λ.

Figure 9.2 shows the predictive density $f(y_{rT} \mid D)$, again under the two assumptions.

As in Figure 9.1, we see that ignoring the length-biased sampling mechanism tends to bias the inference upwards.

As a check on our calculations, which omitted all terms corresponding to values of y_{rT} greater than $M = 100$ (see above), we calculate the predictive mean of y_{rT} under the SRSWR assumption using the formula

$$E(y_{rT} \mid D) \approx \frac{1}{C} \sum_{rT=0}^{M} y_{rT} K(y_{rT})$$

and obtain the value of 14.

This may be compared with the theoretical value, which is exactly
$$E(y_{rT} \mid D) = E\{E(y_{rT} \mid D, \lambda) \mid D\} = E(m\lambda \mid D)$$
$$= m \times \frac{\eta + y_{sT}}{\tau + d} = 7 \times \frac{0 + (3+1)}{0 + 2} = 14.$$

Figure 9.1 Posterior densities of the Poisson mean

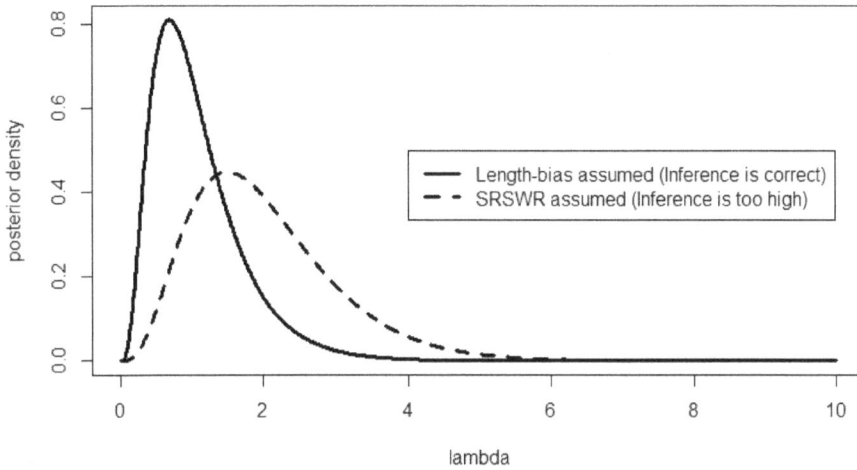

Figure 9.2 Predictive densities of the nonsample total

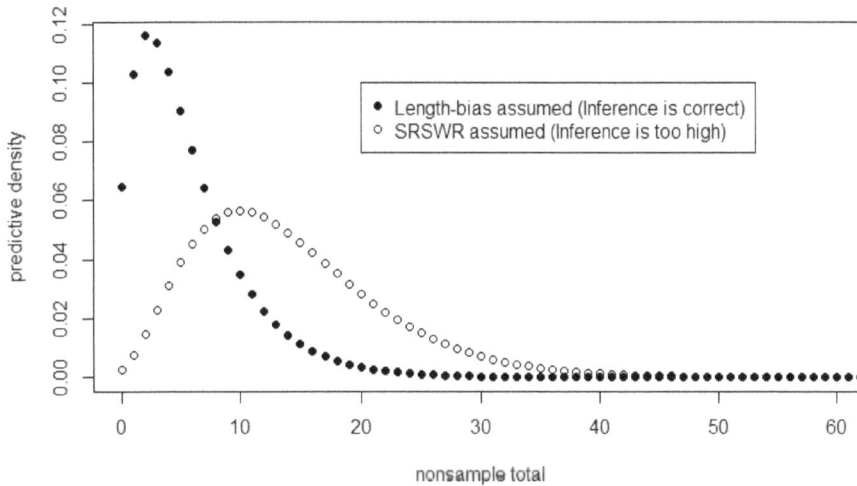

R Code for Exercise 9.4

```
options(digits=5); X11(w=8,h=5); par(mfrow=c(1,1))

N=9; n = 3; ys=c(1,3); ysT = sum(ys); d = 2; m = 7; eta=0; tau=0; yrTv=0:100
kv = ((N-d)/(tau+N))^yrTv *gamma(eta+ysT+yrTv)/
        ( factorial(yrTv) * (ysT+yrTv)^n )
c = sum(kv); fv = kv/c

plot(yrTv,fv,pch=16, xlab="nonsample total",
        ylab="predictive density",xlim=c(0,60), main=" ")
kvigno = ((N-d)/(tau+N))^yrTv *gamma(eta+ysT+yrTv)/( gamma(yrTv+1) * 1)
cigno = sum(kvigno); fvigno = kvigno/cigno
points(yrTv,fvigno,pch=1)
legend(20,0.1,c("Length-bias assumed (Inference is correct)",
        "SRSWR assumed (Inference is too high)"),pch=c(16,1))
c(sum(yrTv*fv), sum(yrTv*fvigno) ) # 5.6302   14.0000
m*(eta+ysT)/(tau+d) # 14

lamv=seq(0,10,0.01); lamfv=lamv
for(i in 1:length(lamv)) lamfv[i]=sum(fv*dgamma(lamv[i],eta+ysT+yrTv,tau+N))
plot(lamv,lamfv,type="l", lty=1, lwd=3,
        xlab="lambda",ylab="posterior density", main=" ")
lamfvigno=lamv
for(i in 1:length(lamv))
        lamfvigno[i]=sum(fvigno*dgamma(lamv[i],eta+ysT+yrTv,tau+N))
# lines(lamv,lamfvigno,lty=2,lwd=1) # Can do as a check on calculations
lines(lamv,dgamma(lamv,eta+ysT,tau+d),lty=2,lwd=3)
legend(4,0.5,c("Length-bias assumed (Inference is correct)",
        "SRSWR assumed (Inference is too high)"),lty=c(1,2),lwd=c(3,3))
```

Exercise 9.5 An exponential finite population model with a biased Poisson sampling scheme

A sample is drawn from a finite population of size $N = 7$ in such a way that unit i has probability of inclusion π_i, independently of all the other units.

The values in the finite population are independent and identically distributed exponentials with mean $\mu = 1/\lambda$, where the prior distribution for λ is given by

$$f(\lambda) \propto 1/\lambda, \lambda > 0.$$

Units **3** and **5** are selected, and their values are **1.6** and **0.4**, respectively.

Find and sketch the posterior density of the superpopulation mean μ and the predictive density of the finite population mean $\bar{y}$ under each of the following specifications:

(a) All the π_i values are equal to 0.3 $(i = 1,...,N)$.

(b) All the π_i values are equal to 0.3 except that:

$$\pi_5 = 0.3 \text{ if } y_5 < 1$$
$$\pi_5 = 0.9 \text{ if } y_5 > 1$$

(thus unit 5 is 3 times as likely to be sampled if its value exceeds 1).

(c) All the π_i values are equal to 0.3 except that:

$$\pi_4 = 0.3 \text{ if } y_4 < 1$$
$$\pi_4 = 0.9 \text{ if } y_4 > 1$$

(thus unit 4 is 3 times as likely to be sampled if its value exceeds 1).

Note: Here, the sample size n is not fixed and is a random variable.

Solution to Exercise 9.5

(a) The relevant Bayesian model is:

$$f(I \mid y, \lambda) = \prod_{i=1}^{N} f(I_i \mid y, \lambda) = \prod_{i=1}^{N} \pi_i^{I_i}(1 - \pi_i)^{1-I_i}, \quad n = \sum_{i=1}^{N} I_i$$

$$f(y \mid \lambda) = \prod_{i=1}^{N} \lambda e^{-\lambda y_i}, \ y_i > 0 \ \forall \ i$$

$$f(\lambda) \propto 1/\lambda, \lambda > 0.$$

Here,

$$\pi_1 = \dots = \pi_N = 0.3,$$

and the data is

$$D = (I, y_s) = ((0,0,1,0,1,0,0), (1.6, 0.4)),$$

with $n = 2$ (the achieved sample size).

The sampling mechanism is ignorable and so

$$f(\lambda \mid D) \propto f(\lambda) f(y_s \mid \lambda) \propto \frac{1}{\lambda} \prod_{i \in s} \lambda e^{-\lambda y_i} = \lambda^{n-1} e^{-\lambda y_{sT}}$$

$$\Rightarrow (\lambda \mid D) \sim G(n, y_{sT})$$

$$\Rightarrow (\mu \mid D) \sim IG(n, y_{sT}).$$

Next,

$$(y_{rT} \mid \lambda) \sim G(m, \lambda),$$

where $m = N - n = 7 - 2 = 5$.

It follows that

$$f(y_{rT} \mid D) = \int f(y_{rT} \mid D, \lambda) f(\lambda \mid D) d\lambda$$

$$\propto \int_0^\infty \lambda^m y_{rT}^{m-1} e^{-\lambda y_{rT}} \times \lambda^{n-1} e^{-\lambda y_{sT}} d\lambda$$

$$= y_{rT}^{m-1} \int_0^\infty \lambda^{n+m-1} e^{-\lambda(y_{rT} + y_{sT})} d\lambda$$

$$= \frac{y_{rT}^{m-1} \Gamma(n + m)}{(y_{rT} + y_{sT})^{n+m}}$$

$$\propto \frac{y_{rT}^{m-1}}{(y_{rT} + y_{sT})^{N}}, y_{rT} > 0.$$

Hence

$$f(\bar{y} \mid D) \propto \frac{\left(\bar{y} - \dfrac{n}{N}\bar{y}_s\right)^{N-n-1}}{\bar{y}^N}, \quad \bar{y} > \frac{n}{N}\bar{y}_s$$

(using the fact that $y_{rT} = N\bar{y} - n\bar{y}_s$).

(b) In this case, inferences will be exactly the same as in (a). This is because, even though the sampling mechanism is *potentially* nonignorable due to $f(I \mid y, \lambda)$ depending on a population value y_5, that value happens to be *known* (since unit 5 is in the sample, i.e. $5 \in s$).

To clarify, we write

$$\pi_5 = \pi_5(y_5) = \begin{cases} 0.3, & y_5 < 1 \\ 0.9, & y_5 > 1 \end{cases} = 0.3 + 0.6I(y_5 > 1).$$

Then, noting that $I_5 = 1$ and $y_5 = 0.4$, we have that

$$f(I_5 \mid y, \lambda) = \pi_5^{I_5}(1 - \pi_5)^{1-I_5} = \pi_5 = 0.3 + 0.6I(y_5 > 1) = 0.3.$$

Thus

$$f(I \mid y, \lambda) = \prod_{i=1}^{N} \pi_i^{I_i}(1 - \pi_i)^{1-I_i}$$

doesn't depend on λ or y_r and is completely known.

Therefore,

$$f(\lambda \mid D) \propto f(\lambda, I, y_s) = \int f(\lambda, I, y_s, y_r) dy_r$$

$$= \int f(\lambda)f(y_s, y_r \mid \lambda)f(I \mid y_s, y_r, \lambda) dy_r$$

$$\propto f(\lambda)f(y_s \mid \lambda)\int f(y_r \mid \lambda) dy_r \propto f(\lambda)f(y_s \mid \lambda) \times 1 \text{ as before in (a).}$$

(c) In this case, the sampling mechanism is **nonignorable** and inferences will be different to those in (a), because $f(I \mid y, \lambda)$ depends on a population value y_4 which is **unknown** (since unit 4 is not in the sample, i.e. $4 \in r$); that is, $f(I \mid y, \lambda)$ is **unknown**. To clarify, we write

$$\pi_4 = \pi_4(y_4) = \begin{cases} 0.3, \ y_4 < 1 \\ 0.9, \ y_4 > 1 \end{cases} = 0.3 + 0.6I(y_4 > 1).$$

Then, noting that $I_4 = 0$ and y_4 is unknown, we have that

$$f(I_4 \mid y, \lambda) = \pi_4^{I_4}(1 - \pi_4)^{1 - I_4} = 1 - \pi_4 = 0.7 - 0.6I(y_4 > 1)$$

(a function of y_4).

So $f(I \mid y, \lambda) = \prod_{i=1}^{N} f(I_i \mid y, \lambda)$ is unknown.

With this in mind, we now write

$$f(\lambda \mid D) \propto f(\lambda, I, y_s) = \int f(\lambda, I, y_s, y_r) dy_r$$

$$= \int f(\lambda) f(y_s, y_r \mid \lambda) f(I \mid y_s, y_r, \lambda) dy_r$$

$$\propto f(\lambda) f(y_s \mid \lambda) W,$$

where

$$W = W(\lambda) = \int f(y_r \mid \lambda) f(I_4 \mid y_4) dy_r$$

$$= \left(\prod_{\substack{i \in r \\ i \neq 4}} \int_0^\infty f(y_i \mid \lambda) dy_i \right) \int f(y_4 \mid \lambda) f(I_4 \mid y_4) dy_4$$

$$= \left(\prod_{\substack{i \in r \\ i \neq 4}} 1 \right) \int_0^\infty \lambda e^{-\lambda y_4} [0.7 - 0.6I(y_4 > 1)] dy_4$$

since $f(y_i \mid \lambda) = \lambda e^{-\lambda y_i} \forall i$

$$= 0.7 \int_0^\infty \lambda e^{-\lambda y_4} dy_4 - 0.6 \int_1^\infty \lambda e^{-\lambda y_4} dy_4$$

$$= 0.7 \times 1 - 0.6 e^{-\lambda}.$$

Thus

$$f(\lambda \mid D) \propto \lambda^{n-1} e^{-\lambda y_{sT}} (7 - 6e^{-\lambda}) = 7\lambda^{n-1} e^{-\lambda y_{sT}} - 6\lambda^{n-1} e^{-\lambda(y_{sT} + 1)}.$$

Thus

$$f(\lambda \mid D) = c \left\{ \frac{7}{y_{sT}^n} \left(\frac{y_{sT}^n \lambda^{n-1} e^{-\lambda y_{sT}}}{\Gamma(n)} \right) - \frac{6}{(y_{sT}+1)^n} \left(\frac{(y_{sT}+1)^n \lambda^{n-1} e^{-\lambda(y_{sT}+1)}}{\Gamma(n)} \right) \right\},$$

where

$$1 = \int f(\lambda \mid D) d\lambda = c \left\{ \frac{7}{y_{sT}^n}(1) - \frac{6}{(y_{sT}+1)^n}(1) \right\}$$

$$\Rightarrow c = \left\{ \frac{7}{y_{sT}^n} - \frac{6}{(y_{sT}+1)^n} \right\}^{-1}.$$

Note 1: The posterior $f(\lambda \mid D)$ is a weighted average of two gamma densities where one of the weights is *negative*.

Note 2: The posterior density of $\mu = 1/\lambda$ is given by

$$f(\mu \mid D) = f(\lambda = 1/\mu \mid D)/\mu^2.$$

We now turn our attention to the predictive distribution of the nonsample total. Observe that

$$f(y_r \mid D) = \int f(y_r \mid D, \lambda) f(\lambda \mid D) d\lambda,$$

where

$$f(y_r \mid D, \lambda) \propto f(y_r, y_s, I, \lambda) \propto \left\{ \left[7 - 6I(y_4 > 1) \right] \lambda e^{-\lambda y_4} \right\} \times \prod_{\substack{i \in r \\ i \neq 4}} \lambda e^{-\lambda y_i}.$$

This suggests that we decompose the nonsample total according to

$$y_{rT} = y_0 + y_4$$

(where y_0 is the total of all values in y_r except for y_4) and think about how we can use the following facts:

$$(y_0 \perp y_4 \mid D, \lambda) \qquad (y_4 \text{ is independent of all other nonsample}$$
$$\text{units, given } D \text{ and } \lambda)$$

$$(y_0 \mid D, \lambda) \sim G(m-1, \lambda) \qquad \text{(a simple distribution)}$$

$$f(y_4 \mid D, \lambda) \propto [7 - 6I(y_4 > 1)]\lambda e^{-\lambda y_4}, \; y_4 > 0$$
$$\text{(a complicated distribution)}.$$

One strategy is to use these facts to obtain the cdf $F(y_4 | D, \lambda)$, hence $F(y_{rT} | D, \lambda)$, hence $f(y_{rT} | D, \lambda) = F'(y_{rT} | D, \lambda)$, hence $f(\bar{y} | D, \lambda)$, and hence ultimately the required $f(\bar{y} | D) = \int f(\bar{y} | D, \lambda) f(\lambda | D) d\lambda$.

First, $F(y_4 | D, \lambda) \propto \begin{cases} 7 \int_0^{y_4} \lambda e^{-\lambda t} dt, & 0 < y_4 < 1 \\ 7 \int_0^{y_4} \lambda e^{-\lambda t} dt - 6 \int_1^{y_4} \lambda e^{-\lambda t} dt, & y_4 > 1 \end{cases}$

$= \begin{cases} 7(1 - e^{-\lambda y_4}), & 0 < y_4 < 1 \\ 7(1 - e^{-\lambda y_4}) - 6(e^{-\lambda 1} - e^{-\lambda y_4}), & y_4 > 1. \end{cases}$

Thus $F(y_4 | D, \lambda) = \begin{cases} k(7 - 7e^{-\lambda y_4}), & 0 < y_4 < 1 \\ k(7 - e^{-\lambda y_4} - 6e^{-\lambda}), & y_4 > 1, \end{cases}$

where $k = k(\lambda) = 1 / (7 - 6e^{-\lambda})$, since $1 = F(y_4 = \infty | D, \lambda) = k(7 - 6e^{-\lambda})$.

Check: Since $(y_4 | D, \lambda)$ is continuous we would expect that
$$F(y_4 = 1^+ | D, \lambda) - F(y_4 = 1^- | D, \lambda) = 0.$$

The left hand side here is
$$k(7 - e^{-\lambda 1} - 6e^{-\lambda}) - k(7 - 7e^{-\lambda 1}) = k(7 - 7e^{-\lambda}) - k(7 - 7e^{-\lambda}) = 0$$
(which is correct).

Next, writing $a \equiv y_{rT}$ for notational convenience, we have that
$$F(a | D, \lambda) = P(y_0 + y_4 \le a | D, \lambda)$$
$$= E\{P(y_0 + y_4 \le a | D, \lambda, y_0) | D, \lambda\}$$
$$= \int_0^a P(y_4 \le a - y_0 | D, \lambda, y_0) f(y_0 | D, \lambda) dy_0$$

(a convolution)

$$= \int_0^a F(y_4 = a - y_0 | D, \lambda) f_{G(m-1, \lambda)}(y_0) dy_0$$

$$= \int_0^a k \left[7 - 7e^{-\lambda(a - y_0)} \right] f_{G(m-1, \lambda)}(y_0) dy_0, \quad 0 < a < 1.$$

For the case $a > 1$ we find that

$$F(a \mid D, \lambda) = \int_{0}^{a-1} k \left[7 - e^{-\lambda(a-y_0)} - 6e^{-\lambda} \right] f_{G(m-1,\lambda)}(y_0) dy_0$$

$$+ \int_{a-1}^{a} k \left[7 - 7e^{-\lambda(a-y_0)} \right] f_{G(m-1,\lambda)}(y_0) dy_0 \,.$$

Note: If $a > 1$ and $0 \le y_0 \le a-1$ then $1 \le a - y_0 \le a$.

If $a > 1$ and $a-1 \le y_0 \le a$ then $0 \le a - y_0 \le 1$.

Check: Since $(a \mid D, \lambda)$ is continuous we would expect that

$$F(a = 1^+ \mid D, \lambda) - F(a = 1^- \mid D, \lambda) = 0 \,.$$

The LHS here is

$$\left\{ \int_{0}^{1-1} k \left[7 - e^{-\lambda(1-y_0)} - 6e^{-\lambda} \right] f_{G(m-1,\lambda)}(y_0) dy_0 \right.$$

$$\left. + \int_{1-1}^{1} k \left[7 - 7e^{-\lambda(1-y_0)} \right] \cancel{f_{G(m-1,\lambda)}(y_0) dy_0} \right\}$$

$$- \int_{0}^{1} k \left[7 - 7e^{-\lambda(1-y_0)} \right] \cancel{f_{G(m-1,\lambda)}(y_0) dy_0} = 0 \quad \text{(which is correct)}.$$

We now consider Leibniz's rule for differentiating an integral:

$$\frac{d}{dy} \int_{a(y)}^{b(y)} f(x,y) dx = \int_{a(y)}^{b(y)} \frac{\partial}{\partial y} f(x,y) dx$$

$$+ b'(y) f(b(y), y) - a'(y) f(a(y), y)$$

(where the symbols here are not directly related to those in this exercise).

Applying this rule for the case $0 < a < 1$, we obtain

$$f(a \mid D, \lambda) = \frac{dF(a \mid D, \lambda)}{da} = \int_{0}^{a} k \left[0 - 7e^{-\lambda(a-y_0)}(-\lambda) \right] f_{G(m-1,\lambda)}(y_0) dy_0$$

$$+ \frac{da}{da} k \left[7 - 7e^{-\lambda(a-a)} \right] f_{G(m-1,\lambda)}(a) \qquad \text{(this is zero)}$$

$$- \frac{d0}{da} k \left[7 - 7e^{-\lambda(a-0)} \right] f_{G(m-1,\lambda)}(0) \qquad \text{(this is zero)}$$

$$= 7k\lambda e^{-\lambda a} \int_0^a e^{\lambda y_0} \left(\frac{\lambda^{m-1} y_0^{m-2} e^{-\lambda y_0}}{\Gamma(m-1)} \right) dy_0 = 7k\lambda e^{-\lambda a} \frac{\lambda^{m-1}}{(m-2)!} \int_0^a y_0^{m-2} dy_0$$

$$= 7k\lambda e^{-\lambda a} \frac{\lambda^{m-1}}{(m-1)!} a^{m-1} \quad = 7k \left(\frac{\lambda^m a^{m-1} e^{-\lambda a}}{(m-1)!} \right) \quad = 7k f_{G(m,\lambda)}(a).$$

Likewise, for the case $a > 1$, we obtain

$$f(a \mid D, \lambda) = \frac{dF(a \mid D, \lambda)}{da} = \int_0^{a-1} k\left[0 - e^{-\lambda(a-y_0)}(-\lambda) - 0 \right] f_{G(m-1,\lambda)}(y_0) dy_0$$

$$+ \frac{d(a-1)}{da} k\left[7 - e^{-\lambda(a-(a-1))} - 6e^{-\lambda} \right] f_{G(m-1,\lambda)}(a-1)$$

$$- \frac{d0}{da} k\left[7 - e^{-\lambda(a-0)} - 6e^{-\lambda} \right] f_{G(m-1,\lambda)}(0) \text{ (this is zero)}$$

$$+ \int_{a-1}^a k\left[0 - 7e^{-\lambda(a-y_0)}(-\lambda) \right] f_{G(m-1,\lambda)}(y_0) dy_0$$

$$+ \frac{da}{da} k\left[7 - 7e^{-\lambda(a-a)} \right] f_{G(m-1,\lambda)}(a) \text{ (this is zero)}$$

$$- \frac{d(a-1)}{da} k\left[7 - 7e^{-\lambda(a-(a-1))} \right] f_{G(m-1,\lambda)}(a-1)$$

$$= k\lambda e^{-\lambda a} \int_0^{a-1} e^{\lambda y_0} \left(\frac{\lambda^{m-1} y_0^{m-2} e^{-\lambda y_0}}{\Gamma(m-1)} \right) dy_0 + k\left[7 - e^{-\lambda} - 6e^{-\lambda} \right] f_{G(m-1,\lambda)}(a-1)$$

$$+ 7k\lambda e^{-\lambda a} \int_{a-1}^a e^{\lambda y_0} \left(\frac{\lambda^{m-1} y_0^{m-2} e^{-\lambda y_0}}{\Gamma(m-1)} \right) dy_0 - 7k\left[1 - e^{-\lambda} \right] f_{G(m-1,\lambda)}(a-1)$$

$$= k\lambda e^{-\lambda a} \frac{\lambda^{m-1}}{(m-1)!} (a-1)^{m-1} + 7k(1 - e^{-\lambda}) f_{G(m-1,\lambda)}(a-1)$$

$$+ 7k\lambda e^{-\lambda a} \frac{\lambda^{m-1}}{(m-1)!} \left[a^{m-1} - (a-1)^{m-1} \right] - 7k(1 - e^{-\lambda}) f_{G(m-1,\lambda)}(a-1)$$

$$= ke^{-\lambda} \left(\frac{\lambda^m (a-1)^{m-1} e^{-\lambda(a-1)}}{(m-1)!} \right) + 7k \left(\frac{\lambda^m a^{m-1} e^{-\lambda a}}{(m-1)!} \right)$$

$$- 7ke^{-\lambda} \left(\frac{\lambda^m (a-1)^{m-1} e^{-\lambda(a-1)}}{(m-1)!} \right)$$

$$= k\left\{ 7 f_{G(m,\lambda)}(a) - 6e^{-\lambda} f_{G(m,\lambda)}(a-1) \right\}.$$

In summary so far,

$$f(a \mid D, \lambda) = k \times \begin{cases} 7 f_{G(m,\lambda)}(a), & 0 < a < 1 \\ 7 f_{G(m,\lambda)}(a) - 6e^{-\lambda} f_{G(m,\lambda)}(a-1), & a > 1. \end{cases}$$

Check: Here,

$$\int f(a \mid D, \lambda) da =$$

$$k \times \left\{ 7 F_{G(m,\lambda)}(1) + 7 \left[1 - F_{G(m,\lambda)}(1) \right] - 6e^{-\lambda} \left[1 - F_{G(m,\lambda)}(1-1) \right] \right\}$$

$$= \frac{1}{7 - 6e^{-\lambda}} \times \left\{ 7 - 6e^{-\lambda} [1 - 0] \right\} = 1$$

(which is correct).

Next, using the relationship $\bar{y} = (n\bar{y}_s + y_{rT}) / N$, we obtain:

$$f(\bar{y} \mid D, \lambda) = f_1(\bar{y}, \lambda)$$

$$\equiv N k(\lambda) 7 f_{G(m,\lambda)}(N\bar{y} - n\bar{y}_s)$$

$$\text{for } \frac{n\bar{y}_s}{N} < \bar{y} < \frac{n\bar{y}_s + 1}{N}$$

$$f(\bar{y} \mid D, \lambda) = f_2(\bar{y}, \lambda)$$

$$\equiv N k(\lambda) \left\{ 7 f_{G(m,\lambda)}(N\bar{y} - n\bar{y}_s) - 6e^{-\lambda} f_{G(m,\lambda)}(N\bar{y} - n\bar{y}_s - 1) \right\}$$

$$\text{for } \bar{y} > \frac{n\bar{y}_s + 1}{N},$$

where:

$$\frac{n\bar{y}_s}{N} = 0.2857$$

$$\frac{n\bar{y}_s + 1}{N} = 0.4286$$

$$k(\lambda) = \frac{1}{7 - 6e^{-\lambda}} \quad \text{(as before)}.$$

Thus we finally obtain the required posterior predictive density:

$$f(\bar{y} \mid D) = g_1(\bar{y}) \equiv \int_0^\infty f_1(\bar{y}, \lambda) f(\lambda \mid D) d\lambda, \quad \frac{n\bar{y}_s}{N} < \bar{y} < \frac{n\bar{y}_s + 1}{N}$$

$$f(\bar{y} \mid D) = g_2(\bar{y}) \equiv \int_0^\infty f_2(\bar{y}, \lambda) f(\lambda \mid D) d\lambda, \quad \bar{y} > \frac{n\bar{y}_s + 1}{N},$$

where $\quad f(\lambda \mid D) = \left\{ \dfrac{7}{y_{sT}^n} - \dfrac{6}{(y_{sT} + 1)^n} \right\}^{-1}$

$$\times \left\{ \frac{7}{y_{sT}^n} f_{G(n, y_{sT})}(\lambda) - \frac{6}{(y_{sT} + 1)^n} f_{G(n, y_{sT} + 1)}(\lambda) \right\}$$

(as obtained earlier).

Figure 9.3 shows the two densities $f(\mu \mid D)$ and $f(\bar{y} \mid D)$ under each of the scenarios in (a) and (c).

We see that inferences under the length-biased sampling scheme in (c) are *lower* than those under SRSWR in (a). This is because, generally speaking, length bias makes larger units more likely to be selected, and not adjusting for that bias naturally leads to inferences that are too high.

These patterns are consistent with the following point estimates as obtained numerically (see the R code below for details of the calculation):

$E(\mu \mid D) = 1.38$ in (c) $\qquad < \qquad E(\mu \mid D) = 2.00$ in (a)

$E(\bar{y} \mid D) = 1.19$ in (c) $\qquad < \qquad E(\bar{y} \mid D) = 1.71$ in (a).

Note 1: In (a),

$$(\mu \mid D) \sim IG(n, y_{sT}),$$

and therefore

$$E(\mu \mid D) = y_{sT} / (n-1) = 2 / (2-1) = 2 \quad \text{(exactly).}$$

Note 2: The posterior predictive mean of $\bar{y}$ in (c) was obtained numerically as follows:

$$\hat{\bar{y}} = E(\bar{y} \mid D) = \int_{\frac{n\bar{y}_s}{N}}^{\frac{n\bar{y}_s + 1}{N}} \bar{y} g_1(\bar{y}) d\bar{y} + \int_{\frac{n\bar{y}_s + 1}{N}}^\infty \bar{y} g_2(\bar{y}) d\bar{y}$$

$$= 0.01140 + 1.17546 = 1.1869.$$

Figure 9.3 Posterior and predictive densities

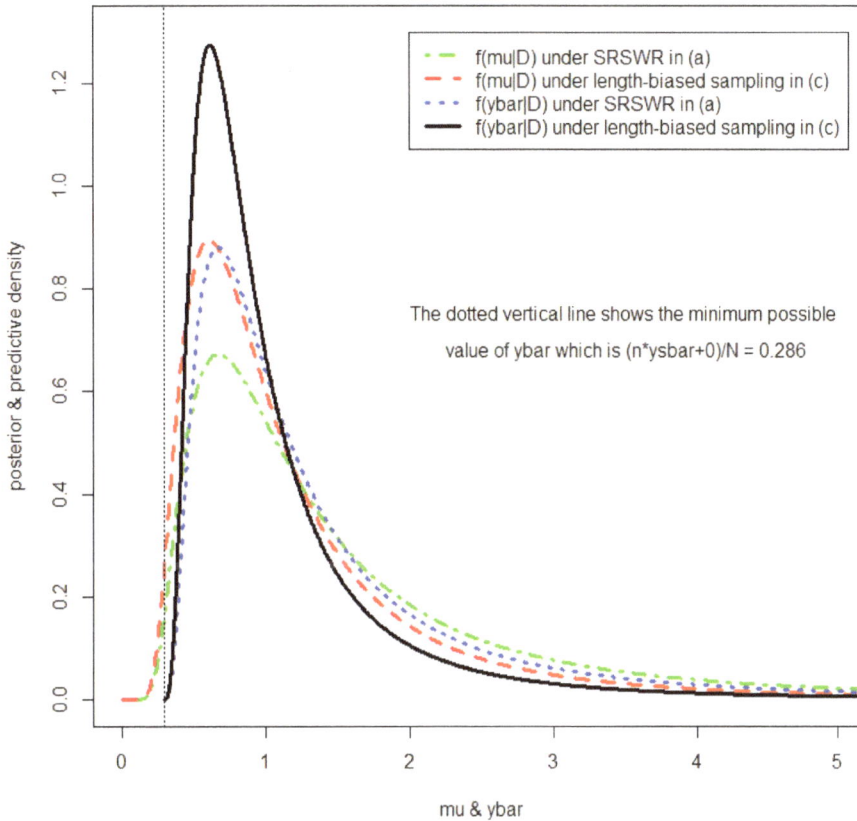

Legend:
- f(mu|D) under SRSWR in (a)
- f(mu|D) under length-biased sampling in (c)
- f(ybar|D) under SRSWR in (a)
- f(ybar|D) under length-biased sampling in (c)

The dotted vertical line shows the minimum possible value of ybar which is (n*ysbar+0)/N = 0.286

y-axis: posterior & predictive density

x-axis: mu & ybar

R Code for Exercise 9.5

```
# (a)
X11(w=8,h=4); par(mfrow=c(1,1))
N=7; ys=c(1.6,0.4); ysT=sum(ys); ysbar=mean(ys); n=length(ys); m=N-n
c(ysT,ysbar,n,m) # 2 1 2 5
fmufun=function(mu,n,ysT) dgamma(1/mu,n,ysT)/mu^2
integrate(fmufun,0, Inf,n=n,ysT=ysT)$value  # 1 check
muv=seq(0.0001,20.0001,0.005); fmuv= fmufun(muv,n=n,ysT=ysT)
plot(muv,fmuv,type="l",xlim=c(0,20)) # check
integrate(function(mu,n,ysT) mu*fmufun(mu,n,ysT),
    0,Inf,n=n,ysT=ysT)$value # 2  check  (posterior mean of mu)
```

```
kybarfun=function(ybar,n,N,ysbar)   (ybar-(n/N)*ysbar)^(N-n-1)   /   ybar^N
const = integrate(kybarfun, (n/N)*ysbar , Inf,n=n,N=N,ysbar=ysbar)$value
const #  0.4083333
ybarv=seq(  (n/N)*ysbar,  (n/N)*ysbar+30,  0.005)
fybarv= kybarfun(ybarv,n=n,N=N,ysbar=ysbar)/const
plot(ybarv,fybarv, type="l",xlim=c(0,20)) # check
(1/const)*integrate(function(ybar,n,N,ysbar) ybar*kybarfun(ybar,n,N,ysbar),
    (n/N)*ysbar,Inf,n=n,N=N,ysbar=ysbar)$value
                          #  1.714286 (predictive mean of ybar)


# (c)
c = 1 / ( 7/ysT^n  -  6/(ysT+1)^n  ); c # 0.9230769
flamfunc=function(lam,n,ysT,c) c*
  ( (7/ysT^n)*dgamma(lam,n,ysT) - (6/(ysT+1)^n)*dgamma(lam,n,ysT+1) )
integrate(flamfunc,0,Inf,n=n,ysT=ysT,c=c)$value # 1   check
lamv=seq(0,20,0.01)
plot(lamv,flamfunc(lamv,n=n,ysT=ysT,c=c),type="l")  # OK
fmufunc=function(mu,n,ysT,c)   c*(1/mu^2)*
  ( (7/ysT^n)*dgamma(1/mu,n,ysT) - (6/(ysT+1)^n)*dgamma(1/mu,n,ysT+1) )

integrate(fmufunc,0,Inf,n=n,ysT=ysT,c=c)$value # 1   check
integrate(function(mu,n,ysT,c) mu*fmufunc(mu,n,ysT,c),
   0,Inf,n=n,ysT=ysT,c=c)$value  # 1.384615    (posterior mean of mu)
fmuvc=fmufunc(mu=muv,n=n,ysT=ysT,c); plot(muv,fmuvc)  # OK

ybarmin=ysT/N; ybarmin # 0.2857143   Minimum possible value of ybar
ybarcut=(ysT+1)/N; ybarcut # 0.4285714   Cut-point for ybar

f1fun=function(ybar,lam,n,N,m,ysT) (N / (7-6*exp(-lam))) *
        7*dgamma(N*ybar-ysT,m,lam)
f2fun=function(ybar,lam,n,N,m,ysT)  (N / (7-6*exp(-lam))) *
  (7*dgamma(N*ybar-ysT,m,lam)-6*exp(-lam)*dgamma(N*ybar-ysT-1,m,lam) )

# Check for particular values of lambda
lam=0.764 # (example in the range ybarmin to ybarcut)
p1 = integrate(f1fun, ybarmin,ybarcut, lam=lam,n=n,N=N,m=m,ysT=ysT)$value
p2 = integrate(f2fun, ybarcut, Inf,      lam=lam,n=n,N=N,m=m,ysT=ysT)$value
c(p1,p2,p1+p2) # 0.001921853 0.998078147 1.000000000  OK

lam=3.214  # (example in the range ybarcut to infinity)
p1 = integrate(f1fun, ybarmin,ybarcut, lam=lam,n=n,N=N,m=m,ysT=ysT)$value
p2 = integrate(f2fun, ybarcut, Inf,      lam=lam,n=n,N=N,m=m,ysT=ysT)$value
c(p1,p2,p1+p2) # 0.2298026 0.7701974 1.0000000   OK
```

448

```
g1fun=function(ybar,n,N,m,ysT,c)
   integrate(function(lam,ybar,n,N,m,ysT,c)
            f1fun(ybar,lam,n,N,m,ysT)*flamfunc(lam,n,ysT,c),
            0,Inf, ybar=ybar, n=n,N=N,m=m,ysT=ysT,c=c)$value
g2fun=function(ybar,n,N,m,ysT,c)
   integrate(function(lam,ybar,n,N,m,ysT,c)
            f2fun(ybar,lam,n,N,m,ysT)*flamfunc(lam,n,ysT,c),
            0,Inf, ybar=ybar, n=n,N=N,m=m,ysT=ysT,c=c)$value

# Check:
g1fun(ybar=0.4,n,N,m,ysT,c)  # 0.4119163   OK
g2fun(ybar=0.6,n,N,m,ysT,c)  # 1.274185    OK


ybarv1=seq(ybarmin,ybarcut,length.out=400); fybarv1=ybarv1
for(j in 1:length(ybarv1)) fybarv1[j] =
     g1fun(ybar=ybarv1[j],n=n,N=N,m=m,ysT=ysT,c=c)


ybarv2=c( seq(ybarcut,1,length.out=200), seq(1,2,length.out=200),
          seq(2,3,length.out=200), seq(3,5,length.out=200),
          seq(5,10,length.out=200), seq(10,50,length.out=200) ,
          seq(50,1000,length.out=200), seq(1000,10000,length.out=200)   )
fybarv2=ybarv2
for(j in 1:length(ybarv2)) fybarv2[j] =
     g2fun(ybar=ybarv2[j],n=n,N=N,m=m,ysT=ysT,c=c)


plot(c(0,5),c(0,1.5),type="n")
lines(ybarv1, fybarv1,lty=1,lwd=2)
lines(ybarv2, fybarv2,lty=1,lwd=2)  # OK


# Check
INTEG <- function(xvec, yvec, a = min(xvec), b = max(xvec)){
# Integrates numerically under a spline through the points given by
# the vectors xvec and yvec, from a to b.
fit <- smooth.spline(xvec, yvec); spline.f <- function(x){predict(fit, x)$y }
integrate(spline.f, a, b)$value   }
INTEG(seq(0,1,0.01),seq(0,1,0.01)^2,0,1)  # 0.3333333    check


prob1=INTEG(ybarv1,fybarv1,ybarmin,ybarcut)
prob2=INTEG(ybarv2,fybarv2,ybarcut,10000)
c(prob1,prob2,prob1+prob2)  # 0.02880659 0.97119399 1.00000058  OK
INTEG(c(ybarv1,ybarv2),c(fybarv1,fybarv2),ybarmin,10000) # 1.000004  OK
```

```
X11(w=8,h=6); par(mfrow=c(2,1))
plot(ybarv1, ybarv1* fybarv1, xlim=c(0,1)) # OK
plot(ybarv2, ybarv2* fybarv2, xlim=c(0,20)) # OK


term1 = INTEG(ybarv1, ybarv1*fybarv1,ybarmin,ybarcut)
term2 = INTEG(ybarv2, ybarv2*fybarv2,ybarcut,10000)
ybarhatc = term1 + term2; c(term1, term2, ybarhatc)
   # 0.01139601 1.17546200 1.18685801   (predictive mean of ybar)


X11(w=8,h=8); par(mfrow=c(1,1))  # Produce final plots
plot(c(0,5),c(0,1.3),type="n",xlab="mu & ybar",
    ylab="posterior & predictive density")
lines(muv,fmuv,lty=4,lwd=3,col="green") # mu under SRS
lines(muv,fmuvc,lty=2,lwd=3, col="red") # mu under length-biased sampling
lines(ybarv,fybarv, lty=3,lwd=3, col="blue") # ybar under SRS
lines(ybarv1, fybarv1,lty=1,lwd=3); lines(ybarv2, fybarv2,lty=1,lwd=3)
abline(v=(n/N)*ysbar,lty=3); (n/N)*ysbar # 0.2857143
legend(2,1.3,c("f(mu|D) under SRSWR in (a)",
        "f(mu|D) under length-biased sampling in (c)",
        "f(ybar|D) under SRSWR in (a)",
        "f(ybar|D) under length-biased sampling in (c)"),
        lty=c(4,2,3,1),lwd=rep(3,4),col=c("green","red","blue","black"))
text(3.5,0.75,"The dotted vertical line shows the minimum possible")
text(3.5,0.68," value of ybar which is (n*ysbar+0)/N = 0.286")
```

Exercise 9.6 A Gibbs sampler for solving a length-biased with-replacement model

Consider the Bayesian model in part (c) of Exercise 9.5, namely:

$$f(I \mid y, \lambda) = \prod_{i=1}^{N} f(I_i \mid y, \lambda) = \prod_{i=1}^{N} \pi_i^{I_i} (1 - \pi_i)^{1 - I_i}, \qquad n = \sum_{i=1}^{N} I_i$$

$$f(y \mid \lambda) = \prod_{i=1}^{N} \lambda e^{-\lambda y_i}, \qquad f(\lambda) \propto 1 / \lambda, \lambda > 0,$$

where: $N = 7$, $\pi_i = 0.3 \ \forall i = 1, 2, 3, 5, 6, ..., N$

$\pi_4 = 0.3$ if $y_4 < 1$ and $\pi_4 = 0.9$ if $y_4 > 1$

$D = (I, y_s) = ((0,0,1,0,1,0,0),(1.6,0.4))$, $n = 2$, $m = N - n = 5$.

Design and implement a suitable Gibbs sampler so as to obtain a random sample from the joint distribution of $\mu = 1 / \lambda$ and $\bar{y}$. Illustrate your results with suitable plots and estimates.

Solution to Exercise 9.6

Motivated by and using results in the previous exercise (Exercise 9.5), define $y_0 = y_{rT} - y_4$ and then note that at the observed value of the data, the Bayesian model implies that:

$$f(I \mid y_s, y_0, y_4, \lambda) \propto 7 - 6I(y_4 > 1)$$

$$f(y_4 \mid y_s, y_0, \lambda) \sim G(1, \lambda)$$

$$f(y_0 \mid y_s, \lambda) \sim G(m-1, \lambda)$$

$$f(y_s \mid \lambda) \propto \prod_{i=1}^{n} \lambda e^{-\lambda y_i}$$

$$f(\lambda) \propto 1/\lambda, \lambda > 0.$$

So

$$f(I, y_s, y_0, y_4, \lambda) \propto \frac{1}{\lambda} \times \left(\prod_{i=1}^{n} \lambda e^{-\lambda y_i} \right) \times \left(\lambda^{m-1} y_0^{m-2} e^{-\lambda y_0} \right)$$

$$\times \lambda e^{-\lambda y_4} \times \left[7 - 6I(y_4 > 1) \right].$$

We see that a suitable Gibbs sampler is defined by the following three conditionals:

1. $f(\lambda \mid I, y_s, y_0, y_4) \propto \lambda^{-1+n+m-1+1} e^{\lambda(y_{sT} + y_0 + y_4)} = \lambda^{N-1} e^{-\lambda y_T}$

$$\Rightarrow (\lambda \mid I, y_s, y_0, y_4) \sim G(N, y_T) \sim G(N, y_{sT} + y_0 + y_4)$$

2. $f(y_0 \mid I, y_s, \lambda, y_4) \propto y_0^{m-2} e^{-\lambda y_0}$

$$\Rightarrow (y_0 \mid I, y_s, \lambda, y_4) \sim G(m-1, \lambda)$$

3. $f(y_4 \mid I, y_s, \lambda, y_0) \propto \left[7 - 6I(y_4 > 1) \right] \lambda e^{-\lambda y_4}, \quad y_4 > 0.$

The first of these three conditionals are straightforward and easy to sample from. The third conditional can be sampled from via the inversion technique as follows.

First, for notational convenience, write the relevant random variable as x with density

$$f(x) \propto \left[7 - 6I(x > 1) \right] \lambda e^{-\lambda x}, \quad x > 0.$$

Then the cdf of x is

$$F(x) = r \begin{cases} 7\int_0^x \lambda e^{-\lambda t}dt, & 0 < x < 1 \\ 7\int_0^x \lambda e^{-\lambda t}dt - 6\int_1^x \lambda e^{-\lambda t}dt, & x > 1 \end{cases}$$

for some constant r

$$= r \begin{cases} 7(1-e^{-\lambda x}), & 0 < x < 1 \\ 7(1-e^{-\lambda x}) - 6(e^{-\lambda 1} - e^{-\lambda x}), & x > 1 \end{cases}$$

$$= r \begin{cases} 7 - 7e^{-\lambda x}, & 0 < x < 1 \\ 7 - e^{-\lambda x} - 6e^{-\lambda}, & x > 1 \end{cases},$$

which equals $1 = 7 - 0 - 6e^{-\lambda}$ in the limit as $x \to \infty$; so $r = 1/(7 - 6e^{-\lambda})$.

Now observe that $F(x=1) = \dfrac{7-7e^{-\lambda}}{7-6e^{-\lambda}}$.

This is a constant in the formula for the quantile function of X, obtained as follows.

First, if $p < \dfrac{7-7e^{-\lambda}}{7-6e^{-\lambda}}$ then we solve $p = r(7 - 7e^{-x\lambda})$

and thereby obtain $x = -\dfrac{1}{\lambda}\log\left(1 - \dfrac{p}{7r}\right)$.

Secondly, if $p > \dfrac{7-7e^{-\lambda}}{7-6e^{-\lambda}}$ then we solve $p = r(7 - e^{-\lambda x} - 6e^{-\lambda})$

and thereby obtain $x = -\dfrac{1}{\lambda}\log\left(7 - 6e^{-\lambda} - \dfrac{p}{r}\right)$.

In summary, the quantile function of x is given by

$$Q(p) = \begin{cases} x = -\dfrac{1}{\lambda}\log\left(1 - \dfrac{p}{7}\left(7-6e^{-\lambda}\right)\right), & p < \dfrac{7-7e^{-\lambda}}{7-6e^{-\lambda}} \\ x = -\dfrac{1}{\lambda}\log\left(7 - 6e^{-\lambda} - p\left(7-6e^{-\lambda}\right)\right), & p > \dfrac{7-7e^{-\lambda}}{7-6e^{-\lambda}} \end{cases}. \quad (9.2)$$

So a procedure for sampling from the third conditional in the Gibbs sampler, namely

$$f(y_4 \mid I, y_s, \lambda, y_0) \propto [7 - 6I(y_4 > 1)] \lambda e^{-\lambda y_4},$$

is to draw $u \sim U(0,1)$ and then return $y_4 = Q(u)$ as per equation (9.2).

Implementing the above Gibbs sampler for 20,000 iterations following a burn-in of 1,000 and then thinning out by a factor of 10 we obtained a random sample of size $J = 2,000$ from the joint posterior/predictive distribution of $f(\lambda, y_0, y_4 \mid I, y_s)$.

Figure 9.4 displays trace plots for the three unknowns, λ, y_0, y_4, sample ACFs for these over the last 20,000 iterations, and the three sample ACFs again over the final samples of size J. Figure 9.5 is a histogram of the J simulated values of $\mu = 1/\lambda$ and Figure 9.6 is a histogram of the J simulated values of $\overline{y} = (y_{sT} + y_0 + y_4)/N$. In each histogram are shown a density estimate as well as three vertical lines for the Monte Carlo point estimate and 95% CI for the mean.

The posterior density of μ, i.e. $f(\mu \mid D)$, was estimated via Rao-Blackwell as

$$\hat{f}(\mu \mid D) = \frac{1}{J} \sum_{j=1}^{J} f_{IG(N, N\overline{y}^{(j)})}(\mu),$$

where

$$\overline{y}^{(j)} = (y_{sT} + y_0^{(j)} + y_4^{(j)})/N,$$

using the fact that

$$(\mu \mid I, y_s, y_0, y_4) \sim IG(N, y_{sT} + y_0 + y_4) \sim IG(N, y_T) \sim IG(N, N\overline{y}).$$

The posterior mean of μ, i.e. $E(\mu \mid D)$, was also estimated via Rao-Blackwell as

$$\hat{\mu} = \frac{1}{J} \sum_{j=1}^{J} \frac{y_{sT} + y_0^{(j)} + y_4^{(j)}}{N-1} = \frac{1}{J} \sum_{j=1}^{J} \frac{N\overline{y}^{(j)}}{N-1} = 1.41,$$

using the fact that

$$E(\mu \mid I, y_s, y_0, y_4) = (y_{sT} + y_0 + y_4)/(N-1),$$

with 95% CI for the posterior mean equal to

$$\left(\hat{\mu} \pm 1.96 \sqrt{\frac{1}{J}} \sqrt{\frac{1}{J-1} \sum_{j=1}^{J} \left(\frac{N\overline{y}^{(j)}}{N-1} - \hat{\mu} \right)^2} \right) = (1.34, 1.47).$$

Note: This is consistent with the exact value, namely $E(\mu\,|\,D) = 1.38$, as obtained in Exercise 9.5.

The predictive density of $\bar{y}$ was estimated by smoothing a probability histogram of the simulated values $\bar{y}^{(j)}$, and the predictive mean of $\bar{y}$, i.e. $E(\bar{y}\,|\,D)$, was estimated by

$$\hat{\bar{y}} = \frac{1}{J}\sum_{j=1}^{J}\bar{y}^{(j)} = 1.21,$$

with 95% CI

$$\left(\hat{\bar{y}}\pm 1.96\sqrt{\frac{1}{J}}\sqrt{\frac{1}{J-1}\sum_{j=1}^{J}\left(\bar{y}^{(j)}-\hat{\bar{y}}\right)^2}\right) = (1.15,\ 1.26).$$

Note 1: This is consistent with the exact value, $E(\bar{y}\,|\,D) = 1.19$, as obtained in Exercise 9.5.

Note 2: We may be able to improve on the above 'histogram' estimation of $E(\bar{y}\,|\,D)$ using Rao-Blackwell methods. For example, observe that

$$E(\bar{y}\,|\,D,\lambda,y_4) = \frac{1}{N}\left(y_{sT} + y_4 + \frac{m-1}{\lambda}\right).$$

So we define

$$e_j = E(\bar{y}\,|\,D,\lambda_j,y_4^{(j)}) = \frac{1}{N}\left(y_{sT} + y_4^{(j)} + \frac{m-1}{\lambda_j}\right).$$

The associated Rao-Blackwell estimate of $E(\bar{y}\,|\,D)$ is

$$\bar{e} = \frac{1}{J}\sum_{j=1}^{J}e_j = 1.21,$$

with 95% CI

$$\left(\bar{e}\pm 1.96\sqrt{\frac{1}{J}}\sqrt{\frac{1}{J-1}\sum_{j=1}^{J}(e_j-\bar{e})^2}\right) = (1.16,\ 1.26).$$

Note 3: In this case, applying Rao-Blackwell methods has only *slightly* narrowed the CI for $E(\bar{y}\,|\,D)$.

Figure 9.4 Trace plots and sample ACFs

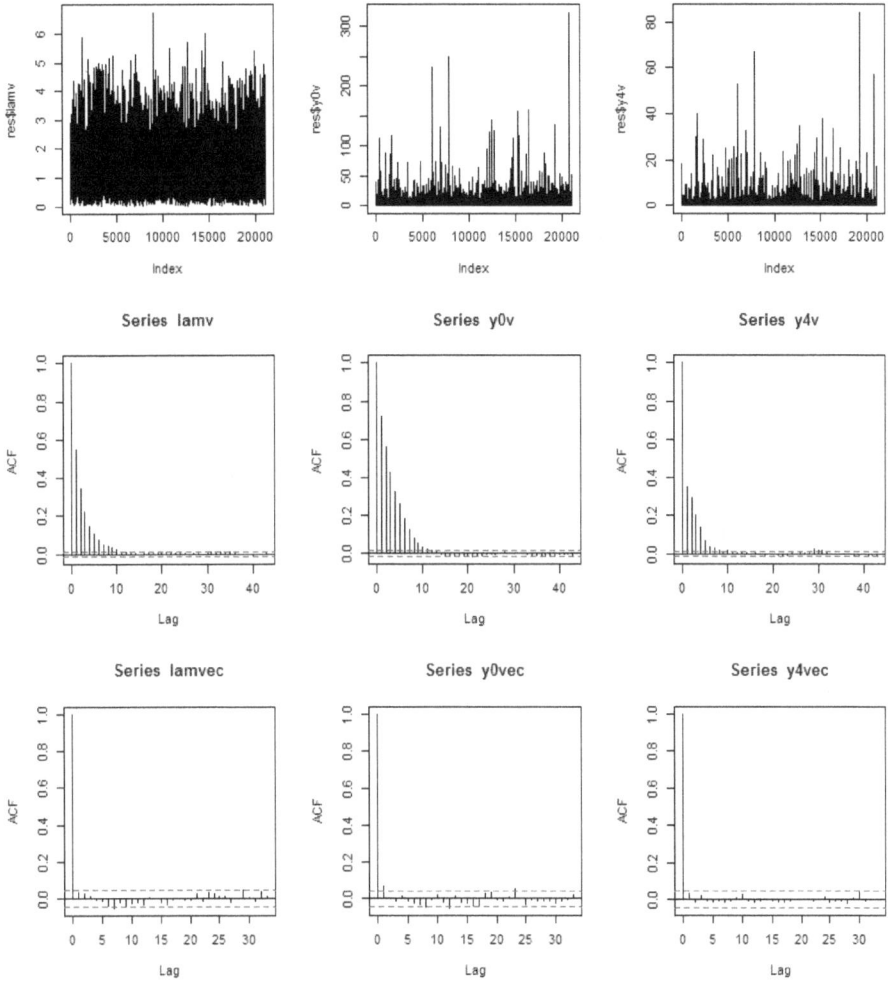

Figure 9.5 Inference on the superpopulation mean

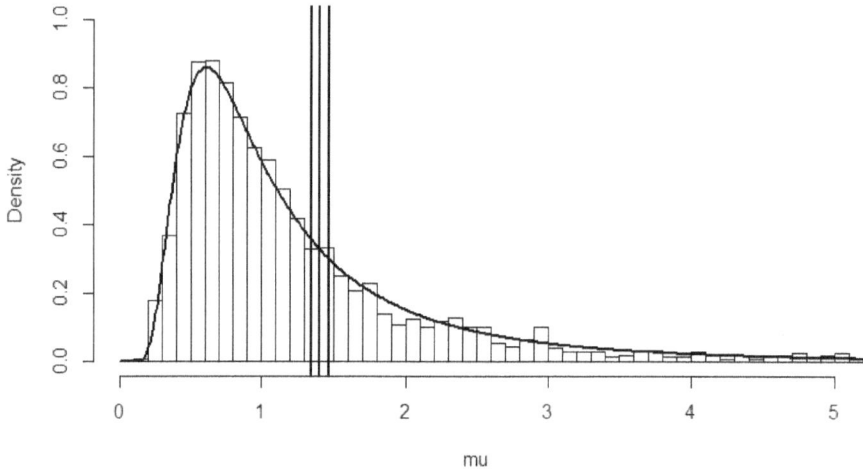

Figure 9.6 Inference on the finite population mean

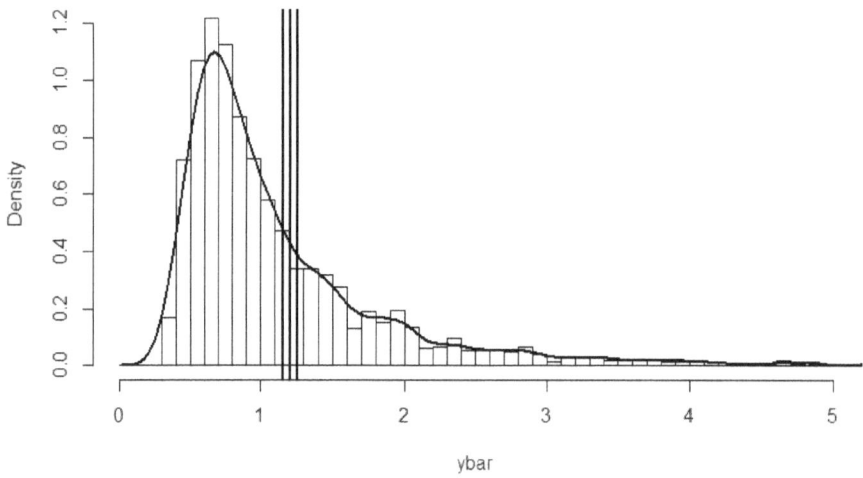

R Code for Exercise 9.6

```
Qfun = function(p=0.5,lam=1){
        c1 = (7-7*exp(-lam))/(7-6*exp(-lam))
        if(p <= c1) c2 = 1- (p/7) * (7-6*exp(-lam))
        if(p > c1) c2 = 7 - 6*exp(-lam) - p*(7-6*exp(-lam))
        -(1/lam)*log(c2)    }
```

```
# Check:
pvec=seq(0,1,0.001); Qvec=pvec
for(i in 1:length(pvec)) Qvec[i] = Qfun(p=pvec[i],lam=1.3)
plot(pvec,Qvec); plot(Qvec,pvec)  # OK
```

```
GS = function(J=1000,N=7,n=2,m=5, ysT=2, lam=1,y0=1,y4=1){
lamv=lam; y0v=y0; y4v=y4;  for(j in 1:J){
                lam=rgamma(1,N,ysT+y0+y4)
                y0=rgamma(1,m-1,lam)
                u=runif(1); y4=Qfun(p=u,lam=lam)
                lamv=c(lamv,lam); y0v=c(y0v,y0); y4v=c(y4v,y4)  }
list(lamv=lamv, y0v=y0v, y4v=y4v)  }
```

```
X11(w=8,h=9); par(mfrow=c(3,3));  set.seed(321); date()
res= GS(J=21000,N=7,n=2,m=5, ysT=2, lam=1,y0=1,y4=1); date()  # took 3 secs
plot(res$lamv,type="l"); plot(res$y0v,type="l"); plot(res$y4v,type="l")  # OK
```

```
lamv=res$lamv[-(1:1001)]; y0v=res$y0v[-(1:1001)]; y4v=res$y4v[-(1:1001)];
acf(lamv); acf(y0v); acf(y4v) # high serial correlation, so need to thin out
inc= seq(10,20000,10); lamvec=lamv[inc]; y0vec=y0v[inc]; y4vec=y4v[inc];
acf(lamvec); acf(y0vec); acf(y4vec)  # OK
J = length(lamvec); J # 2000
```

```
N=7;n=2;m=5; ysT=2; muvec=1/lamvec; ybarvec=(1/N)*(ysT+y0vec+y4vec)
ybarhat=mean(ybarvec);
ybarci=ybarhat+c(-1,1)*qnorm(0.975)*sd(ybarvec)/sqrt(J)
c(ybarhat, ybarci, ybarci[2]-ybarci[1]) # 1.204519 1.151619 1.257419 0.105800
```

```
evec=(1/N)*( ysT+ y4vec + (m-1)/lamvec )
ebar=mean(evec); eci= ebar+c(-1,1)*qnorm(0.975)*sd(evec)/sqrt(J)
c(ebar,eci,eci[2]-eci[1]) # 1.2091236 1.1581903 1.2600569 0.1018666
```

```
muhat=(N/(N-1))*ybarhat
muci=muhat + c(-1,1)*qnorm(0.975)*sd( (N/(N-1))*ybarvec ) / sqrt(J)
c(muhat, muci) # 1.405272 1.343556 1.466989
```

```
mugrid=seq(0.001,10.001,0.01)
fmuhat=mugrid; for(i in 1:length(mugrid))
        fmuhat[i] = mean( dgamma(1/mugrid[i], N, N*ybarvec )/mugrid[i]^2 )

X11(w=8,h=5)

hist(muvec,prob=T, xlim=c(0,5),ylim=c(0,1),breaks=seq(0,80,0.1),
        xlab="mu", main="")
lines(mugrid,fmuhat,lwd=2); abline(v= c(muhat, muci), lwd=2)

hist(ybarvec,prob=T, xlim=c(0,5),ylim=c(0,1.2),breaks=seq(0,80,0.1),
        xlab="ybar", main=" ")
lines(density(ybarvec),lwd=2); abline(v= c(ybarhat, ybarci), lwd=2)
```

Exercise 9.7 Gibbs sampler for a length-biased *without-replacement* sampling model

Earlier we defined $L = (L_1, ..., L_n)$ as the vector of the labels of the selected units *in the order in which they are sampled.*

Now consider the following Bayesian finite population model:

$$f(L \mid y, \lambda) = \prod_{i=1}^{n} \frac{y_{L_i}}{y_T - \sum_{j=1}^{i-1} y_{L_j}}, \quad L = (L_1, ..., L_n) \in \{(a_1, ..., a_n):$$
$$a_i \in \{1, ..., N\} \ \forall \ i \in \{1, ..., n\} \ \& \ a_i \neq a_j \ \forall \ i, j \in \{1, ..., n\}\}$$

$$f(y \mid \lambda) = \prod_{i=1}^{N} \lambda e^{-\lambda y_i}, \ y_i > 0 \ \forall \ i$$

$$f(\lambda) \propto 1 / \lambda, \lambda > 0.$$

Design and implement a suitable Gibbs sampler so as to obtain a random sample from the joint distribution of $\mu = 1 / \lambda$ and $\bar{y}$ in the case where
$$N = 7, \ n = 3, \ m = N - n = 4$$
and when the observed data is
$$D = (L, y_s) = ((4, 3, 6), (1.6, 0.4, 0.7)).$$

Illustrate your results with suitable plots and estimates.

Solution to Exercise 9.7

The sampling mechanism here is defined by the model density of L, which may also be written as

$$f(L_1,...,L_n \mid y, \lambda) = \frac{y_{L_1}}{y_T} \times \frac{y_{L_2}}{y_T - y_{L_1}} \times \frac{y_{L_3}}{y_T - y_{L_1} - y_{L_2}} \times ... \times \frac{y_{L_n}}{y_T - y_{L_1} - y_{L_2} - ... - y_{L_{n-1}}}$$

$$\text{for } L = (1,...,n), (1,3,2,...,n),...,(N, N-1,...,N-n+1).$$

This pdf implies that units are selected from the finite population, one by one and without replacement, in such a way that the probability of selecting a unit on any given draw is its value divided by the sum of the values of all units which have not yet been sampled at that point in time. We call this procedure ***length-biased sampling without replacement***.

Note: This is an example of a sampling mechanism that is nonignorable but ***known***. If $f(L \mid y, \lambda)$ depended on λ, or on some other unknown quantity, then we would say that the sampling mechanism is nonignorable and ***unknown***.

In the present case it is convenient to *relabel* the population units—after sampling—in such a way that $L = (1, 2,...,n)$ and so also $s = (1,...,n)$ and $r = (n+1,...,N)$. Assuming that this is done, we may write the density of the sampling mechanism in various other and simpler ways, for example:

$$f(L \mid y, \lambda) = \frac{y_1}{y_T} \times \frac{y_2}{y_T - y_1} \times \frac{y_3}{y_T - y_1 - y_2} \times ... \times \frac{y_n}{y_T - y_1 - y_2 - ... - y_{n-1}}$$

$$= \frac{y_1}{y_1 + ... + y_N} \times \frac{y_2}{y_2 + ... + y_N} \times \frac{y_3}{y_3 + ... + y_N} \times ... \times \frac{y_n}{y_n + ... + y_N}$$

$$= \prod_{i=1}^{n} \frac{y_i}{y_i + ... + y_N} = \prod_{i=1}^{n} \frac{y_i}{\sum_{j=i}^{N} y_j}, \text{ etc.}$$

Note: We have not previously relabelled population units in this manner because doing so would have provided only marginal notational convenience and may have obscured the nature of the sampling mechanisms we were trying to illustrate. In the next chapter, we will again make use of a convenient relabelling scheme similar to the one applied here.

With the above relabelling in place, and noting that

$$(y_{rT} \mid \lambda) \sim G(m, \lambda),$$

the joint posterior density of λ and y_{rT} (given the data, $D = (L, y_s)$) may now be written as

$$f(\lambda, y_{rT} \mid D) \propto f(\lambda, y_{rT}, y_s, L) = f(\lambda) f(y_s \mid \lambda) f(y_{rT} \mid \lambda) f(L \mid y_s, y_{rT})$$

$$\propto \frac{1}{\lambda} \times \left(\prod_{i=1}^{n} \lambda e^{-\lambda y_i} \right) \times \left(\lambda^m y_{rT}^{m-1} e^{-\lambda y_{rT}} \right) \times \prod_{i=1}^{n} \frac{1}{y_i + \ldots + y_n + y_{rT}}.$$

This joint density suggests a Metropolis-Hastings algorithm with a Gibbs step defined by the conditional posterior distribution

$$(\lambda \mid D, y_{rT}) \sim G(N, y_{sT} + y_{rT})$$

and a Metropolis step defined by a rather complicated conditional predictive density defined by

$$f(y_{rT} \mid D, \lambda) \propto y_{rT}^{m-1} e^{-\lambda y_{rT}} \prod_{i=1}^{n} \frac{1}{(y_i + \ldots + y_n) + y_{rT}}.$$

At this point it is useful to recall a ***data augmentation technique*** based on the identity

$$1 = \int_0^{\infty} x e^{-xw} dw,$$

or equivalently

$$\frac{1}{x} = \int_0^{\infty} e^{-xw} dw,$$

which can be applied here so as to yield the identity

$$\prod_{i=1}^{n} \frac{1}{y_i + \ldots + y_n + y_{rT}} = \prod_{i=1}^{n} \int_0^{\infty} e^{-(y_i + \ldots + y_n + y_{rT}) w_i} dw_i.$$

This suggests that we introduce an artificial or latent random variable $w = (w_1, \ldots, w_n)$ into our model which is defined in such a way that the joint posterior density of λ, y_{rT} and w is given by

$$f(\lambda, y_{rT}, w \mid D) \propto \frac{1}{\lambda} \times \left(\prod_{i=1}^{n} \lambda e^{-\lambda y_i} \right) \times \left(\lambda^m y_{rT}^{m-1} e^{-\lambda y_{rT}} \right) \times \prod_{i=1}^{n} e^{-(y_i + \ldots + y_n + y_{rT}) w_i}.$$

Note: If we integrate this joint density with respect to w then we recover $f(\lambda, y_{rT} \mid D)$ as above.

The above expression for $f(\lambda, y_{rT}, w \mid D)$ now suggests a 'pure' Gibbs sampler defined by the following $n + 2$ conditional distributions:

$$(\lambda \mid D, y_{rT}, w) \sim G(N, y_{sT} + y_{rT})$$
$$(y_{rT} \mid D, \lambda, w) \sim G(m, \lambda + w_T) \quad \text{where} \quad w_T = w_1 + \ldots + w_n$$
$$(w_i \mid D, \lambda, y_{rT}) \sim \perp G(1, y_i + \ldots + y_n + y_{rT}), \quad i = 1, \ldots, n.$$

This Gibbs sampler can be used to generate a random sample
$$(\lambda_j, y_{rT}^{(j)}, w^{(j)}) \sim iid \ f(\lambda, y_{rT}, w \mid D), j = 1, \ldots, J,$$
where
$$w^{(j)} = (w_1^{(j)}, \ldots, w_n^{(j)}).$$

This sample can then be used for Monte Carlo inference on the quantities of interest, namely $\mu = 1/\lambda$ and $\bar{y} = (y_{sT} + y_{rT})/N$.

Applying the above Gibbs sampler (with a suitable burn-in and thinning) we obtained a random sample of size $J = 2,000$ from the joint posterior distribution of λ, y_{rT} and $w = (w_1, \ldots, w_n)$.

The posterior density of μ was estimated via Rao-Blackwell as

$$\hat{f}(\mu \mid D) = \frac{1}{J} \sum_{j=1}^{J} f_{IG(N, N\bar{y}^{(j)})}(\mu),$$

where
$$\bar{y}^{(j)} = (y_{sT} + y_{rT}^{(j)})/N,$$
using the fact that
$$(\mu \mid I, y_s, y_{rT}, w) \sim IG(N, y_{sT} + y_{rT}) \sim IG(N, y_T) \sim IG(N, N\bar{y}).$$

The posterior mean of μ was also estimated via Rao-Blackwell as

$$\hat{\mu} = \frac{1}{J} \sum_{j=1}^{J} \frac{y_{sT} + y_{rT}^{(j)}}{N-1} = \frac{1}{J} \sum_{i=1}^{J} \frac{N\bar{y}^{(j)}}{N-1} = 0.619,$$

using the fact that
$$E(\mu \mid I, y_s, y_{rT}, w) = (y_{sT} + y_{rT})/(N-1),$$
with 95% CI

$$\left(\hat{\mu} \pm 1.96 \sqrt{\frac{1}{J}} \sqrt{\frac{1}{J-1} \sum_{j=1}^{J} \left(\frac{N\bar{y}^{(j)}}{N-1} - \hat{\mu} \right)^2} \right) = (0.614, 0.624).$$

The predictive density of y_{rT} was likewise estimated via Rao-Blackwell as

$$\hat{f}(y_{rT} \mid D) = \frac{1}{J} \sum_{j=1}^{J} f_{G(m, \lambda_j + w_T^{(j)})}(y_{rT}),$$

where

$$w_T^{(j)} = w_1^{(j)} + \dots + w_n^{(j)}.$$

The predictive mean of y_{rT} was also estimated via Rao-Blackwell as

$$\hat{y}_{rT} = \frac{1}{J} \sum_{j=1}^{J} \frac{m}{\lambda_j + w_T^{(j)}} = 1.013,$$

using the fact that

$$E(y_{rT} \mid I, y_s, \lambda, w) = m / (\lambda + w_T),$$

with 95% CI

$$\left(\hat{y}_{rT} \pm 1.96 \sqrt{\frac{1}{J}} \sqrt{\frac{1}{J-1} \sum_{j=1}^{J} \left(\frac{m}{\lambda_j + w_T^{(j)}} - \hat{y}_{rT} \right)^2} \right) = (0.993, \ 1.033).$$

These Rao-Blackwell estimates for y_{rT} were then transformed into estimates for $\bar{y}$ via the equation

$$\bar{y} = (y_{sT} + y_{rT}) / N.$$

In this way, we estimated $\bar{y}$'s posterior mean by 0.530, with 95% CI (0.614, 0.624).

Figure 9.7 shows trace plots for λ, y_{rT} and w_1, sample ACFs for these quantities over the last 10,000 iterations, and these three sample ACFs again but calculated using only the final smaller samples of size $J = 2,000$.

Figures 9.8 and 9.9 (page 464) show two histograms, of the J simulated values of $\mu = 1 / \lambda$, and of the J simulated values of $\bar{y} = (y_{sT} + y_{rT}) / N$. In each histogram are shown a density estimate and three vertical lines representing the Monte Carlo point estimate and 95% CI for the posterior mean.

Note 1: The type of sampling mechanism which features in this exercise has applications in the analysis of oil discovery data. For further details, see West (1996).

Note 2: In this chapter, we have presented several examples of how Bayesian methods can be used to perform inference on an exponential finite population under biased sampling. For another such example, see Puza and O'Neill (2005).

Figure 9.7 Trace plots and sample ACFs for samples obtained via MCMC

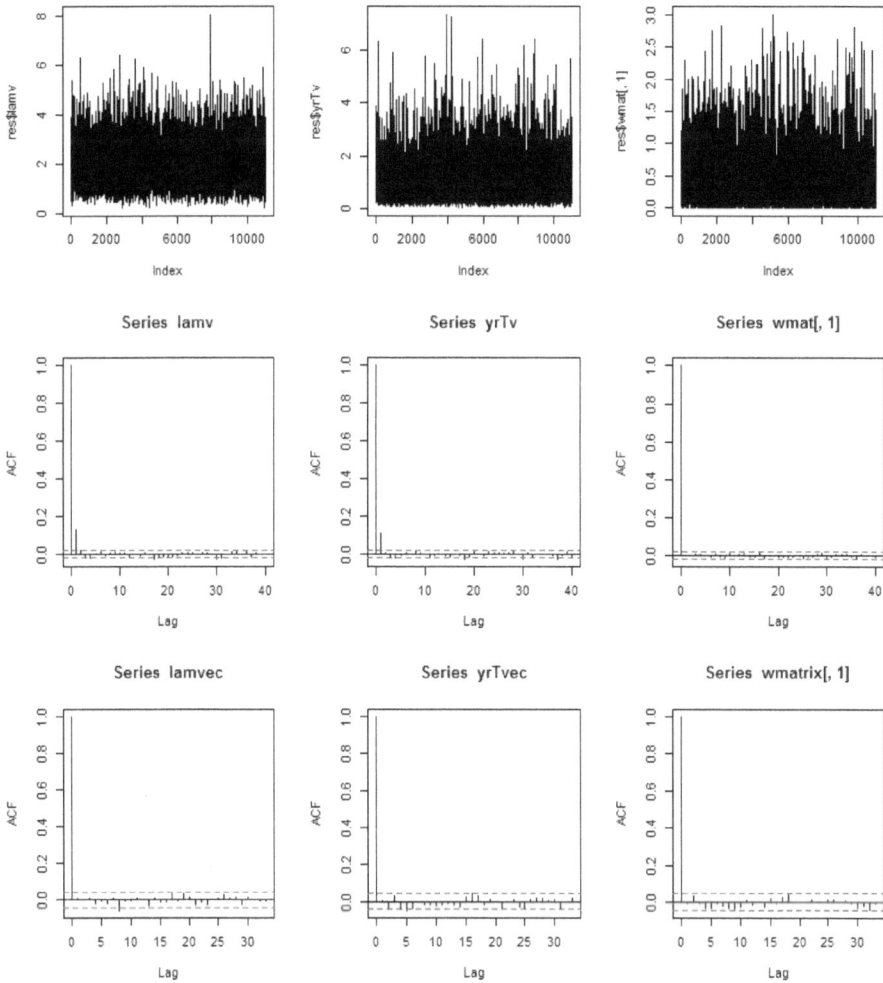

Figure 9.8 Inference on the superpopulation mean via MCMC

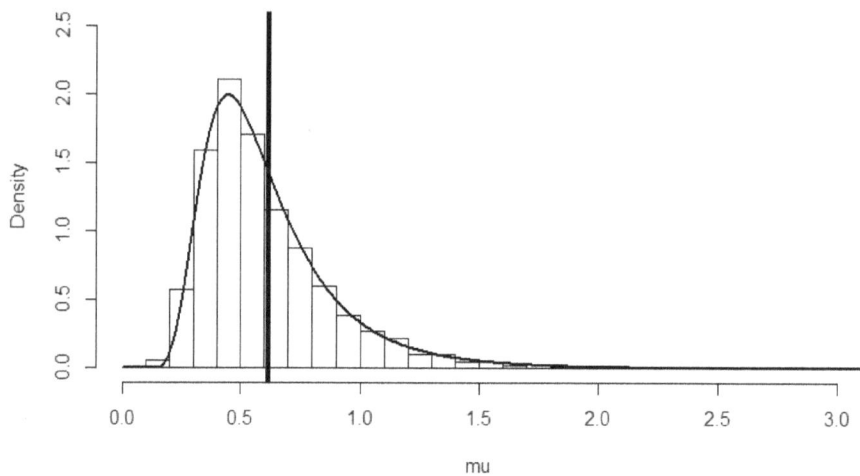

Figure 9.9 Inference on the finite population mean via MCMC

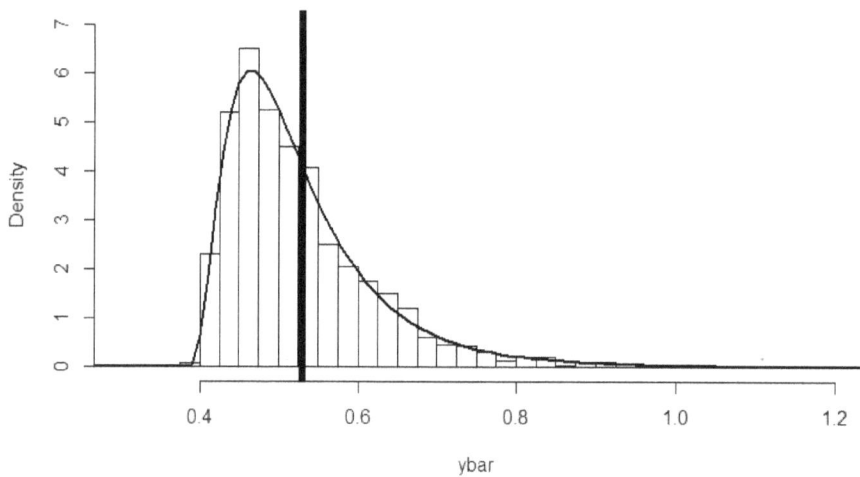

R Code for Exercise 9.7

```
GS = function(J=1000,N=7,n=3,m=4, ys=c(1.6,0.4,0.7),
          lam=1,yrT=1,w=rep(1,3)){
ysT=sum(ys); lamv=lam; yrTv=yrT; wmat=w;  for(j in 1:J){
              lam=rgamma(1,N,ysT+yrT);
              yrT=rgamma(1,m,lam+sum(w))
              for(i in 1:n) w[i] = rgamma(1,1,sum(ys[i:n]))
              lamv=c(lamv,lam); yrTv=c(yrTv,yrT); wmat=rbind(wmat,w)
              }
list(lamv=lamv, yrTv=yrTv, wmat=wmat)
}

set.seed(321); date()
res=GS(J=11000,N=7,n=3,m=4, ys=c(1.6,0.4,0.7), lam=1,yrT=1,w=rep(1,3))
date()  # took 4 secs

X11(w=8,h=9); par(mfrow=c(3,3));

plot(res$lamv,type="l"); plot(res$yrTv,type="l"); plot(res$wmat[,1],type="l")

lamv=res$lamv[-(1:1001)]; yrTv=res$yrTv[-(1:1001)];
wmat=res$wmat[-(1:1001),]
acf(lamv); acf(yrTv); acf(wmat[,1]) #
inc= seq(5,10000,5); lamvec=lamv[inc]; yrTvec=yrTv[inc]; wmatrix=wmat[inc,];
acf(lamvec); acf(yrTvec); acf(wmatrix[,1])  # OK
J = length(lamvec); J # 2000

N=7;n=3;m=4; ys=c(1.6,0.4,0.7); ysT=sum(ys);
muvec=1/lamvec; ybarvec=(1/N)*(ysT+yrTvec)
wTvec=apply(wmatrix,1,sum)
yrThat=mean(m/(lamvec+wTvec))
yrTci=yrThat+c(-1,1)*qnorm(0.975)*sd(m/(lamvec+wTvec))/sqrt(J)
c(yrThat,yrTci) # 1.0131279 0.9930648 1.0331911
ybarhat=(1/N)*(ysT+yrThat)
ybarci=(1/N)*(ysT+yrTci)
c(ybarhat,ybarci) # 0.5304468 0.5275807 0.5333130

muhat=(N/(N-1))*ybarhat
muci=muhat + c(-1,1)*qnorm(0.975)*sd( (N/(N-1))*ybarvec ) / sqrt(J)
c(muhat, muci) # 0.6188547 0.6136692 0.6240401
```

```
mugrid=seq(0.001,10.001,0.01)
fmuhat=mugrid; for(i in 1:length(mugrid))
        fmuhat[i] = mean( dgamma(1/mugrid[i], N, N*ybarvec )/mugrid[i]^2 )

ybargrid=seq(0,10,0.01)
fybarhat= ybargrid; for(i in 1:length(ybargrid))
   fybarhat[i] = mean( dgamma(N*ybargrid[i]-ysT, m, lamvec+wTvec )*N  )

X11(w=8,h=5); par(mfrow=c(1,1))

hist(muvec,prob=T, xlim=c(0,3),ylim=c(0,2.5),breaks=seq(0,80,0.1),
        xlab="mu", main="")
lines(mugrid,fmuhat,lwd=2); abline(v= c(muhat, muci), lwd=2)

hist(ybarvec,prob=T, xlim=c(0.3,1.2),ylim=c(0,7),breaks=seq(0,80,0.025),
        xlab="ybar", main="")
lines(ybargrid, fybarhat,lwd=2); abline(v= c(ybarhat, ybarci), lwd=2)
```

CHAPTER 10

Normal Finite Population Models

10.1 The basic normal-normal finite population model

Consider a finite population of N values $y_1,...,y_N$ from the normal distribution with unknown mean μ and known variance σ^2. Assume we have prior information about μ which may be expressed in terms of a normal distribution with mean μ_0 and variance σ_0^2.

Suppose that we are interested in the finite population mean, namely $\overline{y} = (y_1 + ... + y_N)/N$, and wish to perform inference on $\overline{y}$ based on the observed values in a sample of size n taken from this finite population via simple random sampling without replacement (SRSWOR).

For convenience, we will in what follows label (or rather relabel) the n sample units as $1,...,n$ and the $m = N - n$ nonsample units as $n+1,...,N$. This convention simplifies notation and allows us to write the finite population vector, originally defined by $y = (y_1,,...,y_N)$, as

$$y = ((y_1,...,y_n),(y_{n+1},...,y_N)) = (y_s, y_r).$$

Example: Suppose that we sample units 2, 3 and 5 from a finite population of size 7. Then we change the labels of units 2, 3 and 5 to 1, 2 and 3, respectively, and we change the labels of units 1, 4, 6 and 7 to 4, 5, 6 and 7, respectively.

Thereby, instead of writing $y_s = (y_2, y_3, y_5)$ and $y_r = (y_1, y_4, y_6, y_7)$, we may write $y_s = (y_1, y_2, y_3)$ and $y_r = (y_4, y_5, y_6, y_7)$, respectively.

We will also implicitly condition on $s = (s_1,...,s_n)$ at its fixed value and suppress s from much of the notation. Thus we will sometimes write $f(\overline{y}|s, y_s)$ as $f(\overline{y}|y_s)$, with an understanding that s refers to the particular units which were actually sampled.

Our inferential problem may be thought of as prediction of $\bar{y}_r$ given the data, y_s (and s), since $\bar{y} = (y_{sT} + m\bar{y}_r)/N$. Considering the various distributions that are involved, a suitable Bayesian model is:

$$(\bar{y}_r \mid y_s, \mu) \sim N(\mu, \sigma^2/m)$$

(the model distribution of the nonsample mean)

$$(y_1, ..., y_n \mid \mu) \sim N(\mu, \sigma^2)$$

(the model distribution of the sample values)

$$\mu \sim N(\mu_0, \sigma_0^2) \qquad \text{(the prior distribution)}.$$

This model will be called the *basic normal-normal finite population model*. By results for the normal-normal model reported earlier, we see that the posterior distribution of the superpopulation mean is given by

$$(\mu \mid y_s) \sim N(\mu_*, \sigma_*^2),$$

where: $\mu_* = (1-k)\mu_0 + k\bar{y}_s$ (the posterior mean as a credibility estimate)

$$\sigma_*^2 = k\frac{\sigma^2}{n} \quad \text{(the posterior variance)}, \quad k = \frac{n}{n + \sigma^2/\sigma_0^2}$$

(the credibility factor and weight given to the MLE, $\bar{y}_s$).

It will be recalled that in this context the predictive density of the nonsample mean is

$$f(\bar{y}_r \mid y_s) = \int f(\bar{y}_r \mid y_s, \mu) f(\mu \mid y_s) d\mu.$$

But this is the integral of the exponent of a quadratic equation in μ and $\bar{y}_r$, and so equals the exponent of a quadratic equation in $\bar{y}_r$. It follows that

$$(\bar{y}_r \mid y_s) \sim N(a, b^2),$$

where: $a = E(\bar{y}_r \mid y_s) = E\{E(\bar{y}_r \mid y_s, \mu) \mid y_s\} = E\{\mu \mid y_s\} = \mu_*$

$b^2 = V(\bar{y}_r \mid y_s) = V\{E(\bar{y}_r \mid y_s, \mu) \mid y_s\} + E\{V(\bar{y}_r \mid y_s, \mu) \mid y_s\}$

$$= V\{\mu \mid y_s\} + E\left\{\frac{\sigma^2}{m} \middle| y_s\right\} = \sigma_*^2 + \frac{\sigma^2}{m}.$$

It follows that $(\bar{y} \mid y_s) \sim N(c, d^2)$, where:

$$c = E(\overline{y} \mid y_s) = E\left(\left.\frac{n\overline{y}_s + m\overline{y}_r}{N}\right| y_s\right) = \frac{n\overline{y}_s + mE(\overline{y}_r \mid y_s)}{N}$$

$$= \frac{n\overline{y}_s + ma}{N} = \frac{n\overline{y}_s + m\mu_*}{N}$$

$$d^2 = V(\overline{y} \mid y_s) = V\left(\left.\frac{n\overline{y}_s + m\overline{y}_r}{N}\right| y_s\right) = \left(\frac{m}{N}\right)^2 V(\overline{y}_r \mid y_s)$$

$$= \frac{m^2}{N^2}b^2 = \frac{m^2}{N^2}\left(\sigma_*^2 + \frac{\sigma^2}{m}\right).$$

Then, the $1-\alpha$ central predictive density region (CPDR) for $\overline{y}$ is given by $(c \pm z_{\alpha/2}d)$.

Summary: For the *basic normal-normal finite population model*:

$$(\overline{y}_r \mid y_s, \mu) \sim N\left(\mu, \frac{\sigma^2}{N-n}\right)$$

$$(y_1,...,y_n \mid \mu) \sim iid\ N(\mu, \sigma^2),\quad \mu \sim N(\mu_0, \sigma_0^2),$$

the posterior distribution of the superpopulation mean μ is given by

$$(\mu \mid y_s) \sim N(\mu_*, \sigma_*^2),$$

where: $\mu_* = (1-k)\mu_0 + k\overline{y}_s$, $\sigma_*^2 = k\dfrac{\sigma^2}{n}$, $k = \dfrac{n}{n + \sigma^2/\sigma_0^2}$.

The predictive distribution of the nonsample mean $\overline{y}_r$ is given by

$$(\overline{y}_r \mid y_s) \sim N(a, b^2),$$

where: $a = \mu_*$, $b^2 = \sigma_*^2 + \dfrac{\sigma^2}{m}$, $m = N-n$.

The $1-\alpha$ CPDR for $\overline{y}_r$ is $(a \pm z_{\alpha/2}b)$.

The predictive distribution of the finite population mean $\overline{y}$ is given by

$$(\overline{y} \mid y_s) \sim N(c, d^2),$$

where: $c = \dfrac{n\overline{y}_s + m\mu_*}{N}$, $d^2 = \dfrac{m^2}{N^2}\left(\sigma_*^2 + \dfrac{\sigma^2}{m}\right)$ (with μ_* and σ_*^2 as above).

The $1-\alpha$ CPDR for $\overline{y}$ is $(c \pm z_{\alpha/2}d)$.

Exercise 10.1 Practice with the basic normal-normal finite population model

Consider the Bayesian model given by:

$$(\bar{y}_r \mid y_s, \mu) \sim N(\mu, \sigma^2 / m)$$
$$(y_1, ..., y_n \mid \mu) \sim iid\ N(\mu, \sigma^2)$$
$$\mu \sim N(\mu_0, \sigma_0^2).$$

(a) Express the predictive mean of the finite population mean $\bar{y}$ as a credibility estimate with a suitable credibility factor. Then also express the predictive variance and distribution in terms of that credibility factor. Use your results to answer parts (b) through (e) below.

(b) What is the predictive distribution in the case of *very weak prior information*?

(c) What is the predictive distribution in the case of *very strong prior information*?

(d) What is the predictive distribution in the case of *a very large sample size*?

(e) What is the predictive distribution in *the case of a census*?

(f) Suppose we believe with a priori probability 95% that μ lies between 7.0 and 13.0. We sample the values 5.7, 9.6 and 8.3 from a finite population of seven units. Find the predictive mean and 95% highest predictive density region for the average of all seven values in the finite population if the superpopulation standard deviation is 2.0.

Create a graph showing:
    (i)   the likelihood function for the superpopulation mean
    (ii)   the prior density of the superpopulation mean
    (iii)   the posterior density of the superpopulation mean
    (iv)   the prior density of the nonsample mean
    (v)   the predictive density of the nonsample mean
    (vi)   the prior density of the finite population mean
    (vii)   the predictive density of the finite population mean.

In your graph indicate the predictive mean and 95% highest predictive density region for the average of all seven values in the finite population.

Solution to Exercise 10.1

(a) It is easy to show that the predictive mean of $\bar{y}$, namely

$$c = E(\bar{y} \mid y_s) = \frac{n\bar{y}_s + m\mu_*}{N} = \frac{n\bar{y}_s + (N - n)[(1 - k)\mu_0 + k\bar{y}_s]}{N},$$

may also be written as the credibility estimate

$$c = (1 - q)\mu_0 + q\bar{y}_s,$$

where

$$q = \frac{n + (N - n)k}{N}$$

is the credibility factor, meaning the weight assigned to $\bar{y}_s$ (the direct data estimate of $\bar{y}$), and where $1 - q$ is the weight assigned to μ_0 (the prior estimate of $\bar{y}$).

It can then also be shown that the predictive variance of $\bar{y}$, namely

$$d^2 = V(\bar{y} \mid y_s) = \frac{m^2}{N^2}\left(\sigma_*^2 + \frac{\sigma^2}{m}\right),$$

may be expressed as

$$\frac{(N - n)^2}{N^2}\left(k\frac{\sigma^2}{n} + \frac{\sigma^2}{N - n}\right) = q\frac{\sigma^2}{n}\left(1 - \frac{n}{N}\right).$$

Thus we may also write the predictive distribution of the finite population mean as

$$(\bar{y} \mid y_s) \sim N\left((1 - q)\mu_0 + q\bar{y}_s, q\frac{\sigma^2}{n}\left(1 - \frac{n}{N}\right)\right),$$

where:

$$q = \frac{n + (N - n)k}{N}$$

$$k = \frac{n}{n + \sigma^2 / \sigma_0^2}.$$

Note: If the original credibility factor k equals 1 then the second credibility factor q also equals 1. This then implies that we estimate $\bar{y}$ by

$$c = (1 - 1)\mu_0 + 1\bar{y}_s = \bar{y}_s.$$

This makes sense because if the sample data values are given 'full credibility' then their straight average should intuitively be used to estimate the finite population mean.

On the other hand, if $k = 0$ then $q = n/N$ (the sampling fraction). This then implies that we estimate $\bar{y}$ by

$$c = (1 - n/N)\mu_0 + (n/N)\bar{y}_s = ((N-n)\mu_0 + n\bar{y}_s)/N .$$

This also makes sense because if the sample data are given 'zero credibility' then each of the $N-n$ nonsampled values should intuitively be estimated by the prior mean of the superpopulation mean μ .

(b) In the case of *very weak prior information* we have (in the limit) that $\sigma_0 = \infty$, hence $k = 1$, and hence $q = 1$. Consequently

$$(\bar{y} \mid y_s) \sim N\left((1-1)\mu_0 + 1\bar{y}_s, 1\frac{\sigma^2}{n}\left(1 - \frac{n}{N}\right) \right) \sim N\left(\bar{y}_s, \frac{\sigma^2}{n}\left(1 - \frac{n}{N}\right) \right).$$

This implies a posterior mean and $1 - \alpha$ CPDR for $\bar{y}$ of $\bar{y}_s$ and

$$\left(\bar{y}_s \pm z_{\alpha/2} \frac{\sigma}{\sqrt{n}} \sqrt{1 - \frac{n}{N}} \right).$$

Note: This is the same inference one would make via classical techniques after substituting the sample standard deviation

$$s = \sqrt{\frac{1}{n-1}\sum_{i=1}^{n}(y_i - \bar{y}_s)^2}$$

for σ , assuming that n is 'large'.

(c) In the case of *very strong prior information* we have (in the limit) that $\sigma_0 = 0$, hence $k = 0$, and hence $q = n/N$. Consequently,

$$(\bar{y} \mid y_s) \sim N\left(\left(1 - \frac{n}{N}\right)\mu_0 + \frac{n}{N}\bar{y}_s, \frac{n}{N}\frac{\sigma^2}{n}\left(1 - \frac{n}{N}\right) \right)$$

$$\sim N\left(\frac{(N-n)\mu_0 + n\bar{y}_s}{N}, \frac{\sigma^2}{N}\left(1 - \frac{n}{N}\right) \right).$$

(d) In the case of *a very large sample size* we have (in the limit) that $n = \infty$, hence $k = 1$, and hence $q = 1$. Consequently (just as in (b) for the case of very weak prior information),

$$(\bar{y} \mid y_s) \sim N\left((1-1)\mu_0 + 1\bar{y}_s, 1\frac{\sigma^2}{n}\left(1 - \frac{n}{N}\right) \right)$$

$$\sim N\left(\bar{y}_s, \frac{\sigma^2}{n}\left(1 - \frac{n}{N}\right) \right).$$

(e) In the case of *a census* we have $n = N$, hence

$$q = \frac{N + (N-N)k}{N} = 1,$$

and therefore

$$(\bar{y} \mid y_s) \sim N\left((1-1)\mu_0 + 1\bar{y}_s, 1\frac{\sigma^2}{N}\left(1 - \frac{N}{N}\right) \right)$$

$$\sim N\left(\bar{y}_s, 0 \right),$$

meaning that $\bar{y} = \bar{y}_s$ with posterior probability 1 (obviously).

Note: Some of the equations developed previously implicitly assume that $n < N$.

(f) The given specifications imply that:

$$n = 3, \quad N = 7, \quad m = N - n = 4, \quad \sigma = 2$$

$$\bar{y}_s = \frac{1}{3}(5.7 + 9.6 + 8.3) = 7.8667$$

$$\mu_0 = 10, \quad \sigma_0 = \frac{3}{1.96} = 1.53064$$

$$\mu \sim N(\mu_0, \sigma_0^2)$$

$$k = \frac{n}{n + \sigma^2 / \sigma_0^2} = 0.63731$$

$$\mu_* = (1-k)\mu_0 + k\bar{y}_s = 8.6404, \quad \sigma_* = \sqrt{k\frac{\sigma^2}{n}} = 0.9218141$$

$$(\mu \mid y_s) \sim N(\mu_*, \sigma_*^2)$$

$$a = \mu_* = 8.6404, \quad b = \sqrt{\sigma_*^2 + \sigma^2 / m} = 1.3601$$

$$(\bar{y}_r \mid y_s) \sim N(a, b^2)$$

$$q = \frac{n + (N - n)k}{N} = 0.79275$$

$$c = \frac{n\bar{y}_s + m\mu_*}{N} = (1 - q)\mu_0 + q\bar{y}_s = 8.3088$$

$$d = \sqrt{\frac{m^2}{N^2} b^2} = 0.77717$$

$$(\bar{y} \mid y_s) \sim N(c, d^2).$$

So the predictive mean of $\bar{y}$, the average of all 7 values in the finite population, is $c = 8.3088$, and the 95% highest predictive density region for that average is $(c \pm 1.96d) = (6.7856, 9.8320)$. Figure 10.1 shows:

(i) the likelihood function for the superpopulation mean, $L(\mu)$, equal to the posterior density of μ under a *flat prior*; thus $L(\mu) = f_{N(\bar{y}_s, \sigma^2 / n)}(\mu)$

(ii) the prior density of the superpopulation mean,
$$f(\mu) = f_{N(\mu_0, \sigma_0^2)}(\mu)$$

(iii) the posterior density of the superpopulation mean,
$$f(\mu \mid y_s) = f_{N(\mu_*, \sigma_*^2)}(\mu)$$

(iv) the prior density of the nonsample mean,
$$f(\bar{y}_r) = f_{N(\mu_0, \sigma_0^2 + \sigma^2 / m)}(\bar{y}_r)$$

(v) the predictive density of the nonsample mean,
$$f(\bar{y}_r \mid y_s) = f_{N(a, b^2)}(\bar{y}_r)$$

(vi) the prior density of the finite population mean,
$$f(\bar{y}) = f_{N(\mu_0, \sigma_0^2 + \sigma^2 / N)}(\bar{y})$$

(vii) the predictive density of the finite population mean,
$$f(\bar{y} \mid y_s) = f_{N(c, d^2)}(\bar{y}).$$

Figure 10.1 Various densities in Exercise 10.1

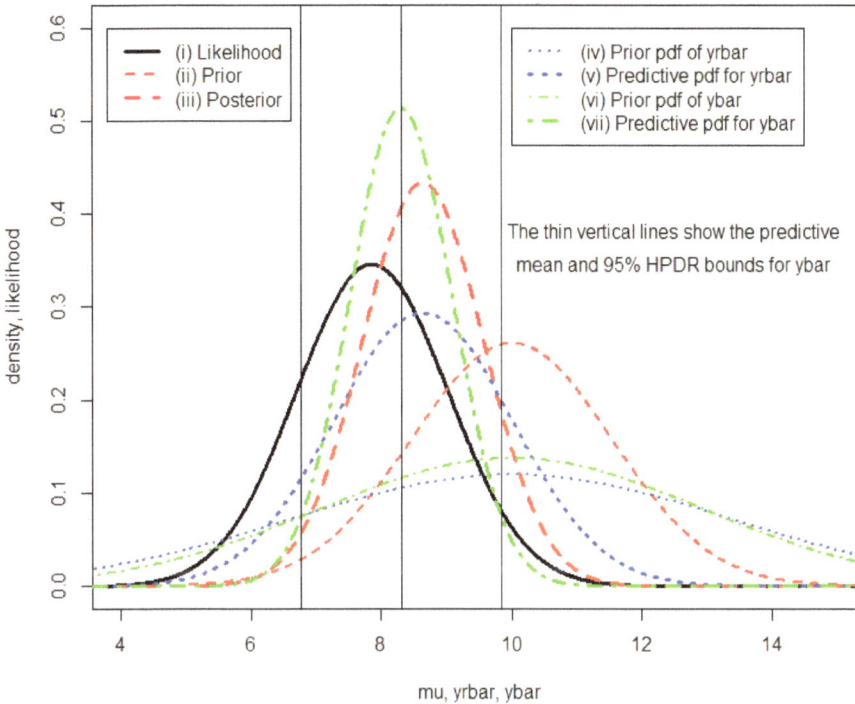

In Figure 10.1, we may observe how the prior densities of μ, $\overline{y}_r$ and $\overline{y}$ are all centred around the prior mean $\mu_0 = 10$. The line for μ is most highly concentrated about 10 because it represents the prior density of the mean of a hypothetically *infinite* number of population values. The line for $\overline{y}_r$ is the least focused about 10 because it represents the prior density of the mean of only 4 such values (compared with the line for $\overline{y}$ which is the prior pdf for the mean of 7 such values).

Each of the posterior/predictive densities for μ, $\overline{y}_r$ and $\overline{y}$ is located somewhere between the corresponding prior density and the likelihood function. The posterior/predictive densities for μ and $\overline{y}_r$ are centred at the same values, namely the posterior mean, $\mu_* = 8.6404$, whereas the predictive density for $\overline{y}$ is centred closer to the likelihood mode, $\overline{y}_s = 7.8667$. This is because the second credibility factor is larger than the first ($q = 0.79275 > k = 0.63731$).

R Code for Exercise 10.1

```
ys=c(5.7,9.6,8.3); ysbar=mean(ys); ysbar # 7.866667
sig=2; n=3; N=7; m=N-n; mu0=10; sig0=3/qnorm(0.975);
k=n/(n+sig^2/sig0^2); q=(n+m*k)/N
c(m,mu0,sig0,k,q) # 4.0000000 10.0000000 1.5306404 0.6373060 0.7927463
mustar=(1-k)*mu0+k*ysbar; sigstar2=k*sig^2/n
c(mustar,sqrt(sigstar2)) # 8.6404139 0.9218141
a=mustar; b2=sigstar2+sig^2/m; c=(n*ysbar+m*a)/N; d2=(m/N)^2*b2
c(a,sqrt(b2),c,sqrt(d2)) # 8.6404139 1.3600519 8.3088080 0.7771725
HPDR=c+c(-1,1)*qnorm(0.975)*sqrt(d2); HPDR # 6.785578 9.832038

X11(w=8,h=7); par(mfrow=c(1,1))
plot(c(4,15),c(0,0.6),type="n",xlab="mu, yrbar, ybar",
        ylab="density, likelihood", main="")
v=seq(0,20,0.01)
lines(v,dnorm(v,ysbar,sig/sqrt(n)),lty=1,lwd=3,col="black")
        # likelihood function (i)

lines(v,dnorm(v,mu0,sig0),lty=2, lwd=2,col="red") # prior (ii)
lines(v,dnorm(v,mustar,sqrt(sigstar2)),lty=2,lwd=3, col="red") # posterior (iii)

lines(v,dnorm(v,mu0,sig0^2+sig^2/m),lty=3,lwd=2, col="blue")
        # prior pdf of yrbar (iv)
lines(v,dnorm(v,a,sqrt(b2)),lty=3,lwd=3, col="blue")
        # predictive pdf of yrbar (v)

lines(v,dnorm(v,mu0,sig0^2+sig^2/N),lty=4,lwd=2, col="green")
        # prior pdf of ybar (vi)
lines(v,dnorm(v,c,sqrt(d2)),lty=4,lwd=3, col="green")
        # predictive pdf of ybar (vii)
abline(v=c(c,HPDR),lty=1,lwd=1)
legend(3.8,0.6,c("(i) Likelihood","(ii) Prior","(iii) Posterior"),
        lty=c(1,2,2), lwd=c(3,2,3), col=c("black","red","red"))
legend(10,0.6,c("(iv) Prior pdf of yrbar","(v) Predictive pdf for yrbar",
                "(vi) Prior pdf of ybar","(vii) Predictive pdf for ybar"),
        lty=c(3,3,4,4), lwd=c(2,3,2,3), col=c("blue","blue","green","green"))
text(12.5,0.38, "The thin vertical lines show the predictive")
text(12.5,0.345,"mean and 95% HPDR bounds for ybar")
```

10.2 The general normal-normal finite population model

The *basic normal-normal finite population model* examined in the previous section assumes that:
- all N values in the finite population are conditionally normal and iid
- we are interested only in the nonsample mean $\bar{y}_r$ and functions of $\bar{y}_r$ (such as the finite population mean $\bar{y}$).

We will now examine a generalisation of this basic model which allows for:
- non-independence of values
- covariate information
- inference on the entire nonsample vector and linear combinations thereof.

We will continue to assume that the values in the population are all (conditionally) normally distributed, and that the (conditional) variance of each value in the finite population is known. We will now also assume that all the covariance terms between these values are known. (These assumptions will be relaxed at a later stage.)

First, define the (finite) population vector in *column* form as

$$y = \begin{pmatrix} y_s \\ y_r \end{pmatrix} = \begin{pmatrix} \begin{pmatrix} y_1 \\ \vdots \\ y_n \end{pmatrix} \\ \begin{pmatrix} y_{n+1} \\ \vdots \\ y_N \end{pmatrix} \end{pmatrix} = \begin{pmatrix} y_1 \\ \vdots \\ y_N \end{pmatrix}.$$

Next, suppose that auxiliary information is available in the form of an N by p matrix

$$X = \begin{pmatrix} x_1' \\ \vdots \\ x_N' \end{pmatrix} = (X_1, ..., X_p) = \begin{pmatrix} x_{11} & \cdots & x_{1p} \\ \vdots & \vdots & \vdots \\ x_{N1} & \cdots & x_{Np} \end{pmatrix},$$

where

$$x_i = \begin{pmatrix} x_{i1} \\ \vdots \\ x_{ip} \end{pmatrix}$$

is the covariate vector for the ith population unit ($i = 1, ..., N$) and

$$X_j = \begin{pmatrix} x_{1j} \\ \vdots \\ x_{Nj} \end{pmatrix}$$

is the population vector for the jth explanatory variable ($j = 1, ..., p$).

Also suppose that the finite population vector y has a known variance-covariance structure in the form of an N by N positive definite matrix

$$\Sigma = \begin{pmatrix} \sigma_{11} & \cdots & \sigma_{1N} \\ \vdots & \ddots & \vdots \\ \sigma_{N1} & \cdots & \sigma_{NN} \end{pmatrix},$$

where: $\sigma_{ij} = C(y_i, y_j) = \sigma_{ji}$

$\sigma_{ii} = Vy_i \equiv \sigma_i^2$,

with the covariance and variance operations here (C and V) implicitly conditional on all model parameters.

In the above context, the Bayesian model we will focus on is:

$$(y \mid \beta) \sim N_N(X\beta, \Sigma)$$
$$\beta \sim N_p(\delta, \Omega).$$

This model will be called the *general normal-normal finite population model*. Here,

$$\beta = \begin{pmatrix} \beta_1 \\ \vdots \\ \beta_p \end{pmatrix}$$

is the vector of regression coefficients, whose prior distribution is multivariate normal with (specified) mean

$$\delta = \begin{pmatrix} \delta_1 \\ \vdots \\ \delta_p \end{pmatrix}$$

and (specified) variance-covariance matrix

$$\Omega = \begin{pmatrix} \omega_{11} & \cdots & \omega_{1p} \\ \vdots & \ddots & \vdots \\ \omega_{p1} & \cdots & \omega_{pp} \end{pmatrix},$$

where: $\omega_{ij} = C(\beta_i, \beta_j) = \omega_{ji}$

$$\omega_{ii} = V\beta_i \equiv \omega_i^2,$$

with the covariance and variance operations here (C and V) implicitly *unconditional*, thereby reflecting prior belief regarding the β_i values.

We will assume interest lies generally in the nonsample vector y_r and functions of that vector, and specifically in the finite population mean $\bar{y}$ (a simple function of y_r and of the known quantities y_s, n and N). Thus the regression coefficient vector β will be treated as a nuisance parameter and inference will be based on the predictive distribution of y_r given y_s.

Note: The *basic normal finite population model* as considered previously is a special case of the just-defined *general normal finite population model* with:

$$p = 1, \qquad \beta = (\beta_1) = \mu, \qquad \delta = (\delta_1) = \mu_0, \qquad \Omega = (\omega_{11}) = \sigma_0^2$$

$$X = 1_N = (1,...,1)' = \begin{pmatrix} 1 \\ \vdots \\ 1 \end{pmatrix} \quad \text{(a column vector of } N \text{ ones)}$$

$$\Sigma = \sigma^2 I_N = \begin{pmatrix} \sigma^2 & 0 & \cdots & 0 \\ 0 & \sigma^2 & & 0 \\ \vdots & & \ddots & \vdots \\ 0 & 0 & 0 & \sigma^2 \end{pmatrix}$$

(where I_N is the N by N identity matrix).

Thus, the previous normal finite population model could also be written as:

$$(y \mid \mu) \sim N_N(\mu 1_N, \sigma^2 I_N)$$

$$\mu \sim N_1(\mu_0, \sigma_0^2).$$

10.3 Derivation of the predictive distribution of the nonsample vector

Observe that the unconditional (or prior) distribution of the entire finite population vector y is given by the density

$$f(y) = \int f(y, \beta) d\beta \ = \int f(y \mid \beta) f(\beta) d\beta .$$

Now, the integrand of this multiple integral is a quadratic in the y_i and β_j values. This implies that the value of the integral has the form of a quadratic in the y_i values alone. This then implies that the prior (or unconditional) distribution of y is also multivariate normal. It then remains to find the mean and covariance vector of that prior distribution, as follows:

$$Ey = EE(y \mid \beta) = E(X\beta) = X\delta$$
$$Vy = EV(y \mid \beta) + VE(y \mid \beta) = E\Sigma + V(X\beta) = \Sigma + X\Omega X' .$$

Thus, $y \sim N_N(X\delta, \Sigma + X\Omega X')$.

This result may also be written as

$$\begin{pmatrix} y_s \\ y_r \end{pmatrix} \sim N_N \left(\begin{pmatrix} X_s \delta \\ X_r \delta \end{pmatrix}, \begin{pmatrix} \Sigma_{ss} + X_s \Omega X_s' & \Sigma_{sr} + X_s \Omega X_r' \\ \Sigma_{rs} + X_r \Omega X_s' & \Sigma_{rr} + X_r \Omega X_r' \end{pmatrix} \right),$$

where we partition X and Σ according to

$$X = \begin{pmatrix} X_s \\ X_r \end{pmatrix} \text{ and } \Sigma = \begin{pmatrix} \Sigma_{ss} & \Sigma_{sr} \\ \Sigma_{rs} & \Sigma_{rr} \end{pmatrix}.$$

Thus, $X_s = \begin{pmatrix} X_1' \\ \vdots \\ X_n' \end{pmatrix}$ is a submatrix consisting of the first n rows of X, etc.

It follows by standard multivariate normal theory (see below) that

$$(y_r \mid y_s) \sim N_m(E_*, V_*),$$

where:

$$E_* = X_r \delta + (\Sigma_{rs} + X_r \Omega X_s')(\Sigma_{ss} + X_s \Omega X_s')^{-1}(y_s - X_s \delta) \qquad (10.1)$$

$$V_* = (\Sigma_{rr} + X_r \Omega X_r') - (\Sigma_{rs} + X_r \Omega X_s')(\Sigma_{ss} + X_s \Omega X_s')^{-1}(\Sigma_{sr} + X_s \Omega X_r') . \qquad (10.2)$$

Note: We have here used the following result (e.g. see equation (81.2.11) in Rao, 1973):

$$\begin{pmatrix} X_1 \\ X_2 \end{pmatrix} \sim N_{n_1+n_2}\left(\begin{pmatrix} \mu_1 \\ \mu_2 \end{pmatrix}, \begin{pmatrix} \Sigma_{11} & \Sigma_{12} \\ \Sigma_{21} & \Sigma_{22} \end{pmatrix} \right)$$

$$\Rightarrow (X_2 \mid X_1) \sim N_{n_2}(\mu_2 + \Sigma_{21}\Sigma_{11}^{-1}(X_1 - \mu_1), \Sigma_{22} - \Sigma_{21}\Sigma_{11}^{-1}\Sigma_{12}).$$

10.4 Alternative formulae for the predictive distribution of the nonsample vector

Another way to obtain the distribution of $(y_r \mid y_s)$ (already derived above) is as follows. First, the posterior density of β is

$$f(\beta \mid y_s) \propto f(\beta)f(y_s \mid \beta)$$

$$\propto \exp\left\{ -\frac{1}{2}(\beta - \delta)'\Omega^{-1}(\beta - \delta) \right\} \exp\left\{ -\frac{1}{2}(y_s - X_s\beta)'\Sigma_{ss}^{-1}(y_s - X_s\beta) \right\}$$

$$= \exp\left(-\frac{1}{2}Q_1 \right),$$

where

$$Q_1 = (\beta - \delta)'\Omega^{-1}(\beta - \delta) + (y_s - X_s\beta)'\Sigma_{ss}^{-1}(y_s - X_s\beta).$$

We see that $f(\beta \mid y_s)$ is proportional to the exponent of a quadratic form in β. This implies that

$$(\beta \mid y_s) \sim N_p(\hat{\beta}, D)$$

for some $\hat{\beta}$ and D to be determined.

Now observe that

$$f(\beta \mid y_s) \propto \exp\left(-\frac{1}{2}Q_2 \right),$$

where

$$Q_2 = (\beta - \hat{\beta})'D^{-1}(\beta - \hat{\beta})$$
$$= \beta'D^{-1}\beta - \beta'D^{-1}\hat{\beta} - \hat{\beta}'D^{-1}\beta + \text{constant} \qquad (10.3)$$

(where the constant does not depend on β).

But $Q_1 = \beta'\Omega^{-1}\beta - \beta'\Omega^{-1}\delta - \delta'\Omega^{-1}\beta - y_s'\Sigma_{ss}^{-1}X_s\beta$

$$-\beta'X_s'\Sigma_{ss}^{-1}y_s + \beta'X_s'\Sigma_{ss}^{-1}X_s\beta + \text{constant}$$

$$= \beta'(\Omega^{-1} + X_s'\Sigma_{ss}^{-1}X_s)\beta - \beta'(\Omega^{-1}\delta + X_s'\Sigma_{ss}^{-1}y_s) - (\delta\Omega^{-1} + y_s'\Sigma_{ss}^{-1}X_s)\beta$$

$$+ \text{constant.} \qquad (10.4)$$

Equating (10.3) and (10.4) we see that:

$$D^{-1} = \Omega^{-1} + X_s'\Sigma_{ss}^{-1}X_s$$

$$D^{-1}\hat{\beta} = \Omega^{-1}\delta + X_s'\Sigma_{ss}^{-1}y_s .$$

It follows that:

$$D = (\Omega^{-1} + X_s'\Sigma_{ss}^{-1}X_s)^{-1}$$

$$\hat{\beta} = D(\Omega^{-1}\delta + X_s'\Sigma_{ss}^{-1}y_s) .$$

We can now use the result

$$(\beta \mid y_s) \sim N_p(\hat{\beta}, D)$$

to find the predictive mean and variance of y_r.

First, observe that

$$(y \mid \beta) \sim N_N(X\beta, \Sigma)$$

may also be written

$$\left(\begin{pmatrix} y_s \\ y_r \end{pmatrix} \middle| \beta \right) \sim N_N \left(\begin{pmatrix} X_s\beta \\ X_r\beta \end{pmatrix}, \begin{pmatrix} \Sigma_{ss} & \Sigma_{sr} \\ \Sigma_{rs} & \Sigma_{rr} \end{pmatrix} \right),$$

which implies that

$$(y_r \mid y_s, \beta) \sim N_m(X_r\beta + \Sigma_{rs}\Sigma_{ss}^{-1}(y_s - X_s\beta), \Sigma_{rr} - \Sigma_{rs}\Sigma_{ss}^{-1}\Sigma_{sr}).$$

It follows that:

$$E(y_r \mid y_s) = E\{E(y_r \mid y_s, \beta) \mid y_s\}$$

$$= E\{X_r\beta + \Sigma_{rs}\Sigma_{ss}^{-1}(y_s - X_s\beta) \mid y_s\}$$

$$= X_r\hat{\beta} + \Sigma_{rs}\Sigma_{ss}^{-1}(y_s - X_s\hat{\beta}) \qquad (10.5)$$

$$V(y_r \mid y_s) = E\{V(y_r \mid y_s, \beta) \mid y_s\} + V\{E(y_r \mid y_s, \beta) \mid y_s\}$$

$$= E\{\Sigma_{rr} - \Sigma_{rs}\Sigma_{ss}^{-1}\Sigma_{sr} \mid y_s\} + V\{X_r\hat{\beta} + \Sigma_{rs}\Sigma_{ss}^{-1}(y_s - X_s\hat{\beta}) \mid y_s\}$$

$$= \Sigma_{rr} - \Sigma_{rs}\Sigma_{ss}^{-1}\Sigma_{sr} + V\{(X_r - \Sigma_{rs}\Sigma_{ss}^{-1}X_s)\hat{\beta} \mid y_s\}$$

$$= \Sigma_{rr} - \Sigma_{rs}\Sigma_{ss}^{-1}\Sigma_{sr} + (X_r - \Sigma_{rs}\Sigma_{ss}^{-1}X_s)D(X_r - \Sigma_{rs}\Sigma_{ss}^{-1}X_s)'. \qquad (10.6)$$

Note: The expression for E_* at (10.1) must be the same as that for $E(y_r \mid y_s)$ at (10.5), and likewise the expression for V_* at (10.2) must be the same as that for $V(y_r \mid y_s)$ at (10.6). This equivalence can also be shown with some algebra by making use of the formula

$$(\Sigma_{ss} + X_s \Omega X_s')^{-1} = \Sigma_{ss}^{-1} \{ I_s - X_s (\Omega^{-1} + X_s \Sigma_{ss}^{-1} X_s)^{-1} X_s' \Sigma_{ss}^{-1} \},$$

which in turn follows from the general matrix identity

$$(A - UW^{-1}V)^{-1} = A^{-1} + A^{-1}U(W - VA^{-1}U)^{-1}VA^{-1}.$$

Here, I_s is the n by n identity matrix and could also be written I_n.

10.5 Prediction of the finite population mean and other linear combinations

We may now write down a general expression for the predictive distribution of the finite population mean. That mean may be expressed as the linear combination

$$\bar{y} = \frac{y_{sT} + m\bar{y}_r}{N} = \frac{1}{N}(y_{sT} + 1_r' y_r).$$

Note: Here, $1_r'$ denotes the row vector with $m = N - n$ ones. This vector could also be written $1_m'$ or $1_{N-n}'$ or $(1, ..., 1)$.

Therefore the predictive distribution of $\bar{y}$ given y_s is normal with mean $e_* = \dfrac{y_{sT} + 1_r' E_*}{N}$ and variance $v_* = \dfrac{1_r' V_* 1_r}{N^2}$.

So the $1 - \alpha$ CPDR for $\bar{y}$ is $(e_* \pm z_{\alpha/2} \sqrt{v_*})$.

More generally, the predictive distribution of the linear combination

$$\psi = c_0 + (c_1 y_1 + ... + c_n y_n) + (c_{n+1} y_{n+1} + ... + c_N y_N)$$

is normal with mean $e_\# = c_0 + c_s' y_s + c_r' E_*$ and variance $v_\# = \dfrac{c_r' V_* c_r}{N^2}$,

where $c_s = (c_1, ..., c_n)'$ and $c_r = (y_{n+1}, ..., c_N)'$.

So the $1 - \alpha$ CPDR for ψ is $(e_\# \pm z_{\alpha/2} \sqrt{v_\#})$.

Summary: For the *general normal-normal finite population model*:

$$(y \mid \beta) \sim N_N(X\beta, \Sigma)$$
$$\beta \sim N_p(\delta, \Omega),$$

the posterior distribution of the regression vector β is given by

$$(\beta \mid y_s) \sim N_p(\hat{\beta}, D),$$

where: $\hat{\beta} = D(\Omega^{-1}\delta + X_s'\Sigma_{ss}^{-1}y_s), \qquad D = (\Omega^{-1} + X_s'\Sigma_{ss}^{-1}X_s)^{-1}.$

The predictive distribution of the nonsample vector y_r is given by

$$(y_r \mid y_s) \sim N_m(E_*, V_*) \qquad (m = N - n),$$

where: $E_* = X_r\delta + (\Sigma_{rs} + X_r\Omega X_s')(\Sigma_{ss} + X_s\Omega X_s')^{-1}(y_s - X_s\delta)$

$$= X_r\hat{\beta} + \Sigma_{rs}\Sigma_{ss}^{-1}(y_s - X_s\hat{\beta})$$

$$V_* = (\Sigma_{rr} + X_r\Omega X_r') - (\Sigma_{rs} + X_r\Omega X_s')(\Sigma_{ss} + X_s\Omega X_s')^{-1}(\Sigma_{sr} + X_s\Omega X_r')$$

$$= \Sigma_{rr} - \Sigma_{rs}\Sigma_{ss}^{-1}\Sigma_{sr} + (X_r - \Sigma_{rs}\Sigma_{ss}^{-1}X_s)D(X_r - \Sigma_{rs}\Sigma_{ss}^{-1}X_s)'.$$

The predictive distribution of the finite population mean $\bar{y}$ is given by

$$(\bar{y} \mid y_s) \sim N(e_*, v_*), \quad \text{where} \quad e_* = \frac{y_{sT} + 1_r'E_*}{N} \quad \text{and} \quad v_* = \frac{1_r'V_*1_r}{N^2},$$

with $1 - \alpha$ CPDR for $\bar{y}$ given by $(e_* \pm z_{\alpha/2}\sqrt{v_*})$.

The predictive distribution of any linear combination of the form $\psi = c_0 + c_s'y_s + c_r'y_r$ is given by

$$(\psi \mid y_s) \sim N(e_\#, v_\#),$$

where $e_\# = c_0 + c_s'y_s + c_r'E_*$ and $v_\# = \dfrac{c_r'V_*c_r}{N^2},$

with $1 - \alpha$ CPDR for ψ given by $(e_\# \pm z_{\alpha/2}\sqrt{v_\#})$.

10.6 Special cases including ratio estimation

In the context of the above general normal-normal finite population model, suppose that $p = 1$ (i.e. there is a single covariate) and the population values are conditionally independent, the ith one having mean $x_i\beta$ and variance $x_i^{2\gamma}\sigma^2$, where $\gamma \in \Re$ and $\sigma^2 > 0$ are known.

Also, suppose that the prior distribution on the single regression coefficient β is normal with mean δ and variance ω^2. Then:

$$(y \mid \beta) \sim N_N(X\beta, \Sigma)$$
$$\beta \sim N_p(\delta, \Omega),$$

where: $p = 1$, $X = \begin{pmatrix} x_1 \\ x_2 \\ \vdots \\ x_N \end{pmatrix}$, $\Sigma = \begin{pmatrix} x_1^{2\gamma} & 0 & \cdots & 0 \\ 0 & x_2^{2\gamma} & & 0 \\ \vdots & \vdots & \ddots & \vdots \\ 0 & 0 & \cdots & x_N^{2\gamma} \end{pmatrix} \sigma^2$, $\Omega = \omega^2$.

The model may also be written in non-matrix form as:

$$(y_i \mid \beta) \sim \perp N(x_i'\beta, x_i^{2\gamma}\sigma^2), i = 1, ..., N$$
$$\beta \sim N(\delta, \omega^2).$$

Under this model it can be shown that the predictive distribution of the finite population mean is given by

$$(\overline{y} \mid y_s) \sim N(A, B^2),$$

where:

$$A = \frac{n}{N}\overline{y}_s + \left(1 - \frac{n}{N}\right)\overline{x}_r\left\{\frac{\delta\sigma^2 + \omega^2 \sum_{i=1}^n y_i x_i^{1-2\gamma}}{\sigma^2 + \omega^2 \sum_{i=1}^n x_i^{2-2\gamma}}\right\}$$

$$B^2 = \frac{\sigma^2}{N^2}\left\{\sum_{i=n+1}^N x_i^{2\gamma} + \frac{m^2\omega^2\overline{x}_r^2}{\sigma^2 + \omega^2 \sum_{i=1}^n x_i^{2-2\gamma}}\right\}$$

$$\overline{x}_r = \frac{1}{m}\sum_{i=n+1}^N x_i \text{ (average of the covariate values in the nonsample).}$$

Now suppose it is believed that the variances of the population values are exactly proportional to the covariate values, i.e. $V(y_i \mid \beta) = x_i\sigma^2$.

Then $\gamma = 1/2$, and we find that:

$$A = \frac{n}{N}\overline{y}_s + \left(1 - \frac{n}{N}\right)\overline{x}_r\left\{\frac{\omega^2\overline{y}_s + \delta\sigma^2/n}{\omega^2\overline{x}_s + \sigma^2/n}\right\}$$

$$B^2 = \frac{\sigma^2}{n}\overline{x}_r\left(1 - \frac{n}{N}\right)\left\{\frac{n}{N} + \left(1 - \frac{n}{N}\right)\frac{\omega^2\overline{x}_r}{\omega^2\overline{x}_s + \sigma^2/n}\right\}$$

$$\overline{x}_s = \frac{1}{n}\sum_{i=1}^n x_i \text{ (the average of the covariate values in the sample).}$$

If there is a priori ignorance regarding β we may further set $\omega = \infty$, and in that case:

$$A = \frac{n}{N}\bar{y}_s + \left(1 - \frac{n}{N}\right)\bar{x}_r\frac{\bar{y}_s}{\bar{x}_s} = \bar{y}_s\left\{\frac{n}{N} + \left(1 - \frac{n}{N}\right)\frac{\bar{x}_r}{\bar{x}_s}\right\}$$

$$= \bar{y}_s\left\{\frac{n\bar{x}_s + (N-n)\bar{x}_r}{N\bar{x}_s}\right\} = \frac{\bar{y}_s}{\bar{x}_s}\bar{x}$$

$$B^2 = \frac{\sigma^2}{n}\bar{x}_r\left(1 - \frac{n}{N}\right)\left\{\frac{n}{N} + \left(1 - \frac{n}{N}\right)\frac{\bar{x}_r}{\bar{x}_s}\right\} = \frac{\sigma^2}{n}\left(1 - \frac{n}{N}\right)\frac{\bar{x}_r}{\bar{x}_s}\bar{x}$$

$$\bar{x} = \frac{1}{N}\sum_{i=1}^{N} x_i \quad \text{(average of covariate values in the finite population)}.$$

As regards this last special case, we see that the predictive mean A is identical to the common design-based ratio estimator.

Also, the predictive variance B^2, although not identical to any design-based formula, is the same as a model-based analogue (e.g. see Brewer, 1963, and Royall, 1970). The formula for B^2 suggests a purposive sampling scheme whereby units with the largest covariate values should be selected.

Note 1: If units with relatively large y-values are selected, then $\bar{x}_s$ will likely be larger than $\bar{x}_r$, so that then $\dfrac{\bar{x}_r}{\bar{x}_s}$ will likely be small, and thereby $B^2 = V(\bar{y} \mid y_s) = \dfrac{\sigma^2}{n}\left(1 - \dfrac{n}{N}\right)\dfrac{\bar{x}_r}{\bar{x}_s}\bar{x}$ will also likely be small.

Note 2: The same formulae as derived in the last special case will also apply approximately when the sample size n is very large. This makes sense because the effect of a very large sample size is the same as that of a very diffuse prior. Note that in the case of a census, $n = N$ and we find that the above formulae correctly yield $A = \bar{y}_s$ and $B^2 = 0$.

In a way similar to the above, it is possible to obtain analogues of other common design-based and model-based results, such as regression and stratified estimators, together with their associated variances (see Ericson, 1969 and 1988).

Exercise 10.2 Derivation of the Bayesian ratio estimator

Consider the Bayesian model given by:

$$(y_i \mid \beta) \sim\!\perp N(x_i\beta, x_i\sigma^2), i = 1,...,N$$
$$f(\beta) \propto 1, \beta \in \Re.$$

Derive the predictive distribution of the finite population mean given data of the form $D = (s, y_s)$.

Solution to Exercise 10.2

The Bayesian model is:

$$(y \mid \beta) \sim N_N(X\beta, \Sigma)$$
$$\beta \sim N_p(\delta, \Omega),$$

where:

$$p = 1, \quad \delta = 0, \quad \Omega = \infty, \quad X = x = \begin{pmatrix} x_1 \\ \vdots \\ x_N \end{pmatrix}$$

$$\Sigma = \sigma^2 diag(x_1,...,x_N) = \sigma^2 \begin{pmatrix} x_1 & & \\ & \ddots & \\ & & x_N \end{pmatrix}.$$

Note: Here, $\sigma^{-2}\Sigma$ is a matrix with zeros everywhere except for $x_1,...,x_N$ along the main diagonal.

Using general results derived previously we first have that

$$(\beta \mid y_s) \sim N(\hat\beta, D),$$

where:

$$D = (\Omega^{-1} + X_s'\Sigma_{ss}^{-1}X_s)^{-1}$$

$$= \left\{ \infty^{-1} + (x_1 \ \cdots \ x_n)\frac{1}{\sigma^2}\begin{pmatrix} x_1^{-1} & & \\ & \ddots & \\ & & x_n^{-1} \end{pmatrix}\begin{pmatrix} x_1 \\ \vdots \\ x_n \end{pmatrix} \right\}^{-1}$$

$$= \sigma^2 \left(\sum_{i=1}^{n} x_i x_i^{-1} x_i \right)^{-1} = \frac{\sigma^2}{x_{sT}}$$

$$\hat{\beta} = D(\Omega^{-1}\delta + X_s'\Sigma_{ss}^{-1}y_s)$$

$$= D\left\{0 + (x_1 \quad \cdots \quad x_n)\frac{1}{\sigma^2}\begin{pmatrix} x_1^{-1} & & \\ & \ddots & \\ & & x_n^{-1} \end{pmatrix}\begin{pmatrix} y_1 \\ \vdots \\ y_n \end{pmatrix}\right\}$$

$$= \frac{\sigma^2}{x_{sT}}\frac{1}{\sigma^2}\sum_{i=1}^{n}x_i x_i^{-1}y_i = \frac{y_{sT}}{x_{sT}} = \frac{\overline{y}_s}{\overline{x}_s}.$$

Next,
$$(y_r \mid y_s) \sim N_m(E_*, V_*),$$
where:

$$m = N - n$$

$$E_* = X_r\hat{\beta} + \Sigma_{rs}\Sigma_{ss}^{-1}(y_s - X_s\hat{\beta}) = \begin{pmatrix} x_{n+1} \\ \vdots \\ x_N \end{pmatrix}\hat{\beta} + 0 = \begin{pmatrix} x_{n+1} \\ \vdots \\ x_N \end{pmatrix}\frac{\overline{y}_s}{\overline{x}_s}$$

$$V_* = \Sigma_{rr} - \Sigma_{rs}\Sigma_{ss}^{-1}\Sigma_{sr} + (X_r - \Sigma_{rs}\Sigma_{ss}^{-1}X_s)D(X_r - \Sigma_{rs}\Sigma_{ss}^{-1}X_s)'$$

$$= \sigma^2\begin{pmatrix} x_{n+1} & & \\ & \ddots & \\ & & x_N \end{pmatrix} - 0 + \left(\begin{pmatrix} x_{n+1} \\ \vdots \\ x_N \end{pmatrix}1_r - 0\right)\frac{\sigma^2}{x_{sT}}\left((x_{n+1} \quad \cdots \quad x_N) - 0\right)$$

$$= \sigma^2\left\{\begin{pmatrix} x_{n+1} & & \\ & \ddots & \\ & & x_N \end{pmatrix} + \frac{1}{x_{sT}}\begin{pmatrix} x_{n+1}x_{n+1} & \cdots & x_{n+1}x_N \\ \vdots & \ddots & \vdots \\ x_N x_{n+1} & \cdots & x_N x_{n+1} \end{pmatrix}\right\}.$$

Thus finally we have that
$$(\overline{y} \mid y_s) \sim N(e_*, v_*),$$
where:

$$e_* = \frac{y_{sT} + 1_r'E_*}{N} = \frac{1}{N}\left\{y_{sT} + (1 \quad \cdots \quad 1)\begin{pmatrix} x_{n+1} \\ \vdots \\ x_N \end{pmatrix}\frac{\overline{y}_s}{\overline{x}_s}\right\}$$

$$= \frac{1}{N}\left\{y_{sT} + x_{rT}\frac{y_{sT}}{x_{sT}}\right\} = \frac{y_{sT}}{N}\left\{\frac{x_{sT} + x_{rT}}{x_{sT}}\right\} = \frac{y_{sT}}{x_{sT}}\frac{x_T}{N} = \frac{\overline{y}_s}{\overline{x}_s}\overline{x}$$

$$v_* = \frac{1'_r V_* 1_r}{N^2} = \frac{1}{N^2} \sigma^2 (1 \quad \cdots \quad 1)$$

$$\times \left\{ \begin{pmatrix} x_{n+1} & & \\ & \ddots & \\ & & x_N \end{pmatrix} + \frac{1}{x_{sT}} \begin{pmatrix} x_{n+1}x_{n+1} & \cdots & x_{n+1}x_N \\ \vdots & \ddots & \vdots \\ x_N x_{n+1} & \cdots & x_N x_{n+1} \end{pmatrix} \right\} \begin{pmatrix} 1 \\ \vdots \\ 1 \end{pmatrix}$$

$$= \frac{1}{N^2} \sigma^2 \left\{ \begin{pmatrix} x_{n+1} & \cdots & x_N \end{pmatrix} + \frac{1}{x_{sT}} \left(\sum_{i=n+1}^{N} x_i x_{n+1} \quad \cdots \quad \sum_{i=n+1}^{N} x_i x_N \right) \right\} \begin{pmatrix} 1 \\ \vdots \\ 1 \end{pmatrix}$$

$$= \frac{1}{N^2} \sigma^2 \left\{ (x_{n+1} + \ldots + x_N) + \frac{1}{x_{sT}} \left(x_{n+1} \sum_{i=n+1}^{N} x_i + \ldots + x_N \sum_{i=n+1}^{N} x_i \right) \right\}$$

$$= \frac{1}{N^2} \sigma^2 \left\{ x_{rT} + \frac{1}{x_{sT}} x_{rT}^2 \right\} = \frac{x_{rT}}{N^2} \sigma^2 \left\{ \frac{x_{sT} + x_{rT}}{x_{sT}} \right\}$$

$$= \sigma^2 \times \frac{1}{N} \times \frac{(N-n)\bar{x}_r}{n\bar{x}_s} \times \frac{x_{sT} + x_{rT}}{N} = \frac{\sigma^2}{n} \left(1 - \frac{n}{N} \right) \frac{\bar{x}_r}{\bar{x}_s} \bar{x} .$$

Exercise 10.3 Practice with the general normal-normal finite population model

Consider a superpopulation model in which all values are independent and normally distributed with mean μ, and where each value y_i has a variance which is either:

σ_0^2 if the corresponding covariate value x_i is 0, or

σ_1^2 if $x_i = 1$ (the only other possibility).

Suppose that σ_0^2, σ_1^2 and all N covariate values x_i are given. Also suppose there is a priori ignorance regarding μ.

Find a simple expression for the predictive distribution of the finite population mean $\bar{y}$. Then calculate the predictive mean and 95% predictive interval for $\bar{y}$ if:

$\sigma_0 = 0.08$, $\sigma_1 = 1.2$, $y_s = (2.1, 4.9, 2.3, 2.0, 0.2)'$

$x = (0,1,0,0,1, \quad 1,1,1,0,0, \quad 1,1,1,1,0, \quad 0,1)'$.

Note: We have here defined a type of *stratification*; the finite population is assumed to consist of two *strata* with different variances but the same underlying mean in both strata.

Solution to Exercise 10.3

Let n_0 denote the number of covariate values x_i in the sample (of size n) which are 0, and let n_1 be the number which are 1. Likewise, let m_0 denote the number of covariate values x_i in the nonsample (of size $m = N - n$) which are 0, and let m_1 be the number which are 1.

(Thus, $n_1 = \sum_{i=1}^{n} x_i$, $n_0 = n - n_1$, $m_1 = \sum_{i=n+1}^{N} x_i$ and $m_0 = m - m_1$.)

Then, without loss of generality, re-order the finite population values in such a way that $x_s = (0,...,0,1,...,1)'$ and $x_r = (0,...,0,1,...,1)'$.

(Thus, in each of the sample and nonsample vectors, place the values with covariate 0 first, and place the values with covariate 1 last.)

With this setup, the Bayesian model is:
$$(y \mid \beta) \sim N_N(X\beta, \Sigma)$$
$$\beta \sim N_p(\delta, \Omega),$$
where:

$p = 1$, $\beta \equiv \mu$, $\delta = 0$, $\Omega = \infty$

$X = 1_N$ (since the covariates do not affect the means)

$\Sigma = diag(\sigma_0^2 1'_{n_0}, \sigma_1^2 1'_{n_1}, \sigma_0^2 1'_{m_0}, \sigma_1^2 1'_{m_1})$

(a matrix with zeros everywhere except for
$$\sigma_0^2,...,\sigma_0^2, \sigma_1^2,...,\sigma_1^2, \sigma_0^2,...,\sigma_0^2, \sigma_1^2,...,\sigma_1^2$$
along the main diagonal).

Then
$$(\beta \mid y_s) \sim N_p(\hat{\beta}, D),$$
where:

$$D = (\Omega^{-1} + X_s' \Sigma_{ss}^{-1} X_s)^{-1} = (\infty^{-1} + 1_s' \Sigma_{ss}^{-1} 1_s)^{-1}$$

$$= \left\{ (1 \quad \cdots \quad 1) \begin{pmatrix} \sigma_0^{-2} & & & & & \\ & \ddots & & & & \\ & & \sigma_0^{-2} & & & \\ & & & \sigma_1^{-2} & & \\ & & & & \ddots & \\ & & & & & \sigma_1^{-2} \end{pmatrix} \begin{pmatrix} 1 \\ \vdots \\ 1 \end{pmatrix} \right\}^{-1}$$

$$= \left\{ (\sigma_0^{-2} \quad \cdots \quad \sigma_0^{-2} \quad \sigma_1^{-2} \quad \cdots \quad \sigma_1^{-2}) \begin{pmatrix} 1 \\ \vdots \\ 1 \end{pmatrix} \right\}^{-1}$$

$$= \frac{1}{n_0 \sigma_0^{-2} + n_1 \sigma_1^{-2}}$$

$$\hat{\beta} = D(\Omega^{-1} \delta + X_s' \Sigma_{ss}^{-1} y_s) = D(\infty^{-1} 0 + 1_s' \Sigma_{ss}^{-1} y_s)$$

$$= D \left\{ (1 \quad \cdots \quad 1) \begin{pmatrix} \sigma_0^{-2} & & & & & \\ & \ddots & & & & \\ & & \sigma_0^{-2} & & & \\ & & & \sigma_1^{-2} & & \\ & & & & \ddots & \\ & & & & & \sigma_1^{-2} \end{pmatrix} \begin{pmatrix} y_1 \\ \vdots \\ y_n \end{pmatrix} \right\}$$

$$= D(\sigma_0^{-2} y_{s0T} + \sigma_1^{-2} y_{s1T}).$$

Note: Here,

$$y_{s0T} = \sum_{i=1}^{n_0} y_i$$

denotes the total of the sample values with covariate $x_i = 0$, and

$$y_{s1T} = \sum_{i=n_0+1}^{n} y_i$$

denotes the total of the sample values with covariate $x_i = 1$.

Next, $(y_r \mid y_s) \sim N_m(E_*, V_*)$, where $m = N - n$ and:

$$E_* = X_r\hat{\beta} + \Sigma_{rs}\Sigma_{ss}^{-1}(y_s - X_s\hat{\beta}) = 1_r\hat{\beta} + 0 = 1_r\hat{\beta}$$

$$V_* = \Sigma_{rr} - \Sigma_{rs}\Sigma_{ss}^{-1}\Sigma_{sr} + (X_r - \Sigma_{rs}\Sigma_{ss}^{-1}X_s)D(X_r - \Sigma_{rs}\Sigma_{ss}^{-1}X_s)'$$

$$= \begin{pmatrix} \sigma_0^2 & & & & & \\ & \ddots & & & & \\ & & \sigma_0^2 & & & \\ & & & \sigma_1^2 & & \\ & & & & \ddots & \\ & & & & & \sigma_1^2 \end{pmatrix} - 0 + (1_r - 0)D(1_r - 0)'$$

$$= \begin{pmatrix} \sigma_0^2 & & & & & \\ & \ddots & & & & \\ & & \sigma_0^2 & & & \\ & & & \sigma_1^2 & & \\ & & & & \ddots & \\ & & & & & \sigma_1^2 \end{pmatrix} - D\begin{pmatrix} 1 & \cdots & 1 \\ \vdots & \ddots & \vdots \\ 1 & \cdots & 1 \end{pmatrix}.$$

Thus $(\bar{y} \mid y_s) \sim N(e_*, v_*)$, where:

$$e_* = \frac{y_{sT} + 1_r'E_*}{N} = \frac{1}{N}\{y_{sT} + (1 \ \cdots \ 1)1_r\hat{\beta}\} = \frac{1}{N}\{y_{sT} + m\hat{\beta}\}$$

$$v_* = \frac{1_r'V_*1_r}{N^2} = \frac{1}{N^2}(1 \ \cdots \ 1)$$

$$\times \left[\begin{pmatrix} \sigma_0^2 & & & & & \\ & \ddots & & & & \\ & & \sigma_0^2 & & & \\ & & & \sigma_1^2 & & \\ & & & & \ddots & \\ & & & & & \sigma_1^2 \end{pmatrix} - D\begin{pmatrix} 1 & \cdots & 1 \\ \vdots & \ddots & \vdots \\ 1 & \cdots & 1 \end{pmatrix} \right]\begin{pmatrix} 1 \\ \vdots \\ 1 \end{pmatrix}$$

$$= \frac{1}{N^2}\left[(\sigma_0^2 \ \cdots \ \sigma_0^2 \ \sigma_1^2 \ \cdots \ \sigma_1^2) - D(m \ \cdots \ m) \right]\begin{pmatrix} 1 \\ \vdots \\ 1 \end{pmatrix}$$

$$= \frac{1}{N^2}(m_0\sigma_0^2 + m_1\sigma_1^2 - Dm^2).$$

In summary, we have that $(\bar{y} \mid y_s) \sim N(e_*, v_*)$, where:

$$e_* = \frac{y_{sT} + m\hat{\beta}}{N}, \quad m = N - n, \quad \hat{\beta} = D(\sigma_0^{-2} y_{s0T} + \sigma_1^{-2} y_{s1T})$$

$$D = \frac{1}{n_0 \sigma_0^{-2} + n_1 \sigma_1^{-2}}, \quad v_* = \frac{m_0 \sigma_0^2 + m_1 \sigma_1^2 - m^2 D}{N^2}.$$

Numerically, we are given:

$$\sigma_0 = 0.08, \quad \sigma_1 = 1.2, \quad y_s = (2.1, 4.9, 2.3, 2.0, 0.2)' \quad (\text{thus } n = 5)$$
$$x = (0,1,0,0,1, \quad 1,1,1,0,0, \quad 1,1,1,1,0, \quad 0,1)'$$
$$(\text{thus } m = 12 \text{ and } N = n + m = 17)$$
$$x_s = (0,1,0,0,1)', \quad x_r = (1,1,1,0,0, \quad 1,1,1,1,0, \quad 0,1)'.$$

We now re-order the sample and nonsample values appropriately and so redefine:

$$y_s = (2.1, 2.0, 2.3, 4.9, 0.2)'$$
$$x_s = (0,0,0,1,1)'$$
$$x_r = (0,0,0,0,1, \quad 1,1,1,1,1, \quad 1,1)'.$$

Note: We have merely swapped units 2 and 4 in both y_s and x_s, respectively, so that all units with covariate 0 appear first and all units with covariate 1 appear last. We have also written the nonsample covariate vector x_r with all four zero values listed at the beginning.

We see that:

$$n_0 = 3, \ n_1 = 2, \quad m_0 = 4, \ m_1 = 8$$
$$y_{s0T} = 2.1 + 2.0 + 2.3 = 6.4, \quad y_{s1T} = 4.9 + 0.1 = 5.1,$$
$$y_{sT} = 6.4 + 5.1 = 11.5$$
$$\bar{y}_{s0} = 6.4 / 3 = 2.1333, \qquad \bar{y}_{s1} = 5.1/2 = 2.55,$$
$$\bar{y}_s = 11.5/5 = 2.3.$$

Thereby we obtain $(\bar{y} \mid y_s) \sim N(e_*, v_*)$, where:

$$D = \frac{1}{n_0 \sigma_0^{-2} + n_1 \sigma_1^{-2}} = \frac{1}{3/0.08^2 + 2/1.2^2} = 0.0021270$$

$$\hat{\beta} = D(\sigma_0^{-2} y_{s0T} + \sigma_1^{-2} y_{s1T}) = 0.0021270(6.4/0.08^2 + 5.1/1.2^2)$$
$$= 2.1345$$

$$e_* = \frac{y_{sT} + m\hat{\beta}}{N} = \frac{11.5 + 12 \times 2.1345}{17} = 2.1832$$

$$v_* = \frac{m_0 \sigma_0^2 + m_1 \sigma_1^2 - m^2 D}{N^2} = \frac{4 \times 0.08^2 + 8 \times 1.2^2 - 12^2 \times 0.0021250}{17^2}$$
$$= 0.038890.$$

Thus the predictive mean of the finite population mean $\bar{y}$ is $\hat{\beta} = 2.13$, and the 95% predictive interval for $\bar{y}$ is $(e_* \pm 1.96\sqrt{v_*}) = (1.80, 2.57)$.

R Code for Exercise 10.3

```
options(digits=4)
sig0=0.08; sig1=1.2; ys = c(2.1,2.0,2.3,4.9,0.2); n=length(ys)
xs=c(0,0,0,1,1); xr = c(0,0,0,0,1,  1,1,1,1,1,  1,1); m=length(xr); N = n+m
n1=sum(xs); n0=n-n1;   m1=sum(xr); m0=m-m1
c(n,n0,n1,  m,m0,m1,  N) # 5 3 2    12 4 8    17
ysT=sum(ys); ys1T=sum(ys*xs); ys0T=ysT-ys1T
ysbar=ysT/n; ys1bar=ys1T/n1;  ys0bar=ys0T/n0
c(ys0T,ys1T,ysT,  ys0bar,ys1bar,ysbar)
        # 6.400  5.100 11.500  2.133  2.550  2.300

D = 1/(  n0/ sig0^2 + n1/ sig1^2   ); betahat = D*(ys0T/ sig0^2 + ys1T/ sig1^2 )
estar=(1/N)*( ysT+m*betahat );
vstar=(1/N^2)*(m0* sig0^2+m1* sig1^2-D*m^2)
c(D,betahat,estar,vstar) # 0.002127 2.134564 2.183222 0.038890
hpdr=estar+c(-1,1)*qnorm(0.975)*sqrt(vstar);   c(hpdr) # 1.797 2.570
```

10.7 The normal-normal-gamma finite population model

For the models so far considered in this chapter, the superpopulation variance σ^2 parameter or variance-covariance matrix parameter Σ has been assumed to be known.

If this parameter were unknown, as might typically be the case in practice, then an estimate could be computed from the data via some method (which need not necessarily be Bayesian) and substituted into the equations derived.

This strategy, which may be considered an example of *empirical Bayes techniques*, may sometimes work well, especially if based on a sufficiently large sample size.

For example, recall that in the case of no covariates, with the superpopulation variance σ^2 known, the $1 - \alpha$ CPDR for $\bar{y}$ is

$$\left(\bar{y}_s \pm z_{\alpha/2} \frac{\sigma}{\sqrt{n}} \sqrt{1 - \frac{n}{N}} \right).$$

Now suppose that n is large and we estimate σ^2 by the sample variance,

$$s^2 = \frac{1}{n-1} \sum_{i=1}^{n} (y_i - \bar{y}_s)^2 .$$

Then the result is the same as the classical design-based CI one would use in the same situation of a large sample size.

However, this strategy will not work well generally. For example, if n is small then it will lead to an interval which has a frequentist coverage well below the intended level of $1 - \alpha$. In such cases, the problem could be addressed to some extent by applying an adjustment which reflects uncertainty regarding the unknown variance parameter. However, the nature of this type of adjustment would be ad hoc and lead to possibly other problems with the inference.

Perhaps the best way to deal with uncertainty regarding the variance parameter is to incorporate it into the finite population model as yet another random variable with its own prior distribution, i.e. to add another level to the hierarchical structure of that model. This is the approach we will now take. Note that parts of the exposition below will be a review of material already covered in previous chapters.

With the above in mind, and with quantities as defined previously, we define the *normal-normal-gamma finite population model* as follows:

$$(y \mid \beta, \lambda) \sim N_N(X\beta, \Sigma / \lambda)$$
$$(\beta \mid \lambda) \sim N_p(\delta, \Omega)$$
$$\lambda \sim G(\eta, \tau) .$$

A problem with this model is that is involves an additional nuisance parameter to deal with relative to the *normal-normal finite population model*, namely λ. This means that the predictive pdf of the nonsample vector cannot be obtained so easily.

That density is now

$$f(y_r \mid y_s) = \int\int f(y_r, \beta, \lambda \mid y_s) d\beta d\lambda \propto \int\int f(y, \beta, \lambda) d\beta d\lambda,$$

(10.7)

where $f(y, \beta, \lambda) = f(\lambda)f(\beta \mid \lambda)f(y \mid \beta, \lambda)$

$$\propto \lambda^{\eta-1}e^{-\tau\lambda} \times \exp\left\{-\frac{1}{2}(\beta - \delta)'\Omega^{-1}(\beta - \delta)\right\}$$

$$\times \lambda^{N/2} \exp\left\{-\frac{1}{2}\lambda(y - X\beta)'\Sigma^{-1}(y - X\beta)\right\}$$

is the joint density of all random variables involved in the model, namely the N finite population values, $y_1, \ldots, y_N$, and the $p+1$ model parameters, namely $\lambda, \beta_1, \ldots, \beta_p$.

In an attempt to perform the second double integral at (10.7) (which is actually a $(p+1)$-fold integral), we may first integrate with respect to λ and obtain

$$f(y_r \mid y_s) \propto \int_{-\infty}^{\infty} \frac{\exp\{-(1/2)(\beta - \delta)'\Omega^{-1}(\beta - \delta)\}}{[\tau + (1/2)(y - X\beta)'\Sigma^{-1}(y - X\beta)]^{\eta+N/2}} d\beta$$

(after recognising a gamma density in λ), or first integrate with respect to β and obtain

$$f(y_r \mid y_s) \propto \int_0^{\infty} \frac{\lambda^{\eta+N/2-1}}{\det(\Omega^{-1} + \lambda X'\Sigma^{-1}X)} \times \exp\left\{-\lambda\left(\tau + \frac{1}{2}y'\Sigma^{-1}y\right)\right.$$

$$+ (\Omega^{-1}\delta + \lambda X'\Sigma^{-1}y)'(\Omega^{-1} + \lambda X'\Sigma^{-1}X)^{-1}(\Omega^{-1}\delta + \lambda X'\Sigma^{-1}y)\bigg\} d\lambda$$

(after recognising a multivariate normal density in β).

Either way, the remaining integral is in general impossible to perform analytically, and the posterior predictive distributions of the nonsample vector and linear combinations of that vector (such as the finite population mean and total) are not normally distributed. However, there is an important special case which simplifies matters considerably.

10.8 Special cases of the normal-normal-gamma finite population model

---

Theorem 10.1: Suppose there is priori ignorance regarding β and it is appropriate to set $\delta = 0$ and $\Omega = \infty$, so that

$$f(\beta \mid \lambda) \propto f(\beta) \propto 1, \, \beta \in \Re .$$

Then the predictive distribution of the finite population mean is given by

$$\left(\left. \frac{\overline{y} - a}{b} \right| y_s \right) \sim t(2\eta + n - p) ,$$

where: $a = \dfrac{y_{sT} + 1'_r [X_r \hat{\beta} + \Sigma_{rs} \Sigma_{ss}^{-1} (y_s - X_s \hat{\beta})]}{N}$

$$b^2 = \frac{1'_r [\Sigma_{rr} - \Sigma_{rs} \Sigma_{ss}^{-1} \Sigma_{sr} + ADA'] 1_r [2\tau + (y_s - X_s \hat{\beta})' \Sigma_{ss}^{-1} (y_s - X_s \hat{\beta})]}{(2\eta + n - p)N^2}$$

$$\hat{\beta} = DX'_s \Sigma_{ss}^{-1} y_s , \qquad D = (X'_s \Sigma_{ss}^{-1} X_s)^{-1} , \qquad A = X_r - \Sigma_{rs} \Sigma_{ss}^{-1} X_s .$$

Note: Here, $\hat{\beta}$ is the MLE of β, and also the posterior mean of β under the simpler *normal-normal finite population model* with improper prior $f(\beta) \propto 1, \beta \in \Re$ (and σ^2 known).

Theorem 10.1 can be proved by first noting that:

(a) $(\lambda \mid y_s)$ is gamma (with parameters that can be obtained by integrating $f(\beta, \lambda \mid y_s)$ with respect to β), and

(b) $(\overline{y} \mid y_s, \lambda)$ is normal (with parameters that can be obtained by examining the normal-normal finite population model above).

Using these two distributions, one can solve for the predictive density of the finite population mean via the identity

$$f(\overline{y} \mid y_s) = \int f(\overline{y}, \lambda \mid y_s) d\lambda = \int f(\overline{y} \mid y_s, \lambda) f(\lambda \mid y_s) d\lambda .$$

A special case of Theorem 10.1 which assumes a priori ignorance of λ by way of setting $\eta = \tau = 0$ can be found in Royall and Pfeffermann (1982).

If we further assume conditional independence (which may expressed by writing $\Sigma = I_N$) and no auxiliary information ($p=1$ and $X = 1_N$), the result in Theorem 10.1 reduces to

$$\left(\left. \frac{\bar{y} - \bar{y}_s}{(s_s / n)\sqrt{1 - n/N}} \right| y_s \right) \sim t(n-1),$$

where $s_s^2 = \dfrac{1}{n-1}\displaystyle\sum_{i=1}^{n}(y_i - \bar{y}_s)^2$ (the sample variance).

This result was already proved in a previous chapter without the involvement of vectors and matrices. Again note that the result leads to point estimates and interval estimates which are identical to those which one might construct using a design-based approach (see Cochran, 1977, Section 2.8).

Exercise 10.4 Proof of Theorem 1

Prove Theorem 10.1 above.

Solution to Exercise 10.4

Using the procedure outlined above, we first derive the unconditional pdf of λ as follows:

$$f(\lambda \mid y_s) = \int f(\beta, \lambda \mid y_s)d\beta \propto \int \lambda^{\eta-1}e^{-\tau\lambda} \times 1 \times \lambda^{\frac{n}{2}}\exp\left(-\frac{\lambda}{2}Q_1\right)d\beta,$$

where

$$Q_1 = (y_s - X_s\beta)'\Sigma_{ss}^{-1}(y_s - X_s\beta)$$
$$= y_s'\Sigma_{ss}^{-1}y_s - y_s'\Sigma_{ss}^{-1}X_s\beta - \beta'X_s'\Sigma_{ss}^{-1}y_s + \beta'X_s'\Sigma_{ss}^{-1}X_s\beta.$$

Now equate Q_1 with

$$Q_2 = (\beta - T)'M(\beta - T)' + R \quad \text{(where } R \text{ stands for 'remainder')}$$
$$= \beta'M\beta - \beta'MT - T'M\beta + T'MT + R.$$

We see that

$$M = X_s'\Sigma_{ss}^{-1}X_s \text{ and } MT = X_s'\Sigma_{ss}^{-1}y_s,$$

so that

$$T = M^{-1}(MT) = (X_s'\Sigma_{ss}^{-1}X_s)^{-1}X_s'\Sigma_{ss}^{-1}y_s.$$

Note: Here, T is the same as $\hat{\beta}$ in Theorem 10.1.

Also, $R = y_s' \Sigma_{ss}^{-1} y_s - T'MT = (y_s - X_s T)' \Sigma_{ss}^{-1} (y_s - X_s T)$.

Note: This is easily proved by noting that the RHS here is
$$y_s' \Sigma_{ss}^{-1} y_s - y_s' \Sigma_{ss}^{-1} X_s T - T'X_s' \Sigma_{ss}^{-1} y_s + T'X_s' \Sigma_{ss}^{-1} X_s T$$
where
$$y_s' \Sigma_{ss}^{-1} X_s T = (y_s' \Sigma_{ss}^{-1} X_s T)' = T'X_s' \Sigma_{ss}^{-1} y_s$$
(since $y_s' \Sigma_{ss}^{-1} X_s T$ is a *scalar* quantity), and where
$$T'X_s' \Sigma_{ss}^{-1} X_s T = T'MT = T' \, \underline{X_s' \Sigma_{ss}^{-1} X_s} \, \underline{(X_s' \Sigma_{ss}^{-1} X_s)^{-1}} X_s' \Sigma_{ss}^{-1} y = T'X_s' \Sigma_{ss}^{-1} y,$$
so that the RHS equals
$$y_s' \Sigma_{ss}^{-1} y_s - T'MT - T'MT + T'MT = y_s' \Sigma_{ss}^{-1} y_s - T'MT .$$

Thus
$$f(\lambda \mid y_s) \propto \int \lambda^{\eta-1} e^{-\tau\lambda} \times 1 \times \lambda^{\frac{n}{2}} \exp\left(-\frac{\lambda}{2}\left[(\beta - T)'M(\beta - T)' + R \right] \right) d\beta$$

$$= \lambda^{\eta+\frac{n}{2}-1} \exp\left(-\lambda\left(\tau + \frac{R}{2} \right) \right) \times I ,$$

where
$$I = \int \exp\left(-\frac{1}{2}(\beta - T)'\left(\frac{M^{-1}}{\lambda} \right)^{-1} (\beta - T)' \right) d\beta$$

$$= (2\pi)^{\frac{p}{2}} \left\{ \det\left(\frac{M^{-1}}{\lambda} \right) \right\}^{-\frac{1}{2}}$$

(using standard multivariate normal theory)

$$\propto \lambda^{\frac{p}{2}} \quad \text{(since } M = X_s' \Sigma_{ss}^{-1} X_s \text{ is a } p \text{ by } p \text{ matrix).}$$

It follows that
$$f(\lambda \mid y_s) \propto \lambda^{\eta+\frac{n}{2}+\frac{p}{2}-1} \exp\left(-\lambda\left(\tau + \frac{R}{2} \right) \right) = \lambda^{\frac{A}{2}-1} \exp\left(-\frac{B}{2}\lambda \right),$$

where:
$$A = 2\eta + n - p, \quad B = 2\tau + R, \quad R = (y_s - X_s T)' \Sigma_{ss}^{-1} (y_s - X_s T).$$

We thereby arrive at the required distribution,

$$(\lambda \mid y_s) \sim G(A/2, B/2),$$

which may also be expressed by writing

$$(B\lambda \mid y_s) \sim G(A/2, 1/2) = \chi^2(A).$$

Having derived the posterior dsn of λ, we now observe that

$$(\bar{y} \mid y_s, \lambda) \sim N(e_0, v_0),$$

where:

$$e_0 = \frac{1}{N}(y_{sT} + 1_r' E_0), \qquad E_0 = X_r T + \Sigma_{rs} \Sigma_{ss}^{-1}(y_s - X_s T)$$

$$v_0 = \frac{1_r' V_0 1_r}{N^2 \lambda} \equiv \frac{w_0}{\lambda}, \qquad w_0 = \frac{1_r' V_0 1_r}{N^2},$$

$$V_0 = G + AM^{-1}A'$$

$$G = \Sigma_{rr} - \Sigma_{rs}\Sigma_{ss}^{-1}\Sigma_{sr}, \qquad A = X_r - \Sigma_{rs}\Sigma_{ss}^{-1}X_s.$$

Note: We have here simply applied the theory of the *normal-normal finite population model* with $\Omega = \infty$ and with quantities such as Σ_{sr} and Σ_{ss} replaced by Σ_{sr}/λ and Σ_{ss}/λ, etc.

Therefore

$$f(\bar{y} \mid y_s) = \int f(\bar{y} \mid y_s, \lambda) f(\lambda \mid y_s) d\lambda$$

$$\propto \int \lambda^{\frac{1}{2}} \exp\left\{-\frac{\lambda}{2w_0}(\bar{y} - e_0)^2\right\} \times \lambda^{\frac{A}{2}-1} \exp\left(-\frac{B}{2}\lambda\right) d\lambda$$

$$= \int \lambda^{\frac{A+1}{2}-1} \exp\left\{-\lambda\left[\frac{B}{2} + \frac{(\bar{y} - e_0)^2}{2w_0}\right]\right\} d\lambda$$

$$\propto \left[\frac{B}{2} + \frac{(\bar{y} - e_0)^2}{2w_0}\right]^{-\left(\frac{A+1}{2}\right)} \qquad \propto \left[1 + \frac{(\bar{y} - e_0)^2}{Bw_0}\right]^{-\left(\frac{A+1}{2}\right)}$$

$$\propto \left[1 + \frac{\left\{\dfrac{(\bar{y} - e_0)^2}{Bw_0/A}\right\}}{A}\right]^{-\left(\frac{A+1}{2}\right)} \qquad \propto \left[1 + \frac{\left(\dfrac{\bar{y} - e_0}{\sqrt{Bw_0/A}}\right)^2}{A}\right]^{-\left(\frac{A+1}{2}\right)}.$$

It follows that $\left(\dfrac{\bar{y}-e_0}{h_0}\middle|\, y_s\right) \sim t(A)$, where $h_0^2 = \dfrac{Bw_0}{A}$.

Here: $A = 2\eta + n - p$ (which is the same as the degrees of freedom in the t distribution in Theorem 10.1)

$$e_0 = \frac{1}{N}(y_{sT} + 1_r' E_0) = \frac{y_{sT} + 1_r'[X_r T + \Sigma_{rs}\Sigma_{ss}^{-1}(y_s - X_s T)]}{N}$$

(which is the same as a in Theorem 10.1).

$$h_0^2 = \frac{B}{A} w_0 = \frac{2\tau + R}{2\eta + n - p} \times \frac{1_r' V_0 1_r}{N^2}$$

$$= \frac{[2\tau + (y_s - X_s T)'\Sigma_{ss}^{-1}(y_s - X_s T)]}{(2\eta + n - p)N^2} 1_r'(G + AM^{-1}A')1_r$$

$$= \frac{[2\tau + (y_s - X_s T)'\Sigma_{ss}^{-1}(y_s - X_s T)]}{(2\eta + n - p)N^2}$$

$$\times 1_r' \left\{ \Sigma_{rr} - \Sigma_{rs}\Sigma_{ss}^{-1}\Sigma_{sr} + (X_r - \Sigma_{rs}\Sigma_{ss}^{-1}X_s)(X_s'\Sigma_{ss}^{-1}X_s)^{-1}(X_r - \Sigma_{rs}\Sigma_{ss}^{-1}X_s)' \right\} 1_r$$

(which is the same as b^2 in Theorem 10.1).

That completes the proof of Theorem 10.1.

10.9 The case of an informative prior on the regression parameter

If there is *some* prior information available regarding the regression parameter β then $\Omega < \infty$ and Theorem 10.1 above cannot be applied. So the problem of inference on the finite population mean $\bar{y}$ becomes much more difficult.

However, that difficulty can be easily 'sidestepped' via Monte Carlo methods based on a random sample from the predictive distribution of $\bar{y}$, namely

$$\bar{y}^{(1)},...,\bar{y}^{(J)} \sim iid\ f(\bar{y}\,|\,y_s).$$

With such a sample we can, for example, estimate $\bar{y}$'s predictive mean, namely $\hat{\bar{y}} = E(\bar{y} \mid y_s)$, by the average of $\bar{y}^{(1)}, ..., \bar{y}^{(J)}$, and estimate $\bar{y}$'s 95% CPDR by the empirical 0.025 and 0.975 quantiles of $\bar{y}^{(1)}, ..., \bar{y}^{(J)}$.

This then raises the question of how the Monte Carlo sample can be obtained. In this context, we may employ the method of composition via the equation

$$f(\bar{y}, \beta, \lambda \mid y_s) = f(\bar{y} \mid y_s, \beta, \lambda) f(\beta, \lambda \mid y_s).$$

Thus, we first generate a sample from the joint posterior distribution the two parameters,

$$(\beta^{(1)}, \lambda^{(1)}), ..., (\beta^{(J)}, \lambda^{(J)}) \sim iid \ f(\beta, \lambda \mid y_s).$$

and then for each $j = 1, ..., J$ we sample

$$\bar{y}^{(j)} \sim f(\bar{y} \mid y_s, \beta^{(j)}, \lambda^{(j)}) \sim N\left(\frac{y_{sT} + 1'_r X_r \beta^{(j)} 1_r}{N}, \frac{1'_r \Sigma_{rr} 1_r}{N^2 \lambda^{(j)}} \right).$$

This in turn raises the question of how to obtain the sample from $f(\beta, \lambda \mid y_s)$. In this case an ideal solution is to apply a Gibbs sampler defined by the following conditional distributions:

1. $(\beta \mid y_s, \lambda) \sim N_p(\tilde{\beta}, D)$,

   where: $\tilde{\beta} = D(\Omega^{-1} \delta + \lambda X'_s \Sigma_{ss}^{-1} y_s)$

   $D = (\Omega^{-1} + \lambda X'_s \Sigma_{ss}^{-1} X_s)^{-1}$

2. $(\lambda \mid y_s, \beta) \sim G\left(\eta + \frac{n}{2}, \tau + \frac{1}{2}(y_s - X_s\beta)' \Sigma_{ss}^{-1}(y_s - X_s\beta) \right).$

Note: The first of these distributions derives directly from the *normal-normal finite population model* with Σ_{sr} and Σ_{ss} replaced by Σ_{sr} / λ and Σ_{ss} / λ, etc.

The second conditional is obtained by noting that

$$f(\lambda \mid y_s, \beta) \propto f(\lambda, \beta \mid y_s)$$
$$\propto f(\lambda, \beta, y_s)$$

$$= f(\lambda)f(\beta \mid \lambda)f(y_s \mid \lambda, \beta)$$

$$\propto \lambda^{\eta-1}e^{-\tau\lambda} \times \exp\left(-\frac{1}{2}(\beta-\delta)'\Omega^{-1}(\beta-\delta)\right)$$

$$\times \lambda^{\frac{n}{2}} \exp\left(-\frac{\lambda}{2}(y_s - X_s\beta)'\Sigma_{ss}^{-1}(y_s - X_s\beta)\right)$$

$$\propto \lambda^{\eta+\frac{n}{2}-1} \exp\left(-\lambda\left\{\tau + \frac{1}{2}(y_s - X_s\beta)'\Sigma_{ss}^{-1}(y_s - X_s\beta)\right\}\right).$$

Exercise 10.5 Practice with the normal-normal-gamma finite population model

In the context of the normal-normal-gamma finite population model, suppose we obtain a sample of size $n = 5$, with values given by

$$y_s = (y_1, ..., y_n)' = (5.6, 2.3, 8.4, 5.1, 4.3)'$$

via SRSWOR from a finite population of size $N = 15$.

Find the predictive mean and 95% central predictive density region for the finite population mean $\bar{y}$ in each of the following scenarios.

(a) There are no covariates, the population values are conditionally *iid* and there is no prior information available regarding the model parameters.

(b) The population values are conditionally *independent*, the ith population value has mean $x_i\beta$ and variance x_i / λ ($i = 1,...,N$), the population covariate vector is

$$x = (x_1, ..., x_N)' = (9.3, 4.6, 15.0, 11.2, 7.8, \quad 2.4, 6.6, 3.0, 2.1, 7.3,$$
$$5.5, 8.0, 2.4, 4.2, 5.5)',$$

and there is no prior information regarding the model parameters.

(c) There are no covariates, the population values are conditionally iid, the prior on the normal mean is normal with mean 10 and variance 2.25, and (independently) the prior on the normal precision parameter (inverse of the normal variance) is gamma with mean 2 and variance 1/2 (or equivalently, gamma with parameters 8 and 4).

Solution to Exercise 10.5

(a) In this case, Theorem 10.1 reduces to

$$\left(\frac{\bar{y} - \bar{y}_s}{(s_s / n)\sqrt{1 - n/N}} \middle| y_s \right) \sim t(n-1),$$

where: $\bar{y}_s = \dfrac{1}{n}(y_1 + ... + y_n) = 5.140$

$$s_s^2 = \frac{1}{n-1}\sum_{i=1}^{n}(y_i - \bar{y})^2 = 4.9030.$$

So the required predictive mean and 95% predictive interval of $\bar{y}$ are

$$\bar{y}_s = 5.140 \quad \text{and} \quad \left(\bar{y}_s \pm t_{\alpha/2}(n-1)\frac{s_s}{n}\sqrt{1 - \frac{n}{N}} \right) = (2.8951, 7.3849).$$

(b) In this case (a variation of Bayesian ratio estimation as discussed earlier) we apply Theorem 10.1 with:

$$p = 1, \ \eta = \tau = 0, \ X = x, \ \Sigma = diag(x) = \begin{pmatrix} x_1 & & \\ & \ddots & \\ & & x_N \end{pmatrix}.$$

Instead of deriving a 'simple' general algebraic expression for the predictive distribution of the finite population mean in this case, we can obtain the specific required result more quickly by directly applying the formulae in Theorem 10.1 using R. An advantage of this approach is that it leads us to write a general algorithm in R which can be straightaway used in other situations requiring Theorem 10.1. Also, the algorithm can be used to check our answer to part (a).

Thereby we obtain the result that

$$\left(\frac{\bar{y} - a}{b} \middle| y_s \right) \sim t(c),$$

where $a = 3.3945$, $b = 0.1159$ and $c = 2\eta + n - p = 4$.

So the required predictive mean and 95% predictive interval of $\bar{y}$ are

$$\bar{y}_s = 3.3945 \text{ and } \left(a \pm t_{\alpha/2}(c)b \right) = (3.0725, 3.7164).$$

Note: This inference is *lower* than that in (a) because the mean of the covariate values in the nonsample is 4.7, which is much lower than their mean in the sample, 9.58. The regression coefficient β in our model is estimated as 0.5365, reflecting the positive linear relationship between the x and y values in the sample.

(c) In this case, a good option is to first employ the Gibbs sampler to generate a random sample from the joint posterior distribution of β and λ, with:

$$p = 1, \ \delta = 10, \ \Omega = 9, \ \eta = 8, \ \tau = 4, \ X = 1_N, \ \Sigma = diag(1_N).$$

The two conditional distributions are:

1. $(\beta \mid y_s, \lambda) \sim N_p(\tilde{\beta}, D)$,

    where:

$$\tilde{\beta} = D(\Omega^{-1}\delta + \lambda X_s' \Sigma_{ss}^{-1} y_s)$$
$$D = (\Omega^{-1} + \lambda X_s' \Sigma_{ss}^{-1} X_s)^{-1}$$

2. $(\lambda \mid y_s, \beta) \sim G\left(\eta + \dfrac{n}{2}, \tau + \dfrac{1}{2}(y_s - X_s\beta)'\Sigma_{ss}^{-1}(y_s - X_s\beta)\right)$.

But, by analogy with the simpler *normal-normal model* and *normal-gamma model*, these conditionals must be equivalent to:

1. $(\beta \mid y_s, \lambda) \sim N(\beta_\lambda, \sigma_\lambda^2)$,

    where:

$$\beta_\lambda = (1 - k_\lambda)\beta_0 + k_\lambda \bar{y}_s$$
$$\sigma_\lambda^2 = \frac{k_\lambda}{n\lambda}, k_\lambda = \frac{n}{n + 1/(\lambda \sigma_0^2)}$$
$$\beta_0 = 10, \quad \sigma_0 = 3$$

2. $(\lambda \mid y_s, \beta) \sim G\left(\eta + \dfrac{n}{2}, \tau + \dfrac{n}{2}s_\beta^2\right)$,

    where

$$s_\beta^2 = \frac{1}{n}\sum_{i=1}^{n}(y_i - \beta)^2.$$

Either way, implementing this Gibbs sampler for 10,100 iterations with a burn-in of 100 we obtain the trace plots and histograms for β and λ in Figure 10.2. (The two subplots on the left are for β, and the two on the right are for λ. The histograms do not include the first 100 iterations.)

Thinning the last 10,000 values of each parameter by a factor of 10 we obtain an approximately random sample of size $J = 1,000$ from the joint posterior distribution of the two parameters, namely
$$(\beta_j, \lambda_j) \sim iid \ f(\beta, \lambda \mid y_s), j = 1,\ldots,J.$$

The sample ACFs over the entire sample of 10,000 and over the thinned sample of 1,000 are shown for each of β and λ in Figure 10.3. (E.g. the top-left subplot is for β over the entire sample of 10,000.) The thinning process has virtually eliminated all signs of autocorrelation.

Figure 10.2 Trace plots and histograms

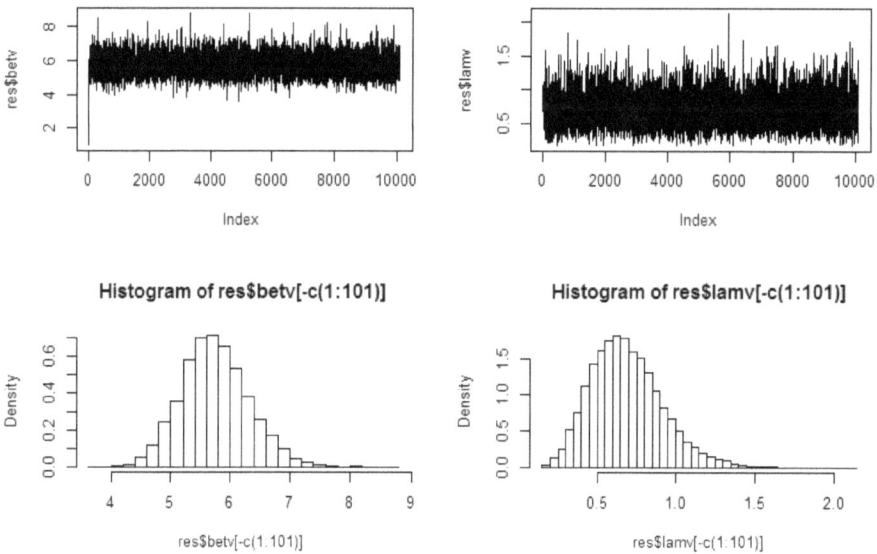

Figure 10.3 Sample ACFs
(Top two: J = 10,000; Bottom two: J = 1,000)

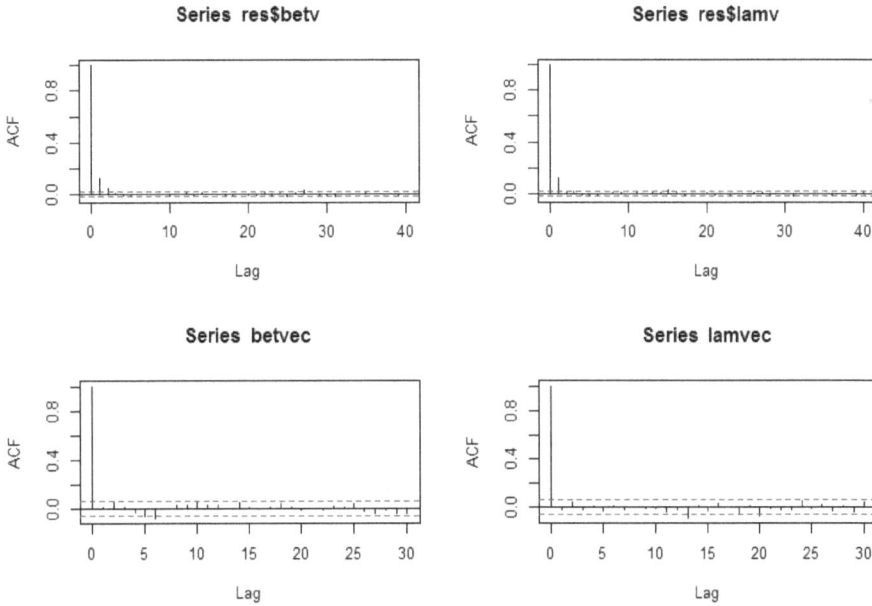

Using our sample from the joint posterior of the two parameters we now generate a sample from the predictive distribution of the nonsample mean by drawing

$$\overline{y}_r^{(j)} \sim f(\overline{y}_r \mid y_s, \beta_j, \lambda_j) \sim N\left(\beta_j, \frac{1}{(N-n)\lambda_j}\right) \text{ for each } j = 1, \ldots, J.$$

Note: The result is

$$\overline{y}_r^{(1)}, \ldots, \overline{y}_r^{(J)} \sim iid \; f(\overline{y}_r \mid y_s),$$

by virtue of the method of composition and the equation

$$f(\overline{y}_r, \beta, \lambda \mid y_s) = f(\overline{y}_r \mid y_s, \beta, \lambda) f(\beta, \lambda \mid y_s).$$

We next form a random sample from the predictive distribution of the finite population mean by calculating

$$\overline{y}^{(j)} = \frac{1}{N}\left(n\overline{y}_s + (N-n)\overline{y}_r^{(j)}\right) \text{ for each } j = 1, \ldots, J.$$

Note: The result is $\overline{y}^{(1)}, \ldots, \overline{y}^{(J)} \sim iid \; f(\overline{y} \mid y_s).$

We now estimate $\bar{y}$ (and $\bar{y}$'s predictive mean, $\hat{\bar{y}} = E(\bar{y} \mid y_s)$) by

$$\bar{\bar{y}} = \frac{1}{J} \sum_{j=1}^{J} \bar{y}^{(J)} = 5.555,$$

with 95% CI for $\hat{\bar{y}}$ equal to

$$\left(\bar{\bar{y}} \pm 1.96 \sqrt{\frac{1}{J(J-1)} \sum_{j=1}^{J} (\bar{\bar{y}} - \bar{y}^{(J)})^2} \right) = (5.526, 5.584).$$

We also estimate the 95% CPDR for $\bar{y}$ by (4.685, 6.633), where the bounds of this interval are the empirical 0.025 and 0.975 quantiles of $\bar{y}^{(1)}, ..., \bar{y}^{(J)}$.

Another approach to performing Monte Carlo inference on $\bar{y}$ is via Rao-Blackwell methods. This approach does not require the sample $\bar{y}_r^{(1)}, ..., \bar{y}_r^{(J)}$ and should provide more accurate Monte Carlo estimates.

The idea is based on the identities:

$$f(\bar{y} \mid y_s) = \int f(\bar{y}, \beta, \lambda \mid y_s) d\beta d\lambda$$

$$= \int f(\bar{y} \mid y_s, \beta, \lambda) f(\beta, \lambda \mid y_s) d\beta d\lambda$$

$$\hat{\bar{y}} = E(\bar{y} \mid y_s) = E_{\beta,\lambda} \left\{ E(\bar{y} \mid y_s, \beta, \lambda) \mid y_s \right\}$$

$$f(\bar{y} \mid y_s) = E_{\beta,\lambda} \left\{ f(\bar{y} \mid y_s, \beta, \lambda) \mid y_s \right\}.$$

Now note once again that:

$$\bar{y} = \frac{1}{N} \left(n\bar{y}_s + (N-n)\bar{y}_r \right)$$

$$(\bar{y}_r \mid y_s, \beta, \lambda) \sim N \left(\beta, \frac{1}{(N-n)\lambda} \right).$$

So we now define:

$$e(\beta, \lambda) = E(\bar{y} \mid y_s, \beta, \lambda)$$

$$= \frac{1}{N} \left(n\bar{y}_s + (N-n)E(\bar{y}_r \mid y_s, \beta, \lambda) \right)$$

$$= \frac{1}{N} \left(n\bar{y}_s + (N-n)\beta \right)$$

$$v(\beta,\lambda) = V(\bar{y} \mid y_s, \beta, \lambda)$$

$$= \frac{(N-n)^2}{N^2} V(\bar{y}_r \mid y_s, \beta, \lambda)$$

$$= \frac{(N-n)^2}{N^2} \times \frac{1}{(N-n)\lambda} = \frac{N-n}{N^2\lambda}$$

$$e_j = e(\beta_j, \lambda_j) = \frac{1}{N}\left(n\bar{y}_s + (N-n)\beta_j\right)$$

$$v_j = v(\beta_j, \lambda_j) = \frac{N-n}{N^2\lambda_j}.$$

Note: Since $e(\beta,\lambda)$ does not depend on λ, we may also write $e(\beta,\lambda)$ as $e(\beta)$. Likewise, since $v(\beta,\lambda)$ does not depend on β, we may also write $v(\beta,\lambda)$ as $v(\lambda)$.

Then the Rao-Blackwell estimate of $\bar{y}$ (and $\hat{\bar{y}} = E(\bar{y} \mid y_s)$) is

$$\bar{e} = \frac{1}{J}\sum_{j=1}^{J} e_j = 5.557,$$

with 95% CI for $\hat{\bar{y}}$ working out as

$$\left(\bar{e} \pm 1.96\sqrt{\frac{1}{J(J-1)}\sum_{j=1}^{J}(\bar{e}-e_j)^2}\right) = (5.534, 5.581).$$

Note: The width of this Rao-Blackwell CI is $5.581 - 5.534 = 0.046$, which (as could be expected) is less than that of the earlier CI, namely $5.584 - 5.526 = 0.058$.

We can now also obtain the Rao-Blackwell estimate of the CPDR for $\bar{y}$.

First, the Rao-Blackwell estimate of the predictive density of $\bar{y}$ (that is, of $f(\bar{y} \mid y_s)$) is

$$\bar{f}(\bar{y} \mid y_s) = \frac{1}{J}\sum_{j=1}^{J} f(\bar{y} \mid y_s, \mu_j, \theta_j)$$

$$= \frac{1}{J}\sum_{j=1}^{J} \frac{1}{\sqrt{v_j}\sqrt{2\pi}} \exp\left\{-\frac{1}{2v_j}(\bar{y}-e_j)^2\right\}.$$

Note: The simplest and most 'basic' estimate of $f(\bar{y}|y_s)$ is the 'histogram' estimate, $\hat{f}(\bar{y}|y_s)$, obtained by smoothing a histogram of the sampled values $\bar{y}^{(1)},...,\bar{y}^{(J)} \sim iid \; f(\bar{y}|y_s)$.

The Rao-Blackwell estimate of the 95% CPDR of $\bar{y}$ is (L,U), where L and U satisfy:

$$\int_{-\infty}^{L} \frac{1}{J}\sum_{j=1}^{J} \frac{1}{\sqrt{v_j}\sqrt{2\pi}} \exp\left\{-\frac{1}{2v_j}(\bar{y}-e_j)^2\right\} d\bar{y} = 0.025$$

$$\int_{-\infty}^{U} \frac{1}{J}\sum_{j=1}^{J} \frac{1}{\sqrt{v_j}\sqrt{2\pi}} \exp\left\{-\frac{1}{2v_j}(\bar{y}-e_j)^2\right\} d\bar{y} = 0.975 \; .$$

To obtain L we rewrite the first of these two equations as

$$\frac{1}{J}\sum_{j=1}^{J} P(X_j < L) = 0.025 \; ,$$

where $X_j \sim N(e_j,v_j)$, or equivalently as

$$\frac{1}{J}\sum_{j=1}^{J} \Phi\left(\frac{L-e_j}{\sqrt{v_j}}\right) = 0.025 \quad \text{(where } \Phi \text{ is the standard normal cdf).}$$

We can now solve this equation in a number of ways, for example by minimising the function

$$g(L) = \left\{\frac{1}{J}\sum_{j=1}^{J}\Phi\left(\frac{L-e_j}{\sqrt{v_j}}\right) - 0.025\right\}^2$$

(whose minimum is 0 at the required L),
e.g. using the optim() function in R.

Likewise we can obtain U by using optim() to minimise

$$h(U) = \left\{\frac{1}{J}\sum_{j=1}^{J}\Phi\left(\frac{L-e_j}{\sqrt{v_j}}\right) - 0.975\right\}^2$$

(whose minimum is 0 at the required U).

Note: We could also obtain L and U using trial and error or the Newton-Raphson algorithm.

Implementing the above procedure we arrive at the required Rao-Blackwell estimate of the central predictive region for the finite population mean: $(L,U) = (4.707, 6.542)$.

Note: This is similar to the previous 'histogram' estimate of the CPDR, $(4.685, 6.633)$.

Figure 10.4 shows a histogram of the $J = 1,000$ simulated values $\overline{y}^{(1)},...,\overline{y}^{(J)} \sim iid\ f(\overline{y}\,|\,y_s)$, together with the histogram estimate $\overline{\overline{y}}$ and the Rao-Blackwell estimate $\overline{e}$ of $\hat{\overline{y}} = E(\overline{y}\,|\,y_s)$. Also shown are the two corresponding 95% CIs for $\hat{\overline{y}}$. The histogram is overlaid with the histogram estimate $\hat{f}(\overline{y}\,|\,y_s)$ and the Rao-Blackwell estimate $\overline{f}(\overline{y}\,|\,y_s)$ of $f(\overline{y}\,|\,y_s)$. It will be observed that the Rao-Blackwell estimate provides the smoother result.

Figure 10.4 Inferences on the finite population mean

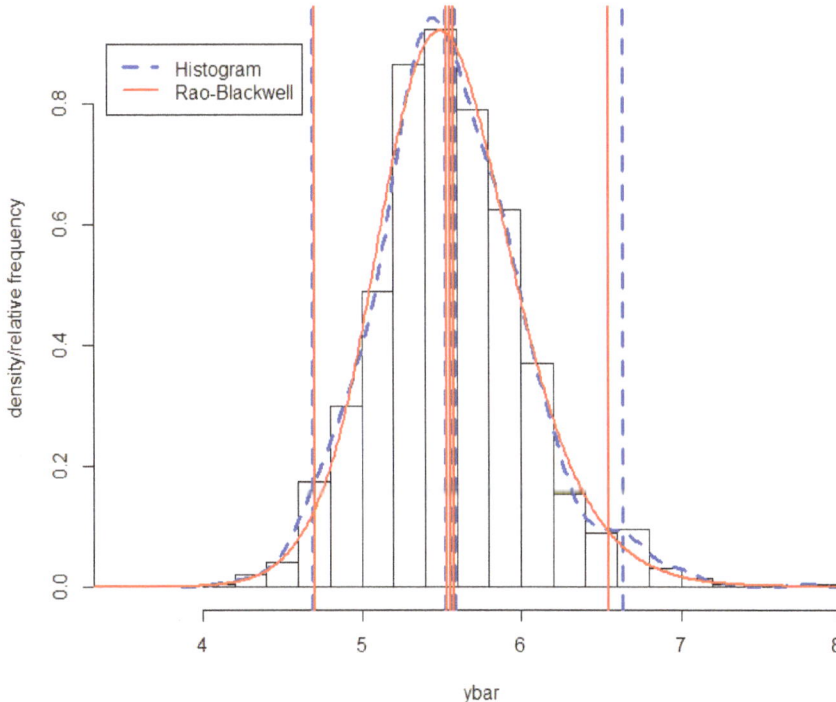

R Code for Exercise 10.5

```
# (a)

options(digits=4); N = 15; ys = c(5.6,2.3,8.4,5.1,4.3); n = length(ys)
est=mean(ys); ss2=var(ys);  varybar=(ss2/n)*(1-n/N);   tval= qt(0.975,n-1)
cpdr=est+c(-1,1)*tval*sqrt(varybar)

c(est,ss2,sqrt(ss2), varybar, sqrt(varybar),  tval,   cpdr)
  # 5.1400 4.9030 2.2143 0.6537 0.8085 2.7764 2.8951 7.3849

# (b)

NNGFPM= function(eta=0, tau=0, alp=0.05,
  ys= c(5.6,2.3,8.4,5.1,4.3), X=rep(1,15) ,  N=15, sigma=diag(rep(1,N))  )
{

# This function performs inference under the normal-normal-gamma
# finite population model.

# Inputs: eta, tau,   alp, ys,  X,  N,  sigma

# Outputs: A list with $a, $b and $c indicating   (ybar-a)/b given ys ~ t(c)

p=ncol(cbind(NA,X))-1;   n = length(ys);   c=2*eta+n-p

ysT=sum(ys); Xs=cbind(NA,X)[1:n,][,-1]; Xr=cbind(NA,X)[(n+1):N,][,-1]
sigmass=sigma[1:n,1:n];   sigmarr=sigma[(n+1):N,(n+1):N]
sigmasr=sigma[1:n,(n+1):N];   sigmars=t(sigmasr)
D=solve(t(Xs)%*%solve(sigmass)%*%Xs)
beta=D%*%t(Xs)%*%solve(sigmass)%*%ys
A=Xr-sigmars%*%solve(sigmass)%*%Xs;     oner=rep(1,N-n)

a=(1/N)*(  ysT   +   t(oner)%*%
  (  Xr%*%beta  +  sigmars%*%solve(sigmass)%*%(ys-Xs%*%beta)   )  )

b2=(1/(c*N^2)) * ( 2*tau + t(ys-Xs%*%beta)%*%solve(sigmass)%*%
      (ys-Xs%*%beta) ) * t(oner)%*%
      ( sigmarr-sigmars%*%solve(sigmass)%*%sigmasr +
      A%*%D%*%t(A)) %*% oner

b=sqrt(b2); cpdr=a+c(-1,1)*qt(1-alp/2,c)*b
list(a=a,b=b,c=c,beta=beta, cpdr=cpdr)
}
```

```
# Test function by using it to check (a):
res= NNGFPM();  c(res$a,res$b,res$c,res$beta, res$cpdr)
  # 5.1400 0.8085 4.0000 5.1400 2.8951 7.3849    Same as in (a) OK


# Apply function with covariate info:
xvec=c(9.3, 4.6, 15.0,11.2, 7.8,    2.4, 6.6, 3.0, 2.1, 7.3,    5.5, 8.0, 2.4, 4.2,
5.5)
res= NNGFPM(X=xvec, sigma=diag(xvec))
c(res$a,res$b,res$c,res$beta,res$cpdr)
        # 3.3945 0.1159 4.0000 0.5365 3.0725 3.7164


c(mean(xvec), mean(xvec[1:5]), mean(xvec[6:15]) ) # 6.327 9.580 4.700


# (c)


ys= c(5.6,2.3,8.4,5.1,4.3); ysbar=mean(ys); n = 5; N = 15; options(digits=4)


GIBBS = function(J=1000,ys= c(5.6,2.3,8.4,5.1,4.3),
        bet=1, lam=1,  bet0=10, sig0=1.5,  eta=8, tau=4)
{
betv=bet; lamv=lam; sig02=sig0^2;  n=length(ys); ysbar=mean(ys);
for(j in 1:J){
        klam=n/(n+1/(lam*sig02));    sig2lam=klam/(n*lam)
        betlam=(1-klam)*bet0+klam*ysbar;
        bet=rnorm(1,betlam,sqrt(sig2lam))
        s2bet=mean((ys-bet)^2); lam=rgamma(1,eta+n/2,tau+n*s2bet/2)
        betv=c(betv,bet); lamv=c(lamv,lam)    }
list(betv=betv,lamv=lamv)
}


set.seed(641);  res=GIBBS(J=10100); X11(w=8,h=5.5); par(mfrow=c(2,2))
plot(res$betv,type="l"); plot(res$lamv,type="l")
hist(res$betv[-c(1:101)],prob=T,nclass=30);
hist(res$lamv[-c(1:101)],prob=T,nclass=30) # Fig. 10.2


betvec=res$betv[-c(1:101)][seq(10,10000,10)]; J = length(betvec); J # 1000
lamvec=res$lamv[-c(1:101)][seq(10,10000,10)]
acf(res$betv); acf(res$lamv);    acf(betvec); acf(lamvec) # Fig. 10.3


betbar=mean(betvec); betci=betbar+c(-1,1)*qnorm(0.975)*sd(betvec)/sqrt(J)
c(betbar,betci) # 5.766 5.731 5.801
```

```
set.seed(121);  yrbarvec=rnorm(J, betvec, 1/sqrt((N-n)*(lamvec))  )
yrbarbar=mean(yrbarvec);
yrbarci= yrbarbar+c(-1,1)*qnorm(0.975)*sd(yrbarvec)/sqrt(J)
yrbarcpdr=quantile(yrbarvec, c(0.025,0.975))
c(yrbarbar,yrbarci,yrbarcpdr) # 5.762 5.718 5.806 4.458 7.380


ybarvec=(1/N)*(  n*ysbar + (N-n)*yrbarvec  )
ybarbar=mean(ybarvec);
ybarci= ybarbar+c(-1,1)*qnorm(0.975)*sd(ybarvec)/sqrt(J)
ybarcpdr=quantile(ybarvec, c(0.025,0.975))
c(ybarbar,ybarci,ybarcpdr) # 5.555 5.526 5.584 4.685 6.633
ybarci[2]-ybarci[1] # 0.05849


evec=(1/N)*(n*ysbar + (N-n)*betvec );   vvec=(N-n)/(N^2*lamvec)
ebar=mean(evec);   eci=ebar+c(-1,1)*qnorm(0.975)*sd(evec)/sqrt(J)


Lfun=function(L){  ( 0.025-mean(pnorm(  (L-evec)/sqrt(vvec)  )  )  )^2 }
   L = optim(par=3,fn=Lfun)$par;  L # 4.707  (ignore warning message)
   mean(   pnorm(  (L-evec)/sqrt(vvec)    ))  # 0.025 OK


Ufun=function(U){  ( 0.975-mean(pnorm(  (U-evec)/sqrt(vvec)  )  )  )^2  }
   U = optim(par=7,fn=Ufun)$par;  U # 6.542   (ignore warning message)
   mean(   pnorm(  (U-evec)/sqrt(vvec)    ))  # 0.975   OK


ecpdr=c(L,U);   c(ebar,eci,ecpdr)  # 5.557 5.534 5.581 4.707 6.542
eci[2]-eci[1] # 0.04642


X11(w=8,h=7); par(mfrow=c(1,1))
hist(ybarvec,prob=T,nclass=20,xlim=c(3.5,8),
        xlab="ybar",ylab="density/relative frequency",main="")
lines(density(ybarvec),lty=2,lwd=3,col="blue")
abline(v=c(ybarbar,ybarci,ybarcpdr),lty=2,lwd=3,col="blue")


ybarv=seq(3,8,0.01); fv=rep(NA,length(ybarv))
for(i in 1:length(ybarv))   fv[i] = mean(dnorm(ybarv[i], evec, sqrt(vvec)))
lines(ybarv,fv,lty=1,lwd=2,col="red")
abline(v=c(ebar,eci,ecpdr),lty=1,lwd=2,col="red")


legend(3.4,0.9,c("Histogram","Rao-Blackwell"),
        lty=c(2,1), lwd=c(3,2),col=c("blue","red"), bg="white")
```

CHAPTER 11

Transformations and Other Topics

11.1 Inference on complicated quantities

So far, in the context of Bayesian finite population models specified by:

$f(\xi \mid y, \theta)$ where ξ is s or I or L (as discussed earlier)

$f(y \mid \theta)$ where

$$y = (y_s, y_r) = ((y_1, ..., y_n), (y_{n+1}, ..., y_N)) = (y_1, ..., y_N)$$

$f(\theta)$ where $\theta = (\theta_1, ..., \theta_q)$,

we have been focusing primarily on two finite population quantities, the finite population total $y_T = y_1 + ... + y_N$ and the finite population mean $\bar{y} = (y_1 + ... + y_N) / N = y_T / N$.

These are special cases of the class of linear combinations of the N population values

$$\tilde{y} = c_0 + c y_1 + ... + c_N y_N,$$

for which inference is often straightforward, such as in the context of the general normal-normal-gamma finite population model.

We will now consider other inferential targets.

Generally, suppose we are interested in the quantity $\psi = g(\theta, y)$, where g is a potentially very complicated function of all q model parameters and all N finite population values. In such cases, we may adopt the following four-step strategy.

Step 1. Obtain a sample from the posterior distribution of $\theta = (\theta_1, ..., \theta_q)$, that is $\theta^{(1)}, ..., \theta^{(J)} \sim iid\ f(\theta \mid D)$, where $\theta^{(j)} = (\theta_1^{(j)}, ..., \theta_q^{(j)})$ and where D is the data, typically defined as (s, y_s) or (I, s) or (L, y_s) as discussed previously, and whichever the case may be.

Make use of special techniques if suitable, e.g. the method of composition and MCMC methods like the Gibbs sampler.

Step 2. Use the sample in Step 1 to generate a random sample from the predictive distribution of the nonsample vector $y_r = (y_{n+1}, ..., y_N)$, that is $y_r^{(1)}, ..., y_r^{(J)} \sim iid \ f(y_r \mid D)$, where $y_r^{(j)} = (y_{n+1}^{(j)}, ..., y_N^{(j)})$.

Make use of special techniques if required.

Often, the sample can be obtained easily via the method of composition and the identity
$$f(y_r, \theta \mid D) = f(y_r \mid D, \theta) f(\theta \mid D),$$
namely by sampling
$$y_r^{(j)} \sim f(y_r \mid D, \theta^{(j)})$$
for each $j = 1, ..., J$.

In many cases, each sampled nonsample vector $y_r^{(j)}$ here can obtained by sampling
$$y_i^{(j)} \sim \perp f(y_i \mid D, \theta^{(j)}), \quad i = n+1, ..., N,$$
and then forming the vector according to
$$y_r^{(j)} = (y_{n+1}^{(j)}, ..., y_N^{(j)}).$$

Step 3. Form the completed population vector
$$y^{(j)} = (y_s, y_r^{(j)}) = (y_1, ..., y_n, y_{n+1}^{(j)}, ..., y_N^{(j)})$$
and then calculate
$$\psi^{(j)} = g(y^{(j)}, \theta^{(j)})$$
for each $j = 1, ..., J$.

The result will be a sample from the posterior/predictive distribution of ψ, namely
$$\psi^{(1)}, ..., \psi^{(J)} \sim iid \ f(\psi \mid D).$$

Step 4. Use the sample obtained in Step 3 to perform Monte Carlo inference on ψ in the usual way. Thus, estimate the posterior/predictive mean of ψ, namely

$$\hat{\psi} = E(\psi \mid D) = \int \psi f(\psi \mid D) d\psi$$

(which may be impossible to obtain analytically), by the Monte Carlo sample mean $\bar{\psi} = \frac{1}{J} \sum_{j=1}^{J} \psi^{(j)}$ (which is unbiased, in that $E(\bar{\psi} \mid D) = \hat{\psi}$).

Also calculate the $1-\alpha$ CI for $\hat{\psi}$ given by

$$\left(\bar{\psi} \pm z_{\alpha/2} \frac{s_{\psi}}{\sqrt{J}} \right), \text{ where } s_{\psi}^2 = \frac{1}{J-1} \sum_{j=1}^{J} (\psi^{(j)} - \bar{\psi})^2.$$

Also, estimate the $1-\alpha$ central posterior/predictive density region (CPDR generally) for ψ by $(Q_{\alpha/2}, Q_{1-\alpha/2})$, where Q_p is the empirical p-quantile of the sample $\psi^{(1)}, ..., \psi^{(J)}$.

Also, estimate the entire posterior/predictive density of ψ, namely $f(\psi \mid D)$, by $\hat{f}(\psi \mid D)$, a smooth of a histogram of $\psi^{(1)}, ..., \psi^{(J)}$ (obtained by adjusting the smooth parameters).

Use Rao-Blackwell methods to improve precision, if possible and practicable. For example, suppose that $q = 2, \theta = (\theta_1, \theta_2)$, $\psi = g(y, \theta_2)$, and $f(\psi \mid D, \theta_1)$ has a simple form. Then, instead of using a 'histogram estimate' $\hat{f}(\psi \mid D)$ to estimate $f(\psi \mid D)$, use the Rao-Blackwell estimate

$$\bar{f}(\psi \mid D) = \frac{1}{J} \sum_{j=1}^{J} f(\psi \mid D, \theta^{(j)}).$$

Exercise 11.1 Estimation of nonstandard target quantities

(a) Suppose that 2.1, 5.2, 3.0, 7.7 and 9.3 constitute a random sample from a normal finite population of size 20 whose mean and variance are unknown. We are interested in the *finite population median*. Estimate this quantity using a suitable Bayesian model.

(b) Repeat (a) but for the quantity:
   average percentage increase between subsequent ordered population values greater than 4.

(c) Repeat (a) but for the quantity:
   sum of finite population values in the upper quartile of the normal superpopulation.

Solution to Exercise 11.1

The Bayesian model here is:

$$f(s \mid y, \mu, \lambda) = \binom{N}{n}^{-1},$$

$$s = (1,...,n), (1,...,n-1,n+1),...,(N-n+1,...,N) \quad \text{(SRSWOR)}$$

$$(y_1,..., y_N \mid \mu, \lambda) \sim iid \ N(\mu, 1/\lambda)$$

$$f(\mu, \lambda) \propto 1/\lambda, \quad \mu \in \Re, \ \lambda > 0,$$

where $N = 20$, $n = 5$, and where the data is

$$D = (s, y_s) = ((1,...,n), (2.1, \ 5.2, \ 3.0, \ 7.7, \ 9.3)).$$

Note 1: This data is presented according to a convenient reordering of population labels, after sampling, so that the sampled values are listed at the beginning of the finite population vector (as discussed earlier).

Note 2: The superpopulation parameter in the model may be thought of as the vector

$$\theta = (\theta_1, \theta_2) = (\mu, \lambda),$$

in which case the model could also be written:

$$(s \mid y, \theta) \sim SRSWOR(N, n)$$

$$(y \mid \theta) \sim N_N(\theta_1 1_N, I_N / \theta_2)$$

$$f(\theta) \propto 1/\theta_2, \quad \theta_1 \in \Re, \ \theta_2 > 0.$$

For the purposes of this exercise, let $y_{(i)}$ denote the *i*th *finite population order statistic*, meaning the *i*th value amongst $y_1,..., y_N$ after these are ordered from smallest to largest. We are interested in three finite population quantities, as follows:

(a) $\psi_1 = g_1(y, \theta) = g_1(y) = \dfrac{y_{(N/2)} + y_{(N/2)+1}}{2}$

(b) $\psi_2 = g_2(y, \theta) = g_2(y) = 100 \dfrac{\displaystyle\sum_{i=2}^{N} \left(\dfrac{y_{(i)} - y_{(i-1)}}{y_{(i-1)}} \right) I(y_{(i-1)} > 4)}{\displaystyle\sum_{i=2}^{N} I(y_{(i-1)} > 4)}$

(c) $\psi_3 = g_3(y, \theta) = \displaystyle\sum_{i=1}^{N} y_i I\left(y_i > \mu + \dfrac{1}{\sqrt{\lambda}} \Phi^{-1}(0.75) \right).$

Note 1: The median ψ_1 is the average of the middle two values, since $N = 20$ is even.

Note 2: In general, ψ_2 is defined only if at least two of the finite population values are greater than 4. For our data, there is no problem with the definition because the observed sample already contains three such values. If there were a problem, then $\psi_2 = g_2(y)$ could be defined as zero (say) in the case where the number of population values is only 0 or 1, i.e. if $\sum_{i=1}^{N} I(y_i > 4) < 2$.

Note 3: As regards ψ_3, if c is the upper quartile of the normal superpopulation then

$$0.75 = P(y_i < c \mid \theta) = P\left(\frac{y_i - \mu}{\sigma} < \frac{c - \mu}{\sigma} \Big| \theta\right)$$

$$\Rightarrow \frac{c - \mu}{\sigma} = \Phi^{-1}(0.75)$$

$$\Rightarrow c = \mu + \sigma\Phi^{-1}(0.75) = \mu + \frac{1}{\sqrt{\lambda}}\Phi^{-1}(0.75).$$

In each case, the inferential target has a posterior/predictive distribution which cannot be obtained analytically. One way to proceed is as follows:

Step 1. Generate $\lambda_1,...,\lambda_J \sim iid \; f(\lambda \mid D) \sim G\left(\frac{n-1}{2}, \frac{n-1}{2}s_s^2\right)$,

where $s_s^2 = \frac{1}{n}\sum_{i=1}^{n}(y_i - \bar{y})^2$.

(This step derives from results for the *normal-normal-gamma model*.)

Step 2. Generate $\mu_j \sim f(\mu \mid D, \lambda_j) \sim N\left(\bar{y}_s, \frac{1}{n\lambda_j}\right)$ for each $j = 1,...,J$.

(This step derives from results for the *normal-normal model*).

Step 3. For each $j = 1,...,J$:

- Generate $y_{n+1}^{(j)},...,y_{n+1}^{(j)} \sim iid\ f(y_i \mid D, \mu_j, \lambda_j) \sim N\left(\mu_j, \dfrac{1}{\lambda_j}\right)$

- Form $y_r^{(j)} = (y_{n+1}^{(j)},...,y_N^{(j)})$ and

  $y^{(j)} = (y_s, y_r^{(j)}) = (y_1,...,y_n, y_{n+1}^{(j)},...,y_N^{(j)})$

- Calculate $\psi^{(j)} = g(y^{(j)}, \theta^{(j)})$, where $\theta^{(j)} = (\mu_j, \lambda_j)$.

Step 4. Use the values $\psi^{(1)},...,\psi^{(J)} \sim iid\ f(\psi \mid D)$ for Monte Carlo inference on ψ in the usual way.

Note 1: Steps 1 and 2 result in the sample
$$(\mu_1, \lambda_1),...,(\mu_J, \lambda_J) \sim iid\ f(\mu, \lambda \mid D).$$

Note 2: In the above, Steps 1 and 2 could be replaced as follows:

Step 1'. Generate $\mu_1,...,\mu_J \sim f(\mu \mid D)$ for each $j = 1,...,J$. Do this by first sampling $w_1,...,w_J \sim iid\ t(n-1)$ and then forming
$$\mu_j = \overline{y}_s + w_j s_s / \sqrt{n} \text{ for each } j = 1,...,J$$
(using results from the *normal-normal-gamma model*).

Step 2'. Generate $\lambda_j \sim\perp f(\lambda \mid D, \mu_j) \sim G\left(\dfrac{n}{2}, \dfrac{n}{2} s_{\mu_j}^2\right)$, where
$$s_{\mu_j}^2 = \frac{1}{n}\sum_{i=1}^{n}(y_i - \mu_j)^2$$
(using results from the *normal-gamma model*).

These modified steps will also result in the sample
$$(\mu_1, \lambda_1),...,(\mu_J, \lambda_J) \sim iid\ f(\mu, \lambda \mid D).$$

Applying the above four-step procedure (using the original Steps 1 and 2) with Monte Carlo sample size $J = 1,000$, we obtain Table 11.1 which shows numerical estimates for the three quantities of interest: $\psi = \psi_1$, ψ_2 and ψ_3, respectively. Figure 11.1 shows histograms which illustrate these inferences.

Table 11.1 and Figure 11.1 also contain analogous results for a fourth quantity of interest which may be defined as

$$\psi_4 = g_4(y,\theta) = (\psi_3 \mid \psi_3 \neq 0)$$

$$= \left\{ \left[\sum_{i=1}^{N} y_i I\left(y_i > \mu + \frac{1}{\sqrt{\lambda}} \Phi^{-1}(0.75) \right) \right] \middle| \left[\sum_{i=1}^{N} I\left(y_i > \mu + \frac{1}{\sqrt{\lambda}} \Phi^{-1}(0.75) \right) \right] > 0 \right\}.$$

The relevant posterior/predictive density may also be written

$$f(\psi_4 \mid D) = f(\psi_3 \mid D, \psi_3 \neq 0).$$

Inferences on ψ_4 were obtained using the 960 values of ψ_3 which were non-zero. It was meaningful to perform this additional inference because there were 40 simulations amongst the 1,000 for which the upper quartile of the normal distribution lay *above* the largest finite population value, resulting in the sum ψ_3 being equal to 0 *exactly*.

Note 1: From the above, we see that ψ_3 is neither a discrete nor a continuous random variable but one with a *mixed distribution*.

The discrete part of this mixed distribution is the probability that $\psi_3 = 0$ *exactly*, and this we estimated via MC as $40/1,000 = 0.04$.

Note 2: We also see that neither ψ_3 nor ψ_4 is necessarily *positive*.

This is because it *might* be the case that the upper quartile of the normal distribution is *negative* and many of the finite population values *happen* (by a very small chance) to lie between that negative quartile and zero.

Table 11.1 Point and interval estimates for four quantities

Quantity of interest:

ψ_1	ψ_2	ψ_3	$\psi_4 = (\psi_3 \mid \psi_3 \neq 0)$

Posterior mean estimate:

| 5.842 | 9.975 | 58.31 | 60.74 |

95% CI for posterior mean:

| (5.790, 5.893) | (9.775, 10.175) | (56.48 60.15) | (58.99, 62.49) |

Posterior mode estimate:

| 5.528 | 8.150 | 62.29 | 62.45 |

Posterior median estimate:

| 5.769 | 9.377 | 59.48 | 60.59 |

95% CPDR estimate:

| (4.308, 7.528) | (5.522, 17.770) | (0.00 114.87) | (11.72, 114.96) |

Figure 11.1 Four histograms and sets of inferences

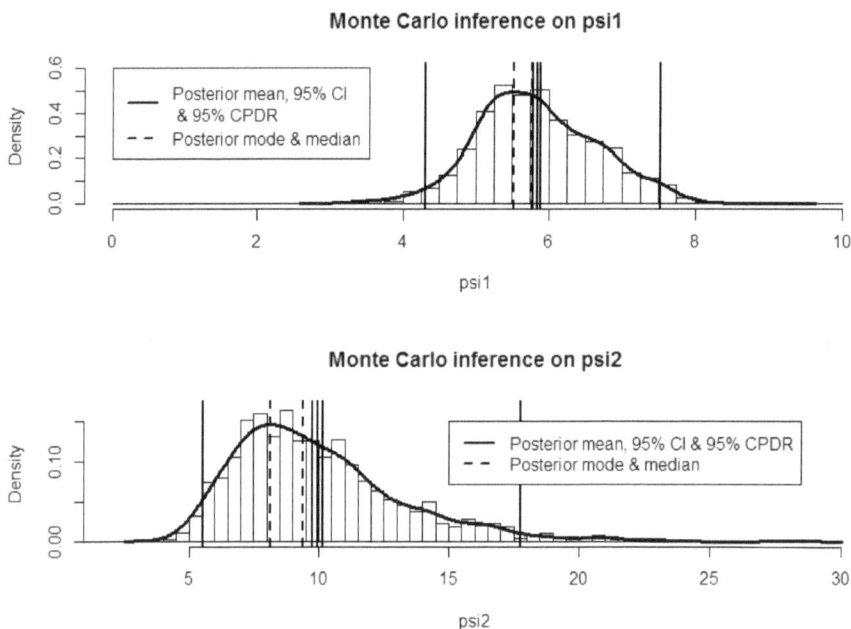

Monte Carlo inference on psi1

— Posterior mean, 95% CI & 95% CPDR
-- Posterior mode & median

Monte Carlo inference on psi2

— Posterior mean, 95% CI & 95% CPDR
-- Posterior mode & median

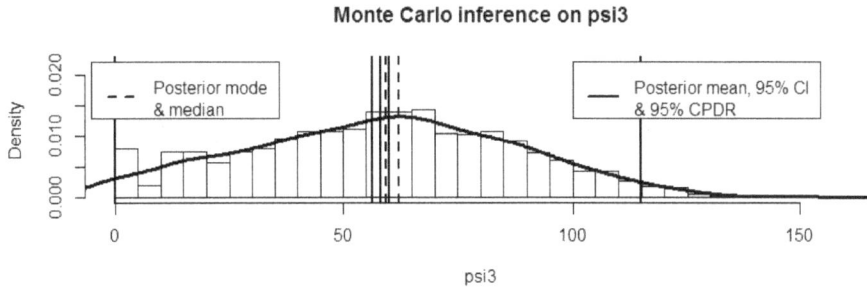

Monte Carlo inference on psi3

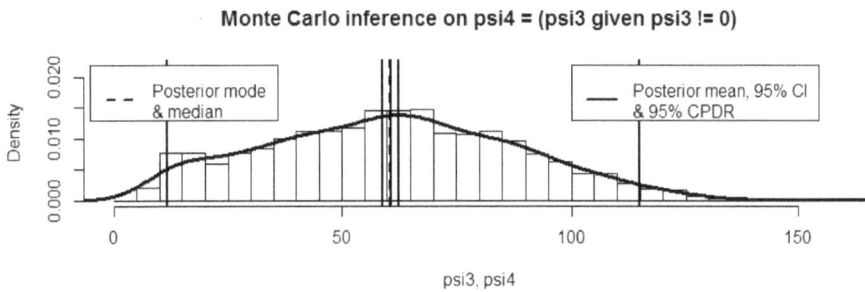

Monte Carlo inference on psi4 = (psi3 given psi3 != 0)

R Code for Exercise 11.1

options(digits=4)

```
# Define 3 psi functions -----------------
PSI1FUN = function(y){ quantile(y,0.5) }
PSI2FUN = function(y){ ynew=sort(y[y>4]); nnew=length(ynew);
  if(nnew<2)   res=NA
  if(nnew>=2)  res = 100*mean(  (ynew[-1]-ynew[-nnew]) / ynew[-nnew] )
  res }
PSI3FUN = function(y,mu,lam){  q = qnorm(0.75); sum(y[y>(mu+q/sqrt(lam))])
}
```

```
# Test 3 psi functions ------------------------
PSI1FUN(y=c(1,2,7)) # 2 OK
PSI1FUN(y=c(1,2,7,8)) # 4.5 OK
PSI2FUN(y=c(5,12,6)) # 60    Correct: 100* (1/2) * ( (6-5)/5 + (12-6)/6 ) = 60
PSI2FUN(y=c(5,3,6)) # 20    Correct: 100* (6-5)/5 = 20
PSI2FUN(y=c(5,2,3)) # NA    Correct
PSI2FUN(y=c(4,4,-3)) # NA    Correct
set.seed(311); PSI3FUN(y=rnorm(100,10,1),mu=10,lam=1) # 267 ~ 25*10,  OK
```

```
# Perform inference on 3 psi functions ----------------------------------------
ys= c(2.1, 5.2, 3.0, 7.7, 9.3); ysbar=mean(ys); n=length(ys); ss2=var(ys); N = 20
options(digits=4); J=1000; set.seed(232)
lamvec=rgamma( J, (n-1)/2, ((n-1)/2) *ss2 )
muvec = rnorm(J,ysbar,1/sqrt(n*lamvec))
yrmat=matrix(NA, nrow=J, ncol=N-n)
for(j in 1:J)   yrmat[j,] = rnorm(N-n,muvec,1/sqrt(lamvec))
psi1vec=rep(NA,J); psi2vec=rep(NA,J); psi3vec=rep(NA,J)
for(j in 1:J){   yrj = yrmat[j,]
        psi1vec[j] = PSI1FUN(y=c(ys, yrj))
        psi2vec[j] = PSI2FUN(y= c(ys, yrj))
        psi3vec[j] = PSI3FUN(y= c(ys, yrj), mu=muvec[j], lam=lamvec[j])   }

cbind(   summary(psi1vec), summary(psi2vec),
         summary(psi3vec), summary(psi3vec[psi3vec!=0])   )
# Min.    3.14 4.44  0.0  9.3
# 1st Qu. 5.28 7.65 37.9 40.3
# Median  5.77 9.38 59.5 60.6
# Mean    5.84 9.97 58.3 60.7
# 3rd Qu. 6.41 11.50 79.6 80.7
# Max.    9.09 28.10 156.0 156.0

X11(w=9,h=6.5); par(mfrow=c(2,1))
psivec=psi1vec; J = length(psivec)
psibar=mean(psivec); psici=psibar+c(-1,1)*qnorm(0.975)*sd(psivec)/sqrt(J)
fpsi=density(psivec);  psimode=fpsi$x[fpsi$y==max(fpsi$y)]
psimedian=quantile(psivec,0.5);  psicpdr=quantile(psivec,c(0.025,0.975))
c(psibar,psici,psimode,psimedian,psicpdr)
# 5.842 5.790 5.893 5.528 5.769 4.308 7.528
hist(psivec, prob=T, xlab="psi1",xlim=c(0,10),ylim=c(0,0.6),
        breaks=seq(0,10,0.25),   main="Monte Carlo inference on psi1")
lines(fpsi,lwd=3)
abline(v= c(psibar, psici, psicpdr, psimedian, psimode) ,
        lty=c(1,1,1,1,2,2), lwd=rep(2,7))
legend(0,0.6,
c("Posterior mean, 95% CI \n & 95% CPDR","Posterior mode & median"),
        lty=c(1,2), lwd=c(2,2), bg="white")

psivec=psi2vec; J = length(psivec)
psibar=mean(psivec); psici=psibar+c(-1,1)*qnorm(0.975)*sd(psivec)/sqrt(J)
fpsi=density(psivec);  psimode=fpsi$x[fpsi$y==max(fpsi$y)]
psimedian=quantile(psivec,0.5);  psicpdr=quantile(psivec,c(0.025,0.975))
c(psibar,psici,psimode,psimedian,psicpdr)
# 9.975 9.775 10.175 8.150 9.377 5.522 17.770
```

524

```
hist(psivec, prob=T, xlab="psi2",xlim=c(2,30),ylim=c(0,0.17),
        breaks=seq(0,30,0.5),main="Monte Carlo inference on psi2")
lines(fpsi,lwd=3)
abline(v= c(psibar, psici, psicpdr, psimedian, psimode) ,
        lty=c(1,1,1,1,1,2,2), lwd=rep(2,7))
legend(15,0.15,
        c("Posterior mean, 95% CI & 95% CPDR","Posterior mode & median"),
        lty=c(1,2), lwd=c(2,2), bg="white")  # End of first 2 graphs

psivec=psi3vec  # Start of next 2 graphs
psibar=mean(psivec); psici=psibar+c(-1,1)*qnorm(0.975)*sd(psivec)/sqrt(J)
fpsi=density(psivec);  psimode=fpsi$x[fpsi$y==max(fpsi$y)]
psimedian=quantile(psivec,0.5);  psicpdr=quantile(psivec,c(0.025,0.975))
c(psibar,psici,psimode,psimedian,psicpdr)
# 58.31  56.48  60.15  62.29  59.48  0.00 114.87

hist(psivec, prob=T, xlab="psi3",xlim=c(0,160),ylim=c(0,0.022),
        breaks=seq(0,200,5), main="Monte Carlo inference on psi3")
lines(fpsi,lwd=3)
abline(v= c(psibar, psici, psicpdr, psimedian, psimode) ,
        lty=c(1,1,1,1,1,2,2), lwd=rep(2,7))
legend(100,0.022,
        c("Posterior mean, 95% CI \n& 95% CPDR"),lty=1,lwd=2,bg="white")
legend(-5,0.022,c("Posterior mode \n& median"), lty=2, lwd=2, bg="white")

length(psi3vec[psi3vec!=0]) # 960
length(psi3vec[psi3vec==0]) # 40     40/1000 = 4%
psivec=psi3vec[psi3vec!=0]; J=length(psivec); J # 960  Condition on psi > 0
psibar=mean(psivec); psici=psibar+c(-1,1)*qnorm(0.975)*sd(psivec)/sqrt(J)
fpsi=density(psivec);  psimode=fpsi$x[fpsi$y==max(fpsi$y)]
psimedian=quantile(psivec,0.5);  psicpdr=quantile(psivec,c(0.025,0.975))
c(psibar,psici,psimode,psimedian,psicpdr)
# 60.74  58.99  62.49  62.45  60.59  11.72 114.96

hist(psivec, prob=T, xlab="psi3, psi4",xlim=c(0,160),ylim=c(0,0.022),
        breaks=seq(0,200,5),
        main="Monte Carlo inference on psi4 = (psi3 given psi3 != 0)")
lines(fpsi,lwd=3)
abline(v= c(psibar, psici, psicpdr, psimedian, psimode),
        lty=c(1,1,1,1,1,2,2), lwd=rep(2,7))
legend(100,0.022,
        c("Posterior mean, 95% CI \n& 95% CPDR"),lty=1,lwd=2,bg="white")
legend(-5,0.022,c("Posterior mode \n& median"), lty=2, lwd=2, bg="white")
```

11.2 Data transformations

In statistical analysis, a common practice is to first transform the data before applying a model. For example, if the data values are strictly positive and highly right skewed, it may be worthwhile taking natural logarithms before applying a normal model.

In the classical setting, e.g. in the design-based survey sampling, this idea may work well for purposes of analytical inference (i.e. estimation of model parameters) but can be problematic for prediction. This is because the quantity requiring prediction (e.g. the nonsample total) does not typically have a simple distribution on the untransformed scale. Although prediction can be performed easily on the *transformed* scale there is no way to translate results back onto the original scale. By contrast, this issue does not create any special problems within the Bayesian framework.

Suppose that we are interested in some finite population quantity which is denoted $\psi = g(y)$, e.g. $\bar{y} = 1'_N y / N$.

Also suppose that there is no convenient superpopulation model for the finite population values y_i, $i = 1,...,N$, but there does exist such a model for some function of those values, say $z_i = h(y_i)$ for a function h.

In that case we may consider a Bayesian model specified in terms of:

$f(\xi \mid z, \theta)$ where ξ is s or I or L (as discussed earlier)

$f(z \mid \theta)$ where $z = (z_s, z_r) = ((z_1,...,z_n),(z_{n+1},...,z_N)) = (z_1,...,z_N)$

$f(\theta)$ where $\theta = (\theta_1,...,\theta_q)$.

We now use Monte Carlo methods (perhaps MCMC methods if needed) to generate a random sample from the predictive distribution of the nonsample vector for the z variable (i.e. z_r), given the data D (for example (s, y_s), (I, s) or (L, y_s)). Let us call this sample

$z_r^{(1)},...,z_r^{(J)} \sim iid\ f(z_r \mid D)$.

We next calculatate $y_i^{(j)} = h^{-1}(z_i^{(j)})$ for each $i = n+1,...,N$ and each $j = 1,...,J$. Thus, we *untransform* the simulated individual data values back to the original scale.

Next, we form the vectors
$$y_r^{(j)} = (y_{n+1}^{(j)}, ..., y_N^{(j)})$$
and
$$y^{(j)} = (y_s, y_r^{(j)})$$
for each $j = 1, ..., J$.

This results in the samples
$$y_r^{(1)}, ..., y_r^{(J)} \sim iid \ f(y_r \mid D)$$
and
$$y^{(1)}, ..., y^{(J)} \sim iid \ f(y \mid D).$$

Finally, we calculate
$$\psi^{(j)} = g(y^{(j)})$$
for each $j = 1, ..., J$.

This results in
$$\psi^{(1)}, ..., \psi^{(J)} \sim iid \ f(\psi \mid D),$$
namely a sample from the predictive distribution of the finite population quantity of interest, on the original scale required for that quantity. This sample can then be used for Monte Carlo inference on ψ in the usual way.

Note: We may think of this topic as an example and special application of the last topic, that is, Bayesian inference on complicated functions of the finite population vector.

Exercise 11.2 Finite population inference using data transformation

Consider the following random sample of size 50 from a finite population of size 200:

28.374, 69.857, 22.721, 57.593, 126.965,   17.816, 16.078, 0.803, 3.164, 3.544,
2.123, 2.353, 184.539, 59.856, 63.701,   585.684, 29.094, 79.245, 18.105, 1.623,
5.513, 1.629, 63.654, 22.060, 187.463,   5.051, 34.299, 27.475, 0.746, 34.016,
8.547, 1.081, 3.151, 55.569, 2.593,   522.377, 1.660, 130.435, 1.246, 169.462,
3.444, 6.376, 18.735, 51.312, 33.920,   350.346, 475.795, 4.972, 24.451, 86.987.

Use Bayesian methods with a suitable transformation to estimate the finite population mean.

Solution to Exercise 11.2

We create a histogram of the sample values and see that the underlying distribution is highly right skewed. However, a histogram of the natural logarithm of the sample values is consistent with a normal superpopulation model. The histograms are shown in Figure 11.2.

Therefore we posit the following Bayesian model involving an uninformative prior and the logarithms of the finite population values, $z_i = h(y_i) = \log y_i$, $i = 1,...,N$ $(N = 200)$:

$$(s \mid z, \mu, \lambda) \sim SRSWOR$$

$$(z_1,...,z_N \mid \mu, \lambda) \sim iid \ N(\mu, 1/\lambda)$$

$$f(\mu, \lambda) \propto 1/\lambda, \ \mu \in \Re, \ \lambda > 0.$$

Figure 11.2 Histograms of the sample data

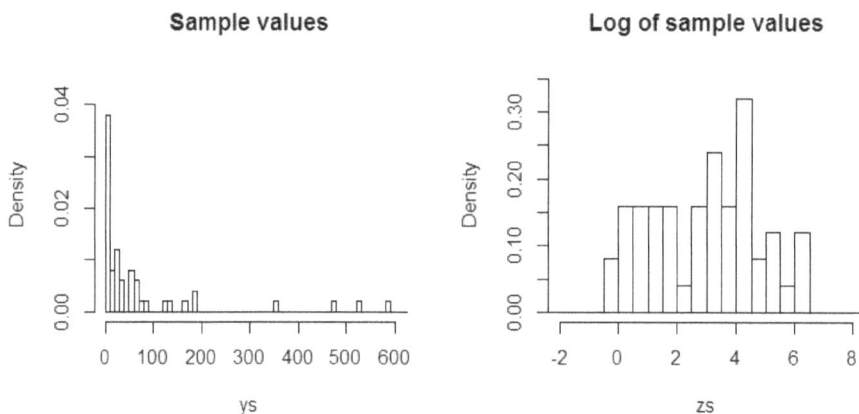

The data is $D = (s, z_s) = ((1,...,50),(28.374,\ 69.857,...,86.987))$ (after a convenient ordering), and the quantity of interest is

$$\bar{y} = \frac{1}{N}\sum_{i=1}^{N} y_i = g(z) = \frac{1}{N}\sum_{i=1}^{N} h^{-1}(z_i) = \frac{1}{N}\sum_{i=1}^{N} \exp(z_i).$$

So we generate

$$(\mu_1, \lambda_1),...,(\mu_J, \lambda_J) \sim iid \ f(\mu, \lambda \mid D)$$

(using methods detailed previously).

Then for each $j = 1, ..., J$ we sample

$$z_{n+1}^{(j)}, ..., z_N^{(j)} \sim iid \ N(\mu_j, 1/\lambda_j)$$

and calculate

$$\bar{y}^{(j)} = \frac{1}{N} \left(\{ y_1 + ... + y_n \} + \left\{ \exp(z_{n+1}^{(j)}) + ... + \exp(z_N^{(j)}) \right\} \right)$$

$$= \frac{1}{N} \left(y_{sT} + \sum_{i=n+1}^{N} \exp(z_i^{(j)}) \right).$$

The result is

$$\bar{y}^{(1)}, ..., \bar{y}^{(J)} \sim iid \ f(\bar{y} \mid D),$$

which can then be used for Monte Carlo inference.

Applying the above procedure with a Monte Carlo sample size of $J = 1,000$ we estimate $\bar{y}$'s posterior mean, $\hat{\bar{y}} = E(\bar{y} \mid D)$, and so also $\bar{y}$ itself, by

$$\bar{\bar{y}} = \frac{1}{J} \sum_{j=1}^{J} \bar{y}^{(j)} = 110.83,$$

with 95% CI for $\hat{\bar{y}}$

$$\left(\bar{\bar{y}} \pm 1.96 \sqrt{\frac{1}{J}} \sqrt{\frac{1}{J-1} \sum_{j=1}^{J} (\bar{y}^{(j)} - \bar{\bar{y}})^2} \right) = (104.64, \ 117.02).$$

We also estimate the bounds of the 95% CPDR for $\bar{y}$ by 49.26 and 302.05, where these are the empirical 0.025 and 0.975 quantiles of $\bar{y}^{(1)}, ..., \bar{y}^{(J)}$.

Figure 11.3 shows a histogram of the simulated values of $\bar{y}$, together with the above five numbers, as well as a 'histogram estimate' of the predictive density $f(\bar{y} \mid D)$. In this histogram the dot shows the true value of the finite population mean, $\bar{y} = 114.2$, which was known prior to the generation of the sample data.

Figure 11.3 Inference on the finite population mean

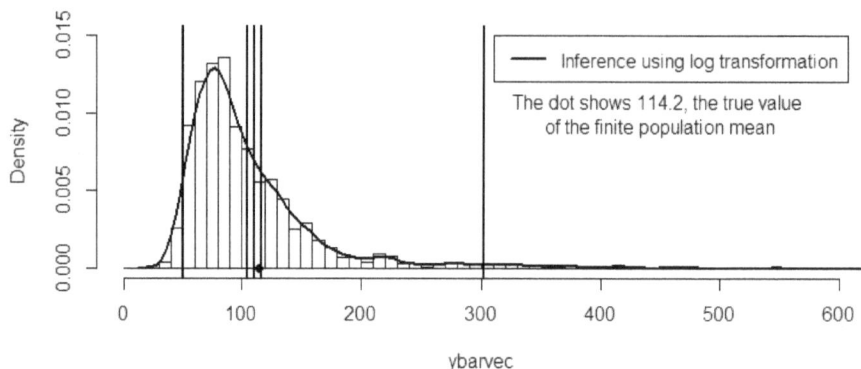

ybarvec

Discussion

Figure 11.4 shows histograms of the values $z_1, ..., z_N$ which were in fact drawn from the normal distribution with mean 3 and standard deviation 2 (left plot), and the values of $y_1 = \exp(z_1), ..., y_N = \exp(z_N)$ (right plot), together with the true underlying superpopulation densities of the variables z_i and y_i.

Figure 11.4 Histogram of all N values of z and y

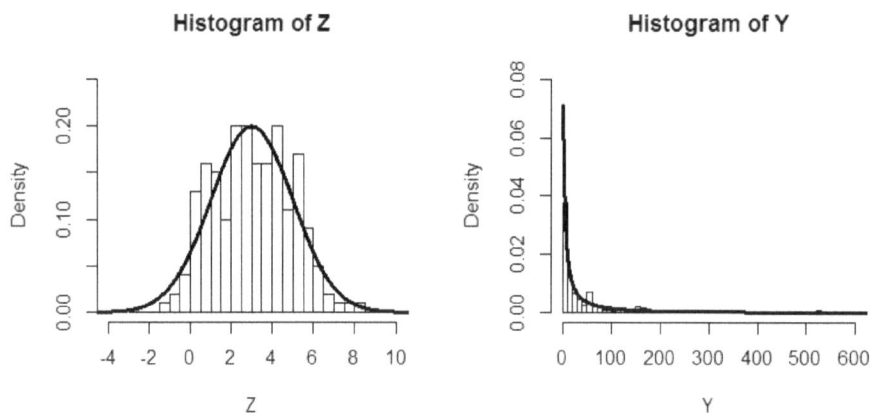

For comparison we repeat the above inference on the *original scale* of the data and 'exactly' (since there is then no need for Monte Carlo methods).

In that case—where we replace z by y in the Bayesian model—we find that the predictive mean of $\bar{y}$ is $\hat{\bar{y}} = E(\bar{y} \mid D) = \bar{y}_s = 74.15$ (the average of the raw data values), and the 95% CPDR for $\bar{y}$ is exactly (41.36, 106.94). We see that this inference does much worse at estimating $\bar{y}$, whose true value is 114.2.

Note: This second set of inference is the same as design-based inference since it is based on the result

$$\left(\frac{\bar{y} - \bar{y}_s}{\frac{s_s}{\sqrt{n}} \sqrt{1 - \frac{n}{N}}} \middle| D \right) \sim t(n-1), \text{ where } s_s^2 = \frac{1}{n-1} \sum_{i=1}^{n} (y_i - \bar{y}_s)^2 .$$

Figure 11.5 shows the original data values (untransformed) and both sets of inferences above. It highlights the value of performing an appropriate prior transformation for purposes of estimating the finite population mean.

Figure 11.5 Comparison of two sets of inference

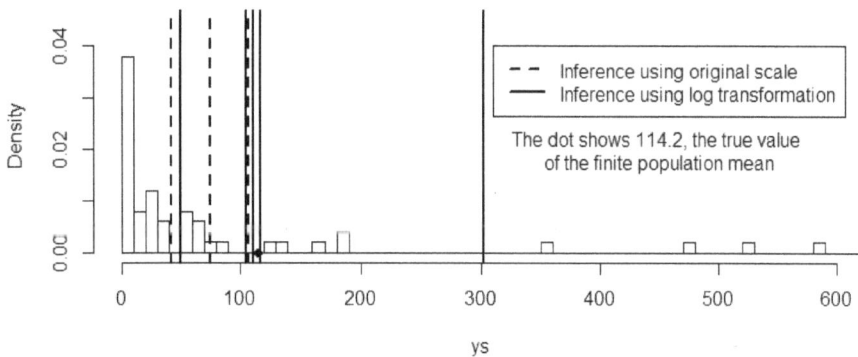

531

For interest, we repeat the above simulations and comparison with a $N(2,1)$ model for the z_is (rather than a $N(3,4)$ model). Figure 11.6 shows the analogue of the last figure above.

We see, of course, that the benefits of applying the log transformation to the data diminishes as the skewness of the sample data decreases.

Figure 11.6 Comparison of two sets of inference with less skewed data

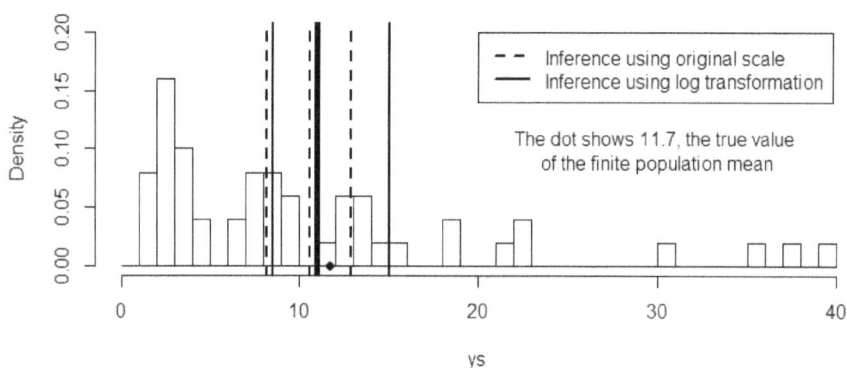

Note 1: Using the formula for sample skewness given by

$$g = \frac{(1/n)\sum_{i=1}^{n}(y_i - \bar{y}_s)^3}{\{(1/n)\sum_{i=1}^{n}(y_i - \bar{y}_s)^2\}^{3/2}},$$

we obtained a value of $g = 2.662$ for the first set of data and a value of $g = 1.549$ for the second set of data.

Note 2: For another example of finite population inference via Bayesian and MCMC methods which involves the logarithmic transformation, see Puza (2002). This other example also features the use of covariate information.

Note 3: It can be shown (mathematically) that $\hat{\bar{y}} = E(\bar{y} \mid D) = \infty$ (exactly). This seems somewhat counterintuitive in light of the fact that our Monte Carlo estimate $\bar{\bar{y}} = 110.83$ is very close to the actual finite population mean, $\bar{y} = 114.2$.

R Code for Exercise 11.2

```
# Data generation used to set up exercise --------------------------------------

options(digits=4); X11(w=8,h=6); par(mfrow=c(2,2))
N=200; n=50; set.seed(432); Z=rnorm(N,3,2); S=sample(1:N,n)
ZS=Z[S]; Y=exp(Z);  YS=exp(ZS); YBAR=mean(Y); YBAR # 114.2
hist(Z,prob=T); hist(Y,prob=T); hist(ZS,prob=T); hist(YS,prob=T)
        # preliminary plots


X11(w=8,h=4); par(mfrow=c(1,2))
hist(Z,prob=T,xlim=c(-4,10), ylim=c(0,0.25),breaks=seq(-3,12,0.5))
        lines(seq(-5,12,0.01),dnorm(seq(-5,12,0.01),3,2),lwd=3)
hist(Y,prob=T,xlim=c(0,600),ylim=c(0,0.08), breaks=seq(0,5000,10));
        yg=seq(0.1,700,0.5);  lines(yg ,dnorm( log(yg),3,2)/yg, lwd=3)


format(list(YS=YS),digits=3) # "28.374, 69.857, 22.721, ...,  24.451, 86.987"

# Look at given data and the log of that data  (load data etc.) ------------------
N = 200; n = 50; m = N-n;  options(digits=4)
ys = c(  28.374, 69.857, 22.721, 57.593, 126.965,
        17.816, 16.078, 0.803, 3.164, 3.544,
        2.123, 2.353, 184.539, 59.856, 63.701,
        585.684, 29.094, 79.245, 18.105, 1.623,
        5.513, 1.629, 63.654, 22.060, 187.463,
        5.051, 34.299, 27.475, 0.746, 34.016,
        8.547, 1.081, 3.151, 55.569, 2.593,
        522.377, 1.660, 130.435, 1.246, 169.462,
        3.444, 6.376, 18.735, 51.312, 33.920,
        350.346, 475.795, 4.972, 24.451, 86.987)

summary(ys)
#   Min. 1st Qu. Median   Mean 3rd Qu.   Max.
#   0.7   3.5   23.6   74.2   63.7   586.0

skewness=mean(  (ys-mean(ys))^3  ) / ( mean((ys-mean(ys))^2) )^(3/2)
skewness # 2.662

zs=log(ys); par(mfrow=c(1,2))
hist(ys,prob=T); hist(zs,prob=T) # preliminary plots
hist(ys,prob=T,xlim=c(0,600),ylim=c(0,0.045),
        breaks=seq(0,700,10), main="Sample values");
hist(zs,prob=T,xlim=c(-2,8), ylim=c(0,0.35),
        breaks=seq(-3,10,0.5),  main="Log of sample values");
```

```
# Finite population inference using original scale and design-based approach
# (same as the 'exact' Bayesian approach without Monte Carlo)  -----------------
ysbar=mean(ys); sy=sd(ys); ybarhat=ysbar
ybarci=ybarhat+c(-1,1)*qt(0.975,n-1)* (sy/sqrt(n)) * sqrt(1-n/N)
inf.original=c(ybarhat,ybarci);
c(inf.original, YBAR) # 74.15   41.36 106.94   114.24


# Finite population inference via Bayesian approach using log transformation
# (and a 'crude' approach which makes no use of Rao-Blackwell ideas etc.) ----

zsbar=mean(zs); sz=sd(zs); J=1000;   set.seed(142);
lamvec=rgamma(J,(n-1)/2,(sz^2)*(n-1)/2)
muvec=rnorm(J,zsbar,1/sqrt(n*lamvec));   yrbarvec=rep(NA,J)

for(j in 1:J){     zr=rnorm(m, muvec[j], 1/sqrt(lamvec[j]) )
                yr=exp(zr);      yrbarvec[j] = mean(yr)    }
ybarvec=(1/N)*(n*ysbar+m*yrbarvec);    ybarhat=mean(ybarvec)
ybarci=ybarhat+c(-1,1)*qnorm(0.975)*sd(ybarvec)/sqrt(J)
ybarcpdr=quantile(ybarvec,c(0.025,0.975))
inf.transform = c(ybarhat,ybarci,ybarcpdr)
c(inf.transform,YBAR) # 110.83 104.64 117.02  49.26 302.05 114.24


summary(ybarvec)
            #  Min. 1st Qu. Median   Mean 3rd Qu.   Max.
            #   37.0   70.6   89.4   111.0  122.0 2080.0

par(mfrow=c(1,1)); hist(ybarvec,prob=T) # preliminary plot
hist(ybarvec,prob=T,xlim=c(0,600),ylim=c(0,0.015),
        breaks=seq(0,3000,10), main=" ");
abline(v=inf.transform,lty=1,lwd=2); points(YBAR,0,pch=16)
legend(310,0.015,c("Inference using log transformation"),lty=c(1),lwd=c(2))
text(450,0.01,
   "The dot shows 114.2, the true value \nof the finite population mean")
lines(density(ybarvec),lwd=2)


par(mfrow=c(1,1)); hist(ys,prob=T) # preliminary plot
hist(ys,prob=T,xlim=c(0,600),ylim=c(0,0.045), breaks=seq(0,700,10), main=" ");
abline(v=inf.original,lty=2,lwd=2); abline(v=inf.transform,lty=1,lwd=2)
points(YBAR,0,pch=16)
legend(310,0.04,c("Inference using original scale",
        "Inference using log transformation"), lty=c(2,1),lwd=c(2,2))
text(450,0.02,
   "The dot shows 114.2, the true value \nof the finite population mean")
```

```
# Repeat with 'less extreme' lognormal data ----------------------------------------

N=200; n=50; set.seed(432); Z=rnorm(N,2,1); S=sample(1:N,n)  # <- difference
ZS=Z[S];  Y=exp(Z);  YS=exp(ZS); YBAR=mean(Y)
X11(w=8,h=6); par(mfrow=c(2,2));
hist(Z,prob=T); hist(Y,prob=T); hist(ZS,prob=T); hist(YS,prob=T)
        # preliminary plots
ys = YS;  zs=log(ys);
skewness=mean(  (ys-mean(ys))^3 ) / ( mean((ys-mean(ys))^2) )^(3/2)
skewness # 1.549
ysbar=mean(ys); sy=sd(ys); ybarhat=ysbar
ybarci=ybarhat+c(-1,1)*qt(0.975,n-1)* (sy/sqrt(n)) * sqrt(1-n/N)
inf.original =c(ybarhat,ybarci);
c(inf.original, YBAR) # 10.541  8.177 12.906 11.698

zsbar=mean(zs); sz=sd(zs); J=1000;   set.seed(142);
lamvec=rgamma(J,(n-1)/2,(sz^2)*(n-1)/2)
muvec=rnorm(J,zsbar,1/sqrt(n*lamvec));   yrbarvec=rep(NA,J)

for(j in 1:J){     zr=rnorm(m,  muvec[j], 1/sqrt(lamvec[j]) )
                yr=exp(zr);     yrbarvec[j] = mean(yr)     }

ybarvec=(1/N)*(n*ysbar+m*yrbarvec);     ybarhat=mean(ybarvec)
ybarci=ybarhat+c(-1,1)*qnorm(0.975)*sd(ybarvec)/sqrt(J)
ybarcpdr=quantile(ybarvec,c(0.025,0.975))
inf.transform = c(ybarhat,ybarci,ybarcpdr)
c(inf.transform,YBAR) # 11.006 10.904 11.108  8.478 15.016 11.698

X11(w=8,h=4); par(mfrow=c(1,1))
hist(ys,prob=T) # preliminary plot
hist(ys,prob=T,xlim=c(0,40),ylim=c(0,0.2), breaks=seq(0,40,1), main=" ");
abline(v=inf.original,lty=2,lwd=2); abline(v=inf.transform,lty=1,lwd=2)
points(YBAR,0,pch=16)
legend(20,0.2,
   c("Inference using original scale", "Inference using log transformation"),
        lty=c(2,1),lwd=c(2,2))
text(30,0.1,
   "The dot shows 11.7, the true value \nof the finite population mean")
```

11.3 Frequentist properties of Bayesian finite population estimators

We have previously studied the frequentist characteristics of Bayesian estimators. That was in the context of analytic inference (i.e. inference on model parameters) and based on a random sample from a hypothetically *infinite* population (e.g. a normal distribution). We will now generalise those ideas in the broader framework of a Bayesian *finite* population model.

As before, we are primarily interested in the frequentist characteristics of Bayesian estimators which are based on uninformative priors and used as proxies for classical or design based estimators. Nevertheless we will consider both types of prior (informative and uninformative).

Consider a Bayesian finite population model specified in terms of:

$f(\xi \mid y, \theta)$ where ξ is s or I or L (as discussed earlier)

$f(y \mid \theta)$ where $y = (y_s, y_r) = ((y_1, ..., y_n), (y_{n+1}, ..., y_N)) = (y_1, ..., y_N)$

$f(\theta)$ where $\theta = (\theta_1, ..., \theta_q)$.

Also suppose that the data is

$D = (s, y_s)$ or (I, s) or (L, y_s)

(as the case may be), and the quantity of interest is

$\psi = g(\theta, y)$

(generally) or $\psi = g(\theta)$ (as considered previously for 'pure' analytic inference) or $\psi = g(y)$ (the case of 'pure' finite population inference).

Now suppose that in the context of this general model, data and quantity of interest, we derive a point estimate for ψ (such as the posterior mean, mode or median) of the form

$\hat{\psi} = \hat{\psi}(D)$

and a $1 - \alpha$ interval estimate for ψ (such as the CPDR or HPDR) of the form

$I = (L, U) = I(D) = (L(D), U(D))$.

Note: If the sampling mechanism is defined in terms of $I = (I_1, ..., I_N)$, the vector of inclusion counters, there is a conflict of notation and one of these quantities needs a different symbol.

In the above context, there may be interest in the frequentist bias of $\hat{\psi}$ and the frequentist coverage probabilities of the interval I, especially if these estimators are intended as proxies for classical ones.

However, because there is now an extra level in the Bayesian model hierarchy relative to previously, in the form of the density defining the sampling mechanism, namely

$$f(\xi \mid y, \theta),$$

there are two ways (at least) of defining the required frequentist characteristics:

- *model-based*, meaning conditional on θ and ξ

- *design-based*, meaning conditional on θ and y.

For definiteness, suppose that the data is $D = (s, y_s)$. Then we define:

- the *model bias* of $\hat{\psi}$ as
$$B_{\theta,s} = E_y\{(\hat{\psi}(s, y_s) - \psi(y, \theta) \mid \theta, s\}$$

- the *relative model bias* of $\hat{\psi}$ as
$$R_{\theta,s} = E_y\left\{\left.\frac{\hat{\psi}(s, y_s) - \psi(y, \theta)}{\psi(y, \theta)}\right| \theta, s\right\}$$

- the *model coverage probability* of I as
$$C_{\theta,s} = P_y\{\psi(y, \theta) \in I(s, y_s) \mid \theta, s\}.$$

Also, we define:

- the *design bias* of $\hat{\psi}$ as
$$B_{\theta,y} = E_s\{(\hat{\psi}(s, y_s) - \psi(y, \theta) \mid \theta, y\}$$

- the *relative design bias* of $\hat{\psi}$ as
$$R_{\theta,y} = E_s\left\{\left.\frac{\hat{\psi}(s, y_s) - \psi(y, \theta)}{\psi(y, \theta)}\right| \theta, y\right\}$$

- the *design coverage probability* of I as
$$C_{\theta,y} = P_s\{\psi(y, \theta) \in I(s, y_s) \mid \theta, y\}.$$

Note 1: Each of the three model-based characteristics is an expectation with respect to the distribution of y given θ and s. Each of the three design-based characteristics is an expectation with respect to the distribution of s given θ and y.

Note 2: Analogous definitions apply if $D = (I, y_s)$ or $D = (L, y_s)$, etc., noting that s is a function of I and L, there is a one-to-one correspondence between I and s under sampling without replacement, etc. For instance, if $D = (I, y_s)$, we define the model bias of $\hat{\psi}$ as

$$B_{\theta, I} = E_y \{ (\hat{\psi}(I, y_s) - \psi(y, \theta) \mid \theta, I \},$$

and when $D = (L, y_s)$, we define the model bias of $\hat{\psi}$ as

$$B_{\theta, L} = E_y \{ (\hat{\psi}(L, y_s) - \psi(y, \theta) \mid \theta, L \}, \text{ etc.}$$

Note 3: If a model-based characteristic such as the model bias $B_{\theta, s}$ is be the same for all possible samples s, then s may be dropped from the subscript; e.g. we may instead write B_θ. Likewise, if a design-based characteristic such as the design bias $B_{\theta, y}$ is the same for all possible values of the model parameter θ, then θ may be dropped; e.g. we may write B_y.

Note 4: If a model-based or design based characteristic cannot be evaluated analytically then it may be possible to estimate via a Monte Carlo simulation. This idea features in the next exercise below.

Note 5: The *model bias* of $\hat{\psi}$ above is a generalisation of the frequentist bias of an estimator as defined earlier and based on a random sample from an *infinite* population (e.g. a normal distribution). The following argument illustrates. Suppose that $\psi = \theta$, $\hat{\psi} = \bar{y}_s$ (the sample mean) and the sampling mechanism is SRSWOR. Then, by the above definitions, the model bias of $\hat{\psi}$ is

$$B_{\theta, s} = E_y \{ (\hat{\psi}(s, y_s) - \psi(y, \theta) \mid \theta, s \} \quad \text{(generally)}$$

$$= E_y (\bar{y}_s - \theta \mid \theta, s) = E_y (\bar{y}_s \mid \theta, s) - \theta,$$

where

$$E_y(\bar{y}_s \mid \theta, s) = \int \bar{y}_s f(y \mid \theta, s) dy .$$

Now, in this case,

$$f(s \mid y, \theta) = f(s) = \binom{N}{n}^{-1} \text{ for all } s = (1, ..., n), ..., (N - n + 1, ..., N),$$

so that

$$f(y \mid \theta, s) \overset{y}{\propto} f(y, \theta, s) = f(s \mid y, \theta) f(y \mid \theta) f(\theta) \overset{y}{\propto} 1 \times f(y \mid \theta) \times 1,$$

and therefore

$$f(y \mid \theta, s) = f(y \mid \theta) = f(y_r, y_s \mid \theta) = f(y_r \mid \theta, y_s) f(y_s \mid \theta),$$

with s fixed at its observed value.

From these observations we see that

$$E_y(\bar{y}_s \mid \theta, s) = \int\int \bar{y}_s f(y_r \mid \theta, y_s) f(y_s \mid \theta) dy_r dy_s$$

$$= \int f(y_r \mid \theta, y_s) dy_r \times \int \bar{y}_s f(y_s \mid \theta) dy_s$$

$$= 1 \times E(\bar{y}_s \mid \theta).$$

Therefore $B_{\theta,s} = E(\bar{y}_s \mid \theta) - \theta = E(\bar{y}_s - \theta \mid \theta).$

We have shown that the model bias here is the same as the bias of $\bar{y}_s$ in the earlier non-finite population context (where s did not feature in the notation).

This is an example of where s could be dropped from the subscript in $B_{\theta,s}$, i.e. where this could also be written B_θ.

If the sampling mechanism in this illustration were *nonignorable*, with $f(s \mid y, \theta)$ depending on y in some way, then the simplifications above might not be possible and the bias might need to be evaluated, with more difficulty, according to the formula

$$B_{\theta,s} = -\theta + \int \bar{y}_s f(y \mid \theta, s) dy = -\theta + \int \bar{y}_s \frac{f(y, \theta, s)}{f(\theta, s)} dy$$

where: $f(\theta, s) = \int f(y, \theta, s) dy$

$$f(y, \theta, s) = f(s \mid y, \theta) f(y \mid \theta) f(\theta), \text{ etc.}$$

Note 6: The *design bias* of $\hat{\psi}$ above is a generalisation of the bias of an estimator in the classical survey sampling context where a sample is drawn from a finite population of values which are thought of as constants. The following argument illustrates. Suppose that $\psi = \bar{y}$ (the finite population mean), $\hat{\psi} = \bar{y}_s$ (the sample mean) and the sampling mechanism is SRSWOR. Then, by the above definitions, the design bias of $\hat{\psi}$ is

$$B_{\theta,y} = E_s\{(\hat{\psi}(s, y_s) - \psi(y, \theta) \mid \theta, y\} \quad \text{(generally)}$$
$$= E_s\{\bar{y}_s - \bar{y} \mid \theta, y\} = E_s(\bar{y}_s \mid \theta, y) - \bar{y}.$$

Now, $\quad E_s(\bar{y}_s \mid \theta, y) = \sum_s \bar{y}_s f(s \mid y, \theta) = \dfrac{1}{kn} \sum_s (y_{s_1} + \ldots + y_{s_n})$

$$\text{where } f(s \mid y, \theta) = \frac{1}{k} \quad \text{and } k = \binom{N}{n}$$

$$= \frac{1}{kn}\{(y_1 + \ldots + y_n) + \ldots (y_{N-n+1} + \ldots + y_N)\}.$$

Here, expression $\{\ \}$ contains a total of kn terms, with each of $y_1, \ldots, y_N$ is represented equally often and therefore kn/N times.

We see that $\{\ \} = \dfrac{kn}{N}(y_1 + \ldots + y_N) = kn\bar{y}.$

Thus $E_s(\bar{y}_s \mid \theta, y) = \dfrac{1}{kn} kn \bar{y} = \bar{y},$

and so $B_{\theta,y} = E_s(\bar{y}_s \mid \theta, y) - \bar{y} = \bar{y} - \bar{y} = 0.$

We have here simply followed through with our general definitions and notation to show that under SRSWOR the sample mean is unbiased for the population mean.

If the sampling mechanism were *nonignorable*, with $f(s \mid y, \theta)$ depending on y in some way, then the bias of the sample mean might need to be evaluated, with more difficulty, according to the formula

$$B_{\theta,y} = -\bar{y} + \sum_s \bar{y}_s f(s \mid y, \theta) = -\bar{y} + \sum_s \bar{y}_s \frac{f(s, y, \theta)}{f(y, \theta)},$$

where $f(y, \theta) = \sum_s f(s, y, \theta)$, $f(s, y, \theta) = f(s \mid y, \theta) f(y \mid \theta) f(\theta)$, etc.

Exercise 11.3 Frequentist properties of Bayesian estimators in a normal finite population model

Consider a sample of size $n = 20$ taken from a finite population of size $N = 100$ according to SRSWOR, where the population values are normal with mean $\mu = 10$ and variance $\sigma^2 = 1/\lambda = 4$, with prior given by
$$f(\mu, \lambda) \propto 1/\lambda, \quad \mu \in \Re, \lambda > 0 \text{ (uninformative)}.$$

(a) Using these specifications, generate a finite population vector $y = (y_1, ..., y_N)$, take the sample vector as $y_s = (y_1, ..., y_n)$, and then use Monte Carlo (MC) methods with a sample size of $J = 1,000$ to estimate the *superpopulation signal to noise ratio* defined by $\gamma = \mu / \sigma$.

Report a point estimate of γ in the form of a MC estimate of the posterior mean $\hat{\gamma} = E(\gamma | D)$ where $D = (s, y_s)$ is the data, and an interval estimate in the form of a MC estimate of the 95% CPDR for γ. (Do not bother to calculate a 95% CI for $\hat{\gamma}$.)

What is the difference between your point estimate and γ? Does γ lie inside the interval? Calculate $\tilde{\gamma}$, the MLE of γ and report the difference between $\tilde{\gamma}$ and γ.

Illustrate your inferences by drawing a suitable histogram of the simulated values of γ, marked over with the various estimates.

(b) Perform the procedure in (a) $K = 100$ times independently, with K different finite populations but the sample always consisting of the first n values in that finite population.

Based on your results, estimate the model bias and relative model bias of your point estimator, and the model coverage of your interval estimator. Also estimate the model bias and relative model bias of the MLE $\tilde{\gamma}$.

Illustrate your results by drawing a suitable histogram of the K simulated MC estimates, marked over with the various relevant quantities.

(c) Repeat (b) but with $K = 5,000$ and discuss.

(d) Generate a finite population, vector $y = (y_1, ..., y_N)$, and then take a sample from the finite population via SRSWOR. Then use MC methods with sample size $J = 1,000$ to estimate the *finite population ratio of largest value to median*, which is given by the formula

$$\psi = \frac{y_{(100)}}{(y_{(50)} + y_{(51)})/2},$$

where $y_{(i)}$ is the ith order statistic for the N population values $y_1, ..., y_N$.

Report a point estimate of ψ in the form of a MC estimate of the posterior mean $\hat{\psi} = E(\psi \mid D)$ and an interval estimate in the form of a MC estimate of the 95% CPDR for ψ. (Do not bother to calculate a 95% CI for $\hat{\psi}$.)

What is the difference between your point estimate and ψ? Does ψ lie inside the interval?

Illustrate your inferences by drawing a suitable histogram of the simulated values of ψ, marked over with the various estimates.

(e) Perform the procedure in (d) $K = 100$ times independently, with K different samples taken from the same finite population.

Based on your results, estimate the design bias and relative design bias of your point estimator, and the design coverage of your interval estimator.

Illustrate your results by drawing a suitable histogram of the K simulated MC estimates, marked over with the various relevant quantities.

(f) Repeat (e) using two other point estimators, respectively.

Solution to Exercise 11.3

(a) A finite population of size $N = 100$ from the $N(\mu = 10, \sigma^2 = 4)$ distribution was generated. The sample mean and standard deviation of the 100 finite population values were $\bar{y} = 9.932$ and $s_y = 1.907$. Figure 11.7 shows a histogram of these values.

Figure 11.7 Histogram of N = 100 finite population values

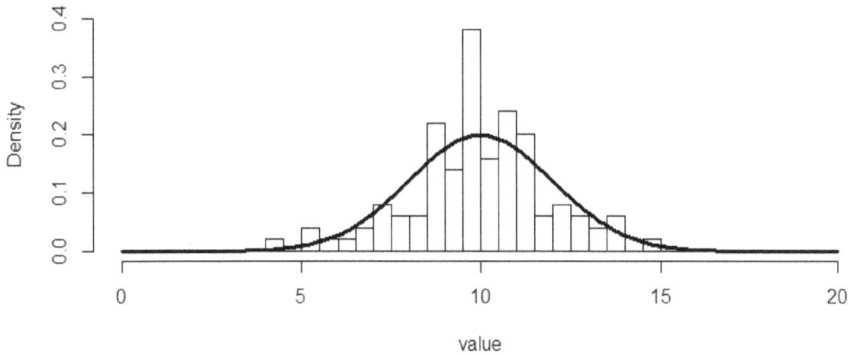

Then the first $n = 20$ values were taken as a sample from the finite population. Figure 11.8 shows a histogram of these sample values. The sample mean and standard deviation of the sample values were $\overline{y}_s = 10.516$ and $s_s = 1.749$. So the MLE of $\gamma = \mu / \sigma$ was calculated as $\tilde{\gamma} = \tilde{\mu} / \tilde{\sigma} = \overline{y}_s / s_s = 6.011$.

Figure 11.8 Histogram of n = 20 sample values

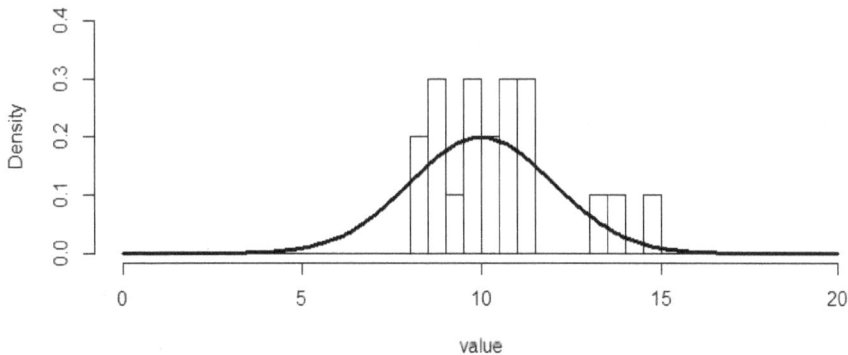

Then a Monte Carlo sample of size $J = 1,000$ was taken from the joint posterior distribution of μ and $\lambda = 1/\sigma^2$, i.e. from $f(\mu, \lambda \mid D)$ where $D = (s, y_s)$. Hence a MC sample of size J was obtained from the posterior distribution of γ, namely $\gamma_1, ..., \gamma_J \sim iid \ f(\gamma \mid D)$.

Note: As explained in previous exercises, this was done by:

- first sampling $\lambda_1, ..., \lambda_J \sim iid \ G\left(\dfrac{n-1}{2}, \dfrac{n-1}{2} s_s^2\right)$
- then sampling $w_1, ..., w_J \sim iid \ t(n-1)$
- next forming $\mu_j = \bar{y}_s + w_j s / \sqrt{n}$
- finally calculating $\gamma_j = \mu_j \sqrt{\lambda_j}$.

The MC sample from γ's posterior was used to calculate the point estimate

$$\bar{\gamma} = \frac{1}{J} \sum_{j=1}^{J} \gamma_j = 5.925$$

(the MC estimate of γ's posterior mean) and the interval estimate

$$I = (4.115, 7.963)$$

(formed by the empirical 0.025 and 0.975 quantiles of $\gamma_1, ..., \gamma_J$).

Figure 11.9 shows a histogram of the simulated values $\gamma_1, ..., \gamma_J$ overlaid by an estimate of γ's posterior density $f(\gamma \mid D)$. Also shown in the figure are the Bayesian estimates (3 vertical lines), the MLE $\hat{\gamma} = 6.011$, and the true value of γ, namely $\gamma = \mu / \sigma = 10/2 = 5$. We see that the true value of γ lies in the Bayesian interval estimate, and the difference between the Bayesian estimate and the true value is $5.925 - 5 = 0.925$. Likewise, the MLE is in 'error' by $6.011 - 5 = 1.011$.

Figure 11.9 Inference on γ based on a MC sample ($J = 1,000$)

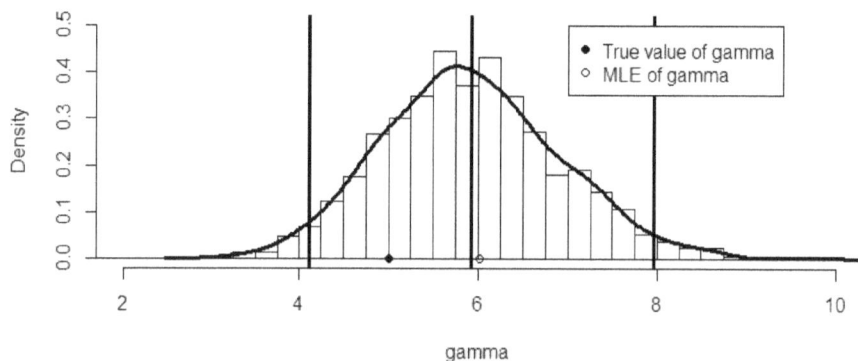

(b) The procedure in (a) was repeated so as to yield a total of $K = 100$ Bayesian estimates $\bar{\gamma}_1,...,\bar{\gamma}_K$, as well as K Bayesian interval estimates $I_1,...,I_K$ and K MLEs $\tilde{\gamma}_1,...,\tilde{\gamma}_K$.

From these results we estimated the model mean of the Bayesian estimate $\bar{\gamma}$ by

$$\bar{\bar{\gamma}} = \frac{1}{K}\sum_{k=1}^{K}\bar{\gamma}_k = 5.2226,$$

with 95% CI (for that mean) of

$$\left(\bar{\bar{\gamma}} \pm 1.96\sqrt{\frac{1}{K(K-1)}\sum_{k=1}^{K}(\bar{\gamma}_k - \bar{\bar{\gamma}})^2}\right) = (4.9986, 5.4466).$$

Hence we estimated the model bias of $\bar{\gamma}$ by $\bar{\bar{\gamma}} - \gamma = 0.2226$ with 95% CI $(-0.0014, 0.4466)$.

Likewise, we estimated the model mean of the MLE $\tilde{\gamma}$ by

$$\bar{\tilde{\gamma}} = \frac{1}{K}\sum_{k=1}^{K}\tilde{\gamma}_k = 5.298,$$

with 95% CI (for that mean) of

$$\left(\bar{\tilde{\gamma}} \pm 1.96\sqrt{\frac{1}{K(K-1)}\sum_{k=1}^{K}(\tilde{\gamma}_k - \bar{\tilde{\gamma}})^2}\right) = (5.070, 5.526).$$

Hence we estimate the model bias of $\tilde{\gamma}$ by $\bar{\tilde{\gamma}} - \gamma = 0.298$ with 95% CI $(0.0705, 0.5255)$.

Thus we also estimate the relative model biases of $\bar{\gamma}$ and $\tilde{\gamma}$ by

$$(\bar{\bar{\gamma}} - \gamma)/\gamma = 0.0445 \quad \text{with 95\% CI} \quad (-0.0003, 0.0893)$$
$$(\bar{\tilde{\gamma}} - \gamma)/\gamma = 0.0596 \quad \text{with 95\% CI} \quad (0.0141, 0.1051).$$

Note: These could also be reported as the percentages (%):

$$(\bar{\bar{\gamma}} - \gamma)/\gamma = 4.5 \quad \text{with 95\% CI} \quad (-0.03, 8.9)$$
$$(\bar{\tilde{\gamma}} - \gamma)/\gamma = 6.0 \quad \text{with 95\% CI} \quad (1.4, 10.5).$$

Also, exactly 91 of the 100 Bayesian interval estimates $I_1,...,I_K$ actually contained the true value $\gamma = 5$.

So we estimate the model coverage of the 95% CPDR estimate of $\bar{\gamma}$ (based on a MC sampled size of specifically $J = 1,000$) as 0.91, with 95% CI (for that coverage)

$$(0.91 \pm 1.96\sqrt{0.91(1-0.91)/100}) = (0.854, 0.966).$$

Figure 11.10 shows a histogram of the K simulated values of $\bar{\gamma}_1,...,\bar{\gamma}_K$ and related quantities.

We see that the Bayesian inference appears to have slightly outperformed the MLE as regards model bias.

Note that this applies in a very particular situation, namely one with $N = 100$, $n = 20$, $\mu = 10$, $\sigma = 2$, and a MC estimation scheme as described above with specifically $J = 1,000$.

Note: If we were to use a different common sample from each finite population (e.g. $y_s = (y_2, y_{14}, y_{15},..., y_{87})$), or a different sample each time, the results would be the same, subject to Monte Carlo variation. This might not be the case in a situation where the sampling mechanism is nonignorable or where there are covariate values. But as a matter of form when calculating model-based properties, we must condition on the sample being taken, i.e. on s.

Figure 11.10 Distribution of $K = 100$ estimates

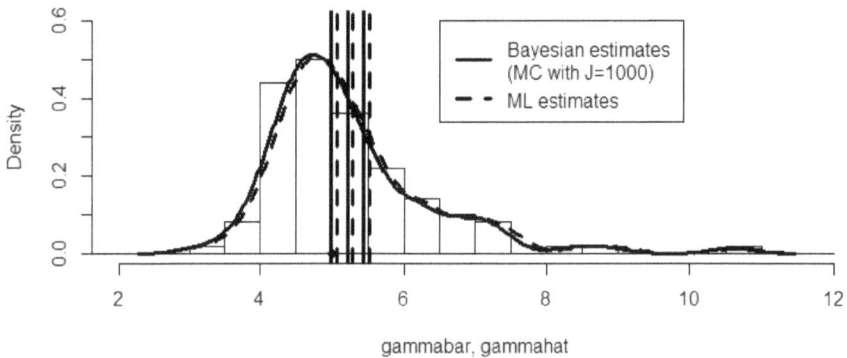

gammabar, gammahat

(c) Repeating (a) and (b) with $K = 5,000$, we obtained the following results:

Estimate of model bias of $\bar{\gamma}$ is 0.1616 with 95% CI (0.1359, 0.1872)

Estimate of model bias of $\tilde{\gamma}$ is 0.2301 with 95% CI (0.2041, 0.2561)

Estimate of relative model bias of $\bar{\gamma}$ is 3.2 with 95% CI (2.7, 3.7) (%)

Estimate of relative model bias of $\tilde{\gamma}$ is 4.6 with 95% CI (4.1, 5.1) (%).

Exactly 4,755 of the 5,000 Bayesian interval estimates $I_1, ..., I_K$ actually contained the true value $\gamma = 5$.

So we estimate the model coverage of the 95% CPDR estimate of $\bar{\gamma}$ (based on a MC sample of size $J = 1,000$) as $4,755/5,000 = 0.951$, with 95% CI (for that coverage),

$$(0.951 \pm 1.96\sqrt{0.951(1-0.951)/5,000}) = (0.945, 0.957).$$

From these results it appears that both the Bayesian and ML estimators are indeed positively biased by several percent, with the Bayesian estimator slightly outperforming the MLE.

It also appears that the model coverage of the Bayesian interval estimate is very close to the nominal 95%.

Figure 11.11 shows a histogram of the 5,000 simulated Bayesian estimates and related information. A detail in this figure is shown as Figure 11.12.

Figure 11.11 Distribution of $K = 5,000$ estimates

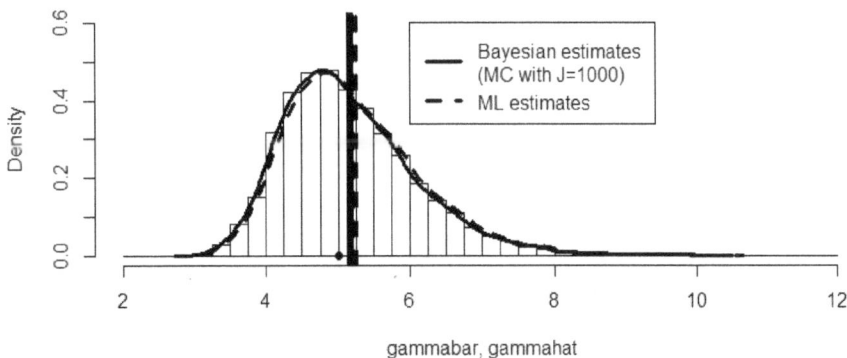

So we estimate the model coverage of the 95% CPDR estimate

Figure 11.12 Detail in Figure 11.11

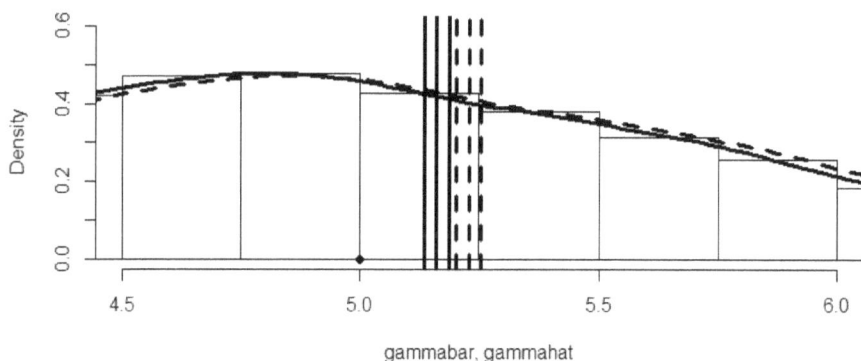

gammabar, gammahat

(d) A finite population of size $N = 100$ from the $N(\mu = 10, \sigma^2 = 4)$ distribution was generated. The sample mean and standard deviation of the 100 finite population values were $\bar{y} = 9.675$ and $s_y = 2.159$.

A histogram of the values is shown in Figure 11.13. The true value of the ratio requiring inference was in this case calculated as

$$\psi = \frac{y_{(100)}}{(y_{(50)} + y_{(51)})/2} = \frac{15.622}{10.171} = 1.536.$$

Figure 11.13 Histogram of N = 100 finite population values

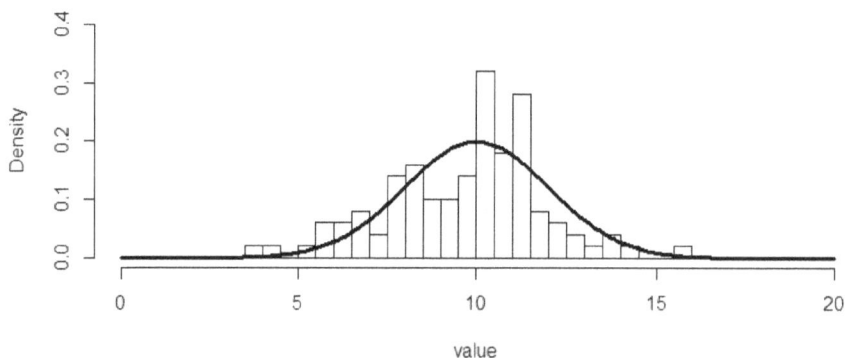

value

Then a sample of size $n = 20$ values was taken from the finite population. The sample mean and standard deviation of the sampled values were $\bar{y}_s = 9.438$ and $s_s = 2.448$. A histogram of the sample values is shown in Figure 11.14.

Figure 11.14 Histogram of $n = 20$ sample values

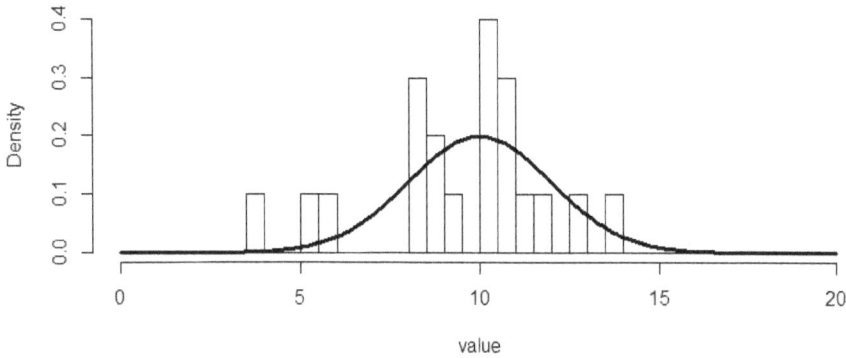

Then a MC sample of size $J = 1,000$ was taken from the joint posterior distribution of μ and $\lambda = 1/\sigma^2$, i.e. from $f(\mu, \lambda \mid D)$ with $D = (s, y_s)$. Hence a MC sample of size J was obtained from the predictive distribution of ψ, namely $\psi_1, ..., \psi_J \sim iid\ f(\psi \mid D)$.

Note: As explained in previous exercises, this was done by doing the following for each $j = 1, ..., J$:

- first sample $y_i^{(j)} \sim iid\ N\left(\mu_j, 1/\lambda_j\right), i \in r$
- then form $y_r^{(j)} = (y_{r_1}^{(j)}, ..., y_{r_{N-n}}^{(j)})$
- finally calculate ψ_j from $(y_s, y_r^{(j)})$.

The MC sample from ψ's predictive distribution was used to calculate the point estimate

$$\bar{\psi} = \frac{1}{J}\sum_{j=1}^{J}\psi_j = 1.715$$

(the MC estimate of ψ's predictive mean) and the interval $I = (1.456, 2.078)$ formed by the empirical 0.025 and 0.975 quantiles of $\psi_1, ..., \psi_J$.

Figure 11.15 shows a probability histogram of the simulated values $\psi_1, ..., \psi_J$ overlaid by an estimate of ψ's predictive density $f(\psi \mid D)$. Also shown are the Bayesian estimates (represented by three vertical lines), and the true value of ψ, which is 1.536 (represented by the dot).

We note that the true value of ψ lies in the Bayesian interval estimate, and the difference between the Bayesian estimate and the true value is $1.715 - 1.536 = 0.179$.

Figure 11.15 Inference on ψ based on a MC sample ($J = 1,000$)

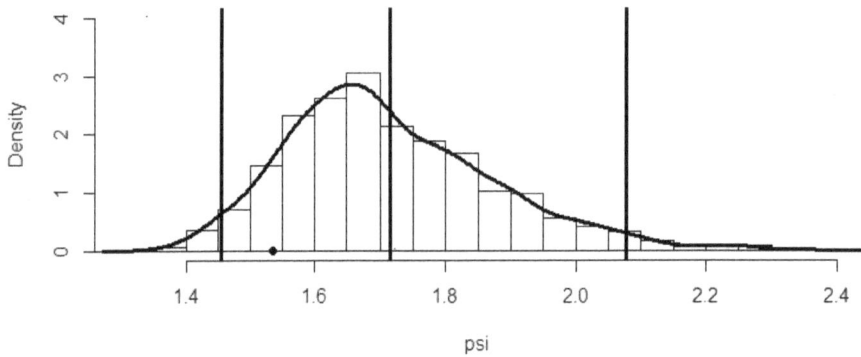

(e) The procedure in (d) was repeated so as to yield a total of $K = 100$ Bayesian estimates $\bar{\psi}_1, ..., \bar{\psi}_K$ and K corresponding Bayesian interval estimates $I_1, ..., I_K$. From these results we estimate the design mean of the Bayesian predictive mean estimate $\bar{\psi}$ by

$$\bar{\bar{\psi}} = \frac{1}{K} \sum_{k=1}^{K} \bar{\psi}_k = 1.6168,$$

with 95% CI (for that mean)

$$\left(\bar{\bar{\psi}} \pm 1.96 \sqrt{\frac{1}{K(K-1)} \sum_{k=1}^{K} (\bar{\psi}_k - \bar{\bar{\psi}})^2} \right) = (1.5962, 1.6374).$$

Hence we estimate the design bias of $\bar{\psi}$ by $\bar{\bar{\psi}} - \psi = 0.0808$, with 95% CI (0.0602, 0.1014). Thus we also estimate the relative design bias of $\bar{\psi}$ by $(\bar{\bar{\psi}} - \psi)/\psi = 5.3$, with 95% CI (3.9, 6.6) (%).

Also, 91 of the 100 Bayesian interval estimates $I_1, ..., I_K$ contained the true value, $\psi = 1.536$. So we estimate the design coverage of the 95% CPDR estimate of $\bar{\psi}$ (based on a MC sample with size $J = 1,000$) as 0.91, with 95% CI $(0.91 \pm 1.96\sqrt{0.91(1-0.91)/100}) = (0.8539, 0.9661)$.

Figure 11.16 shows a probability histogram of the K simulated values $\bar{\psi}_1, ..., \bar{\psi}_K$ and related quantities.

Figure 11.16 Distribution of $K = 100$ estimates of ψ

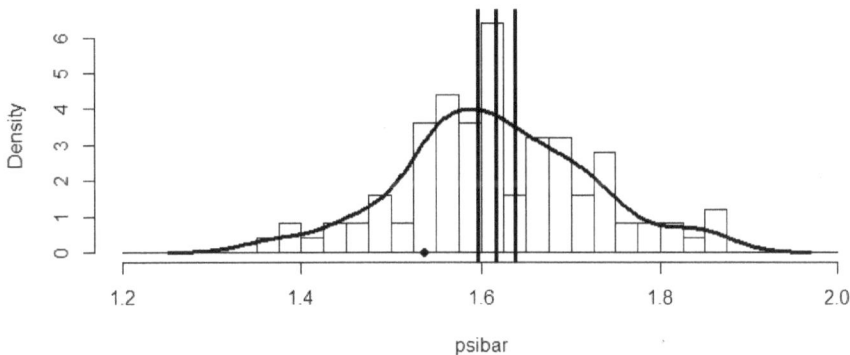

(f) Figure 11.17 is an analogue of Figure 11.16 but obtained by replacing the Monte Carlo sample *mean* estimate $\bar{\psi} = (\psi_1 + ... + \psi_J)/J$ by the *empirical median* of $\psi_1, ..., \psi_J$.

Likewise, Figure 11.18 is an analogue of Figure 11.16 but obtained by replacing the posterior mean estimate by the *empirical mode* of $\psi_1, ..., \psi_J$.

Note: The empirical mode was obtained using the R function density().

We see that the design bias of the empirical mode appears to be smaller than that of the empirical median, which in turn is smaller than that of the posterior mean. The biases of the Monte Carlo predictive mean, median and mode estimates (based on a Monte Carlo sample size of $J = 1,000$) are estimated as +5.3%, +3.8% and +1.4%.

Note: From Figure 11.15 in (d) we may have already guessed that the posterior mode is better than the posterior mean as an estimate of ψ (whose true value is 1.536, as shown by the dot in Figures 11.15–18).

Figure 11.17 Distribution of $K = 100$ estimates of ψ

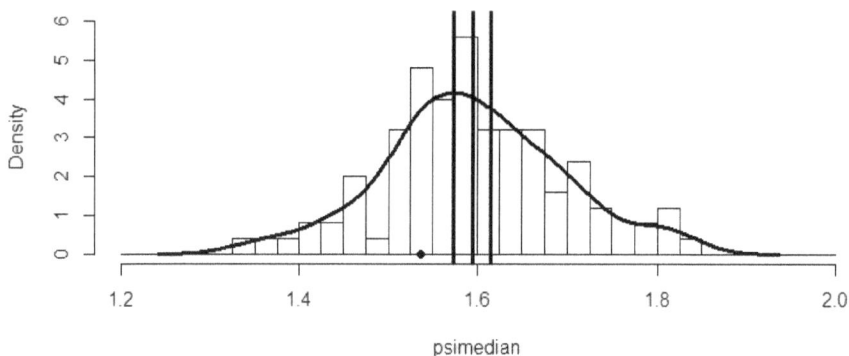

psimedian

Figure 11.18 Distribution of $K = 100$ estimates of ψ

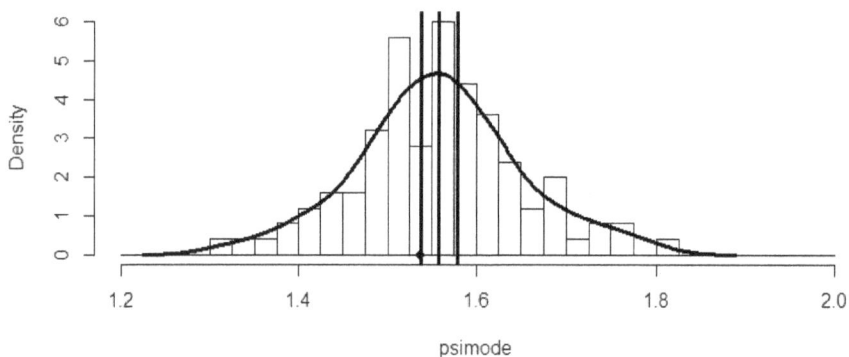

psimode

R Code for Exercise 11.3

```
# (a)

X11(w=8,h=4); par(mfrow=c(1,1)); options(digits=4)

N=100; n=20; mu=10; sig=2; lam=1/sig^2; gam=mu/sig
set.seed(332); y=rnorm(N,mu,sig); # hist(y,prob=T)
hist(y,prob=T,xlab="value", xlim=c(0,20), ylim=c(0,0.4), breaks=seq(0,20,0.5),
        main=" ")
lines(seq(0,20,0.1),dnorm(seq(0,20,0.1),mu,sig),lwd=3)

ys=y[1:n]
hist(ys,prob=T,xlab="value", xlim=c(0,20), ylim=c(0,0.4), breaks=seq(0,20,0.5),
        main=" ")
lines(seq(0,20,0.1),dnorm(seq(0,20,0.1),mu,sig),lwd=3)

ysbar=mean(ys); sys=sd(ys); gammle=ysbar/sys
ybar=mean(y); sy=sd(y); ygam=ybar/sy; c(ybar,sy,ygam) # 9.932 1.907 5.207
c(lam,ysbar,sys, gam, gammle) # 0.250 10.516  1.749  5.000  6.011

J=1000;  set.seed(171);
lamv=rgamma(J,(n-1)/2,sys^2*(n-1)/2); muv=rnorm(J,ysbar,1/sqrt((n*lamv)))
gamv=muv*sqrt(lamv)
gambar=mean(gamv);  gamint=quantile(gamv,c(0.025,0.975))
c(gambar, gamint) # 5.925 4.115 7.963

hist(gamv,prob=T,xlab="gamma", xlim=c(2,10), ylim=c(0,0.5),
        breaks=seq(0,12,0.25), main=" ")
abline(v=c(gambar, gamint),lwd=3); lines(density(gamv),lwd=3)
points(c(gam,gammle),c(0,0),pch=c(16,1))
legend(7,0.5,c("True value of gamma","MLE of gamma"),
        pch=c(16,1),bg="white")

# (b) Follows on from (a)

K = 100; J=1000; gambarvec=rep(NA,K); gammlevec=rep(NA,K);
gamlie=rep(0,K);
```

```
set.seed(143);  for(k in 1:K){
        y=rnorm(N,mu,sig); s=1:n; ys=y[s];  ysbar=mean(ys); sys=sd(ys)
        lamv=rgamma(J,(n-1)/2,sys^2*(n-1)/2);
        muv=rnorm(J,ysbar,1/sqrt((n*lamv)))
        gamv=muv*sqrt(lamv);  gambar=mean(gamv);
        gammlevec[k]=ysbar/sys
        gamint=quantile(gamv,c(0.025,0.975));   gambarvec[k]=gambar
        if((gamint[1]<=gam)&&(gam<=gamint[2])) gamlie[k]=1   }

Eest=mean(gambarvec);
Eci=Eest+c(-1,1)*qnorm(0.975)*sd(gambarvec)/sqrt(K)
Cest=mean(gamlie); Cci=Cest+c(-1,1)*qnorm(0.975)*sqrt(Cest*(1-Cest)/K)
c(Eest,Eci,Cest,Cci) # 5.2226 4.9986 5.4466 0.9100 0.8539 0.9661
Emleest=mean(gammlevec)
Emleci=Emleest+c(-1,1)*qnorm(0.975)*sd(gammlevec)/sqrt(K)
c(Emleest,Emleci) # 5.298 5.070 5.526

Biasest=Eest-gam; Biasci=Eci-gam
Biasmleest=Emleest-gam; Biasmleci=Emleci-gam
c(Biasest,Biasci, Biasmleest,Biasmleci)
        # 0.222583 -0.001418  0.446583  0.298019  0.070493  0.525544
c(Biasest,Biasci, Biasmleest,Biasmleci)/gam
        # 0.0445165 -0.0002836  0.0893166  0.0596037  0.0140986  0.1051088

# hist(gambarvec,prob=T)
hist(gambarvec,prob=T,xlab="gammabar, gammahat", xlim=c(2,12),
        ylim=c(0,0.6), breaks=seq(0,12,0.5), main= "")
abline(v=c(Eest,Eci), lty=1, lwd=3); abline(v=c(Emleest,Emleci), lty=2, lwd=3)
lines(density(gambarvec),lty=1,lwd=3); lines(density(gammlevec),lty=2,lwd=3)
points(gam,0,pch=16)
legend(6.5,0.6,c("Bayesian estimates \n(MC with J=1000)", "ML estimates"),
        lty=c(1,2), lwd=c(3,3))

# (c)

K = 5000; J=1000; gambarvec=rep(NA,K);
gammlevec=rep(NA,K); gamlie=rep(0,K);
```

```
set.seed(213); for(k in 1:K){ # Takes a few seconds
        y=rnorm(N,mu,sig); s=1:n; ys=y[s]; ysbar=mean(ys); sys=sd(ys)
        lamv=rgamma(J,(n-1)/2,sys^2*(n-1)/2);
        muv=rnorm(J,ysbar,1/sqrt((n*lamv)))
        gamv=muv*sqrt(lamv);
        gambar=mean(gamv); gammlevec[k]=ysbar/sys
        gamint=quantile(gamv,c(0.025,0.975)); gambarvec[k]=gambar
        if((gamint[1]<=gam)&&(gam<=gamint[2])) gamlie[k]=1   }


Eest=mean(gambarvec);
Eci=Eest+c(-1,1)*qnorm(0.975)*sd(gambarvec)/sqrt(K)
Cest=mean(gamlie); Cci=Cest+c(-1,1)*qnorm(0.975)*sqrt(Cest*(1-Cest)/K)
c(Eest,Eci,Cest,Cci) # 5.162 5.136 5.187 0.951 0.945 0.957
Emleest=mean(gammlevec)
Emleci=Emleest+c(-1,1)*qnorm(0.975)*sd(gammlevec)/sqrt(K)
c(Emleest,Emleci) # 5.230 5.204 5.256


Biasest=Eest-gam; Biasci=Eci-gam
Biasmleest=Emleest-gam; Biasmleci=Emleci-gam
c(Biasest,Biasci, Biasmleest,Biasmleci)
        # 0.1616 0.1359 0.1872 0.2301 0.2041 0.2561
c(Biasest,Biasci, Biasmleest,Biasmleci)/gam
        # 0.03231 0.02718 0.03745 0.04602 0.04081 0.05122


#   hist(gambarvec,prob=T)
hist(gambarvec,prob=T,xlab="gammabar, gammahat", xlim=c(2,12),
        ylim=c(0,0.6), breaks=seq(2,12,0.25), main= "")
abline(v=c(Eest,Eci), lty=1, lwd=3); abline(v=c(Emleest,Emleci), lty=2, lwd=3)
lines(density(gambarvec),lty=1,lwd=3); lines(density(gammlevec),lty=2,lwd=3)
points(gam,0,pch=16)
legend(6,0.6,c("Bayesian estimates \n(MC with J=1000)", "ML estimates"),
        lty=c(1,2), lwd=c(3,3))


hist(gambarvec,prob=T,xlab="gammabar, gammahat", xlim=c(4.5,6),
        ylim=c(0,0.6), breaks=seq(2,12,0.25), main= "")
abline(v=c(Eest,Eci), lty=1, lwd=3); abline(v=c(Emleest,Emleci), lty=2, lwd=3)
lines(density(gambarvec),lty=1,lwd=3); lines(density(gammlevec),lty=2,lwd=3)
points(gam,0,pch=16)
```

(d)

```
psifun=function(y){ max(y)/median(y) } # Function for the quantity of interest
N=100; n=20; mu=10; sig=2; set.seed(119); y=rnorm(N,mu,sig)
ybar=mean(y); sy=sd(y); psi=psifun(y=y)
c(ybar,sy,min(y),max(y), median(y), psi)
        # 9.675  2.159  3.678 15.622 10.171  1.536


hist(y,prob=T,xlab="value", xlim=c(0,20), ylim=c(0,0.4), breaks=seq(0,20,0.5),
        main="")
lines(seq(0,20,0.1),dnorm(seq(0,20,0.1),mu,sig),lwd=3)


set.seed(421); ys=sample(y,n)
ys=y[s]; ysbar=mean(ys); sy=sd(ys); sy2=var(ys)
c(ysbar,sy, sy2) # 9.438 2.448 5.994


hist(ys,prob=T,xlab="value", xlim=c(0,20), ylim=c(0,0.4), breaks=seq(0,20,0.5),
        main="")
lines(seq(0,20,0.1),dnorm(seq(0,20,0.1),mu,sig),lwd=3)


set.seed(323); J=1000;
lamv=rgamma(J,(n-1)/2,sy2*(n-1)/2); muv=rnorm(J,ysbar,1/sqrt((n*lamv)))
psiv=rep(NA,J);
for(j in 1:J){    yrsim=rnorm(N-n,muv,1/sqrt(lamv));   ysim=c(ys,yrsim);
                  psiv[j]=psifun(y=ysim)         }


psibar=mean(psiv);  psiint=quantile(psiv,c(0.025,0.975))
c(psibar,psiint)  # 1.715 1.456 2.078
summary(psiv)
#   Min. 1st Qu. Median   Mean 3rd Qu.   Max.
#   1.37   1.60   1.69   1.72   1.81   2.34


#       hist(psiv,prob=T)
hist(psiv,prob=T,xlab="psi", xlim=c(1.3,2.4), ylim=c(0,4),breaks=seq(1,2.5,0.05),
        main="")
abline(v=c(psibar,psiint),lwd=3); den=density(psiv)
lines(den,lwd=3); points(psi,0,pch=16)


psimedian=median(psiv)
psimode=den$x[den$y==max(den$y)]
c(psibar,psimedian,psimode) # 1.715 1.688 1.659
```

```
# (e)  Follows on from (d)

K = 100; J=1000; psibarvec=rep(NA,K); LBvec= psibarvec; UBvec=LBvec;
alp=0.05
set.seed(411);

date() #
for(k in 1:K){
        ys=sample(y,n);  ysbar=mean(ys); sy2=var(ys)
        lamv=rgamma(J,(n-1)/2,sy2*(n-1)/2);
        muv=rnorm(J,ysbar,1/sqrt((n*lamv)))
        psiv=rep(NA,J); for(j in 1:J){
                yrsim=rnorm(N-n,muv,1/sqrt(lamv))
                ysim=c(ys,yrsim)
                psiv[j]=psifun(y=ysim)
                }
        psibarvec[k] = mean(psiv);
        LBvec[k]=quantile(psiv,alp/2);  UBvec[k]=quantile(psiv,1-alp/2)
        };
date() #    Simulation with K=100 & J=1000  takes 12 seconds

ct=0; for(k in 1:K) if((LBvec[k]<=psi)&&(psi<=UBvec[k])) ct=ct+1

# hist(psibarvec,prob=T)
hist(psibarvec,prob=T,xlab="psibar", xlim=c(1.2,2), ylim=c(0,6.5),
        breaks=seq(1.2,2,0.025), main= "")
points(psi,0,pch=16)

# Characteristics of posterior mean estimate --------------
Eest=mean(psibarvec); Eci=Eest+c(-1,1)*qnorm(0.975)*sd(psibarvec)/sqrt(K)
Cest=ct/K; Cci=Cest+c(-1,1)*qnorm(0.975)*sqrt(Cest*(1-Cest)/K)
c(Eest,Eci,Cest,Cci) # 1.6168 1.5962 1.6374 0.9100 0.8539 0.9661
Biasest=Eest-psi; Biasci=Eci-psi; c(Biasest,Biasci)  # 0.08084 0.06024 0.10144
c(Biasest,Biasci)/psi  # 0.05263 0.03922 0.06604
abline(v=c(Eest,Eci), lty=1, lwd=3);   lines(density(psibarvec),lty=1,lwd=3)

# (f)  Follows on from (e)

K = 100; J=1000; LBvec= rep(NA,K); UBvec=LBvec;  alp=0.05
psimodevec= LBvec; psimedianvec= LBvec; set.seed(411);

date() #
```

```
for(k in 1:K){
        ys=sample(y,n); ysbar=mean(ys); sy2=var(ys)
        lamv=rgamma(J,(n-1)/2,sy2*(n-1)/2);
        muv=rnorm(J,ysbar,1/sqrt((n*lamv)))
        psiv=rep(NA,J); for(j in 1:J){
                yrsim=rnorm(N-n,muv,1/sqrt(lamv))
                ysim=c(ys,yrsim)
                psiv[j]=psifun(y=ysim)
                }
        psimedianvec[k] = median(psiv)
        den=density(psiv); psimodevec[k]=den$x[den$y==max(den$y)]
        LBvec[k]=quantile(psiv,alp/2); UBvec[k]=quantile(psiv,1-alp/2)
        }
date() #   Simulation with K=100 & J=1000  takes 12 seconds
ct=0; for(k in 1:K) if((LBvec[k]<=psi)&&(psi<=UBvec[k])) ct=ct+1

# hist(psimedianvec,prob=T)
hist(psimedianvec,prob=T,xlab="psimedian", xlim=c(1.2,2),
        ylim=c(0,6),breaks=seq(1.2,2,0.025), main= "")
points(psi,0,pch=16)

# Characteristics of posterior median estimate ------------------
Eest=mean(psimedianvec);
Eci=Eest+c(-1,1)*qnorm(0.975)*sd(psibarvec)/sqrt(K)
Cest=ct/K; Cci=Cest+c(-1,1)*qnorm(0.975)*sqrt(Cest*(1-Cest)/K)
c(Eest,Eci,Cest,Cci) # 1.5947 1.5741 1.6153 0.9100 0.8539 0.9661
Biasest=Eest-psi; Biasci=Eci-psi; c(Biasest,Biasci)  # 0.05873 0.03813 0.07934
c(Biasest,Biasci)/psi  # 0.03824 0.02483 0.05165
abline(v=c(Eest,Eci), lty=1, lwd=3);   lines(density(psimedianvec),lty=1,lwd=3)

# hist(psimodevec,prob=T)
hist(psimodevec,prob=T,xlab="psimode", xlim=c(1.2,2),
        ylim=c(0,6),breaks=seq(1.2,2,0.025), main= "")
points(psi,0,pch=16)

# Characteristics of posterior mode estimate --------------------
Eest=mean(psimodevec); Eci=Eest+c(-1,1)*qnorm(0.975)*sd(psibarvec)/sqrt(K)
Cest=ct/K; Cci=Cest+c(-1,1)*qnorm(0.975)*sqrt(Cest*(1-Cest)/K)
c(Eest,Eci,Cest,Cci) # 1.5579 1.5373 1.5785 0.9100 0.8539 0.9661
Biasest=Eest-psi; Biasci=Eci-psi; c(Biasest,Biasci)
        # 0.021933 0.001332 0.042534
c(Biasest,Biasci)/psi  # 0.0142795 0.0008672 0.0276917
abline(v=c(Eest,Eci), lty=1, lwd=3);   lines(density(psimodevec),lty=1,lwd=3)
```

CHAPTER 12

Biased Sampling and Nonresponse

12.1 Review of sampling mechanisms

We have already discussed the topic of ignorable and nonignorable sampling in the context of Bayesian finite population models. To be definite, let us now focus on the model defined by:

$f(s \mid y, \theta)$ (the probability of obtaining sample s for given values of y and θ)

$f(y \mid \theta)$ (the model density of the finite population vector)

$f(\theta)$ (the prior density of the parameter),

where the data is $D = (s, y_s)$ and the quantity of interest is some functional $\psi = g(\theta, y)$, e.g. a function of two components of θ or a function of y only, etc.

We say that the sampling mechanism is *ignorable* if

$$f(\psi \mid s, y_s) = f(\psi \mid y_s)$$

for all values of ψ, where s is fixed at its observed value, or equivalently, if the posterior distribution of ψ is exactly the same when it is calculated solely on the basis of the 'reduced model' as given by:

$f(y \mid \theta)$ (same as before)

$f(\theta)$ (same as before),

that is, with $f(s \mid y, \theta)$ effectively being 'ignored'. Otherwise, we say that the sampling mechanism is *nonignorable* (or *biased*).

Equivalently, the sampling mechanism is *ignorable* if

$$f(\psi \mid s, y_s) = f(\psi \mid y_s)$$

for all ψ, and the sampling mechanism is *nonignorable* if

$$f(\psi \mid s, y_s) \neq f(\psi \mid y_s)$$

for at least one value of ψ.

Recall that in some situations, whether the sampling mechanism is ignorable may depend on which particular units happen to be sampled.

For example if $f(s|y,\theta)$ is a function of only N, n and y_3 (say), then (typically) the sampling mechanism is ignorable if and only if unit 3 is sampled (and thereby observed).

Also, recall that analogous definitions apply if the sampling mechanism is alternatively specified in terms of
$$f(I|y,\theta)$$
or in terms of
$$f(L|y,\theta),$$
rather than in terms of
$$f(s|y,\theta).$$

Here, as previously, $I = (I_1,...,I_N)$ denotes the vector of inclusion counters, i.e. the numbers of times units $1,...,N$ are sampled (possibly more than once in the case of sampling with replacement), and $L = (L_1,...,L_n)$ is the vector of the labels of the units sampled in the temporal order in which they are sampled.

12.2 Nonresponse mechanisms

An issue related to nonignorable sampling is nonignorable nonresponse. Once a sample has been taken, some of the units may then *fail to respond*. This may be for whatever reason, but the underlying issue is that the values of the nonresponding units will then be unobserved, with possibly serious consequences to the resulting inference.

This issue can be addressed by introducing another variable and level into the modelling equation. Let R_i denote the ith *response indicator*, meaning the indicator variable for the ith population unit responding.

Thus $R_i = 1$ if unit i responds, and $R_i = 0$ otherwise ($i = 1,...,N$).

Now let $R = (R_1,...,R_N)$ (or the transpose of this) be called the *population response vector*, and likewise, define:
$$R_s = (R_{s_1},...,R_{s_n}) \text{ as the } \textit{sample response vector}$$
$$R_r = (R_{r_1},...,R_{r_{N-n}}) \text{ as the } \textit{nonsample response vector}.$$

With these definitions we may now augment our 'base model' above with a new level in the hierarchy, typically in between y and s, as follows:

$$f(s \mid R, y, \theta) \quad \text{(the probability of obtaining sample } s \text{ for given}$$
$$\text{values of } R, y \text{ and } \theta\text{)}$$

$$f(R \mid y, \theta) \quad \text{(the probability of units responding as indicated}$$
$$\text{by } R, \text{ given } y \text{ and } \theta\text{)}$$

$$f(y \mid \theta) \quad \text{(same as before)}$$

$$f(\theta) \quad \text{(same as before)}. \tag{12.1}$$

Note 1: This general formulation, with $f(s \mid R, y, \theta)$ a function of R, means that which units are sampled could *potentially* depend on which units *would* respond *if* sampled. However, typically it will be reasonable to assume that the sampling and response mechanisms are independent in the model, meaning that $f(s \mid R, y, \theta) = f(s \mid y, \theta)$.

Note 2: The statistical literature contains many different and sometimes inconsistent treatments of nonignorable nonresponse. For a review of the term 'missing at random', which relates to but does not feature in the exposition here, see Seaman et al. (2013).

In the context of this model, let
$$n_o = R_{s_1} + \ldots + R_{s_n} = 1'_n R_s$$
be the number of values in the sample that respond (have a value that is *observed*), and let
$$n_u = n - n_o$$
be the number of units in the sample that do not respond (have a value that is *unobserved*).

Then define
$$o = (o_1, \ldots, o_{n_o})$$
as the *observed vector* (the vector of the labels of the units sampled and observed), and define
$$u = (u_1, \ldots, u_{n_u})$$
as the *unobserved vector* (the vector of the labels of the units sampled and unobserved).

Note: In each of these vectors, the values (labels) are assumed to be in increasing order.

Then define the *observed sample vector* as
$$y_o = (y_{o_1}, ..., y_{o_{n_o}})$$
and the *unobserved sample vector* as
$$y_u = (y_{u_1}, ..., y_{u_{n_u}}).$$

With these definitions, the data has the general form
$$D = (s, R_s, y_o)$$
and also the quantity of interest has the general form
$$\psi = g(\theta, y, R).$$

Note 1: The function g defining ψ takes into account the possibility there may be interest in whether some of the nonsampled units *would have responded* had they been sampled.

Note 2: As mentioned previously, it is often convenient to re-label the N finite population values in such a way that
$$y = (y_s, y_r) = (y_o, y_u, y_r)$$
$$= ((y_1, ..., y_{n_o}), (y_{n_o+1}, ..., y_n), (y_{n+1}, ..., y_N))$$
$$= (y_1, ..., y_N).$$

In the context of the general four-level Bayesian finite population model given by (12.1) above (which involves s, R, y and θ), we may make the following definitions:

- The *sampling mechanism* is *ignorable* if
$$f(\psi \mid s, R_s, y_o) = f(\psi \mid R_s, y_o) \ \forall \psi$$
with s fixed at its observed value (note that o is a function of s and R_s); otherwise the sampling mechanism is *nonignorable.*

- The *response mechanism* is *ignorable* if
$$f(\psi \mid s, R_s, y_o) = f(\psi \mid s, y_o) \ \forall \psi$$
with o fixed at its observed value; otherwise the response mechanism is *nonignorable.*

562

These two basic definitions then lead to four general cases, defined as follows:

- The *sampling mechanism* and *response mechanism* are both *ignorable* if
$$f(\psi \mid s, R_s, y_o) = f(\psi \mid y_o) \; \forall \psi$$
with o fixed at its observed values.

- The *sampling mechanism* is *ignorable* and the *response mechanism* is *nonignorable* if
$$f(\psi \mid s, R_s, y_o) = f(\psi \mid R_s, y_o) \; \forall \psi$$
with s fixed at its observed value, and
$$f(\psi \mid R_s, y_o) \neq f(\psi \mid y_o)$$
for at least one value of ψ.

- The *response mechanism* is *ignorable* and the *sampling mechanism* is *nonignorable* if
$$f(\psi \mid s, R_s, y_o) = f(\psi \mid s, y_o) \; \forall \psi$$
with o fixed at its observed value and
$$f(\psi \mid s, y_o) \neq f(\psi \mid y_o)$$
for at least one value of ψ.

- The *sampling mechanism* and *response mechanism* are both *nonignorable* if
$$f(\psi \mid s, R_s, y_o) \neq f(\psi \mid R_s, y_o)$$
for at least one value of ψ and
$$f(\psi \mid s, R_s, y_o) \neq f(\psi \mid s, y_o)$$
for at least one value of ψ.

Exercise 12.1 A model with sampling and response mechanisms that are both ignorable

Consider a Bayesian finite population model defined by:

$$f(s \mid R, y, \theta)$$
$$f(R \mid y, \theta)$$
$$f(y \mid \theta)$$
$$f(\theta),$$

where the data is

$$D = (s, R_s, y_o)$$

and the quantity of interest is

$$\psi = g(\theta, y, R) = 1'_N y = \sum_{i=1}^{N} y_i = y_T \quad \text{(finite population total)}.$$

Suppose that in this context:

- the sample of n units is taken from the N in the population via SRSWOR

- each unit in the population has the same probability of response, π

- the population values in the model are iid, each with a distribution which depends only on a single parameter μ

- the model parameter vector is

$$\theta = (\mu, \pi)$$

  with $\mu \perp \pi$ (thus the two model parameters are independent, a priori).

Show that the sampling mechanism and response mechanism are both ignorable, and that this is true for all possible values of the data.

Solution to Exercise 12.1

Observe that for all s, R, y and θ:

$$f(s \mid R, y, \theta) = f(s) = \binom{N}{n}^{-1}$$

$$f(R \mid y, \theta) = f(R) = \prod_{i=1}^{N} \pi^{R_i} (1 - \pi)^{1 - R_i}$$

$$y_T = y_{oT} + y_{uT} + y_{rT},$$

where:

$\quad y_{oT} = 1_o' y_o = \sum_{i \in o} y_i$ is the total of the observed sample values

$\quad y_{uT} = 1_u' y_u = \sum_{i \in u} y_i$ is the total of the unobserved sample values

$\quad y_{rT} = 1_r' y_r = \sum_{i \in r} y_i$ is the total of the nonsample values.

Note: Here, $1_o'$ denotes a column vector of n_o ones, etc.

Consequently, the relevant predictive density of the quantity of interest, namely

$$f(\psi \mid D) = f(y_T \mid s, R_s, y_o),$$

is derived from the joint predictive density of all unobserved and nonsampled values, namely

$$f(y_u, y_r \mid s, R_s, y_o).$$

We will now proceed to show that

$$f(y_u, y_r \mid s, R_s, y_o) = f(y_u, y_r \mid y_o)$$

with o fixed at its observed value, and that this is true for all possible values of y_u, y_r, s, R_s and y_o.

If this can be shown then also

$$f(y_T \mid s, R_s, y_o) = f(y_T \mid y_o),$$

for all possible values of y_T, s, R_s and y_o.

It will thereby be established that the sampling mechanism and response mechanism are both ignorable, and that this is true for all possible values of the data $D = (s, R_s, y_o)$.

Observe that for any y_u, y_r, s, R_s and y_o, it is true that

$$f(y_u, y_r \mid s, R_s, y_o) \propto f(y_u, y_r, s, R_s, y_o)$$

$$= \sum_{R_r} \iint f(y_u, y_r, s, R_s, y_o, R_r, \mu, \pi) d\mu d\pi$$

$$= \sum_{R_r} \iint f(\mu) f(\pi) f(y_o \mid \mu) f(y_u, y_r \mid \mu, y_o)$$
$$\times f(R_s \mid \pi) f(R_r \mid \pi) f(s) d\mu d\pi$$

$$= f(s) \times \int f(y_u, y_r \mid \mu, y_o) \Big\{ f(\mu) f(y_o \mid \mu) \Big\} d\mu$$
$$\times \left[\int f(\pi) f(R_s \mid \pi) \sum_{R_r} f(R_r \mid \pi) \, d\pi \right]$$
$$\text{where } [\,\bullet\,] = \int f(\pi, R_s) \times 1 d\pi = f(R_s)$$

$$\overset{y_u, y_r}{\propto} 1 \times \int f(y_u, y_r \mid \mu, y_o) \left\{ \frac{f(\mu) f(y_o \mid \mu)}{f(y_o)} \right\} d\mu \times 1$$

$$= \int f(y_u, y_r \mid \mu, y_o) f(\mu \mid y_o) d\mu$$

$$= \int f(y_u, y_r, \mu \mid y_o) d\mu$$

$$= f(y_u, y_r \mid y_o).$$

That is,

$$f(y_u, y_r \mid s, R_s, y_o) = f(y_u, y_r \mid y_o),$$

as required.

566

Exercise 12.2 An *ignorable* sampling mechanism with a *nonignorable* response mechanism

A finite population consists of $N = 500$ values that are modelled as normally distributed with unknown mean μ and unknown variance $\sigma^2 = 1/\lambda$. A sample of size $n = 100$ is taken from this population via SRSWOR. We find that only $n_o = 34$ values are observed, with values:

   12.57, 13.35, 11.47, 14.81, 13.25,   14.09, 11.55, 11.32, 13.2, 11.28,
   9.7, 12.18, 11.49, 10.52, 9.93,   11.84, 12.2, 10.57, 11.9, 14.75,
   10.34, 14.37, 12.13, 8.56, 11.91,   11.79, 11.45, 14.98, 10.57, 12.28,
   9.91, 10.94, 13.28, 11.43.

(a) Assuming that the response mechanism is ignorable, estimate the finite population mean.

(b) A follow-up sample of size $n_f = 15$ is taken from the $n_u = 66$ non-responding units via SRSWOR, and these n_f units are observed (by 'force'), yielding the values:

   5.4, 9.41, 7.03, 8.88, 11.47,   7, 9.44, 8.58, 9.27, 8.18,
   8.62, 8.73, 7.33, 9.81, 9.88.

Thus there remain $n - n_o - n_f = n_u - n_f = 51$ nonresponding sample units with unknown values.

Assuming that the response mechanism is ignorable, use all of the available data to re-estimate the finite population mean.

(c) Repeat (b) but using a suitable Bayesian model which takes into account the response mechanism and appropriately incorporates it into the inferential procedure.

Solution to Exercise 12.2

(a) We estimate $\bar{y}$ by the average of the $n_o = 34$ observed values, which is $\bar{y}_o = 11.94$. The sample standard deviation of these n_o values is equal to $s_o = 1.552$. So a 95% CPDR for $\bar{y}$ is

$$\left(\bar{y}_o \pm t_{0.025}(n_o - 1) \frac{s_o}{\sqrt{n_o}} \sqrt{1 - \frac{n_o}{N}} \right) = (11.42, 12.46).$$

(b) We estimate $\bar{y}$ by the average of all $n_{of} = n_o + n_f = 34 + 15 = 49$ observed values, which is equal to $\bar{y}_{of} = 10.92$. The sample standard deviation of these n_{of} values is $s_{of} = 2.168$. So a 95% CPDR for $\bar{y}$ is

$$\left(\bar{y}_{of} \pm t_{0.025}(n_{of} - 1) \frac{s_{of}}{\sqrt{n_{of}}} \sqrt{1 - \frac{n_{of}}{N}} \right) = (10.33,\ 11.51).$$

(c) Figure 12.1 is a histograms of the $n_o = 34$ initially observed values and the $n_f = 15$ follow-up values, respectively. We see that the 'forced' follow-up values which initially failed to respond seem to be smaller on average than the values of the units which responded. This suggests a biased or nonignorable nonresponse mechanism whereby units with *large* values are *more likely to respond* than units with small values.

Figure 12.1 Initially observed and and follow-up sample values

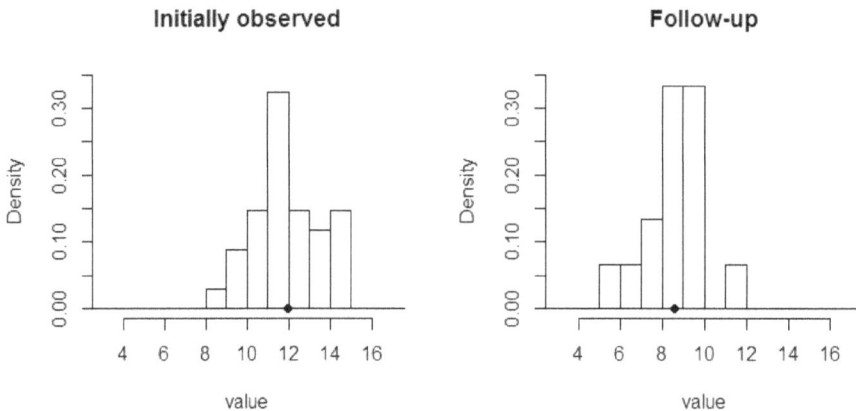

One way (amongst several) to model such a response mechanism is via the formulation

$$(R_i \mid y, \mu, \lambda) \sim \perp Bernoulli(p_i), \quad i = 1, ..., N,$$

where

$$\log\left(\frac{p_i}{1 - p_i} \right) = a + by_i$$

is the logit of the probability of unit i responding.

Noting that the sampling mechanism is ignorable, and that the response mechanism *would be ignorable* if all n sample values were known, we posit a suitable Bayesian model as follows:

$$(\overline{y} \mid y_{rT}, R_s, y_s, \mu, \lambda) \sim (\overline{y} \mid y_{rT}, y_s) = \frac{1}{N}(y_{sT} + y_{rT})$$

$$(y_{rT} \mid R_s, y_s, \mu, \lambda) \sim (y_{rT} \mid \mu, \lambda) \sim N\left((N-n)\mu, \frac{N-n}{\lambda}\right)$$

$$f(R_s \mid y_s, \mu, \lambda) = f(R_s \mid y_s) = \prod_{i \in s} p_i^{R_i}(1-p_i)^{1-R_i}$$

$$\text{where } p_i = \frac{1}{1+e^{-(a+by_i)}}$$

$$f(y_s \mid \mu, \lambda) = \prod_{i \in s} \frac{\sqrt{\lambda}}{\sqrt{2\pi}} e^{-\frac{\lambda}{2}(y_i-\mu)^2}$$

$$f(\mu, \lambda) \propto 1/\lambda, \ \mu \in \Re, \lambda > 0.$$

Note: There is no need to include the nonsample response vector R_r in the model.

Let $m = s - o - f = u - f$ be the vector of labels for the units which are sampled but still 'missing' after the follow-up sample has been observed.

Then the joint posterior/predictive density of all the relevant unknowns in the model may be written

$$f(y_{rT}, \mu, \lambda, y_m \mid R_s, y_o, y_f) \propto f(y_{rT}, \mu, \lambda, y_m, R_s, y_o, y_f)$$

$$= f(\mu, \lambda) \times \left\{ f(y_o \mid \mu, \lambda) f(y_f \mid \mu, \lambda) f(y_m \mid \mu, \lambda \right\}$$

$$\times \left\{ f(R_o \mid y_s) f(R_f \mid y_s) f(R_m \mid y_s) \right\} \times f(y_{rT} \mid \mu, \lambda)$$

$$\propto \frac{1}{\lambda} \times \prod_{i \in o} \frac{\sqrt{\lambda}}{\sqrt{2\pi}} e^{-\frac{\lambda}{2}(y_i-\mu)^2} \prod_{i \in f} \frac{\sqrt{\lambda}}{\sqrt{2\pi}} e^{-\frac{\lambda}{2}(y_i-\mu)^2} \prod_{i \in m} \frac{\sqrt{\lambda}}{\sqrt{2\pi}} e^{-\frac{\lambda}{2}(y_i-\mu)^2}$$

$$\times \prod_{i \in o} p_i^1 (1-p_i)^{1-1} \prod_{i \in f} p_i^0 (1-p_i)^{1-0} \prod_{i \in m} p_i^0 (1-p_i)^{1-0}$$

$$\times \frac{\sqrt{\lambda}}{\sqrt{N-n}\sqrt{2\pi}} e^{-\frac{\lambda}{2(N-n)}(y_i-(N-n)\mu)^2}.$$

This joint density defines a suitable Metropolis-Hastings algorithm with Gibbs steps that could be run to obtain a Monte Carlo sample from the predictive distribution of the finite population mean $\bar{y}$.

One way to proceed is to implement this algorithm using WinBUGS and the code shown below (underneath the R Code below). Some of the results are as shown in Table 12.1. These inferences are based on $J = 10,000$ iterations of a WinBUGS run, following an initial burn-in of size 1,000.

Table 12.1 Results of WinBUGS analysis

node	mean	sd	MC error	2.5%	median	97.5%
a	-17.86	4.582	0.4184	-26.96	-17.79	-10.31
b	1.676	0.4535	0.04136	0.9301	1.672	2.586
lam	0.1921	0.04236	0.001112	0.118	0.189	0.2828
mu	9.688	0.3508	0.01358	8.976	9.693	10.35
ps[1]	0.9348	0.07378	0.006256	0.7572	0.959	0.997
ps[2]	0.9721	0.0535	0.004619	0.8664	0.9886	0.9996
....						
ps[99]	0.1417	0.2097	0.003545	1.224E-5	0.04017	0.7787
ps[100]	0.1423	0.2101	0.003883	1.12E-5	0.03954	0.7731
ybar	9.687	0.3353	0.01329	9.013	9.696	10.32
yrT	3874.0	147.9	5.408	3573.0	3878.0	4156.0

From Table 12.1, we estimate the posterior mean of $\bar{y}$ by 9.69 and we estimate the 95% CPDR for $\bar{y}$ as (9.01, 10.32). It will be noted that this inference is significantly *lower* than the inferences in (a) and (b) where the response mechanism was taken as ignorable. Some of the graphical output from the WinBUGS run are shown in Figure 12.2

Figure 12.2 Graphical output from WinBUGS

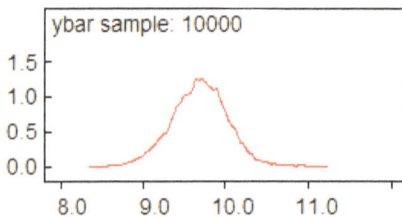

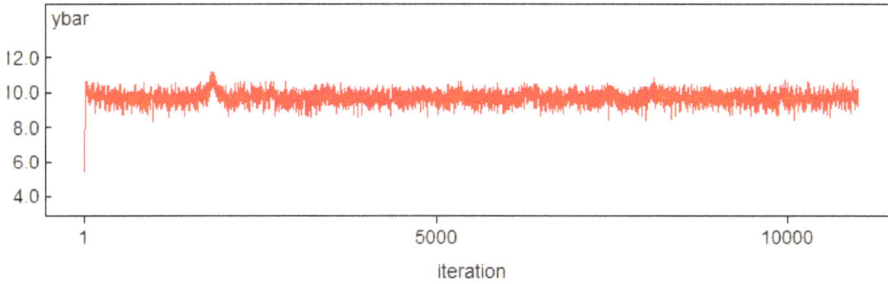

Discussion

It is instructive to now reveal that the data values in this exercise were in fact generated as follows.

First, a finite population of size $N = 500$ was generated from the normal distribution with mean $\mu = 10$ and standard deviation $\sigma = 2$. The mean of the finite population values was calculated as $\bar{y} = 10.10$.

Note: We see that the CPDR in (c), (9.013, 10.32), contains this true value of $\bar{y}$, whereas the CPDRs in (a) and (b), (11.42, 12.46) and (10.33, 11.51), do not. This suggests the analysis in (c) was on the right track.

Then a random sample of size $n = 100$ was taken from the finite population according to SRSWOR. The sample mean was calculated as $\bar{y}_s = 9.91$.

Note: Thus, if there had been no nonresponse then the finite population mean (with true value 10.10) would have been estimated by 9.91.

Figure 12.3 shows histograms of the population and sample values, each overlaid by the superpopulation density. The dots in the two subplots show $\bar{y} = 10.10$ and $\bar{y}_s = 9.91$, respectively.

Figure 12.3 Histograms of the population and sample values

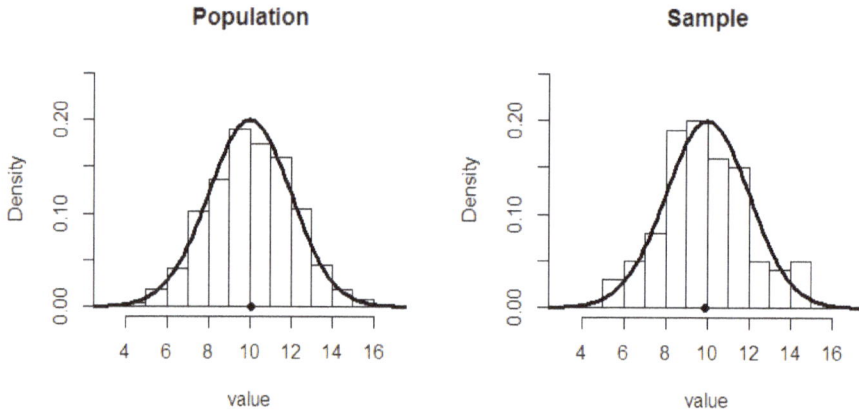

Then the probabilities of response were calculated as

$$p_i = \frac{1}{1 + e^{-(a + by_i)}}$$

with $a = -15$ and $b = 1.4$ (set in advance).

Using these probabilities, it was next determined which units would respond, by sampling

$$R_i \sim Bernoulli(p_i)$$

for each $i = 1, \ldots, N$.

Thereby it was established which *sample* units would respond and which would not. Figure 12.4 shows histograms of these two groups (of size $n_o = 34$ and $n_u = 66$), each overlaid by the superpopulation density. The dots in the left and right subplots show $\bar{y}_o$ and $\bar{y}_u$, respectively, and each histogram is overlaid by the superpopulation density.

We see how the *respondent* values are systematically *larger* than the *nonrespondent* values. This reflects the fact that units with larger values were more likely to respond.

Figure 12.4 Observed and unobserved (non-responding) sample values

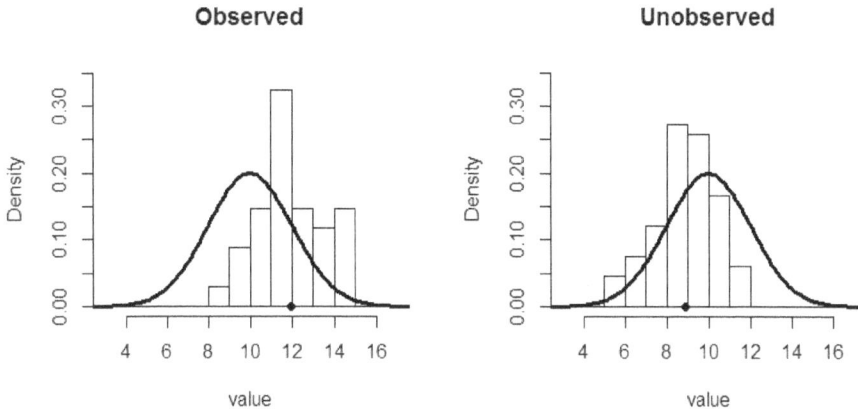

Figure 12.5 shows all N probabilities of response $p_1, ..., p_N$ plotted against the population values $y_1, ..., y_N$. The crosses indicate population units which would not respond if sampled, and these naturally tend to be the units with the smallest values.

Figure 12.5 Probabilities of response in the population

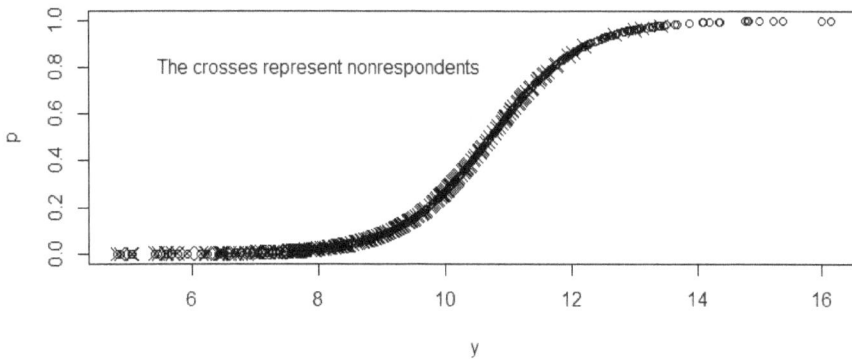

Likewise, Figure 12.6 shows the n probabilities of response in the *sample* plotted against the sample values. The crosses indicate sample units which did not respond in actuality, and these tend to be the units with the smallest values. The solid dots indicate the 15 units which were selected for 'forced' follow-up according to SRSWOR (from the 66 non-responding sample units). Without these 15 'representative' follow-up values it would have been impossible to appropriately address the nonignorable nonresponse problem and correct the biased inference in (a) and (b) downward.

Figure 12.6 Probabilities of response in the sample

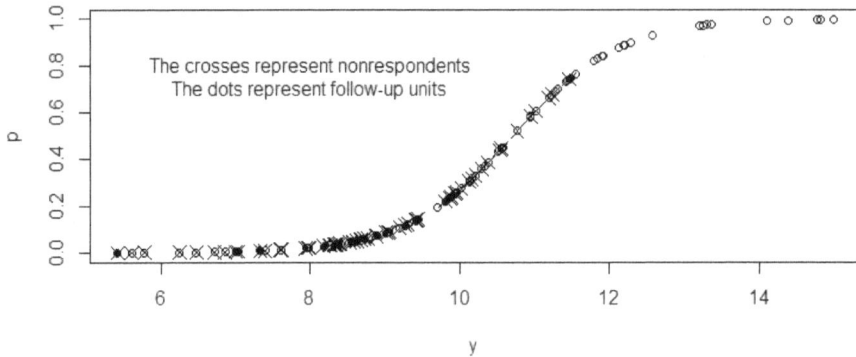

R Code for Exercise 12.2

```
# Preliminary: Data generation and description ===========

X11(w=8,h=4);  par(mfrow=c(1,1));    options(digits=4);
N=500; n=100; mu=10; sig=2; a=-15;  b=1.4;
set.seed(421); y=rnorm(N,mu,sig)  # N finite population values
p=1/(1+exp(-(a+b*y)))  # N probabilities of response (logistic)
plot(y,p) # OK

set.seed(123); R=rbinom(N,1,p) # N response indicators
set.seed(421); s=sort(sample(1:N,n))     # n sample labels

r = (1:N)[-s]      # N-n nonsample labels
ys=y[s]            # n sample values
yr=y[r]            # N-n nonsample values
Rs = R[s]          # n sample response indicators
Rr = R[r]          # N-n nonsample response indicators

no = sum(Rs); nu = n-no; c(no,nu)
          #  34 66    numbers of observed and unobserved units
o = s[Rs==1]            # labels of observed sample values
u = s[Rs==0]            # labels of unobserved sample values
```

```
rbind(s[1:10],Rs[1:10])
# [1,]   6   7  14  17  22  37  39  48  66   69
# [2,]   0   0   1   0   1   0   0   0   1   1
o[1:5] # 14 22 66 69 78      Correct
u[1:5] # 6  7 17 37 39       Correct


yo = y[o]; yu = y[u]
ybar=mean(y); ysbar=mean(ys); yrbar=mean(yr);
yobar=mean(yo); yubar=mean(yu)
c(ybar,ysbar,yrbar,yobar,yubar) #   10.095  9.907 10.143 11.938  8.860


# Plot population and sample values ------------------------------
par(mfrow=c(1,2))
hist(y,prob=T,xlab="value", main="Population",
        xlim=c(3,17),ylim=c(0,0.25), breaks=seq(0,20,1))
lines(seq(0,20,0.1),dnorm(seq(0,20,0.1),mu,sig),lwd=3)
points(ybar,0,pch=16)
hist(ys,prob=T,xlab="value", main="Sample",
        xlim=c(3,17),ylim=c(0,0.25), breaks=seq(0,20,1))
lines(seq(0,20,0.1),dnorm(seq(0,20,0.1),mu,sig),lwd=3)
points(ysbar,0,pch=16)


# Plot observed and unobserved sample values ------------------------------
par(mfrow=c(1,2))
hist(yo,prob=T,xlab="value", main="Observed",
        xlim=c(3,17),ylim=c(0,0.35), breaks=seq(0,20,1))
lines(seq(0,20,0.1),dnorm(seq(0,20,0.1),mu,sig),lwd=3)
points(yobar,0,pch=16)
hist(yu,prob=T,xlab="value", main="Unobserved",
        xlim=c(3,17),ylim=c(0,0.35), breaks=seq(0,20,1))
lines(seq(0,20,0.1),dnorm(seq(0,20,0.1),mu,sig),lwd=3)
points(yubar,0,pch=16)


# Plot probabilities of response in population --------------
par(mfrow=c(1,1))
plot(y,p,xlab="y",ylab="p",main="")
points(y[R==0], p[R==0],pch=4,cex=1.5)
text(8,0.8,"The crosses represent nonrespondents")


# Plot probabilities of response in sample and follow-up subsample --------------
par(mfrow=c(1,1)); plot(ys,p[s],xlab="y",ylab="p",main="")
points(ys[Rs==0], p[s][Rs==0],pch=4,cex=1.5)
```

```
nf=15; set.seed(112); followup = sort(sample(1:nu, nf))   # Follow up sample
f=u[followup]  # pop. labels of follow-up units
yf=y[f]   # The follow-up sample vector
yfbar=mean(yf); yfbar   # 8.601    mean of follow-up values
points(yf, p[f], pch=16) # OK
text(8,0.8,"The crosses represent nonrespondents")
text(8,0.7,"The dots represent follow-up units")


# Print data ---------------------------------------------------
s  # [1]  6  7 14 17 22 37 39 48 66 69 73  77 78 103 105 106 117.........
o  # [1] 14 22 66 69 78 141 152 156 172 228 230 232 ......
f  # [1]  17 73 77 128 145 163 187 196 253 271 318 357 436 438 481


paste(as.character(round(yo,2)), collapse=", ")
# 12.57, 13.35, 11.47, 14.81, 13.25, 14.09, 11.55, 11.32,13.2,11.28,9.7,12.18,
# 11.49, 10.52, 9.93, 11.84, 12.2, 10.57, 11.9, 14.75, 10.34, 14.37, 12.13, 8.56,
# 11.91, 11.79, 11.45, 14.98, 10.57, 12.28, 9.91, 10.94, 13.28, 11.43
paste(as.character(round(yf,2)), collapse=", ")
# 5.4, 9.41, 7.03, 8.88, 11.47, 7, 9.44, 8.58,9.27,8.18,8.62,8.73,7.33,9.81, 9.88


# (a) ===================================
yo = c(12.57, 13.35, 11.47, 14.81, 13.25, 14.09, 11.55, 11.32, 13.2, 11.28, 9.7,
12.18, 11.49, 10.52, 9.93, 11.84, 12.2, 10.57, 11.9, 14.75, 10.34, 14.37, 12.13,
8.56, 11.91, 11.79, 11.45, 14.98, 10.57, 12.28, 9.91, 10.94, 13.28, 11.43)
no=length(yo); N=500; ybarhata = mean(yo); so=sd(yo)
ybarcpdra=ybarhata+c(-1,1)*qt(0.975,no-1)*(so/sqrt(no))*sqrt(1-no/N)
c(no,so,ybarhata, ybarcpdra) # 34.000  1.552 11.939 11.416 12.461


# (b) ===================================
yf = c(5.4,9.41,7.03,8.88,11.47,7,9.44,8.58,9.27,8.18,8.62,8.73,7.33, 9.81,9.88)
yof=c(yo,yf); nof=no+nf; ybarhatb = mean(yof);sof=sd(yof)
ybarcpdrb=ybarhatb+c(-1,1)*qt(0.975,nof-1)*(sof/sqrt(nof))*sqrt(1-nof/N)
c(nof,sof,ybarhatb, ybarcpdrb) # 49.000  2.168 10.917 10.326 11.509


# (c) ============================================
# Plot observed and follow-up sample values separately
par(mfrow=c(1,2))
hist(yo,prob=T,xlab="value", main="Initially observed",
       xlim=c(3,17),ylim=c(0,0.35), breaks=seq(0,20,1));
points(mean(yo),0,pch=16);
hist(yf,prob=T,xlab="value", main="Follow-up",
       xlim=c(3,17),ylim=c(0,0.35), breaks=seq(0,20,1));
points(mean(yf),0,pch=16)
```

WinBUGS code for Exercise 12.2

```
model
{
for(i in 1:n){
  zs[i] <- a + b*ys[i]
  logit(ps[i])<- zs[i]
  rs[i] ~ dbern(ps[i])
  ys[i] ~ dnorm(mu,lam)
  }
a ~ dnorm(0.0,0.001)
b ~ dnorm(0.0,0.001)
mu ~ dnorm(0.0,0.001)
lam ~ dgamma(0.001,0.001)
ysT <- sum(ys[])
meanyrT <- nr*mu
precyrT <- lam/nr
yrT ~ dnorm(meanyrT,precyrT)
ybar <- (ysT+yrT)/(n+nr)
}

# data
list( n=100, nr=400,
        rs=c(   1,1,1,1,1,1,1,1,1,1,  1,1,1,1,1,1,1,1,1,1,
                1,1,1,1,1,1,1,1,1,1,  1,1,1,1,0,0,0,0,0,0,
                0,0,0,0,0,0,0,0,0,0,  0,0,0,0,0,0,0,0,0,0,
                0,0,0,0,0,0,0,0,0,0,  0,0,0,0,0,0,0,0,0,0,
                0,0,0,0,0,0,0,0,0,0,  0,0,0,0,0,0,0,0,0,0),

ys=c(
12.57, 13.35, 11.47, 14.81, 13.25,      14.09, 11.55, 11.32, 13.2, 11.28,
9.7, 12.18, 11.49, 10.52, 9.93,         11.84, 12.2, 10.57, 11.9, 14.75,
10.34, 14.37, 12.13, 8.56, 11.91,       11.79, 11.45, 14.98, 10.57, 12.28,
9.91, 10.94, 13.28, 11.43, 5.4,         9.41, 7.03, 8.88, 11.47, 7,
9.44, 8.58, 9.27, 8.18, 8.62,           8.73, 7.33, 9.81, 9.88, NA,
NA, NA, NA, NA, NA,                     NA, NA, NA, NA, NA,
NA, NA, NA, NA, NA,                     NA, NA, NA, NA, NA,
NA, NA, NA, NA, NA,                     NA, NA, NA, NA, NA,
NA, NA, NA, NA, NA,                     NA, NA, NA, NA, NA,
NA, NA, NA, NA, NA,                     NA, NA, NA, NA, NA)   )

# inits
list(a=0,b=0,mu=0,lam=1)
```

12.3 Selection bias in volunteer surveys

Volunteer surveys are common nowadays, with the main mediums being the telephone and Internet. However, they can be misleading on account of selection bias, and this has been known for a long time. For example, in 1983 a major television network in the US conducted a phone-in (or dial-in) poll. Viewers were invited to phone the network and answer the following question:

Should the United Nations continue to be based in the United States?

Of the 185,000 phones calls subsequently registered, 33% were from persons answering yes, and 67% from persons answering no. The question then arose as to how reliable these figures are when applied to the American population as a whole. Many factors could affect said reliability, for example whether some people phoned in more than once.

A key concern is that maybe yes-respondents were more, or less, likely to phone in than no-respondents. For example, if yes-respondents were *less* likely to phone in, then the sample almost certainly contained an unrepresentatively low proportion of yes-responses. Consequently, the figure 33% is biased and *too low* when taken as an estimate of the percentage of all Americans in favour of the UN being based in the US.

Concerned about the accuracy of its phone-in polls generally, the TV network conducted an independent survey of the entire American population using proper probability sampling techniques. A SRSWOR of 1,000 persons yielded 72% yes-responses to the same question and 28% no-responses.

From these results, we may suspect that yes-respondents were indeed less likely to phone in than no-respondents. This prompts us to now study the issue in more depth, starting with the following model. This model and parts of the subsequent exposition can also be found in Puza and O'Neill (2006).

12.4 A classical model for self-selection bias

Suppose that there are a large number N units in the population (e.g. persons in the US) and each unit has the same probability p of having a particular characteristic in question (e.g. being in favour of the UN being based in the US).

Then define:

y_i as the indicator for pop. unit i having the characteristic (0 or 1)

π_i as the probability that unit i will be sampled (e.g. phone in to answer the question)

I_i as the indicator that population unit i is sampled.

In this context the data is $D = (n, y_{sT})$, where:

$n = I_1 + \ldots + I_N$ is the observed sample size

$y_{sT} = y_{s_1} + \ldots + y_{s_n}$ is the number of yes-respondents in the sample.

Now, a 'naïve' or 'base' model here is

$$y_{sT} \sim Bin(n, p),$$

and this leads to the straight sample proportion

$$\overline{y}_s = y_{sT} / n$$

as an estimate of p.

We now wish to generalise this model to account for the possibility that $\overline{y}_s$ may be biased. To this end, suppose each π_i can be one of two values:

ϕ_1 if that unit **has** the characteristic in question, i.e. if $y_i = 1$

ϕ_0 if that unit **does not have** the characteristic, i.e. if $y_i = 0$.

Note: We may then write $\pi_i = \phi_{y_i}$.

Next, suppose that a unit *with* the characteristic in question is λ times as likely to respond as a unit *without* the characteristic. Thus

$$\phi_1 = \lambda\phi_0.$$

Also, write ϕ_0 simply as ϕ Then,

$$\pi_i = \begin{cases} \phi & \text{if } y_i = 0 \\ \lambda\phi & \text{if } y_i = 1 \end{cases} = \phi\lambda^{y_i}.$$

With the above definitions, we now consider the probability of a *respondent* having the characteristic (as distinct from the probability of a *nonrespondent* having the characteristic):

$$P(y_i = 1 | I_i = 1) = \frac{P(y_i = 1)P(I_i = 1 | y_i = 1)}{P(I_i = 1)}$$

(note that we are applying Bayes' rule here)

$$= \frac{P(y_i = 1)P(I_i = 1 | y_i = 1)}{P(y_i = 0)P(I_i = 1 | y_i = 0) + P(y_i = 1)P(I_i = 1 | y_i = 1)}$$

$$= \frac{p\phi_1}{(1-p)\phi_0 + p\phi_1} = \frac{p\phi\lambda}{(1-p)\phi + p\phi\lambda} = \frac{p\lambda}{1-p+p\lambda}.$$

Note: Observe how one of the parameters, namely ϕ, cancels out here.

We may now write $y_{sT} \sim Bin(n, \omega)$, where $\omega = \dfrac{p\lambda}{1-p+p\lambda}$.

Next, the MLE and method of moments estimator of ω is $\bar{y}_s = y_{sT} / n$.

Also, solving $\omega = \dfrac{p\lambda}{1-p+p\lambda}$ for p yields $p = \dfrac{\omega}{\lambda - \lambda\omega + \omega}$.

It follows that the MLE and MOME of p is $\hat{p} = \dfrac{\bar{y}_s}{\lambda - \lambda\bar{y}_s + \bar{y}_s}$.

Also, $(L, U) = \left(\bar{y}_s \pm z_{\alpha/2} \sqrt{\dfrac{\bar{y}_s(1-\bar{y}_s)}{n}} \right)$ is a $1-\alpha$ CI for ω.

Therefore, a $1-\alpha$ CI for p is $\left(\dfrac{L}{\lambda - \lambda L + L}, \dfrac{U}{\lambda - \lambda U + U} \right)$.

It is of interest to now discuss the biases of the two estimators mentioned above. First, the bias of $\bar{y}_s$ is

$$B(\bar{y}_s) = \omega - p = p\left(1 - \frac{\lambda}{1-p+p\lambda} \right) = p\frac{(1-p)(1-\lambda)}{1-p(1-\lambda)}.$$

This is not zero but reduces to zero when $\lambda = 1$, i.e. when $\pi_1 = \pi_0$.

Also, the bias of $\hat{p}$ is $B(\hat{p}) = E\left(\dfrac{\overline{y}_s}{\lambda - \lambda\overline{y}_s + \overline{y}_s}\right) - p$.

Just like $B(\overline{y}_s)$, $B(\hat{p})$ is nonzero but reduces to zero when $\lambda = 1$. But unlike $B(\overline{y}_s)$, $B(\hat{p})$ converges to zero as the sample size n tends to infinity, this being true for all λ.

That is, $\hat{p} = \dfrac{\overline{y}_s}{\lambda - \lambda\overline{y}_s + \overline{y}_s}$ is asymptotically unbiased for p as $n \to \infty$.

Note: This is obvious by construction. But just to check, we note that

$E\overline{y}_s = \dfrac{p\lambda}{1 - p + p\lambda}$ and $V\overline{y}_s < \infty$. Therefore

$$B(\hat{p}) \to \left(\dfrac{\left[\dfrac{p\lambda}{1 - p + p\lambda}\right]}{\lambda - \lambda\left[\dfrac{p\lambda}{1 - p + p\lambda}\right] + \left[\dfrac{p\lambda}{1 - p + p\lambda}\right]}\right) - p = 0 \text{ as } n \to \infty.$$

Example 12.1 Application to the US TV network scenario (a classical analysis)

Observe that $\omega = \dfrac{p\lambda}{1 - p + p\lambda}$ implies $\lambda = \dfrac{\omega(1 - p)}{p(1 - \omega)}$.

Then recall that the phone-in poll conducted by the TV network yielded an estimate of 0.33, and that the parallel scientifically designed (and 'proper') survey yielded an estimate of 0.72.

Thus we may estimate $\lambda = \pi_1 / \pi_0$ by

$$\hat{\lambda} = \dfrac{\hat{\omega}(1 - \hat{p})}{\hat{p}(1 - \hat{\omega})} = \dfrac{0.33(1 - 0.72)}{0.72(1 - 0.33)} = 0.19.$$

This estimate being less than unity is consistent with our earlier intuition that the phone-in poll estimate might be too low due to yes-respondents being *less* likely to phone in than no-respondents.

Example 12.2 Inference on p in a flag poll (a classical analysis)

On 28 January 2000 an Internet poll was conducted by the Nine TV Network in Australia with the question:
Should the Australian flag be replaced by a new one?

To this poll there were 4,941 yes-responses and 4,512 no-responses, thus a proportion of
$$4,941/(4,941 + 4,512) = 4,941/9,453 = 0.523 \text{ yeses.}$$

A similar question was asked in the Australian Constitutional Referendum Study, 1999 (Gow et al., 2000), and this proper survey yielded 829 yes-responses and 1,394 no-responses, thus a proportion of
$$829/(829 + 1,394) = 829/2,223 = 0.373 \text{ yeses.}$$

Hence, for the 28 January Internet poll we may estimate $\lambda = \pi_1 / \pi_0$ by
$$\hat{\lambda} = \frac{\hat{\omega}(1 - \hat{p})}{\hat{p}(1 - \hat{\omega})} = \frac{0.523(1 - 0.373)}{0.373(1 - 0.523)} = 1.84.$$

This suggests that persons who wanted the flag replaced were *almost twice as likely* to register their opinion via the Internet poll as persons who were happy with the old flag.

Example 12.3 Inference on p in a currency poll (a classical analysis)

On 4 June 2000 an Internet poll was conducted by the Nine TV Network with the question:
Should the Queen's image be removed from our currency?

To this there were 2,544 yes-responses and 1,755 no-responses, thus a proportion of
$$2,544/(2,544 + 1,755) = 2,544/4,299 = 0.592 \text{ yeses.}$$

Now recall Example 12.2. Clearly there is some similarity between the two polls. Both were conducted on the Internet by the same organisation within the same half-year, and the two questions asked both relate to changing something about Australia's heritage. This similarity suggests that 1.84 *may* be a plausible value of $\lambda = \pi_1 / \pi_0$ to be used in the 4 June poll here.

If so, we may estimate the true proportion of Australians in favour of removing the Queen's image from our currency as

$$\hat{p} = \frac{\overline{y}_s}{\lambda - \lambda \overline{y}_s + \overline{y}_s} = \frac{0.592}{1.84 - 1.84 \times 0.592 + 0.592} = 0.441.$$

Then, a 95% CI for $\omega = \dfrac{p\lambda}{1-p+p\lambda}$ (the probability of a yes-response for a respondent) is

$$(L,U) = \left(\overline{y}_s \pm z_{\alpha/2} \sqrt{\frac{\overline{y}_s(1-\overline{y}_s)}{n}} \right) = \left(0.592 \pm 1.96 \sqrt{\frac{0.592(1-0.592)}{4,299}} \right)$$
$$= (0.577, 0.607).$$

Therefore, a $1-\alpha$ CI for p is

$$\left(\frac{L}{\lambda - \lambda L + L}, \frac{U}{\lambda - \lambda U + U} \right)$$
$$= \left(\frac{0.577}{1.84 - 1.84 \times 0.577 + 0.577}, \frac{0.607}{1.84 - 1.84 \times 0.607 + 0.607} \right)$$
$$= (0.426, 0.456).$$

12.5 Uncertainty regarding the sampling mechanism

In Example 12.3 above, the value of λ was taken to be exactly 1.84. However, there is in fact uncertainty about λ which ought to be taken into account and perhaps lead to a wider CI for p than the one reported.

With this in mind we now postulate the following Bayesian model:

$$(y_{sT} \mid p, \lambda) \sim Bin(n, \omega) \quad \text{where} \quad \omega = \frac{p\lambda}{1-p+p\lambda} \quad \text{(as before)}$$

$$(p \mid \lambda) \sim Beta(\alpha, \beta)$$

$$\lambda \sim Gamma(\eta, \tau). \tag{12.2}$$

Note: This model implicitly conditions on the sample size n.

Example 12.4 Bayesian re-analysis of poll data Example 12.2

Recall the 28 January 2000 Internet poll yielding 4,941 yeses out of 9,453 responses and the related properly conducted probability survey yielding 829 yeses and 1,394 nos.

This suggests we apply the Bayesian model (12.2) in WinBUGS to estimate λ, with:

$$\eta = \tau = 0.000001 \quad \text{(implying an uninformative prior on } \lambda \text{)}$$

$$\alpha = 829 + 1 = 830, \ \beta = 1,394 + 1 = 1,395$$
   (the posterior of p implied by the proper survey in a
   binomial-beta model and then fed here as the prior for p)

$$n = 9,453, \ y_{sT} = 4,941$$
   (the observed data in the self-selected sample).

Using suitable WinBUGS code (see below) and a sample size of 10,000 after a burn-in of 1,000, we obtained results shown in Table 12.2. Figure 12.7 shows some of the graphical output from WinBUGS.

Table 12.2 Results of WinBUGS analysis

node	mean	sd	MC error	2.5%	median	97.5%
lam	1.843	0.08879	0.00271	1.677	1.841	2.026
p	0.373	0.01022	3.15E-4	0.3529	0.373	0.393

We see that λ has been estimated as 1.84 again, but now with some measure of uncertainty: the 95% posterior interval estimate for λ is (1.68, 2.03).

Figure 12.7 Graphical output from WinBUGS

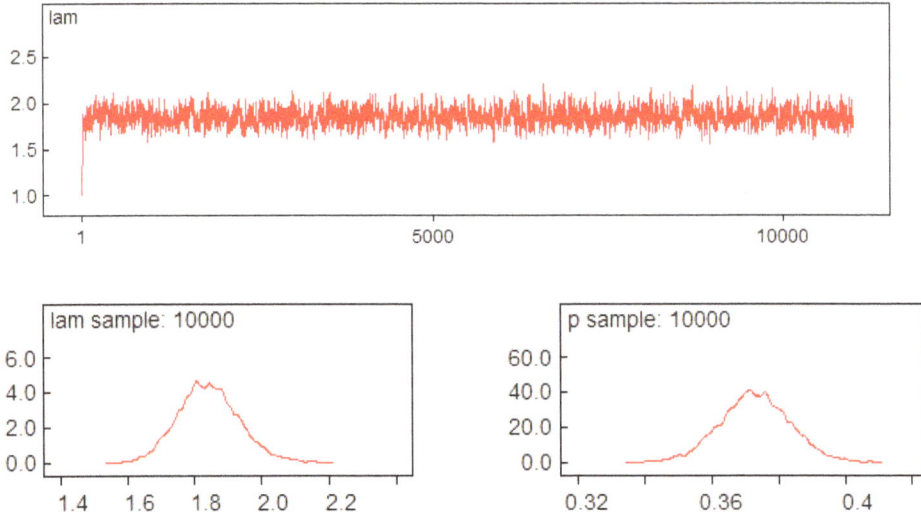

Equating the sample mean and sample variance of the 10,000 simulated values with the theoretical mean and variance of the $Gamma(\eta, \tau)$, namely η / τ and η / τ^2, respectively, we may *approximate* the posterior distribution of λ as $Gamma(\eta, \tau)$ with $\eta = 431$ and $\tau = 234$.

Figure 12.8 shows a histogram of the simulated values overlaid by the gamma density defined by these parameters. We see that the gamma posterior approximation fits quite well.

Figure 12.8 Histogram of simulated values and fitted gamma density

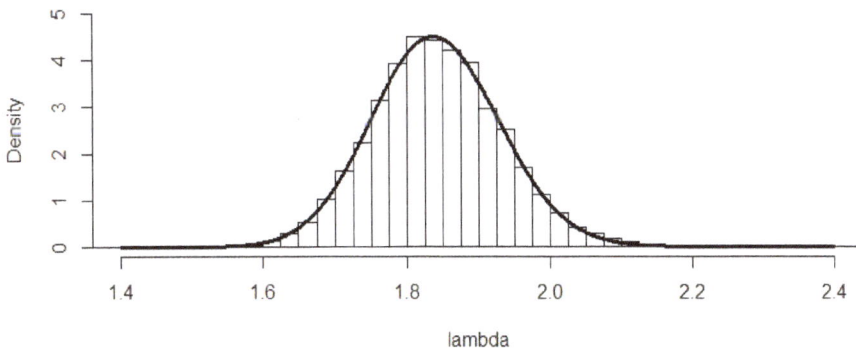

WinBUGS Code for Example 12.4

```
model;
{
  ysT~ dbin(omega,n)
  omega <- (p*lam)/(1-p+lam*p)
  lam ~ dgamma(eta,tau)
  p ~ dbeta(alpha,beta)
}
```

```
# data
list(ysT=4941,n=9453,eta=0.000001,
tau=0.000001,alpha=830,beta=1395)
```

```
# inits
list(p=0.5,lam=1)
```

R Code for Example 12.4

```
# Need to run BUGS code above first, using coda to create output in data.txt

options(digits=3);   0.33*0.28/(0.72*0.67) # 0.192
0.523*(1-0.373)/(0.373*(1-0.523)) # 1.84
0.592/(1.84-1.84*0.592+0.592) # 0.441
CIomega = 0.592+c(-1,1)*1.96*sqrt(0.592*(1-0.592)/4299)
CIp = (CIomega/(1.84-1.84*CIomega+CIomega))
c(CIomega, CIp) # 0.577 0.607 0.426 0.456

out=read.table(file=file.choose()) # choose data.txt from BUGS run
lamvec = out[1:10000,2]; options(digits=5)
lambar=mean(lamvec); lamvar=var(lamvec)
taufit=lambar/lamvar;  etafit=lambar*taufit
c(lambar, lamvar, etafit, taufit)
        # 1.8432e+00 7.8849e-03 4.3087e+02 2.3376e+02
summary(lamvec)
#   Min. 1st Qu. Median   Mean 3rd Qu.   Max.
#   1.55   1.78   1.84   1.84   1.90   2.20
X11(w=8,h=4); par(mfrow=c(1,1))
lamv <- seq(1.4,2.4,0.001)
fv <- dgamma(lamv,431,234)
hist(lamvec,prob=T,xlim=c(1.4,2.4),ylim=c(0,5),xlab="lambda",cex=1.5,
        breaks=seq(1,3,0.025), main="")
lines(lamv,fv,lwd=3)
```

Example 12.5 Bayesian re-analysis of poll data in Example 12.3 using results in Example 12.4

Recall the 4 June 2000 poll yielding 2,544 yeses out of 4,299 responses, leading to 0.441 as an estimate of p, with 95% CI (0.426, 0.456), based on λ being *exactly* equal to 1.84. This suggests we apply our Bayesian model in WinBUGS to estimate p with:

$$\eta = 431, \quad \tau = 234$$

(using the posterior for λ in Example 4 as the prior)

$$\alpha = \beta = 1 \qquad \text{(implying an uninformative prior for } p\text{)}$$

$$n = 4{,}299, \quad y_{sT} = 2{,}544$$

(the observed data in the self-selected sample).

Using suitable WinBUGS code (see below), we obtained the results shown in Table 12.3. Some of the graphical output is shown in Figure 12.9.

Table 12.3 Results of WinBUGS analysis

node	mean	sd	MC error	2.5%	median	97.5%
lam	1.841	0.08801	0.001991	1.67	1.84	2.014
p	0.4409	0.01408	3.18E-4	0.414	0.4406	0.4698

We see that p has been estimated as 0.441 again, with 95% interval estimate (0.414, 0.470). It will be noted that this interval is *wider* than the one in Example 12.3; this may be attributed to the fact that in Example 12.3 uncertainty regarding λ was not properly taken into account. For more information on the topic in this section, see Puza and O'Neill (2006).

Note: The posterior for λ is virtually the same as the prior for λ. This was to be expected, since—unlike in Example 12.4—the data here does not contain any structure which could tell us anything about the relationship between the sampling propensities π_0 and π_1.

Figure 12.9 Graphical output from WinBUGS

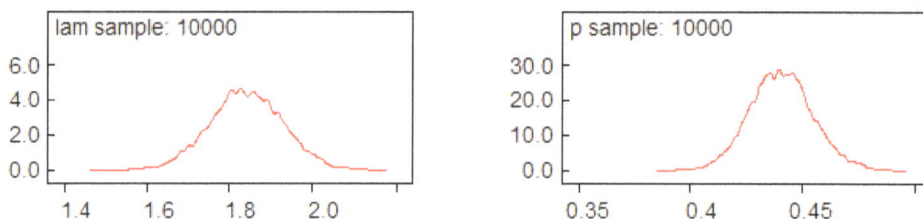

WinBUGS Code for Example 12.5

```
model;
{
  ysT~ dbin(omega,n)
  omega <- (p*lam)/(1-p+lam*p)
  lam ~ dgamma(eta,tau)
  p ~ dbeta(alpha,beta)
}

# data
list(ysT=2544,n=4299,eta=431,
tau=234,alpha=1,beta=1)

# inits
list(p=0.5,lam=1)
```

12.6 Finite population inference under selection bias in volunteer surveys

In the last section on selection bias in volunteer surveys, the finite population size N was introduced at the beginning, but then seemed to disappear from the notation. The Bayesian model subsequently developed did not feature N at all.

This is a clue to the fact that the Bayesian model in that section is only useful for *infinite* population inference, in particular on the superpopulation parameter p, and cannot be used for inference on *finite* population quantities, in particular the finite population mean

$$\bar{y} = (y_1 + \ldots + y_N) / N .$$

This is not an issue when N is very large (as it was assumed there), since in that case inference on $\bar{y}$ is, by the law of large numbers, virtually identical to inference on the superpopulation mean p.

The following exercise develops a 'true' Bayesian *finite* population model in the same setting, one which could be useful in scenarios where N is not so large as to be effectively infinity.

Exercise 12.3 A Bayesian finite population self-selection model

Consider a finite population of N units, where each unit has common probability p of having some characteristic, independently of all the other units, and where our prior beliefs regarding p can be represented by way of a beta distribution with parameters α and β.

A sample is selected from the finite population in such a way that every unit *without* the characteristic has probability ϕ of being sampled, and every unit *with* the characteristic has probability $\lambda\phi$ of being selected. Every unit that is sampled has its value fully observed.

The prior on ϕ is beta with parameters δ and γ but evenly spread over the interval $(0, c)$, where $c < 1$ is a specified constant representing an absolute upper bound for what the value of ϕ could possibly be. (Examples of a potentially suitable values of c are 0.1, 0.2 and 0.5.)

Also, the prior on λ is beta with parameters η and τ but evenly spread over the interval $(0, 1/c)$, so as to permit a suitably wide range of possible values for the ratio of sampling propensities $\pi_1 = \lambda\phi$ to $\pi_0 = \phi$. (For example, if $c = 0.2$ then that ratio could be anything from 0 to 5.)

(a) Write down a Bayesian model which comprehensively represents the above situation. Assume that all of the model parameters are independent a priori. Clearly identify the data.

(b) Suppose we are interested in both the *superpopulation mean* (i.e. the common probability of a unit having the characteristic, p) and the *finite population mean* (i.e. proportion of the N finite population units which have the characteristic, $\bar{y}$). Write down a formula for the joint posterior (and predictive) density of all quantities which are relevant to and could used be as a basis for the desired inference.

(c) Use the density in (a) to construct a suitable Metropolis-Hastings algorithm. Then run the algorithm in R so as to redo the analyses in Examples 12.4 and 12.5. Perform each new analysis thrice, assuming the finite population size N is 200,000, 400,000 and 40,000, respectively.

(d) Modify the MH algorithm in (c) so that its output features only the three model parameters and none of the nonsample values. (NB: The idea here is to design a superior MH algorithm, one with better 'mixing' than the one in (c).)

(e) Describe a procedure whereby the output from the algorithm in (d) could be used to obtain a sample from the predictive distribution of the nonsample mean. Then run that algorithm and implement the procedure so as to produce results intended to be equivalent to those in the reanalysis of Example 5 in (c) with $N = 200,000$.

Solution to Exercise 12.3

(a) With $y = (y_1,..., y_N)$ and $I = (I_1,..., I_N)$, the Bayesian model may be written as follows:

$$(I_i \mid y, p, \lambda, \phi) \sim\perp Bernoulli(\phi\lambda^{y_i}), \quad i = 1,..., N$$

$$(y_1,..., y_N \mid p, \lambda, \phi) \sim iid \ Bernoulli(p)$$

$$(p \mid \lambda, \phi) \sim Beta(\alpha, \beta), \quad (\lambda \mid \phi) \sim (1/c) \times Beta(\eta, \tau)$$

$$\phi \sim c \times Beta(\delta, \gamma) \quad (0 < c < 1).$$

Note: The sampling mechanism here is nonignorable and *unknown*, since $f(I \mid y, p, \lambda, \phi)$ depends on the unknown quantities ϕ and λ. If λ were equal to 1 then the sampling mechanism would again be unknown but in that case *ignorable*, since $\phi \perp p$ a priori.

The data here may be written as $D = (n, y_{sT})$, where:

$$n = \sum_{i=1}^{N} I_i \quad \text{is the sample size}$$

$$y_{sT} = \sum_{i \in s} y_i \quad \text{is the number of sampled units with the characteristic.}$$

Since the data is a function of (I, y_s), the relevant joint posterior/predictive density is

$$f(\phi, \lambda, p, y_r \mid I, y_s) \propto f(\phi, \lambda, p, \lambda, y_r, I, y_s)$$

$$= f(\phi, \lambda, p, y_r, I_s, I_r, y_s)$$

$$= f(\phi) f(\lambda) f(p) \times f(y_s \mid p) f(y_r \mid p)$$
$$\times f(I_s \mid y_s, \phi, \lambda) f(I_r \mid y_r, \phi, \lambda) \qquad (12.3)$$

$$= \frac{(\phi/c)^{\delta-1}(1-\phi/c)^{\gamma-1}}{cB(\delta, \gamma)} \times \frac{c(c\lambda)^{\eta-1}(1-c\lambda)^{\tau-1}}{B(\eta, \tau)} \times \frac{p^{\alpha-1}(1-p)^{\beta-1}}{B(\alpha, \beta)}$$

$$\times \left(\prod_{i \in s} p^{y_i}(1-p)^{1-y_i} \right) \left(\prod_{i \in r} p^{y_i}(1-p)^{1-y_i} \right) \times \left(\prod_{i \in s} \left(\phi \lambda^{y_i} \right)^{I_i} \left(1 - \phi \lambda^{y_i} \right)^{1-I_i} \right)$$

$$\times \left(\prod_{i \in r} \left(\phi \lambda^{y_i} \right)^{I_i} \left(1 - \phi \lambda^{y_i} \right)^{1-I_i} \right) \qquad (12.4)$$

$$\propto \phi^{\delta-1}(1-\phi/c)^{\gamma-1} \times \lambda^{\eta-1}(1-c\lambda)^{\tau-1} \times p^{\alpha-1}(1-p)^{\beta-1}$$
$$\times p^{y_{sT}}(1-p)^{n-y_{sT}} p^{y_{rT}}(1-p)^{N-n-y_{rT}}$$
$$\times \left(\prod_{i \in s} \left(\phi \lambda^{y_i} \right)^{1} \left(1 - \phi \lambda^{y_i} \right)^{1-1} \right) \left(\prod_{i \in r} \left(\phi \lambda^{y_i} \right)^{0} \left(1 - \phi \lambda^{y_i} \right)^{1-0} \right) \qquad (12.5)$$

$$= \phi^{\delta-1}(1-\phi/c)^{\gamma-1} \times \lambda^{\eta-1}(1-c\lambda)^{\tau-1} \times p^{\alpha-1}(1-p)\tau^{\beta-1}$$
$$\times p^{y_{sT}+y_{rT}}(1-p)^{N-y_{sT}-y_{rT}} \times \phi^n \lambda^{y_{sT}}(1-\phi\lambda)^{y_{rT}}(1-\phi)^{N-n-y_{rT}} . \qquad (12.6)$$

Note 1: In all of the above e.g. (12.3), s and r are fixed at their observed values.

Note 2: In the step from (12.4) to (12.5), be aware that $I_i = 1 \, \forall \, i \in s$ and $I_i = 0 \, \forall \, i \in r$.

Note 3: In the step form (12.3) to (12.4), $f(\phi)$ is derived as follows.

If $w \equiv \dfrac{\phi}{c} \sim Beta(\delta, \gamma)$ then $f(w) = \dfrac{w^{\delta-1}(1-w)^{\gamma-1}}{B(\delta, \gamma)}$.

Therefore

$$f(\phi) = f(w)\left|\frac{dw}{d\phi}\right| = \frac{(\phi/c)^{\delta-1}(1-\phi/c)^{\gamma-1}}{cB(\delta, \gamma)}.$$

A similar logic can be used to derive the density

$$f(\lambda) = \frac{c(c\lambda)^{\eta-1}(1-c\lambda)^{\tau-1}}{B(\eta, \tau)}.$$

(b) Examining the density in (a), in particular (12.6), we see that:

$$f(y_{rT} \mid D, \phi, \lambda, p) \propto \left[p(1-\phi\lambda)\right]^{y_{rT}} \left[(1-p)(1-\phi)\right]^{N-n-y_{rT}}$$

$$\Rightarrow (y_{rT} \mid D, \phi, \lambda, p) \sim Bin(N-n, q),$$

$$\text{where} \quad q = \frac{p(1-\phi\lambda)}{p(1-\phi\lambda)+(1-p)(1-\phi)} \tag{12.7}$$

$$f(p \mid D, \phi, \lambda, y_{rT}) = p^{\alpha+y_{sT}+y_{rT}-1}(1-p)\tau^{\beta+N-n-y_{sT}-y_{rT}-1}$$

$$\Rightarrow (p \mid D, \phi, \lambda, y_{rT}) \sim Beta(\alpha + y_{sT} + y_{rT}, \beta + N - y_{sT} - y_{rT}). \tag{12.8}$$

Also:

$$f(\phi \mid D, y_{rT}, \lambda, p) \propto \phi^{\delta+n-1}(1-\phi/c)^{\gamma-1}(1-\phi\lambda)^{y_{rT}}(1-\phi)^{N-n-y_{rT}} \tag{12.9}$$

$$f(\lambda \mid D, y_{rT}, \phi, p) \propto \lambda^{\eta+y_{sT}-1}(1-c\lambda)^{\tau-1}(1-\phi\lambda)^{y_{rT}}. \tag{12.10}$$

The above implies a suitable MH algorithm with two Gibbs steps as defined at (12.7) and (12.8) and two Metropolis steps as defined by (12.9) and (12.10).

(c) The MH algorithm in (b) was applied with the following specifications so as to redo the analysis in Example 12.4:

$$N = 200{,}000, \quad n = 9453, \quad y_{sT} = 4941, \quad c = 0.2$$
$$\alpha = 830, \quad \beta = 1395, \quad \eta = \tau = 1, \quad \delta = \gamma = 1.$$

A run with burn-in 2,000 followed by $J = 10{,}000$ iterations for inference was performed. Numerical results from this run are shown in Table 12.4.

Table 12.4 Monte Carlo inferences using N = 200,000

phi, ϕ	lam, λ	p	ybar, $\bar{y}$	
0.03597	1.84686	0.37259	0.37259	mean of simulated values
0.08789	0.08789	0.01017	0.01022	sample standard deviation
0.03449	1.68272	0.35266	0.35250	LB of 95% CPDR estimate
0.03749	2.02311	0.39190	0.39202	UB of 95% CPDR estimate

Our point and interval estimates for λ are 1.85 and (1.68, 2.02), which are very similar to 1.84 and (1.68, 2.03) in Example 12.4.

Note: The primary object here is estimation of λ, not of p or $\bar{y}$. But it will be noted that the estimates of these other two quantities (p or $\bar{y}$) are very alike, which is as one might expect.

Repeating the above but with finite population sizes 400,000 and 40,000, respectively, we obtain the corresponding results shown in Tables 12.5.

Table 12.5 Inferences using different N (same details as in Table 12.4)

	N = 400,000				N = 40,000		
phi, ϕ	lam, λ	p	ybar, $\bar{y}$	phi, ϕ	lam, λ	p	ybar, $\bar{y}$
0.01803	1.83548	0.37394	0.373948	0.18123	1.81588	0.375693	0.375834
0.08546	0.08546	0.00981	0.009832	0.07579	0.07579	0.009203	0.009399
0.01731	1.68407	0.35413	0.354113	0.17492	1.66922	0.357356	0.357050
0.01878	2.00923	0.39122	0.391193	0.18813	1.97208	0.393969	0.394500

Note: The three sets of inferences in Tables 12.4 and 12.5 have yielded different estimates of ϕ but very similar results for the other three quantities, in particular the object of this study, λ.

Figure 12.10 shows graphical output from the first of the three Metropolis-Hastings algorithms (i.e. the one with $N = 200,000$).

Figure 12.10 Graphical output from run with $N = 200,000$

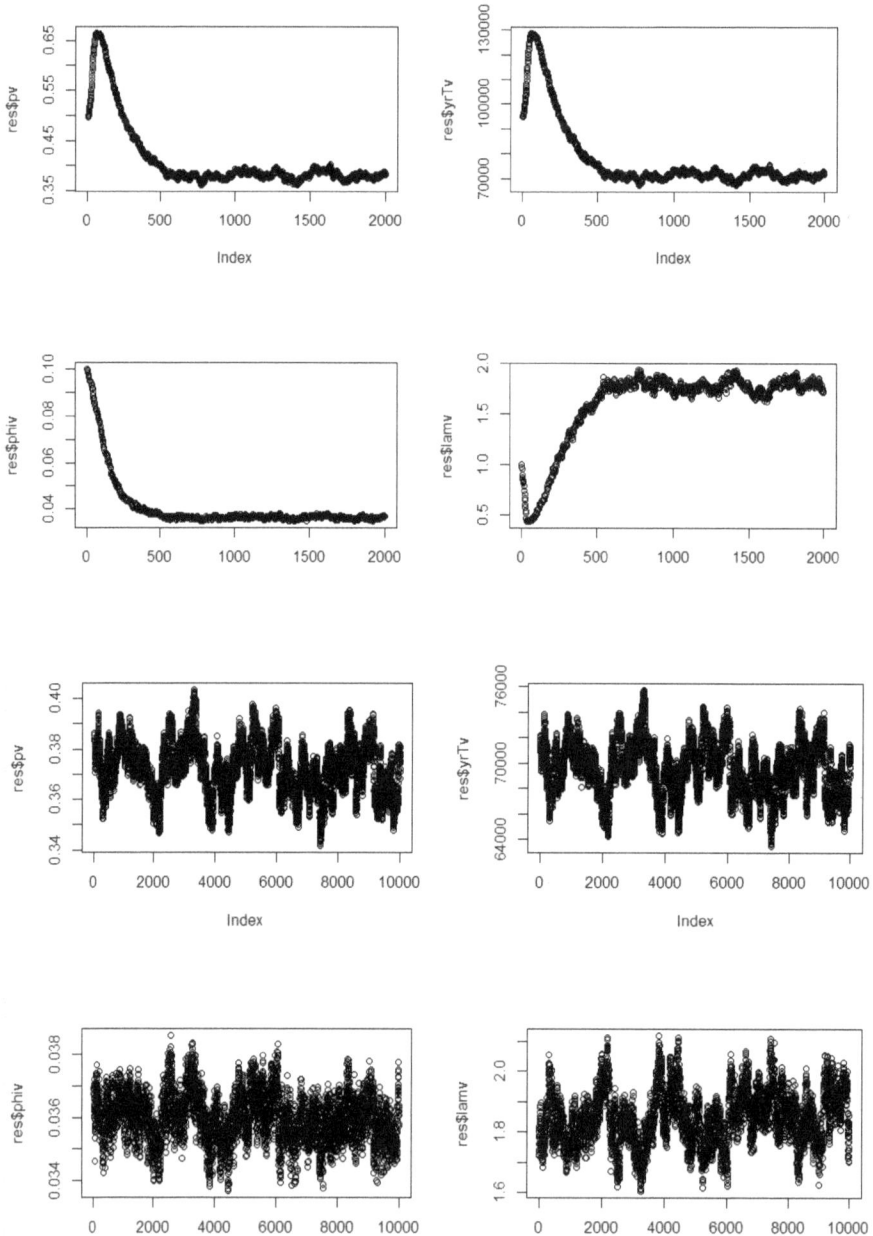

Histogram of pv

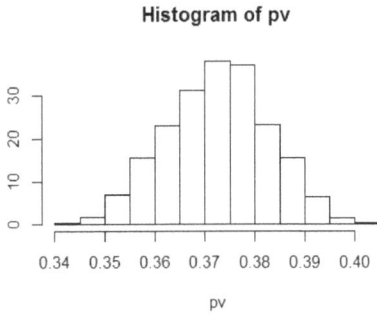

Histogram of yrTv

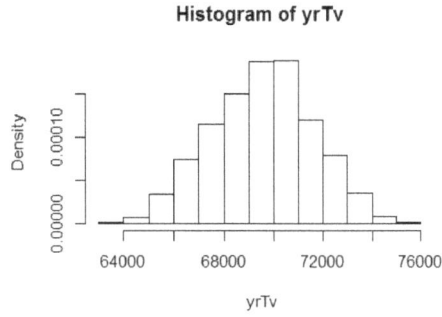

Histogram of phiv

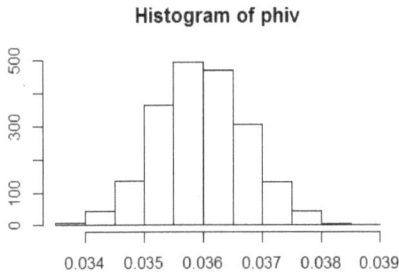

Histogram of lamv

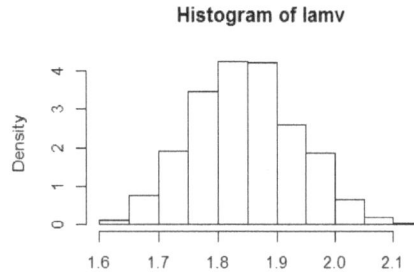

Next, a beta distribution was fitted to the 10,000 simulated values of λ above (taken from the run with $N = 200,000$) so as to define the approximate posterior given by

$$(\lambda \mid D) \sim (1/c) \times Beta(\eta', \tau'),$$

where $\eta' = 278.1$ and $\tau' = 474.8$ (with $c = 0.2$ as before).

This *posterior* for λ was then fed in as the *prior* for λ so as to redo the analysis in Example 12.5.

Accordingly, the MH algorithm in (b) was next applied once again but with the following specifications:

$$N = 200,000, \quad n = 4299, \quad y_{sT} = 2544, \quad c = 0.2$$
$$\alpha = 1, \quad \beta = 1, \quad \eta = 278.1, \quad \tau = 474.8, \quad \delta = \gamma = 1.$$

The relevant numerical estimates are as shown in Table 12.6.

Table 12.6: Inferences using N = 200,000 and a fitted beta prior

phi, ϕ	lam, λ	p	ybar, $\bar{y}$	
0.01570	1.84272	0.44049	0.45248	mean of simulated values
0.08792	0.08792	0.01408	0.01403	sample standard deviation
0.01495	1.67553	0.41344	0.42555	LB of 95% CPDR estimate
0.01656	2.01139	0.46602	0.47799	UB of 95% CPDR estimate

Thus point and interval estimates for p are 0.440 and (0.413, 0.466), which we note are similar to 0.441 and (0.414, 0.470) in Example 12.5.

Also point and 95% interval estimates for $\bar{y}$ are 0.452 and (0.426, 0.478).

Note 1: The inference on $\bar{y}$ here was not possible using the theory in the section just above the present exercise, i.e. using the *infinite* population models developed in that section.

Note 2: The posterior for λ is very similar to its prior, which is as one might expect, since the data now has no structure which could tell us anything further about that parameter.

Repeating the above but with finite population sizes 400,000 and 40,000, respectively, we obtain the corresponding results shown in Tables 12.7.

Table 12.7 Inferences using different N (same details as in Table 12.6)

N = 400,000				N = 40,000			
phi, ϕ	lam, λ	p	ybar, $\bar{y}$	phi, ϕ	lam, λ	p	ybar, $\bar{y}$
0.007863	1.83516	0.44193	0.44792	0.07888	1.82895	0.44228	0.50220
0.087755	0.08776	0.01375	0.01372	0.08162	0.08162	0.01359	0.01337
0.007482	1.66809	0.41563	0.42160	0.07538	1.66402	0.41490	0.47517
0.008299	2.00048	0.46819	0.47409	0.08278	1.99275	0.47007	0.52985

Discussion

Something to be noted above is that estimation of $\bar{y}$ appears to increase slightly as N decreases, whereas estimation of p remains about the same.

Estimation of ϕ also increases as N decreases. This could present a 'problem' if N is 'too small'. Figures 12.11, 12.12 and 12.13 (pages 598 and 599) show histograms of the simulated values when $N = 200,000$, 20,000 and 15,000, respectively.

We see no problem in the first two of these three cases. But for $N = 15,000$, the estimation of ϕ appears to be artificially restricted by our arbitrary choice of c as 0.2. (Observe that the simulated values are strongly 'bunched up' at just below 0.2.)

Repeating the MCMC run with $N = 15,000$ but with c also changed to 0.5 appears to solve this problem. Results are shown in Figure 12.14 (page 599). We note that estimation of λ has changed from about 2 to less than 1. This suggests that we might get very similar results with c even larger, e.g. $c = 1$.

But when we do this, we get very different results (not shown). Why?

Because when we changed c from 0.2 to 0.5, we forgot to reconfigure the prior for λ, which also involves c.

Note: The prior for ϕ also involves c but does not need reconfiguring (because that prior is uniform for all values of c, since $\delta = \gamma = 1$).

Thus, Figure 12.14 (the case of $N = 15,000$ and $c = 0.5$) in fact illustrates output which is 'flawed' (in this sense) and so should be disregarded.

Although these technical issues could satisfactorily be resolved with some effort, we will leave that task as an avenue of investigation for further research and move on to answering part (d).

Figure 12.11 Histograms using *N* = 200,000 and *c* = 0.2

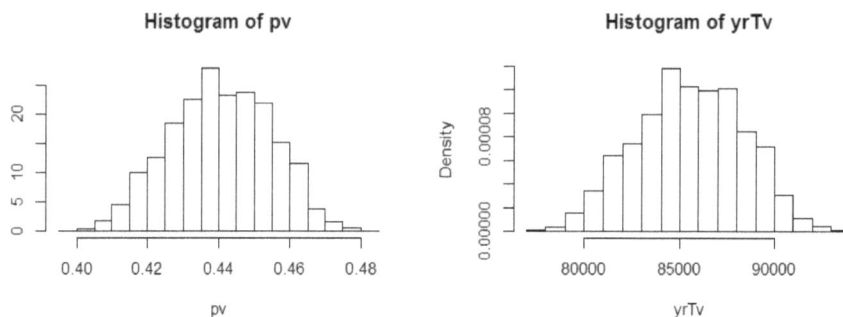

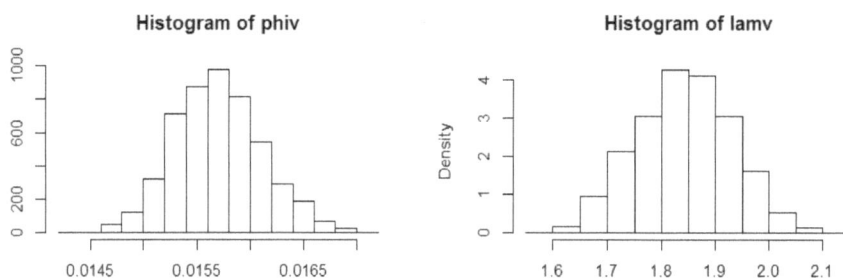

Histogram of pv

Histogram of yrTv

Histogram of phiv

Histogram of lamv

Figure 12.12 Histograms using *N* = 20,000 and *c* = 0.2

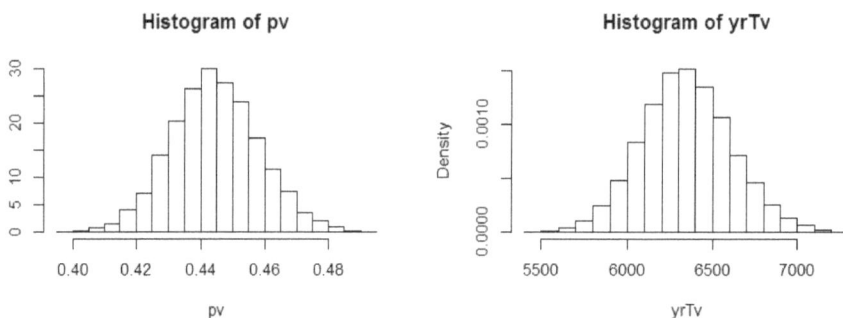

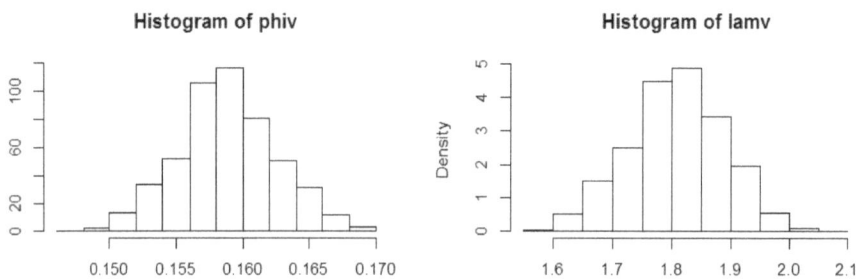

Histogram of pv

Histogram of yrTv

Histogram of phiv

Histogram of lamv

Figure 12.13 Histograms using N = 15,000 and c = 0.2

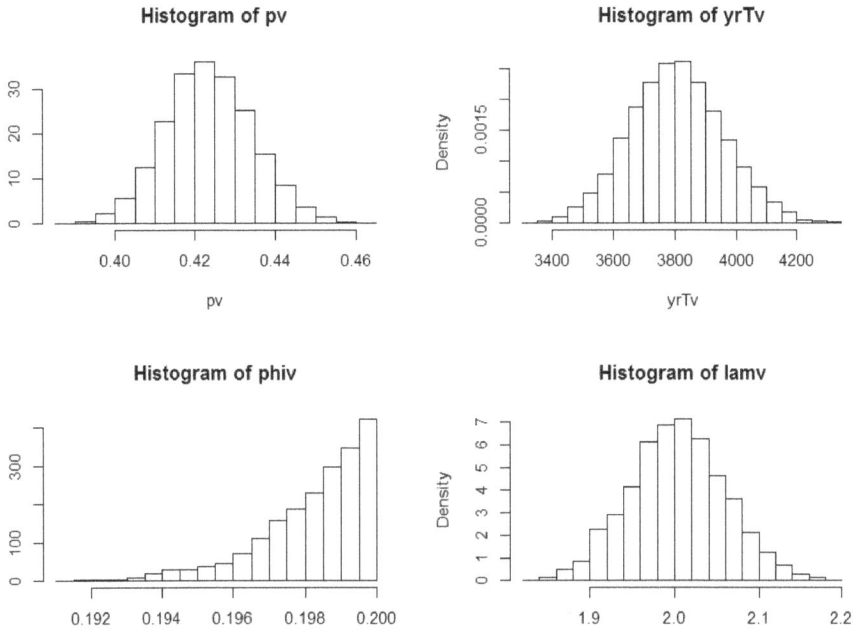

Histogram of pv

Histogram of yrTv

Histogram of phiv

Histogram of lamv

Figure 12.14 Histograms using N = 15,000 and c = 0.5

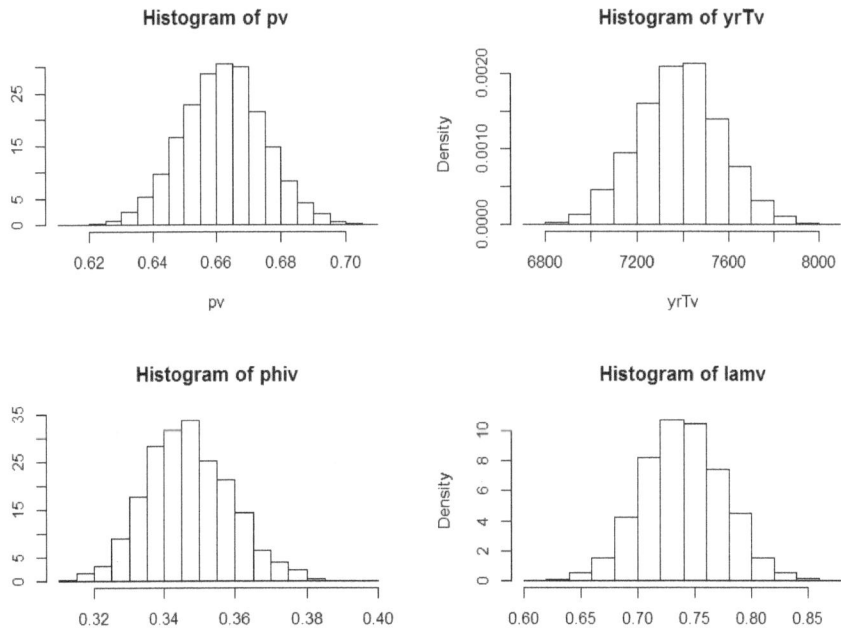

Histogram of pv

Histogram of yrTv

Histogram of phiv

Histogram of lamv

(d) Recall the joint density (12.6). This density may also be written as:

$$f(\phi, \lambda, p, y_r \mid I, y_s) \propto f(\phi, \lambda, p) p^{y_{sT} + y_{rT}} (1-p)^{N - y_{sT} - y_{rT}}$$
$$\times \phi^n \lambda^{y_{sT}} (1 - \phi\lambda)^{y_{rT}} (1-\phi)^{N - n - y_{rT}},$$

where $f(\phi, \lambda, p) \propto \phi^{\delta-1}(1 - \phi/c)^{\gamma-1} \times \lambda^{\eta-1}(1 - c\lambda)^{\tau-1} \times p^{\alpha-1}(1-p)^{\beta-1}$.

Now observe that

$$f(\phi, \lambda, p, y_r \mid I, y_s) \propto f(\phi, \lambda, p) \times \left\{ p^{y_{sT}} (1-p)^{n - y_{sT}} \phi^n \lambda^{y_{sT}} \right\} \times \xi,$$

where: $\xi = \left[p(1 - \phi\lambda) \right]^{y_{rT}} \left[(1-p)(1-\phi) \right]^{N - n - y_{rT}}$

$$= \left[p(1 - \phi\lambda) + (1-p)(1-\phi) \right]^{N-n} \times \prod_{i \in r} z^{y_i} (1 - z)^{1 - y_i}$$

$$z = \frac{p(1 - \phi\lambda)}{p(1 - \phi\lambda) + (1-p)(1-\phi)}.$$

Further observe that

$$\sum_{y_r} \prod_{i \in r} z^{y_i} (1 - z)^{1 - y_i} = \prod_{i \in r} \sum_{y_i = 0}^{1} z^{y_i} (1 - z)^{1 - y_i} = 1$$

(since the first product is the joint pdf of $N - n$ iid *Bernoulli(z)* variables).

It follows that

$$f(\phi, \lambda, p \mid I, y_s) = \sum_{y_r} f(\phi, \lambda, p, y_r \mid I, y_s)$$

$$\propto f(\phi, \lambda, p) \times \left\{ p^{y_{sT}} (1-p)^{n - y_{sT}} \phi^n \lambda^{y_{sT}} \right\}$$

$$\times \left[p(1 - \phi\lambda) + (1-p)(1-\phi) \right]^{N-n}.$$

The above defines a MH algorithm with three steps based on the following conditionals:

$$f(\phi \mid D, \lambda, p) \propto \phi^{\delta+n-1}(1 - \phi/c)^{\gamma-1} \left[p(1 - \phi\lambda) + (1-p)(1-\phi) \right]^{N-n}$$

$$f(\lambda \mid D, \phi, p) \propto \lambda^{\eta + y_{sT} - 1}(1 - c\lambda)^{\tau-1} \left[p(1 - \phi\lambda) + (1-p)(1-\phi) \right]^{N-n}$$

$$f(p \mid D, \phi, \lambda) \propto p^{\alpha + y_{sT} - 1}(1-p)^{\beta + n - y_{sT} - 1} \left[p(1 - \phi\lambda) + (1-p)(1-\phi) \right]^{N-n}.$$

(e) From the working in (d) we see that
$$(y_{rT} \mid I, y_s, \phi, \lambda, p) \sim Bin(N - n, z),$$
where
$$z = \frac{p(1 - \phi\lambda)}{p(1 - \phi\lambda) + (1 - p)(1 - \phi)}. \qquad (12.11)$$

So, to get a sample from the predictive distribution of $\bar{y}$ we do as follows:

1. Obtain $(\phi_j, \lambda_j, p_j) \sim iid\ f(\phi, \lambda, p \mid I, y_s)$, $j = 1, \ldots, J$
   using the MH algorithm in (d)

2. Sample $y_{rT}^{(j)} \sim Bin(N - n, z_j)$, where
   $$z_j = \frac{p_j(1 - \phi_j\lambda_j)}{p_j(1 - \phi_j\lambda_j) + (1 - p_j)(1 - \phi_j)}, j = 1, \ldots, J \ \text{(from (12.11))}$$

3. Calculate $\bar{y}^{(j)} = \dfrac{1}{N}(y_{sT} + y_{rT}^{(j)})$, $j = 1, \ldots, J$.

We now perform the MH algorithm in (d) and the above procedure with:
$$N = 200{,}000, \quad n = 4299, \quad y_{sT} = 2544, \quad c = 0.2$$
$$\alpha = 1, \quad \beta = 1, \quad \eta = 278.1, \quad \tau = 474.8, \quad \delta = \gamma = 1.$$

We thereby obtain the inferences shown in Table 12.8.

Table 12.8 Results obtained in part (e)

phi, ϕ	lam, λ	p	ybar, $\bar{y}$	
0.01567	1.8491	0.43973	0.43973	mean of simulated values
0.08660	0.0866	0.01387	0.01382	sample standard deviation
0.01491	1.6844	0.41331	0.41346	LB of 95% CPDR estimate
0.01650	2.0278	0.46689	0.46673	UB of 95% CPDR estimate

We see that inferences are very similar to those in the reanalysis of Example 12.5 in (c) with $N = 200{,}000$ (where $\bar{y}$ was estimated as 0.45248). But the results here should in fact be considered more accurate because they are based on a MH algorithm with fewer components.

Note 1: The inferences on $\bar{y}$ could be further improved via Rao-Blackwell arguments which obviate the need to sample values of y_{rT} at all. In particular, the Rao-Blackwell estimate of the predictive mean of the finite population mean, $\hat{\bar{y}} = E(\bar{y} \mid D)$, is

$$\bar{z} = \frac{1}{J} \sum_{j=1}^{J} z_j = 0.4364,$$

with 95% CI for $\hat{\bar{y}}$

$$\left(\bar{z} \pm 1.96 \sqrt{\frac{1}{J(J-1)} \sum_{j=1}^{J} (z_j - \bar{z})^2} \right) = (0.4361,\, 0.4367).$$

Actually, this is not quite right, since $\bar{z}$ is the Rao-Blackwell estimate of $\hat{\bar{y}}_r = E(\bar{y}_r \mid D)$, and the 95% CI is for $\hat{\bar{y}}_r$. To see this, refer to (12.3).

Thus, since

$$\bar{y} = \frac{1}{N}(y_{sT} + (N-n)\bar{y}_r),$$

the RB estimate of $\hat{\bar{y}}$ is actually

$$\frac{1}{N}(y_{sT} + (N-n)\bar{z}) = 0.440,$$

with a 95% confidence interval for $\hat{\bar{y}}$ equal to

$$\left(\frac{1}{N}(y_{sT} + (N-n)0.4361,\, \frac{1}{N}(y_{sT} + (N-n)0.4367 \right) = (0.439,\, 0.440).$$

Note 2: The Monte Carlo 95% confidence intervals reported here are unduly narrow (i.e. will have less than 95% actual coverage). This is because we did not address the problem of the very strong serial correlation amongst the values outputted from the Metropolis-Hastings algorithm, for example by way of thinning or the batch means method.

But this remark only applies to confidence intervals for mean estimates and not to posterior or predictive interval estimates, such as (0.413, 0.467) for $\bar{y}$ in Table 12.8.

R Code for Exercise 12.3

```
MH = function(J=100,  n=9453, ysT=4941,  alp=830, bet=1395,
       p=0.5,   phi0=0.1,  lam0=1,    phisd=0.1, lamsd=0.1,
       eta=1, tau=1,   del=1, gam=1,   c=0.2,  N=200000 ){
phi=phi0; lam=lam0;  phiv=phi; lamv=lam; phict=0; lamct=0; pv=NA; yrTv=NA
for(j in 1:J){
       q=p*(1-phi*lam)/(  p*(1-phi*lam) + (1-p)*(1-phi)  )
       yrT=rbinom(1,N-n,q); yT=ysT+yrT;  p=rbeta(1,alp+yT,bet+N-yT)
       phinew=rnorm(1,phi,phisd)
       if((phinew>0)&&(phinew<c)){
              logprobnum=(del-1)*log(phinew)+(gam-1)*log(1- phinew/c)+
          n*log(phinew) +yrT*log(1- phinew*lam)+(N-n-yrT)*log(1-phinew)
              logprobden=(del-1)*log(phi)+(gam-1)*log(1-phi/c)+
                     n*log(phi) +yrT*log(1-phi*lam)+(N-n-yrT)*log(1-phi)
              logprob= logprobnum- logprobden; prob=exp(logprob)
              u=runif(1); if(u<=prob){ phict=phict+1; phi=phinew }    }
       lamnew=rnorm(1,lam,lamsd)
       if((lamnew>0)&&(lamnew<(1/c))){
         logprobnum=  (eta-1)*log(lamnew)+(tau-1)*log(1- lamnew*c)+
                     ysT*log(lamnew)+yrT*log(1-phi*lamnew)
              logprobden=  (eta-1)*log(lam)+(tau-1)*log(1-lam*c)+
                     ysT*log(lam)+yrT*log(1-phi*lam)
              logprob= logprobnum- logprobden; prob=exp(logprob)
              u=runif(1); if(u<=prob){ lamct=lamct+1; lam=lamnew }    }
       phiv=c(phiv,phi); lamv=c(lamv,lam); pv=c(pv,p); yrTv=c(yrTv,yrT)  }
phiar=phict/J; lamar=lamct/J
list(pv=pv, yrTv=yrTv, phiv=phiv, lamv=lamv, phiar=phiar, lamar=lamar)  }
    # end fn
X11(w=8,h=6);    par(mfrow=c(2,2));    options(digits=5); N=200000


# A --------------------------------


set.seed(531); res=MH(J=2000,  n=9453, ysT=4941,  alp=830, bet=1395,
       p=0.5,    phi0=0.1,  lam0=1,     phisd=0.0007, lamsd=0.04,
       eta=1, tau=1,    del=1, gam=1,    c=0.2,  N=N )
c(res$phiar,res$lamar) # 0.513 0.536   OK
plot(res$pv); plot(res$yrTv); plot(res$phiv); plot(res$lamv)  # Has burnt in OK
p0=res$pv[2001]; lam0=res$lamv[2001]; phi0=res$phiv[2001]
   # record last values
```

```
set.seed(131); K=10000; date() #
res=MH(J=K,  n=9453, ysT=4941,  alp=830, bet=1395,
        p=p0,   phi0=phi0,  lam0=lam0,  phisd=0.0006, lamsd=0.04,
        eta=1, tau=1,   del=1, gam=1,   c=0.2,   N=N  ); date() #
c(res$phiar,res$lamar) # 0.5548 0.5707    OK
plot(res$pv); plot(res$yrTv); plot(res$phiv); plot(res$lamv)   # OK

# Example of optional thinning to reduce serial correlation:
# acf(res$pv[-1]); acf (res$yrTv[-1]); acf (res$phiv[-1]); acf (res$lamv[-1])
# skip=10; inc=1+seq(skip,K,skip); J=length(inc); J # 1000
# pv= res$pv[inc]; yrTv= res$yrTv[inc]; phiv=res$phiv[inc]; lamv=res$lamv[inc]
# acf(pv); acf(yrTv); acf(phiv); acf(lamv)  # better

skip=1;  inc=1+seq(skip,K,skip);  J=length(inc);  J # 10000    (Just use whole
sample)
pv= res$pv[inc]; yrTv= res$yrTv[inc]; phiv=res$phiv[inc]; lamv=res$lamv[inc]
hist(pv,prob=T); hist(yrTv,prob=T); hist(phiv,prob=T); hist(lamv,prob=T);  # OK

# Calculate estimates (Note we could improve these via Rao-Blackwell):
phat=mean(pv); pcpdr=quantile(pv,c(0.025,0.975)); pse=sd(pv)
lamhat=mean(lamv); lamcpdr=quantile(lamv,c(0.025,0.975)); lamse=sd(lamv)
phihat=mean(phiv); phicpdr=quantile(phiv,c(0.025,0.975)); phise=sd(lamv)
n= 9453; ysT=4941;  ybarv=(1/N)*(ysT+yrTv);
ybarhat=mean(ybarv);  ybarcpdr=quantile(ybarv,c(0.025,0.975));
ybarse=sd(ybarv)
print(cbind(c(phihat, phise ,phicpdr), c(lamhat, lamse ,lamcpdr),
        c(phat, pse,pcpdr), c(ybarhat,ybarse, ybarcpdr)), digits=4)

# B --------------------------------


#       phi           lam           p             ybar
#       0.03597       1.84686       0.37259       0.37259 mean
#       0.08789       0.08789       0.01017       0.01022  se
# 2.5%  0.03449       1.68272       0.35266       0.35250  LB
# 97.5% 0.03749       2.02311       0.39190       0.39202  UB

# Repeat above exactly from A to B but after setting  N=400000. Results:
#       0.01803       1.83548       0.37394       0.373948
#       0.08546       0.08546       0.00981       0.009832
# 2.5%  0.01731       1.68407       0.35413       0.354113
# 97.5% 0.01878       2.00923       0.39122       0.391193
```

```
# Repeat above exactly from A to B but after setting N=40000. Results:
#        0.18123        1.81588        0.375693        0.375834
#        0.07579        0.07579        0.009203        0.009399
# 2.5%  0.17492        1.66922        0.357356        0.357050
# 97.5% 0.18813        1.97208        0.393969        0.394500


# Now calculate new prior from posterior of lambda (based on 1st run above):
c(lamhat,lamse) # 1.846864 0.087889
fun=function(etatau, c=0.2, est=lamhat, se=lamse){
        (est-(1/c)*etatau[1]/sum(etatau))^2+
    ( se^2 - (1/c^2)*prod(etatau)/( sum(etatau)^2*(1 + sum(etatau)) )  )^2  }
etataunew0 = optim(par=c(2,5), fn=fun)$par
etataunew = optim(par= etataunew0, fn=fun)$par


etanew=etataunew[1]; taunew=etataunew[2]
c(etanew, taunew) # 278.10 474.79
(1/0.2)*etanew/(etanew+taunew) # 1.8469
sqrt((1/0.2^2)*etanew*taunew/((etanew+taunew)^2*(etanew+taunew+1)))
    # 0.087889 OK


# Now run MCMC with new prior and data: -----------------------------
par(mfrow=c(2,2));   N=200000


# C ----------------------------------------------------------


set.seed(531); res=MH(J=2000,  n=4299, ysT=2544,  alp=1, bet=1,
        p=0.5,   phi0=0.1,  lam0=1,     phisd=0.0007, lamsd=0.04,
        eta=etanew, tau=taunew,   del=1, gam=1,   c=0.2,  N=N )
c(res$phiar,res$lamar) # 0.4295 0.5485   OK
plot(res$pv); plot(res$yrTv); plot(res$phiv); plot(res$lamv)  # Has burnt in OK
p0=res$pv[2001]; lam0=res$lamv[2001]; phi0=res$phiv[2001]
    # record last values


set.seed(131); K=10000; date() #
res=MH(J=K,  n=4299, ysT=2544,  alp=1, bet=1,
        p=p0,    phi0=phi0,  lam0=lam0,  phisd=0.0004, lamsd=0.05,
        eta= etanew, tau= taunew,    del=1, gam=1,   c=0.2,  N=N ); date() #
c(res$phiar,res$lamar) # 0.5473 0.5908   OK
plot(res$pv); plot(res$yrTv); plot(res$phiv); plot(res$lamv)  # OK


skip=1;  inc=1+seq(skip,K,skip);  J=length(inc); J #  10000    (Just use whole
sample)
pv= res$pv[inc]; yrTv= res$yrTv[inc]; phiv=res$phiv[inc]; lamv=res$lamv[inc]
hist(pv,prob=T); hist(yrTv,prob=T); hist(phiv,prob=T); hist(lamv,prob=T);  # OK
```

Calculate estimates (Note we could improve these via Rao-Blackwell):

```
phat=mean(pv); pcpdr=quantile(pv,c(0.025,0.975)); pse=sd(pv)
lamhat=mean(lamv); lamcpdr=quantile(lamv,c(0.025,0.975)); lamse=sd(lamv)
phihat=mean(phiv); phicpdr=quantile(phiv,c(0.025,0.975)); phise=sd(lamv)
n= 9453; ysT=4941; ybarv=(1/N)*(ysT+yrTv);
ybarhat=mean(ybarv); ybarcpdr=quantile(ybarv,c(0.025,0.975));
ybarse=sd(ybarv)
print(cbind(c(phihat, phise ,phicpdr), c(lamhat, lamse ,lamcpdr),
        c(phat, pse,pcpdr), c(ybarhat,ybarse, ybarcpdr)), digits=4)
```

D --

```
#        phi        lam         p          ybar
#        0.01570    1.84272     0.44049    0.45248 mean
#        0.08792    0.08792     0.01408    0.01403 se
# 2.5%  0.01495    1.67553     0.41344    0.42555 LB
# 97.5% 0.01656    2.01139     0.46602    0.47799 UB
```

Repeat above exactly from C to D but with N=400000. Results:
```
#        0.007863   1.83516     0.44193    0.44792
#        0.087755   0.08776     0.01375    0.01372
# 2.5%  0.007482   1.66809     0.41563    0.42160
# 97.5% 0.008299   2.00048     0.46819    0.47409
```

Repeat above exactly from C to D but with N=40000. Results:
```
#        0.07888    1.82895     0.44228    0.50220
#        0.08162    0.08162     0.01359    0.01337
# 2.5%  0.07538    1.66402     0.41490    0.47517
# 97.5% 0.08278    1.99275     0.47007    0.52985
```

Repeat above exactly from C to D but with N=20000 and 15000 to produce
extra graphs. We omit the code for the case N = 15000, c=0.5 and the case
N = 15000, c = 1

```
# (e)
MH2 = function(J=100,  n=9453, ysT=4941,  alp=830, bet=1395,
        p0=0.5, phi0=0.1, lam0=1,    psd=0.1, phisd=0.1, lamsd=0.1,
        eta=1, tau=1,   del=1, gam=1,  c=0.2,  N=200000 ){
p=p0; phi=phi0; lam=lam0;   pv=p; phiv=phi; lamv=lam; pct=0; phict=0;
lamct=0;
```

```
for(j in 1:J){
       pnew=rnorm(1,p,psd)
       if((pnew >0)&&(pnew <1)){
        logprobnum=(alp-1+ysT)*log(pnew)+(bet-1+n-ysT)*log(1-pnew) +
                  (N-n)*log((1-pnew)*(1-phi)+pnew*(1-phi*lam))
            logprobden=(alp-1+ysT)*log(p)+(bet-1+n-ysT)*log(1-p) +
                  (N-n)*log((1-p)*(1-phi)+p*(1-phi*lam))
            logprob= logprobnum- logprobden; prob=exp(logprob)
            u=runif(1); if(u<=prob){ pct=pct+1; p=pnew }     }
       phinew=rnorm(1,phi,phisd)
       if((phinew>0)&&(phinew<c)){
         logprobnum=(del-1+n)*log(phinew)+(gam-1)*log(1- phinew/c)+
                  (N-n)*log((1-p)*(1-phinew)+p*(1-phinew*lam))
            logprobden=(del-1+n)*log(phi)+(gam-1)*log(1-phi/c)+
                  (N-n)*log((1-p)*(1-phi)+p*(1-phi*lam))
            logprob= logprobnum- logprobden; prob=exp(logprob)
            u=runif(1); if(u<=prob){ phict=phict+1; phi=phinew }     }
       lamnew=rnorm(1,lam,lamsd)
       if((lamnew>0)&&(lamnew<(1/c))){
         logprobnum= (eta-1+ysT)*log(lamnew)+(tau-1)*log(1- lamnew*c)+
                  (N-n)*log((1-p)*(1-phi)+p*(1-phi*lamnew))
            logprobden= (eta-1+ysT)*log(lam)+(tau-1)*log(1- lam*c)+
                  (N-n)*log((1-p)*(1-phi)+p*(1-phi*lam))
            logprob= logprobnum- logprobden; prob=exp(logprob)
            u=runif(1); if(u<=prob){ lamct=lamct+1; lam=lamnew }     }
       pv=c(pv,p); phiv=c(phiv,phi); lamv=c(lamv,lam) }
par=pct/J; phiar=phict/J; lamar=lamct/J
list(pv=pv, phiv=phiv, lamv=lamv, par=par, phiar=phiar, lamar=lamar)  }
       # end fn


X11(w=8,h=6);    par(mfrow=c(2,2))
N=200000; n = 4299; ysT=2544; K=2000
set.seed(531); res=MH2(J=K,  n=4299, ysT=2544,  alp=1, bet=1,
       p0=0.5, phi0=0.1, lam0=1,    psd=0.008, phisd=0.0007, lamsd=0.04,
       eta= etanew, tau= taunew,    del=1, gam=1,   c=0.2, N=N )
c(res$par, res$phiar,res$lamar) # 0.6580 0.4135 0.6045  OK
plot(res$pv); plot(res$phiv); plot(res$lamv)  # Has burnt in OK
p0=res$pv[2001]; lam0=res$lamv[2001]; phi0=res$phiv[2001]
       # record last values
```

```
set.seed(131); K=10000; par(mfrow=c(2,2)); date() #
res=MH2(J=K,  n=4299, ysT=2544,  alp=1, bet=1,
        p0=p0,  phi0=phi0,  lam0=lam0, psd=0.008,  phisd=0.0006,
        lamsd=0.04,
        eta= etanew, tau= taunew,   del=1, gam=1,   c=0.2,  N=N ); date() #
c(res$par, res$phiar,res$lamar) # 0.6825 0.4315 0.6643   OK
plot(res$pv); plot(res$phiv); plot(res$lamv)  # OK

skip=1; inc=1+seq(skip,K,skip); J=length(inc); J
        # 10000  (Just use whole sample)
pv= res$pv[inc]; phiv=res$phiv[inc]; lamv=res$lamv[inc]
par(mfrow=c(2,2)); hist(pv,prob=T); hist(phiv,prob=T); hist(lamv,prob=T);
        # OK

# Calculate estimates
phat=mean(pv); pcpdr=quantile(pv,c(0.025,0.975)); pse=sd(pv)
lamhat=mean(lamv); lamcpdr=quantile(lamv,c(0.025,0.975)); lamse=sd(lamv)
phihat=mean(phiv); phicpdr=quantile(phiv,c(0.025,0.975)); phise=sd(lamv)

# Generate sample from predictive dsn of finite population mean
zv=pv*(1-phiv*lamv)/( pv*(1-phiv*lamv) + (1-pv)*(1-phiv)  )
set.seed(331); yrTv = rbinom(J, N-n, zv);   ybarv=(1/N)*(ysT+yrTv)
ybarhat=mean(ybarv); ybarcpdr=quantile(ybarv,c(0.025,0.975));
ybarse=sd(ybarv)

# Print out inferences
print(cbind(c(phihat, phise ,phicpdr), c(lamhat, lamse ,lamcpdr),
        c(phat, pse,pcpdr), c(ybarhat,ybarse, ybarcpdr)), digits=4)
#         0.01567     1.8491        0.43973        0.43973
#         0.08660     0.0866        0.01387        0.01382
# 2.5% 0.01491     1.6844        0.41331        0.41346
# 97.5% 0.01650    2.0278        0.46689        0.46673

RBest=mean(zv); RBci=RBest+c(-1,1)*qnorm(0.975)*sd(zv)/sqrt(J)
c(RBest,RBci)  #  0.43639 0.43612 0.43667
(1/N)*(ysT+(N-n)*RBest) # 0.43973
(1/N)*(ysT+(N-n)*RBci)  # 0.43946 0.44000
```

APPENDIX A

Additional Exercises

Exercise A.1 Practice with the Metropolis algorithm

(a) Sample a value m from the standard exponential distribution. Then randomly sample $n = 100$ values from the normal distribution with mean m and variance $v = m^2$.

Then design and implement a Metropolis algorithm so as to obtain a random sample of size $J = 1{,}000$ from the posterior of m.

Use this sample to perform Monte Carlo inference on m. Be sure to provide a 95% CI for the posterior mean of m, an estimate of the 95% central posterior density region for m, and an estimate of the entire marginal posterior density of m.

Then predict c, the average of a future independent sample of size $k = 10$ from the normal distribution with the same mean m and variance v.

Be sure to provide a 95% CI for the predictive mean of c, an estimate of the 95% central predictive density region for c, and an estimate of the entire posterior predictive density of c.

Illustrate your results with suitable figures (for example, trace plots and histograms).

(b) Consider the following values in a sample obtained via SRSWOR from a finite population of size $N = 50$:

$$3.4, 6.3, 1.0, 2.9, 1.8, \quad 2.0, 0.5, 7.9, 4.8, 6.5.$$

Suppose we model the finite population values as normal with (unknown) mean m and variance $v = m^2$, with a standard exponential prior on m.

Using MCMC methods, estimate the finite population mean and provide a suitable 95% interval estimate.

Solution to Exercise A.1

(a) The sampled value of m was 0.7071. A histogram of the 100 sampled normal values is shown in Figure A.1(a) (page 612). This histogram is overlaid by the (known) normal distribution with mean m and variance $v = m^2 = 0.5$.

The posterior density of m is

$$f(m \mid y) \propto f(m) f(y \mid m)$$

$$\propto e^{-m} \prod_{i=1}^{n} \frac{1}{m} \exp\left\{ -\frac{1}{2m^2} (y_i - m)^2 \right\}$$

$$= e^{-m} m^{-n} \exp\left\{ -\frac{1}{2m^2} \sum_{i=1}^{n} (y_i - m)^2 \right\}.$$

So the log-posterior is

$$l(m) = \log f(m \mid y) = -m - n \log m - \frac{1}{2m^2} \sum_{i=1}^{n} (y_i - m)^2.$$

A suitable Metropolis algorithm is one which at each iteration proposes a value

$$m' \sim U(m - \delta, m + \delta),$$

where δ is a tuning constant, and accepts this value with probability

$$p = e^q,$$

where

$$q = l(m') - l(m).$$

Implementing this algorithm we obtained the 10,100 values of m, whose trace is shown in Figure A.1(b) (page 612). Stochastic convergence appears to have been attained immediately, and so the burn-in was conservatively taken to be 100.

The last 10,000 of these 10,100 values are highly autocorrelated, as evidenced by the sample ACF in Figure A.1(c) (page 612). However, thinning out by a factor of 10 removes almost all of the autocorrelation, as seen in the sample ACF in Figure A.1(d) (page 612), and yields the required random sample

$$m_1, \ldots, m_J \sim iid \ f(m \mid y),$$

where $J = 1,000$.

A histogram of these 1,000 values of m is shown in Figure A.1(e).

The dashed line in this subplot is a histogram estimate of $f(m \mid y)$, and the solid line is the true posterior density. The vertical lines show the posterior mean estimate, $\bar{m} = 0.7377$, the 95% CI for the posterior mean, (0.7350, 0.7404), and the 95% CPDR estimate for m, (0.6620, 0.8298).

The dots show the true posterior mean, $\hat{m} = E(m \mid y) = 0.7393$, and the true 95% CPDR for m. The cross shows the true value of m, 0.7071.

The Monte Carlo sample was used to generate a random sample from the predictive distribution of
$$c = (y_{n+1} + \ldots + y_{n+10})/10$$
by sampling
$$c_j \sim N(m_j, m_j^2/10), j = 1, \ldots, J.$$
A histogram of these c-values is shown in Figure A.1(f).

The dashed line in this subplot is a histogram estimate of $f(c \mid y)$, and the solid line is the Rao-Blackwell estimate
$$\bar{f}(c \mid y) = \frac{1}{J} \sum_{j=1}^{J} \frac{1}{m_j \sqrt{2\pi}} \exp\left\{-\frac{1}{2m_j^2}(c - m_j)^2\right\}.$$

The vertical lines show the predictive mean estimate, $\bar{c} = 0.741$, the 95% CI for the predictive mean, (0.7270, 0.7549), and the 95% CPDR estimate for c, (0.3063, 1.1893).

The dot shows the Rao-Blackwell estimate of $\hat{c} = E(c \mid y)$, which is the same as $\bar{m} = 0.7377$.

The Rao-Blackwell 95% CI for $\hat{c}$ is the same as the 95% CI (0.7350, 0.7404) reported earlier.

Figure A.1 Graphical results for part (a)

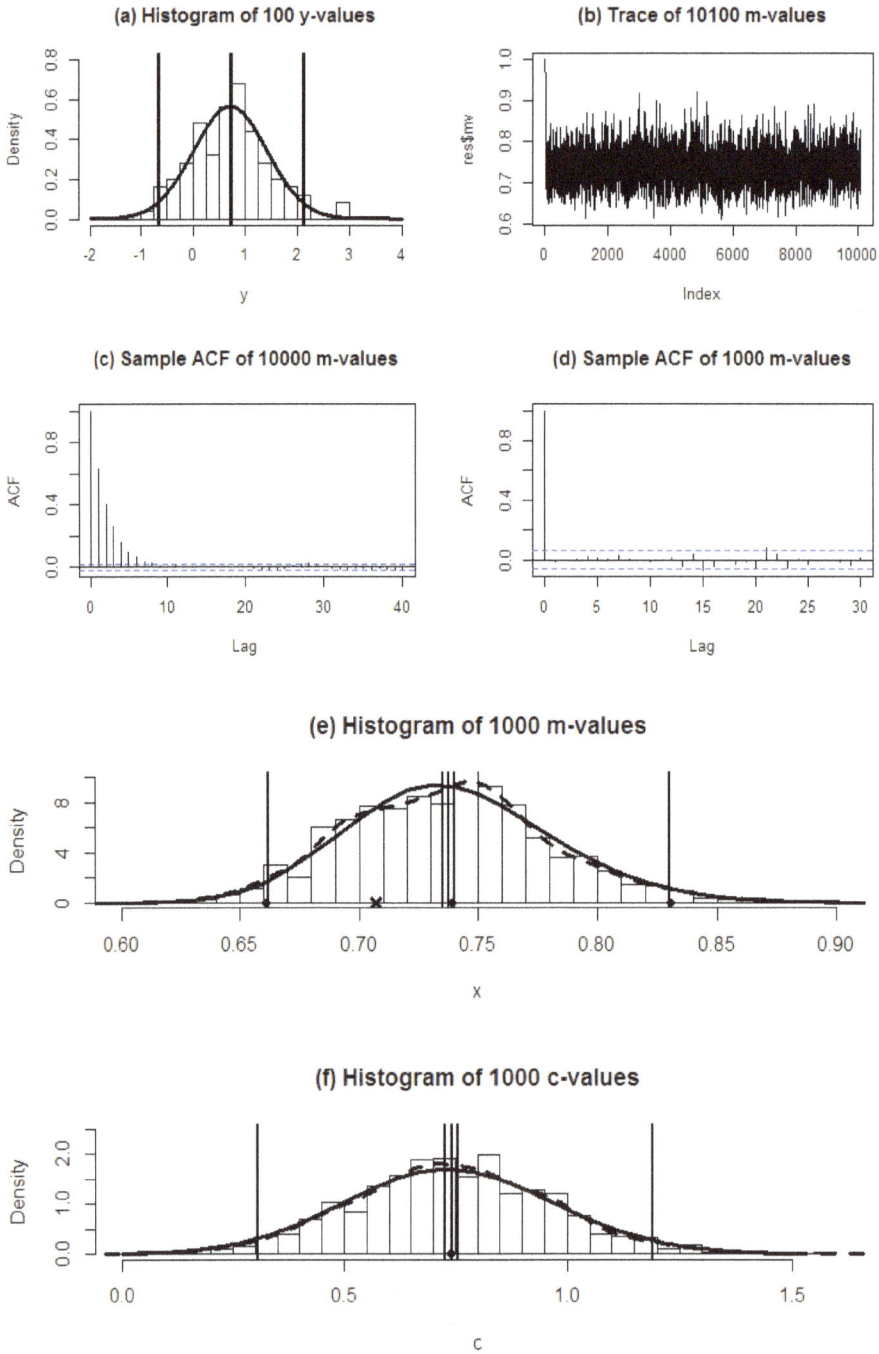

(a) Histogram of 100 y-values

(b) Trace of 10100 m-values

(c) Sample ACF of 10000 m-values

(d) Sample ACF of 1000 m-values

(e) Histogram of 1000 m-values

(f) Histogram of 1000 c-values

(b) Here we repeat the procedure in part (a), but:

- with $n = 10$ (rather than 100)

- using the 10 given sample values, whose mean is 3.71
  (instead of the 100 generated values, as previously)

- with $c = \dfrac{1}{40}(y_{11} + \ldots + y_{50})$ (instead of $c = \dfrac{1}{10}(y_{101} + \ldots + y_{110})$).

Figure A.2 is an analogue of Figure A.1, except that subplot (a) does not have a normal density overlaid, and there is an extra subplot (g) that shows inference on the finite population mean, which may be denoted here by

$$a = \frac{1}{50}(10 \times 3.71 + 40c).$$

Figure A.2 Graphical results for part (b)

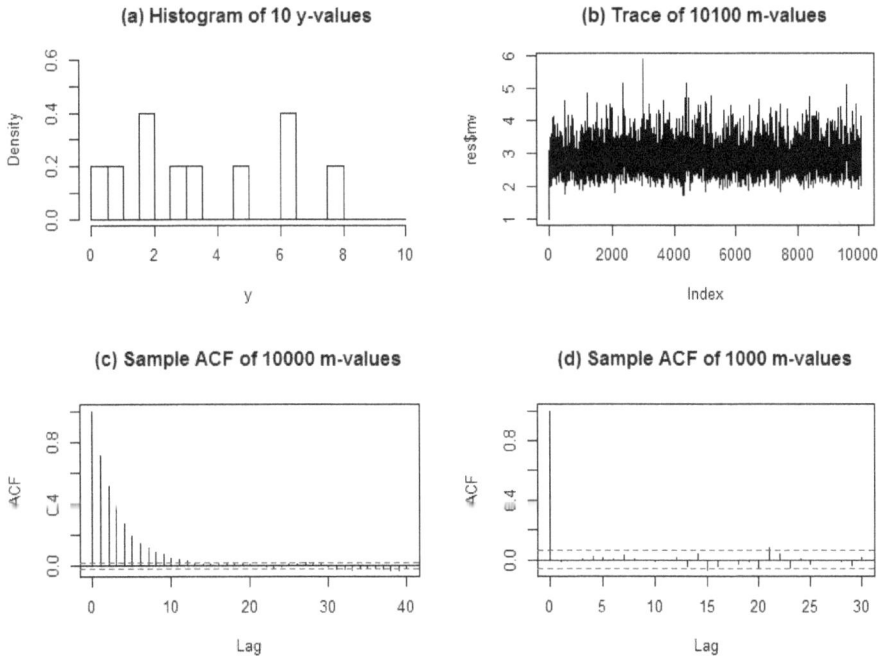

(a) Histogram of 10 y-values

(b) Trace of 10100 m-values

(c) Sample ACF of 10000 m-values

(d) Sample ACF of 1000 m-values

(e) Histogram of 1000 m-values

(f) Histogram of 1000 c-values

(g) Histogram of 1000 a-values (finite population mean)

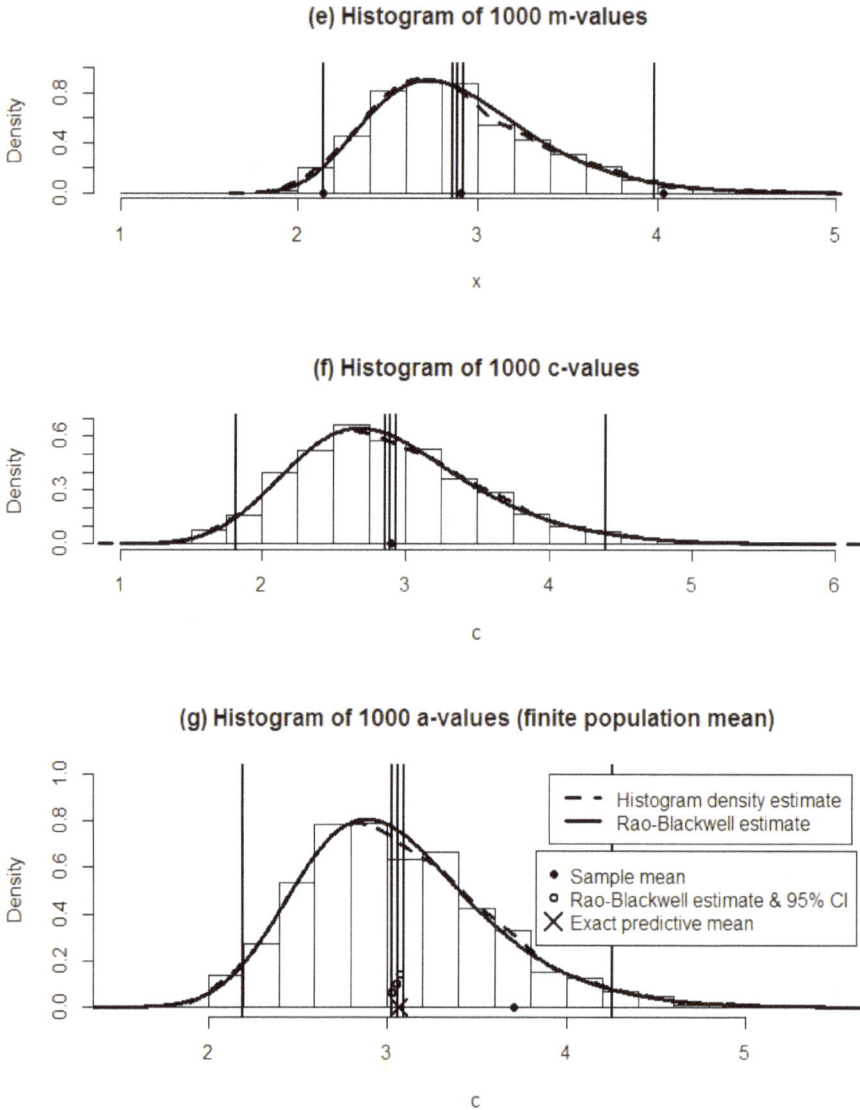

Some of the estimates and quantities shown in the last subplot (g) are as follows. The histogram estimate of a's predictive mean is $\bar{a} = 3.061$ with 95% CI (3.028, 3.094). The Rao-Blackwell estimate of a's predictive mean is $(10 \times 3.71 + 40\bar{m}) / 50 = 3.055$, with 95% CI (3.031, 3.078). The exact predictive mean of a is the same as the posterior mean of m and equal to 3.068. The 95% CPDR estimate for a is 2.190 4.256.

R Code for Exercise A.1

```
# (a)
options(digits=4)
INTEG <- function(xvec, yvec, a = min(xvec), b = max(xvec)){
# Integrates numerically under a spline through the points given by
# the vectors xvec and yvec, from a to b.
fit <- smooth.spline(xvec, yvec)
spline.f <- function(x){predict(fit, x)$y }
integrate(spline.f, a, b)$value   }
INTEG(seq(0,1,0.01), seq(0,1,0.01)^2, 0,1)   # 0.3333 correct

X11(w=8,h=6);    par(mfrow=c(2,2));
set.seed(221); m=rgamma(1,1,1); v=m^2; n=100; y=rnorm(n,m,m); c(m,v)
          # 0.7071 0.5000
hist(y,prob=T,xlim=c(-2,4),ylim=c(0,0.8), breaks=seq(-2,4,0.25),
          main="(a) Histogram of 100 y-values")
yvec=seq(-2,4,0.01); lines(yvec,dnorm(yvec,m,m),lwd=3)
abline(v=c(m,m+c(-1,1)*qnorm(0.975)*m), lwd=3)

LOGPOST=function(m=2,n=10,y=c(2,1)){
          -m-n*log(m)-(1/(2*m^2))*sum((y-m)^2)   }
LOGPOST() # -9.056   OK

METALG = function(J=1000,y,m0=1,mdel=0.4){
m=m0; mv=m; mct=0; n=length(y); for(j in 1:J){
          mcand=runif(1,m-mdel,m+mdel)
          if(mcand>0){    logprob=LOGPOST(m= mcand,n=n,y=y)-
                                    LOGPOST(m=m,n=n,y=y)
                          prob=exp(logprob)
                          u=runif(1); if(u<=prob){ mct=mct+1; m= mcand }
                          }
          mv=c(mv,m)
          }
list(mv=mv,mar=mct/J)    }

set.seed(312); res=METALG(J=10100,y=y,m0=1,mdel=0.11); res$mar # 0.5528
plot(res$mv,type="l",main="(b) Trace of 10100 m-values");

acf(res$mv, main="(c) Sample ACF of 10000 m-values")
acf(res$mv, plot=F)[1:5]  # 0.628 0.404 0.259 0.157 0.100
mv=res$mv[-(1:101)][seq(10,10000,10)];
acf(mv, main="(d) Sample ACF of 1000 m-values")
acf(mv,plot=F)[1:5] # -0.014 -0.001  0.006  0.018  0.014
```

```
J=length(mv); J # 1000

mbar=mean(mv); mci=mbar+c(-1,1)*qnorm(0.975)*sd(mv)/sqrt(J)
mcpdr=quantile(mv,c(0.025,0.975));
mvec=seq(0.5,1,0.01); kvec=mvec;
        for(i in 1:length(mvec)) kvec[i] = exp(LOGPOST(m=mvec[i],n=n,y=y))
k0=INTEG(mvec,kvec); postvec=kvec/k0;   k0 # 6.269e-11
mhat=INTEG(mvec,mvec*postvec);
c(mbar,sd(mv),mhat,mci,mcpdr)
        # 0.73769 0.04305 0.73935 0.73502 0.74036 0.66197 0.82984

fun=function(q,p=0.025){ (INTEG(mvec,postvec,0,q)-p)^2 }
LB0 = optim(par=0.5,fn=fun)$par;  LB = optim(par= LB0,fn=fun)$par
fun=function(q,p=0.975){ (INTEG(mvec,postvec,0,q)-p)^2 }
UB0 = optim(par=0.8,fn=fun)$par;  UB = optim(par= UB0,fn=fun)$par
c(LB,UB) # 0.6609 0.8305
INTEG(mvec,postvec,0,LB) # 0.025
INTEG(mvec,postvec,UB,1) # 0.025 OK     (Ignore all the warnings)

par(mfrow=c(2,1))
hist(mv,prob=T,xlim=c(0.6,0.9),ylim=c(0,10), breaks=seq(0.5,1,0.01),
        xlab="x",main="(e) Histogram of 1000 m-values")
lines(mvec,postvec,lty=1,lwd=3)
lines(density(mv),lty=2,lwd=3)
abline(v=c(mbar,mci,mcpdr),lwd=2)
points(c(mhat,LB,UB),c(0,0,0),pch=16)
points(m,0,pch=4,lwd=3)

# Prediction of c ----------------------
set.seed(332); cv=rnorm(J,mv,mv/sqrt(10))
cbar=mean(cv); cci=cbar+c(-1,1)*qnorm(0.975)*sd(cv)/sqrt(J)
ccpdr=quantile(cv,c(0.025,0.975))
c(cbar,sd(cv),cci,ccpdr)  # 0.7410 0.2253 0.7270 0.7549 0.3063 1.1893

hist(cv,prob=T,xlim=c(0,1.6),ylim=c(0,2.5), breaks=seq(0,1.6,0.05),
        xlab="c",main="(f) Histogram of 1000 c-values")
cvec=seq(0,1.5,0.01); fcvec=seq(0,1.5,0.01);  for(i in 1:length(cvec))
        fcvec[i]=mean(dnorm(cvec[i],mv,mv/sqrt(10)))
lines(cvec,fcvec,lty=1,lwd=3)
lines(density(cv),lty=2,lwd=3)
abline(v=c(cbar,cci,ccpdr),lwd=2)
points(mhat,0,pch=16)
```

```
# (b)
X11(w=8,h=6);    par(mfrow=c(2,2));
y = c(3.4, 6.3, 1.0, 2.9, 1.8,   2.0, 0.5, 7.9, 4.8, 6.5); n = 10; ybar=mean(y);
ybar # 3.71
hist(y,prob=T,xlim=c(0,10),ylim=c(0,0.6), breaks=seq(0,10,0.5),
        main="(a) Histogram of 10 y-values")

set.seed(312); res=METALG(J=10100,y=y,m0=1,mdel=1); res$mar # 0.5954
plot(res$mv,type="l",main="(b) Trace of 10100 m-values");
acf(res$mv, main="(c) Sample ACF of 10000 m-values")
acf(res$mv,plot=F)[1:5]  # 0.710 0.513 0.374 0.270 0.195
acf(mv, main="(d) Sample ACF of 1000 m-values")
mv=res$mv[-(1:101)][seq(10,10000,10)];
acf(mv,plot=F)[1:5]  # 0.056  0.001 -0.006 -0.027  0.035
J=length(mv); J # 1000

mbar=mean(mv); mci=mbar+c(-1,1)*qnorm(0.975)*sd(mv)/sqrt(J)
mcpdr=quantile(mv,c(0.025,0.975));
mvec=seq(1.8,5,0.01); kvec=mvec;
        for(i in 1:length(mvec)) kvec[i] = exp(LOGPOST(m=mvec[i],n=n,y=y))
k0=INTEG(mvec,kvec); postvec=kvec/k0;   k0 # 3.317e-08
mhat=INTEG(mvec,mvec*postvec);
c(mbar,sd(mv),mhat,mci,mcpdr)
        #  2.8907 0.4823 2.9071 2.8608 2.9206 2.1456 3.9827

fun=function(q,p=0.025){ (INTEG(mvec,postvec,1.8,q)-p)^2 }
LB0 = optim(par=2.1,fn=fun)$par;  LB = optim(par= LB0,fn=fun)$par
fun=function(q,p=0.975){ (INTEG(mvec,postvec,1.8,q)-p)^2 }
UB0 = optim(par=4.1,fn=fun)$par;  UB = optim(par= UB0,fn=fun)$par
c(LB,UB) # 2.143 4.033
INTEG(mvec,postvec,1.8,LB) # 0.025
INTEG(mvec,postvec,UB,5) # 0.025 OK     (Ignore all the warnings)

par(mfrow=c(2,1))
hist(mv,prob=T,xlim=c(1,5),ylim=c(0,1), breaks=seq(1,5,0.2),
        xlab="x",main="(e) Histogram of 1000 m-values")
lines(mvec,postvec,lty=1,lwd=3)
lines(density(mv),lty=2,lwd=3)
abline(v=c(mbar,mci,mcpdr),lwd=2)
points(c(mhat,LB,UB),c(0,0,0),pch=16)
points(m,0,pch=4,lwd=3)
```

```
# Prediction of c = (1/40)(y11+...+y50)   (new definition) ----------------------
set.seed(332); cv=rnorm(J,mv,mv/sqrt(40))
cbar=mean(cv); cci=cbar+c(-1,1)*qnorm(0.975)*sd(cv)/sqrt(J)
ccpdr=quantile(cv,c(0.025,0.975))
c(cbar,sd(cv),cci,ccpdr) # 2.8985 0.6594 2.8577 2.9394 1.8105 4.3925

hist(cv,prob=T,xlim=c(1,6), ylim=c(0,0.7), breaks=seq(1,6,0.25),
        xlab="c",main="(f) Histogram of 1000 c-values")
cvec=seq(1,6,0.01); fcvec=seq(1,6,0.01); for(i in 1:length(cvec))
        fcvec[i]=mean(dnorm(cvec[i],mv,mv/sqrt(40)))
lines(cvec,fcvec,lty=1,lwd=3)
lines(density(cv),lty=2,lwd=3)
abline(v=c(cbar,cci,ccpdr),lwd=2)
points(mhat,0,pch=16)


# Now perform inference on the finite population mean,
# a=(1/50)*(10*ybar +40*c)
av=(1/50)*(10*ybar+40*cv)
abar=mean(av); aci=abar+c(-1,1)*qnorm(0.975)*sd(av)/sqrt(J)
acpdr=quantile(av,c(0.025,0.975))
c(abar,sd(av),aci,acpdr) # 3.0608 0.5276 3.0281 3.0935 2.1904 4.2560
 (1/50)*(10*ybar+40*mbar) # 3.055  RB estimate of predictive mean of a
 (1/50)*(10*ybar+40*mci)  # 3.031 3.078  RB CI for predictive mean of a
 (1/50)*(10*ybar+40*mhat)  # 3.068   Exact predictive mean of a

X11(w=8,h=4); par(mfrow=c(1,1))
hist(av,prob=T,xlim=c(1.5,5.5), ylim=c(0,1), breaks=seq(1,6,0.2), xlab="c",
        main="(g) Histogram of 1000 a-values (finite population mean)")
avec=seq(1,6,0.01); favec=seq(1,6,0.01); for(i in 1:length(avec))
        favec[i]=
    mean( dnorm( avec[i], (1/50)*( 10*ybar+40*mv), mv*sqrt(40)/50 ) )
lines(avec,favec,lty=1,lwd=3); lines(density(av),lty=2,lwd=3)
abline(v=c(abar,aci,acpdr),lwd=2)
points( (1/50)*(10*ybar+40*mbar) ,0.1,pch=1,cex=1, lwd=2)
points( (1/50)*(10*ybar+40*mci) ,c(0.06,0.14), pch=1,cex=1, lwd=2)
points( (1/50)*(10*ybar+40*mhat) ,0,pch=4,lwd=2,cex=2)
points(ybar,0,cex=1,lwd=2,pch=16)
legend(3.9,1, c("Histogram density estimate","Rao-Blackwell estimate"),
        lty=c(2,1), lwd=c(3,3), bg="white")
legend(3.83,0.67,c("Sample mean","Rao-Blackwell estimate & 95% CI",
        "Exact predictive mean"),
        pch=c(16,1,4), pt.cex=c(1,1,2), pt.lwd= c(2,2,2),   bg="white")
```

Exercise A.2 Practice with the MH algorithm

(a) Sample a value a from the standard exponential distribution and a value b from the uniform distribution between 0 and 10 (independently).

Then randomly sample $n = 100$ values from the gamma distribution with mean $m = a/b$ and variance $v = a/b^2$.

Then design and implement a Metropolis-Hastings algorithm so as to generate a random sample of size $J = 1,000$ from the joint posterior distribution of a and b.

Use this sample to perform Monte Carlo inference on m.

Be sure to provide a 95% CI for the posterior mean of m, an estimate of the 95% central posterior density region for m, and an estimate of the entire marginal posterior density of m.

Then predict c, the average of a future independent sample of size $k = 10$ from the gamma distribution with the same mean m and variance v.

Be sure to provide a 95% CI for the predictive mean of c, an estimate of the 95% central predictive density region for c, and an estimate of the entire posterior predictive density of c.

Illustrate your results with suitable figures (e.g. trace plots and histograms).

(b) Consider the following values in a sample obtained via SRSWOR from a finite population of size $N = 30$:
        0.4, 3.3, 1.0, 2.9, 1.8, 4.1.

Suppose we model the finite population values as gamma with mean $m = a/b$ and variance $v = a/b^2$, with a standard exponential prior on m and a uniform prior on b between 0 and 10.

Using MCMC methods, estimate/predict the *finite population mean absolute deviation about the superpopulation mean*, equivalently referred to as the MAD for short, and defined by

$$\psi = \frac{1}{N} \sum_{i=1}^{N} |y_i - m|.$$

Solution to Exercise A.2

The sampled values of a and b were 1.463 and 5.528. So the value of m was $a/b = 0.2647$. The 100 sampled gamma values are shown in Figure A.3(a) (page 621).

Next, the posterior density of the two parameters a and b is

$$f(a,b \mid y) \propto f(a,b) f(y \mid a,b)$$

$$\propto e^{-a} \prod_{i=1}^{n} \frac{b^a y_i^{a-1} e^{-by_i}}{\Gamma(a)} = \frac{e^{-a} b^{na} (\prod_{i=1}^{n} y_i)^{a-1} e^{-by_T}}{\Gamma(a)^n}.$$

So the log-posterior is

$$l(a,b) = \log f(a,b \mid y)$$

$$= -a + na \log b + (a-1) \sum_{i=1}^{n} \log y_i - by_T - n \log \Gamma(a).$$

A suitable Metropolis algorithm is one which at each iteration:

1. Proposes a value
   $$a' \sim U(a - \delta_a, a + \delta_a),$$
   where δ_a is a tuning constant, and accepts this value with
   probability $p = e^q$, where $q = l(a',b) - l(a,b)$

2. Proposes a value
   $$b' \sim U(b - \delta_b, b + \delta_b),$$
   where δ_b is a tuning constant, and accepts this value with
   probability $p = e^q$, where $q = l(a,b') - l(a,b)$.

Implementing this algorithm we obtained the required $J = 1,000$ values
$$(a_1, b_1), ..., (a_J, b_J) \sim iid \ f(a,b \mid y)$$
and hence
$$m_1, ..., m_J \sim iid \ f(m \mid y)$$
by calculating $m_j = a_j / b_j$ for each $j = 1, ..., J$.

A histogram of these simulated m-values is shown in Figure A.3(b) (page 622).

The dashed line is a histogram estimate of $f(m\,|\,y)$. The vertical lines show the posterior mean estimate, $\bar{m} = 0.3017$, the 95% CI for the posterior mean, $(0.3001, 0.3033)$, and the 95% CPDR estimate for m, $(0.2566, 0.3570)$. The cross shows the true value of m, 0.7071.

The Monte Carlo sample was then used to generate a random sample from the predictive distribution of

$$c = (y_{n+1} + ... + y_{n+10})/10.$$

This was done by sampling

$$y_{n+1}^{(j)},..., y_{n+10}^{(j)} \sim iid\ G(a_j, b_j)$$

and forming

$$c_j = (y_{n+1}^{(j)},..., y_{n+10}^{(j)})/10,\ j = 1,...,J.$$

A histogram of the c-values is shown in Figure A.3(c). The dashed line is a histogram estimate of $f(c\,|\,y)$. The vertical lines are the predictive mean estimate, $\bar{c} = 0.2981$, the 95% CI for the predictive mean, $(0.2929, 0.3033)$, and the 95% CPDR estimate for c, $(0.1584, 0.4878)$.

Figure A.3 Graphical results for part (a)

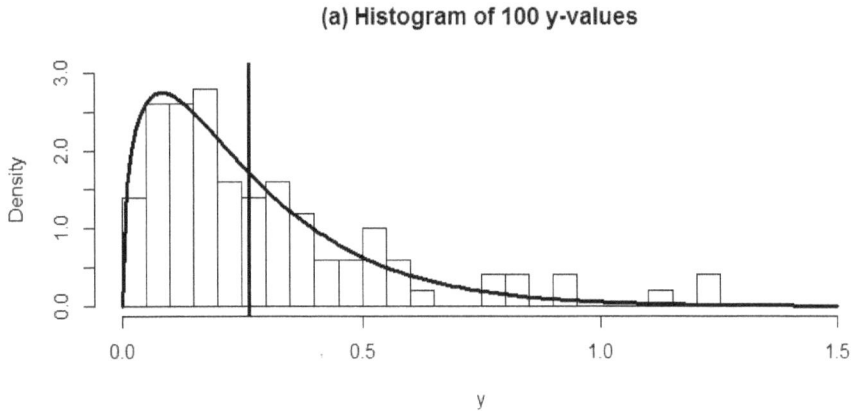

(a) Histogram of 100 y-values

(b) Histogram of 1000 m-values

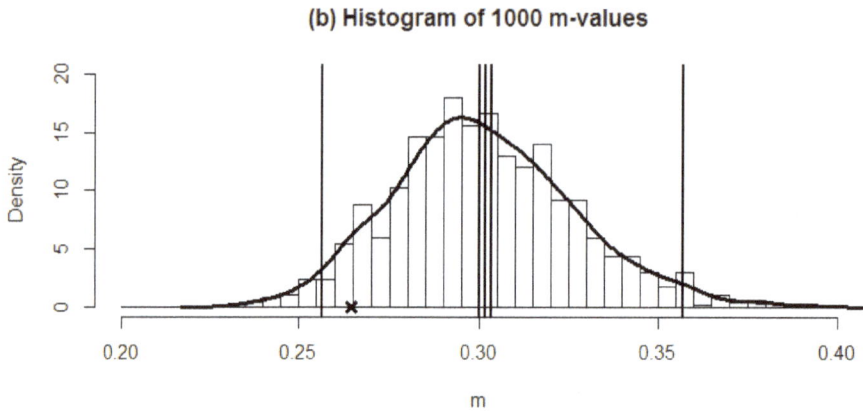

(c) Histogram of 1000 c-values

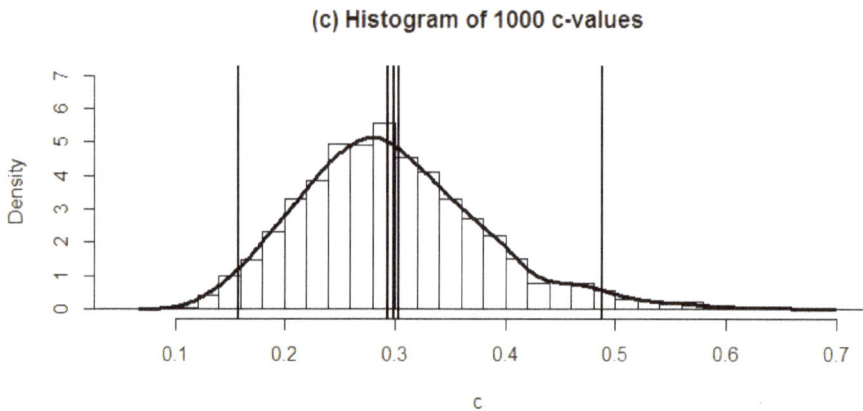

(b) Here we repeat the procedure in (a) but using $n = 6$ (rather than 100), and the 6 given sample values whose mean is 2.25 (instead of the 100 generated values as before), so as to generate a Monte Carlo sample of size $J = 1,000$ from the posterior distribution of a and b.

We then use each pair of values, a_j and b_j, to generate 24 values which are iid from the gamma distribution with parameters a_j and b_j.

Then for each j we calculate the associated value of the MAD, namely

$$\psi_j = \frac{1}{N} \sum_{i=1}^{N} \left| y_i - \frac{a_j}{b_j} \right|.$$

We then use the resulting J values of the MAD, i.e. $\psi_1, ..., \psi_J$, for Monte Carlo inference in the usual way.

Figure A.4 shows a histogram of these J values and related information.

Numerically, we estimate ψ's posterior/predictive mean by 1.307 with 95% CI (1.27, 1.34), and we estimate ψ's CPDR by (0.75, 2.73).

Figure A.4 Histogram of 1,000 MAD values

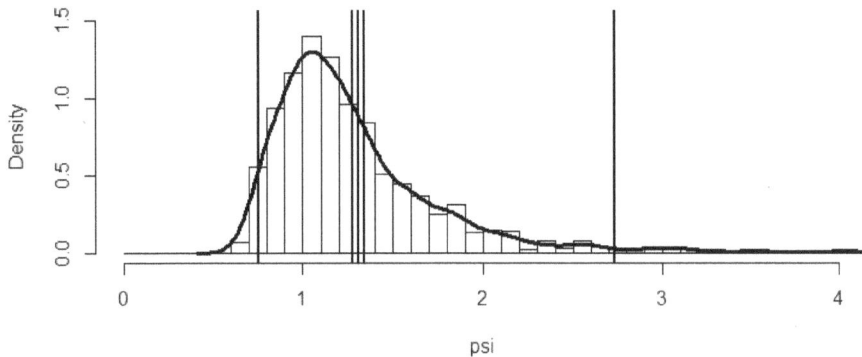

R Code for Exercise A.2

```
# (a)
options(digits=4); n = 100; X11(w=8,h=4);    par(mfrow=c(1,1));
set.seed(192); a=rgamma(1,1,1); b=runif(1,0,10); y=rgamma(n,a,b);
m=a/b; v=a/b^2; c(a,b,m,v) # 1.46321 5.52763 0.26471 0.04789

hist(y,prob=T,xlim=c(0,1.5),ylim=c(0,3), breaks=seq(0,1.5,0.05),
        main="(a) Histogram of 100 y-values")
yvec=seq(0,1.5,0.01); lines(yvec,dgamma(yvec,a,b),lwd=3)
abline(v=m,lwd=3)

sumlogy=sum(log(y)); sumy=sum(y)  # sufficient statistics
LOGPOST=function(a=1,b=1,n=3,sumlogy=2,sumy=2){
        -a+n*a*log(b)+(a-1)*sumlogy-b*sumy-n*lgamma(a)      }
LOGPOST() # -3    OK
```

```
MHALG = function(J=1000,y,a0=1,b0=1,adel=1,bdel=1){
a=a0; b=b0;  av=a; bv=b;   act=0; bct=0;  n=length(y);
sumlogy=sum(log(y)); sumy=sum(y)  # sufficient statistics
for(j in 1:J){
        acand=runif(1,a-adel,a+adel)
        if(acand>0){
                logprob=
        LOGPOST (a=acand,b=b,n=n,sumlogy=sumlogy,sumy=sumy)-
        LOGPOST (a=a,b=b,n=n,sumlogy=sumlogy,sumy=sumy)
                prob=exp(logprob)
                u=runif(1); if(u<=prob){ act=act+1; a= acand }     }
        bcand=runif(1,b-bdel,b+bdel)
        if((bcand>0)&&(bcand<10)){
                logprob=
        LOGPOST (a=a,b=bcand,n=n,sumlogy=sumlogy,sumy=sumy)-
         LOGPOST (a=a,b=b,n=n,sumlogy=sumlogy,sumy=sumy)
                prob=exp(logprob)
                u=runif(1); if(u<=prob){ bct=bct+1; b= bcand }
                }
        av=c(av,a); bv=c(bv,b)
        }
list(av=av,bv=bv,aar=act/J,bar=bct/J)
}

set.seed(312); res=MHALG(J=10100,y=y,a0=1,b0=1,adel=0.3,bdel=1)
X11(w=8,h=6); par(mfrow=c(2,1));
plot(res$av); plot(res$bv); c(res$aar,res$bar) # 0.5055 0.5611

av=res$av[-(1:101)][seq(10,10000,10)]; J=length(av); J # 1000
bv=res$bv[-(1:101)][seq(10,10000,10)]; mv=av/bv
mbar=mean(mv); mci=mbar+c(-1,1)*qnorm(0.975)*sd(mv)/sqrt(J)
mcpdr=quantile(mv,c(0.025,0.975));
c(mbar,mci,mcpdr) #  0.3017 0.3001 0.3033 0.2566 0.3570

X11(w=8,h=4); par(mfrow=c(1,1));
hist(mv,prob=T,xlim=c(0.2,0.4),ylim=c(0,20), breaks=seq(0.2,0.4,0.005),
        xlab="m",main="(b) Histogram of 1000 m-values")
lines(density(mv),lty=1,lwd=3)
abline(v=c(mbar,mci,mcpdr),lwd=2)
points(m,0,pch=4,lwd=3)
```

```
# Prediction of c ----------------------
set.seed(332); cv=rep(NA,J); for(j in 1:J) cv[j]=mean(rgamma(10,av[j],bv[j]))
cbar=mean(cv); cci=cbar+c(-1,1)*qnorm(0.975)*sd(cv)/sqrt(J)
ccpdr=quantile(cv,c(0.025,0.975))
c(cbar,sd(cv),cci,ccpdr)  # 0.29812 0.08356 0.29294 0.30329 0.15843 0.48783
hist(cv,prob=T,xlim=c(0.05,0.7),ylim=c(0,7), breaks=seq(0,1.6,0.02),
       xlab="c",main="(c) Histogram of 1000 c-values")
lines(density(cv),lty=1,lwd=3); abline(v=c(cbar,cci,ccpdr),lwd=2)

# (b)
y=c( 0.4, 3.3, 1.0, 2.9, 1.8, 4.1); X11(w=8,h=6); par(mfrow=c(2,1));
n=length(y); sumlogy=sum(log(y)); sumy=sum(y)  # sufficient statistics
set.seed(312); res=MHALG(J=10100,y=y,a0=1,b0=1,adel=1.3,bdel=0.7)
plot(res$av); plot(res$bv); c(res$aar,res$bar) # 0.5129 0.5094

av=res$av[-(1:101)][seq(10,10000,10)]; J=length(av); J # 1000
bv=res$bv[-(1:101)][seq(10,10000,10)]; mv=av/bv
mbar=mean(mv); mci=mbar+c(-1,1)*qnorm(0.975)*sd(mv)/sqrt(J)
mcpdr=quantile(mv,c(0.025,0.975));
c(mbar,mci,mcpdr) #  2.256 2.208 2.305 1.148 4.188

X11(w=8,h=4); par(mfrow=c(1,1));
hist(mv,prob=T,xlim=c(0,7),ylim=c(0,0.8), breaks=seq(0,10,0.5),
       xlab="x",main="Histogram of 1000 simulated m-values")
lines(density(mv),lty=2,lwd=3);  abline(v=c(mbar,mci,mcpdr),lwd=2)

# Prediction of psi ----------------------
set.seed(332); psiv=rep(NA,J);
for(j in 1:J){      yrem=rgamma(24,av[j],bv[j])
        yall = c(y,yrem); psiv[j]=mean((abs(yall-mv[j])) )  }
psibar=mean(psiv); psici =psibar+c(-1,1)*qnorm(0.975)*sd(psiv)/sqrt(J)
psicpdr=quantile(psiv,c(0.025,0.975))
c(psibar,sd(psiv),psici,psicpdr)  # 1.3068 0.5411 1.2732 1.3403 0.7497 2.7349

hist(psiv,prob=T,xlim=c(0,4),ylim=c(0,1.5), breaks=seq(0,7,0.1),
       xlab="psi",main="")
lines(density(psiv),lty=1,lwd=3); abline(v=c(psibar,psici,psicpdr),lwd=2)
```

Exercise A.3 Practice with a Bayesian finite population regression model

(a) Generate a population of covariates
$$x_1, ..., x_N \sim iid \; U(10, 20),$$
where $N = 100$.

Then generate a population of values
$$y_i \sim N(a + bx_i, \sigma^2), \; i = 1, ..., N,$$
where $a = 3$, $b = 0.5$, $\sigma = 2$.

Then select a random sample of size $n = 20$ from the N units in the finite population, without replacement.

Plot the y values against the x values, over the population and over the sample, respectively. Draw the true regression line $y = a + bx$ and the two least squares regression lines estimated using the population data and sample data, respectively.

(b) Consider the following Bayesian model:
$$(y_i \mid a, b, \lambda) \sim \perp N(a + bx_i, 1/\lambda), \quad i = 1, ..., N$$
$$f(a, b, \lambda) \propto 1/\lambda; \quad a, b \in \Re; \quad \lambda > 0.$$

Generate a random sample of size $J = 1,000$ from the joint posterior distribution of a, b and λ, given the sample data generated in (a).

Then use this sample and R to estimate each of the following quantities:

$m = a + 16b$ (average of a hypothetically infinite number of values with covariate 16)

$$\bar{y} = \frac{y_1 + ... + y_N}{N}$$ (the finite population mean)

$$\psi = \frac{2y_{(100)}}{y_{(50)} + y_{(51)}}$$ (ratio of maximum to median of the 100 finite population values).

Assume that all N covariate values in the population are known.

(c) Repeat the inferences in (b) but using WinBUGS and a sample size of $J = 10,000$.

Solution to Exercise A.3

(a) The required plot and regression lines are shown in the Figure A.5.

Figure A.5 Graphical results for part (a)

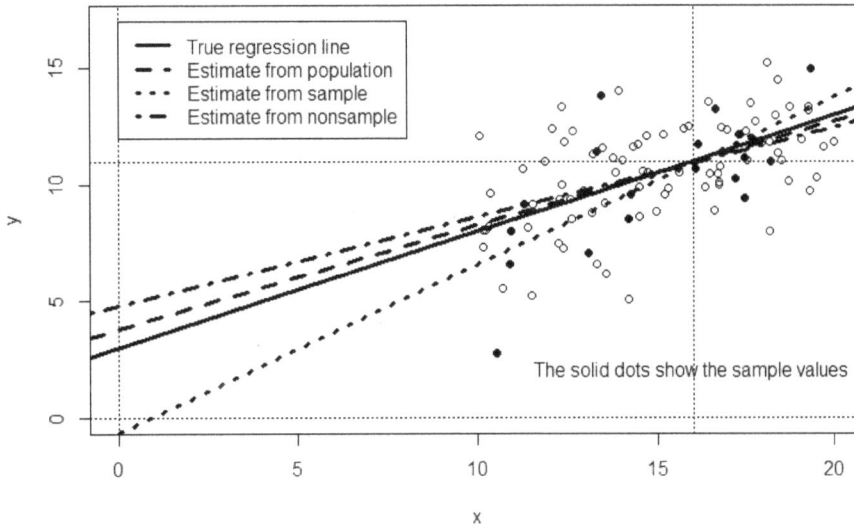

(b) Denote the sample values by $s_1, ..., s_n \in \{1, ..., N\}$, where $s_1 < ... < s_n$, and define $s = (s_1, ..., s_n)$.

Then define the population vector as $y = (y_1, ..., y_N)'$ and the sample vector as $y_s = (y_{s_1}, ..., y_{s_n})'$.

Also define $r = (r_1, ..., r_{N-n}) = \{1, ..., N\} - s$ in such a way that $r_1 < ... < r_{N-n}$, and define the nonsample vector as $y_r = (y_{r_1}, ..., y_{r_{N-n}})'$.

Likewise, define the population covariate vector as $x = (x_1, ..., x_N)'$, the sample covariate vector as $x_s = (x_{s_1}, ..., x_{s_n})'$, and the nonsample covariate vector as $x_r = (x_{r_1}, ..., x_{r_{N-n}})'$.

Also consider all of $x_1, \ldots, x_N$ as known constants, and define $D = (s, y_s)$ as the data. Also let:

$$\beta = \begin{pmatrix} a \\ b \end{pmatrix}, \quad X_s = (1_n, x_s), \quad X_r = (1_{N-n}, x_r), \quad \Sigma_{ss} = I_n, \quad \Sigma_{rr} = I_{N-n}.$$

Then, from the theory of the normal-normal-gamma finite population model, we have that:

$$(y_r \mid D, \beta, \lambda) \sim N_{N-n}(X_r \beta, \Sigma_{rr} / \lambda)$$

$$(\beta \mid D, \lambda) \sim N_2(T, D / \lambda),$$

where $D = (X_s' \Sigma_{ss}^{-1} X_s)^{-1}$ and $T = (X_s' \Sigma_{ss}^{-1} X_s)^{-1} X_s' \Sigma_{ss}^{-1} y_s$

$$(\lambda \mid D) \sim G(A/2, B/2),$$

where $A = n - 2$ and $B = (y_s - X_s T)' \Sigma_{ss}^{-1}(y_s - X_s T).$

Thus, to do the required inference, first carry out the following steps:

1. Relabel the population units so that $y_s = (y_1, \ldots, y_n)'$, $x_s = (x_1, \ldots, x_n)'$, $y_r = (y_{n+1}, \ldots, y_N)'$, $x_r = (x_{n+1}, \ldots, x_N)'$, etc., so that $y = (y_s', y_r')'$, etc.

2. Calculate A, B, D and T as per the above

3. Generate $\lambda_1, \ldots, \lambda_J \sim iid\ G(A/2, B/2)$ (easy)

4. Generate $\beta^{(j)} \sim \perp N_2(T, D / \lambda_j)$, for $j = 1, \ldots, J$ (easy)

5. Generate $y_r^{(1)}, \ldots, y_r^{(J)} \sim N_{N-n}(X_r \beta^{(j)}, \Sigma_{rr} / \lambda_j)$, for $j = 1, \ldots, J$

   (e.g. for each j, generate $y_i^{(j)} \sim \perp N(a_j + b_j x_i, 1 / \lambda_j)$, $i = n+1, \ldots, N$, and form $y_r^{(j)} = (y_{n+1}^{(j)}, \ldots, y_N^{(j)})'$

6. Form $y^{(j)} = (y_s', y_r^{(j)'})'$ for each $j = 1, \ldots, J$.

Now calculate

$$m_j = a_j + 16 b_j$$

and perform Monte Carlo inference on m, using the fact that

$$m_1, \ldots, m_J \sim iid\ f(m \mid D).$$

(For example, estimate m by $\bar{m} = J^{-1} \sum_{j=1}^{J} m_j$.)

Likewise, calculate $\bar{y}^{(j)} = 1_N' y^{(j)} / N$ and perform Monte Carlo inference on $\bar{y}$ in the usual way, using the fact that $\bar{y}^{(1)}, \ldots, \bar{y}^{(J)} \sim iid\ f(\bar{y} \mid D)$.

Finally, calculate

$$\psi_j = \frac{2y_{(100)}^{(j)}}{y_{(50)}^{(j)} + y_{(51)}^{(j)}}$$

and perform Monte Carlo inference on ψ, using the fact that

$$\psi_1,...,\psi_J \sim iid \ f(\psi \mid D).$$

Optionally, we may improve on some of the above 'basic' inferences by considering Rao-Blackwell techniques, e.g. estimate m by its exact posterior mean, $\hat{m} = E(m \mid D) = (1,16)T$.

Figure A.6 shows histograms of the simulated values of m (subplot (a)), $\bar{y}$ (subplot (b)) and ψ (subplot (c)), with each subplot overlaid by various points, interval and density estimates.

Subplot (d) (page 631) illustrates 'exact' inference on $\bar{y}$ based on the theory of the normal-normal-gamma finite population model, and subplot (e) (page 631) is a detail in subplot (d). Each plot features a cross showing the true value of the quantity being estimated.

Figure A.6 Graphical results for part (b)

(a) Histogram of 1000 m-values

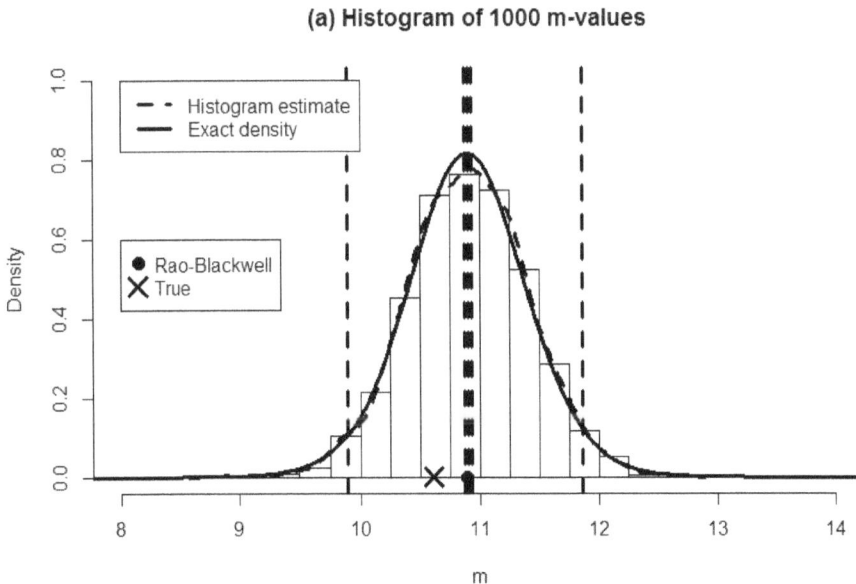

(b) Histogram of 1000 ybar-values

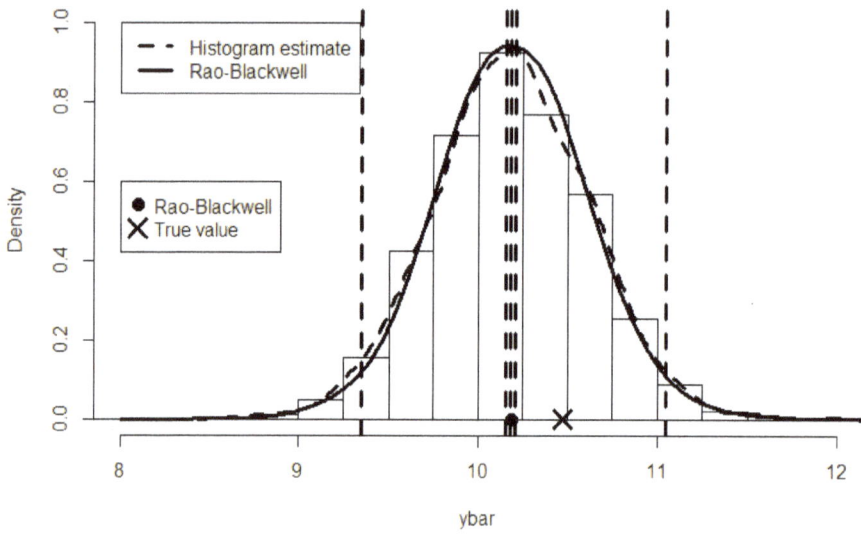

(c) Histogram of 1000 psi-values

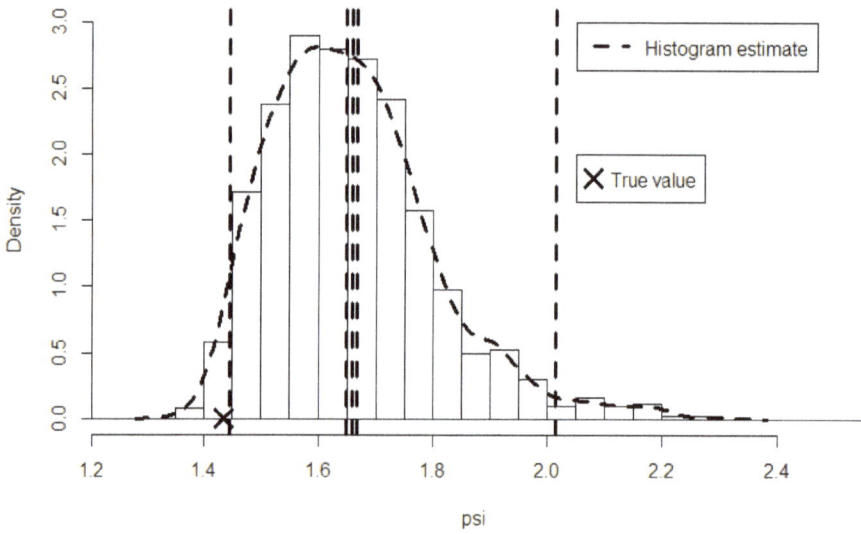

(d) Histogram of 1000 ybar-values

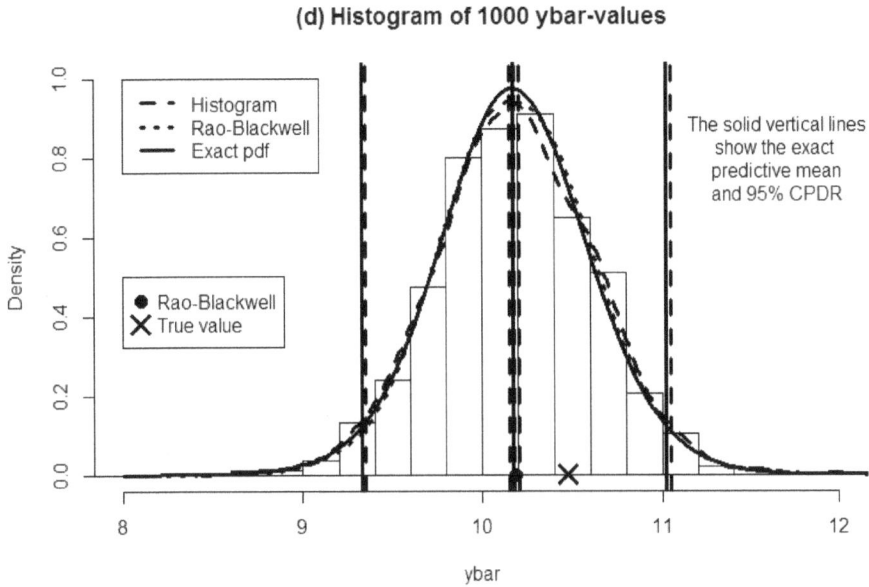

(e) Detail in subplot (d)

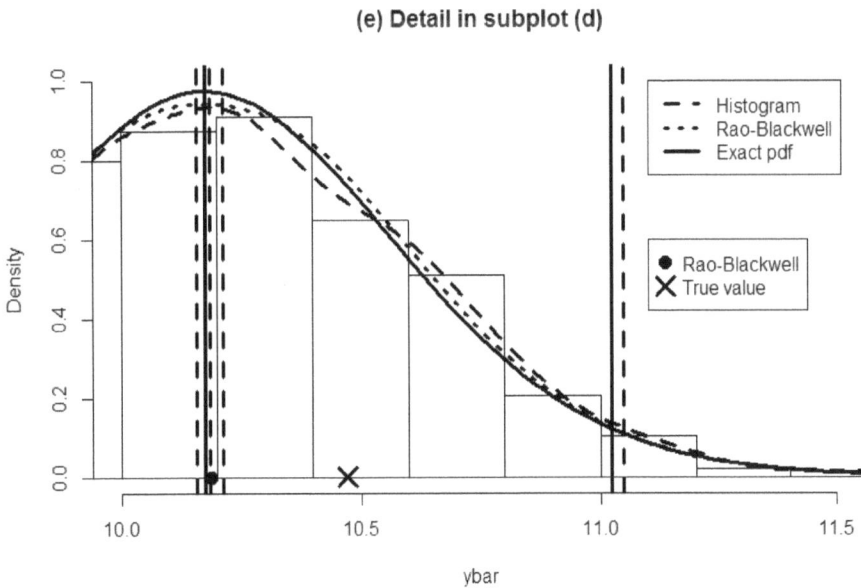

Table A.1 shows some of the true values and corresponding numerical estimates featuring in Figure A.6.

(c) Using the WinBUGS code below we obtained results as shown in Figure A.7. It will be noted that these are consistent with those in Table A.1

Table A.1 Numerical results for part (b)

Quantity	True value	Posterior mean	MC estimate	95% CI for post. mean	MC estimate of 95% CPDR
m	11.000	10.895	10.906	(10.875, 10.937)	(9.893, 11.863)
$\bar{y}$	10.473	10.174	10.185	(10.158, 10.211)	(9.353, 11.049)
ψ	1.435	NA	1.659	(1.650, 1.668)	(1.444, 2.014)

Figure A.7 Output from WinBUGS run

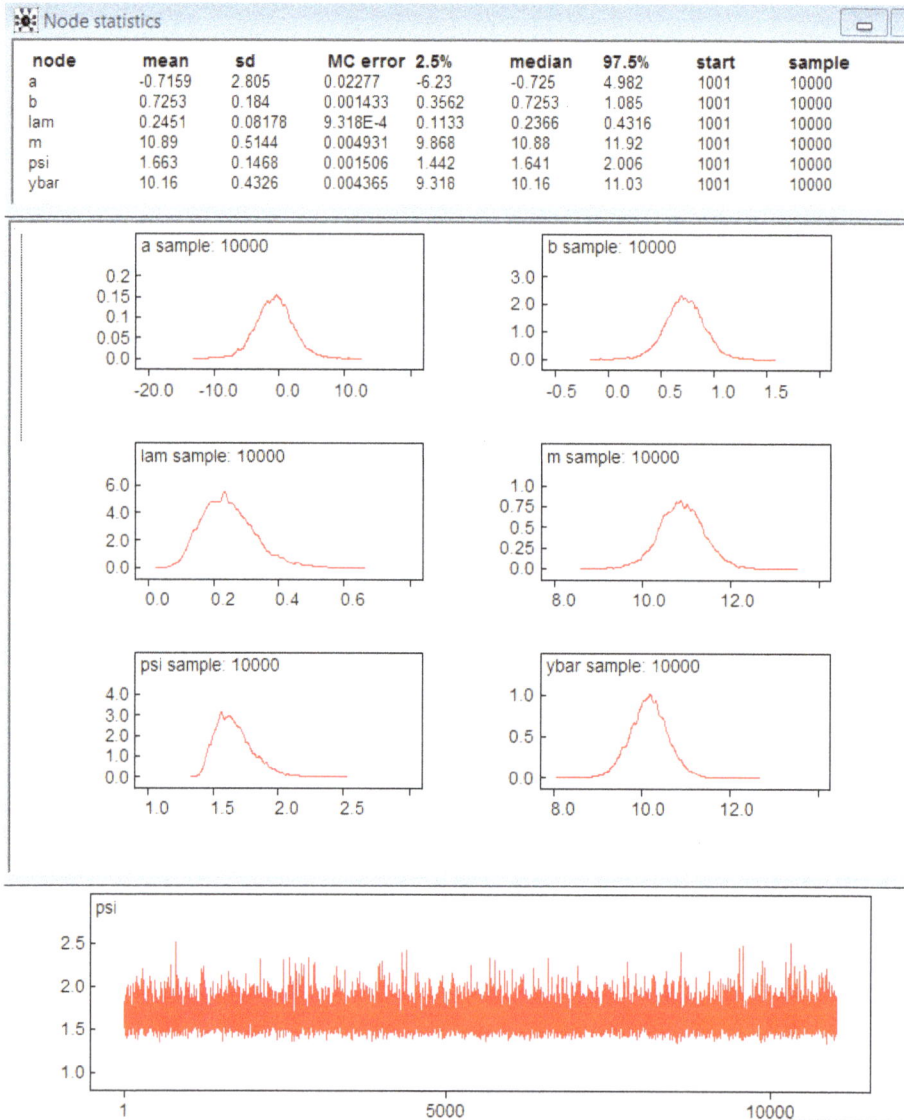

Node statistics

node	mean	sd	MC error	2.5%	median	97.5%	start	sample
a	-0.7159	2.805	0.02277	-6.23	-0.725	4.982	1001	10000
b	0.7253	0.184	0.001433	0.3562	0.7253	1.085	1001	10000
lam	0.2451	0.08178	9.318E-4	0.1133	0.2366	0.4316	1001	10000
m	10.89	0.5144	0.004931	9.868	10.88	11.92	1001	10000
psi	1.663	0.1468	0.001506	1.442	1.641	2.006	1001	10000
ybar	10.16	0.4326	0.004365	9.318	10.16	11.03	1001	10000

R Code for Exercise A.3

```
# (a)
X11(w=8,h=5.5);    par(mfrow=c(1,1)); options(digits=4)
N=100; n=20; a=3; b=0.5; sig=2; set.seed(312); x=runif(N,10,20);
y=rnorm(N,a+b*x,sig); s=sort(sample(1:N,n)); xs=x[s]; ys=y[s];
r=(1:N)[-s];   xr=x[r];   yr=y[r];   yT=sum(y); ysT=sum(ys); yrT=sum(yr)
ybar=mean(y); ysbar=mean(ys); yrbar=mean(yr);
xT=sum(x);  xsT=sum(xs); xrT=sum(xr)
xbar=mean(x); xsbar=mean(xs); xrbar=mean(xr);

m=a+16*b; psi=max(y)/median(y)
c(m, ybar,max(y),median(y),psi)  # 11.000 10.473    15.234 10.616  1.435

plot(x,y,xlim=c(0,20),ylim=c(0,17));
points(xs,ys,pch=16); abline(v=0,lty=3); abline(h=0,lty=3); abline(v=16,lty=3);
abline(h=a+16*b,lty=3);
abline(a,b,lwd=3);
abline(lm(y~x),lty=2,lwd=3); abline(lm(ys~xs),lty=3,lwd=3);
abline(lm(yr~xr),lty=4,lwd=3)
legend(0,17,bg="white", c("True regression line","Estimate from population",
        "Estimate from sample","Estimate from nonsample"),
        lty=1:4,lwd=rep(3,4) )
text(16,2,"The solid dots show the sample values")

# (b) Follows on from (a)....
# Packages, Load package, MASS      (for use further down)

eta=0; tau=0;  sigma=diag(rep(1,N)); sigmass=diag(rep(1,n));
sigmarr=diag(rep(1,N-n));
p=2;  c=2*eta+n-p; Xs=cbind(1,xs);   Xr=cbind(1,xr); X=rbind(Xs,Xr)
D=solve(t(Xs)%*%solve(sigmass)%*%Xs)
T=D%*%t(Xs)%*%solve(sigmass)%*%ys; t(T) #  -0.6637  0.7224
A=2*eta+n-p;  B= 2*tau + t(ys-Xs%*%T) %*% solve(sigmass) %*% (ys-Xs%*%T)

J=1000; set.seed(5);   lamvec=rgamma(J,A/2,B/2);
betamat=matrix(NA,nrow=2,ncol=J)
for(j in 1:J) betamat[,j] = mvrnorm( n=1, mu=T, Sigma=D/lamvec[j] )

avec=betamat[1,]; bvec=betamat[2,]
ahat=mean(avec); bhat=mean(bvec); c(ahat,bhat) # -0.5742  0.7175
yrmat=matrix(NA,nrow=N-n,ncol=J)
set.seed(334); for(j in 1:J)
        yrmat[,j]= rnorm(N-n,avec[j]+bvec[j]*xr,1/sqrt(lamvec[j]))
```

```
# Use simulated values of beta and yr to do inference
mvec=avec+16*bvec; ybarvec=rep(NA,J); psivec=rep(NA,J)
for(j in 1:J){       ysim = c(ys, yrmat[,j])
                     ybarvec[j]=mean(ysim)
                     psivec[j] = max(ysim)/median(ysim)   }
mhat=mean(mvec); mci= mhat +c(-1,1)*qnorm(0.975)*sd(mvec)/sqrt(J)
        mcpdr=quantile(mvec,c(0.025,0.975))
ybarhat=mean(ybarvec);
ybarci = ybarhat +c(-1,1)*qnorm(0.975)*sd(ybarvec)/sqrt(J)
                ybarcpdr=quantile(ybarvec,c(0.025,0.975))
psihat=mean(psivec); psici = psihat +c(-1,1)*qnorm(0.975)*sd(psivec)/sqrt(J)
                psicpdr=quantile(psivec,c(0.025,0.975))

hist(mvec,prob=T,xlim=c(8,14),ylim=c(0,1), breaks=seq(7,14,0.25),
        xlab="m",main="(a) Histogram of 1000 m-values")  # Ignore warnings
lines(density(mvec),lty=2,lwd=3) # Histogram estimate
abline(v=c(mhat,mci,mcpdr),lty=2,lwd=3) # Histogram estimates
mhat2=c(1,16)%*%T; points(mhat2,0, pch=16,cex=1.5) # Exact posterior mean
mvarterm2=c(1,16)%*%D%*%c(1,16); msdterm2=sqrt(mvarterm2)
mv=seq(6,16,0.05); fmv2=mv
for(k in 1:length(mv))
    fmv2[k]=mean(dnorm(mv[k],mhat2,msdterm2/sqrt(lamvec)))
lines(mv,fmv2,lwd=3);  # Exact posterior density of m
points(median(y),0, pch=4,cex=2,lwd=3  ) # True value of m
legend(8,1,c("Histogram estimate","Exact density"), lty=c(2,1),lwd=c(3,3),
        bg="white")
legend(8,0.6,c("Rao-Blackwell","True"),pch=c(16,4),
          pt.cex=c(1.5,2), pt.lwd=c(1,3), bg="white")

hist(ybarvec,prob=T,xlim=c(8,12),ylim=c(0,1), breaks=seq(3,18,0.25),
        xlab="ybar",main="(b) Histogram of 1000 ybar-values")
lines(density(ybarvec),lty=2,lwd=3)  # Histogram estimate
abline(v=c(ybarhat, ybarci, ybarcpdr),lty=2,lwd=3) # Histogram estimates
ybarv=seq(8,13,0.02); fybarhatv=ybarv;
meanvalvec = (1/N)*( ysT+(N-n)*(avec+bvec*xrbar)  )
varvalvec = (N-n)/(lamvec*N^2)
for(k in 1:length(ybarv)){
        fybarhatv[k]= mean( dnorm(ybarv[k], meanvalvec, sqrt(varvalvec) ) )  }
lines(ybarv, fybarhatv,lty=1,lwd=3)  # Rao-Blackwell
points(mean(meanvalvec),0,pch=16,cex=1.5) # Rao-Blackwell
points(ybar, 0, pch=4,cex=2,lwd=3  )  # True value of ybar

legend(8,1,c("Histogram estimate","Rao-Blackwell"),
        lty=c(2,1),lwd=c(3,3), bg="white")
```

```
legend(8,0.6,c("Rao-Blackwell","True value"),pch=c(16,4),
        pt.cex=c(1.5,2), pt.lwd=c(1,3), bg="white")

hist(psivec,prob=T,xlim=c(1.25,2.5),ylim=c(0, 3), breaks=seq(0,10,0.05),
        xlab="psi",main="(c) Histogram of 1000 psi-values")
den=density(psivec); lines(den, lty=2,lwd=3)
abline(v=c(psihat, psici, psicpdr),lty=2,lwd=3)

# psimode=den$x[(1:length(den$x))[den$y==max(den$y)]] # optional extras....
# psimedian=median(psivec); abline(v=c(psimode,psimedian),lty=1,lwd=3)

points(psi, 0, pch=4,cex=2,lwd=3 )  # True value of psi
legend(2.05,3,c("Histogram estimate"), lty=c(2),lwd=c(3), bg="white")
legend(2.05,2,c("True value"),pch=c(4), pt.cex=c(2), pt.lwd=c(3), bg="white")

# Perform exact inference on ybar using a function from a previous exercise:
NNGFPM= function(eta=0, tau=0, alp=0.05,
  ys= c(5.6,2.3,8.4,5.1,4.3), X=rep(1,15) , N=15, sigma=diag(rep(1,N))  )  {
# This function performs inference under the normal-normal-gamma
# finite population model.
# Inputs: eta, tau,    alp, ys, X, N, sigma
# Outputs: A list with $a, $b and $c indicating   (ybar-a)/b given ys ~ t(c)
p=ncol(cbind(NA,X))-1;   n = length(ys);   c=2*eta+n-p
ysT=sum(ys); Xs=cbind(NA,X)[1:n,][,-1]; Xr=cbind(NA,X)[(n+1):N,][,-1]
sigmass=sigma[1:n,1:n];      sigmarr=sigma[(n+1):N,(n+1):N]
sigmasr=sigma[1:n,(n+1):N];   sigmars=t(sigmasr)
D=solve(t(Xs)%*%solve(sigmass)%*%Xs)
beta=D%*%t(Xs)%*%solve(sigmass)%*%ys
A=Xr-sigmars%*%solve(sigmass)%*%Xs;      oner=rep(1,N-n)
a=(1/N)*(   ysT   +   t(oner)%*%
  (  Xr%*%beta  + sigmars%*%solve(sigmass)%*%(ys-Xs%*%beta)   )  )
b2=(1/(c*N^2)) * ( 2*tau + t(ys-Xs%*%beta)%*%solve(sigmass)%*%
  (ys-Xs%*%beta) ) *    t(oner)%*%
  (sigmarr-sigmars%*%solve(sigmass)%*%sigmasr +A%*%D%*%t(A))  %*%
  oner
b=sqrt(b2); cpdr=a+c(-1,1)*qt(1-alp/2,c)*b
list(a=a,b=b,c=c,beta=beta, cpdr=cpdr) }

res= NNGFPM(  eta=0, tau=0, alp=0.05, ys=ys,X=X,N=N, sigma=sigma  )
c(res$a,res$b,res$c, res$cpdr) # 10.1744  0.4035 18.0000  9.3267 11.0221

# Plot for inference on ybar again
hist(ybarvec,prob=T,xlim=c(8,12),ylim=c(0,1), breaks=seq(3,18,0.2),
        xlab="ybar",main="(d) Histogram of 1000 ybar-values")
```

```
abline(v=c(ybarhat, ybarci, ybarcpdr),lty=2,lwd=3) # Histogram point estimates
points(mean(meanvalvec),0,pch=16,cex=1.5)
        # Rao-Blackwell estimate of predictive mean
abline(v=c(res$a,res$cpdr), lty=1, lwd=3) # # Exact point estimates
points(ybar, 0, pch=4,cex=2,lwd=3  )   # True value of ybar
lines(density(ybarvec),lty=2,lwd=3)  # Histogram estimate of predictive pdf
lines(ybarv, fybarhatv,lty=3,lwd=3)   # Rao-Blackwell estimate of pdf
lines(ybarv, dt((ybarv-res$a)/res$b,c)/res$b,lty=1,lwd=3) # Exact predictive pdf
legend(8,1,c("Histogram","Rao-Blackwell","Exact pdf"),
        lty=c(2,3,1),lwd=c(3,3,3))
legend(8,0.5,c("Rao-Blackwell","True value"),
        pch=c(16,4),pt.cex=c(1.5,2), pt.lwd=c(1,3))
text(11.65,0.8,
"The solid vertical lines\nshow the exact \npredictive mean\nand 95% CPDR")


# Detail in last figure
hist(ybarvec,prob=T,xlim=c(10,11.5),ylim=c(0,1), breaks=seq(3,18,0.2),
        xlab="ybar",main="(e) Detail in subplot (d)")
abline(v=c(ybarhat, ybarci, ybarcpdr),lty=2,lwd=3) # Histogram point estimates
points(mean(meanvalvec),0,pch=16,cex=1.5)
        # Rao-Blackwell estimate of predictive mean
abline(v=c(res$a,res$cpdr), lty=1, lwd=3) # # Exact point estimates
points(ybar, 0, pch=4,cex=2,lwd=3  )   # True value of ybar
lines(density(ybarvec),lty=2,lwd=3)  # Histogram estimate of predictive pdf
lines(ybarv, fybarhatv,lty=3,lwd=3)   # Rao-Blackwell estimate of pdf
lines(ybarv, dt((ybarv-res$a)/res$b,c)/res$b,lty=1,lwd=3) # Exact predictive pdf
legend(11.1,1,c("Histogram","Rao-Blackwell",
        "Exact pdf"),lty=c(2,3,1),lwd=c(3,3,3))
legend(11.1,0.6,c("Rao-Blackwell","True value"),
        pch=c(16,4),pt.cex=c(1.5,2), pt.lwd=c(1,3))


# Exact values of the quantities of interest and summary estimates ------------
c(m,mhat2,mhat,mci,mcpdr)
        # 11.000 10.895 10.906 10.875 10.937  9.893 11.863
c(ybar,res$a,ybarhat,ybarci,ybarcpdr)
        # 10.473 10.174 10.185 10.158 10.211  9.353 11.049
c(psi,psihat,psici,psicpdr) # 1.435 1.659 1.650 1.668 1.444 2.014


#  Preparation of data for input to WinBUGS -------------------------------------
paste(as.character(round(ys,2)), collapse=",")
# 14.98,10.99,9.58,6.56,13.83,......., 10.66,10.41"
paste(as.character(round(c(xs,xr),2)), collapse=",")
# 19.34,18.2,14.27,10.91,13.45,.....,12.57,10.36,19.49
```

WinBUGS Code for Exercise A.3

```
model
{
for(i in 1:100){
  mu[i] <- a + b*x[i]
  y[i] ~ dnorm(mu[i],lam)
  }
a ~ dnorm(0.0,0.0001)
b ~ dnorm(0.0,0.0001)
lam ~ dgamma(0.0001,0.0001)
m <- a+16*b
ybar <- mean(y[])
max <- ranked(y[],100)
medL <- ranked(y[],50)
medU <- ranked(y[],51)
med <- (medL + medU)/2
psi <- max/med
}

# data
list(y=c(
        14.98,10.99,9.58,6.56,13.83,    11.38,9.13,13.25,7.03,11.14,
        2.74,11.97,12.15,9.39,11.71,    10.25,7.98,8.54,10.66,10.41,
NA,NA,NA,NA,NA,  NA,NA,NA,NA,NA,  NA,NA,NA,NA,NA,  NA,NA,NA,NA,NA,
NA,NA,NA,NA,NA,  NA,NA,NA,NA,NA,  NA,NA,NA,NA,NA,  NA,NA,NA,NA,NA,
NA,NA,NA,NA,NA,  NA,NA,NA,NA,NA,  NA,NA,NA,NA,NA,  NA,NA,NA,NA,NA,
NA,NA,NA,NA,NA,  NA,NA,NA,NA,NA,  NA,NA,NA,NA,NA,  NA,NA,NA,NA,NA),

x=c(19.34,18.2,14.27,10.91,13.45,13.3,11.31,16.62,13.07,17.45,10.55,
17.66,17.34,17.46,16.14,17.19,10.96,14.19,16.08,14.83,17.92,16.61,
14.52,16.7,12.28,14.61,14.51,11.5,15.17,16.72,11.27,15.21,16.34,
10.36,12.62,19.27,19.7,12.26,10.07,18.74,11.86,12.35,16.79,13.18,
14.05,17.52,18.17,18.7,18.1,10.17,10.26,12.95,12.64,12.35,18.39,
12.08,17.48,13.47,14.47,16.76,17.64,14.32,19.07,17.29,15.87,14.2,
18.49,14.69,13.57,14.74,12.41,19.99,18.39,16.43,15.6,15.74,18.33,
16.98,16.72,19.3,13.92,11.4,11.55,13.83,12.36,13.3,15.3,19.26,18.15,
17.75,10.72,13.78,13.2,14.98,13.53,10.19,16.46,12.57,10.36,19.49))

# inits
list(a=0,b=0,lam=1)
```

Exercise A.4 Case study in Bayesian finite population models with biased sampling

A finite population of size $N = 4$ consists of values $y_1, ..., y_4$ that are iid Bernoulli with parameter θ.

A priori, θ is equally likely to be 1/4 or 3/4 (with no other possibilities).

We are interested in two quantities:

the *superpopulation mean* $\qquad \theta = E(y_i \mid \theta)$

the *finite population mean* $\qquad \bar{y} = \dfrac{y_1 + ... + y_N}{N}$.

We sample $n = 2$ units from the finite population without replacement in such a way that

every sample is equally likely to be selected, apart from one exception, as follows:

if the value of unit 1 is 1 then each sample with unit 1 is twice as likely to be selected as each sample without unit 1.

We observe the values of the two sampled units (each being 0 or 1) as well as the labels identifying them (each being 1, 2, 3 or 4).

(a) Write down a suitable Bayesian model for the above scenario in terms of the densities of the parameter θ, the finite population vector, $y = (y_1, ..., y_N)$, and the sample, $s = (s_1, ..., s_n)$.

Your formulae may involve only these variables, as well as n, N, and the vector of inclusion counters, $I = (I_1, ..., I_N)$, where $I_i = 1$ if the ith unit is in the sample, and $I_i = 0$ otherwise. (Note that there is a one-to-one correspondence between s and I in this exercise.)

(b) Identify a condition which determines whether the sampling mechanism is ignorable or nonignorable. Then write down an expression for the density of s in each of these two cases.

(c) Derive the posterior density and mean of θ generally.

(d) Find the *model bias* of the *posterior mean* of θ if:

    (i) $\theta = 1/4$ and $s = (1,3)$

    (ii) $\theta = 1/4$ and $s = (2,3)$.

(e) Find the *design bias* of the *posterior mean* of θ if:

    (i) $\theta = 1/4$ and $y = (0,0,1,1)$

    (ii) $\theta = 1/4$ and $y = (1,0,1,1)$.

(f) Derive the predictive mean of $\bar{y}$ generally.

(g) Find the *model bias* of the *predictive mean* of $\bar{y}$ if:

    (i) $\theta = 1/4$ and $s = (1,3)$

    (ii) $\theta = 1/4$ and $s = (2,3)$.

(h) Find the *design bias* of the *predictive mean* of $\bar{y}$ if:

    (i) $\theta = 1/4$ and $y = (0,0,1,1)$

    (ii) $\theta = 1/4$ and $y = (1,0,1,1)$.

(i) Design and run a Gibbs sampler to check the posterior mean of θ in (c) and the predictive mean of $\bar{y}$ in (f).

(j) Use Monte Carlo methods to check the two design biases in (h).

(k) Find the mean of the predictive mean of the finite population mean. Then apply Monte Carlo methods to check your answer.

Solution to Exercise A.4

(a) Part of the Bayesian model is:

$$f(y \mid \theta) = \prod_{i=1}^{N} \theta^{y_i} (1-\theta)^{1-y_i}$$

$$f(\theta) = 1/2, \ \theta = 1/4, 3/4 .$$

As regards the sampling mechanism, if $y_1 = 0$ then

$$f(s \mid y, \theta) = f(s) = \binom{N}{n}^{-1}, \quad s = (1,2),(1,3),(1,4),(2,3),(2,4),(3,4).$$

Also, if $y_1 = 1$ then

$$f(s \mid y, \theta) = f(s \mid y_1) = \begin{cases} c, & i \notin s \\ 2c, & i \in s \end{cases}$$

$$= \begin{cases} c, & s = (1,2),(1,3),(1,4) \\ 2c, & s = (2,3),(2,4),(3,4) \end{cases}.$$

To find the value of c, we may equate

$$1 = \sum_s f(s \mid y_1) = c \times 3 + (2c) \times 3 = 9c .$$

We thereby obtain $c = 1/9$.

Note 1: Alternatively, we may observe that

$$f(s \mid y_1) = c(1 + I_i),$$

where

$$I_i = I(i \in s).$$

Hence

$$1 = \sum_s f(s \mid y_1) = c \sum_s (1 + I_i) = c \left\{ \sum_s 1 + \sum_s I(i \in s) \right\}$$

$$= c \left\{ \sum_s 1 + \sum_{s:i \in s} 1 \right\} = c \left\{ \binom{N}{n} + \binom{N-1}{n-1} \right\} = c \left\{ \binom{4}{2} + \binom{3}{1} \right\}$$

$$= c(6 + 3) = 9c$$

$$\Rightarrow c = 1/9 .$$

Note 2: There are a total of $\binom{N-1}{n-1}$ samples s which contain any given particular unit i. So if $y_1 = 1$ then

$$f(s \mid y, \theta) = f(s \mid y_1) = \begin{cases} 1/9, & s = (1,2),(1,3),(1,4) \\ 2/9, & s = (2,3),(2,4),(3,4) \end{cases}.$$

Putting together the two cases above ($y_1 = 0$ and 1), we see that the sampling mechanism is given generally by

$$f(s \mid y, \theta) = f(s \mid y_1)$$

$$= \frac{1 + I_1 y_1}{\binom{N}{n} + \binom{N-1}{n-1} y_1}$$

$$= \frac{1 + I_1 y_1}{6 + 3 y_1}, \quad s = (1,2),(1,3),(1,4),(2,3),(2,4),(3,4),$$

where of course $I_1 = I(s \in \{(1,2),(1,3),(1,4)\})$.

As a check, it is useful to list all of the values produced by this formula. These values are as shown in Table A.2. Observe that the sum of $f(s \mid y_1)$ over all values of s is equal to 1, both when $y_1 = 0$ and when $y_1 = 1$.

From Table A.2 we may also confirm that, as specified in the problem:

every sample is equally likely to be selected, apart from one exception, as follows: if the value of unit 1 is 1 then each sample with unit 1 is twice as likely to be selected as each sample without unit 1.

Table A.2 All possible values of s and their probabilities

Sample, s:	(1,2)	(1,3)	(1,4)	(2,3)	(2,4)	(3,4)
$I_1 = I(1 \in s)$:	1	1	1	0	0	0
$f(s \mid y_1 = 0)$:	1/6	1/6	1/6	1/6	1/6	1/6
$f(s \mid y_1 = 1)$:	2/9	2/9	2/9	1/9	1/9	1/9

(b) If unit 1 is selected ($1 \in s$, $I_i = 1$) then $y_1 = 0$ or 1 is *known* and so the sampling mechanism is *ignorable*. In that case,

$$f(s \mid y, \theta) = \frac{1 + y_1}{6 + 3y_1} = \begin{cases} 1/6, y_1 = 0 \\ 2/9, y_1 = 1 \end{cases} = \begin{cases} 3/18, y_1 = 0 \\ 4/18, y_1 = 1 \end{cases} = \frac{3 + y_1}{18},$$

$$s = (1,2), (1,3), (1,4).$$

Conversely, if unit 1 is not selected ($1 \notin s$, $I_i = 0$) then $y_1 = 0$ or 1 is *unknown* and so the sampling mechanism is *nonignorable*.

In that case:

$$f(s \mid y, \theta) = \frac{1}{6 + 3y_1} = \begin{cases} 1/6, y_1 = 0 \\ 1/9, y_1 = 1 \end{cases} = \begin{cases} 3/18, y_1 = 0 \\ 2/18, y_1 = 1 \end{cases} = \frac{3 - y_1}{18},$$

$$s = (2,3), (2,4), (3,4).$$

(c) The posterior distribution of θ given data $D = (s, y_s)$ can now be derived by considering the two cases in the note above.

First, if unit 1 happens to be sampled then the value of the sampling density $f(s \mid y, \theta)$ is known, and so the sampling mechanism is ignorable.

Explicitly, we find in that case,

$$f(\theta \mid D) = f(\theta \mid s, y_s) \propto f(\theta, s, y_s) = \sum_{y_r} f(\theta, s, y_s, y_r)$$

$$= \sum_{y_r} f(\theta) f(y_s, y_r \mid \theta) f(s \mid y_s, y_r, \theta)$$

$$= \sum_{y_r} f(\theta) f(y_s \mid \theta) f(y_r \mid \theta) f(s \mid y_1)$$

$$= f(\theta) f(y_s \mid \theta) f(s \mid y_1) \sum_{y_r} f(y_r \mid \theta)$$

since $f(s \mid y, \theta) = f(s \mid y_1)$, where s is fixed at its observed value, $s = (s_1, s_2) = (1,2)$, $(1,3)$ or $(1,4)$

$$\overset{\theta}{\propto} f(\theta) f(y_s \mid \theta) \times 1 \times 1$$

since $f(s \mid y_1)$ does not depend on θ.

Note: This is the point at which $f(s \mid y, \theta)$ can be 'ignored'.

Thus we have that

$$f(\theta \mid D) \propto 1 \times \prod_{i \in s} \theta^{y_i} (1-\theta)^{1-y_i}$$

$$= \theta^{y_{sT}} (1-\theta)^{2-y_{sT}}$$

$$= \begin{cases} \left(\dfrac{1}{4}\right)^{y_{sT}} \left(\dfrac{3}{4}\right)^{y_{sT} \, 2-y_{sT}}, & \theta = 1/4 \\[4mm] \left(\dfrac{3}{4}\right)^{y_{sT}} \left(\dfrac{1}{4}\right)^{y_{sT} \, 2-y_{sT}}, & \theta = 3/4 \end{cases}$$

$$\propto \begin{cases} 3^{2-y_{sT}}, & \theta = 1/4 \\ 3^{y_{sT}}, & \theta = 3/4 \end{cases}$$

$$\propto \begin{cases} 3^2, & \theta = 1/4 \\ 3^{y_{sT}+y_{sT}}, & \theta = 3/4 \end{cases}$$

$$= \begin{cases} 9, & \theta = 1/4 \\ 9^{y_{sT}}, & \theta = 3/4 \end{cases}.$$

That is (if $1 \in s$),

$$f(\theta \mid D) = \begin{cases} \dfrac{9}{9+9^{y_{sT}}}, & \theta = 1/4 \\[4mm] \dfrac{9^{y_{sT}}}{9+9^{y_{sT}}}, & \theta = 3/4 \end{cases} = \begin{cases} \begin{cases} 9/10, & \theta = 1/4 \\ 1/10, & \theta = 3/4 \end{cases}, & y_{sT} = 0 \\[4mm] \begin{cases} 1/2, & \theta = 1/4 \\ 1/2, & \theta = 3/4 \end{cases}, & y_{sT} = 1 \\[4mm] \begin{cases} 1/10, & \theta = 1/4 \\ 9/10, & \theta = 3/4 \end{cases}, & y_{sT} = 2. \end{cases}$$

So then also (if $1 \in s$) the posterior mean of θ is

$$\hat{\theta} = E(\theta \mid D) = \begin{cases} \dfrac{1}{4}\left(\dfrac{9}{10}\right) + \dfrac{3}{4}\left(\dfrac{1}{10}\right) = \dfrac{3}{10}, & y_{sT} = 0 \\[4mm] \dfrac{1}{4}\left(\dfrac{1}{2}\right) + \dfrac{3}{4}\left(\dfrac{1}{2}\right) = \dfrac{1}{2}, & y_{sT} = 1 \\[4mm] \dfrac{1}{4}\left(\dfrac{1}{10}\right) + \dfrac{3}{4}\left(\dfrac{9}{10}\right) = \dfrac{7}{10}, & y_{sT} = 2. \end{cases}$$

Note: This could also be written as $\hat{\theta} = \dfrac{3+2y_{sT}}{10}$ (if $1 \in s$).

Next, suppose that unit 1 is *not* sampled. Then the value of unit 1 is unknown and so the sampling mechanism is nonignorable.

In that case, we see from (b) that

$$f(s \mid y, \theta) = f(s \mid y_1) = \frac{3 - y_1}{18} \propto 3 - y_1, \quad s = (2,3),(2,4),(3,4),$$

where y_1 is an unknown value in the nonsample vector $y_r = (y_1, y_k)$ where $k = 2$, 3 or 4.

Working through as before,

$$f(\theta \mid D) = f(\theta \mid s, y_s)$$
$$\propto f(\theta, s, y_s)$$
$$= \sum_{y_r} f(\theta, s, y_s, y_r)$$
$$= \sum_{y_r} f(\theta) f(y_s, y_r \mid \theta) f(s \mid y_s, y_r, \theta)$$
$$= \sum_{y_r} f(\theta) f(y_s \mid \theta) f(y_r \mid \theta) f(s \mid y_1)$$
$$= f(\theta) f(y_s \mid \theta) \sum_{y_r} f(s \mid y_1) f(y_r \mid \theta)$$
$$= f(\theta) f(y_s \mid \theta) q(\theta),$$

where

$$q(\theta) \propto \sum_{y_r} (3 - \theta) f(y_r \mid \theta)$$
$$= E_{y_r}(3 - \theta \mid \theta)$$
$$= 3 - \theta.$$

Note: We could also have written

$$q(\theta) \propto \sum_{y_1=0}^{1} \sum_{y_k=0}^{1} (3 - y_1)\left[\theta^{y_1}(1-\theta)^{1-y_1}\right]\left[\theta^{y_k}(1-\theta)^{1-y_k}\right]$$

$$= \left(\sum_{y_k=0}^{1} \theta^{y_k}(1-\theta)^{1-y_k}\right)\sum_{y_1=0}^{1}(3 - y_1)\theta^{y_1}(1-\theta)^{1-y_1}$$

$$= 1 \times \left\{(3-0)\theta^0(1-\theta)^{1-0} + (3-1)\theta^1(1-\theta)^{1-1}\right\}$$

$$= 3(1-\theta) + 2\theta$$

$$= 3 - \theta.$$

Having shown (in the case $1 \notin s$) that
$$f(\theta \mid D) \propto f(\theta) f(y_s \mid \theta)(3 - \theta),$$
it now follows that

$$f(\theta \mid D) \propto \begin{cases} \left(\dfrac{1}{4}\right)^{y_{sT}} \left(\dfrac{3}{4}\right)^{y_{sT}\,2-y_{sT}} \left(3 - \dfrac{1}{4}\right), & \theta = 1/4 \\[2mm] \left(\dfrac{3}{4}\right)^{y_{sT}} \left(\dfrac{1}{4}\right)^{y_{sT}\,2-y_{sT}} \left(3 - \dfrac{3}{4}\right), & \theta = 3/4 \end{cases}$$

$$\propto \begin{cases} 3^{2-y_{sT}} \times 11, & \theta = 1/4 \\ 3^{y_{sT}} \times 9, & \theta = 3/4 \end{cases}$$

$$\propto \begin{cases} 3^2 \times 11, & \theta = 1/4 \\ 3^{y_{sT}+y_{sT}} \times 9, & \theta = 3/4 \end{cases}$$

$$= \begin{cases} 11, & \theta = 1/4 \\ 9^{y_{sT}}, & \theta = 3/4 \end{cases}.$$

Thus (if $1 \notin s$), we have that

$$f(\theta \mid D) = \begin{cases} \dfrac{11}{11 + 9^{y_{sT}}}, & \theta = 1/4 \\[2mm] \dfrac{9^{y_{sT}}}{11 + 9^{y_{sT}}}, & \theta = 3/4 \end{cases} = \begin{cases} \begin{cases} 11/12, & \theta = 1/4 \\ 1/12, & \theta = 3/4 \end{cases}, & y_{sT} = 0 \\[3mm] \begin{cases} 11/20, & \theta = 1/4 \\ 9/20, & \theta = 3/4 \end{cases}, & y_{sT} = 1 \\[3mm] \begin{cases} 11/92, & \theta = 1/4 \\ 81/92, & \theta = 3/4 \end{cases}, & y_{sT} = 2. \end{cases}$$

So then also (if $1 \notin s$) the posterior mean of θ is

$$\hat{\theta} = E(\theta \mid D)$$

$$= \begin{cases} \dfrac{1}{4}\left(\dfrac{11}{12}\right) + \dfrac{3}{4}\left(\dfrac{1}{12}\right) = \dfrac{14}{48} = \dfrac{7}{24} = \dfrac{805}{2760} = 0.2917, & y_{sT} = 0 \\[3mm] \dfrac{1}{4}\left(\dfrac{11}{20}\right) + \dfrac{3}{4}\left(\dfrac{9}{20}\right) = \dfrac{38}{80} = \dfrac{19}{40} = \dfrac{1311}{2760} = 0.4750, & y_{sT} = 1 \\[3mm] \dfrac{1}{4}\left(\dfrac{11}{92}\right) + \dfrac{3}{4}\left(\dfrac{81}{92}\right) = \dfrac{254}{368} = \dfrac{127}{184} = \dfrac{1905}{2760} = 0.6902, & y_{sT} = 2. \end{cases}$$

645

Note: This mean may also be written as

$$\hat{\theta} = \frac{805 + 462 y_{sT} + 44 y_{sT}^2}{2760} \quad \text{(if } 1 \notin s \text{)}.$$

This alternative formula was obtained by solving the equation

$$a + bx + cx^2 = \begin{cases} 805, & x = 0 \\ 1311, & x = 1 \\ 1905, & x = 2 \end{cases}$$

for a, b and c.

Putting the two cases together we find that the posterior mean of θ is given generally by:

$$\hat{\theta} = E(\theta \mid D) = \hat{\theta}(D) = \hat{\theta}(s, y_s)$$

$$= \begin{cases} 3/10 = 0.3000 & \text{if } 1 \in s \text{ and } y_{sT} = 0 \\ 1/2 = 0.5000 & \text{if } 1 \in s \text{ and } y_{sT} = 1 \\ 7/10 = 0.7000 & \text{if } 1 \in s \text{ and } y_{sT} = 2 \\ 7/24 = 0.2917 & \text{if } 1 \notin s \text{ and } y_{sT} = 0 \\ 19/40 = 0.4750 & \text{if } 1 \notin s \text{ and } y_{sT} = 1 \\ 127/184 = 0.6902 & \text{if } 1 \notin s \text{ and } y_{sT} = 2, \end{cases}$$

or equivalently, by

$$\hat{\theta} = \left(\frac{3 + 2 y_{sT}}{10} \right) I_1 + \left(\frac{805 + 462 y_{sT} + 44 y_{sT}^2}{2760} \right) (1 - I_1).$$

Note: Here:

$$1 \in s \quad \Leftrightarrow \quad I_1 = 1 \quad \Leftrightarrow \quad s = (1,2), (1,3) \text{ or } (1,4)$$
$$1 \notin s \quad \Leftrightarrow \quad I_1 = 0 \quad \Leftrightarrow \quad s = (2,3), (2,4) \text{ or } (3,4).$$

Also:

$y_{sT} = 0$ iff both sampled values are 0

$y_{sT} = 1$ iff one sampled value is 0 and the other is 1

$y_{sT} = 2$ iff both sampled values are 1.

(d)(i) If $\theta = 1/4$ and $s = (1,3)$ then $1 \in s$ and $I_1 = 1$, and so

$$\hat{\theta} = \left(\frac{3 + 2y_{sT}}{10} \right) 1 + \left(\frac{805 + 462 y_{sT} + 44 y_{sT}^2}{2760} \right)(1-1) = \frac{3 + 2y_{sT}}{10}.$$

So the model mean of $\hat{\theta}$ is

$$E(\hat{\theta} \mid \theta, s) = \frac{1}{10} \{ 3 + 2E(y_{sT} \mid \theta, s) \}.$$

Now,

$$f(y \mid \theta, s) = \frac{f(y, s \mid \theta)}{f(s \mid \theta)},$$

where:

$$f(y, s \mid \theta) = f(s \mid y, \theta) f(y \mid \theta) = \frac{3 + y_1}{18} \prod_{i=1}^{4} \theta^{y_i} (1-\theta)^{1-y_i}$$

$$\text{(using the result in (b) that } f(s \mid y, \theta) = \frac{3 + y_1}{18} \text{ if } 1 \in s \text{)}$$

$$f(s \mid \theta) = \sum_y f(y, s \mid \theta) = \sum_y f(s \mid y, \theta) f(y \mid \theta) = E_y \{ f(s \mid y, \theta) \mid \theta \}$$

$$= E_y \left(\frac{3 + y_1}{18} \, \middle| \, \theta \right) = \frac{3 + \theta}{18}.$$

Therefore

$$f(y \mid \theta, s) = \frac{\left(\dfrac{3 + y_1}{18} \displaystyle\prod_{i=1}^{4} \theta^{y_i} (1-\theta)^{1-y_i} \right)}{\left(\dfrac{3 + \theta}{18} \right)}$$

$$= \left(\frac{3 + y_1}{3 + \theta} \theta^{y_1} (1-\theta)^{1-y_1} \right) \prod_{i=2}^{4} \theta^{y_i} (1-\theta)^{1-y_i}.$$

We see that

$$(y_i \mid \theta, s) \sim \perp Bernoulli(\pi_i), \quad i = 1, 2, 3, 4,$$

where:

$$\pi_2 = \pi_3 = \pi_4 = \theta$$

$$\pi_1 = \frac{3+1}{3+\theta} \theta^1 (1-\theta)^{1-1} = \frac{4\theta}{3+\theta}.$$

Check: $\dfrac{3+0}{3+\theta}\theta^{0}(1-\theta)^{1-0} = \dfrac{3(1-\theta)}{3+\theta} = 1 - \dfrac{4\theta}{3+\theta} = 1 - \pi_1.$

It follows that

$$E(y_{sT} \mid \theta, s) = E(y_1 \mid \theta, s) + E(y_3 \mid \theta, s)$$

$$= \pi_1 + \pi_3 = \theta + \dfrac{4\theta}{3+\theta} = \dfrac{\theta(7+\theta)}{3+\theta} = \dfrac{\dfrac{1}{4}\left(7+\dfrac{1}{4}\right)}{\left(3+\dfrac{1}{4}\right)} = \dfrac{29/16}{13/4} = \dfrac{29}{52}.$$

Hence

$$E(\hat{\theta} \mid \theta, s) = \dfrac{1}{10}\left\{3 + 2\left(\dfrac{29}{52}\right)\right\} = \dfrac{107}{260} = 0.4115.$$

So, if $\theta = 1/4$ and $s = (1,3)$, then the model bias of $\hat{\theta}$ is

$$E(\hat{\theta} - \theta \mid \theta, s) = \dfrac{107}{260} - \theta = \dfrac{107}{260} - \dfrac{1}{4} = \dfrac{21}{130} = 0.1615.$$

Note: We can also report the relative model bias of $\hat{\theta}$ as

$$E\left(\left.\dfrac{\hat{\theta} - \theta}{\theta}\right| \theta, s\right) = \dfrac{21/130}{1/4} = \dfrac{42}{65} = +64.6\%.$$

(d)(ii) If $\theta = 1/4$ and $s = (2,3)$ then $1 \in r$ and $I_1 = 0$, and so

$$\hat{\theta} = \left(\dfrac{3+2y_{sT}}{10}\right)0 + \left(\dfrac{805 + 462 y_{sT} + 44 y_{sT}^2}{2760}\right)(1-0)$$

$$= \dfrac{805 + 462 y_{sT} + 44 y_{sT}^2}{2760}.$$

So the model mean of $\hat{\theta}$ is

$$E(\hat{\theta} \mid \theta, s) = \dfrac{805 + 462 E(y_{sT} \mid \theta, s) + 44 E(y_{sT}^2 \mid \theta, s)}{2760}.$$

In this case,

$$f(y \mid \theta, s) = \frac{f(y, s \mid \theta)}{f(s \mid \theta)},$$

as before, but with

$$f(y, s \mid \theta) = f(s \mid y, \theta) f(y \mid \theta) = \frac{3 - y_1}{18} \prod_{i=1}^{4} \theta^{y_i} (1 - \theta)^{1 - y_i}$$

(using the result in (b) that $f(s \mid y, \theta) = \dfrac{3 - y_1}{18}$ if $1 \notin s$).

Thus,

$$f(s \mid \theta) = \sum_{y} f(y, s \mid \theta) = \sum_{y} f(s \mid y, \theta) f(y \mid \theta)$$

$$= E_y \left\{ f(s \mid y, \theta) \mid \theta \right\} = E_y \left(\frac{3 - y_1}{18} \middle| \theta \right)$$

$$= \frac{3 - \theta}{18}.$$

So

$$f(y \mid \theta, s) = \frac{\left(\dfrac{3 - y_1}{18} \displaystyle\prod_{i=1}^{4} \theta^{y_i} (1 - \theta)^{1 - y_i} \right)}{\left(\dfrac{3 - \theta}{18} \right)}$$

$$= \left(\frac{3 - y_1}{3 - \theta} \theta^{y_1} (1 - \theta)^{1 - y_1} \right) \prod_{i=2}^{4} \theta^{y_i} (1 - \theta)^{1 - y_i}.$$

We see that

$$(y_i \mid \theta, s) \sim \perp Bernoulli(\pi_i), \quad i = 1, 2, 3, 4,$$

where:

$$\pi_2 = \pi_3 = \pi_4 = \theta$$

$$\pi_1 = \frac{3 - 1}{3 - \theta} \theta^1 (1 - \theta)^{1 - 1} = \frac{2\theta}{3 - \theta}.$$

Check: $\dfrac{3 - 0}{3 - \theta} \theta^0 (1 - \theta)^{1 - 0} = \dfrac{3(1 - \theta)}{3 - \theta} = 1 - \dfrac{2\theta}{3 - \theta} = 1 - \pi_1.$

It follows that
$$E(y_{sT} \mid \theta, s) = E(y_2 \mid \theta, s) + E(y_3 \mid \theta, s)$$
$$= \pi_2' + \pi_3' = \theta + \theta = \frac{1}{4} + \frac{1}{4} = \frac{1}{2}.$$

Equivalently,
$$(y_{sT} \mid \theta, s) \sim Bin(2, \theta),$$
and so
$$E(y_{sT} \mid \theta, s) = 2\theta.$$

By the same token,
$$E(y_{sT}^2 \mid \theta, s) = V(y_{sT} \mid s, \theta) + \{E(y_{sT} \mid s, \theta)\}^2$$
$$= 2\theta(1-\theta) + (2\theta)^2 = 2\theta(1+\theta) = 2 \times \frac{1}{4}\left(1 + \frac{1}{4}\right) = \frac{5}{8}.$$

Hence
$$E(\hat{\theta} \mid \theta, s) = \frac{805 + 462\left(\dfrac{1}{2}\right) + 44\left(\dfrac{5}{8}\right)}{2760} = \frac{2127}{5520} = 0.3853.$$

So, if $\theta = 1/4$ and $s = (2,3)$, then the model bias of $\hat{\theta}$ is
$$E(\hat{\theta} - \theta \mid \theta, s) = E(\hat{\theta} \mid \theta, s) - \theta = \frac{2127}{5520} - \frac{1}{4} = \frac{747}{5520} = 0.1353.$$

Note: As regards the model bias of $\hat{\theta}$, there are a total of 4 cases, corresponding to whether $1 \in s$ or $1 \notin s$, and to whether $\theta = 1/4$ or $\theta = 3/4$. We have covered two of these four cases.

(e)(i) If $\theta = 1/4$ and $y = (0,0,1,1)$ then $y_1 = 0$. So in that particular case the sampling mechanism is definitely SRSWOR and ignorable. Without further thought, the posterior density of θ can be obtained as follows:
$$f(\theta \mid D) = f(\theta \mid s, y_s) = f(\theta \mid y_s)$$

$$\propto f(\theta) f(y_s \mid \theta)$$

$$\propto 1 \times \prod_{i \in s} \theta^{y_i} (1-\theta)^{1-y_i}.$$

Recalling (c), note that

$$f(\theta \mid D) = \begin{cases} \begin{cases} 9/10, & \theta = 1/4 \\ 1/10, & \theta = 3/4 \end{cases}, \ y_{sT} = 0 \\[1em] \begin{cases} 1/2, & \theta = 1/4 \\ 1/2, & \theta = 3/4 \end{cases}, \ y_{sT} = 1 \\[1em] \begin{cases} 1/10, & \theta = 1/4 \\ 9/10, & \theta = 3/4 \end{cases}, \ y_{sT} = 2, \end{cases}$$

and

$$\hat{\theta} = E(\theta \mid D) = \begin{cases} 3/10, & y_{sT} = 0 \\ 1/2, & y_{sT} = 1 \\ 7/10, & y_{sT} = 2 \end{cases} = \frac{3 + 2y_{sT}}{10}.$$

The design mean of $\hat{\theta}$ is therefore

$$E(\hat{\theta} \mid \theta, y) = \frac{3 + 2E(y_{sT} \mid \theta, y)}{10},$$

where

$$E(y_{sT} \mid \theta, y) = nE(\bar{y}_s \mid \theta, y) = n\sum_s \bar{y}_s f(s \mid \theta, y) = n\bar{y},$$

since (making use of basic results in the classical theory)

$$f(s \mid \theta, y) = f(s) = \binom{N}{n}^{-1} = 2 \times \frac{0 + 0 + 1 + 1}{4} = 1.$$

Therefore the design mean of $\hat{\theta}$ is

$$E(\hat{\theta} \mid \theta, y) = \frac{3 + 2 \times 1}{10} = \frac{1}{2}.$$

So the design bias of $\hat{\theta}$ is

$$E(\hat{\theta} - \theta \mid \theta, y) = \frac{1}{2} - \theta = \frac{1}{2} - \frac{1}{4} = 0.25.$$

Note: In the above, $E(\hat{\theta} \mid \theta, y)$ does not depend on θ. So, for the case $\theta = 3/4$ and $y = (0,0,1,1)$, the design bias of $\hat{\theta}$ is $\frac{1}{2} - \frac{3}{4} = -0.25$.

(e)(ii) If $\theta = 1/4$ and $y = (1,0,1,1)$ then $y_1 = 1$, and so the sampling mechanism is potentially nonignorable (depending on which sample s happens to be drawn).

Recall from (c) that the posterior mean of θ is a function of the data given generally by

$$\hat{\theta} = \hat{\theta}(s, y_s) = \begin{cases} 3/10 = 0.3000 & \text{if } 1 \in s \text{ and } y_{sT} = 0 \\ 1/2 = 0.5000 & \text{if } 1 \in s \text{ and } y_{sT} = 1 \\ 7/10 = 0.7000 & \text{if } 1 \in s \text{ and } y_{sT} = 2 \\ 7/24 = 0.2917 & \text{if } 1 \notin s \text{ and } y_{sT} = 0 \\ 19/40 = 0.4750 & \text{if } 1 \notin s \text{ and } y_{sT} = 1 \\ 127/184 = 0.6902 & \text{if } 1 \notin s \text{ and } y_{sT} = 2. \end{cases}$$

Also recall from (b) that

$$f(s \mid y, \theta) = \begin{cases} \dfrac{3+y_1}{18}, & s = (1,2),(1,3),(1,4) \\ \dfrac{3-y_1}{18}, & s = (2,3),(2,4),(3,4) \end{cases}.$$

The design bias of $\hat{\theta}$ can now be worked out according to

$$E(\hat{\theta} \mid \theta, y) = \sum_s \hat{\theta}(s, y_s) f(s \mid \theta, y).$$

Now, suppose that we draw the sample $s = (1,2)$.

Then $y_s = (y_1, y_2) = (1,0)$.

Thus $1 \in s$ and $y_{sT} = 1$, and so by the above,

$$\hat{\theta}(s, y_s) f(s \mid \theta, y) = \frac{1}{2} \times \frac{3+1}{18} = \frac{1}{9}.$$

Likewise:

If $s = (1,3)$ then $y_s = (y_1, y_3) = (1,1)$ and so

$$\hat{\theta}(s, y_s) f(s \mid \theta, y) = \frac{7}{10} \times \frac{3+1}{18} = \frac{7}{45}.$$

If $s = (1,4)$ then $y_s = (y_1, y_4) = (1,1)$ and so

$$\hat{\theta}(s, y_s) f(s \mid \theta, y) = \frac{7}{10} \times \frac{3+1}{18} = \frac{7}{45}.$$

If $s = (2,3)$ then $y_s = (y_2, y_3) = (0,1)$ and so

$$\hat{\theta}(s, y_s) f(s \mid \theta, y) = \frac{19}{40} \times \frac{3-1}{18} = \frac{19}{360}.$$

If $s = (2,4)$ then $y_s = (y_2, y_4) = (0,1)$ and so

$$\hat{\theta}(s, y_s) f(s \mid \theta, y) = \frac{19}{40} \times \frac{3-1}{18} = \frac{19}{360}.$$

If $s = (3,4)$ then $y_s = (y_3, y_4) = (1,1)$ and so

$$\hat{\theta}(s, y_s) f(s \mid \theta, y) = \frac{127}{184} \times \frac{3-1}{18} = \frac{127}{1656}.$$

It follows that

$$E(\hat{\theta} \mid \theta, y) = \sum_s \hat{\theta}(s, y_s) f(s \mid \theta, y)$$
$$= (1/9) + (7/45) + (7/45) + (19/360) + (19/360) + (127/1656)$$
$$= 0.6045.$$

Thus, if $\theta = 1/4$ and $y = (1,0,1,1)$, then the design bias of $\hat{\theta}$ is

$$E(\hat{\theta} - \theta \mid \theta, y) = 0.6045 - \frac{1}{4} = 0.3545.$$

Note 1: Also, if $\theta = 3/4$ and $y = (1,0,1,1)$, then the design bias of $\hat{\theta}$ is

$$0.6045 - \frac{3}{4} = -0.1455.$$

Note 2: As regards the design bias of $\hat{\theta}$, there are a total of $2 \times 4 \times 2 = 16$ cases to be considered, corresponding to:

y_1	being either 0 or 1	(2 possibilities)
$y_T - y_1$	being 0 or 1 or 2 or 3	(4 possibilities)
θ	being either 1/4 or 3/4	(2 possibilities).

We have covered four of these 16 cases.

(f) Recall from (c) that

$$(y_i \mid \theta, s) \sim\perp Bernoulli(\pi_i), \quad i = 1, 2, 3, 4,$$

where:

$$\pi_2 = \pi_3 = \pi_4 = \theta$$

$$\pi_1 = \left\{ \begin{array}{ll} \dfrac{4\theta}{3+\theta}, & 1 \in s \\[2mm] \dfrac{2\theta}{3-\theta}, & 1 \notin s \end{array} \right\}.$$

Therefore

$$E(y_{rT} \mid \theta, s, y_s) = E(y_{rT} \mid \theta, s) = \left\{ \begin{array}{ll} \theta + \theta, & 1 \in s \\ \theta + \phi, & 1 \notin s \end{array} \right\},$$

where

$$\phi = \frac{2\theta}{3-\theta}.$$

So

$$E(y_{rT} \mid s, y_s) = E\{E(y_{rT} \mid \theta, s, y_s) \mid s, y_s\}$$

$$= \left\{ \begin{array}{ll} E(2\theta \mid D), & 1 \in s \\ E(\theta \mid D) + E(\phi \mid D), & 1 \notin s \end{array} \right\}$$

$$= \left\{ \begin{array}{ll} 2\hat{\theta}, & 1 \in s \\ \hat{\theta} + \hat{\phi}, & 1 \notin s \end{array} \right\},$$

where

$$\hat{\phi} = E(\phi \mid D) = E_\theta \left(\frac{2\theta}{3-\theta} \bigg| D \right)$$

$$= \sum_{\theta = 1/4, 3/4} \left(\frac{2\theta}{3-\theta} \right) f(\theta \mid D).$$

The finite population mean is

$$\bar{y} = \frac{1}{4}(y_{sT} + y_{rT}),$$

and so the predictive mean of $\bar{y}$ may be expressed as

$$\hat{\bar{y}} = E(\bar{y} \mid s, y_s) = \frac{1}{4}(y_{sT} + E(y_{rT} \mid s, y_s)).$$

Using suitable R functions, we find that $\hat{\phi}$ and $\hat{\bar{y}}$ are as follows:

If $1 \in s$ and $y_{sT} = 0$ then $\hat{\phi} = 0.2303030$ and $\hat{\bar{y}} = 0.1500000$

If $1 \in s$ and $y_{sT} = 1$ then $\hat{\phi} = 0.4242424$ and $\hat{\bar{y}} = 0.5000000$

If $1 \in s$ and $y_{sT} = 2$ then $\hat{\phi} = 0.6181818$ and $\hat{\bar{y}} = 0.8500000$

If $1 \notin s$ and $y_{sT} = 0$ then $\hat{\phi} = 0.2222222$ and $\hat{\bar{y}} = 0.1284722$

If $1 \notin s$ and $y_{sT} = 1$ then $\hat{\phi} = 0.4000000$ and $\hat{\bar{y}} = 0.4687500$

If $1 \notin s$ and $y_{sT} = 2$ then $\hat{\phi} = 0.6086957$ and $\hat{\bar{y}} = 0.8247283$.

Note: Working through the above equation using exact fractions, it can be shown that

$$\hat{\bar{y}} = \hat{\bar{y}}(s, y_s) = \begin{cases} 3/20, & 1 \in s, y_{sT} = 0 \\ 1/2, & 1 \in s, y_{sT} = 1 \\ 17/20, & 1 \in s, y_{sT} = 2 \\ 37/288, & 1 \notin s, y_{sT} = 0 \\ 15/32, & 1 \notin s, y_{sT} = 1 \\ 607/736, & 1 \notin s, y_{sT} = 2. \end{cases}$$

The following are details of the working for 37/288, 15/32 and 607/736.

Observe that

$$E(y_{rT} \mid \theta, s, y_s) = \theta + \frac{2\theta}{3-\theta} = \frac{\theta(5-\theta)}{3-\theta}.$$

Therefore

$$\hat{y}_{rT} = E\{E(y_{rT} \mid s, y_s, \theta) \mid s, y_s\} = E\left\{ \frac{\theta(5-\theta)}{3-\theta} \middle| s, y_s \right\}.$$

So, if $y_{sT} = 0$ then

$$\hat{y}_{rT} = E\left\{ \frac{\theta(5-\theta)}{3-\theta} \middle| D \right\} = \frac{\frac{1}{4}\left(5-\frac{1}{4}\right)}{3-\frac{1}{4}} \frac{11}{12} + \frac{\frac{3}{4}\left(5-\frac{3}{4}\right)}{3-\frac{3}{4}} \frac{1}{12}.$$

$$= \frac{\frac{1}{\cancel{A}}\binom{19}{4}\cancel{N}}{\frac{\cancel{N}}{\cancel{A}}}\frac{1}{12} + \frac{\frac{3}{\cancel{A}}\binom{17}{4}}{\frac{9}{\cancel{A}}}\frac{1}{12} = \frac{1}{48}\left\{19 + \frac{17}{3}\right\}$$

$$= \frac{1}{48}\left\{\frac{57+17}{3}\right\} = \frac{74}{48\times 3} = \frac{37}{72}.$$

Also, if $y_{sT} = 1$ then

$$\hat{y}_{rT} = E\left\{\frac{\theta(5-\theta)}{3-\theta}\Big| D\right\} = \frac{\frac{1}{4}\left(5-\frac{1}{4}\right)}{3-\frac{1}{4}}\frac{11}{20} + \frac{\frac{3}{4}\left(5-\frac{3}{4}\right)}{3-\frac{3}{4}}\frac{9}{20}$$

$$= \frac{\frac{1}{\cancel{A}}\binom{19}{4}\cancel{N}}{\frac{\cancel{N}}{\cancel{A}}}\frac{9}{20} + \frac{\frac{3}{\cancel{A}}\binom{17}{4}}{\frac{9}{\cancel{A}}}\frac{9}{20} = \frac{1}{80}\{19+51\} = \frac{7}{8}.$$

And if $y_{sT} = 2$ then

$$\hat{y}_{rT} = E\left\{\frac{\theta(5-\theta)}{3-\theta}\Big| D\right\} = \frac{\frac{1}{4}\left(5-\frac{1}{4}\right)}{3-\frac{1}{4}}\frac{11}{92} + \frac{\frac{3}{4}\left(5-\frac{3}{4}\right)}{3-\frac{3}{4}}\frac{81}{92}$$

$$= \frac{\frac{1}{\cancel{A}}\binom{19}{4}\cancel{N}}{\frac{\cancel{N}}{\cancel{A}}}\frac{81}{92} + \frac{\frac{3}{\cancel{A}}\binom{17}{4}}{\frac{9}{\cancel{A}}}\frac{81}{92}$$

$$= \frac{1}{368}\{19+27\times 17\} = \frac{478}{368} = \frac{239}{184}.$$

Thus (for $1 \notin s$) we have that

$$\hat{y}_{rT} = \begin{cases} 37/72, & y_{sT} = 0 \\ 7/8, & y_{sT} = 1 \\ 239/184, & y_{sT} = 2. \end{cases}$$

Hence

$$\hat{y}_T = y_{sT} + \hat{y}_{rT} = \hat{y}_{rT} = \begin{cases} 0 + 37/72 = 37/72, & y_{sT} = 0 \\ 1 + 7/8 = 15/8, & y_{sT} = 1 \\ 2 + 239/184 = 607/184, & y_{sT} = 2. \end{cases}$$

Thus, finally (for $1 \notin s$), we obtain

$$\hat{\bar{y}} = \frac{\hat{y}_T}{4} = \begin{cases} 37/288, & y_{sT} = 0 \\ 15/32, & y_{sT} = 1 \\ 607/736, & y_{sT} = 2. \end{cases}$$

A similar logic can be used to obtain the fractions 3/20, 1/2 and 17/20.

(g)(i) Suppose that $\theta = 1/4$ and $s = (1,3)$. Then $1 \in s$ and so

$$(y_1, \dots, y_4 \mid \theta, s) \sim\perp Bernoulli(\pi_i),$$

where:

$$\pi_1 = \frac{4\theta}{3+\theta} = \frac{4\left(\frac{1}{4}\right)}{3+\left(\frac{1}{4}\right)} = \frac{4}{13}$$

$$\pi_i = \theta = \frac{1}{4}, i > 1.$$

In this case,

$$y_{sT} = y_1 + y_3,$$

and so:

$$P(y_{sT} = 0 \mid \theta, s) = \frac{9}{13} \times \frac{3}{4} = \frac{27}{52}$$

$$P(y_{sT} = 2 \mid \theta, s) = \frac{4}{13} \times \frac{1}{4} = \frac{4}{52}$$

$$P(y_{sT} = 1 \mid \theta, s) = 1 - \frac{27}{52} - \frac{4}{52} = \frac{21}{52}.$$

So the model mean of $\hat{\bar{y}}$ is

$$
\begin{aligned}
E(\hat{\bar{y}} \mid \theta, s) &= E\{E(\hat{\bar{y}} \mid \theta, s, y_{sT}) \mid \theta, s\} \\
&= E\{\hat{\bar{y}}(s, y_{sT}) \mid \theta, s\} \\
&= \sum_{y_{sT}=0}^{2} \hat{\bar{y}}(s, y_{sT}) f(y_{sT} \mid \theta, s) \\
&= 0.15(27/52) + 0.5(21/52) + 0.85(4/52) \\
&= 0.3451923.
\end{aligned}
$$

Also, the model mean of $\bar{y}$ is

$$
E(\bar{y} \mid \theta, s) = \frac{1}{4}(\pi_1 + \ldots + \pi_4) = \frac{1}{4}\left(\frac{4}{13} + \frac{1}{4} + \frac{1}{4} + \frac{1}{4}\right)
$$
$$
= 55/208 = 0.2644231.
$$

So the model bias of $\hat{\bar{y}}$ is

$$
E(\hat{\bar{y}} - \bar{y} \mid \theta, s) = 0.3451923 - 0.2644231 = 0.08077.
$$

(g)(ii) Suppose that $\theta = 1/4$ and $s = (2,3)$. Then $1 \notin s$ and so
$$(y_1, \ldots, y_4 \mid \theta, s) \sim\perp Bernoulli(\pi_i),$$
where:

$$
\pi_1 = \frac{2\theta}{3-\theta} = \frac{2\left(\dfrac{1}{4}\right)}{3 - \left(\dfrac{1}{4}\right)} = \frac{2}{11}
$$

$$
\pi_i = \theta = \frac{1}{4}, i > 1.
$$

In this case,
$$y_{sT} = y_2 + y_3,$$
and so:

$$
P(y_{sT} = 0 \mid \theta, s) = \frac{3}{4} \times \frac{3}{4} = \frac{9}{16}
$$

$$
P(y_{sT} = 2 \mid \theta, s) = \frac{1}{4} \times \frac{1}{4} = \frac{1}{16}
$$

$$
P(y_{sT} = 1 \mid \theta, s) = 1 - \frac{9}{16} - \frac{1}{16} = \frac{6}{16}.
$$

So (using results in (g)(i)) the model mean of $\hat{\bar{y}}$ is

$0.1284722(9/16) + 0.46875(6/16) + 0.8247283(1/16) = 0.2995924.$

Also, the model mean of $\bar{y}$ is

$$E(\bar{y}\,|\,\theta,s) = \frac{1}{4}(\pi_1 + ... + \pi_4) = \frac{1}{4}\left(\frac{2}{11} + \frac{1}{4} + \frac{1}{4} + \frac{1}{4}\right)$$
$$= 41/176 = 0.2329545.$$

So the model bias of $\hat{\bar{y}}$ is

$$E(\hat{\bar{y}} - \bar{y}\,|\,\theta,s) = 0.2995924 - 0.2329545 = 0.06664.$$

(h)(i) Suppose that $\theta = 1/4$ and $y = (0,0,1,1)$. Then $y_1 = 0$ and so the sampling mechansim is definitely SRSWOR and ignorable.

Explicitly, we have that
$$f(s\,|\,\theta,y) = f(s) = 1/6.$$

So the design mean of $\hat{\bar{y}}$ is

$$E(\hat{\bar{y}}\,|\,\theta,y) = E\{E(\hat{\bar{y}}\,|\,\theta,y,s)\,|\,\theta,y\}$$
$$= \sum_s \hat{\bar{y}}(s,y_s)f(s\,|\,\theta,y) = \frac{1}{6}\sum_s \hat{\bar{y}}(s,y_s)$$
$$= \frac{1}{6}\{\hat{\bar{y}}((1,2),(0,0)) + \hat{\bar{y}}((1,3),(0,1)) + \hat{\bar{y}}((1,4),(0,1))$$
$$+ \hat{\bar{y}}((2,3),(0,1)) + \hat{\bar{y}}((2,4),(0,1)) + \hat{\bar{y}}((3,4),(1,1))\}$$
$$= (1/6)(0.15 + 0.5 + 0.5 + 0.46875 + 0.46875 + 0.8247283)$$
$$= 0.4853714.$$

Also, the design mean of $\bar{y}$ is
$$E(\bar{y}\,|\,\theta,y) = (0 + 0 + 1 + 1)/4 = 0.5.$$

So the design bias of $\hat{\bar{y}}$ is
$$E(\hat{\bar{y}} - \bar{y}\,|\,\theta,y) = 0.4853714 - 0.5 = -0.01463.$$

Note: The derivation of this result did not involve θ. So for the case $\theta = 3/4$ and $y = (0,0,1,1)$, the design bias of $\hat{\bar{y}}$ is also -0.01463.

(h)(ii) Suppose that $\theta = 1/4$ and $y = (1, 0, 1, 1)$. Then $y_1 = 1$ and so the sampling mechansim is possibly nonignorable, with

$$f(s \mid y, \theta) = \begin{cases} \dfrac{3 + y_1}{18} = \dfrac{3 + 1}{18} = \dfrac{2}{9}, & s = (1, 2), (1, 3), (1, 4) \\[2mm] \dfrac{3 - y_1}{18} = \dfrac{3 - 1}{18} = \dfrac{1}{9}, & s = (2, 3), (2, 4), (3, 4) \end{cases}.$$

So the design mean of $\hat{\bar{y}}$ is

$$E(\hat{\bar{y}} \mid \theta, y) = E\{E(\hat{\bar{y}} \mid \theta, y, s) \mid \theta, y\} = \sum_s \hat{\bar{y}}(s, y_s) f(s \mid \theta, y)$$

$$= \hat{\bar{y}}((1, 2), (1, 0)) \frac{2}{9} + \hat{\bar{y}}((1, 3), (1, 1)) \frac{2}{9} + \hat{\bar{y}}((1, 4), (1, 1)) \frac{2}{9}$$

$$+ \hat{\bar{y}}((2, 3), (0, 1)) \frac{1}{9} + \hat{\bar{y}}((2, 4), (0, 1)) \frac{1}{9} + \hat{\bar{y}}((3, 4), (1, 1)) \frac{1}{9}$$

$$= (2/9)(0.5 + 0.85 + 0.85) + (1/9)(0.46875 + 0.46875 + 0.8247283)$$

$$= 0.684692.$$

Also, the design mean of $\bar{y}$ is $E(\bar{y} \mid \theta, y) = (1 + 0 + 1 + 1)/4 = 0.75$.

So the design bias of $\hat{\bar{y}}$ is $E(\hat{\bar{y}} - \bar{y} \mid \theta, y) = 0.684692 - 0.75 = -0.06531$.

Note: The derivation of this result did not involve θ. So for the case $\theta = 3/4$ and $y = (1, 0, 1, 1)$, the design bias of $\hat{\bar{y}}$ is also -0.06531.

(i) A suitable Gibbs sampler is based on the joint density

$$f(s, y, \theta) = f(\theta) f(y \mid \theta) f(s \mid y, \theta) \propto 1 \times \prod_{i=1}^{4} \theta^{y_i} (1 - \theta)^{1 - y_i} \times \frac{1 + I_1 y_1}{6 + 3 y_1}.$$

We can identify three conditional distributions here. First observe that

$$f(\theta \mid s, y) \propto \prod_{i=1}^{4} \theta^{y_i} (1 - \theta)^{1 - y_i} = \theta^{y_T} (1 - \theta)^{1 - y_T}, \quad \theta = 1/4, 3/4$$

$$= \begin{cases} (3/4)^{y_T} (1 - 1/4)^{1 - y_T}, & \theta = 1/4 \\ (1/4)^{y_T} (1 - 3/4)^{1 - y_T}, & \theta = 3/4. \end{cases} \tag{A.1}$$

Next, recall from (d)(ii) that

$$(y_i \mid \theta, s) \sim \perp Bernoulli(\pi_i), \quad i = 1, 2, 3, 4,$$

where: $\pi_2 = \pi_3 = \pi_4 = \theta$

$$\pi_1 = \frac{3-1}{3-\theta}\theta^1(1-\theta)^{1-1} = \frac{2\theta}{3-\theta}.$$

Now, the second component of $r = (r_1, r_2)$ must be 2, 3 or 4.

Therefore

$$(y_{r_2} \mid \theta, s, y_s, y_{r_1}) \sim Bernoulli(\theta). \tag{A.2}$$

However, there are two possibilities for y_{r_1}. If the data is such that $s_1 = 1$ then

$$(y_{r_1} \mid \theta, s, y_s, y_{r_2}) \sim Bernoulli(\theta). \tag{A.3}$$

On the other hand, if the data is such that $s_1 > 1$ then $r_1 = 1$, and this implies that

$$(y_{r_1} \mid \theta, s, y_s, y_{r_2}) \sim Bernoulli\left(\frac{2\theta}{3-\theta}\right). \tag{A.4}$$

Equations (A.1), (A.2), (A.3) and (A.4) imply three conditional distributions which define a suitable Gibbs sampler (for θ, y_{r_1} and y_{r_2}).

Note: At (15.4), the ratio of probabilities of $y_{r_1} = 0$ to $y_{r_1} = 1$ is

$$\frac{\left(1 - \dfrac{2\theta}{3-\theta}\right)}{\left(\dfrac{2\theta}{3-\theta}\right)} = \frac{3(1-\theta)}{2\theta} = \frac{3}{2} \times \left(\frac{1-\theta}{\theta}\right),$$

which is exactly 3/2 times the ratio of the probabilities of $y_{r_1} = 0$ to $y_{r_1} = 1$ at (A.3). (This observation provided some assistance when formulating the required R code, as detailed below.)

Implementing the above Gibbs sampler, we obtained a random sample
$$(\theta_1, \bar{y}^{(1)}), ..., (\theta_{10000}, \bar{y}^{(10000)}) \sim iid \ f(\theta, \bar{y} \mid D)$$
for each of the six possible data configurations in (c) and (f).

The respective sample means for θ were:
    0.3007, 0.4924, 0.6997, 0.2952, 0.4764, 0.6925.

It will be observed that these numbers are very close to the corresponding values obtained in (c), namely

$$\hat{\theta} = \begin{cases} 3/10 = 0.3000 & \text{if } 1 \in s \text{ and } y_{sT} = 0 \\ 1/2 = 0.5000 & \text{if } 1 \in s \text{ and } y_{sT} = 1 \\ 7/10 = 0.7000 & \text{if } 1 \in s \text{ and } y_{sT} = 2 \\ 7/24 = 0.2917 & \text{if } 1 \notin s \text{ and } y_{sT} = 0 \\ 19/40 = 0.4750 & \text{if } 1 \notin s \text{ and } y_{sT} = 1 \\ 127/184 = 0.6902 & \text{if } 1 \notin s \text{ and } y_{sT} = 2. \end{cases}$$

The respective sample means for $\bar{y}$ were:

0.1518, 0.4929, 0.8485, 0.1308, 0.4719, 0.8269.

It will be noted that these are very close to the corresponding values obtained in (f), namely:

0.15, 0.5, 0.85, 0.1284722, 0.4687500, 0.8247283.

(j) To check the design bias in (h)(i) we note that for $y = (0,0,1,1)$ the sampling mechanism is *ignorable*.

So proceed as follows. Simply select one of the 6 possible samples randomly. Then calculate the corresponding value of $\hat{\bar{y}}$. Repeat another $J - 1$ times, independently. Then take the mean of the simulated $\hat{\bar{y}}$ values and subtract $\bar{y} = 2/4$.

Implementing this procedure with $J = 10,000$ yielded a point estimate of -0.01562 with 95% CI $(-0.01945, -0.01179)$. This is consistent with the result -0.01463 in (h)(i).

To check the design bias in (h)(ii) we note that for $y = (1,0,1,1)$ the sampling mechanism is *nonignorable* with each sample containing unit 1 twice as likely as each unit not containing unit 1.

So, select a sample s from $(1,2), (1,3), (1,4), (2,3), (2,4), (3,4)$, in such a way that each of the first three of these has probability 2/9 and each of the last three has probability 1/9. Then calculate the corresponding value of $\hat{\bar{y}}$. Repeat another $J - 1$ times, independently. Then take the mean of the simulated $\hat{\bar{y}}$ values and subtract $\bar{y} = 3/4$.

Implementing this procedure with $J = 10{,}000$ yielded a point estimate of -0.06592 with 95% CI $(-0.06944, -0.06239)$. This is consistent with the result -0.06531 in (h)(ii).

(k) The mean of the predictive mean of the finite population mean is the same as the unconditional mean of the finite population mean, which is the same as the prior mean of the superpopulation mean, which in our case equals 1/2. Mathematically,

$$E\hat{\bar{y}} = EE(\bar{y} \mid s, y_s) \quad \text{by the definition of } \hat{\bar{y}}$$
$$= E\bar{y} \qquad\qquad \text{by the law of conditional expectation}$$
$$= EE(\bar{y} \mid \theta) \qquad \text{by the law of conditional expectation}$$
$$= E\theta \qquad\qquad \text{since } E(\bar{y} \mid \theta) = \frac{1}{4}\sum_{i=1}^{4} E(y_i \mid \theta) = \frac{1}{4}\sum_{i=1}^{4}\theta = \theta$$
$$= \sum_{\theta} \theta f(\theta) = \frac{1}{4}\times\frac{1}{2} + \frac{3}{4}\times\frac{1}{2} = \frac{1}{2}.$$

To verify this obvious result via Monte Carlo is a good final check on previous calculations.

To this end, simulate θ, then simulate y, then simulate s, hence obtain the data (s, y_s), then calculate the associated $\hat{\bar{y}}$. Then repeat all of the above independently another $J - 1$ times.

Implementing this procedure with $J = 10{,}000$ yielded a point estimate of 0.4992 with 95% CI $(0.4938, 0.5047)$. This is consistent with the answer of 1/2 above.

R Code for Exercise A.4

```
# (g)
postfun = function(s=c(1,2), ys=c(0,1)){   ysT=sum(ys)
        if(any(s==1)==T){        if(ysT==0) probs=c(0.9,0.1)
                                 if(ysT==1) probs=c(0.5,0.5)
                                 if(ysT==2) probs=c(0.1,0.9)  }
        if(any(s==1)==F){        if(ysT==0) probs=c(11/12,1/12)
                                 if(ysT==1) probs=c(11/20,9/20)
                                 if(ysT==2) probs=c(11/92,81/92)  }

        probs  }

postfun() # 0.5 0.5   Just testing
```

```
postfun(s=c(2,4),ys=c(1,1)) # 0.1195652 0.8804348

thetahatfun=function(s=c(1,2), ys=c(0,1)){  probs= postfun(s=s,ys=ys);
        thetavals=c(1,3)/4;   sum( thetavals * probs ) }
thetahatfun() # 0.5    Just testing
thetahatfun(s=c(2,4),ys=c(1,1)) # 0.6902174
phihatfun=function(s=c(1,2), ys=c(0,1)){ probs=postfun(s=s,ys=ys);
        thetavals=c(1,3)/4; phivals=2*thetavals/(3-thetavals)
        sum( phivals * probs )   }

phihatfun() # 0.4242424    Just testing
phihatfun(s=c(2,4),ys=c(1,1)) # 0.6086957

yrThatfun=function(s=c(1,2), ys=c(0,1)){  thetahat=thetahatfun(s=s,ys=ys)
        if(any(s==1)==T){        res=2*thetahat }
        if(any(s==1)==F){
                phihat=phihatfun(s=s,ys=ys); res = thetahat + phihat  }
        res   }

yrThatfun() # 1    Just testing
yrThatfun (s=c(2,4),ys=c(1,1)) # 1.298913

ybarhatfun=function(s=c(1,2), ys=c(0,1)){  EyrT= yrThatfun (s=s,ys=ys)
        (sum(ys)+EyrT)/4  }

ybarhatfun() # 0.5    Just testing
ybarhatfun(s=c(2,4),ys=c(1,1)) # 0.8247283

smat=matrix(c(1,2, 1,2,  1,2,  1,2,  2,3,  2,3,  2,3,  2,3), byrow=T,nrow=8, ncol=2)
ysmat= matrix(c(0,0, 0,1, 1,0, 1,1,   0,0,   0,1,  1,0,  1,1),
        byrow=T,nrow=8, ncol=2)
thetahatvec=rep(NA,8); phihatvec=rep(NA,8); ybarhatvec=rep(NA,8);

for(k in 1:8){   thetahatvec[k]= thetahatfun(s=smat[k,],ys=ysmat[k,])
                phihatvec[k]= phihatfun(s=smat[k,],ys=ysmat[k,])
                ybarhatvec[k]= ybarhatfun(s=smat[k,],ys=ysmat[k,])  }

cbind(smat,NA,ysmat,NA,thetahatvec, NA, phihatvec, NA, ybarhatvec)
#            thetahatvec  phihatvec   ybarhatvec
# [1,] 1 2 NA 0 0 NA  0.3000000 NA 0.2303030 NA 0.1500000
# [2,] 1 2 NA 0 1 NA  0.5000000 NA 0.4242424 NA 0.5000000
# [3,] 1 2 NA 1 0 NA  0.5000000 NA 0.4242424 NA 0.5000000  repeat OK
# [4,] 1 2 NA 1 1 NA  0.7000000 NA 0.6181818 NA 0.8500000
# [5,] 2 3 NA 0 0 NA  0.2916667 NA 0.2222222 NA 0.1284722
```

```
# [6,] 2 3 NA 0 1 NA   0.4750000 NA 0.4000000 NA 0.4687500
# [7,] 2 3 NA 1 0 NA   0.4750000 NA 0.4000000 NA 0.4687500  repeat OK
# [8,] 2 3 NA 1 1 NA   0.6902174 NA 0.6086957 NA 0.8247283

0.15*(27/52) + 0.5*(21/52) +  0.85*(4/52) #  0.3451923
0.1284722*(9/16) +  0.46875*(6/16) + 0.8247283*(1/16)  # 0.2995924

# (h)
(1/6)*(0.15 + 0.5 + 0.5 + 0.46875+ 0.46875 + 0.8247283)  # 0.4853714
(2/9)*(0.5 + 0.85 + 0.85) + (1/9)*(0.46875+ 0.46875 + 0.8247283) # 0.684692

# (i) Check posterior means and predcitive means via Gibbs sampler
options(digits=4)
GS=function(J=1000,  s=c(1,2),ys=c(1,0),   theta=1/4 ){
thetav=rep(NA,J); yrTv=rep(NA,J); yTv=rep(NA,J)
yrmat=matrix(NA,nrow=J,ncol=2); ysT=sum(ys)

for(j in 1:J){
   probsyi = c(1-theta, theta)
   yr2=sample(x=c(0,1),size=1,prob=probsyi)
   if(s[1]==1) yr1=sample(x=c(0,1),size=1,prob=probsyi)  else
            yr1=sample(x=c(0,1),size=1,prob=c(3,2)*probsyi)
   yr=c(yr1,yr2); yrT=sum(yr); yT=ysT+yrT
   probstheta=c( (1/4)^yT *(3/4)^(4-yT), (3/4)^yT *(1/4)^(4-yT) )
   theta = sample( x=c(1/4,3/4),  size=1, prob= probstheta)
   thetav[j]=theta; yrTv[j]=yrT;  yTv[j]=yT; yrmat[j,]=yr
   }
list(thetav=thetav, yrTv=yrTv, yTv=yTv, ybarv=yTv/4, yrmat=yrmat)   }

set.seed(111); J = 10000; thetahatvec=rep(NA,6); ybarhatvec=rep(NA,6)
res=GS(J=J,s=c(1,2),ys=c(0,0))
   thetahatvec[1] = mean(res$thetav); ybarhatvec[1] = mean(res$ybarv);
res= GS(J=J,s=c(1,2),ys=c(0,1))
   thetahatvec[2] = mean(res$thetav); ybarhatvec[2] = mean(res$ybarv);
res= GS(J=J,s=c(1,2),ys=c(1,1))
   thetahatvec[3] = mean(res$thetav); ybarhatvec[3] = mean(res$ybarv);
res=GS(J=J,s=c(2,3),ys=c(0,0))
   thetahatvec[4] = mean(res$thetav); ybarhatvec[4] = mean(res$ybarv);
res= GS(J=J,s=c(2,3),ys=c(0,1))
   thetahatvec[5] = mean(res$thetav); ybarhatvec[5] = mean(res$ybarv);
res= GS(J=J,s=c(2,3),ys=c(1,1))
   thetahatvec[6] = mean(res$thetav); ybarhatvec[6] = mean(res$ybarv);
thetahatvec # 0.3007 0.4924 0.6997 0.2952 0.4764 0.6925
          # All very close to results in (c)
```

```
ybarhatvec  # 0.1518 0.4929 0.8485 0.1308 0.4719 0.8269
            # All very close to results in (f)


# (j) Check design bias of predictive mean of ybar if theta=1/4 and y=(0,0,1,1)
smatrix=matrix(c(1,2, 1,3,  1,4,  2,3,  2,4,  3,4), byrow=T,nrow=6, ncol=2)
y=c(0,0,1,1); J = 10000; ybarhatsimv=rep(NA,J); set.seed(413)

for(j in 1:J){    indexsim=sample(1:6,1,prob=c(1,1,1,1,1,1))
                  ssim=smatrix[indexsim,];  yssim= y[ssim]
                  ybarhatsimv[j] = ybarhatfun(s=ssim,ys=yssim)       }

est=mean(ybarhatsimv)-0.5;
ci=est+c(-1,1)*qnorm(0.975)*sd(ybarhatsimv-0.5)/sqrt(J)
c(est,ci) # -0.01562 -0.01945 -0.01179    Consistent with -0.01463 in (h)(i)


# Check design bias of predictive mean of ybar if theta=1/4 and y=(1,0,1,1)
y=c(1,0,1,1); J = 10000; ybarhatsimv=rep(NA,J); set.seed(442)

for(j in 1:J){    indexsim=sample(1:6,1,prob=c(2,2,2,1,1,1))
                  ssim=smatrix[indexsim,];  yssim= y[ssim]
                  ybarhatsimv[j] = ybarhatfun(s=ssim,ys=yssim)       }

est=mean(ybarhatsimv)-0.75;
ci=est+c(-1,1)*qnorm(0.975)*sd(ybarhatsimv-0.5)/sqrt(J)
c(est,ci) #  -0.06592 -0.06944 -0.06239  Consistent with -0.06531 in (h)(ii)


# (k)  Check mean of predictive mean of finite population mean
smatrix=matrix(c(1,2, 1,3,  1,4,  2,3,  2,4,  3,4), byrow=T,nrow=6, ncol=2)
J = 10000; ybarhatsimv=rep(NA,J); set.seed(102);

for(j in 1:J){
    thetasim=sample(c(1/4,3/4),1);   ysim=rbinom(4,1,thetasim)
    if(ysim[1]==0) indexsim = sample(1:6,1,prob=c(1,1,1,1,1,1))
    if(ysim[1]==1) indexsim = sample(1:6,1,prob=c(2,2,2,1,1,1))
    ssim=smatrix[indexsim,];  yssim= ysim[ssim];
    ybarhatsimv[j]= ybarhatfun(s=ssim,ys=yssim)   }

est = mean(ybarhatsimv);
ci = est+c(-1,1)*qnorm(0.975)*sd(ybarhatsimv)/sqrt(J)
c(est,ci) # 0.4992 0.4938 0.5047    Consistent with 0.5
```

APPENDIX B

Distributions and Notation

Below are several probability distributions which feature in this book. The purpose of this appendix is to provide a brief guide to the style of notation and terminology used throughout. It is not intended to be a comprehensive listing. Some of the notation introduced here is repeated in Appendix C.

B.1 The normal distribution

A random variable (rv) X has the *normal distribution* with parameters μ and σ^2 if its probability density function (pdf), or density, has the form

$$f(x) = \frac{1}{\sigma\sqrt{2\pi}} \exp\left\{-\frac{1}{2\sigma^2}(x-\mu)^2\right\}, \mu \in \Re.$$

We then write $X \sim N(\mu, \sigma^2)$. To be more explicit, we will sometimes write $f(x)$ as $f_X(x)$ or $f_{N(\mu,\sigma^2)}(x)$. To avoid subscripting notation and so aid legibility, $f_{N(\mu,\sigma^2)}(x)$ may sometimes be written as $f(x, N(\mu, \sigma^2))$. Likewise for other functions and expressions which contain subscripts.

If $X \sim N(\mu, \sigma^2)$ then $EX = Mode(X) = Median(X) = \mu$ and $VX = \sigma^2$.

The cumulative distribution function (cdf) of X is

$$F(x) = P(X \leq x) = F_{N(\mu,\sigma^2)}(x) = F(x, N(\mu, \sigma^2)) = \int_{-\infty}^{x} f_{N(\mu,\sigma^2)}(t)dt .$$

The (lower) p-quantile of X is the value of x such that $F(x) = p$.

Thus the p-quantile of X is the inverse cdf of X. This may also be written
$$F^{-1}(p) = F_X^{-1}(p) = F_{N(\mu,\sigma^2)}^{-1}(p) = FInv(p, N(\mu, \sigma^2)) .$$

If $Z \sim N(0,1)$, we say that Z has the *standard normal distribution*. The pdf, cdf, (lower) p-quantile and upper p-quantile of Z may be denoted by $\phi(z)$, $\Phi(z)$, $\Phi^{-1}(p)$, and $z_p = \Phi^{-1}(1-p)$, respectively.

This notation means that if $X \sim N(\mu, \sigma^2)$, then we may write:

$$f(x) = \frac{1}{\sigma} \phi\left(\frac{x - \mu}{\sigma}\right), \quad F(x) = \Phi\left(\frac{x - \mu}{\sigma}\right), \quad F_X^{-1}(p) = \mu + z_{1-p}\sigma.$$

Note: We sometimes use upper and lower case letters interchangeably. Thus $X \sim N(\mu, \sigma^2)$ may also be written $x \sim N(\mu, \sigma^2)$. The pdf of a rv X when evaluated at c may also be denoted by $f(x = c)$.

B.2 The gamma distribution

A random variable X has the *gamma distribution* with parameters a and b if its pdf has the form

$$f_X(x) = \frac{b^a x^{a-1} e^{-bx}}{\Gamma(a)}, \, x > 0.$$

We then write $X \sim Gamma(a,b)$ or $X \sim Gam(a,b)$ or $X \sim G(a,b)$. We may also write $f_X(x)$ as $f(x)$ or $f_{G(a,b)}(x)$ or $f(x, G(a,b))$.

The cdf of X may be written $F_X(x) = F_{G(a,b)}(x) = F(x, G(a,b))$, and X's p-quantile is $F_X^{-1}(p) = F_{G(a,b)}^{-1}(p) = F^{-1}(p, G(a,b)) = FInv(p, G(a,b))$.

If $X \sim G(a,b)$ then:
$$Mode(X) = (a-1)/b \text{ if } a > 1$$
$$Mode(X) = 0 \text{ if } a \leq 1$$
$$EX = a/b, \quad VX = a/b^2$$
$$EX^k = \frac{\Gamma(a+k)}{b^k \Gamma(a)} \quad \text{(the } k\text{th raw moment of } X\text{).}$$

The last result may be proved by writing

$$EX^k = \int_0^\infty x^k \frac{b^a x^{a-1} e^{-bx}}{\Gamma(a)} dx = \frac{b^a \Gamma(a+k)}{b^{a+k} \Gamma(a)} \int_0^\infty \frac{b^{a+k} x^{a+k-1} e^{-bx}}{\Gamma(a+k)} dx$$

and noting that the last integral is equal to unity.

The definition of the gamma distribution involves the *gamma function*,

$$\Gamma(k) = \int_0^\infty t^{k-1} e^{-t} dt.$$

Some properties of the gamma function are as follows:

$\Gamma(k) \to \infty$ as $k \to \infty$ or $k \to 0$

$\Gamma(k) = (k-1)\Gamma(k-1)$ for $k > 1$

$\Gamma(k) = (k-1)!$ if $k \in \{1, 2, 3, ...\}$ (with $0! = 1$)

$\Gamma(1/2) = \sqrt{\pi}$.

Note: There is an alternative definition of the gamma distribution, whereby $X \sim G(a,b)$ means $f(x) = b^{-a}x^{a-1}e^{-x/b} / \Gamma(a)$, $x > 0$, so that $EX = ab$. This alternative definition is not used in this book.

B.3 The exponential distribution

If $X \sim G(1,b)$ then X has the *exponential distribution* with parameter b, and we write $X \sim Exponential(b)$ or $X \sim Expo(b)$.

Note: We do not write $X \sim Exp(b)$ because this could more easily be confused with $X = \exp(b) = e^b$ (where exp is the exponential function).

The pdf of X, namely $f(x) = be^{-bx}$, $x > 0$, may also be written as $f_{Expo(b)}(x)$ or $f(x, Expo(b))$.

If $X \sim Expo(1)$, we say that X has the *standard exponential distribution*.

B.4 The chi-squared distribution

If $X \sim G(m/2, 1/2)$ then X has the *chi-squared distribution* with parameter m (called the *degrees of freedom*, abbreviated dof).

We then write $X \sim \chi^2(m)$ or $X \sim Chisq(m)$, and denote the pdf of X by $f_{\chi^2(m)}(x)$ or $f(x, Chisq(m))$.

The upper p-quantile of the $\chi^2(m)$ distribution may be written

$$\chi_p^2(m) = F_{\chi^2(m)}^{-1}(1-p) = FInv(1-p, Chisq(m)).$$

A useful result is that if $Y = rX$, where $X \sim Gamma(m/2, r/2)$, then $Y \sim G(m/2, 1/2) \sim \chi^2(m)$. This result can be proved easily using the transformation rule, as follows:

$$f(y) = f(x) \left| \frac{dx}{dy} \right| \overset{y}{\propto} \left(\frac{y}{r} \right)^{\frac{m}{2}-1} e^{-\frac{r}{2} \times \frac{y}{r}} \left| \frac{1}{r} \right| \overset{y}{\propto} y^{\frac{m}{2}-1} e^{-\frac{1}{2}y}.$$

Note: The symbol $\overset{y}{\propto}$ here denotes 'proportionality with respect to y'. The statement $g \overset{t}{\propto} h$ means $g = c \times h$, where c is a constant that does not depend on t. E.g. if $g = 5t^2 r^3$, we may write: $g \overset{t}{\propto} t^2$, $g \overset{r}{\propto} r^3$, $g \overset{t,r}{\propto} t^2 r^3$, $g \overset{r}{\not\propto} t$, $g \overset{r}{\not\propto} r^4$, etc. By default, $g(t) \propto t^5$ means $g(t) \overset{t}{\propto} t^5$, and $g(t|u) \propto t^5$ means $g(t|u) \overset{t}{\propto} t^5$ (not $g(t|u) \overset{t,u}{\propto} t^5$).

B.5 The inverse gamma distribution

If $X \sim G(a,b)$, then $Y = 1/X$ has the *inverse gamma distribution* with parameters a and b. In that case, we write $Y \sim InverseGamma(a,b)$ or $Y \sim IGam(a,b)$ or $Y \sim IG(a,b)$.

By the transformation rule, the pdf of Y is

$$f(y) = f(x) \left| \frac{dx}{dy} \right| = \frac{b^a (1/y)^{a-1} e^{-b(1/y)}}{\Gamma(a)} \left| -\frac{1}{y^2} \right| = \frac{b^a y^{-(a+1)} e^{-b/y}}{\Gamma(a)}, \; y > 0,$$

which may also be written $f_{IG(a,b)}(y)$ or $f(y, IG(a,b))$.

Some other properties of Y are as follows:

$EY = b/(a-1)$ if $a > 1$, $\qquad\qquad EY = \infty$ if $a \le 1$

$VY = b^2 / \{(a-1)^2 (a-2)\}$ if $a > 2$, $\quad Mode(Y) = b/(a+1)$.

B.6 The t distribution

A random variable X has the *t distribution* with parameter m if

$$f(x) = \frac{\Gamma((m+1)/2)}{\Gamma(m/2)\sqrt{m\pi}} \left(1 + \frac{x^2}{m} \right)^{-\frac{1}{2}(m+1)}, \; -\infty < x < \infty.$$

In that case, we write $X \sim t(m)$ and denote the density of X by $f_{t(m)}(x)$ or $f(x, t(m))$. The cdf of X is denoted $F_{t(m)}(x)$ or $F(x, t(m))$, and the

upper p-quantile may be written $t_p(m) = F_{t(m)}^{-1}(1-p) = FInv(1-p, t(m))$.
We call m the *degrees of freedom* parameter.

An equivalent definition of the t distribution is as follows. If $Z \sim N(0,1)$,
$Y \sim \chi^2(m)$ and $Z \perp Y$, then $X = Z / \sqrt{Y/m} \sim t(m)$.

Note: The symbol $\perp$ here denotes independence. Thus, the statement
$A \perp B$ means that A and B are independent random variables. Likewise,
$(A \perp B \mid C)$ means that A and B are independent conditional on C.

B.7 The F distribution

Suppose that $U \sim \chi^2(a)$, $W \sim \chi^2(b)$ and $U \perp W$. Then $X = \dfrac{U/a}{W/b}$ has
the *F distribution* with parameters a and b. We then write $X \sim F(a,b)$.
The pdf and cdf of X (both omitted here) may be denoted $f_{F(a,b)}(x)$ and
$F_{F(a,b)}(x)$, respectively. We call a the *numerator degrees of freedom* and
b the *denominator degrees of freedom*. The upper p-quantile of X may be
denoted as $F_p(a,b)$ or $F_{F(a,b)}^{-1}(1-p)$ or $Finv(1-p, F(a,b))$.

B.8 The (continuous) uniform distribution

A random variable X has the *(continuous) uniform distribution* with
parameters a and b if its pdf is $f(x) = 1/(b-a)$, $a < x < b$.

We then write $X \sim U(a,b)$ and $f(x) = f_{U(a,b)}(x) = f(x, U(a,b))$.
The cdf of X is $F_{U(a,b)}(x) = F(x, U(a,b)) = (x-a)/(b-a)$, $a < x < b$.
The mean and variance of X are $(a+b)/2$ and $(b-a)^2/12$.

B.9 The discrete uniform distribution

A random variable X has the *discrete uniform distribution* with parameters
$a_1, ..., a_K$ if its density is $f(x) = 1/K$, $x = a_1, ..., a_K$.

We then write $X \sim DU(a_1, ..., a_K)$. The density $f(x)$ may also be written
as $f_{DU(a_1, ..., a_K)}(x)$ or $f(x, DU(a_1, ..., a_K))$.

Equivalently, we may describe X as having the discrete uniform distribution with parameter $a = (a_1, ..., a_K)$ (a vector). In that case, we may write $X \sim DU(a)$ and denote $f(x)$ by $f_{DU(a)}(x)$ or $f(x, DU(a))$.

Note: Because X here is discrete, $f(x)$ may more aptly be called the *probability mass function* (pmf) of X. But for simplicity, we usually use the term *probability density function* (pdf) or *density* in reference to any type of random variable (continuous, discrete or mixed).

B.10 The binomial distribution

A rv X has the *binomial distribution* with parameters n and p if its density has the form

$$f(x) = \binom{n}{x} p^x (1-p)^{n-x}, \ x = 0, 1, ..., n.$$

We then write $X \sim Bin(n, p)$. The density $f(x)$ may also be denoted by $f_{Bin(n,p)}(x)$ or $f(x, Bin(n, p))$. The mean and variance of X are np and $np(1-p)$. We call n the *number of trials* and p the *probability of success* (equivalently, the *binomial parameter* or the *binomial proportion*).

B.11 The Bernoulli distribution

If $X \sim Bin(1, p)$ then we say that X has the *Benoulli distribution* with parameter p. We then write $X \sim Bernoulli(p)$ or $X \sim Bern(p)$.

B.12 The geometric distribution

A random variable is said to have the *geometric distribution* with parameter p if its pdf has the form

$$f(x) = (1-p)^{x-1} p, \ x = 1, 2, 3, ...$$

We then write $X \sim Geo(p)$. The pdf of X may be denoted by $f_{Geo(p)}(x)$ or $f(x, Geo(p))$. The mean and variance of X are $1/p$ and $(1-p)/p^2$. The cdf of X is given by

$$F_{Geo(p)}(x) = F(x, Geo(p)) = P(X \leq x) = 1 - (1-p)^x, \ x = 1, 2, 3, ...$$

APPENDIX C

Abbreviations and Acronyms

Below are some of the abbreviations and acronyms used in this book. The list may not be comprehensive. Some of the expressions listed have more than one meaning, depending on the context.

ACF	autocorrelation function
AELF	absolute error loss function
AR	autoregressive (process); acceptance rate
ARMA	autoregressive moving average (process)
B	beta function; bias
Bern	Bernoulli distribution
Beta	beta distribution
BF	Bayes factor
Bin, Binom	binomial distribution
BUGS	Bayesian inference Using Gibbs Sampling (software environment for performing MCMC)
C, Cov	covariance operator
cdf	cumulative distribution function (same as df)
CDR	central density region
Chisq	chi-squared distribution (equivalent to χ^2)
CI	confidence interval
CNR	conditional Newton-Raphson (algorithm)
CPDI	central posterior (or predictive) density interval
CPDR	central posterior (or predictive) density region
cts	continuous
D	data
DA	data augmentation (algorithm)
df	distribution function (same as cdf)
dof	degrees of freedom
dsn	distribution
DU	discrete uniform distribution

E	expectation operator
e	Euler's number (2.71828)
ECM	Expectation-Conditional-Maximisation (algorithm)
ELF	error loss function
EM	Expectation-Maximisation (algorithm)
E-Step	Expectation Step (in EM algorithm)
exp	exponential function (*e* raised to a power)
Expo	exponential distribution
F	F distribution; (cumulative) distribution function
f	pdf or pmf (same as *p*); finite population correction factor
FCP	frequentist coverage probability
FInv	inverse distribution function (equivalent to F^{-1})
fpc	finite population correction (factor)
G, Gam	gamma distribution (not to be confused with the gamma function, which is denoted by the Greek letter Γ)
Geo	geometric distribution
GLM	generalised linear model
GS	Gibbs sampler/sampling
HPDI	highest posterior (or predictive) density interval
HPDR	highest posterior (or predictive) density region
Hyp	hypergeometric distribution
I	standard indicator function; vector of sample inclusion indicators (or counters); Fisher information
id	identically distributed (not necessarily independent)
IELF	indicator error loss function
IG, IGam	inverse gamma distribution
iid	independent and identically distributed (as)
ind, indep	independent (not necessarily identically distributed)
J	Monte Carlo sample size
L	loss function; lower bound; ordered sample (vector of the labels of selected units in the order that they are sampled)
LIC	law of iterated covariance: $$C(X,Y) = EC(X,Y \mid Z) + C\{E(X \mid Z), E(Y \mid Z)\}$$
LIE	law of iterated expectation: $EX = EE(X \mid Z)$
LIV	law of iterated variance: $VX = EV(X \mid Z) + VE(X \mid Z)$
ln, log	natural logarithm (to base *e*)

m	nonsample size ($m = N - n$)
MA	moving average (process); Metropolis algorithm
MAD	mean absolute deviation; finite population mean absolute deviation about the superpopulation mean
max	maximum/maximise
MC	Monte Carlo (method); Markov chain
MCMC	Markov chain Monte Carlo (method)
MH	Metropolis-Hastings (algorithm)
min	minimum/minimise
ML	maximum likelihood (method)
MLE	maximum likelihood estimate/estimator/estimation
MOME	method of moments estimate/estimator/estimation
M-Step	Maximisation Step (in EM algorithm)
N	normal (or Gaussian) distribution; finite population size
n	sample size
NG	normal-gamma (Bayesian model)
NN	normal-normal (Bayesian model)
NNG	normal-normal-gamma (Bayesian model)
NR	Newton-Raphson (algorithm)
$P, Pr, Prob$	probability function
p	binomial proportion; pdf or pmf (same as f)
PACF	partial autocorrelation function
PDF	portable document format (file)
pdf	probability density function (used for all types of rvs: continuous, discrete and mixed); used instead of pmf
PEL	posterior expected loss (function)
pmf	probability mass function (rarely used; see pdf)
Poi	Poisson distribution
POO	posterior odds
pop	population
post	posterior
ppp-value	posterior predictive p-value
pr, prob	probability
pred	predictive/prediction/predictor
PRO	prior odds
pt	point
Q	quantity of interest; quantile function; Q-function (in the EM algorithm)
QELF	quadratic error loss function

R	R (software environment for statistical computing)
R	relative bias; risk function (not to be confused with $\Re$, which denotes the whole real line)
r	Bayes risk; nonsample (vector of the labels of the units that are not sampled)
RB	Rao-Blackwell (estimate/estimator/estimation or method)
rv	random variable
s	sample standard deviation; sample (vector of the labels of the units that are sampled)
SD, sd	standard deviation
SE, se	standard error (estimate of standard deviation)
SMA	seasonal moving average (process)
SRS	simple random sampling (with or without replacement)
SRSWOR	simple random sampling without replacement
SRSWR	simple random sampling with replacement
st	such that
T	random variable with the t distribution
t	t distribution; upper quantile of the t distribution
TIAP	Total International Airline Passengers (time series)
U	(continuous) uniform distribution; random variable with the standard uniform distribution; upper bound
V, Var	variance operator
WinBUGS	BUGS for Microsoft Windows (see BUGS)
wrt	with respect to
X	finite population covariate vector (of N values)
x	sample covariate vector (of n values)
Y	random variable or vector of random variables; finite population vector (of N values)
y	realised value of a random variable or vector of random variables; sample vector (of n values); sometimes used interchangeably with Y
Z	standard normal random variable
z	upper quantile of the standard normal distribution

Bibliography

Albert, J. (2009). *Bayesian Computation with R, 2nd Edition*. New York: Springer.

Bolstad, W.M. (2009). *Computational Bayesian Statistics*. Hoboken NJ: Wiley.

Box, G.E.P, and Tiao, G.C. (1992). *Bayesian Inference in Statistical Analysis* by Box and Tiao (1973). Reading: Addison-Wesley.

Brooks, S., Gelman, A., Jones, G.L., and Meng, X.-L. (Eds.) (2011). *Handbook of Markov Chain Monte Carlo*. London: Chapman & Hall/CRC.

Bühlmann, H. (1967). Experience rating and credibility. *ASTIN Bulletin*. Website: www.casact.org/library/astin/vol4no3/199.pdf

Byrne, A.P., and Dracoulis, G.D. (1985). Monte Carlo calculations for asymmetric NaI(Tl) and BGO Compton suppression shields, *Nuclear Instruments and Methods in Physics Research*, **A234**: 281–287.

Cochran, W.G. (1977). *Sampling Techniques, 3rd Edition*. New York: Wiley.

Ericson, W.A. (1969). Subjective Bayesian models in sampling finite populations. *Journal of the Royal Statisticial Society, Series B*, **31**: 195–224.

Ericson, W.A. (1988). Bayesian inference in finite populations. In *Handbook of Statistics, Vol. 6*, P.R. Krishnaiah and C.R. Rao (Eds.), pp 213–246. Amsterdam: North Holland.

Gelman, A., Carlin, J.B., Stern, H.S., and Rubin, D.B. (2004). *Bayesian Data Analysis, 2nd Edition*. New York: Chapman and Hall.

Gilks, W.R., Richardson, S., and Spiegelhalter, D.J. (1996). *Markov Chain Monte Carlo in Practice*. New York: Chapman & Hall.

Hobert, J.P. and Casella, G. (1996). The effect of improper priors on Gibbs sampling in hierarchical linear mixed models. *Journal of the American Statistical Association*, **91**: 1461–1473.

Jeffreys, H. (1961). *Theory of Probability, 3rd Edition*. Oxford: Oxford University Press.

Kéry, M., and Schaub, M. (2012). *Bayesian Population Analysis Using WinBUGS*. New York: Elsevier.

Lachlan, G.J., and and Krishnan, T. (2008). *The EM Algorithm and Extensions*. Hoboken, NJ: John Wiley & Sons.

Leonard, T., and Hsu, J.S.J. (1999). *Bayesian Methods: An Analysis for Statisticians and Interdisciplinary Researchers*. Cambridge: Cambridge University Press.

Lee, P. (1997). *Bayesian Statistics: An Introduction*. New York: Oxford University Press.

Lunn, D.J., Thomas, A., Best., N., and Spiegelhalter, D. (2000). WinBUGS – A Bayesian modelling framework: Concepts, structure, and extensibility. *Statistics and Computing*, **10**: 325–337.

Maindonald, J., and Braun, W.J. (2010). *Data Analysis and Graphics Using R: An Example-Based Approach, 3rd Edition*. Cambridge: Cambridge University Press.

Meng, X.-L. (1994). Posterior predictive *p*-values. *The Annals of Statistics*, **22**: 1142–1160.

Ntzoufras, I. (2009). *Bayesian Modeling Using WinBUGS*. Hoboken NJ: Wiley.

O'Hagan, A, and Forster, J. (2004). *Kendall's Advanced Theory of Statistics, Second Edition, Volume 2B, Bayesian Inference*. London: Arnold.

Puza, B. (1995). *Monte Carlo Methods for Finite Population Inference*. Internal document. Canberra: Australian Bureau of Statistics.

Puza, B.D. (2002). 'Postscript: Bayesian methods for estimation' and 'Appendix C: Details of calculations in the Postscript'. In *Combined Survey Sampling Inference: Weighing Basu's Elephants*, by K. Brewer, London: Arnold, 2002, pp 293–296 and 299–302.

Puza, B.D., and O'Neill, T.J. (2005). Length-biased, with-replacement sampling from an exponential finite population. *Journal of Statistical Computation and Simulation*, **75**: 159–174.

Puza, B. and O'Neill, T.J. (2006). Selection bias in binary data from volunteer surveys. *The Mathematical Scientist*, **31**: 85–94.

Rao, C.R. (1973). *Linear Statistical Inference and its Applications, 2nd Edition*. New York: Wiley.

Rao, J.N.K. (2011). Impact of frequentist and Bayesian methods on survey sampling practice: a selective appraisal. *Statistical Science*, **26**: 240–256.

Särndal, C.-E., Swensson, B., and Wretman, J. (1992). *Model Assisted Survey Sampling*. New York: Springer.

Seaman, S., Galati, J., Jackson, D., and Carlin, J. (2013). What is meant by 'Missing at Random'? *Statistical Science*, **28(2)**: 257–268.

Shaw D, (1988). On-site samples' regression: Problems of non-negative integers, truncation, and endogenous stratification. *Journal of Econometrics*, **37**: 211–223.

Smith, A.F.M., and Gelfand, A.E. (1992). Bayesian statistics without tears: A sampling-resampling perspective. *The American Statistician*, **46(2)**: 84–88.

Wackerly, D.D., Mendenhall III, W., and Scheaffer, R.L. (2008). *Mathematical Statistics with Applications, 7th edition.* Duxbury: Thomson, Brooks/Cole.

West, M. (1996). Inference in successive sampling discovery models. *Journal of Econometrics*, **75**: 217–238.

www.ingramcontent.com/pod-product-compliance
Lightning Source LLC
Chambersburg PA
CBHW042031220326

41598CB00074BA/7403